Sixth Edition

Intermediate Algebra: An Applied Approach

Richard N. Aufmann
Palomar College, California

Vernon C. Barker
Palomar College, California

Joanne S. Lockwood
Plymouth State College, New Hampshire

HOUGHTON MIFFLIN COMPANY
Boston   New York

Editor-in-Chief: *Jack Shira*
Sponsoring Editor: *Lynn Cox*
Senior Development Editor: *Dawn Nuttall*
Editorial Assistant: *Melissa Parkin*
Senior Project Editor: *Nancy Blodget*
Editorial Assistant: *Kristin Penta*
Senior Production/Design Coordinator: *Carol Merrigan*
Manufacturing Manager: *Florence Cadran*
Marketing Manager: *Ben Rivera*

Cover photographer: Harold Burch, Harold Burch Design/NYC.

Printed in the U.S.A.

Library of Congress Control Number: 2001099899

ISBN
Student Text: 0-618-203044
Instructor's Annotated Edition: 0-618-203052

23456789-VH-06 05 04 03 02

Contents

Preface

The sixth edition of *Intermediate Algebra: An Applied Approach* provides comprehensive, mathematically sound coverage of the topics considered essential in an intermediate algebra course. The text has been designed not only to meet the needs of the traditional college student, but also to serve the needs of returning students whose mathematical proficiency may have declined during years away from formal education.

In this new edition of *Intermediate Algebra: An Applied Approach*, we have continued to integrate some of the approaches suggested by AMATYC. Each chapter opens with an illustration and a reference to a mathematical application within the chapter. At the end of each section, there are "Applying the Concepts" exercises, which include writing, synthesis, critical thinking, and challenge problems. At the end of each chapter, there is a "Focus on Problem Solving," which introduces students to various problem-solving strategies. This is followed by "Projects and Group Activities," which can be used for cooperative-learning activities.

NEW! Changes to This Edition

In response to user requests, we have expanded the material on functions to include the mapping of functions and illustrations of function machines. See, for example, Section 2 of Chapter 3, *Linear Functions and Inequalities in Two Variables*. This chapter also now includes graphing exercises that require critical thinking. See the exercises in Section 3.4.

In Chapter 4, *Systems of Linear Equations and Inequalities*, the discussion of solutions of systems of equations has been expanded and improved. This development should lead to greater student understanding of inconsistent, dependent, and independent systems of equations.

In Chapter 7, *Exponents and Radicals*, nth roots are simplified by using perfect powers rather than by using prime factorization. Therefore, rather than rewriting the square root of 20 as the square root of 2^2 times 5, 20 is written as the product of 4 and 5; the cube root of 16 is rewritten as the cube root of 8 times 2, not 2^4 and then 2^3 times 2. This approach better mirrors what the student is thinking mentally when simplifying radical expressions.

Concept-based writing exercises have been added at the beginning of many of the objective-specific exercise sets. These exercises require the student to verbalize basic concepts presented in the lesson. The student must have an understanding of these concepts before attempting any exercises. For example, in Chapter 10, *Exponential and Logarithmic Functions*, the first exercise for Objective 10.2A asks, "What is a common logarithm? How is the common logarithm of $4x$ written?"

Many of the exercise sets now include developmental exercises, which are intended to reinforce the concepts underlying the skills presented in the lesson. For example, in Chapter 8, *Quadratic Equations*, exercises require students to write quadratic equations in standard form and then to name the values of a, b, and c prior to having them solve any quadratic equations.

Throughout the text, data problems have been updated to reflect current data and trends. These application problems will demonstrate to students the variety of problems that require mathematical analysis. Instructors will find that many of these problems may lead to interesting class discussions.

Another new feature of this edition is *AIM for Success*, which explains what is required of a student to be successful and how this text has been designed to foster student success. *AIM for Success* can be used as a lesson on the first day of class or as a project for students to complete in order to strengthen their study skills. There are suggestions for teaching this lesson in the *Instructor's Resource Manual*.

Related to *AIM for Success* are *Prep Tests,* which occur at the beginning of each chapter. These tests focus on the particular prerequisite skills that will be used in the upcoming chapter. The answers to these questions can be found in the Answer Appendix, along with a reference (except for Chapter 1) to the objective from which each question was taken. Students who miss a question are encouraged to review the objective from which the question was taken.

The *Go Figure* problem that follows the *Prep Test* is a puzzle problem designed to engage students in problem solving.

Chapter Opening Features

NEW! Chapter Opener

New, motivating chapter opener photos and captions have been added, illustrating and referencing specific applications from the chapter.

The WWW at the bottom of the page lets students know about additional on-line resources at math.college.hmco.com/students.

Objective-Specific Approach

Each chapter begins with a list of learning objectives that form the framework for a complete learning system. The objectives are woven throughout the text (i.e., Exercises, Prep Tests, Chapter Reviews, Chapter Tests, Cumulative Reviews) as well as throughout the print and multimedia ancillaries. This results in a seamless learning system delivered in one consistent voice.

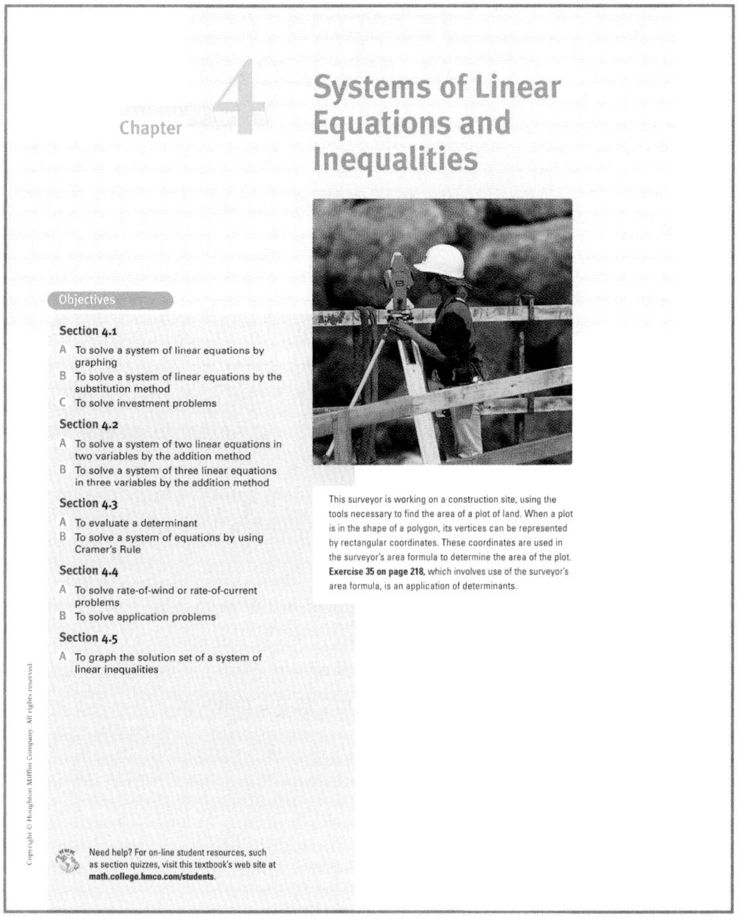

Chapter 4 Systems of Linear Equations and Inequalities

Objectives

Section 4.1
A To solve a system of linear equations by graphing
B To solve a system of linear equations by the substitution method
C To solve investment problems

Section 4.2
A To solve a system of two linear equations in two variables by the addition method
B To solve a system of three linear equations in three variables by the addition method

Section 4.3
A To evaluate a determinant
B To solve a system of equations by using Cramer's Rule

Section 4.4
A To solve rate-of-wind or rate-of-current problems
B To solve application problems

Section 4.5
A To graph the solution set of a system of linear inequalities

This surveyor is working on a construction site, using the tools necessary to find the area of a plot of land. When a plot is in the shape of a polygon, its vertices can be represented by rectangular coordinates. These coordinates are used in the surveyor's area formula to determine the area of the plot. **Exercise 35 on page 218**, which involves use of the surveyor's area formula, is an application of determinants.

Need help? For on-line student resources, such as section quizzes, visit this textbook's web site at **math.college.hmco.com/students**.

Page 185

NEW! Prep Test and Go Figure

Prep Tests occur at the beginning of each chapter and test students on previously covered concepts that are required in the upcoming chapter. Answers are provided in the Answer Appendix. Objective references are also provided if the student needs to review specific concepts.

The **Go Figure** problem that follows the *Prep Test* is a playful puzzle problem designed to engage students in problem solving.

Page 466

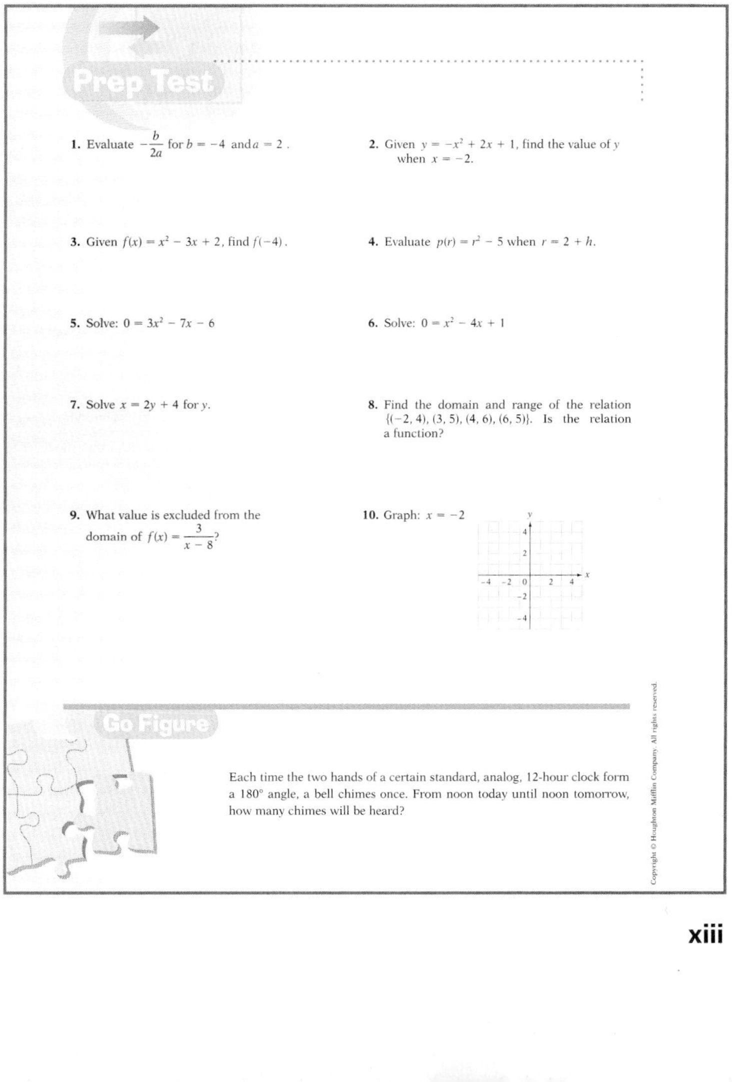

Prep Test

1. Evaluate $-\frac{b}{2a}$ for $b = -4$ and $a = 2$.
2. Given $y = -x^2 + 2x + 1$, find the value of y when $x = -2$.
3. Given $f(x) = x^2 - 3x + 2$, find $f(-4)$.
4. Evaluate $p(r) = r^2 - 5$ when $r = 2 + h$.
5. Solve: $0 = 3x^2 - 7x - 6$
6. Solve: $0 = x^2 - 4x + 1$
7. Solve $x = 2y + 4$ for y.
8. Find the domain and range of the relation $\{(-2, 4), (3, 5), (4, 6), (6, 5)\}$. Is the relation a function?
9. What value is excluded from the domain of $f(x) = \frac{3}{x-8}$?
10. Graph: $x = -2$

Go Figure

Each time the two hands of a certain standard, analog, 12-hour clock form a 180° angle, a bell chimes once. From noon today until noon tomorrow, how many chimes will be heard?

Aufmann Interactive Method (AIM)

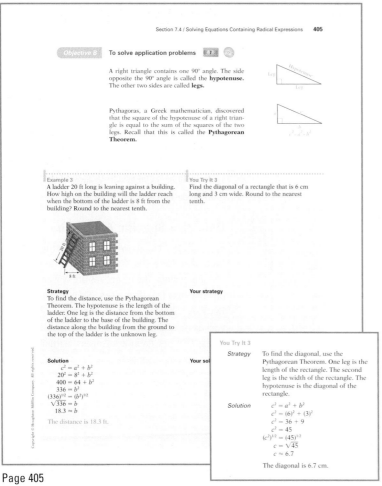

Page 405

Page S23

An Interactive Approach

Intermediate Algebra: An Applied Approach uses an interactive style that provides the student with the opportunity to try each skill as it is presented. Each section is divided into objectives, and every objective contains one or more sets of matched-pair examples. The first example in each set is worked out; the second example, called "You Try It," is for the student to work. By solving this problem, the student actively practices concepts as they are presented in the text.

There are complete worked-out solutions to these examples in the appendix. By comparing their own solutions to the solution in the appendix, students obtain immediate feedback on, and reinforcement of, the concept.

Page xxv

NEW! AIM for Success Student Preface

This new student "how to use this book" preface explains what is required of the student to be successful and how this text has been designed to foster student success, including the Aufmann Interactive Method (AIM). *AIM for Success* can be used as a lesson on the first day of class or as a project for students to complete to strengthen their study skills. There are suggestions for teaching this lesson in the *Instructor's Resource Manual.*

AIM for Success

INSTRUCTOR NOTE
See the *Instructor's Resource Manual* or *Class Prep* CD for suggestions on how to teach this lesson.

Welcome to *Intermediate Algebra: An Applied Approach*. As you begin this course, we know two important facts: (1) We want you to succeed. (2) You want to succeed. To do that requires an effort from each of us. For the next few pages, we are going to show you what is required of you to achieve that success and how you can use the features of this text to be successful.

Motivation

One of the most important keys to success is motivation. We can try to motivate you by offering interesting or important ways mathematics can benefit you. But, in the end, the motivation must come from you. On the first day of class, it is easy to be motivated. Eight weeks into the term, it is harder to keep that motivation.

Problem Solving

Focus on Problem Solving

At the end of each chapter is a Focus on Problem Solving feature, which introduces the student to various successful problem-solving strategies. Strategies such as drawing a diagram, applying solutions to other problems, working backwards, inductive reasoning, and trial and error are some of the techniques that are demonstrated.

Page 194

Page 43

Focus on Problem Solving

Polya's Four-Step Process

Your success in mathematics and your success in the workplace are heavily dependent on your ability to solve problems. One of the foremost mathematicians to study problem solving was George Polya (1887–1985). The basic structure that Polya advocated for problem solving has four steps, as outlined below.

Point of Interest

George Polya was born in Hungary and moved to the United States in 1940. He lived in Providence, Rhode Island, where he taught at Brown University until 1942, when he moved to California. There he taught at Stanford University until his retirement. While at Stanford, he published 10 books and a number of articles for mathematics journals. Of the books Polya published, *How To Solve It* (1945) is one of his best known. In this book, Polya outlines a strategy for solving problems. This strategy, although frequently applied to mathematics, can be used to solve problems from virtually any discipline.

1. Understand the Problem

You must have a clear understanding of the problem. To help you focus on understanding the problem, here are some questions to think about.

- Can you restate the problem in your own words?
- Can you determine what is known about this type of problem?
- Is there missing information that you need in order to solve the problem?
- Is there information given that is not needed?
- What is the goal?

2. Devise a Plan

Successful problem solvers use a variety of techniques when they attempt to solve a problem. Here are some frequently used strategies.

- Make a list of the known information.
- Make a list of information that is needed to solve the problem.
- Make a table or draw a diagram.
- Work backwards.
- Try to solve a similar but simpler problem.
- Research the problem to determine whether there are known techniques for solving problems of its kind.
- Try to determine whether some pattern exists.
- Write an equation.

3. Carry Out the Plan

Once you have devised a plan, you must carry it out.

- Work carefully.
- Keep an accurate and neat record of all your attempts.
- Realize that some of your initial plans will not work and that you may have to return to Step 2 and devise another plan or modify your existing plan.

4. Review Your Solution

Once you have found a solution, check the solution against the known facts.

- Make sure that the solution is consistent with the facts of the problem.
- Interpret the solution in the context of the problem.
- Ask yourself whether there are generalizations of the solution that could apply to other problems.
- Determine the strengths and weaknesses of your solution. For instance, is your solution only an approximation to the actual solution?
- Consider the possibility of alternative solutions.

Focus on Problem Solving 43

Problem-Solving Strategies

The text features a carefully developed approach to problem solving that emphasizes the importance of *strategy* when solving problems. Students are encouraged to develop their own strategies—to draw diagrams, to write out the solution steps in words—as part of their solution to a problem. In each case, model strategies are presented as guides for students to follow as they attempt the "You Try It" problem. Having students provide strategies is a natural way to incorporate writing into the math curriculum.

---

194 Chapter 4 / Systems of Linear Equations and Inequalities

Example 7

An investment of $4000 is made at an annual simple interest rate of 4.9%. How much additional money must be invested at an annual simple interest rate of 7.4% so that the total interest earned is 6.4% of the total investment?

Strategy

- Amount invested at 4.9%: 4000
  Amount invested at 7.4%: x
  Amount invested at 6.4%: y

	Principal	Rate	Interest
Amount at 4.9%	4000	0.049	0.049(4000)
Amount at 7.4%	x	0.074	0.074x
Amount at 6.4%	y	0.064	0.064y

- The amount invested at 6.4% (y) is $4000 more than the amount invested at 7.4% (x):
  $y = x + 4000$
- The sum of the interest earned at 4.9% and the interest earned at 7.4% equals the interest earned at 6.4%:
  $0.049(4000) + 0.074x = 0.064y$

Solution

$$y = x + 4000 \quad (1)$$
$$0.049(4000) + 0.074x = 0.064y \quad (2)$$

Replace y in Equation (2) by $x + 4000$ from Equation (1). Then solve for x.

$$0.049(4000) + 0.074x = 0.064(x + 4000)$$
$$196 + 0.074x = 0.064x + 256$$
$$0.01x = 60$$
$$x = 6000$$

$6000 must be invested at an annual simple interest rate of 7.4%.

You Try It 7

An investment club invested $13,600 into two simple interest accounts. On one account, the annual simple interest rate is 4.2%. On the other, the annual simple interest rate is 6%. How much should be invested in each account so that both accounts earn the same annual interest?

Your strategy

Your solution

Solution on p. S11

xv

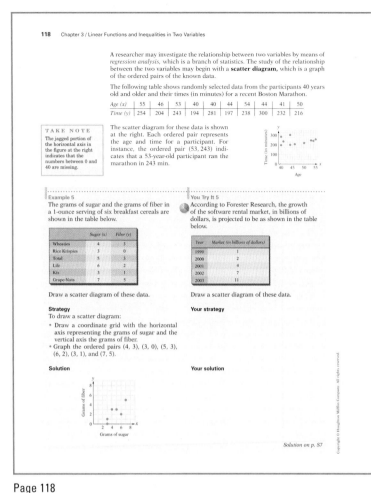

Page 118

Real Data and Applications

Applications

One way to motivate interest in mathematics is through applications. Wherever appropriate, the last objective of a section presents applications that require the student to use problem-solving strategies, along with the skills covered in that section, to solve practical problems. This carefully integrated, applied approach generates student awareness of the value of algebra as a real-life tool.

Applications are taken from many disciplines, including architecture, business, carpentry, chemistry, construction, Earth science, education, manufacturing, nutrition, real estate, and sports.

Page 154

Real Data

Real data examples and exercises, identified by ⬤, ask students to analyze and solve problems taken from actual situations. Students are often required to work with tables, graphs, and charts drawn from a variety of disciplines.

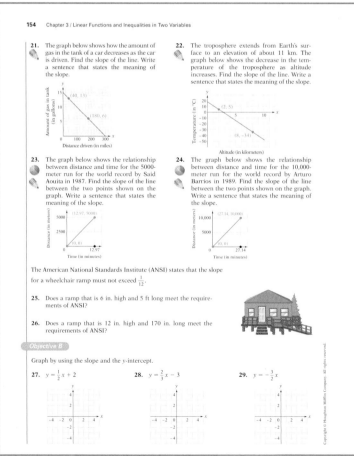

Student Pedagogy

Icons

The at each objective head remind students that both a video and a tutorial lesson are available for that objective.

Key Terms and Concepts

Key terms, in bold, emphasize important terms. The key terms are also provided in a **Glossary** at the back of the text.

Key concepts are presented in purple boxes in order to highlight these important concepts and to provide easy reference.

Point of Interest

These margin notes contain interesting sidelights about mathematics, its history, and its application.

7.2 Operations on Radical Expressions

Objective A To simplify radical expressions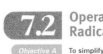

Point of Interest
The Latin expression for irrational numbers was *numerus surdus,* which literally means "inaudible number." A prominent 16th-century mathematician wrote of irrational numbers, ". . . just as an infinite number is not a number, so an irrational number is not a true number, but lies hidden in some sort of cloud of infinity." In 1872, Richard Dedekind wrote a paper that established the first logical treatment of irrational numbers.

If a number is not a perfect power, its root can only be approximated; examples include $\sqrt{5}$ and $\sqrt[3]{3}$. These numbers are **irrational numbers.** Their decimal representations never terminate or repeat.

$$\sqrt{5} = 2.2360679\ldots \qquad \sqrt[3]{3} = 1.4422495\ldots$$

A radical expression is in simplest form when the radicand contains no factor that is a perfect power. The Product Property of Radicals is used to simplify radical expressions whose radicands are not perfect powers.

The Product Property of Radicals
If $\sqrt[n]{a}$ and $\sqrt[n]{b}$ are positive real numbers, then $\sqrt[n]{ab} = \sqrt[n]{a} \cdot \sqrt[n]{b}$ and $\sqrt[n]{a} \cdot \sqrt[n]{b} = \sqrt[n]{ab}$.

Page 385

Take Note

These margin notes are used to amplify the concept under discussion or to alert students to a point requiring special attention.

Annotated Examples

Examples indicated by ⇒ use blue annotations to explain what is happening in key steps of the complete, worked-out solutions.

TAKE NOTE
When a quadratic equation has two solutions that are the same number, the solution is called a **double root** of the equation. 3 is a double root of $x^2 - 6x = -9$.

⇒ Solve by factoring: $x^2 - 6x = -9$

$$x^2 - 6x = -9$$
$$x^2 - 6x + 9 = 0$$
$$(x - 3)(x - 3) = 0$$
$$x - 3 = 0 \qquad x - 3 = 0$$
$$x = 3 \qquad x = 3$$

- Write the equation in standard form.
- Factor.
- Use the Principle of Zero Products.
- Solve each equation.

3 checks as a solution. The solution is 3.

Page 421

Calculator Note

These margin notes provide suggestions for using a calculator in certain situations.

Page 524

Logarithms base 10 are called **common logarithms.** Usually the base, 10, is omitted when writing the common logarithm of a number. Therefore, $\log_{10} x$ is written $\log x$. To find the common logarithm of most numbers, a calculator is necessary. A calculator was used to find the value of log 384, shown below.

CALCULATOR NOTE
The logarithms of most numbers are irrational numbers. Therefore, the value displayed by a calculator is an approximation.

$$\log 384 \approx 2.5843312$$

Mantissa / Characteristic

Exercises and Projects

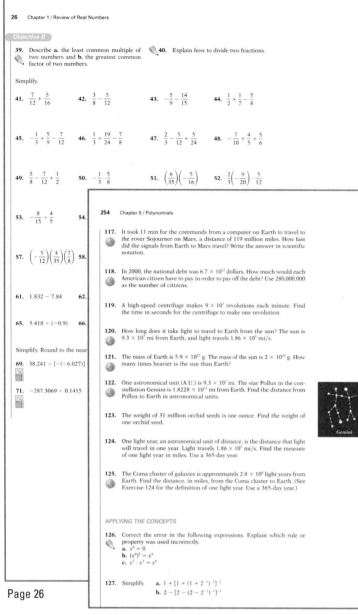

Objective B

39. Describe **a.** the least common multiple of two numbers and **b.** the greatest common factor of two numbers.

40. Explain how to divide two fractions.

Simplify.

41. $\dfrac{7}{12} + \dfrac{5}{16}$ **42.** $\dfrac{3}{8} - \dfrac{5}{12}$ **43.** $-\dfrac{5}{9} - \dfrac{14}{15}$ **44.** $\dfrac{1}{2} + \dfrac{1}{7} - \dfrac{5}{8}$

45. $-\dfrac{1}{3} + \dfrac{5}{9} - \dfrac{7}{12}$ **46.** $\dfrac{1}{3} + \dfrac{19}{24} - \dfrac{7}{8}$ **47.** $\dfrac{2}{3} - \dfrac{5}{12} + \dfrac{5}{24}$ **48.** $-\dfrac{7}{10} + \dfrac{4}{5} + \dfrac{5}{6}$

49. $\dfrac{5}{8} - \dfrac{7}{12} + \dfrac{1}{2}$ **50.** $-\dfrac{1}{3} \cdot \dfrac{5}{8}$ **51.** $\left(\dfrac{6}{35}\right)\left(-\dfrac{5}{16}\right)$ **52.** $\dfrac{2}{3}\left(-\dfrac{9}{20}\right) \cdot \dfrac{5}{12}$

53. $-\dfrac{8}{15} \div \dfrac{4}{5}$ **54.**

57. $\left(-\dfrac{5}{12}\right)\left(\dfrac{4}{35}\right)\left(\dfrac{7}{8}\right)$ **58.**

61. $1.832 - 7.84$ **62.**

65. $5.418 \div (-0.9)$ **66.**

Simplify. Round to the near

69. $38.241 \div [-(-6.027)]$

71. $-287.3069 \div 0.1415$

Page 26

117. It took 11 min for the commands from a computer on Earth to travel to the rover Sojourner on Mars, a distance of 119 million miles. How fast did the signals from Earth to Mars travel? Write the answer in scientific notation.

118. In 2000, the national debt was 6.7×10^{12} dollars. How much would each American citizen have to pay in order to pay off the debt? Use 280,000,000 as the number of citizens.

119. A high-speed centrifuge makes 9×10^{7} revolutions each minute. Find the time in seconds for the centrifuge to make one revolution.

120. How long does it take light to travel to Earth from the sun? The sun is 9.3×10^{7} mi from Earth, and light travels 1.86×10^{5} mi/s.

121. The mass of Earth is 5.9×10^{27} g. The mass of the sun is 2×10^{33} g. How many times heavier is the sun than Earth?

122. One astronomical unit (A.U.) is 9.3×10^{7} mi. The star Pollux in the constellation Gemini is 1.8228×10^{12} mi from Earth. Find the distance from Pollux to Earth in astronomical units.

123. The weight of 31 million orchid seeds is one ounce. Find the weight of one orchid seed.

124. One light year, an astronomical unit of distance, is the distance that light will travel in one year. Light travels 1.86×10^{5} mi/s. Find the measure of one light year in miles. Use a 365-day year.

125. The Coma cluster of galaxies is approximately 2.8×10^{8} light years from Earth. Find the distance, in miles, from the Coma cluster to Earth. (See Exercise 124 for the definition of one light year. Use a 365-day year.)

APPLYING THE CONCEPTS

126. Correct the error in the following expressions. Explain which rule or property was used incorrectly.
a. $x^{0} = 0$
b. $(x^{4})^{5} = x^{9}$
c. $x^{2} \cdot x^{3} = x^{6}$

127. Simplify.   **a.** $1 + [1 + (1 + 2^{-1})^{-1}]^{-1}$
b. $2 - [2 - (2 - 2^{-1})^{-1}]^{-1}$

Gemini

Page 254

Exercises

The exercise sets of *Intermediate Algebra: An Applied Approach* emphasize skill building, skill maintenance, and applications.

Concept-based writing or developmental exercises have been integrated with the exercise sets.

Icons identify appropriate writing , data analysis , and calculator exercises.

Included in each exercise set is a section called **Applying the Concepts,** which presents extensions of topics, requires analysis, or offers challenge problems. The writing exercises ask students to explain answers, write about a topic in the section, or research and report on a related topic.

Page 552

Projects and Group Activities

The Projects and Group Activities feature at the end of each chapter can be used as extra credit or for cooperative-learning activities. The projects cover various aspects of mathematics, including the use of calculators, Internet data collection, data analysis, and extended applications.

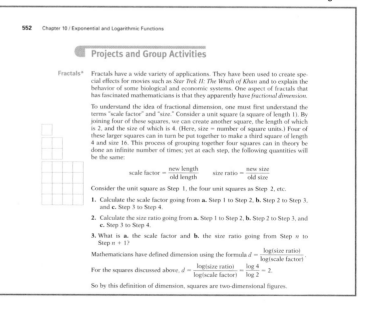

Projects and Group Activities

Fractals*   Fractals have a wide variety of applications. They have been used to create special effects for movies such as *Star Trek II: The Wrath of Khan* and to explain the behavior of some biological and economic systems. One aspect of fractals that has fascinated mathematicians is that they apparently have *fractional dimension*.

To understand the idea of fractional dimension, one must first understand the terms "scale factor" and "size." Consider a unit square (a square of length 1). By joining four of these squares, we can create another square, the length of which is 2, and the size of which is 4. (Here, size = number of square units.) Four of these larger squares can in turn be put together to make a third square of length 4 and size 16. This process of grouping together four squares can in theory be done an infinite number of times; yet at each step, the following quantities will be the same:

$$\text{scale factor} = \frac{\text{new length}}{\text{old length}} \qquad \text{size ratio} = \frac{\text{new size}}{\text{old size}}$$

Consider the unit square as Step 1, the four unit squares as Step 2, etc.

1. Calculate the scale factor going from **a.** Step 1 to Step 2, **b.** Step 2 to Step 3, and **c.** Step 3 to Step 4.

2. Calculate the size ratio going from **a.** Step 1 to Step 2, **b.** Step 2 to Step 3, and **c.** Step 3 to Step 4.

3. What is **a.** the scale factor and **b.** the size ratio going from Step n to Step $n + 1$?

Mathematicians have defined dimension using the formula $d = \dfrac{\log(\text{size ratio})}{\log(\text{scale factor})}$.

For the squares discussed above, $d = \dfrac{\log(\text{size ratio})}{\log(\text{scale factor})} = \dfrac{\log 4}{\log 2} = 2$.

So by this definition of dimension, squares are two-dimensional figures.

End of Chapter

506   Chapter 9 / Functions and Relations

Chapter Summary

Key Words   A *quadratic function* is one of the form $f(x) = ax^2 + bx + c$, where $a \neq 0$. The graph of a quadratic function is a *parabola*. The *vertex* of a parabola is the point with the smallest y-coordinate or the largest y-coordinate. When $a > 0$, the parabola opens up and the vertex of the parabola is the point with the smallest y-coordinate. The value of the function at this point is a *minimum*. When $a < 0$, the parabola opens down and the vertex is the point with the largest y-coordinate. The value of the function at this point is a *maximum*. [pp. 467, 468, 473]

The *axis of symmetry* of a parabola is the vertical line that passes through the vertex of the parabola and is parallel to the y-axis. [p. 468]

A point at which a graph crosses the x-axis is called an *x-intercept* of the graph. [p. 470]

The *inverse* of a function is the set of ordered pairs formed by reversing the coordinates of each ordered pair of the function. A function f has an inverse function if and only if f is a 1–1 function. [pp. 496–497]

Essential Rules   Vertex and Axis of Symmetry of a Parabola
Let $f(x) = ax^2 + bx + c$ be the equation of a parabola. The coordinates of the vertex are $\left(-\frac{b}{2a}, f\left(-\frac{b}{2a}\right)\right)$. The equation of the axis of symmetry is $x = -\frac{b}{2a}$. [p. 468]

The Effect of the Discriminant on the Number of x-Intercepts of a Parabola
1. If $b^2 - 4ac = 0$, the parabola has one x-intercept.
2. If $b^2 - 4ac > 0$, the parabola has two x-intercepts.
3. If $b^2 - 4ac < 0$, the parabola has no x-intercepts. [p. 471]

To Find the Minimum or Maximum Value of a Quadratic Function:
Find the x-coordinate of the vertex. Then evaluate the function at that value. [p. 473]

Vertical-Line Test
A graph defines the graph of a function if any vertical line intersects the graph at no more than one point. [p. 484]

Page 506

Chapter Summary

At the end of each chapter is a Chapter Summary that includes Key Words and Essential Rules that were covered in the chapter. These chapter summaries provide a single point of reference as the student prepares for a test. Each concept references the page number of the lesson in which the concept was introduced.

Page 507

Chapter Review

Review exercises are found at the end of each chapter. These exercises are selected to help the student integrate all of the topics presented in the chapter.

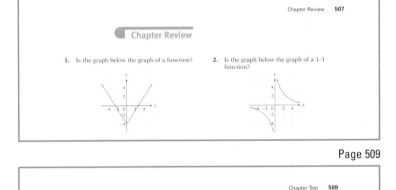

Page 509

Chapter Test

The Chapter Test exercises are designed to simulate a possible test of the material in the chapter.

Page 511

Cumulative Review

Cumulative Review exercises, which appear at the end of each chapter (beginning with Chapter 2), help students maintain skills learned in previous chapters.

The answers to all Chapter Review exercises, Chapter Test exercises, and Cumulative Review exercises are given in the Answer Section. Along with the answer, there is a reference to the objective that pertains to each exercise.

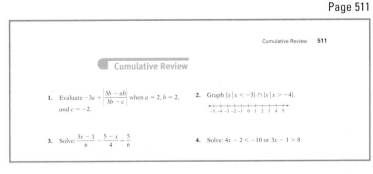

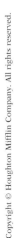

xix

Supplements for the Instructor

Intermediate Algebra: An Applied Approach has a complete set of teaching aids for the instructor.

Instructor's Annotated Edition   This edition contains a replica of the student text in reduced format and additional items just for the instructor. These include: *Instructor Notes, Transparency Master icons, In-Class Examples, Concept Checks, Discuss the Concepts, New Vocabulary/Symbols, etc., Vocabulary/Symbols, etc. to Review, Optional Student Activities, Quick Quizzes, Answers to Writing Exercises,* and *Suggested Assignments.* Answers to all exercises are also provided.

Instructor's Solutions Manual   The *Instructor's Solutions Manual* contains worked-out solutions for all exercises in the text.

Instructor's Resource Manual with Chapter Tests   The *Instructor's Resource Manual* contains ready-to-use, printed Chapter Tests, which are the first of three sources of testing material. Eight printed tests (in two formats—free-response and multiple-choice) are provided for each chapter. Cumulative tests and final exams are also provided. These tests are available on the *Class Prep* CD or can be downloaded from our web site at math.college.hmco.com/instructors. The tests are in Microsoft Word® format and can be edited to suit the needs of the instructor. The *Instructor's Resource Manual* also includes a lesson plan for the *AIM for Success.*

Printed Test Bank   The *Printed Test Bank,* the second component of the testing material, offers a printout of the items in the *HM Testing* computerized testing software, including an example of each algorithmic item. All multiple-choice items are carefully worded to make them appropriate to use as free-response items also.

NEW! ***HM Testing***   *HM Testing,* our computerized test generator, is our third source of testing material. The database contains test items and algorithms and is designed to produce an unlimited number of tests for each chapter of the text, including cumulative tests and final exams. These questions are unique to *HM Testing* and do not repeat items provided in the Chapter Tests of the *Instructor's Resource Manual.* It is available for Microsoft Windows® and the Macintosh. Both versions provide **on-line testing** and **gradebook** functions.

HM³ Tutorial (Instructor version)   This state-of-the-art tutorial CD software package for Microsoft Windows® was written and developed by the authors specifically for use with this text. Every objective is supported by the HM³ Tutor; exercises and quizzes are algorithmically generated; solution steps are animated; lessons and exercises are presented in a colorful, lively manner; and an integrated classroom management system tracks student performance.

NEW! ***WebCT ePacks***   *WebCT ePacks* provide instructors with a flexible, Internet-based education platform providing multiple ways in which to present learning materials. The *WebCT ePacks* come with a full array of features to enrich the on-line learning experience.

NEW! ***Blackboard Cartridges***   The *Houghton Mifflin Blackboard cartridge* allows flexible, efficient, and creative ways to present learning materials and opportunities. In addition to course-management benefits, instructors may make use of an electronic grade book, receive papers from students enrolled in the course via the Internet, and track the students' use of the communication and collaboration functions.

NEW! *HMClassPrep*   These CD-ROMs contain a multitude of text-specific resources for instructors to use to enhance the classroom experience. The resources, or assets, are available as pdf and/or customizable Microsoft Word® files and include transparency masters and Chapter Tests from the IRM, to name only a few. These resources can be easily accessed from the CD-ROM by chapter or resource type. The CD can also link you to the text's web site.

NEW! *Instructor Text-Specific Web Site*   The resources available on the *Class Prep* CD are also available on the instructor web site at math.college.hmco.com/instructors. Appropriate items will be password-protected. Instructors also have access to the student part of the text's web site.

Supplements for the Student

Student Solutions Manual   The *Student Solutions Manual* contains complete solutions to all odd-numbered exercises in the text.

Math Study Skills Workbook *by Paul D. Nolting*   This workbook is designed to reinforce skills and minimize frustration for students in any math class, lab, or study skills course. It offers a wealth of study tips and sound advice on note taking, time management, and reducing math anxiety. In addition, numerous opportunities for self assessment enable students to track their own progress.

NEW! *HM eduSpace*   eduSpace is a new content delivery system, combining an algorithmic tutorial program, online delivery of course materials, and classroom management functions. The interactive on-line content correlates directly to this text and can be accessed 24 hours a day.

HM³ Tutorial (Student version)   This state-of-the-art tutorial CD software package for Microsoft Windows® was written and developed by the authors specifically for use with this text. HM³ is an interactive tutorial containing lessons, exercises, and quizzes for every section of the text. Lessons are presented in a colorful, lively manner, and solution steps are animated. Exercises and quizzes are algorithmically generated. The lessons provide additional instruction and practice and can be used in several ways: (1) to cover material that was missed because of absence from class; (2) to reinforce instruction on a concept that has not yet been mastered; and (3) to review material in preparation for an examination. Following each lesson are practice exercises for the student to try. These exercises are created by carefully constructed algorithms that enable a student to practice a variety of exercise types. Because the exercises are created algorithmically, each time the student uses the tutorial, the student is presented with different problems. Following the practice exercises is an algorithmically generated quiz. Next to every objective head in the text, the ⊙ serves as a reminder that there is an HM³ tutorial lesson that corresponds to that objective.

NEW! *SMARTHINKING™ Live, On-Line Tutoring*   Houghton Mifflin has partnered with SMARTHINKING to provide an easy-to-use, effective, on-line tutorial service. **Whiteboard Simulations** and **Practice Area** promote real-time visual interaction.

Three levels of service are offered.

- **Text-Specific Tutoring** provides real-time, one-on-one instruction with a specially qualified "e-structor."
- **Questions Any Time** allows students to submit questions to the tutor outside the scheduled hours and receive a reply within 24 hours.

- **Independent Study Resources** connect students to additional educational services, including interactive web sites, diagnostic tests, and Frequently Asked Questions posed to SMARTHINKING e-structors. These services are available 24 hours a day, 7 days a week.

NEW! *Videos* This edition offers brand new text-specific videos, hosted by Dana Mosely, covering all sections of the text. These videos, professionally produced specifically for this text, offer a valuable resource for further instruction and review. Next to every objective head in the text, the [   ] serves as a reminder that the objective is covered in a video lesson.

NEW! *Student Text-Specific Web Site* On-line student resources, such as section quizzes, can be found at this text's web site at math.college.hmco.com/students.

Acknowledgments

The authors would like to thank the people who have reviewed this manuscript and provided many valuable suggestions.

Mary Kay Best, *Coastal Bend College, TX*
Jeremy Carr, *Pensacola Junior College, FL*
Sally Copeland, *Johnson County Community College, KS*
Barry Dayton, *Northeastern Illinois University*
William C. Hoston
Marcella Jones, *Minneapolis Community and Technical College*
Richard A. Langlie, *North Hennepin Community College, MN*
Mike McCoy, *Vernon Regional Junior College, TX*
Lauri Semarne
Susan Sharkey, *Waukesha County Technical College, WI*
Rick Simon, *University of La Verne, CA*
John Wardigo, *Bloomsburg University, PA*

AIM for Success

Welcome to *Intermediate Algebra: An Applied Approach*. As you begin this course, we know two important facts: (1) We want you to succeed. (2) You want to succeed. To do that requires an effort from each of us. For the next few pages, we are going to show you what is required of you to achieve that success and how you can use the features of this text to be successful.

Motivation

One of the most important keys to success is motivation. We can try to motivate you by offering interesting or important ways mathematics can benefit you. But, in the end, the motivation must come from you. On the first day of class, it is easy to be motivated. Eight weeks into the term, it is harder to keep that motivation.

To stay motivated, there must be outcomes from this course that are worth your time, money, and energy.

List some reasons you are taking this course.

_____

_____

_____

Although we hope that one of the reasons you listed was an interest in mathematics, we know that many of you are taking this course because it is required to graduate, it is a prerequisite for a course you must take, or because it is required for your major. Although you may not agree that this course is necessary, it is! If you are motivated to graduate or complete the requirements for your major, then use that motivation to succeed in this course. Do not become distracted from your goal to complete your education!

Commitment

To be successful, you must make a committment to succeed. This means devoting time to math so that you achieve a better understanding of the subject.

List some activities (sports, hobbies, talents such as dance, art, or music) that you enjoy and at which you would like to become better.

ACTIVITY	TIME SPENT	TIME WISHED SPENT

Thinking about these activities, put the number of hours that you spend each week practicing these activities next to the activity. Next to that number, indicate the number of hours per week you would like to spend on these activities.

Whether you listed surfing or sailing, aerobics or restoring cars, or any other activity you enjoy, note how many hours a week you spend doing it. To succeed in math, you must be willing to commit the same amount of time. Success requires some sacrifice.

The "I Can't Do Math" Syndrome

There may be things you cannot do, such as lift a two-ton boulder. You can, however, do math. It is much easier than lifting the two-ton boulder. When you first learned the activities you listed above, you probably could not do them well. With practice, you got better. With practice, you will be better at math. Stay focused, motivated, and committed to success.

It is difficult for us to emphasize how important it is to overcome the "I Can't Do Math Syndrome." If you listen to interviews of very successful atheletes after a particularly bad performance, you will note that they focus on the positive aspect of what they did, not the negative. Sports psychologists encourage athletes to always be positive—to have a "Can Do" attitude. Develop this attitude toward math.

Strategies for Success

Textbook Reconnaissance Right now, do a 15-minute "textbook reconnaissance" of this book. Here's how:

First, read the table of contents. Do it in three minutes or less. Next, look through the entire book, page by page. Move quickly. Scan titles, look at pictures, notice diagrams.

A textbook reconnaissance shows you where a course is going. It gives you the big picture. That's useful because brains work best when going from the general to the specific. Getting the big picture before you start makes details easier to recall and understand later on.

Your textbook reconnaissance will work even better if, as you scan, you look for ideas or topics that are interesting to you. List three facts, topics, or problems that you found interesting during your textbook reconnaissance.

_____

_____

_____

The idea behind this technique is simple: It's easier to work at learning material if you know it's going to be useful to you.

Not all the topics in this book will be "interesting" to you. But that is true of any subject. Surfers find that on some days the waves are better than others, musicians find some music more appealing than other music, computer gamers find some computer games more interesting than others, car enthusiasts find some cars more exciting than others. Some car enthusiasts would rather have a completely restored 1957 Chevrolet than a new Ferrari.

Know the Course Requirements To do your best in this course, you must know exactly what your instructor requires. Course requirements may be stated in a *syllabus*, which is a printed outline of the main topics of the course, or they may be presented orally. When they are listed in a syllabus or on other printed pages, keep them in a safe place. When they are presented orally, make sure to take complete notes. In either case, it is important that you understand them completely and follow them exactly. Be sure you know the answer to each of the following questions.

1. What is your instructor's name?
2. Where is your instructor's office?
3. At what times does your instructor hold office hours?
4. Besides the textbook, what other materials does your instructor require?
5. What is your instructor's attendance policy?
6. If you must be absent from a class meeting, what should you do before returning to class? What should you do when you return to class?

7. What is the instructor's policy regarding collection or grading of homework assignments?

8. What options are available if you are having difficulty with an assignment? Is there a math tutoring center?

9. If there is a math lab at your school, where is it located? What hours is it open?

10. What is the instructor's policy if you miss a quiz?

11. What is the instructor's policy if you miss an exam?

12. Where can you get help when studying for an exam?

Remember: Your instructor wants to see you succeed. If you need help, ask! Do not fall behind. If you are running a race and fall behind by 100 yards, you may be able to catch up but it will require more effort than had you not fallen behind.

> **TAKE NOTE**
>
> Besides time management, there must be realistic ideas of how much time is available. There are very few people who can *successfully* work full-time and go to school full-time. If you work 40 hours a week, take 15 units, spend the recommended study time given at the right, and sleep 8 hours a day, you will use over 80% of the available hours in a week. That leaves less than 20% of the hours in a week for family, friends, eating, recreation, and other activities.

Time Management We know that there are demands on your time. Family, work, friends, and entertainment all compete for your time. We do not want to see you receive poor job evaluations because you are studying math. However, it is also true that we do not want to see you receive poor math test scores because you devoted too much time to work. When several competing and important tasks require your time and energy, the only way to manage the stress of being successful at both is to manage your time efficiently.

Instructors often advise students to spend twice the amount of time outside of class studying as they spend in the classroom. Time management is important if you are to accomplish this goal and succeed in school. The following activity is intended to help you structure your time more efficiently.

List the name of each course you are taking this term, the number of class hours each course meets, and the number of hours you should spend studying each subject outside of class. Then fill in a weekly schedule like the one printed below. Begin by writing in the hours spent in your classes, the hours spent at work (if you have a job), and any other commitments that are not flexible with respect to the time that you do them. Then begin to write down commitments that are more flexible, including hours spent studying. Remember to reserve time for activities such as meals and exercise. You should also schedule free time.

	Monday	Tuesday	Wednesday	Thursday	Friday	Saturday	Sunday
7–8 a.m.							
8–9 a.m.							
9–10 a.m.							
10–11 a.m.							
11–12 p.m.							
12–1 p.m.							
1–2 p.m.							
2–3 p.m.							
3–4 p.m.							
4–5 p.m.							
5–6 p.m.							
6–7 p.m.							
7–8 p.m.							
8–9 p.m.							
9–10 p.m.							
10–11 p.m.							
11–12 a.m.							

We know that many of you must work. If that is the case, realize that working 10 hours a week at a part-time job is equivalent to taking a three-unit class. If you must work, consider letting your education progress at a slower rate to allow you to be successful at both work and school. There is no rule that says you must finish school in a certain time frame.

Schedule Study Time   As we encouraged you to do by filling out the time management form above, schedule a certain time to study. You should think of this time the way you would the time for work or class—that is, reasons for missing study time should be as compelling as reasons for missing work or class. "I just didn't feel like it" is not a good reason to miss your scheduled study time.

Although this may seem like an obvious exercise, list a few reasons you might want to study.

_____

_____

_____

Of course we have no way of knowing the reasons you listed, but from our experience one reason given quite frequently is "To pass the course." There is nothing wrong with that reason. If that is the most important reason for you to study, then use it to stay focused.

One method of keeping to a study schedule is to form a ***study group***. Look for people who are committed to learning, who pay attention in class, and who are punctual. Ask them to join your group. Choose people with similar educational goals but different methods of learning. You can gain insight from seeing the material from a new perspective. Limit groups to four or five people; larger groups are unwieldy.

There are many ways to conduct a study group. Begin with the following suggestions and see what works best for your group.

1. Test each other by asking questions. Each group member might bring two or three sample test questions to each meeting.
2. Practice teaching each other. Many of us who are teachers learned a lot about our subject when we had to explain it to someone else.
3. Compare class notes. You might ask other students about material in your notes that is difficult for you to understand.
4. Brainstorm test questions.
5. Set an agenda for each meeting. Set approximate time limits for each agenda item and determine a quitting time.

And finally, probably the most important aspect of studying is that it should be done in relatively small chunks. If you can only study three hours a week for this course (probably not enough for most people), do it in blocks of one hour on three separate days, preferably after class. Three hours of studying on a Sunday is not as productive as three hours of paced study.

Text Features That Promote Success   There are 12 chapters in this text. Each chapter is divided into sections, and each section is subdivided into learning objectives. Each learning objective is labeled with a letter from A to D.

Preparing for a Chapter   Before you begin a new chapter, you should take some time to review previously learned skills. There are two ways to do this. The first is to complete the **Cumulative Review**, which occurs after every chapter (except Chapter 1). For instance, turn to page 239. The questions in this review are taken from the previous chapters. The answers for all these exercises can be found on page A14. Turn to that page now and locate the answers for the Chapter 4 Cumulative Review. After the answer to the first exercise, which is $-\dfrac{11}{28}$, you will see the objective reference [2.1C]. This means that this question was taken from Chapter 2, Section 1, Objective C. If you missed this question, you should return to that objective and restudy the material.

A second way of preparing for a new chapter is to complete the **Prep Test**. This test focuses on the particular skills that will be required for the new chapter. Turn to page 186 to see a Prep Test. The answers for the Prep Test are the first set of answers in the answer section for a chapter. Turn to page A11 to see the answers for the Chapter 4 Prep Test. Note that an objective reference is given for each question. If you answer a question incorrectly, restudy the objective from which the question was taken.

Before the class meeting in which your professor begins a new section, you should read each objective statement for that section. Next, browse through the objective material, being sure to note each word in bold type. These words indicate important concepts that you must know in order to learn the material. Do not worry about trying to understand all the material. Your professor is there to assist you with that endeavor. The purpose of browsing through the material is so that your brain will be prepared to accept and organize the new information when it is presented to you.

Turn to page 3. Write down the title of the first objective in Section 1.1. Under the title of the objective, write down the words in the objective that are in bold print. It is not necessary for you to understand the meaning of these words. You are in this class to learn their meaning.

_____   _____   _____   _____

_____   _____   _____   _____

_____   _____   _____   _____

_____   _____   _____   _____

_____   _____   _____   _____

Math is Not a Spectator Sport   To learn mathematics you must be an active participant. Listening and watching your professor do mathematics is not enough. Mathematics requires that you interact with the lesson you are studying. If you filled in the blanks above, you were being interactive. There are other ways this textbook has been designed to help you be an active learner.

Annotated Examples   An orange arrow indicates an example with explanatory remarks to the right of the work. Using paper and pencil, you should work along as you go through the example.

$$3x + 2 < -4$$
$$3x < -6$$
$$\frac{3x}{3} < \frac{-6}{3}$$
$$x < -2$$

The solution set is
$\{x\,|\,x < -2\}$.

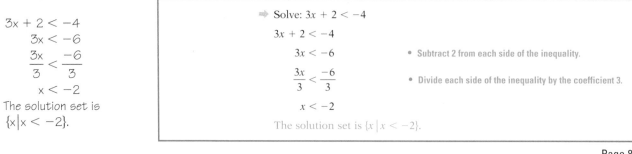

Solve: $3x + 2 < -4$

$$3x + 2 < -4$$
$$3x < -6$$ • Subtract 2 from each side of the inequality.
$$\frac{3x}{3} < \frac{-6}{3}$$ • Divide each side of the inequality by the coefficient 3.
$$x < -2$$

The solution set is $\{x\,|\,x < -2\}$.

When you complete the example, get a clean sheet of paper. Write down the problem and then try to complete the solution without referring to your notes or the book. When you can do that, move on to the next part of the objective.

Leaf through the book now and write down the page numbers of two other occurrences of an arrowed example.

You Try Its   One of the key instructional features of this text is the paired examples. Notice that in each example box, the example on the left is completely worked out and the "You Try It" example on the right is not. Study the worked-out example carefully by working through each step. Then work the You Try It. If you get stuck, refer to the page number at the end of the example, which directs you to the place where the You Try It is solved—a complete worked-out solution is provided. Try to use the given solution to get a hint for the step you are stuck on. Then try to complete your solution.

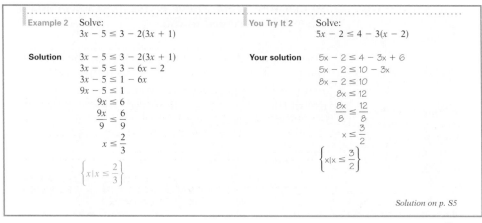

Example 2   Solve:
$$3x - 5 \le 3 - 2(3x + 1)$$

Solution
$$3x - 5 \le 3 - 2(3x + 1)$$
$$3x - 5 \le 3 - 6x - 2$$
$$3x - 5 \le 1 - 6x$$
$$9x - 5 \le 1$$
$$9x \le 6$$
$$\frac{9x}{9} \le \frac{6}{9}$$
$$x \le \frac{2}{3}$$
$$\left\{x\,|\,x \le \frac{2}{3}\right\}$$

You Try It 2   Solve:
$$5x - 2 \le 4 - 3(x - 2)$$

Your solution
$$5x - 2 \le 4 - 3x + 6$$
$$5x - 2 \le 10 - 3x$$
$$8x - 2 \le 10$$
$$8x \le 12$$
$$\frac{8x}{8} \le \frac{12}{8}$$
$$x \le \frac{3}{2}$$
$$\left\{x\,|\,x \le \frac{3}{2}\right\}$$

Solution on p. S5

When you have completed your solution, check your work against the solution we provided. (Turn to page S5 to see the solution of You Try It 2.) Be aware that frequently there is more than one way to solve a problem. Your answer, however, should be the same as the given answer. If you have any question as to whether your method will "always work," check with your instructor or with someone in the math center.

Browse through the textbook and write down the page numbers where two other paired example features occur.

Remember: Be an active participant in your learning process. When you are sitting in class watching and listening to an explanation, you may think that you understand. However, until you actually try to do it, you will have no confirmation of the new knowledge or skill. Most of us have had the experience of sitting in class thinking we knew how to do something only to get home and realize that we didn't.

TAKE NOTE

There is a strong connection between reading and being a successful student in math or any other subject. If you have difficulty reading, consider taking a reading course. Reading is much like other skills. There are certain things you can learn that will make you a better reader.

Word Problems   Word problems are difficult because we must read the problem, determine the quantity we must find, think of a method to do that, and then actually solve the problem. In short, we must formulate a *strategy* to solve the problem and then devise a *solution*.

Note in the paired example below that part of every word problem is a strategy and part is a solution. The strategy is a written description of how we will solve the problem. In the corresponding You Try It, you are asked to formulate a strategy. Do not skip this step, and be sure to write it out.

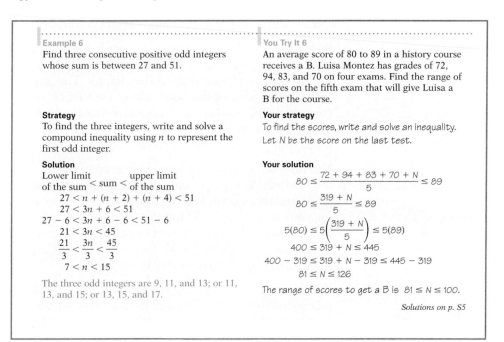

Example 6
Find three consecutive positive odd integers whose sum is between 27 and 51.

Strategy
To find the three integers, write and solve a compound inequality using n to represent the first odd integer.

Solution
$$\text{Lower limit} \atop \text{of the sum} < \text{sum} < {\text{upper limit} \atop \text{of the sum}}$$
$$27 < n + (n + 2) + (n + 4) < 51$$
$$27 < 3n + 6 < 51$$
$$27 - 6 < 3n + 6 - 6 < 51 - 6$$
$$21 < 3n < 45$$
$$\frac{21}{3} < \frac{3n}{3} < \frac{45}{3}$$
$$7 < n < 15$$

The three odd integers are 9, 11, and 13; or 11, 13, and 15; or 13, 15, and 17.

You Try It 6
An average score of 80 to 89 in a history course receives a B. Luisa Montez has grades of 72, 94, 83, and 70 on four exams. Find the range of scores on the fifth exam that will give Luisa a B for the course.

Your strategy
To find the scores, write and solve an inequality. Let N be the score on the last test.

Your solution
$$80 \le \frac{72 + 94 + 83 + 70 + N}{5} \le 89$$
$$80 \le \frac{319 + N}{5} \le 89$$
$$5(80) \le 5\left(\frac{319 + N}{5}\right) \le 5(89)$$
$$400 \le 319 + N \le 445$$
$$400 - 319 \le 319 + N - 319 \le 445 - 319$$
$$81 \le N \le 126$$

The range of scores to get a B is $81 \le N \le 100$.

Solutions on p. S5

Page 84

TAKE NOTE
If a rule has more than one part, be sure to make a notation to that effect.

I can add the same number to both sides of an inequality and not change the solution set.

TAKE NOTE
If you are working at home and need assistance, there is on-line help available at math.college.hmco.com/students, at this text's web site.

Rule Boxes   Pay special attention to rules placed in boxes. These rules give you the reasons certain types of problems are solved the way they are. When you see a rule, try to rewrite the rule in your own words.

When solving an inequality, we use the **Addition and Multiplication Properties of Inequalities** to rewrite the inequality in the form *variable* < *constant* or in the form *variable* > *constant*.

The Addition Property of Inequalities
If $a > b$, then $a + c > b + c$.
If $a < b$, then $a + c < b + c$.

Page 79

Chapter Exercises   When you have completed studying an objective, do the exercises in the exercise set that correspond with that objective. The exercises are labeled with the same letter as the objective. Math is a subject that needs to be learned in small sections and practiced continually in order to be mastered. Doing all of the exercises in each exercise set will help you master the problem-solving techniques necessary for success. As you work through the exercises for an objective, check your answers to the odd-numbered exercises with those in the back of the book.

Preparing for a Test   There are important features of this text that can be used to prepare for a test.

- Chapter Summary
- Chapter Review
- Chapter Test

After completing a chapter, read the Chapter Summary. (See page 234 for the Chapter 4 Summary.) This summary highlights the important topics covered in the chapter. The page number following each topic refers you to the page in the text on which you can find more information about the concept.

Following the Chapter Summary are a Chapter Review (see page 235) and a Chapter Test (see page 237). Doing the review exercises is an important way of testing your understanding of the chapter. The answer to each review exercise is given at the back of the book, along with its objective reference. After checking your answers, restudy any objective from which a question you missed was taken. It may be helpful to retry some of the exercises for that objective to reinforce your problem-solving techniques.

The Chapter Test should be used to prepare for an exam. We suggest that you try the Chapter Test a few days before your actual exam. Take the test in a quiet place and try to complete the test in the same amount of time you will be allowed for your exam. When taking the Chapter Test, practice the strategies of successful test takers: (1) scan the entire test to get a feel for the questions; (2) read the directions carefully; (3) work the problems that are easiest for you first; and perhaps most importantly, (4) try to stay calm.

When you have completed the Chapter Test, check your answers. If you missed a question, review the material in that objective and rework some of the exercises from that objective. This will strengthen your ability to perform the skills in that objective.

Is it difficult to be successful? YES! Successful music groups, artists, professional athletes, chefs, and <u>Write your major here</u> have to work very hard to achieve their goals. They focus on their goals and ignore distractions. The things we ask you to do to achieve success take time and commitment. We are confident that if you follow our suggestions, you will succeed.

Index of Applications

Review of Real Numbers

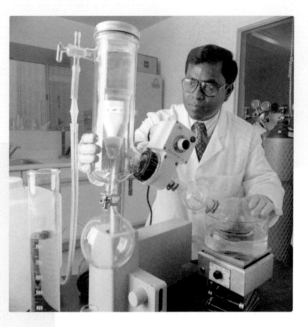

The entomologist in this photo is performing research in an agricultural laboratory. Doing research involves problem solving, as does the job you will hold after you graduate. The basic steps involved in problem solving are outlined in the **Focus on Problem Solving on page 43** and then integrated throughout the text. Your success in the workplace is heavily dependent on your ability to solve problems. It is a skill that improves with study and practice.

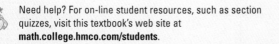 Need help? For on-line student resources, such as section quizzes, visit this textbook's web site at **math.college.hmco.com/students**.

For Exercises 1 to 8, add, subtract, multiply, or divide.

1. $\dfrac{5}{12} + \dfrac{7}{30}$

2. $\dfrac{8}{15} - \dfrac{7}{20}$

3. $\dfrac{5}{6} \cdot \dfrac{4}{15}$

4. $\dfrac{4}{15} \div \dfrac{2}{5}$

5. $8 + 29.34 + 7.065$

6. $92 - 18.37$

7. $2.19(3.4)$

8. $32.436 \div 0.6$

9. Which of the following numbers are greater than -8?
   a. -6 **b.** -10 **c.** 0 **d.** 8

10. Match the fraction with its decimal equivalent.
   a. $\dfrac{1}{2}$ **A.** 0.75

   b. $\dfrac{7}{10}$ **B.** 0.89

   c. $\dfrac{3}{4}$ **C.** 0.5

   d. $\dfrac{89}{100}$ **D.** 0.7

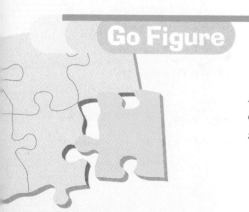

Go Figure

A pair of perpendicular lines are drawn through the interior of a rectangle, dividing it into four smaller rectangles. The area of the smaller rectangles are x, 2, 3, and 6. Find the possible values of x.

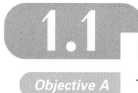 Introduction to Real Numbers

Objective A   **To use inequality and absolute value symbols with real numbers**

Point of Interest

The Big Dipper, known to the Greeks as *Ursa Major*, the great bear, is a constellation that can be seen from northern latitudes. The stars of the Big Dipper are Alkaid, Mizar, Alioth, Megrez, Phecda, Merak, and Dubhe. The star at the bend of the handle, Mizar, is actually two stars, Mizar and Alcor. An imaginary line from Merak through Dubhe passes through Polaris, the north star.

Point of Interest

The concept of zero developed very gradually over many centuries. It has been variously denoted by leaving a blank space, a dot, and finally as 0. Negative numbers, although evident in Chinese manuscripts dating from 200 B.C., were not fully integrated into mathematics until late in the 14th century.

It seems to be a human characteristic to put similar items in the same place. For instance, an astronomer places stars in *constellations* and a geologist divides the history of Earth into *eras*.

Mathematicians likewise place objects with similar properties in *sets*. A **set** is a collection of objects. The objects are called **elements** of the set. Sets are denoted by placing braces around the elements in the set.

The numbers that we use to count things, such as the number of books in a library or the number of CDs sold by a record store, have similar characteristics. These numbers are called the *natural numbers.*

> **Natural numbers** = {1, 2, 3, 4, 5, 6, 7, 8, 9, 10, 11,…}

Each natural number greater than 1 is a *prime* number or a *composite* number. A **prime number** is a natural number greater than 1 that is divisible (evenly) only by itself and 1. For example, 2, 3, 5, 7, 11, and 13 are the first six prime numbers. A natural number that is not a prime number is a **composite number.** The numbers 4, 6, 8, and 9 are the first four composite numbers.

The natural numbers do not have a symbol to denote the concept of none, for instance, the number of trees taller than 1000 feet. The *whole numbers* include zero and the natural numbers.

> **Whole numbers** = {0, 1, 2, 3, 4, 5, 6, 7, 8,…}

The whole numbers alone do not provide all the numbers that are useful in applications. For instance, a meteorologist needs numbers below zero and above zero.

> **Integers** = {…,−5, −4, −3, −2, −1, 0, 1, 2, 3, 4, 5,…}

The integers …,−5, −4, −3, −2, −1 are **negative integers.** The integers 1, 2, 3, 4, 5,… are **positive integers.** Note that the natural numbers and the positive integers are the same set of numbers. The integer zero is neither a positive nor a negative integer.

Still other numbers are necessary to solve the variety of application problems that exist. For instance, a landscape architect may need to purchase irrigation pipe that has a diameter of $\frac{5}{8}$ in. The numbers that include fractions are called *rational numbers.*

> **Rational numbers** = $\left\{ \dfrac{p}{q}, \text{ where } p \text{ and } q \text{ are integers and } q \neq 0 \right\}$

Examples of rational numbers include $\frac{2}{3}$, $-\frac{9}{2}$, and $\frac{5}{1}$. Note that $\frac{5}{1} = 5$, so all integers are rational numbers. The number $\frac{4}{\pi}$ is not a rational number because π is not an integer.

A rational number written as a fraction can be written in decimal notation by dividing the numerator by the denominator.

⟹ Write $\frac{3}{8}$ as a decimal.

Divide 3 by 8.

$$
\begin{array}{r}
0.375 \\
8\overline{)3.000} \\
-2\,4 \\
\hline
60 \\
-56 \\
\hline
40 \\
-40 \\
\hline
0
\end{array}
$$

← This is a terminating decimal.

← The remainder is zero.

$\frac{3}{8} = 0.375$

⟹ Write $\frac{2}{15}$ as a decimal.

Divide 2 by 15.

$$
\begin{array}{r}
0.133 \\
15\overline{)2.000} \\
-1\,5 \\
\hline
50 \\
-45 \\
\hline
50 \\
-45 \\
\hline
5
\end{array}
$$

← This is a repeating decimal.

← The remainder is never zero.

$\frac{2}{15} = 0.1\overline{3}$

• The bar over 3 indicates this digit repeats.

Some numbers cannot be written as terminating or repeating decimals, for example, $0.01001000100001\ldots$, $\sqrt{7} \approx 2.6457513$, and $\pi \approx 3.1415927$. These numbers have decimal representations that neither terminate nor repeat. They are called **irrational numbers.** The rational numbers and the irrational numbers taken together are the *real numbers*.

Real numbers = {rational numbers and irrational numbers}

The relationship among sets of numbers is shown in the figure below.

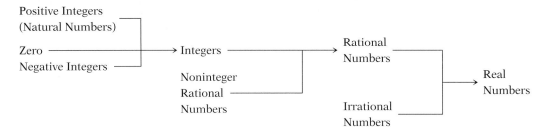

The **graph of a real number** is made by placing a heavy dot on a number line directly above the number. The graphs of some real numbers are shown below.

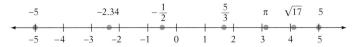

Consider the following sentences:

A restaurant's chef prepared a dinner and served *it* to the customers.
A maple tree was planted and *it* grew two feet in one year.

In the first sentence, "it" means dinner; in the second sentence, "it" means tree. In language, this word can stand for many different objects. Similarly, in mathematics, a letter of the alphabet can be used to stand for some number. A letter used in this way is called a **variable**.

It is convenient to use a variable to represent or stand for any one of the elements of a set. For instance, the statement "*x* is an *element of* the set {0, 2, 4, 6}" means that *x* can be replaced by 0, 2, 4, or 6. The set {0, 2, 4, 6} is called the **domain** of the variable.

The symbol for "is an element of" is ∈; the symbol for "is not an element of" is ∉. For example,

$$2 \in \{0, 2, 4, 6\} \qquad 6 \in \{0, 2, 4, 6\} \qquad 7 \notin \{0, 2, 4, 6\}$$

Variables are used in the next definition.

> **Definition of Inequality Symbols**
>
> If a and b are two real numbers and a is to the left of b on the number line, then a is **less than** b. This is written $a < b$.
>
> If a and b are two real numbers and a is to the right of b on the number line, then a is **greater than** b. This is written $a > b$.

Here are some examples.

$$5 < 9 \qquad -4 > -10 \qquad \pi < \sqrt{17} \qquad 0 > -\frac{2}{3}$$

The inequality symbols ≤ (is less than or equal to) and ≥ (is greater than or equal to) are also important. Note the examples below.

$$4 \le 5 \quad \text{is a true statement because } 4 < 5.$$
$$5 \le 5 \quad \text{is a true statement because } 5 = 5.$$

The numbers 5 and −5 are the same distance from zero but on opposite sides of zero. The numbers 5 and −5 are called **additive inverses** or **opposites** of each other.

The additive inverse (or opposite) of 5 is −5. The additive inverse of −5 is 5. The symbol for additive inverse is −.

−(2) means the additive inverse of *positive* 2. $-(2) = -2$

−(−5) means the additive inverse of *negative* 5. $-(-5) = 5$

The **absolute value** of a number is its distance from zero on the number line. The symbol for absolute value is | |.

Note from the figure above that the distance from 0 to 5 is 5. Therefore, $|5| = 5$. That figure also shows that the distance from 0 to −5 is 5. Therefore, $|-5| = 5$.

> **Absolute Value**
>
> The absolute value of a positive number is the number itself.
> The absolute value of a negative number is the opposite of the negative number. The absolute value of zero is zero.

⇒ Evaluate: $-|-12|$

$$-|-12| = -12$$

• The absolute value sign does not affect the negative sign *in front of* the absolute value sign.

Example 1

Let $y \in \{-7, 0, 6\}$. For which values of y is the inequality $y < 4$ a true statement?

Solution

Replace y by each of the elements of the set and determine whether the inequality is true.

$$y < 4$$
$$-7 < 4 \quad \text{True}$$
$$0 < 4 \quad \text{True}$$
$$6 < 4 \quad \text{False}$$

The inequality is true for -7 and 0.

You Try It 1

Let $z \in \{-10, -5, 6\}$. For which values of z is the inequality $z > -5$ a true statement?

Your solution

Example 2

Let $y \in \{-12, 0, 4\}$.
a. Determine $-y$, the additive inverse of y, for each element of the set.
b. Evaluate $|y|$ for each element of the set.

Solution

a. Replace y in $-y$ by each element of the set and determine the value of the expression.

$$-y$$
$$-(-12) = 12$$
$$-(0) = 0 \quad \bullet \text{ 0 is neither positive nor negative.}$$
$$-(4) = -4$$

b. Replace y in $|y|$ by each element of the set and determine the value of the expression.

$$|y|$$
$$|-12| = 12$$
$$|0| = 0$$
$$|4| = 4$$

You Try It 2

Let $d \in \{-11, 0, 8\}$.
a. Determine $-d$, the additive inverse of d, for each element of the set.
b. Evaluate $|d|$ for each element of the set.

Your solution

Solutions on p. S1

Objective B   **To write sets using the roster method and set-builder notation**

The **roster method** of writing a set encloses the list of the elements of the set in braces. The set of whole numbers, written $\{0, 1, 2, 3, 4, \ldots\}$, and the set of natural numbers, written $\{1, 2, 3, 4, \ldots\}$, are **infinite sets.** The pattern of numbers continues without end. It is impossible to list all the elements of an infinite set.

The set of even natural numbers less than 10 is written {2, 4, 6, 8}. This is an example of a **finite set;** all the elements of the set can be listed.

The set that contains no elements is called the **empty set,** or **null set,** and is symbolized by ∅ or { }. The set of trees over 1000 feet tall is the empty set.

⇒ Use the roster method to write the set of whole numbers less than 5.

{0, 1, 2, 3, 4}   • **Recall that the whole numbers include 0.**

A second method of representing a set is **set-builder notation.** Set-builder notation can be used to describe almost any set, but it is especially useful when writing infinite sets. In set-builder notation, the set of integers greater than −3 is written

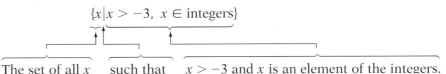

The set of all x such that $x > -3$ and x is an element of the integers.

This is an infinite set. It is impossible to list all the elements of the set, but the set can be described using set-builder notation.

The set of real numbers less than 5 is written

$$\{x|x < 5, x \in \text{real numbers}\}$$

and is read "the set of all x such that x is less than 5 and x is an element of the real numbers."

⇒ Use set-builder notation to write the set of integers greater than 20.

$\{x|x > 20, x \in \text{integers}\}$

Example 3

Use the roster method to write the set of positive integers less than or equal to 7.

Solution

{1, 2, 3, 4, 5, 6, 7}

You Try It 3

Use the roster method to write the set of negative integers greater than −6.

Your solution

Example 4

Use set-builder notation to write the set of integers less than −9.

Solution

$\{x|x < -9, x \in \text{integers}\}$

You Try It 4

Use set-builder notation to write the set of whole numbers greater than or equal to 15.

Your solution

Solutions on p. S1

Objective C

To perform operations on sets and write sets in interval notation

Just as operations such as addition and multiplication are performed on real numbers, operations are performed on sets. Two operations performed on sets are *union* and *intersection*.

Union of Two Sets

The **union** of two sets, written $A \cup B$, is the set of all elements that belong to either set A **or** set B. In set-builder notation, this is written

$$A \cup B = \{x \mid x \in A \quad \text{or} \quad x \in B\}$$

Given $A = \{2, 3, 4\}$ and $B = \{0, 1, 2, 3\}$, the union of A and B contains all the elements that belong to either A or B. The elements that belong to both sets are listed only once.

$$A \cup B = \{0, 1, 2, 3, 4\}$$

Intersection of Two Sets

The **intersection** of two sets, written $A \cap B$, is the set of all elements that are common to both set A **and** set B. In set-builder notation, this is written

$$A \cap B = \{x \mid x \in A \quad \text{and} \quad x \in B\}$$

Point of Interest

The symbols $\in$, $\cup$, and $\cap$ were first used by Giuseppe Peano in *Arithmetices Principia, Nova Exposita* (The Principle of Mathematics, a New Method of Exposition), published in 1889. The purpose of this book was to deduce the principles of mathematics from pure logic.

Given $A = \{2, 3, 4\}$ and $B = \{0, 1, 2, 3\}$, the intersection of A and B contains all the elements that are common to both A and B.

$$A \cap B = \{2, 3\}$$

➡ Given $A = \{2, 3, 5, 7\}$ and $B = \{0, 1, 2, 3, 4\}$, find $A \cup B$ and $A \cap B$.

$A \cup B = \{0, 1, 2, 3, 4, 5, 7\}$ • List the elements of each set. The elements that belong to both sets are listed only once.

$A \cap B = \{2, 3\}$ • List the elements that are common to both A and B.

Set-builder notation and the inequality symbols $>$, $<$, $\geq$, and $\leq$ are used to describe infinite sets of real numbers. These sets can also be graphed on the real number line.

The graph of $\{x|x > -2, x \in \text{real numbers}\}$ is shown below. The set is the real numbers greater than -2. The parenthesis on the graph indicates that -2 is not included in the set.

The graph of $\{x|x \geq -2, x \in \text{real numbers}\}$ is shown below. The set is the real numbers greater than or equal to -2. The bracket at -2 indicates that -2 is included in the set.

In many cases, we will assume that real numbers are being used and omit "$x \in \text{real numbers}$" from set-builder notation. For instance, the above set is written $\{x|x \geq -2\}$.

➡ Graph $\{x|x \leq 3\}$.

• Draw a bracket at 3 to indicate that 3 is in the set. Draw a solid line to the left of 3.

The union of two sets is the set of all elements belonging to either one or the other of the two sets. The set $\{x|x \leq -1\} \cup \{x|x > 3\}$ is the set of real numbers that are either less than or equal to -1 or greater than 3.

The set is written $\{x|x \leq -1 \text{ or } x > 3\}$.

The set $\{x|x > 2\} \cup \{x|x > 4\}$ is the set of real numbers that are either greater than 2 or greater than 4.

The set is written $\{x|x > 2\}$.

➡ Graph $\{x|x > 1\} \cup \{x|x < -4\}$.

• The graph includes all the numbers that are either greater than 1 or less than -4.

The intersection of two sets is the set that contains the elements common to both sets. The set $\{x|x > -2\} \cap \{x|x < 5\}$ is the set of real numbers that are greater than -2 and less than 5.

The set can be written $\{x|x > -2 \text{ and } x < 5\}$. However, it is more commonly written $\{x|-2 < x < 5\}$, which is read "the set of all x such that x is greater than -2 and less than 5."

The set $\{x|x < 4\} \cap \{x|x < 5\}$ is the set of real numbers that are less than 4 and less than 5.

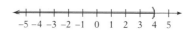

The set is written $\{x|x < 4\}$.

➡ Graph $\{x|x < 0\} \cap \{x|x > -3\}$.

● The set is $\{x|-3 < x < 0\}$.

Some sets can also be expressed using **interval notation.** For example, the interval notation $(-3, 2]$ indicates the interval of all real numbers greater than -3 and less than or equal to 2. As on the graph of a set, the left parenthesis indicates that -3 is not included in the set. The right bracket indicates that 2 is included in the set.

An interval is said to be **closed** if it includes both endpoints; it is **open** if it does not include either endpoint. An interval is **half-open** if one endpoint is included and the other is not. In each example given below, -3 and 2 are the **endpoints** of the interval. In each case, the set notation, the interval notation, and the graph of the set are shown.

$\{x|-3 < x < 2\}$ $(-3, 2)$
                       Open interval

$\{x|-3 \le x \le 2\}$ $[-3, 2]$
                       Closed interval

$\{x|-3 \le x < 2\}$ $[-3, 2)$
                       Half-open interval

$\{x|-3 < x \le 2\}$ $(-3, 2]$
                       Half-open interval

To indicate an interval that extends forever in one or both directions using interval notation, we use the **infinity symbol** ∞ or the **negative infinity symbol** $-\infty$. The infinity symbol is not a number; it is simply a notation to indicate that the interval is unlimited. In interval notation, a parenthesis is always used to the right of an infinity symbol or to the left of a negative infinity symbol, as shown in the following examples.

$\{x|x > 1\}$ $(1, \infty)$

$\{x|x \ge 1\}$ $[1, \infty)$

$\{x|x < 1\}$ $(-\infty, 1)$

$\{x|x \le 1\}$ $(-\infty, 1]$

$\{x|-\infty < x < \infty\}$ $(-\infty, \infty)$

⇒ Graph $(-2, 5)$.

• The graph is all numbers that are greater than -2 and less than 5.

Example 5

Given $A = \{0, 2, 4, 6, 8, 10\}$ and $B = \{0, 3, 6, 9\}$, find $A \cup B$.

Solution

$A \cup B = \{0, 2, 3, 4, 6, 8, 9, 10\}$

You Try It 5

Given $C = \{1, 5, 9, 13, 17\}$ and $D = \{3, 5, 7, 9, 11\}$, find $C \cup D$.

Your solution

Example 6

Given $A = \{x | x \in \text{natural numbers}\}$ and $B = \{x | x \in \text{negative integers}\}$, find $A \cap B$.

Solution

$A \cap B = \varnothing$

• There are no natural numbers that are also negative integers.

You Try It 6

Given $E = \{x | x \in \text{odd integers}\}$ and $F = \{x | x \in \text{even integers}\}$, find $E \cap F$.

Your solution

Example 7

Graph $\{x | x \geq 3\}$.

Solution

Draw a bracket at 3 to indicate that 3 is in the set. Draw a solid line to the right of 3.

You Try It 7

Graph $\{x | x < 0\}$.

Your solution

Example 8

Graph $\{x | x > -1\} \cup \{x | x < 2\}$.

Solution

This is the set of real numbers greater than -1 *or* less than 2. Any real number satisfies this condition. The graph is the entire real number line.

You Try It 8

Graph $\{x | x < -2\} \cup \{x | x > -1\}$.

Your solution

Solutions on p. S1

Example 9

Graph $\{x|x < 3\} \cap \{x|x \geq -1\}$.

Solution

This is the set of real numbers greater than or equal to -1 *and* less than 3.

Example 10

Write each set in interval notation.

a. $\{x|x > 3\}$

b. $\{x|-2 < x \leq 4\}$

Solution

a. The set $\{x|x > 3\}$ is the numbers greater than 3. In interval notation, this is written $(3, \infty)$.

b. The set $\{x|-2 < x \leq 4\}$ is the numbers greater than -2 and less than or equal to 4. In interval notation, this is written $(-2, 4]$.

Example 11

Write each set in set-builder notation.

a. $(-\infty, 4]$

b. $[-3, 0]$

Solution

a. $(-\infty, 4]$ is the numbers less than or equal to 4. In set-builder notation, this is written $\{x|x \leq 4\}$.

b. $[-3, 0]$ is the numbers greater than or equal to -3 and less than or equal to 0. In set-builder notation, this is written $\{x|-3 \leq x \leq 0\}$.

Example 12

Graph $(-2, 2]$.

Solution

Draw a parenthesis at -2 to show it is not in the set. Draw a bracket at 2 to show it is in the set. Draw a solid line between -2 and 2.

You Try It 9

Graph $\{x|x < 1\} \cap \{x|x > -3\}$.

Your solution

You Try It 10

Write each set in interval notation.

a. $\{x|x < -1\}$

b. $\{x|-2 \leq x < 4\}$

Your solution

You Try It 11

Write each set in set-builder notation.

a. $(3, \infty)$

b. $(-4, 1]$

Your solution

You Try It 12

Graph $[-2, \infty)$.

Your solution

Solutions on p. S1

1.1 Exercises

Objective A

Determine which of the numbers are **a.** integers, **b.** rational numbers, **c.** irrational numbers, **d.** real numbers. List all that apply.

1. $-\dfrac{15}{2}, 0, -3, \pi, 2.\overline{33}, 4.232232223\ldots, \dfrac{\sqrt{5}}{4}, \sqrt{7}$

2. $-17, 0.3412, \dfrac{3}{\pi}, -1.010010001\ldots, \dfrac{27}{91}, 6.1\overline{2}$

Find the additive inverse of each of the following.

3. 27

4. -3

5. $\dfrac{3}{4}$

6. $\sqrt{17}$

7. 0

8. $-\pi$

9. $-\sqrt{33}$

10. -1.23

11. -91

12. $-\dfrac{2}{3}$

Solve.

13. Let $x \in \{-3, 0, 7\}$. For which values of x is $x < 5$ true?

14. Let $z \in \{-4, -1, 4\}$. For which values of z is $z > -2$ true?

15. Let $y \in \{-6, -4, 7\}$. For which values of y is $y > -4$ true?

16. Let $x \in \{-6, -3, 3\}$. For which values of x is $x < -3$ true?

17. Let $w \in \{-2, -1, 0, 1\}$. For which values of w is $w \le -1$ true?

18. Let $p \in \{-10, -5, 0, 5\}$. For which values of p is $p \ge 0$ true?

19. Let $b \in \{-9, 0, 9\}$. Evaluate $-b$ for each element of the set.

20. Let $a \in \{-3, -2, 0\}$. Evaluate $-a$ for each element of the set.

21. Let $c \in \{-4, 0, 4\}$. Evaluate $|c|$ for each element of the set.

22. Let $q \in \{-3, 0, 7\}$. Evaluate $|q|$ for each element of the set.

23. Let $m \in \{-6, -2, 0, 1, 4\}$. Evaluate $-|m|$ for each element of the set.

24. Let $x \in \{-5, -3, 0, 2, 5\}$. Evaluate $-|x|$ for each element of the set.

Objective B

Use the roster method to write the set.

25. the integers between −3 and 5

26. the integers between −4 and 0

27. the even natural numbers less than 14

28. the odd natural numbers less than 14

29. the positive-integer multiples of 3 that are less than or equal to 30

30. the negative-integer multiples of 4 that are greater than or equal to −20

31. the negative-integer multiples of 5 that are greater than or equal to −35

32. the positive-integer multiples of 6 that are less than or equal to 36

Use set-builder notation to write the set.

33. the integers greater than 4

34. the integers less than −2

35. the real numbers greater than or equal to −2

36. the real numbers less than or equal to 2

37. the real numbers between 0 and 1

38. the real numbers between −2 and 5

39. the real numbers between 1 and 4, inclusive

40. the real numbers between 0 and 2, inclusive

Objective C

Find $A \cup B$.

41. $A = \{1, 4, 9\}, B = \{2, 4, 6\}$

42. $A = \{-1, 0, 1\}, B = \{0, 1, 2\}$

43. $A = \{2, 3, 5, 8\}, B = \{9, 10\}$

44. $A = \{1, 3, 5, 7\}, B = \{2, 4, 6, 8\}$

45. $A = \{-4, -2, 0, 2, 4\}, B = \{0, 4, 8\}$

46. $A = \{-3, -2, -1\}, B = \{-2, -1, 0, 1\}$

47. $A = \{1, 2, 3, 4, 5\}, B = \{3, 4, 5\}$

48. $A = \{2, 4\}, B = \{0, 1, 2, 3, 4, 5\}$

Find $A \cap B$.

49. $A = \{6, 12, 18\}, B = \{3, 6, 9\}$

50. $A = \{-4, 0, 4\}, B = \{-2, 0, 2\}$

51. $A = \{1, 5, 10, 20\}, B = \{5, 10, 15, 20\}$

52. $A = \{1, 3, 5, 7, 9\}, B = \{1, 9\}$

53. $A = \{1, 2, 4, 8\}, B = \{3, 5, 6, 7\}$

54. $A = \{-3, -2, -1, 0\}, B = \{1, 2, 3, 4\}$

55. $A = \{2, 4, 6, 8, 10\}, B = \{4, 6\}$

56. $A = \{-9, -5, 0, 7\}, B = \{-7, -5, 0, 5, 7\}$

Graph.

57. $\{x \mid x < 2\}$

58. $\{x \mid x < -1\}$

59. $\{x \mid x \geq 1\}$

60. $\{x \mid x \leq -2\}$

61. $\{x \mid -1 < x < 5\}$

62. $\{x \mid 1 < x < 3\}$

63. $\{x \mid 0 \leq x \leq 3\}$

64. $\{x \mid -1 \leq x \leq 1\}$

65. $\{x \mid x > 1\} \cup \{x \mid x < -1\}$

66. $\{x \mid x \leq 2\} \cup \{x \mid x > 4\}$

67. $\{x \mid x \leq 2\} \cap \{x \mid x \geq 0\}$

68. $\{x \mid x > -1\} \cap \{x \mid x \leq 4\}$

69. $\{x \mid x > 1\} \cap \{x \mid x \geq -2\}$

70. $\{x \mid x < 4\} \cap \{x \mid x \leq 0\}$

71. $\{x \mid x > 2\} \cup \{x \mid x > 1\}$

72. $\{x \mid x < -2\} \cup \{x \mid x < -4\}$

Write each interval in set-builder notation.

73. $(0, 8)$ **74.** $(-2, 4)$ **75.** $[-5, 7]$ **76.** $[3, 4]$ **77.** $[-3, 6)$

78. $(4, 5]$ **79.** $(-\infty, 4]$ **80.** $(-\infty, -2)$ **81.** $(5, \infty)$ **82.** $[-2, \infty)$

Write each set of real numbers in interval notation.

83. $\{x \mid -2 < x < 4\}$ **84.** $\{x \mid 0 < x < 3\}$ **85.** $\{x \mid -1 \le x \le 5\}$ **86.** $\{x \mid 0 \le x \le 3\}$

87. $\{x \mid -\infty < x < 1\}$ **88.** $\{x \mid -\infty < x \le 6\}$ **89.** $\{x \mid -2 \le x < \infty\}$ **90.** $\{x \mid 3 \le x < \infty\}$

Graph.

91. $(-2, 5)$

92. $(0, 3)$

93. $[-1, 2]$

94. $[-3, 2]$

95. $(-\infty, 3]$

96. $(-\infty, -1)$

97. $[3, \infty)$

98. $[-2, \infty)$

APPLYING THE CONCEPTS

Let $R = \{\text{real numbers}\}$, $A = \{x \mid -1 \le x \le 1\}$, $B = \{x \mid 0 \le x \le 1\}$, $C = \{x \mid -1 \le x \le 0\}$, and $\varnothing$ be the empty set. Answer the following using R, A, B, C, or $\varnothing$.

99. $A \cup B$ **100.** $A \cup A$ **101.** $B \cap B$ **102.** $A \cup C$ **103.** $A \cap R$

104. $C \cap R$ **105.** $B \cup R$ **106.** $A \cup R$ **107.** $R \cup R$ **108.** $R \cap \varnothing$

109. The set $B \cap C$ cannot be expressed using R, A, B, C, or $\varnothing$. What real number is represented by $B \cap C$?

110. A student wrote $-3 > x > 5$ as the inequality that represents the real numbers less than -3 or greater than 5. Explain why this is incorrect.

1.2 Operations on Rational Numbers

Objective A   **To add, subtract, multiply, and divide integers**

An understanding of the operations on integers is necessary to succeed in algebra. Let's review those properties, beginning with the sign rules for addition.

> **Rules for Addition of Real Numbers**
>
> **To add numbers with the same sign,** add the absolute values of the numbers. Then attach the sign of the addends.
>
> **To add numbers with different signs,** find the absolute value of each number. Subtract the smaller of the two numbers from the larger. Then attach the sign of the number with the larger absolute value.

⇒ Add: **a.** $-65 + (-48)$ **b.** $27 + (-53)$

a. $-65 + (-48) = -113$

- The signs are the same. Add the absolute values of the numbers. Then attach the sign of the addends.

b. $27 + (-53)$
$|27| = 27$ $|-53| = 53$
$53 - 27 = 26$

$27 + (-53) = -26$

- The signs are different. Find the absolute value of each number.
- Subtract the smaller number from the larger.
- Because $|-53| > |27|$, attach the sign of -53.

Subtraction is defined as addition of the additive inverse.

> **Rule for Subtraction of Real Numbers**
>
> If a and b are real numbers, then $a - b = a + (-b)$.

⇒ Subtract: **a.** $48 - (-22)$ **b.** $-31 - 18$

Change to +

Change to +

a. $48 - (-22) = 48 + 22 = 70$ **b.** $-31 - 18 = -31 + (-18) = -49$

Opposite of -22

Opposite of 18

⇒ Simplify: $-3 - (-16) + (-12)$

$-3 - (-16) + (-12) = -3 + 16 + (-12)$

- Write subtraction as addition of the opposite.

$= 13 + (-12) = 1$

- Add from left to right.

The sign rules for multiplying real numbers are given below.

Sign Rules for Multiplication of Real Numbers

The product of two numbers with the same sign is positive.

The product of two numbers with different signs is negative.

➡ Multiply: **a.** $-4(-9)$ **b.** $84(-4)$

a. $-4(-9) = 36$ • The product of two numbers with
                                       the same sign is positive.

b. $84(-4) = -336$ • The product of two numbers with
                                       different signs is negative.

The **multiplicative inverse** of a nonzero real number a is $\frac{1}{a}$. This number is also called the **reciprocal** of a. For instance, the reciprocal of 2 is $\frac{1}{2}$ and the reciprocal of $-\frac{3}{4}$ is $-\frac{4}{3}$. Division of real numbers is defined in terms of multiplication by the multiplicative inverse.

Rule for Division of Real Numbers

If a and b are real numbers and $b \neq 0$, then $a \div b = a \cdot \frac{1}{b}$.

Because division is defined in terms of multiplication, the sign rules for dividing real numbers are the same as the sign rules for multiplying.

➡ Divide: **a.** $\dfrac{-54}{9}$ **b.** $(-21) \div (-7)$ **c.** $-\dfrac{-63}{-9}$

a. $\dfrac{-54}{9} = -6$ • The quotient of two numbers with
                                              different signs is negative.

b. $(-21) \div (-7) = 3$ • The quotient of two numbers with
                                              the same sign is positive.

c. $-\dfrac{-63}{-9} = -7$

Note that $\dfrac{-12}{3} = -4$, $\dfrac{12}{-3} = -4$, and $-\dfrac{12}{3} = -4$. This suggests the following rule.

If a and b are real numbers and $b \neq 0$, then $\dfrac{-a}{b} = \dfrac{a}{-b} = -\dfrac{a}{b}$.

Properties of Zero and One in Division

• Zero divided by any number other than zero is zero.

$\dfrac{0}{a} = 0, \ a \neq 0$

- Division by zero is not defined.

  To understand that division by zero is not permitted, suppose $\frac{4}{0} = n$, where n is some number. Because each division problem has a related multiplication problem, $\frac{4}{0} = n$ means $n \cdot 0 = 4$. But $n \cdot 0 = 4$ is impossible because any number times 0 is 0. Therefore, division by 0 is not defined.
- Any number other than zero divided by itself is 1.

  $\frac{a}{a} = 1, a \neq 0$
- Any number divided by 1 is the number.

  $\frac{a}{1} = a$

Example 1 Simplify: $-3 - (-5) - 9$

Solution $-3 - (-5) - 9 = -3 + 5 + (-9)$
$= 2 + (-9) = -7$

You Try It 1 Simplify: $6 - (-8) - 10$

Your solution

Example 2 Simplify: $-6|-5|(-15)$

Solution $-6|-5|(-15) = -6(5)(-15)$
$= -30(-15) = 450$

You Try It 2 Simplify: $12(-3)|-6|$

Your solution

Example 3 Simplify: $-\frac{36}{-3}$

Solution $-\frac{36}{-3} = -(-12) = 12$

You Try It 3 Simplify: $-\frac{28}{|-14|}$

Your solution

Solutions on p. S1

Objective B   **To add, subtract, multiply, and divide rational numbers**

Recall that a rational number is one that can be written in the form $\frac{p}{q}$, where p and q are integers and $q \neq 0$. Examples of rational numbers are $-\frac{5}{9}$ and $\frac{12}{5}$. The number $\frac{9}{\sqrt{7}}$ is not a rational number because $\sqrt{7}$ is not an integer. All integers are rational numbers. Terminating and repeating decimals are also rational numbers.

➡ Add: $-12.34 + 9.059$

$12.340 - 9.059 = 3.281$
- The signs are different. Subtract the absolute values of the numbers.

$-12.34 + 9.059 = -3.281$
- Attach the sign of the number with the larger absolute value.

➡ Multiply: $(-0.23)(0.04)$

$(-0.23)(0.04) = -0.0092$

- The signs are different. The product is negative.

➡ Divide: $(-4.0764) \div (-1.72)$

$(-4.0764) \div (-1.72) = 2.37$

- The signs are the same. The quotient is positive.

To add or subtract rational numbers written as fractions, first rewrite the fractions as equivalent fractions with a common denominator. The common denominator is the **least common multiple** (LCM) of the denominators.

➡ Add: $\dfrac{5}{6} + \left(-\dfrac{7}{8}\right)$

$$\dfrac{5}{6} + \left(-\dfrac{7}{8}\right) = \dfrac{5}{6} \cdot \dfrac{4}{4} + \left(\dfrac{-7}{8} \cdot \dfrac{3}{3}\right)$$

$$= \dfrac{20}{24} + \left(\dfrac{-21}{24}\right) = \dfrac{20 + (-21)}{24}$$

$$= \dfrac{-1}{24} = -\dfrac{1}{24}$$

- The common denominator is 24. Write each fraction in terms of the common denominator.
- Add the numerators. Place the sum over the common denominator.

> **TAKE NOTE**
> Although the sum could have been left as $\dfrac{-1}{24}$, all answers in this text are written with the negative sign in front of the fraction.

➡ Subtract: $-\dfrac{7}{15} - \left(-\dfrac{1}{6}\right)$

$$-\dfrac{7}{15} - \left(-\dfrac{1}{6}\right) = -\dfrac{7}{15} + \dfrac{1}{6}$$

$$= \dfrac{-7}{15} \cdot \dfrac{2}{2} + \dfrac{1}{6} \cdot \dfrac{5}{5} = \dfrac{-14}{30} + \dfrac{5}{30}$$

$$= \dfrac{-9}{30} = -\dfrac{3}{10}$$

- Write subtraction as addition of the opposite.
- Write each fraction in terms of the common denominator, 30.
- Add the numerators. Write the answer in simplest form.

The product of two fractions is the product of the numerators over the product of the denominators.

➡ Multiply: $-\dfrac{5}{12} \cdot \dfrac{8}{15}$

$$-\dfrac{5}{12} \cdot \dfrac{8}{15} = -\dfrac{5 \cdot 8}{12 \cdot 15} = -\dfrac{40}{180}$$

$$= -\dfrac{20 \cdot 2}{20 \cdot 9} = -\dfrac{2}{9}$$

- The signs are different. The product is negative. Multiply the numerators and multiply the denominators.
- Write the answer in simplest form.

In the last problem, the fraction was written in simplest form by dividing the numerator and denominator by 20, which is the largest integer that divides evenly into both 40 and 180. The number 20 is called the **greatest common factor** (GCF) of 40 and 180. To write a fraction in simplest form, divide the numerator and denominator by the GCF. If you have difficulty finding the GCF, try finding the prime factorization of the numerator and denominator and then divide by the common prime factors. For instance,

$$-\dfrac{5}{12} \cdot \dfrac{8}{15} = -\dfrac{\overset{1}{\cancel{5}} \cdot (\overset{1}{\cancel{2}} \cdot \overset{1}{\cancel{2}} \cdot 2)}{(\underset{1}{\cancel{2}} \cdot \underset{1}{\cancel{2}} \cdot 3) \cdot (3 \cdot \underset{1}{\cancel{5}})} = -\dfrac{2}{9}$$

Division of Fractions

To divide two fractions, multiply by the reciprocal of the divisor.

$$\frac{a}{b} \div \frac{c}{d} = \frac{a}{b} \cdot \frac{d}{c}$$

➡ Divide: $\dfrac{3}{8} \div \dfrac{9}{16}$

$$\frac{3}{8} \div \frac{9}{16} = \frac{3}{8} \cdot \frac{16}{9} = \frac{48}{72} = \frac{2}{3}$$

• Multiply by the reciprocal of the divisor. Write the answer in simplest form.

Example 4

Simplify: $-\dfrac{3}{8} - \dfrac{5}{12} - \left(-\dfrac{9}{16}\right)$

Solution

$$-\frac{3}{8} - \frac{5}{12} - \left(-\frac{9}{16}\right) = \frac{-3}{8} + \frac{-5}{12} + \frac{9}{16}$$

$$= \frac{-3}{8} \cdot \frac{6}{6} + \frac{-5}{12} \cdot \frac{4}{4} + \frac{9}{16} \cdot \frac{3}{3}$$

$$= \frac{-18 + (-20) + 27}{48}$$

$$= \frac{-11}{48} = -\frac{11}{48}$$

You Try It 4

Simplify: $\dfrac{5}{6} - \dfrac{3}{8} - \dfrac{7}{9}$

Your solution

Example 5

Simplify: $\left(-\dfrac{5}{12}\right)\left(\dfrac{4}{25}\right)$

Solution $\left(-\dfrac{5}{12}\right)\left(\dfrac{4}{25}\right) = \dfrac{-5 \cdot 4}{12 \cdot 25} = -\dfrac{1}{15}$

You Try It 5

Simplify: $\dfrac{5}{8} \div \left(-\dfrac{15}{40}\right)$

Your solution

Solutions on pp. S1–S2

Objective C   **To evaluate exponential expressions**   ◀ 1 ▶ ◉

Repeated multiplication of the same factor can be written using an exponent.

$2 \cdot 2 \cdot 2 \cdot 2 \cdot 2 \cdot 2 = 2^6 \leftarrow$ **Exponent** $b \cdot b \cdot b \cdot b \cdot b = b^5 \leftarrow$ **Exponent**

　　　　　　　↑——**Base**   ↑——**Base**

The **exponent** indicates how many times the factor, called the **base,** occurs in the multiplication. The multiplication $2 \cdot 2 \cdot 2 \cdot 2 \cdot 2 \cdot 2$ is in **factored form.** The exponential expression 2^6 is in **exponential form.**

2^1 is read "the first power of two" or just "two."
2^2 is read "the second power of two" or "two squared."
2^3 is read "the third power of two" or "two cubed."
2^4 is read "the fourth power of two."
2^5 is read "the fifth power of two."
b^5 is read "the fifth power of b."

• Usually the exponent 1 is not written.

> **nth Power of a**
>
> If a is a real number and n is a positive integer, the nth power of a is the product of n factors of a.
>
> $$a^n = \underbrace{a \cdot a \cdot a \cdot \cdots \cdot a}_{a \text{ as a factor } n \text{ times}}$$

TAKE NOTE

Examine the results of $(-3)^4$ and -3^4 very carefully. As another example, $(-2)^6 = 64$ but $-2^6 = -64$.

$5^3 = 5 \cdot 5 \cdot 5 = 125$

$(-3)^4 = (-3)(-3)(-3)(-3) = 81$

$-3^4 = -(3)^4 = -(3 \cdot 3 \cdot 3 \cdot 3) = -81$

Note the difference between $(-3)^4$ and -3^4. The placement of the parentheses is very important.

Example 6   Evaluate $(-2)^5$ and -5^2.

Solution

$(-2)^5 = (-2)(-2)(-2)(-2)(-2)$
$= -32$
$-5^2 = -(5 \cdot 5) = -25$

You Try It 6   Evaluate -2^5 and $(-5)^2$.

Your solution

Example 7

Evaluate $\left(-\dfrac{3}{4}\right)^3$.

Solution

$\left(-\dfrac{3}{4}\right)^3 = \left(-\dfrac{3}{4}\right)\left(-\dfrac{3}{4}\right)\left(-\dfrac{3}{4}\right)$

$= -\dfrac{3 \cdot 3 \cdot 3}{4 \cdot 4 \cdot 4} = -\dfrac{27}{64}$

You Try It 7

Evaluate $\left(\dfrac{2}{3}\right)^4$.

Your solution

Solutions on p. S2

Objective D   **To use the Order of Operations Agreement**   1

Suppose we wish to evaluate $16 + 4 \cdot 2$. There are two operations, addition and multiplication. The operations could be performed in different orders.

Multiply first. $16 + \underbrace{4 \cdot 2}$ Add first. $\underbrace{16 + 4} \cdot 2$

Then add. $\underbrace{16 + 8}$ Then multiply. $\underbrace{20 \cdot 2}$

           24   40

Note that the answers are different. To avoid possibly getting more than one answer to the same problem, an *Order of Operations Agreement* is followed.

> **Order of Operations Agreement**
>
> **Step 1** Perform operations inside grouping symbols. Grouping symbols include parentheses ( ), brackets [ ], braces { }, the absolute value symbol, and the fraction bar.
> **Step 2** Simplify exponential expressions.
> **Step 3** Do multiplication and division as they occur from left to right.
> **Step 4** Do addition and subtraction as they occur from left to right.

⇨ Simplify: $8 - \dfrac{2 - 22}{4 + 1} \cdot 2^2$

$$8 - \frac{2 - 22}{4 + 1} \cdot 2^2 = 8 - \frac{-20}{5} \cdot 2^2$$

$$= 8 - \frac{-20}{5} \cdot 4$$

$$= 8 - (-4) \cdot 4$$

$$= 8 - (-16)$$

$$= 24$$

- The fraction bar is a grouping symbol. Perform the operations above and below the fraction bar.
- Simplify exponential expressions.

- Do multiplication and division as they occur from left to right.

- Do addition and subtraction as they occur from left to right.

One or more of the steps in the Order of Operations Agreement may not be needed. In that case, just proceed to the next step.

⇨ Simplify: $14 - [(25 - 9) \div 2]^2$

$$14 - [(25 - 9) \div 2]^2 = 14 - [16 \div 2]^2$$

$$= 14 - [8]^2$$

$$= 14 - 64$$

$$= -50$$

- Perform the operations inside grouping symbols.

- Simplify exponential expressions.

- Do addition and subtraction as they occur from left to right.

A **complex fraction** is a fraction whose numerator or denominator contains one or more fractions. Examples of complex fractions are shown at the right.

$$\dfrac{\dfrac{2}{3}}{-\dfrac{5}{2}}, \quad \dfrac{\dfrac{1}{7} + 5}{\dfrac{3}{4} - \dfrac{7}{8}} \quad \leftarrow \text{Main fraction bar}$$

When simplifying complex fractions, recall that $\dfrac{\dfrac{a}{b}}{\dfrac{c}{d}} = \dfrac{a}{b} \div \dfrac{c}{d} = \dfrac{a}{b} \cdot \dfrac{d}{c}$.

⇨ Simplify: $\dfrac{\dfrac{3}{4} - \dfrac{1}{3}}{\dfrac{1}{5} - 2}$

$$\frac{\dfrac{3}{4} - \dfrac{1}{3}}{\dfrac{1}{5} - 2} = \frac{\dfrac{9}{12} - \dfrac{4}{12}}{\dfrac{1}{5} - \dfrac{10}{5}} = \frac{\dfrac{5}{12}}{-\dfrac{9}{5}}$$

$$= \frac{5}{12}\left(-\frac{5}{9}\right)$$

$$= -\frac{25}{108}$$

- Perform the operations above and below the main fraction bar.

- Multiply the numerator of the complex fraction by the reciprocal of the denominator of the complex fraction.

Example 8

Simplify: $-3 \cdot 4^2 + [2 - 5(8 \div 2)]$

Solution

$$-3 \cdot 4^2 + [2 - 5(8 \div 2)]$$
$$= -3 \cdot 4^2 + [2 - 5(4)]$$
$$= -3 \cdot 4^2 + [2 - 20]$$
$$= -3 \cdot 4^2 + [-18]$$
$$= -3 \cdot 16 + [-18]$$
$$= -48 + [-18]$$
$$= -66$$

You Try It 8

Simplify: $-2\left[3 - 3^3 + \dfrac{16 + 2}{1 - 10}\right]$

Your solution

Example 9

Simplify: $\left(\dfrac{1}{2}\right)^3 - \left[\left(\dfrac{2}{3} + \dfrac{1}{4}\right) \div \dfrac{5}{6}\right]$

Solution

$$\left(\dfrac{1}{2}\right)^3 - \left[\left(\dfrac{2}{3} + \dfrac{1}{4}\right) \div \dfrac{5}{6}\right] = \left(\dfrac{1}{2}\right)^3 - \left[\dfrac{11}{12} \div \dfrac{5}{6}\right]$$
$$= \left(\dfrac{1}{2}\right)^3 - \left[\dfrac{11}{12} \cdot \dfrac{6}{5}\right]$$
$$= \left(\dfrac{1}{2}\right)^3 - \dfrac{11}{10}$$
$$= \dfrac{1}{8} - \dfrac{11}{10}$$
$$= -\dfrac{39}{40}$$

You Try It 9

Simplify: $\dfrac{1}{3} + \dfrac{5}{8} \div \dfrac{15}{16} - \dfrac{7}{12}$

Your solution

Example 10

Simplify: $9 \cdot \dfrac{\dfrac{5}{6} - 2}{\dfrac{3}{8}} \div \dfrac{7}{6}$

Solution

$$9 \cdot \dfrac{\dfrac{5}{6} - 2}{\dfrac{3}{8}} \div \dfrac{7}{6} = 9 \cdot \dfrac{-\dfrac{7}{6}}{\dfrac{3}{8}} \div \dfrac{7}{6}$$
$$= 9 \cdot \left(-\dfrac{7}{6} \cdot \dfrac{8}{3}\right) \div \dfrac{7}{6}$$
$$= 9 \cdot \left(-\dfrac{28}{9}\right) \div \dfrac{7}{6}$$
$$= -28 \div \dfrac{7}{6}$$
$$= -28 \cdot \dfrac{6}{7}$$
$$= -24$$

You Try It 10

Simplify: $\dfrac{11}{12} - \dfrac{\dfrac{5}{4}}{2 - \dfrac{7}{2}} \cdot \dfrac{3}{4}$

Your solution

Solutions on p. S2

1.2 Exercises

Objective A

1. Explain how to add **a.** two integers with the same sign and **b.** two integers with different signs.

2. Explain how to rewrite $8 - (-12)$ as addition of the opposite.

Simplify.

3. $-18 + (-12)$

4. $-18 - 7$

5. $5 - 22$

6. $16 \cdot (-60)$

7. $3 \cdot 4 \cdot (-8)$

8. $18 \cdot 0 \cdot (-7)$

9. $18 \div (-3)$

10. $25 \div (-5)$

11. $-60 \div (-12)$

12. $(-9)(-2)(-3)(10)$

13. $-20(35)(-16)$

14. $54(19)(-82)$

15. $-271(-365)$

16. $|(-16)(10)|$

17. $|12(-8)|$

18. $|7 - 18|$

19. $|15 - (-8)|$

20. $|-16 - (-20)|$

21. $|-56 \div 8|$

22. $|81 \div (-9)|$

23. $|-153 \div (-9)|$

24. $|-4| - |-2|$

25. $-|-8| + |-4|$

26. $-|-16| - |24|$

27. $-30 + (-16) - 14 - 2$

28. $3 - (-2) + (-8) - 11$

29. $-2 + (-19) - 16 + 12$

30. $-6 + (-9) - 18 + 32$

31. $13 - |6 - 12|$

32. $-9 - |-7 - (-15)|$

33. $738 - 46 + (-105) - 219$

34. $-871 - (-387) - 132 - 46$

35. $-442 \div (-17)$

36. $621 \div (-23)$

37. $-4897 \div 59$

38. $-17|-5|$

39. Describe **a.** the least common multiple of two numbers and **b.** the greatest common factor of two numbers.

40. Explain how to divide two fractions.

Simplify.

41. $\dfrac{7}{12} + \dfrac{5}{16}$

42. $\dfrac{3}{8} - \dfrac{5}{12}$

43. $-\dfrac{5}{9} - \dfrac{14}{15}$

44. $\dfrac{1}{2} + \dfrac{1}{7} - \dfrac{5}{8}$

45. $-\dfrac{1}{3} + \dfrac{5}{9} - \dfrac{7}{12}$

46. $\dfrac{1}{3} + \dfrac{19}{24} - \dfrac{7}{8}$

47. $\dfrac{2}{3} - \dfrac{5}{12} + \dfrac{5}{24}$

48. $-\dfrac{7}{10} + \dfrac{4}{5} + \dfrac{5}{6}$

49. $\dfrac{5}{8} - \dfrac{7}{12} + \dfrac{1}{2}$

50. $-\dfrac{1}{3} \cdot \dfrac{5}{8}$

51. $\left(\dfrac{6}{35}\right)\left(-\dfrac{5}{16}\right)$

52. $\dfrac{2}{3}\left(-\dfrac{9}{20}\right) \cdot \dfrac{5}{12}$

53. $-\dfrac{8}{15} \div \dfrac{4}{5}$

54. $-\dfrac{2}{3} \div \left(-\dfrac{6}{7}\right)$

55. $-\dfrac{11}{24} \div \dfrac{7}{12}$

56. $\dfrac{7}{9} \div \left(-\dfrac{14}{27}\right)$

57. $\left(-\dfrac{5}{12}\right)\left(\dfrac{4}{35}\right)\left(\dfrac{7}{8}\right)$

58. $\dfrac{6}{35}\left(-\dfrac{7}{40}\right)\left(-\dfrac{8}{21}\right)$

59. $-14.27 + 1.296$

60. $-0.4355 + 172.5$

61. $1.832 - 7.84$

62. $(3.52)(4.7)$

63. $(0.03)(10.5)(6.1)$

64. $(1.2)(3.1)(-6.4)$

65. $5.418 \div (-0.9)$

66. $-0.2645 \div (-0.023)$

67. $-0.4355 \div 0.065$

68. $-6.58 - 3.97 + 0.875$

Simplify. Round to the nearest hundredth.

69. $38.241 \div [-(-6.027)] - 7.453$

70. $-9.0508 - (-3.177) + 24.77$

71. $-287.3069 \div 0.1415$

72. $6472.3018 \div (-3.59)$

Objective C

Simplify.

73. 5^3

74. 3^4

75. -2^3

76. -4^3

77. $(-5)^3$

78. $(-8)^2$

79. $2^2 \cdot 3^4$

80. $4^2 \cdot 3^3$

81. $-2^2 \cdot 3^2$

82. $-3^2 \cdot 5^3$

83. $(-2)^3(-3)^2$

84. $(-4)^3(-2)^3$

85. $4 \cdot 2^3 \cdot 3^3$

86. $-4(-3)^2(4^2)$

87. $2^2 \cdot (-10)(-2)^2$

88. $-3(-2)^2(-5)$

89. $\left(-\dfrac{2}{3}\right)^2 \cdot 3^3$

90. $-\left(\dfrac{2}{5}\right)^3 \cdot 5^2$

91. $2^5 \cdot (-3)^4 \cdot 4^5$

92. $4^4(-3)^5(6)^2$

Objective D

93. Why do we need an Order of Operations Agreement?

94. Describe each step in the Order of Operations Agreement.

Simplify.

95. $5 - 3(8 \div 4)^2$

96. $4^2 - (5 - 2)^2 \cdot 3$

97. $16 - \dfrac{2^2 - 5}{3^2 + 2}$

98. $\dfrac{4(5 - 2)}{4^2 - 2^2} \div 4$

99. $\dfrac{3 + \dfrac{2}{3}}{\dfrac{11}{16}}$

100. $\dfrac{\dfrac{11}{14}}{4 - \dfrac{6}{7}}$

101. $5[(2 - 4) \cdot 3 - 2]$

102. $2[(16 \div 8) - (-2)] + 4$

103. $16 - 4\left(\dfrac{8 - 2}{3 - 6}\right) \div \dfrac{1}{2}$

104. $25 \div 5\left(\dfrac{16 + 8}{-2^2 + 8}\right) - 5$

105. $6[3 - (-4 + 2) \div 2]$

106. $12 - 4[2 - (-3 + 5) - 8]$

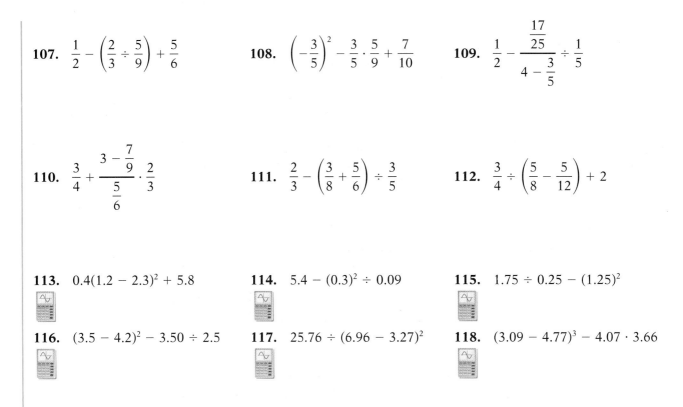

107. $\dfrac{1}{2} - \left(\dfrac{2}{3} \div \dfrac{5}{9}\right) + \dfrac{5}{6}$

108. $\left(-\dfrac{3}{5}\right)^2 - \dfrac{3}{5} \cdot \dfrac{5}{9} + \dfrac{7}{10}$

109. $\dfrac{1}{2} - \dfrac{\frac{17}{25}}{4 - \frac{3}{5}} \div \dfrac{1}{5}$

110. $\dfrac{3}{4} + \dfrac{3 - \frac{7}{9}}{\frac{5}{6}} \cdot \dfrac{2}{3}$

111. $\dfrac{2}{3} - \left(\dfrac{3}{8} + \dfrac{5}{6}\right) \div \dfrac{3}{5}$

112. $\dfrac{3}{4} \div \left(\dfrac{5}{8} - \dfrac{5}{12}\right) + 2$

113. $0.4(1.2 - 2.3)^2 + 5.8$

114. $5.4 - (0.3)^2 \div 0.09$

115. $1.75 \div 0.25 - (1.25)^2$

116. $(3.5 - 4.2)^2 - 3.50 \div 2.5$

117. $25.76 \div (6.96 - 3.27)^2$

118. $(3.09 - 4.77)^3 - 4.07 \cdot 3.66$

APPLYING THE CONCEPTS

119. A number that is its own additive inverse is _____ .

120. Which two numbers are their own multiplicative inverse?

121. Do all real numbers have a multiplicative inverse? If not, which ones do not have a multiplicative inverse?

122. What is the tens digit of 11^{22}?

123. What is the ones digit of 7^{18}?

124. What are the last two digits of 5^{33}?

125. What are the last three digits of 5^{234}?

126. Does $(2^3)^4 = 2^{(3^4)}$? If not, which expression is larger?

127. What is the Order of Operations Agreement for a^{b^c}? *Note:* Even calculators that normally follow the Order of Operations Agreement may not do so for this expression.

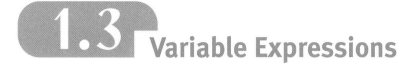

Variable Expressions

Objective A   **To use and identify the properties of the real numbers**

The properties of the real numbers describe the way operations on numbers can be performed. Following is a list of some of the real-number properties and an example of each property.

PROPERTIES OF THE REAL NUMBERS

The Commutative Property of Addition

$$a + b = b + a$$

$3 + 2 = 2 + 3$
$5 = 5$

The Commutative Property of Multiplication

$$a \cdot b = b \cdot a$$

$(3)(-2) = (-2)(3)$
$-6 = -6$

The Associative Property of Addition

$$(a + b) + c = a + (b + c)$$

$(3 + 4) + 5 = 3 + (4 + 5)$
$7 + 5 = 3 + 9$
$12 = 12$

The Associative Property of Multiplication

$$(a \cdot b) \cdot c = a \cdot (b \cdot c)$$

$(3 \cdot 4) \cdot 5 = 3 \cdot (4 \cdot 5)$
$12 \cdot 5 = 3 \cdot 20$
$60 = 60$

The Addition Property of Zero

$$a + 0 = 0 + a = a$$

$3 + 0 = 0 + 3 = 3$

The Multiplication Property of Zero

$$a \cdot 0 = 0 \cdot a = 0$$

$3 \cdot 0 = 0 \cdot 3 = 0$

> **The Multiplication Property of One**
> $$a \cdot 1 = 1 \cdot a = a$$

$$5 \cdot 1 = 1 \cdot 5 = 5$$

> **The Inverse Property of Addition**
> $$a + (-a) = (-a) + a = 0$$

$$4 + (-4) = (-4) + 4 = 0$$

$-a$ is called the **additive inverse** of a. a is the additive inverse of $-a$. The sum of a number and its additive inverse is 0.

> **The Inverse Property of Multiplication**
> $$a \cdot \frac{1}{a} = \frac{1}{a} \cdot a = 1, \quad a \neq 0$$

$$(4)\left(\frac{1}{4}\right) = \left(\frac{1}{4}\right)(4) = 1$$

$\frac{1}{a}$ is called the **multiplicative inverse** of a. $\frac{1}{a}$ is also called the **reciprocal** of a. The product of a number and its multiplicative inverse is 1.

> **The Distributive Property**
> $$a(b + c) = ab + ac$$

$$3(4 + 5) = 3 \cdot 4 + 3 \cdot 5$$
$$3 \cdot 9 = 12 + 15$$
$$27 = 27$$

Example 1

Complete the statement by using the Commutative Property of Multiplication.

$$(x)\left(\frac{1}{4}\right) = (?)(x)$$

Solution

$$(x)\left(\frac{1}{4}\right) = \left(\frac{1}{4}\right)(x)$$

You Try It 1

Complete the statement by using the Inverse Property of Addition.

$$3x + ? = 0$$

Your solution

Example 2

Identify the property that justifies the statement: $3(x + 4) = 3x + 12$

Solution

The Distributive Property

You Try It 2

Identify the property that justifies the statement: $(a + 3b) + c = a + (3b + c)$

Your solution

Solutions on p. S2

Objective B   **To evaluate a variable expression**

An expression that contains one or more variables is a **variable expression.** The variable expression $6x^2y - 7x - z + 2$ contains four **terms:** $6x^2y$, $-7x$, $-z$, and 2. The first three terms are **variable terms.** The 2 is a **constant term.**

Each variable term is composed of a **numerical coefficient** and a **variable part.**

Variable Term	Numerical Coefficient	Variable Part
$6x^2y$	6	x^2y
$-7x$	-7	x
$-z$	-1	z

- When the coefficient is 1 or -1, the 1 is usually not written.

⇒ Evaluate $a^2 - (a - b^2c)$ when $a = -2$, $b = 3$, and $c = -4$.

Replace each variable with its value. Then use the Order of Operations Agreement to simplify the resulting numerical expression.

See the Appendix: Guidelines for Using Graphing Calculators for instructions on using a graphing calculator to evaluate variable expressions.

$$a^2 - (a - b^2c)$$

$$(-2)^2 - [(-2) - 3^2(-4)] = (-2)^2 - [(-2) - 9(-4)] \quad • \ a = -2, b = 3, c = -4$$

$$= (-2)^2 - [(-2) - (-36)]$$

$$= (-2)^2 - [34]$$

$$= 4 - 34$$

$$= -30$$

. . . .

Example 3

Evaluate $2x^3 - 4(2y - 3z)$ when $x = 2$, $y = 3$, and $z = -2$.

Solution

$2x^3 - 4(2y - 3z)$

$2(2)^3 - 4[2(3) - 3(-2)] = 2(2)^3 - 4[6 - (-6)]$

$\qquad\qquad\qquad\qquad\quad = 2(2)^3 - 4[12]$

$\qquad\qquad\qquad\qquad\quad = 2(8) - 4[12]$

$\qquad\qquad\qquad\qquad\quad = 16 - 48 = -32$

You Try It 3

Evaluate $-2x^2 + 3(4xy - z)$ when $x = -3$, $y = -1$, and $z = 2$.

Your solution

. . . .

Example 4

Evaluate $3 - 2|3x - 2y^2|$ when $x = -1$ and $y = -2$.

Solution

$3 - 2|3x - 2y^2|$

$3 - 2|3(-1) - 2(-2)^2| = 3 - 2|3(-1) - 2(4)|$

$\qquad\qquad\qquad\qquad\quad = 3 - 2|(-3) - 8|$

$\qquad\qquad\qquad\qquad\quad = 3 - 2|-11| = 3 - 2(11)$

$\qquad\qquad\qquad\qquad\quad = 3 - 22 = -19$

You Try It 4

Evaluate $2x - y|4x^2 - y^2|$ when $x = -2$ and $y = -6$.

Your solution

Solutions on p. S2

Objective C **To simplify a variable expression** ◀ 1 ▶

Like terms of a variable expression are the terms with the same variable part.

Constant terms are like terms.

$$\overbrace{4x \quad \underbrace{- \ 5 \quad + \quad 7x^2 \quad + \ 3x}_{\text{like terms}} \quad - \ 9}^{\text{like terms}}$$

To simplify a variable expression, combine the like terms by using the Distributive Property. For instance,

$$7x + 4x = (7 + 4)x = 11x$$

Adding the coefficients of like terms is called **combining like terms.**

The Distributive Property is also used to remove parentheses from a variable expression so that like terms can be combined.

➡ Simplify: $3(x + 2y) + 2(4x - 3y)$

$3(x + 2y) + 2(4x - 3y)$

$= 3x + 6y + 8x - 6y$ • Use the Distributive Property.

$= (3x + 8x) + (6y - 6y)$ • Use the Commutative and Associative Properties of Addition to rearrange and group like terms.

$= 11x$ • Combine like terms.

➡ Simplify: $2 - 4[3x - 2(5x - 3)]$

$2 - 4[3x - 2(5x - 3)] = 2 - 4[3x - 10x + 6]$ • Use the Distributive Property to remove the inner parentheses.

$= 2 - 4[-7x + 6]$ • Combine like terms.

$= 2 + 28x - 24$ • Use the Distributive Property to remove the brackets.

$= 28x - 22$ • Combine like terms.

Example 5
Simplify: $7 + 3(4x - 7)$

Solution
$7 + 3(4x - 7)$
$= 7 + 12x - 21$ • The Order of Operations Agreement requires multiplication before addition.
$= 12x - 14$

Example 6
Simplify: $5 - 2(3x + y) - (x - 4)$

Solution
$5 - 2(3x + y) - (x - 4)$
$= 5 - 6x - 2y - x + 4$
$= -7x - 2y + 9$

You Try It 5
Simplify: $9 - 2(3y - 4) + 5y$

Your solution

You Try It 6
Simplify: $6z - 3(5 - 3z) + 5(2z - 3)$

Your solution

Solutions on p. S2

1.3 Exercises

Objective A

Use the given property of the real numbers to complete the statement.

1. The Commutative Property of Multiplication
$3 \cdot 4 = 4 \cdot ?$

2. The Commutative Property of Addition
$7 + 15 = ? + 7$

3. The Associative Property of Addition
$(3 + 4) + 5 = ? + (4 + 5)$

4. The Associative Property of Multiplication
$(3 \cdot 4) \cdot 5 = 3 \cdot (? \cdot 5)$

5. The Division Properties of Zero
$\dfrac{5}{?}$ is undefined.

6. The Multiplication Property of Zero
$5 \cdot ? = 0$

7. The Distributive Property
$3(x + 2) = 3x + ?$

8. The Distributive Property
$5(y + 4) = ? \cdot y + 20$

9. The Division Properties of Zero
$\dfrac{?}{-6} = 0$

10. The Inverse Property of Addition
$(x + y) + ? = 0$

11. The Inverse Property of Multiplication
$\dfrac{1}{mn} (mn) = ?$

12. The Multiplication Property of One
$? \cdot 1 = x$

13. The Associative Property of Multiplication
$2(3x) = ? \cdot x$

14. The Commutative Property of Addition
$ab + bc = bc + ?$

Identify the property that justifies the statement.

15. $\dfrac{0}{-5} = 0$

16. $-8 + 8 = 0$

17. $(-12)\left(-\dfrac{1}{12}\right) = 1$

18. $(3 \cdot 4) \cdot 2 = 2 \cdot (3 \cdot 4)$

19. $y + 0 = y$

20. $2x + (5y + 8) = (2x + 5y) + 8$

21. $\dfrac{-9}{0}$ is undefined.

22. $(x + y)z = xz + yz$

23. $6(x + y) = 6x + 6y$

24. $(-12y)(0) = 0$

25. $(ab)c = a(bc)$

26. $(x + y) + z = (y + x) + z$

Objective B

Evaluate the variable expression when $a = 2$, $b = 3$, $c = -1$, and $d = -4$.

27. $ab + dc$

28. $2ab - 3dc$

29. $4cd \div a^2$

30. $b^2 - (d - c)^2$

31. $(b - 2a)^2 + c$

32. $(b - d)^2 \div (b - d)$

33. $(bc + a)^2 \div (d - b)$

34. $\dfrac{1}{3}b^3 - \dfrac{1}{4}d^3$

35. $\dfrac{1}{4}a^4 - \dfrac{1}{6}bc$

36. $2b^2 \div \dfrac{ad}{2}$

37. $\dfrac{3ac}{-4} - c^2$

38. $\dfrac{2d - 2a}{2bc}$

39. $\dfrac{3b - 5c}{3a - c}$

40. $\dfrac{2d - a}{b - 2c}$

41. $\dfrac{a - d}{b + c}$

42. $|a^2 + d|$

43. $-a|a + 2d|$

44. $d|b - 2d|$

45. $\dfrac{2a - 4d}{3b - c}$

46. $\dfrac{3d - b}{b - 2c}$

47. $-3d \div \left|\dfrac{ab - 4c}{2b + c}\right|$

48. $-2bc + \left|\dfrac{bc + d}{ab - c}\right|$

49. $2(d - b) \div (3a - c)$

50. $(d - 4a)^2 \div c^3$

51. $-d^2 - c^3 a$

52. $a^2 c - d^3$

53. $-d^3 + 4ac$

54. b^a

55. $4^{(a^2)}$

56. a^b

Objective C

Simplify.

57. $5x + 7x$

58. $3x + 10x$

59. $-8ab - 5ab$

60. $-2x + 5x - 7x$

61. $3x - 5x + 9x$

62. $-2a + 7b + 9a$

63. $5b - 8a - 12b$

64. $12\left(\dfrac{1}{12}x\right)$

65. $\dfrac{1}{3}(3y)$

66. $-3(x - 2)$

67. $-5(x - 9)$

68. $(x + 2)5$

69. $-(x + y)$

70. $-(-x - y)$

71. $3(a - 5)$

72. $3(x - 2y) - 5$

73. $4x - 3(2y - 5)$

74. $-2a - 3(3a - 7)$

75. $3x - 2(5x - 7)$

76. $2x - 3(x - 2y)$

77. $3[a - 5(5 - 3a)]$

78. $5[-2 - 6(a - 5)]$

79. $3[x - 2(x + 2y)]$

80. $5[y - 3(y - 2x)]$

81. $-2(x - 3y) + 2(3y - 5x)$

82. $4(-a - 2b) - 2(3a - 5b)$

83. $5(3a - 2b) - 3(-6a + 5b)$

84. $-7(2a - b) + 2(-3b + a)$

85. $3x - 2[y - 2(x + 3[2x + 3y])]$

86. $2x - 4[x - 4(y - 2[5y + 3])]$

87. $4 - 2(7x - 2y) - 3(-2x + 3y)$

88. $3x + 8(x - 4) - 3(2x - y)$

89. $\frac{1}{3}[8x - 2(x - 12) + 3]$

90. $\frac{1}{4}[14x - 3(x - 8) - 7x]$

APPLYING THE CONCEPTS

In each of the following, it is possible that at least one of the properties of real numbers has been applied incorrectly. If the statement is incorrect, state the incorrect application of the properties of real numbers and correct the answer. If the statement is correct, state the property of real numbers that is being used.

91. $4(3y + 1) = 12y + 4$

92. $-4(5x - y) = -20x + 4y$

93. $2 + 3x = (2 + 3)x = 5x$

94. $6 - 6x = 0x = 0$

95. $2(3y) = (2 \cdot 3)(2y) = 12y$

96. $3a - 4b = 4b - 3a$

97. $-x^2 + y^2 = y^2 - x^2$

98. $x^4 \cdot \frac{1}{x^4} = 1, x \neq 0$

1.4 Verbal Expressions and Variable Expressions

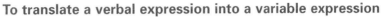

Objective A   **To translate a verbal expression into a variable expression**

One of the major skills required in applied mathematics is translating a verbal expression into a mathematical expression. Doing so requires recognizing the verbal phrases that translate into mathematical operations. Following is a partial list of the verbal phrases used to indicate the different mathematical operations.

Addition	more than	8 more than w	$w + 8$
	added to	x added to 9	$9 + x$
	the sum of	the sum of z and 9	$z + 9$
	the total of	the total of r and s	$r + s$
	increased by	x increased by 7	$x + 7$
Subtraction	less than	12 less than b	$b - 12$
	the difference between	the difference between x and 1	$x - 1$
	minus	z minus 7	$z - 7$
	decreased by	17 decreased by a	$17 - a$
Multiplication	times	negative 2 times c	$-2c$
	the product of	the product of x and y	xy
	multiplied by	3 multiplied by n	$3n$
	of	three-fourths of m	$\dfrac{3}{4}m$
	twice	twice d	$2d$
Division	divided by	v divided by 15	$\dfrac{v}{15}$
	the quotient of	the quotient of y and 3	$\dfrac{y}{3}$
	the ratio of	the ratio of x to 7	$\dfrac{x}{7}$
Power	the square of or the second power of	the square of x	x^2
	the cube of or the third power of	the cube of r	r^3
	the fifth power of	the fifth power of a	a^5

Translating a phrase that contains the word *sum, difference, product,* or *quotient* can sometimes cause a problem. In the examples at the right, note where the operation symbol is placed.

the *sum* of x and y $x + y$

the *difference* between x and y $x - y$

the *product* of x and y $x \cdot y$

the *quotient* of x and y $\dfrac{x}{y}$

⇒ Translate "three times the sum of c and five" into a variable expression.

Identify words that indicate the mathematical operations.	3 <u>times</u> the <u>sum of</u> c and 5
Use the identified words to write the variable expression. Note that the phrase <u>times the sum of</u> requires parentheses.	$3(c + 5)$

⇒ The sum of two numbers is thirty-seven. If x represents the smaller number, translate "twice the larger number" into a variable expression.

Write an expression for the larger number by subtracting the smaller number, x, from 37.	larger number: $37 - x$
Identify the words that indicate the mathematical operations.	<u>twice</u> the larger number
Use the identified words to write a variable expression.	$2(37 - x)$

⇒ Translate "five less than twice the difference between a number and seven" into a variable expression. Then simplify.

Identify words that indicate the mathematical operations.	the unknown number: x <u>5 less than</u> <u>twice</u> the <u>difference between</u> x and 7
Use the identified words to write the variable expression.	$2(x - 7) - 5$
Simplify the expression.	$2x - 14 - 5 = 2x - 19$

Example 1

Translate "the quotient of r and the sum of r and four" into a variable expression.

Solution

the quotient of r and the sum of r and four

$$\frac{r}{r + 4}$$

You Try It 1

Translate "twice x divided by the difference between x and seven" into a variable expression.

Your solution

Example 2

Translate "the sum of the square of y and six" into a variable expression.

Solution

the sum of the square of y and six

$y^2 + 6$

You Try It 2

Translate "the product of negative three and the square of d" into a variable expression.

Your solution

Example 3

The sum of two numbers is twenty-eight. Using x to represent the smaller number, translate "the sum of three times the larger number and the smaller number" into a variable expression. Then simplify.

Solution

The smaller number is x.
The larger number is $28 - x$.
the sum of three times the larger number and the smaller number

$3(28 - x) + x$ • **This is the variable expression.**
$84 - 3x + x$ • **Simplify.**
$84 - 2x$

You Try It 3

The sum of two numbers is sixteen. Using x to represent the smaller number, translate "the difference between twice the smaller number and the larger number" into a variable expression. Then simplify.

Your solution

Example 4

Translate "eight more than the product of four and the total of a number and twelve" into a variable expression. Then simplify.

Solution

Let the unknown number be x.
8 more than the product of 4 and the total of x and 12

$4(x + 12) + 8$ • **This is the variable expression.**
$4x + 48 + 8$ • **Simplify.**
$4x + 56$

You Try It 4

Translate "the difference between fourteen and the sum of a number and seven" into a variable expression. Then simplify.

Your solution

Solutions on p. S2

Objective B   **To solve application problems**

Many of the applications of mathematics require that you identify the unknown quantity, assign a variable to that quantity, and then attempt to express other unknowns in terms of that quantity.

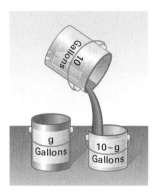

⇒ Ten gallons of paint were poured into two containers of different sizes. Express the amount of paint poured into the smaller container in terms of the amount poured into the larger container.

Assign a variable to the amount of paint poured into the larger container.	the number of gallons of paint poured into the larger container: g
Express the amount of paint in the smaller container in terms of g. (g gallons of paint were poured into the larger container.)	the number of gallons of paint poured into the smaller container: $10 - g$

Example 5

A cyclist is riding at a rate that is twice the speed of a runner. Express the speed of the cyclist in terms of the speed of the runner.

Solution

the speed of the runner: r
the speed of the cyclist is twice r: $2r$

You Try It 5

The length of the Carnival cruise ship *Destiny* is 56 ft more than the height of the Empire State Building. Express the length of *Destiny* in terms of the height of the Empire State Building.

Your solution

Example 6

The length of a rectangle is 2 ft more than 3 times the width. Express the length of the rectangle in terms of the width.

Solution

the width of the rectangle: w
the length is 2 more than 3 times w: $3w + 2$

You Try It 6

The depth of the deep end of a swimming pool is 2 ft more than twice the depth of the shallow end. Express the depth of the deep end in terms of the depth of the shallow end.

Your solution

Example 7

In a survey of listener preferences for AM or FM radio stations, one-third of the number of people surveyed preferred AM stations. Express the number of people who preferred AM stations in terms of the number of people surveyed.

Solution

the number of people surveyed: x
the number of people who preferred AM stations: $\frac{1}{3}x$

You Try It 7

A customer's recent credit card bill showed that one-fourth of the total bill was for the cost of restaurant meals. Express the cost of restaurant meals in terms of the total credit card bill.

Your solution

Solutions on p. S3

1.4 Exercises

Objective A

Translate into a variable expression.

1. eight less than a number

2. the product of negative six and a number

3. four-fifths of a number

4. the difference between a number and twenty

5. the quotient of a number and fourteen

6. a number increased by two hundred

Translate into a variable expression. Then simplify.

7. a number minus the sum of the number and two

8. a number decreased by the difference between five and the number

9. five times the product of eight and a number

10. a number increased by two-thirds of the number

11. the difference between seventeen times a number and twice the number

12. one-half of the total of six times a number and twenty-two

13. the difference between the square of a number and the total of twelve and the square of the number

14. eleven more than the square of a number added to the difference between the number and seventeen

15. the sum of five times a number and twelve added to the product of fifteen and the number

16. four less than twice the sum of a number and eleven

17. The sum of two numbers is fifteen. Using x to represent the smaller of the two numbers, translate "the sum of two more than the larger number and twice the smaller number" into a variable expression. Then simplify.

18. The sum of two numbers is twenty. Using x to represent the smaller of the two numbers, translate "the difference between two more than the larger number and twice the smaller number" into a variable expression. Then simplify.

19. The sum of two numbers is thirty-four. Using x to represent the larger of the two numbers, translate "the quotient of five times the smaller number and the difference between the larger number and three" into a variable expression.

20. The sum of two numbers is thirty-three. Using x to represent the larger of the two numbers, translate "the difference between six more than twice the larger number and the sum of the smaller number and three" into a variable expression. Then simplify.

Objective B *Application Problems*

21. The distance from Earth to the sun is approximately 390 times the distance from Earth to the moon. Express the distance from Earth to the sun in terms of the distance from Earth to the moon.

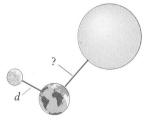

22. The length of an infrared ray is twice the length of an ultraviolet ray. Express the length of the infrared ray in terms of the ultraviolet ray.

23. A mixture of candy contains 3 lb more of milk chocolate than of caramel. Express the amount of milk chocolate in the mixture in terms of the amount of caramel in the mixture.

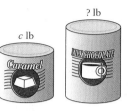

24. The length of a rectangle is three more than twice the width. Express the length of the rectangle in terms of the width.

25. A financial advisor has invested $10,000 in two accounts. If one account contains x dollars, express the amount in the second account in terms of x.

26. A fishing line 3 ft long is cut into two pieces, one shorter than the other. Express the length of the shorter piece in terms of the length of the longer piece.

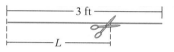

27. The measure of angle A of a triangle is twice the measure of angle B. The measure of angle C is twice the measure of angle A. Write expressions for angle A and angle C in terms of angle B.

28. A 12-foot board is cut into two pieces of different lengths. Express the length of the longer piece in terms of the length of the shorter piece.

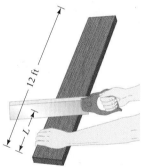

APPLYING THE CONCEPTS

For each of the following, write a phrase that would translate into the given expression.

29. $2x + 3$ **30.** $5y - 4$ **31.** $2(x + 3)$ **32.** $5(y - 4)$

33. Translate each of the following into a variable expression. Each expression is part of a formula from the sciences, and its translation requires more than one variable.
  a. the product of one-half the acceleration due to gravity (g) and the time (t) squared
  b. the product of mass (m) and acceleration (a)
  c. the product of the area (A) and the square of the velocity (v)

Focus on Problem Solving

Polya's Four-Step Process

Your success in mathematics and your success in the workplace are heavily dependent on your ability to solve problems. One of the foremost mathematicians to study problem solving was George Polya (1887–1985). The basic structure that Polya advocated for problem solving has four steps, as outlined below.

1. Understand the Problem

You must have a clear understanding of the problem. To help you focus on understanding the problem, here are some questions to think about.

- Can you restate the problem in your own words?
- Can you determine what is known about this type of problem?
- Is there missing information that you need in order to solve the problem?
- Is there information given that is not needed?
- What is the goal?

2. Devise a Plan

Successful problem solvers use a variety of techniques when they attempt to solve a problem. Here are some frequently used strategies.

- Make a list of the known information.
- Make a list of information that is needed to solve the problem.
- Make a table or draw a diagram.
- Work backwards.
- Try to solve a similar but simpler problem.
- Research the problem to determine whether there are known techniques for solving problems of its kind.
- Try to determine whether some pattern exists.
- Write an equation.

3. Carry Out the Plan

Once you have devised a plan, you must carry it out.

- Work carefully.
- Keep an accurate and neat record of all your attempts.
- Realize that some of your initial plans will not work and that you may have to return to Step 2 and devise another plan or modify your existing plan.

4. Review Your Solution

Once you have found a solution, check the solution against the known facts.

- Make sure that the solution is consistent with the facts of the problem.
- Interpret the solution in the context of the problem.
- Ask yourself whether there are generalizations of the solution that could apply to other problems.
- Determine the strengths and weaknesses of your solution. For instance, is your solution only an approximation to the actual solution?
- Consider the possibility of alternative solutions.

We will use Polya's four-step process to solve the following problem.

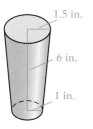

1.5 in.

6 in.

1 in.

A large soft drink costs $1.25 at a college cafeteria. The dimensions of the cup are shown at the left. Suppose you don't put any ice in the cup. Determine the cost per ounce for the soft drink.

1. *Understand the problem.* We must determine the cost per ounce for the soft drink. To do this, we need the dimensions of the cup (which are given), the cost of the drink (given), and a formula for the volume of the cup (unknown). Also, because the dimensions are given in inches, the volume will be in cubic inches. We need a conversion factor that will convert cubic inches to fluid ounces.

2. *Devise a plan.* Consult a resource book that gives an equation for the volume of the figure, which is called a **frustrum.** The formula for the volume is

$$V = \frac{\pi h}{3}(r^2 + rR + R^2)$$

where h is the height, r is the radius of the base, and R is the radius of the top. Also from a reference book, $1 \text{ in}^3 \approx 0.55 \text{ fl oz}$. The general plan is to calculate the volume, convert the answer to fluid ounces, and then divide the cost by the number of fluid ounces.

3. *Carry out the plan.* Using the information from the drawing, evaluate the formula for the volume.

$$V = \frac{6\pi}{3}[1^2 + 1(1.5) + 1.5^2] = 9.5\pi \approx 29.8451 \text{ in}^3$$

$V \approx 29.8451(0.55) \approx 16.4148 \text{ fl oz}$ • **Convert to fluid ounces.**

$\text{Cost per ounce} \approx \dfrac{1.25}{16.4148} \approx 0.07615$ • **Divide the cost by the volume.**

The cost of the soft drink is approximately 7.62 cents per ounce.

4. *Review the solution.* The cost of a 12-ounce can of soda from a vending machine is generally about 75¢. Therefore, the cost of canned soda is 75¢ ÷ 12 = 6.25¢ per ounce. This is consistent with our solution. This does not mean our solution is correct, but it does indicate that it is at least reasonable. Why might soda from a cafeteria be more expensive per ounce than soda from a vending machine?

Is there an alternative way to obtain the solution? There are probably many, but one possibility is to get a measuring cup, pour the soft drink into it, and read the number of ounces. Name an advantage and a disadvantage of this method.

Use the four-step solution process to solve the following problems.

1. A cup dispenser next to a water cooler holds cups that have the shape of a right circular cone. The height of the cone is 4 in. and the radius of the circular top is 1.5 in. How many ounces of water can the cup hold?

2. Soft drink manufacturers research the preferences of consumers with regard to the look, feel, and size of a soft drink can. Suppose a manufacturer has determined that people want to have their hands reach around approximately 75% of the can. If this preference is to be achieved, how tall should the can be if it contains 12 oz of fluid? Assume the can is a right circular cylinder.

Projects and Group Activities

Water Displacement

When an object is placed in water, the object displaces an amount of water that is equal to the volume of the object.

⇒ A sphere with a diameter of 4 in. is placed in a rectangular tank of water that is 6 in. long and 5 in. wide. How much does the water level rise? Round to the nearest hundredth.

$$V = \frac{4}{3}\pi r^3$$
• Use the formula for the volume of a sphere.

$$V = \frac{4}{3}\pi (2^3) = \frac{32}{3}\pi$$
• $r = \frac{1}{2}d = \frac{1}{2}(4) = 2$

Let x represent the amount of the rise in water level. The volume of the sphere will equal the volume displaced by the water. As shown at the left, this volume is the rectangular solid with width 5 in., length 6 in., and height x in.

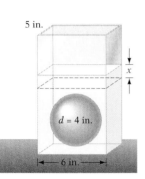

$$V = LWH$$
• Use the formula for the volume of a rectangular solid.

$$\frac{32}{3}\pi = (6)(5)x$$
• Substitute $\frac{32}{3}\pi$ for V, 5 for W, and 6 for L.

$$\frac{32}{90}\pi = x$$
• The exact height that the water will fill is $\frac{32}{90}\pi$.

$$1.12 \approx x$$
• Use a calculator to find an approximation.

The water will rise approximately 1.12 in.

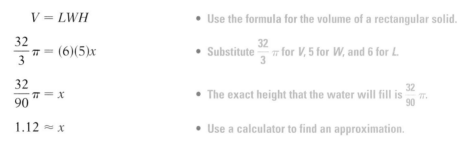

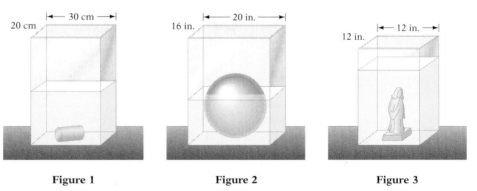

Figure 1 **Figure 2** **Figure 3**

1. A cylinder with a 2-centimeter radius and a height of 10 cm is submerged in a tank of water that is 20 cm wide and 30 cm long (see Figure 1). How much does the water level rise? Round to the nearest hundredth.

2. A sphere with a radius of 6 in. is placed in a rectangular tank of water that is 16 in. wide and 20 in. long (see Figure 2). The sphere displaces water until two-thirds of the sphere is submerged. How much does the water level rise? Round to the nearest hundredth.

3. A chemist wants to know the density of a statue that weighs 15 lb. The statue is placed in a rectangular tank of water that is 12 in. long and 12 in. wide (see Figure 3). The water level rises 0.42 in. Find the density of the statue. Round to the nearest hundredth. (*Hint:* density = weight ÷ volume)

Chapter Summary

Key Words
The *integers* are $\ldots -4, -3, -2, -1, 0, 1, 2, 3, 4, \ldots$. The *negative integers* are $\ldots -4, -3, -2, -1$. The *positive integers*, or *natural numbers*, are $1, 2, 3, 4, \ldots$ [p. 3]

A *rational number* can be written in the form $\frac{a}{b}$, where a and b are integers and $b \neq 0$. An *irrational number* has a decimal representation that never terminates or repeats. The rational numbers and the irrational numbers taken together are the *real numbers*. [pp. 3, 4]

The *additive inverse* of a number is the opposite of the number. The *multiplicative inverse* of a number is the reciprocal of the number. The *absolute value* of a number is its distance from zero on the number line. [pp. 5, 18]

A *set* is a collection of objects. The objects are called the *elements* of the set. The *roster method* of writing sets encloses a list of the elements in braces. *Set-builder notation* makes use of a variable and a certain property that only elements of that set possess. Sets of real numbers are also expressed using *interval notation*. [pp. 3, 6, 7, 10]

The *union* of two sets, written $A \cup B$, is the set that contains all the elements of A and all the elements of B. The *intersection* of two sets, written $A \cap B$, is the set that contains the elements that are common to both A and B. [p. 8]

The expression a^n is in *exponential form*, where a is the *base* and n is the *exponent*. [p. 21]

A *variable expression* is an expression that contains one or more variables. The *terms* of a variable expression are the addends of the expression. A *variable term* is composed of a numerical coefficient and a variable part. [p. 31]

Essential Rules
To add numbers with the same sign, add the absolute values of the numbers; then attach the sign of the addends. **To add numbers with different signs,** find the difference between the absolute values of the numbers; then attach the sign of the addend with the larger absolute value. [p. 17]

To subtract two integers, add the opposite of the second number to the first number. [p. 17]

The product or quotient of two numbers with the same sign is positive. **The product or quotient of two numbers with different signs** is negative. [p. 18]

Commutative Property of Addition	$a + b = b + a$
Commutative Property of Multiplication	$a \cdot b = b \cdot a$
Associative Property of Addition	$(a + b) + c = a + (b + c)$
Associative Property of Multiplication	$(a \cdot b) \cdot c = a \cdot (b \cdot c)$
Addition Property of Zero	$a + 0 = a$ or $0 + a = a$
Multiplication Property of Zero	$a \cdot 0 = 0$ or $0 \cdot a = 0$
Multiplication Property of One	$a \cdot 1 = a$ or $1 \cdot a = a$
Inverse Property of Addition	$a + (-a) = -a + a = 0$
Inverse Property of Multiplication	If $a \neq 0$, $a \cdot \dfrac{1}{a} = \dfrac{1}{a} \cdot a = 1$.
Distributive Property	$a(b + c) = ab + ac$
Division Properties of Zero and One [pp. 18–19, 29–30]	If $a \neq 0$, $0 \div a = 0$ and $a \div a = 1$. $a \div 1 = a$ $a \div 0$ is undefined.

The Order of Operations Agreement [p. 22]

Step 1 Perform all operations inside grouping symbols.
Step 2 Simplify exponential expressions.
Step 3 Do multiplication and division as they occur from left to right.
Step 4 Do addition and subtraction as they occur from left to right.

Chapter Review

1. Use the roster method to write the set of integers between -3 and 4.

2. Find $A \cap B$ given $A = \{0, 1, 2, 3\}$ and $B = \{2, 3, 4, 5\}$.

3. Graph $(-2, 4]$.

4. Identify the property that justifies the statement.
$$2(3x) = (2 \cdot 3)x$$

5. Simplify: $-4.07 + 2.3 - 1.07$

6. Evaluate $(a - 2b^2) \div ab$ when $a = 4$ and $b = -3$.

7. Simplify: $-2 \cdot (4^2) \cdot (-3)^2$

8. Simplify: $4y - 3[x - 2(3 - 2x) - 4y]$

9. Find the additive inverse of $-\dfrac{3}{4}$.

10. Use set-builder notation to write the set of real numbers less than -3.

11. Graph $\{x \mid x < 1\}$.

12. Simplify: $-10 - (-3) - 8$

13. Simplify: $-\dfrac{2}{3} + \dfrac{3}{5} - \dfrac{1}{6}$

14. Use the Associative Property of Addition to complete the statement.
$$3 + (4 + y) = (3 + ?) + y$$

15. Simplify: $-\dfrac{3}{8} \div \dfrac{3}{5}$

16. Let $x \in \{-4, -2, 0, 2\}$. For what values of x is $x > -1$ true?

17. Evaluate $2a^2 - \dfrac{3b}{a}$ when $a = -3$ and $b = 2$.

18. Simplify: $18 - |-12 + 8|$

19. Simplify: $20 \div \dfrac{3^2 - 2^2}{3^2 + 2^2}$

20. Graph $[-3, \infty)$.

21. Find $A \cup B$ given $A = \{1, 3, 5, 7\}$ and $B = \{2, 4, 6, 8\}$.

22. Simplify: $-204 \div (-17)$

23. Write $[-2, 3]$ in set-builder notation.

24. Simplify: $\dfrac{3}{5}\left(-\dfrac{10}{21}\right)\left(-\dfrac{7}{15}\right)$

25. Use the Distributive Property to complete the statement.
$6x - 21y = ?(2x - 7y)$

26. Graph $\{x \,|\, x \le -3\} \cup \{x \,|\, x > 0\}$.

27. Simplify: $-2(x - 3) + 4(2 - x)$

28. Let $p \in \{-4, 0, 7\}$. Evaluate $-|p|$ for each element of the set.

29. Identify the property that justifies the statement.
$-4 + 4 = 0$

30. Simplify: $-3.286 \div (-1.06)$

31. Translate and simplify "twelve minus the quotient of three more than a number and the number."

32. The sum of two numbers is forty. Using x to represent the smaller of the two numbers, translate "the sum of twice the smaller number and five more than the larger number" into a variable expression. Then simplify.

33. The length of a rectangle is three feet less than three times the width. Express the length of the rectangle in terms of the width.

34. Jimmy Carter used the presidential veto 47 fewer times than Ronald Reagan. Express the number of vetos used by Jimmy Carter in terms of the number of vetos used by Ronald Reagan. (*Source:* U.S. Senate Library, Congressional Research Service)

Chapter Test

1. Simplify: $(-2)(-3)(-5)$

2. Find $A \cap B$ given $A = \{1, 3, 5, 7\}$ and $B = \{5, 7, 9, 11\}$.

3. Simplify: $(-2)^3(-3)^2$

4. Graph $(-\infty, 1]$.

$$\xleftarrow{\hspace{0.3cm}}\!\!\!\overset{\displaystyle +\!\!+\!\!+\!\!+\!\!+\!\!+\!\!+\!\!+\!\!+\!\!+\!\!+}{\underset{\text{-5 -4 -3 -2 -1 \ 0 \ 1 \ 2 \ 3 \ 4 \ 5}}{}}\!\!\!\xrightarrow{\hspace{0.3cm}}$$

5. Find $A \cap B$ given $A = \{-3, -2, -1, 0, 1, 2, 3\}$ and $B = \{-1, 0, 1\}$.

6. Evaluate $(a - b)^2 \div (2b + 1)$ when $a = 2$ and $b = -3$.

7. Simplify: $\left| -3 - (-5) \right|$

8. Simplify: $2x - 4[2 - 3(x + 4y) - 2]$

9. Find the additive inverse of -12.

10. Simplify: $-5^2 \cdot 4$

11. Graph $\{x \mid x < 3\} \cap \{x \mid x > -2\}$.

$$\xleftarrow{\hspace{0.3cm}}\!\!\!\overset{\displaystyle +\!\!+\!\!+\!\!+\!\!+\!\!+\!\!+\!\!+\!\!+\!\!+\!\!+}{\underset{\text{-5 -4 -3 -2 -1 \ 0 \ 1 \ 2 \ 3 \ 4 \ 5}}{}}\!\!\!\xrightarrow{\hspace{0.3cm}}$$

12. Simplify: $2 - (-12) + 3 - 5$

13. Simplify: $\dfrac{2}{3} - \dfrac{5}{12} + \dfrac{4}{9}$

14. Use the Commutative Property of Addition to complete the statement.
$(3 + 4) + 2 = (? + 3) + 2$

15. Simplify: $-\dfrac{2}{3}\left(\dfrac{9}{15}\right)\left(\dfrac{10}{27}\right)$

16. Let $x \in \{-5, 3, 7\}$. For what values of x is $x < -1$ true?

17. Evaluate $\dfrac{b^2 - c^2}{a - 2c}$ when $a = 2$, $b = 3$, and $c = -1$.

18. Simplify: $-180 \div 12$

19. Simplify: $12 - 4\left(\dfrac{5^2 - 1}{3}\right) \div 16$

20. Graph $(3, \infty)$.

21. Find $A \cup B$ given $A = \{1, 3, 5, 7\}$ and $B = \{2, 3, 4, 5\}$.

22. Simplify: $3x - 2(x - y) - 3(y - 4x)$

23. Simplify: $8 - 4(2 - 3)^2 \div 2$

24. Simplify: $\dfrac{3}{5}\left(-\dfrac{10}{21}\right)\left(-\dfrac{7}{15}\right)$

25. Identify the property that justifies the statement.
$-2(x + y) = -2x - 2y$

26. Graph $\{x \mid x \le 3\} \cup \{x \mid x < -2\}$.

27. Simplify: $4.27 - 6.98 + 1.3$

28. Find $A \cup B$ given $A = \{-2, -1, 0, 1, 2, 3\}$ and $B = \{-1, 0, 1\}$.

29. Translate and simplify "thirteen decreased by the product of three less than a number and nine."

30. The sum of two numbers is nine. Using x to represent the larger of the two numbers, translate "the difference between one more than the larger number and twice the smaller number" into a variable expression. Then simplify.

Chapter 2

First-Degree Equations and Inequalities

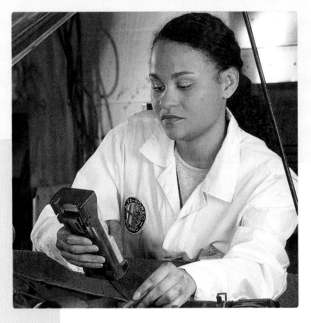

Auto mechanics rely on readings from voltage meters to keep cars running efficiently. The meters provide exact numbers since precision is key in measuring certain aspects of an engine's performance. Precision is also crucial in the manufacture of engine parts. **Example 8 on page 96** illustrates how an absolute value inequality is used to determine the lower and upper limits of the diameter of the piston.

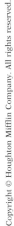

Need help? For on-line student resources, such as section quizzes, visit this textbook's web site at **math.college.hmco.com/student**.

For Exercises 1 to 5, add, subtract, multiply, or divide.

1. $8 - 12$

2. $-9 + 3$

3. $\dfrac{-18}{-6}$

4. $-\dfrac{3}{4}\left(-\dfrac{4}{3}\right)$

5. $-\dfrac{5}{8}\left(\dfrac{4}{5}\right)$

For Exercises 6 to 9, simplify.

6. $3x - 5 + 7x$

7. $6(x - 2) + 3$

8. $n + (n + 2) + (n + 4)$

9. $0.08x + 0.05(400 - x)$

10. A twenty-ounce snack mixture contains nuts and pretzels. Let n represent the number of ounces of nuts in the mixture. Express the number of ounces of pretzels in the mixture in terms of n.

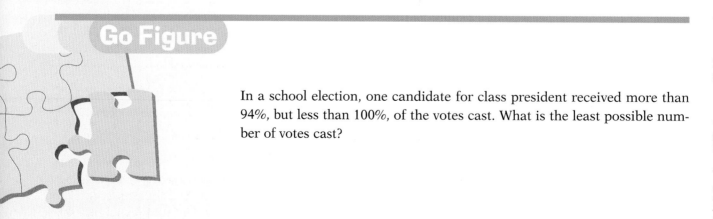

Go Figure

In a school election, one candidate for class president received more than 94%, but less than 100%, of the votes cast. What is the least possible number of votes cast?

2.1   Solving First-Degree Equations

Objective A   **To solve an equation using the Addition or the Multiplication Property of Equations**

An **equation** expresses the equality of two mathematical expressions. The expressions can be either numerical or variable expressions.

$$\left. \begin{array}{l} 2 + 8 = 10 \\ x + 8 = 11 \\ x^2 + 2y = 7 \end{array} \right\} \quad \text{Equations}$$

The equation at the right is a **conditional equation.** The equation is true if the variable is replaced by 3. The equation is false if the variable is replaced by 4. A conditional equation is true for at least one value of the variable.

$x + 2 = 5$ A conditional equation

$3 + 2 = 5$ A true equation

$4 + 2 = 5$ A false equation

The replacement values of the variable that will make an equation true are called the **roots** or the **solutions** of the equation. The solution of the equation $x + 2 = 5$ is 3 because $3 + 2 = 5$ is a true equation.

The equation at the right is an **identity.** Any replacement for x will result in a true equation.

$x + 2 = x + 2$

The equation at the right has no solution because there is no number that equals itself plus one. Any replacement value for x will result in a false equation. This equation is a **contradiction.**

$x = x + 1$

Each of the equations at the right is a **first-degree equation in one variable.** All variables have an exponent of 1.

$x + 2 = 12$

$3y - 2 = 5y$

$3(a + 2) = 14a$

To **solve an equation** means to find a root or solution of the equation. The simplest equation to solve is an equation of the form *variable = constant*, because the constant is the solution. If $x = 3$, then 3 is the solution of the equation, because $3 = 3$ is a true equation.

Equivalent equations are equations that have the same solution. For instance, $x + 4 = 6$ and $x = 2$ are equivalent equations because the solution of each equation is 2. In solving an equation, the goal is to produce *simpler* but equivalent equations until you reach the goal of *variable = constant*. The **Addition Property of Equations** can be used to rewrite an equation in this form.

The Addition Property of Equations

If a, b, and c are algebraic expressions, then the equation $a = b$ has the same solutions as the equation $a + c = b + c$.

The Addition Property of Equations states that the same quantity can be added to each side of an equation without changing the solution of the equation. This property is used to remove a term from one side of the equation by adding the opposite of that term to each side of the equation.

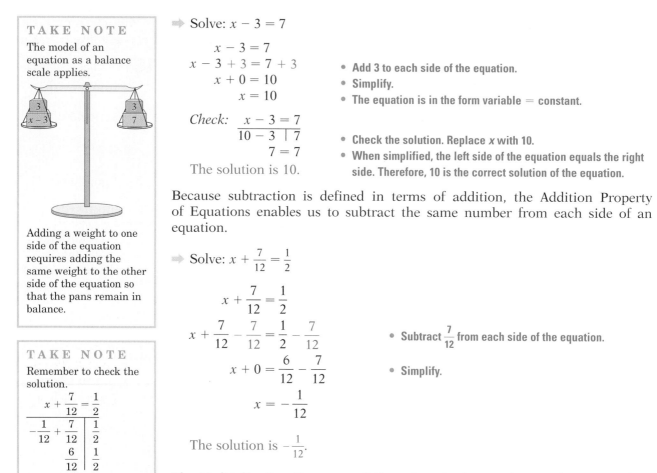

TAKE NOTE

The model of an equation as a balance scale applies.

Adding a weight to one side of the equation requires adding the same weight to the other side of the equation so that the pans remain in balance.

TAKE NOTE

Remember to check the solution.

$$\begin{array}{c|c} x + \dfrac{7}{12} = \dfrac{1}{2} \\ \hline -\dfrac{1}{12} + \dfrac{7}{12} & \dfrac{1}{2} \\ \dfrac{6}{12} & \dfrac{1}{2} \\ \dfrac{1}{2} = \dfrac{1}{2} \end{array}$$

Solve: $x - 3 = 7$

$$\begin{aligned} x - 3 &= 7 \\ x - 3 + 3 &= 7 + 3 \\ x + 0 &= 10 \\ x &= 10 \end{aligned}$$

- Add 3 to each side of the equation.
- Simplify.
- The equation is in the form variable = constant.

Check: $\dfrac{x - 3 = 7}{\,10 - 3 \mid 7\,}$

$7 = 7$

- Check the solution. Replace x with 10.
- When simplified, the left side of the equation equals the right side of the equation. Therefore, 10 is the correct solution of the equation.

The solution is 10.

Because subtraction is defined in terms of addition, the Addition Property of Equations enables us to subtract the same number from each side of an equation.

Solve: $x + \dfrac{7}{12} = \dfrac{1}{2}$

$$\begin{aligned} x + \dfrac{7}{12} &= \dfrac{1}{2} \\ x + \dfrac{7}{12} - \dfrac{7}{12} &= \dfrac{1}{2} - \dfrac{7}{12} \\ x + 0 &= \dfrac{6}{12} - \dfrac{7}{12} \\ x &= -\dfrac{1}{12} \end{aligned}$$

- Subtract $\dfrac{7}{12}$ from each side of the equation.
- Simplify.

The solution is $-\dfrac{1}{12}$.

The **Multiplication Property of Equations** is also used to produce equivalent equations.

> **The Multiplication Property of Equations**
>
> If a, b, and c are algebraic expressions and $c \neq 0$, then the equation $a = b$ has the same solution as the equation $ac = bc$.

This property states that each side of an equation can be multiplied by the same nonzero number without changing the solution of the equation.

Recall that the goal of solving an equation is to rewrite the equation in the form *variable = constant*. The Multiplication Property of Equations is used to rewrite an equation in this form by multiplying each side by the reciprocal of the coefficient.

Solve: $-\dfrac{3}{4}x = 12$

TAKE NOTE

Remember to check the solution.

$$\begin{array}{c|c} -\dfrac{3}{4}x = 12 \\ \hline -\dfrac{3}{4}(-16) & 12 \\ 12 = 12 \end{array}$$

$$\begin{aligned} -\dfrac{3}{4}x &= 12 \\ \left(-\dfrac{4}{3}\right)\left(-\dfrac{3}{4}\right)x &= \left(-\dfrac{4}{3}\right)12 \\ 1x &= -16 \\ x &= -16 \end{aligned}$$

- Multiply each side of the equation by $-\dfrac{4}{3}$, the reciprocal of $-\dfrac{3}{4}$.
- Simplify.

The solution is -16.

Because division is defined in terms of multiplication, the Multiplication Property of Equations enables us to divide each side of an equation by the same nonzero quantity.

⟹ Solve: $-5x = 9$

Multiplying each side of the equation by the reciprocal of -5 is equivalent to dividing each side of the equation by -5.

$$-5x = 9$$
$$\frac{-5x}{-5} = \frac{9}{-5}$$ • **Divide each side of the equation by -5.**
$$1x = -\frac{9}{5}$$ • **Simplify.**
$$x = -\frac{9}{5}$$

The solution is $-\frac{9}{5}$. You should check the solution.

When using the Multiplication Property of Equations, it is usually easier to multiply each side of the equation by the reciprocal of the coefficient when the coefficient is a fraction. Divide each side of the equation by the coefficient when the coefficient is an integer or a decimal.

....

Example 1

Solve: $x - 7 = -12$

Solution
$$x - 7 = -12$$
$$x - 7 + 7 = -12 + 7$$
$$x = -5$$

The solution is -5.

You Try It 1

Solve: $x + 4 = -3$

Your solution

....

Example 2

Solve: $-\frac{2x}{7} = \frac{4}{21}$

Solution
$$-\frac{2x}{7} = \frac{4}{21}$$
$$\left(-\frac{7}{2}\right)\left(-\frac{2}{7}x\right) = \left(-\frac{7}{2}\right)\left(\frac{4}{21}\right)$$
$$x = -\frac{2}{3}$$

The solution is $-\frac{2}{3}$.

You Try It 2

Solve: $-3x = 18$

Your solution

Solutions on p. S3

Objective B

To solve an equation using both the Addition and the Multiplication Properties of Equations

In solving an equation, it is often necessary to apply both the Addition and the Multiplication Properties of Equations.

➡ Solve: $4x - 3 + 2x = 8 + 9x - 12$

$$4x - 3 + 2x = 8 + 9x - 12$$
$$6x - 3 = -4 + 9x$$

- Simplify each side of the equation by combining like terms.

$$6x - 3 - 9x = -4 + 9x - 9x$$
$$-3x - 3 = -4$$

- Subtract $9x$ from each side of the equation. Then simplify.

$$-3x - 3 + 3 = -4 + 3$$
$$-3x = -1$$

- Add 3 to each side of the equation. Then simplify.

$$\frac{-3x}{-3} = \frac{-1}{-3}$$
$$x = \frac{1}{3}$$

- Divide each side of the equation by the coefficient -3. Then simplify.

TAKE NOTE

You should always check your work by substituting your answer into the original equation and simplifying the resulting numerical expressions. If the left and right sides are equal, your answer is the solution of the equation.

Check:

$$\frac{4x - 3 + 2x = 8 + 9x - 12}{}$$

$$4\left(\frac{1}{3}\right) - 3 + 2\left(\frac{1}{3}\right) \,\bigg|\, 8 + 9\left(\frac{1}{3}\right) - 12$$

$$\frac{4}{3} - 3 + \frac{2}{3} \,\bigg|\, 8 + 3 - 12$$

$$-\frac{5}{3} + \frac{2}{3} \,\bigg|\, 11 - 12$$

$$-1 = -1$$

The solution is $\frac{1}{3}$.

- -

Example 3

Solve: $3x - 5 = x + 2 - 7x$

Solution

$$3x - 5 = x + 2 - 7x$$
$$3x - 5 = -6x + 2$$
$$3x - 5 + 6x = -6x + 2 + 6x$$
$$9x - 5 = 2$$
$$9x - 5 + 5 = 2 + 5$$
$$9x = 7$$
$$\frac{9x}{9} = \frac{7}{9}$$
$$x = \frac{7}{9}$$

The solution is $\frac{7}{9}$.

You Try It 3

Solve: $6x - 5 - 3x = 14 - 5x$

Your solution

Solution on p. S3

Objective C   **To solve an equation containing parentheses**

When an equation contains parentheses, one of the steps required to solve the equation involves using the Distributive Property.

➡ Solve: $3(x - 2) + 3 = 2(6 - x)$

$$3(x - 2) + 3 = 2(6 - x)$$
$$3x - 6 + 3 = 12 - 2x$$
$$3x - 3 = 12 - 2x$$

- Use the Distributive Property to remove parentheses. Then simplify.

$$3x - 3 + 2x = 12 - 2x + 2x$$
$$5x - 3 = 12$$

- Add $2x$ to each side of the equation.

$$5x - 3 + 3 = 12 + 3$$
$$5x = 15$$

- Add 3 to each side of the equation.

$$\frac{5x}{5} = \frac{15}{5}$$
$$x = 3$$

- Divide each side of the equation by the coefficient 5.

The solution is 3.

To solve an equation containing fractions, first clear the denominators by multiplying each side of the equation by the least common multiple (LCM) of the denominators.

➡ Solve: $\frac{x}{2} - \frac{7}{9} = \frac{x}{6} + \frac{2}{3}$

Find the LCM of the denominators. The LCM of 2, 9, 6, and 3 is 18.

$$\frac{x}{2} - \frac{7}{9} = \frac{x}{6} + \frac{2}{3}$$

$$18\left(\frac{x}{2} - \frac{7}{9}\right) = 18\left(\frac{x}{6} + \frac{2}{3}\right)$$

- Multiply each side of the equation by the LCM of the denominators.

$$\frac{18x}{2} - \frac{18 \cdot 7}{9} = \frac{18x}{6} + \frac{18 \cdot 2}{3}$$

- Use the Distributive Property to remove parentheses. Then simplify.

$$9x - 14 = 3x + 12$$

$$6x - 14 = 12$$

- Subtract $3x$ from each side of the equation. Then simplify.

$$6x = 26$$

- Add 14 to each side of the equation. Then simplify.

$$\frac{6x}{6} = \frac{26}{6}$$

- Divide each side of the equation by the coefficient of x. Then simplify.

$$x = \frac{13}{3}$$

The solution is $\frac{13}{3}$.

Example 4 Solve:
$$5(2x - 7) + 2 = 3(4 - x) - 12$$

Solution
$$5(2x - 7) + 2 = 3(4 - x) - 12$$
$$10x - 35 + 2 = 12 - 3x - 12$$
$$10x - 33 = -3x$$
$$-33 = -13x$$
$$\frac{-33}{-13} = \frac{-13x}{-13}$$
$$\frac{33}{13} = x$$

The solution is $\frac{33}{13}$.

You Try It 4 Solve:
$$6(5 - x) - 12 = 2x - 3(4 + x)$$

Your solution

Solution on p. S3

Objective D **To solve a literal equation for one of the variables** ‹ 2 ›

A **literal equation** is an equation that contains more than one variable. Some examples are shown at the right.

$$3x - 2y = 4$$
$$v^2 = v_0^2 + 2as$$

Formulas are used to express a relationship among physical quantities. A **formula** is a literal equation that states rules about measurement. Examples are shown at the right.

$$s = vt - 16t^2 \quad \text{(Physics)}$$
$$c^2 = a^2 + b^2 \quad \text{(Geometry)}$$
$$I = P(1 + r)^n \quad \text{(Business)}$$

The Addition and Multiplication Properties of Equations can be used to solve a literal equation for one of the variables. The goal is to rewrite the equation so that the variable being solved for is alone on one side of the equation and all the other numbers and variables are on the other side.

⇒ Solve $A = P + Prt$ for t.

$$A = P + Prt$$
$$A - P = Prt$$
$$\frac{A - P}{Pr} = \frac{Prt}{Pr}$$
$$\frac{A - P}{Pr} = t$$

- Subtract P from each side of the equation.

- Divide each side of the equation by Pr.

Example 5

Solve $C = \frac{5}{9}(F - 32)$ for F.

Solution

$$C = \frac{5}{9}(F - 32)$$
$$\frac{9}{5}C = \frac{9}{5} \cdot \frac{5}{9}(F - 32)$$
$$\frac{9}{5}C = F - 32$$
$$\frac{9}{5}C + 32 = F$$

You Try It 5

Solve $S = C - rC$ for r.

Your solution

Solution on p. S3

2.1 Exercises

Objective A

1. How does an equation differ from an expression?

2. What is the solution of an equation?

3. What is the Addition Property of Equations and how is it used?

4. What is the Multiplication Property of Equations and how is it used?

5. Is 1 a solution of $7 - 3m = 4$?

6. Is 5 a solution of $4y - 5 = 3y$?

7. Is -2 a solution of $6x - 1 = 7x + 1$?

8. Is 3 a solution of $x^2 = 4x - 5$?

Solve and check.

9. $x - 2 = 7$

10. $x - 8 = 4$

11. $a + 3 = -7$

12. $a + 5 = -2$

13. $b - 3 = -5$

14. $10 + m = 2$

15. $-7 = x + 8$

16. $-12 = x - 3$

17. $3x = 12$

18. $8x = 4$

19. $-3x = 2$

20. $-5a = 7$

21. $-\dfrac{3}{2} + x = \dfrac{4}{3}$

22. $\dfrac{2}{7} + x = \dfrac{17}{21}$

23. $x + \dfrac{2}{3} = \dfrac{5}{6}$

24. $\dfrac{5}{8} - y = \dfrac{3}{4}$

25. $\dfrac{2}{3}y = 5$

26. $\dfrac{3}{5}y = 12$

27. $-\dfrac{5}{8}x = \dfrac{4}{5}$

28. $-\dfrac{5}{12}x = \dfrac{7}{16}$

29. $-\dfrac{3b}{5} = -\dfrac{3}{5}$

30. $-\dfrac{7b}{12} = \dfrac{7}{8}$

31. $-\dfrac{2}{3}x = -\dfrac{5}{8}$

32. $-\dfrac{3}{4}x = -\dfrac{4}{7}$

33. $-\dfrac{5}{8}x = 40$

34. $\dfrac{2}{7}y = -8$

35. $-\dfrac{5}{6}y = -\dfrac{25}{36}$

36. $-\dfrac{15}{24}x = -\dfrac{10}{27}$

Objective B

Solve and check.

37. $3x + 5x = 12$

38. $2x - 7x = 15$

39. $2x - 4 = 12$

40. $5x + 9 = 6$ **41.** $3y - 5y = 0$ **42.** $2y - 9 = -9$

43. $4x - 6 = 3x$ **44.** $2a - 7 = 5a$ **45.** $7x + 12 = 9x$

46. $3x - 12 = 5x$ **47.** $4x + 2 = 4x$ **48.** $3m - 7 = 3m$

49. $2x + 2 = 3x + 5$ **50.** $7x - 9 = 3 - 4x$ **51.** $2 - 3t = 3t - 4$

52. $7 - 5t = 2t - 9$ **53.** $3b - 2b = 4 - 2b$ **54.** $3x - 5 = 6 - 9x$

55. $3x + 7 = 3 + 7x$ **56.** $\dfrac{5}{8}b - 3 = 12$ **57.** $\dfrac{1}{3} - 2b = 3$

Objective C

Solve and check.

58. $2x + 2(x + 1) = 10$ **59.** $2x + 3(x - 5) = 15$

60. $2(a - 3) = 2(4 - 2a)$ **61.** $5(2 - b) = -3(b - 3)$

62. $3 - 2(y - 3) = 4y - 7$ **63.** $3(y - 5) - 5y = 2y + 9$

64. $2(3x - 2) - 5x = 3 - 2x$ **65.** $4 - 3x = 7x - 2(3 - x)$

66. $8 - 5(4 - 3x) = 2(4 - x) - 8x$ **67.** $-3x - 2(4 + 5x) = 14 - 3(2x - 3)$

68. $3[2 - 3(y - 2)] = 12$ **69.** $3y = 2[5 - 3(2 - y)]$

70. $4[3 + 5(3 - x) + 2x] = 6 - 2x$ **71.** $2[4 + 2(5 - x) - 2x] = 4x - 7$

72. $3[4 - 2(a - 2)] = 3(2 - 4a)$

73. $2[3 - 2(z + 4)] = 3(4 - z)$

74. $-3(x - 2) = 2[x - 4(x - 2) + x]$

75. $3[x - (2 - x) - 2x] = 3(4 - x)$

76. $\dfrac{2}{9}t - \dfrac{5}{6} = \dfrac{1}{12}t$

77. $\dfrac{3}{4}t - \dfrac{7}{12}t = 1$

78. $\dfrac{2}{3}x - \dfrac{5}{6}x - 3 = \dfrac{1}{2}x - 5$

79. $\dfrac{1}{2}x - \dfrac{3}{4}x + \dfrac{5}{8} = \dfrac{3}{2}x - \dfrac{5}{2}$

80. $\dfrac{5 - 2x}{5} + \dfrac{x - 4}{10} = \dfrac{3}{10}$

81. $\dfrac{2x - 5}{12} - \dfrac{3 - x}{6} = \dfrac{11}{12}$

82. $\dfrac{x - 2}{4} - \dfrac{x + 5}{6} = \dfrac{5x - 2}{9}$

83. $\dfrac{2x - 1}{4} + \dfrac{3x + 4}{8} = \dfrac{1 - 4x}{12}$

84. If $5x - 7 = 2x + 2$, evaluate $7x + 4$.

85. If $3x - 5 = 9x + 4$, evaluate $6x - 3$.

86. If $2x - 5(x + 1) = 7$, evaluate $x^2 - 1$.

87. If $3(2x + 1) = 5 - 2(x - 2)$, evaluate $2x^2 + 1$.

88. If $4 - 3(2x + 3) = 5 - 4x$, evaluate $x^2 - 2x$.

89. If $5 - 2(4x - 1) = 3x + 7$, evaluate $x^4 - x^2$.

Objective D

Solve the formula for the given variable.

90. $I = Prt$; r (Business)

91. $C = 2\pi r$; r (Geometry)

92. $PV = nRT$; R (Chemistry)

93. $A = \dfrac{1}{2}bh$; h (Geometry)

94. $V = \dfrac{1}{3}\pi r^2 h; h$ (Geometry)

95. $I = \dfrac{100M}{C}; M$ (Intelligence Quotient)

96. $P = 2L + 2W; W$ (Geometry)

97. $A = P + Prt; r$ (Business)

98. $s = V_0 t - 16t^2; V_0$ (Physics)

99. $s = \dfrac{1}{2}(a + b + c); c$ (Geometry)

100. $F = \dfrac{9}{5}C + 32; C$ (Temperature Conversion)

101. $S = 2\pi r^2 + 2\pi rh; h$ (Geometry)

102. $A = \dfrac{1}{2}h(b_1 + b_2); b_2$ (Geometry)

103. $P = \dfrac{R - C}{n}; R$ (Business)

104. $a_n = a_1 + (n - 1)d; d$ (Mathematics)

105. $S = 2WH + 2WL + 2LH; H$ (Geometry)

APPLYING THE CONCEPTS

106. The following is offered as the solution of $5x + 15 = 2x + 3(2x + 5)$.

$$5x + 15 = 2x + 3(2x + 5)$$
$$5x + 15 = 2x + 6x + 15$$
$$5x + 15 = 8x + 15$$
$$5x + 15 - 15 = 8x + 15 - 15$$

- Use the Distributive Property.
- Combine like terms.
- Subtract 15 from each side of the equation.

$$5x = 8x$$
$$\frac{5x}{x} = \frac{8x}{x}$$

- Divide each side of the equation by x.

$$5 = 8$$

Because $5 = 8$ is not a true equation, the equation has no solution.

If this is correct, so state. If not, explain why it is not correct and supply the correct answer.

107. Why, when the Multiplication Property of Equations is used, must the quantity that multiplies each side of the equation not be zero?

2.2 Applications: Puzzle Problems

Objective A   **To solve integer problems**

An equation states that two mathematical expressions are equal. Therefore, to translate a sentence into an equation requires recognition of the words or phrases that mean "equals." A partial list of these phrases includes "is," "is equal to," "amounts to," and "represents."

Once the sentence is translated into an equation, the equation can be solved by rewriting it in the form *variable = constant*.

Recall that an **even integer** is an integer that is divisible by 2. An **odd integer** is an integer that is not divisible by 2.

Consecutive integers are integers that follow one another in order. Examples of consecutive integers are shown at the right. (Assume that the variable n represents an integer.)

$8, 9, 10$
$-3, -2, -1$
$n, n + 1, n + 2$

Examples of **consecutive even integers** are shown at the right. (Assume that the variable n represents an even integer.)

$16, 18, 20$
$-6, -4, -2$
$n, n + 2, n + 4$

Examples of **consecutive odd integers** are shown at the right. (Assume that the variable n represents an odd integer.)

$11, 13, 15$
$-23, -21, -19$
$n, n + 2, n + 4$

➡ The sum of three consecutive even integers is seventy-eight. Find the integers.

Strategy for Solving an Integer Problem

1. Let a variable represent one of the integers. Express each of the other integers in terms of that variable. Remember that for consecutive integer problems, consecutive integers differ by 1. Consecutive even or consecutive odd integers differ by 2.

First even integer: n
Second even integer: $n + 2$
Third even integer: $n + 4$

• **Represent three consecutive even integers.**

2. Determine the relationship among the integers.

$n + (n + 2) + (n + 4) = 78$
$3n + 6 = 78$
$3n = 72$
$n = 24$

• **The sum of the three even integers is 78.**

• **The first integer is 24.**

$n + 2 = 24 + 2 = 26$
$n + 4 = 24 + 4 = 28$

• **Find the second and third integers.**

The three consecutive even integers are 24, 26, and 28.

Example 1

One number is four more than another number. The sum of the two numbers is sixty-six. Find the two numbers.

Strategy

- The smaller number: n
  The larger number: $n + 4$
- The sum of the numbers is 66.

$$n + (n + 4) = 66$$

Solution

$$\begin{aligned} n + (n + 4) &= 66 \\ 2n + 4 &= 66 \\ 2n &= 62 \\ n &= 31 \end{aligned}$$

$$n + 4 = 31 + 4 = 35$$

The numbers are 31 and 35.

You Try It 1

The sum of three numbers is eighty-one. The second number is twice the first number, and the third number is three less than four times the first number. Find the numbers.

Your strategy

Your solution

Example 2

Five times the first of three consecutive even integers is five more than the product of four and the third integer. Find the integers.

Strategy

- First even integer: n
  Second even integer: $n + 2$
  Third even integer: $n + 4$
- Five times the first integer equals five more than the product of four and the third integer.

$$5n = 4(n + 4) + 5$$

Solution

$$\begin{aligned} 5n &= 4(n + 4) + 5 \\ 5n &= 4n + 16 + 5 \\ 5n &= 4n + 21 \\ n &= 21 \end{aligned}$$

Because 21 is not an even integer, there is no solution.

You Try It 2

Find three consecutive odd integers such that three times the sum of the first two integers is ten more than the product of the third integer and four.

Your strategy

Your solution

Solutions on p. S3

Objective B   **To solve coin and stamp problems**   ◖ 2 ◗   ◉

In solving problems dealing with coins or stamps of different values, it is necessary to represent the value of the coins or stamps in the same unit of money. The unit of money is frequently cents. For example,

The value of five 8¢ stamps is 5 · 8, or 40 cents.
The value of four 20¢ stamps is 4 · 20, or 80 cents.
The value of n 10¢ stamps is n · 10, or $10n$ cents.

➡ A collection of stamps consists of 5¢, 13¢, and 18¢ stamps. The number of 13¢ stamps is two more than three times the number of 5¢ stamps. The number of 18¢ stamps is five less than the number of 13¢ stamps. The total value of all the stamps is $1.68. Find the number of 18¢ stamps.

> **Strategy for Solving a Stamp Problem**
>
> **1.** For each denomination of stamp, write a numerical or variable expression for the number of stamps, the value of the stamp, and the total value of the stamps in cents. The results can be recorded in a table.

The number of 5¢ stamps: x
The number of 13¢ stamps: $3x + 2$
The number of 18¢ stamps: $(3x + 2) - 5 = 3x - 3$

Stamp	Number of Stamps	·	Value of Stamp in Cents	=	Total Value in Cents
5¢	x	·	5	=	$5x$
13¢	$3x + 2$	·	13	=	$13(3x + 2)$
18¢	$3x - 3$	·	18	=	$18(3x - 3)$

> **2.** Determine the relationship between the total values of the stamps. Use the fact that the sum of the total values of each denomination of stamp is equal to the total value of all the stamps.

The sum of the total values of each denomination of stamp is equal to the total value of all the stamps (168 cents).

$$5x + 13(3x + 2) + 18(3x - 3) = 168$$
$$5x + 39x + 26 + 54x - 54 = 168$$
$$98x - 28 = 168$$
$$98x = 196$$
$$x = 2$$

• The sum of the total values equals 168.

The number of 18¢ stamps is $3x - 3$. Replace x by 2 and evaluate.

$$3x - 3 = 3(2) - 3 = 3$$

There are three 18¢ stamps in the collection.

Some of the problems in Section 4 of the chapter "Review of Real Numbers" involved using one variable to describe two numbers whose sum was known. For example, given that the sum of two numbers is 12, we let one of the two numbers be x. Then the other number is $12 - x$. Note that the sum of these two numbers, $x + 12 - x$, equals 12.

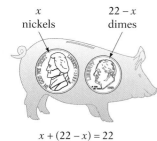

x
nickels

$22 - x$
dimes

$x + (22 - x) = 22$

In Example 3 below, we are told that there are only nickels and dimes in a coin bank, and that there is a total of twenty-two coins. This means that the sum of the number of nickels and the number of dimes is 22. Let the number of nickels be x. Then the number of dimes is $22 - x$. (If you let the number of dimes be x and the number of nickels be $22 - x$, the solution to the problem will be the same.)

Example 3

A coin bank contains $1.80 in nickels and dimes; in all, there are twenty-two coins in the bank. Find the number of nickels and the number of dimes in the bank.

You Try It 3

A collection of stamps contains 3¢, 10¢, and 15¢ stamps. The number of 10¢ stamps is two more than twice the number of 3¢ stamps. There are three times as many 15¢ stamps as there are 3¢ stamps. The total value of the stamps is $1.56. Find the number of 15¢ stamps.

Strategy

• Number of nickels: x
  Number of dimes: $22 - x$

Your strategy

Coin	Number	Value	Total Value
Nickel	x	5	$5x$
Dime	$22 - x$	10	$10(22 - x)$

• The sum of the total values of each denomination of coin equals the total value of all the coins (180 cents).

$5x + 10(22 - x) = 180$

Solution

$$5x + 10(22 - x) = 180$$
$$5x + 220 - 10x = 180$$
$$-5x + 220 = 180$$
$$-5x = -40$$
$$x = 8$$

$22 - x = 22 - 8 = 14$

The bank contains 8 nickels and 14 dimes.

Your solution

Solution on p. S4

2.2 Exercises

Application Problems

1. What number must be added to the numerator of $\frac{3}{10}$ to produce the fraction $\frac{4}{5}$?

2. What number must be added to the numerator of $\frac{5}{12}$ to produce the fraction $\frac{2}{3}$?

3. The sum of two integers is ten. Three times the larger integer is three less than eight times the smaller integer. Find the integers.

4. The sum of two integers is thirty. Eight times the smaller integer is six more than five times the larger integer. Find the integers.

5. One integer is eight less than another integer. The sum of the two integers is fifty. Find the integers.

6. One integer is four more than another integer. The sum of the integers is twenty-six. Find the integers.

7. The sum of three numbers is one hundred twenty-three. The second number is two more than twice the first number. The third number is five less than the product of three and the first number. Find the three numbers.

8. The sum of three numbers is forty-two. The second number is twice the first number, and the third number is three less than the second number. Find the three numbers.

9. The sum of three consecutive integers is negative fifty-seven. Find the integers.

10. The sum of three consecutive integers is one hundred twenty-nine. Find the integers.

11. Five times the smallest of three consecutive odd integers is ten more than twice the largest. Find the integers.

12. Find three consecutive even integers such that twice the sum of the first and third integers is twenty-one more than the second integer.

13. Find three consecutive odd integers such that three times the middle integer is seven more than the sum of the first and third integers.

14. Find three consecutive even integers such that four times the sum of the first and third integers is twenty less than six times the middle integer.

15. A collection of fifty-three coins has a value of $3.70. The collection contains only nickels and dimes. Find the number of dimes in the collection.

16. A collection of twenty-two coins has a value of $4.75. The collection contains dimes and quarters. Find the number of quarters in the collection.

17. A coin bank contains twenty-two coins in nickels, dimes, and quarters. There are four times as many dimes as quarters. The value of the coins is $2.30. How many dimes are in the bank?

18. A coin collection contains nickels, dimes, and quarters. There are twice as many dimes as quarters and seven more nickels than dimes. The total value of all the coins is $2.00. How many quarters are in the collection?

19. A stamp collector has some 15¢ stamps and some 20¢ stamps. The number of 15¢ stamps is eight less than three times the number of 20¢ stamps. The total value is $4. Find the number of each type of stamp in the collection.

20. An office has some 20¢ stamps and some 28¢ stamps. All together the office has 140 stamps for a total value of $31.20. How many of each type of stamp does the office have?

21. A stamp collection consists of 3¢, 8¢, and 13¢ stamps. The number of 8¢ stamps is three less than twice the number of 3¢ stamps. The number of 13¢ stamps is twice the number of 8¢ stamps. The total value of all the stamps is $2.53. Find the number of 3¢ stamps in the collection.

22. An account executive bought 330 stamps for $79.50. The purchase included 15¢ stamps, 20¢ stamps, and 40¢ stamps. The number of 20¢ stamps is four times the number of 15¢ stamps. How many 40¢ stamps were purchased?

23. A stamp collector has 8¢, 13¢, and 18¢ stamps. The collector has twice as many 8¢ stamps as 18¢ stamps. There are three more 13¢ than 18¢ stamps. The total value of the stamps in the collection is $3.68. Find the number of 18¢ stamps in the collection.

24. A stamp collection consists of 3¢, 12¢, and 15¢ stamps. The number of 3¢ stamps is five times the number of 12¢ stamps. The number of 15¢ stamps is four less than the number of 12¢ stamps. The total value of the stamps in the collection is $3.18. Find the number of 15¢ stamps in the collection.

APPLYING THE CONCEPTS

25. Find three consecutive odd integers such that the product of the second and third minus the product of the first and second is 42.

26. The sum of the digits of a three-digit number is six. The tens digit is one less than the units digit, and the number is twelve more than one hundred times the hundreds digit. Find the number.

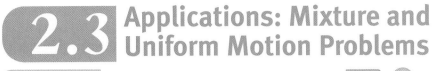

Applications: Mixture and Uniform Motion Problems

Objective A   **To solve value mixture problems**

A **value mixture problem** involves combining two ingredients that have different prices into a single blend. For example, a coffee manufacturer may blend two types of coffee into a single blend.

The solution of a value mixture problem is based on the equation $AC = V$, where A is the amount of the ingredient, C is the cost per unit of the ingredient, and V is the value of the ingredient.

➡ How many pounds of peanuts that cost $2.25 per pound must be mixed with 40 lb of cashews that cost $6.00 per pound to make a mixture that costs $3.50 per pound?

> **Strategy for Solving a Value Mixture Problem**
>
> **1.** For each ingredient in the mixture, write a numerical or variable expression for the amount of the ingredient used, the unit cost of the ingredient, and the value of the amount used. For the mixture, write a numerical or variable expression for the amount, the unit cost of the mixture, and the value of the amount. The results can be recorded in a table.

Pounds of peanuts: x
Pounds of cashews: 40
Pounds of mixture: $x + 40$

	Amount (A)	·	Unit Cost (C)	=	Value (V)
Peanuts	x	·	2.25	=	2.25x
Cashews	40	·	6.00	=	6.00(40)
Mixture	$x + 40$	·	3.50	=	3.50($x + 40$)

> **2.** Determine how the values of the ingredients are related. Use the fact that the sum of the values of all the ingredients taken separately is equal to the value of the mixture.

The sum of the values of the peanuts and the cashews is equal to the value of the mixture.

$$2.25x + 6.00(40) = 3.50(x + 40)$$ • Value of cashews plus value of peanuts equals value of mixture.
$$2.25x + 240 = 3.50x + 140$$
$$-1.25x + 240 = 140$$
$$-1.25x = -100$$
$$x = 80$$

The mixture must contain 80 lb of peanuts.

Example 1

How many ounces of a gold alloy that costs $320 per ounce must be mixed with 100 oz of an alloy that costs $100 per ounce to make a mixture that costs $160 per ounce?

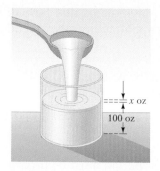

You Try It 1

A butcher combined hamburger that costs $3.00 per pound with hamburger that costs $1.80 per pound. How many pounds of each were used to make a 75-pound mixture costing $2.20 per pound?

Strategy

• Ounces of $320 gold alloy: x
 Ounces of $100 gold alloy: 100
 Ounces of $160 mixture: $x + 100$

Your strategy

	Amount	*Cost*	*Value*
$320 alloy	x	320	$320x$
$100 alloy	100	100	100(100)
Mixture	$x + 100$	160	$160(x + 100)$

• The sum of the values before mixing equals the value after mixing.

$$320x + 100(100) = 160(x + 100)$$

Solution

$$320x + 100(100) = 160(x + 100)$$
$$320x + 10{,}000 = 160x + 16{,}000$$
$$160x + 10{,}000 = 16{,}000$$
$$160x = 6000$$
$$x = 37.5$$

The mixture must contain 37.5 oz of the $320 gold alloy.

Your solution

Solution on p. S4

Objective B   **To solve percent mixture problems**   ‹ 2 ›

The amount of a substance in a solution or alloy can be given as a percent of the total solution or alloy. For example, in a 10% hydrogen peroxide solution, 10% of the total solution is hydrogen peroxide. The remaining 90% is water.

The solution of a **percent mixture problem** is based on the equation $Ar = Q$, where A is the amount of solution or alloy, r is the percent of concentration, and Q is the quantity of a substance in the solution or alloy.

> **TAKE NOTE**
>
> The equation $Ar = Q$ is used to find the amount of a substance in a mixture. For example, the number of grams of silver in 50 g of a 40% alloy is:
>
> $$Ar = Q$$
> $$(50 \, g)(0.40) = Q$$
> $$20 \, g = Q$$

➡ A chemist mixes an 11% acid solution with a 4% acid solution. How many milliliters of each solution should the chemist use to make a 700-milliliter solution that is 6% acid?

> **Strategy for Solving a Percent Mixture Problem**
>
> **1.** For each solution, use the equation $Ar = Q$. Write a numerical or variable expression for the amount of solution, the percent of concentration, and the quantity of the substance in the solution. The results can be recorded in a table.

Amount of 11% solution: x
Amount of 4% solution: $700 - x$
Amount of 6% mixture: 700

	Amount of Solution (A)	·	Percent of Concentration (r)	=	Quantity of Substance (Q)
11% solution	x	·	0.11	=	$0.11x$
4% solution	$700 - x$	·	0.04	=	$0.04(700 - x)$
6% solution	700	·	0.06	=	$0.06(700)$

> **2.** Determine how the quantities of the substance in each solution are related. Use the fact that the sum of the quantities of the substances being mixed is equal to the quantity of the substance after mixing.

The sum of the quantities of the substance in the 11% solution and the 4% solution is equal to the quantity of the substance in the 6% solution.

$$0.11x + 0.04(700 - x) = 0.06(700)$$ • Quantity in 11% solution plus quantity in 4% solution equals quantity in 6% solution.

$$0.11x + 28 - 0.04x = 42$$
$$0.07x + 28 = 42$$
$$0.07x = 14$$
$$x = 200$$

The amount of 4% solution is $700 - x$. Replace x by 200 and evaluate.

$$700 - x = 700 - 200 = 500$$ • $x = 200$

The chemist should use 200 ml of the 11% solution and 500 ml of the 4% solution.

Example 2

How many grams of pure acid must be added to 60 g of an 8% acid solution to make a 20% acid solution?

You Try It 2

A butcher has some hamburger that is 22% fat and some that is 12% fat. How many pounds of each should be mixed to make 80 lb of hamburger that is 18% fat?

Strategy

• Grams of pure acid: x

Your strategy

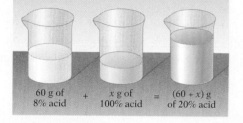

60 g of 8% acid + x g of 100% acid = $(60 + x)$ g of 20% acid

	Amount	Percent	Quantity
Pure Acid (100%)	x	1.00	x
8%	60	0.08	0.08(60)
20%	$60 + x$	0.20	0.20(60 + x)

• The sum of the quantities before mixing equals the quantity after mixing.

$x + 0.08(60) = 0.20(60 + x)$

Solution

$x + 0.08(60) = 0.20(60 + x)$
$x + 4.8 = 12 + 0.20x$
$0.8x + 4.8 = 12$
$0.8x = 7.2$
$x = 9$

To make the 20% acid solution, 9 g of pure acid must be used.

Your solution

Solution on p. S4

Objective C   **To solve uniform motion problems**

A car that travels constantly in a straight line at 55 mph is in uniform motion. **Uniform motion** means that the speed of an object does not change.

The solution of a uniform motion problem is based on the equation $rt = d$, where r is the rate of travel, t is the time spent traveling, and d is the distance traveled.

➡ An executive has an appointment 785 mi from the office. The executive takes a helicopter from the office to the airport and a plane from the airport to the business appointment. The helicopter averages 70 mph and the plane averages 500 mph. The total time spent traveling is 2 h. Find the distance from the executive's office to the airport.

> **Strategy for Solving a Uniform Motion Problem**
>
> **1.** For each object, write a numerical or variable expression for the distance, rate, and time. The results can be recorded in a table. It may also help to draw a diagram.

Unknown time in the helicopter: t
Time in the plane: $2 - t$

	Rate (r)	·	Time (t)	=	Distance (d)
Helicopter	70	·	t	=	$70t$
Plane	500	·	$2 - t$	=	$500(2 - t)$

> **2.** Determine how the distances traveled by each object are related. For example, the total distance traveled by both objects may be known, or it may be known that the two objects traveled the same distance.

The total distance traveled is 785 mi.

$$70t + 500(2 - t) = 785$$ • Distance by helicopter plus distance by plane equals 785.

$$70t + 1000 - 500t = 785$$
$$-430t + 1000 = 785$$
$$-430t = -215$$
$$t = 0.5$$

The time spent traveling from the office to the airport in the helicopter is 0.5 h. To find the distance between these two points, substitute the values of r and t into the equation $rt = d$.

$$rt = d$$
$$70 \cdot 0.5 = d$$ • $r = 70$; $t = 0.5$
$$35 = d$$

The distance from the office to the airport is 35 mi.

Office   Airport   Appointment
⊢— 70 t —⊢—— 500 (2 – t) ——⊣
⊢———————— 785 ————————⊣

Example 3

A long-distance runner started a course running at an average speed of 6 mph. One and one-half hours later, a cyclist traveled the same course at an average speed of 12 mph. How long after the runner started did the cyclist overtake the runner?

You Try It 3

Two small planes start from the same point and fly in opposite directions. The first plane is flying 30 mph faster than the second plane. In 4 h the planes are 1160 mi apart. Find the rate of each plane.

Strategy

Your strategy

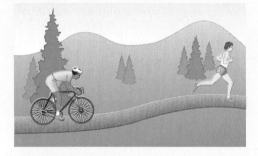

- Unknown time for the cyclist: t
  Time for the runner: $t + 1.5$

	Rate	Time	Distance
Runner	6	$t + 1.5$	$6(t + 1.5)$
Cyclist	12	t	$12t$

- The runner and the cyclist travel the same distance. Therefore the distances are equal.

$$6(t + 1.5) = 12t$$

Solution

Your solution

$$6(t + 1.5) = 12t$$
$$6t + 9 = 12t$$
$$9 = 6t$$
$$\frac{3}{2} = t$$

The cyclist traveled for 1.5 h.

$$t + 1.5 = 1.5 + 1.5 = 3$$

The cyclist overtook the runner 3 h after the runner started.

Solution on p. S4

2.3 Exercises

Objective A   *Application Problems*

1. Forty pounds of cashews costing $5.60 per pound were mixed with 100 lb of peanuts costing $1.89 per pound. Find the cost of the resulting mixture.

2. A coffee merchant combines coffee costing $6 per pound with coffee costing $3.50 per pound. How many pounds of each should be used to make 25 lb of a blend costing $5.25 per pound?

3. Adult tickets for a play cost $5.00 and children's tickets cost $2.00. For one performance, 460 tickets were sold. Receipts for the performance were $1880. Find the number of adult tickets sold.

4. Tickets for a school play sold for $2.50 for each adult and $1.00 for each child. The total receipts for 113 tickets sold were $221. Find the number of adult tickets sold.

5. A restaurant manager mixes 5 L of pure maple syrup that costs $9.50 per liter with imitation maple syrup that costs $4.00 per liter. How much imitation maple syrup is needed to make a mixture that costs $5.00 per liter?

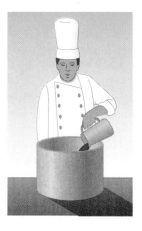

6. To make a flour mixture, a miller combined soybeans that cost $8.50 per bushel with wheat that cost $4.50 per bushel. How many bushels of each were used to make a mixture of 1000 bushels costing $5.50 per bushel?

7. A goldsmith combined pure gold that cost $400 per ounce with an alloy of gold that cost $150 per ounce. How many ounces of each were used to make 50 oz of gold alloy costing $250 per ounce?

8. A silversmith combined pure silver that cost $5.20 per ounce with 50 oz of a silver alloy that cost $2.80 per ounce. How many ounces of the pure silver were used to make an alloy of silver costing $4.40 per ounce?

9. A tea mixture was made from 40 lb of tea costing $5.40 per pound and 60 lb of tea costing $3.25 per pound. Find the cost of the tea mixture.

10. Find the cost per ounce of a sunscreen made from 100 oz of lotion that cost $3.46 per ounce and 60 oz of lotion that cost $12.50 per ounce.

11. The owner of a fruit stand combined cranberry juice that cost $4.60 per gallon with 50 gal of apple juice that cost $2.24 per gallon. How much cranberry juice was used to make the cranapple juice if the mixture cost $3.00 per gallon? Round to the nearest tenth.

12. Walnuts that cost $4.05 per kilogram were mixed with cashews that cost $7.25 per kilogram. How many kilograms of each were used to make a 50-kilogram mixture costing $6.25 per kilogram? Round to the nearest tenth.

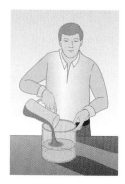

Objective B   *Application Problems*

13. How many pounds of a 15% aluminum alloy must be mixed with 500 lb of a 22% aluminum alloy to make a 20% aluminum alloy?

14. A hospital staff mixed a 75% disinfectant solution with a 25% disinfectant solution. How many liters of each were used to make 20 L of a 40% disinfectant solution?

15. Rubbing alcohol is typically diluted with water to 70% strength. If you need 3.5 oz of 45% rubbing alcohol, how many ounces of 70% rubbing alcohol and how much water should you combine?

16. A silversmith mixed 25 g of a 70% silver alloy with 50 g of a 15% silver alloy. What is the percent concentration of the resulting alloy?

17. How many ounces of pure water must be added to 75 oz of an 8% salt solution to make a 5% salt solution?

18. How many quarts of water must be added to 5 qt of an 80% antifreeze solution to make a 50% antifreeze solution?

19. How many milliliters of alcohol must be added to 200 ml of a 25% iodine solution to make a 10% iodine solution?

20. A butcher has some hamburger that is 21% fat and some that is 15% fat. How many pounds of each should be mixed to make 84 lb of hamburger that is 17% fat?

21. Many fruit drinks are actually only 5% real fruit juice. If you let 2 oz of water evaporate from 12 oz of a drink that is 5% fruit juice, what is the percent concentration of the result?

22. How much water must be evaporated from 6 qt of a 50% antifreeze solution to produce a 75% solution?

23. A car radiator contains 12 qt of a 40% antifreeze solution. How many quarts will have to be replaced with pure antifreeze if the resulting solution is to be 60% antifreeze?

Objective C   *Application Problems*

24. A car traveling at 56 mph overtakes a cyclist who, traveling at 14 mph, had a 1.5-hour head start. How far from the starting point does the car overtake the cyclist?

25. A helicopter traveling 130 mph overtakes a speeding car traveling 80 mph. The car had a 0.5-hour head start. How far from the starting point does the helicopter overtake the car?

26. Two planes are 1620 mi apart and are traveling toward each other. One plane is traveling 120 mph faster than the other plane. The planes meet in 1.5 h. Find the speed of each plane.

27. Two cars are 310 mi apart and are traveling toward each other. One car travels 8 mph faster than the other car. The cars meet in 2.5 h. Find the speed of each car.

28. A ferry leaves a harbor and travels to a resort island at an average speed of 20 mph. On the return trip, the ferry travels at an average speed of 12 mph because of fog. The total time for the trip is 5 h. How far is the island from the harbor?

29. A commuter plane provides transportation from an international airport to the surrounding cities. One commuter plane averaged 250 mph flying to a city and 150 mph returning to the international airport. The total flying time was 4 h. Find the distance between the two airports.

30. A student rode a bicycle to the repair shop and then walked home. The student averaged 12 mph riding to the shop and 3 mph walking home. The round trip took 1 h. How far is it from the student's home to the repair shop?

31. A passenger train leaves a depot 1.5 h after a freight train leaves the same depot. The passenger train is traveling 18 mph faster than the freight train. Find the rate of each train if the passenger train overtakes the freight train in 2.5 h.

32. A plane leaves an airport at 3 P.M. At 4 P.M. another plane leaves the same airport traveling in the same direction at a speed 150 mph faster than that of the first plane. Four hours after the first plane takes off, the second plane is 250 mi ahead of the first plane. How far does the second plane travel?

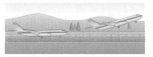

33. A jogger and a cyclist set out at 9 A.M. from the same point headed in the same direction. The average speed of the cyclist is four times the average speed of the jogger. In 2 h, the cyclist is 33 mi ahead of the jogger. How far did the cyclist ride?

APPLYING THE CONCEPTS

34. a. If a parade 2 mi long is proceeding at 3 mph, how long will it take a runner jogging at 6 mph to travel from the front of the parade to the end of the parade?

   b. If a parade 2 mi long is proceeding at 3 mph, how long will it take a runner jogging at 6 mph to travel from the end of the parade to the start of the parade?

35. The concentration of gold in an alloy is measured in karats, which indicate how many parts out of 24 are pure gold. For example, 1 karat is $\frac{1}{24}$ pure gold. What amount of 12-karat gold should be mixed with 3 oz of 24-karat gold to create 14-karat gold, the most commonly used alloy?

36. A student jogs 1 mi at a rate of 8 mph and jogs back at a rate of 6 mph. Does it seem reasonable that the average rate is 7 mph? Why or why not? Support your answer.

37. Two cars are headed directly toward each other at rates of 40 mph and 60 mph. How many miles apart are they 2 min before impact?

38. a. A radiator contains 6 qt of a 25% antifreeze solution. How much should be removed and replaced with pure antifreeze to yield a 33% solution?

   b. A radiator contains 6 qt of a 25% antifreeze solution. How much should be removed and replaced with pure antifreeze to yield a 60% solution?

39. Two birds start flying, at the same time and at the same rate, from the tops of two towers that are 50 ft apart. One tower is 30 ft high, and the other tower is 40 ft high. At exactly the same time, the two birds reach a grass seed on the ground. How far is the grass seed from the base of the 40-foot tower? (This problem appeared in a math text written around A.D. 1200.)

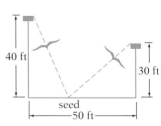

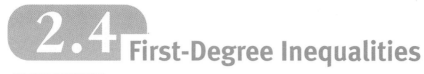

2.4 First-Degree Inequalities

Objective A   **To solve an inequality in one variable**   〈 2 〉 ⊙

The **solution set of an inequality** is a set of numbers, each element of which, when substituted for the variable, results in a true inequality.

The inequality at the right is true if the variable is replaced by (for instance) 3, -1.98, or $\frac{2}{3}$.

$$x - 1 < 4$$
$$3 - 1 < 4$$
$$-1.98 - 1 < 4$$
$$\frac{2}{3} - 1 < 4$$

There are many values of the variable x that will make the inequality $x - 1 < 4$ true. The solution set of the inequality is any number less than 5. The solution set can be written in set-builder notation as $\{x \mid x < 5\}$.

The graph of the solution set of $x - 1 < 4$ is shown at the right.

⟵—+——+——+——+——+——+——+——+——+——+——)——→
   −5 −4 −3 −2 −1 0 1 2 3 4 5

When solving an inequality, we use the **Addition and Multiplication Properties of Inequalities** to rewrite the inequality in the form *variable < constant* or in the form *variable > constant*.

The Addition Property of Inequalities

If $a > b$, then $a + c > b + c$.
If $a < b$, then $a + c < b + c$.

The Addition Property of Inequalities states that the same number can be added to each side of an inequality without changing the solution set of the inequality. This property is also true for an inequality that contains the symbol $\leq$ or $\geq$.

The Addition Property of Inequalities is used to remove a term from one side of an inequality by adding the additive inverse of that term to each side of the inequality. Because subtraction is defined in terms of addition, the same number can be subtracted from each side of an inequality without changing the solution set of the inequality.

⟹ Solve: $x + 2 \geq 4$

$$x + 2 \geq 4$$
$$x + 2 - 2 \geq 4 - 2$$ • Subtract 2 from each side of the inequality.
$$x \geq 2$$ • Simplify.

The solution set is $\{x \mid x \geq 2\}$.

⇒ Solve: $3x - 4 < 2x - 1$

$$3x - 4 < 2x - 1$$

$$3x - 4 - 2x < 2x - 1 - 2x$$

$$x - 4 < -1$$

$$x - 4 + 4 < -1 + 4$$

$$x < 3$$

- Subtract $2x$ from each side of the inequality.

- Add 4 to each side of the inequality.

The solution set is $\{x \mid x < 3\}$.

The Multiplication Property of Inequalities is used to remove a coefficient from one side of an inequality by multiplying each side of the inequality by the reciprocal of the coefficient.

TAKE NOTE

$c > 0$ means c is a positive number. Note that the inequality symbols do not change.

$c < 0$ means c is a negative number. Note that the inequality symbols are reversed.

The Multiplication Property of Inequalities

Rule 1   If $a > b$ and $c > 0$, then $ac > bc$.
         If $a < b$ and $c > 0$, then $ac < bc$.

Rule 2   If $a > b$ and $c < 0$, then $ac < bc$.
         If $a < b$ and $c < 0$, then $ac > bc$.

Here are some examples of this property.

Rule 1		**Rule 2**	
$3 > 2$	$2 < 5$	$3 > 2$	$2 < 5$
$3(4) > 2(4)$	$2(4) < 5(4)$	$3(-4) < 2(-4)$	$2(-4) > 5(-4)$
$12 > 8$	$8 < 20$	$-12 < -8$	$-8 > -20$

Rule 1 states that when each side of an inequality is multiplied by a positive number, the inequality symbol remains the same. However, Rule 2 states that when each side of an inequality is multiplied by a negative number, the inequality symbol must be reversed. Because division is defined in terms of multiplication, when each side of an inequality is divided by a *positive* number, the inequality symbol remains the same. But when each side of an inequality is divided by a *negative* number, the inequality symbol must be reversed.

The Multiplication Property of Inequalities is also true for the symbols $\leq$ and $\geq$.

TAKE NOTE

Each side of the inequality is divided *by* a negative number; the inequality symbol must be reversed.

⇒ Solve: $-3x > 9$

$$-3x > 9$$

$$\frac{-3x}{-3} < \frac{9}{-3}$$

- Divide each side of the inequality by the coefficient -3. Because -3 is a negative number, the inequality symbol must be reversed.

$$x < -3$$

The solution set is $\{x \mid x < -3\}$.

➡ Solve: $3x + 2 < -4$

$3x + 2 < -4$

$3x < -6$ • Subtract 2 from each side of the inequality.

$\dfrac{3x}{3} < \dfrac{-6}{3}$ • Divide each side of the inequality by the coefficient 3.

$x < -2$

The solution set is $\{x \,|\, x < -2\}$.

➡ Solve: $2x - 9 > 4x + 5$

$2x - 9 > 4x + 5$

$-2x - 9 > 5$ • Subtract 4x from each side of the inequality.

$-2x > 14$ • Add 9 to each side of the inequality.

$\dfrac{-2x}{-2} < \dfrac{14}{-2}$ • Divide each side of the inequality by the coefficient −2. Reverse the inequality symbol.

$x < -7$

The solution set is $\{x \,|\, x < -7\}$.

➡ Solve: $5(x - 2) \ge 9x - 3(2x - 4)$

$5(x - 2) \ge 9x - 3(2x - 4)$

$5x - 10 \ge 9x - 6x + 12$ • Use the Distributive Property to remove parentheses.

$5x - 10 \ge 3x + 12$

$2x - 10 \ge 12$ • Subtract 3x from each side of the inequality.

$2x \ge 22$ • Add 10 to each side of the inequality.

$\dfrac{2x}{2} \ge \dfrac{22}{2}$ • Divide each side of the inequality by the coefficient 2.

$x \ge 11$

The solution set is $\{x \,|\, x \ge 11\}$.

Example 1
Solve: $x + 3 > 4x + 6$

Solution

$x + 3 > 4x + 6$
$-3x + 3 > 6$ • Subtract 4x from each side.
$-3x > 3$ • Subtract 3 from each side.
$\dfrac{-3x}{-3} < \dfrac{3}{-3}$ • Divide each side by −3.
$x < -1$

The solution set is $\{x \,|\, x < -1\}$.

You Try It 1
Solve: $2x - 1 < 6x + 7$

Your solution

Solution on p. S5

Example 2   Solve:
$3x - 5 \leq 3 - 2(3x + 1)$

You Try It 2   Solve:
$5x - 2 \leq 4 - 3(x - 2)$

Solution

$$3x - 5 \leq 3 - 2(3x + 1)$$
$$3x - 5 \leq 3 - 6x - 2$$
$$3x - 5 \leq 1 - 6x$$
$$9x - 5 \leq 1$$
$$9x \leq 6$$
$$\frac{9x}{9} \leq \frac{6}{9}$$
$$x \leq \frac{2}{3}$$

$$\left\{ x \,\middle|\, x \leq \frac{2}{3} \right\}$$

Your solution

Solution on p. S5

Objective B   **To solve a compound inequality** ⟨ 2 ⟩

A **compound inequality** is formed by joining two inequalities with a connective word such as "and" or "or." The inequalities at the right are compound inequalities.

$2x < 4$ and $3x - 2 > -8$

$2x + 3 > 5$ or $x + 2 < 5$

The solution set of a compound inequality with the connective word *and* is the set of all elements that appear in the solution sets of both inequalities. Therefore, it is the intersection of the solution sets of the two inequalities.

➡ Solve: $2x < 6$ and $3x + 2 > -4$

$2x < 6$	and	$3x + 2 > -4$	
$x < 3$		$3x > -6$	• Solve each inequality.
$\{x \mid x < 3\}$		$x > -2$	
		$\{x \mid x > -2\}$	

The solution set of a compound inequality with *and* is the intersection of the solution sets of the two inequalities.

$$\{x \mid x < 3\} \cap \{x \mid x > -2\} = \{x \mid -2 < x < 3\}$$

➡ Solve: $-3 < 2x + 1 < 5$

This inequality is equivalent to the compound inequality $-3 < 2x + 1$ and $2x + 1 < 5$.

$-3 < 2x + 1$	and $2x + 1 < 5$	
$-4 < 2x$	$2x < 4$	• Solve each inequality.
$-2 < x$	$x < 2$	
$\{x \mid x > -2\}$	$\{x \mid x < 2\}$	

The solution set of a compound inequality with *and* is the intersection of the solution sets of the two inequalities.

$$\{x \mid x > -2\} \cap \{x \mid x < 2\} = \{x \mid -2 < x < 2\}$$

There is an alternative method for solving the inequality in the last example.

➡ Solve: $-3 < 2x + 1 < 5$

$$-3 < 2x + 1 < 5$$
$$-3 - 1 < 2x + 1 - 1 < 5 - 1$$
$$-4 < 2x < 4$$
$$\frac{-4}{2} < \frac{2x}{2} < \frac{4}{2}$$
$$-2 < x < 2$$

- Subtract 1 from each of the three parts of the inequality.

- Divide each of the three parts of the inequality by the coefficient 2.

The solution set is $\{x \mid -2 < x < 2\}$.

The solution set of a compound inequality with the connective word *or* is the union of the solution sets of the two inequalities.

➡ Solve: $2x + 3 > 7$ or $4x - 1 < 3$

$$
\begin{array}{ll}
2x + 3 > 7 & \text{or} \quad 4x - 1 < 3 \\
2x > 4 & \qquad 4x < 4 \\
x > 2 & \qquad x < 1 \\
\{x \mid x > 2\} & \qquad \{x \mid x < 1\}
\end{array}
$$

- Solve each inequality.

Find the union of the solution sets.

$$\{x \mid x > 2\} \cup \{x \mid x < 1\} = \{x \mid x > 2 \text{ or } x < 1\}$$

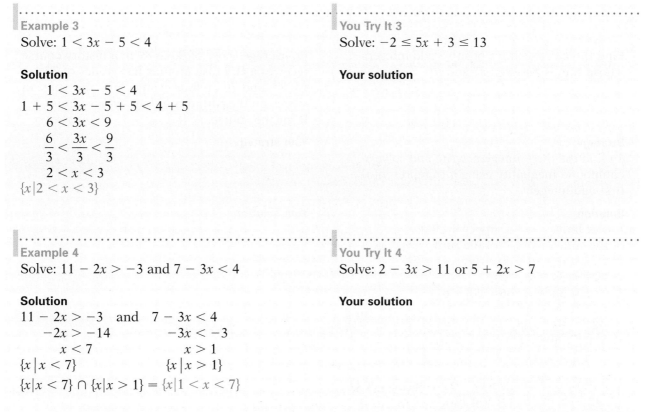

Example 3
Solve: $1 < 3x - 5 < 4$

Solution
$$1 < 3x - 5 < 4$$
$$1 + 5 < 3x - 5 + 5 < 4 + 5$$
$$6 < 3x < 9$$
$$\frac{6}{3} < \frac{3x}{3} < \frac{9}{3}$$
$$2 < x < 3$$
$$\{x \mid 2 < x < 3\}$$

You Try It 3
Solve: $-2 \le 5x + 3 \le 13$

Your solution

Example 4
Solve: $11 - 2x > -3$ and $7 - 3x < 4$

Solution
$$
\begin{array}{ll}
11 - 2x > -3 & \text{and} \quad 7 - 3x < 4 \\
-2x > -14 & \qquad -3x < -3 \\
x < 7 & \qquad x > 1 \\
\{x \mid x < 7\} & \qquad \{x \mid x > 1\}
\end{array}
$$
$$\{x \mid x < 7\} \cap \{x \mid x > 1\} = \{x \mid 1 < x < 7\}$$

You Try It 4
Solve: $2 - 3x > 11$ or $5 + 2x > 7$

Your solution

Solutions on p. S5

Objective C   **To solve application problems**

Example 5

A cellular phone company advertises two pricing plans. The first is $19.95 per month with 20 free minutes and $.39 per minute thereafter. The second is $23.95 per month with 20 free minutes and $.30 per minute thereafter. How many minutes can you talk per month for the first plan to cost less than the second?

Strategy

To find the number of minutes, write and solve an inequality using N to represent the number of minutes. Then $N - 20$ is the number of minutes for which you are charged after the first free 20 min.

Solution

Cost of first plan $<$ cost of second plan
$$19.95 + 0.39(N - 20) < 23.95 + 0.30(N - 20)$$
$$19.95 + 0.39N - 7.8 < 23.95 + 0.30N - 6$$
$$12.15 + 0.39N < 17.95 + 0.30N$$
$$12.15 + 0.09N < 17.95$$
$$0.09N < 5.8$$
$$N < 64.\overline{4}$$

The first plan costs less if you talk less than 65 min.

You Try It 5

The base of a triangle is 12 in. and the height is $(x + 2)$ in. Express as an integer the maximum height of the triangle when the area is less than 50 in².

Your strategy

Your solution

Example 6

Find three consecutive positive odd integers whose sum is between 27 and 51.

Strategy

To find the three integers, write and solve a compound inequality using n to represent the first odd integer.

Solution

Lower limit $<$ sum $<$ upper limit
of the sum   of the sum
$$27 < n + (n + 2) + (n + 4) < 51$$
$$27 < 3n + 6 < 51$$
$$27 - 6 < 3n + 6 - 6 < 51 - 6$$
$$21 < 3n < 45$$
$$\frac{21}{3} < \frac{3n}{3} < \frac{45}{3}$$
$$7 < n < 15$$

The three odd integers are 9, 11, and 13; or 11, 13, and 15; or 13, 15, and 17.

You Try It 6

An average score of 80 to 89 in a history course receives a B. Luisa Montez has grades of 72, 94, 83, and 70 on four exams. Find the range of scores on the fifth exam that will give Luisa a B for the course.

Your strategy

Your solution

Solutions on p. S5

2.4 Exercises

Objective A

1. State the Addition Property of Inequalities and give numerical examples of its use.

2. State the Multiplication Property of Inequalities and give numerical examples of its use.

3. Which numbers are solutions of the inequality $x + 7 \leq -3$?
   a. -17 **b.** 8 **c.** -10 **d.** 0

4. Which numbers are solutions of the inequality $2x - 1 > 5$?
   a. 6 **b.** -4 **c.** 3 **d.** 5

Solve.

5. $x - 3 < 2$

6. $x + 4 \geq 2$

7. $4x \leq 8$

8. $6x > 12$

9. $-2x > 8$

10. $-3x \leq -9$

11. $3x - 1 > 2x + 2$

12. $5x + 2 \geq 4x - 1$

13. $2x - 1 > 7$

14. $3x + 2 < 8$

15. $5x - 2 \leq 8$

16. $4x + 3 \leq -1$

17. $6x + 3 > 4x - 1$

18. $7x + 4 < 2x - 6$

19. $8x + 1 \geq 2x + 13$

20. $5x - 4 < 2x + 5$

21. $4 - 3x < 10$

22. $2 - 5x > 7$

23. $7 - 2x \geq 1$

24. $3 - 5x \leq 18$

25. $-3 - 4x > -11$

26. $-2 - x < 7$

27. $4x - 2 < x - 11$

28. $6x + 5 \leq x - 10$

29. $x + 7 \geq 4x - 8$

30. $3x + 1 \leq 7x - 15$

31. $3x + 2 \leq 7x + 4$

32. $3x - 5 \geq -2x + 5$

33. $\frac{3}{5}x - 2 < \frac{3}{10} - x$

34. $\frac{5}{6}x - \frac{1}{6} < x - 4$

35. $\frac{2}{3}x - \frac{3}{2} < \frac{7}{6} - \frac{1}{3}x$

36. $\frac{7}{12}x - \frac{3}{2} < \frac{2}{3}x + \frac{5}{6}$

37. $\frac{1}{2}x - \frac{3}{4} < \frac{7}{4}x - 2$

38. $6 - 2(x - 4) \leq 2x + 10$

39. $4(2x - 1) > 3x - 2(3x - 5)$

40. $2(1 - 3x) - 4 > 10 + 3(1 - x)$

41. $2 - 5(x + 1) \geq 3(x - 1) - 8$

42. $2 - 2(7 - 2x) < 3(3 - x)$

43. $3 + 2(x + 5) \geq x + 5(x + 1) + 1$

44. $10 - 13(2 - x) < 5(3x - 2)$

45. $3 - 4(x + 2) \leq 6 + 4(2x + 1)$

46. $3x - 2(3x - 5) \leq 2 - 5(x - 4)$

47. $12 - 2(3x - 2) \geq 5x - 2(5 - x)$

Objective B

48. **a.** Which set operation is used when a compound inequality is combined with *or*?
  b. Which set operation is used when a compound inequality is combined with *and*?

49. Explain why writing $-3 < x > 4$ does not make sense.

Solve.

50. $3x < 6$ and $x + 2 > 1$

51. $x - 3 \leq 1$ and $2x \geq -4$

52. $x + 2 \geq 5$ or $3x \leq 3$

53. $2x < 6$ or $x - 4 > 1$

54. $-2x > -8$ and $-3x < 6$

55. $\frac{1}{2}x > -2$ and $5x < 10$

56. $\frac{1}{3}x < -1$ or $2x > 0$

57. $\frac{2}{3}x > 4$ or $2x < -8$

58. $x + 4 \geq 5$ and $2x \geq 6$

59. $3x < -9$ and $x - 2 < 2$

60. $-5x > 10$ and $x + 1 > 6$

61. $7x < 14$ and $1 - x < 4$

62. $2x - 3 > 1$ and $3x - 1 < 2$

63. $4x + 1 < 5$ and $4x + 7 > -1$

64. $3x + 7 < 10$ or $2x - 1 > 5$

65. $6x - 2 < -14$ or $5x + 1 > 11$

66. $-5 < 3x + 4 < 16$

67. $5 < 4x - 3 < 21$

68. $0 < 2x - 6 < 4$

69. $-2 < 3x + 7 < 1$

70. $4x - 1 > 11$ or $4x - 1 \leq -11$

71. $3x - 5 > 10$ or $3x - 5 < -10$

72. $9x - 2 < 7$ and $3x - 5 > 10$

73. $8x + 2 \leq -14$ and $4x - 2 > 10$

74. $3x - 11 < 4$ or $4x + 9 \geq 1$

75. $5x + 12 \geq 2$ or $7x - 1 \leq 13$

76. $-6 \leq 5x + 14 \leq 24$

77. $3 \leq 7x - 14 \leq 31$

78. $3 - 2x > 7$ and $5x + 2 > -18$

79. $1 - 3x < 16$ and $1 - 3x > -16$

80. $5 - 4x > 21$ or $7x - 2 > 19$

81. $6x + 5 < -1$ or $1 - 2x < 7$

82. $3 - 7x \leq 31$ and $5 - 4x > 1$

83. $9 - x \geq 7$ and $9 - 2x < 3$

Objective C   *Application Problems*

84. Five times the difference between a number and two is greater than the quotient of two times the number and three. Find the smallest integer that will satisfy the inequality.

85. Two times the difference between a number and eight is less than or equal to five times the sum of the number and four. Find the smallest number that will satisfy the inequality.

86. The length of a rectangle is two feet more than four times the width. Express as an integer the maximum width of the rectangle when the perimeter is less than thirty-four feet.

87. The length of a rectangle is five centimeters less than twice the width. Express as an integer the maximum width of the rectangle when the perimeter is less than sixty centimeters.

88. In 1997, the computer service America Online offered its customers a rate of $19.95 per month for unlimited use or $4.95 per month with 3 free hours plus $2.50 for each hour thereafter. How many hours can you use this service per month if the second plan is to cost you less than the first?

89. TopPage advertises local paging service for $6.95 per month for up to 400 pages, and $.10 per page thereafter. A competitor advertises service for $3.95 per month for up to 400 pages and $.15 per page thereafter. For what number of pages per month is the TopPage plan less expensive?

90. Suppose PayRite Rental Cars rents compact cars for $32 per day with unlimited mileage and Otto Rentals offers compact cars for $19.99 per day but charges $.19 for each mile beyond 100 mi driven per day. You want to rent a car for one week. How many miles can you drive during the week if Otto Rentals is to be less expensive than PayRite?

91. During a weekday, to call a city 40 mi away from a certain pay phone costs $.70 for the first 3 min and $.15 for each additional minute. If you use a calling card, there is a $.35 fee and then the rates are $.196 for the first minute and $.126 for each additional minute. How long must a call be if it is cheaper to pay with coins rather than a calling card?

92. The temperature range for a week was between 14°F and 77°F. Find the temperature range in Celsius degrees. $F = \dfrac{9}{5}C + 32$

93. The temperature range for a week in a mountain town was between 0°C and 30°C. Find the temperature range in Fahrenheit degrees. $C = \dfrac{5(F - 32)}{9}$

94. You are a sales account executive earning $1200 per month plus 6% commission on the amount of sales. Your goal is to earn a minimum of $6000 per month. What amount of sales will enable you to earn $6000 or more per month?

95. George Stoia earns $1000 per month plus 5% commission on the amount of sales. George's goal is to earn a minimum of $3200 per month. What amount of sales will enable George to earn $3200 or more per month?

96. Heritage National Bank offers two different checking accounts. The first charges $3 per month, and $.50 per check after the first 10 checks. The second account charges $8 per month with unlimited check writing. How many checks can be written per month if the first account is to be less expensive than the second account?

97. Glendale Federal Bank offers a checking account to small businesses. The charge is $8 per month plus $.12 per check after the first 100 checks. A competitor is offering an account for $5 per month plus $.15 per check after the first 100 checks. If a business chooses the first account, how many checks does the business write monthly if it is assumed that the first account will cost less than the competitor's account?

98. An average score of 90 or above in a history class receives an A grade. You have scores of 95, 89, and 81 on three exams. Find the range of scores on the fourth exam that will give you an A grade for the course.

99. An average of 70 to 79 in a mathematics class receives a C grade. A student has scores of 56, 91, 83, and 62 on four tests. Find the range of scores on the fifth test that will give the student a C for the course.

100. Find four consecutive integers whose sum is between 62 and 78.

101. Find three consecutive even integers whose sum is between 30 and 51.

APPLYING THE CONCEPTS

102. Determine whether the following statements are always true, sometimes true, or never true.
   a. If $a > b$, then $-a < -b$.
   b. If $a < b$ and $a \neq 0, b \neq 0$, then $\frac{1}{a} < \frac{1}{b}$.
   c. When dividing both sides of an inequality by an integer, we must reverse the inequality symbol.
   d. If $a < 1$, then $a^2 < a$.
   e. If $a < b < 0$ and $c < d < 0$, then $ac > bd$.

103. The following is offered as the solution of $2 + 3(2x - 4) < 6x + 5$.

$$2 + 3(2x - 4) < 6x + 5$$
$$2 + 6x - 12 < 6x + 5 \qquad \bullet \text{ Use the Distributive Property.}$$
$$6x - 10 < 6x + 5 \qquad \bullet \text{ Simplify.}$$
$$6x - 6x - 10 < 6x - 6x + 5 \qquad \bullet \text{ Subtract } 6x \text{ from each side.}$$
$$-10 < 5$$

Because $-10 < 5$ is a true inequality, the solution set is all real numbers.

If this is correct, so state. If it is not correct, explain the incorrect step and supply the correct answer.

2.5 Absolute Value Equations and Inequalities

Objective A To solve an absolute value equation

The **absolute value** of a number is its distance from zero on the number line. Distance is always a positive number or zero. Therefore, the absolute value of a number is always a positive number or zero.

The distance from 0 to 3 or from 0 to -3 is 3 units.

$$|3| = 3 \qquad |-3| = 3$$

Absolute value can be used to represent the distance between any two points on the number line. The **distance between two points** on the number line is the absolute value of the difference between the coordinates of the two points.

The distance between point a and point b is given by $|b - a|$.

The distance between 4 and -3 on the number line is 7 units. Note that the order in which the coordinates are subtracted does not affect the distance.

$$\begin{aligned} \text{Distance} &= |-3 - 4| & \text{Distance} &= |4 - (-3)| \\ &= |-7| & &= |7| \\ &= 7 & &= 7 \end{aligned}$$

For any two numbers a and b, $|b - a| = |a - b|$.

An equation containing an absolute value symbol is called an **absolute value equation.** Here are three examples.

$$|x| = 3 \qquad |x + 2| = 8 \qquad |3x - 4| = 5x - 9$$

Solution of an Absolute Value Equation

If $a \geq 0$ and $|x| = a$, then $x = a$ or $x = -a$.

For instance, given $|x| = 3$, then $x = 3$ or $x = -3$ because $|3| = 3$ and $|-3| = 3$. We can solve this equation as follows:

$$|x| = 3$$
$$x = 3 \qquad x = -3$$

• Remove the absolute value sign from $|x|$ and let x equal 3 and the opposite of 3.

Check:

$$\begin{array}{c|c} |x| = 3 & |x| = 3 \\ \hline |3| \ | \ 3 & |-3| \ | \ 3 \\ 3 = 3 & 3 = 3 \end{array}$$

The solutions are 3 and -3.

➡ Solve: $|x + 2| = 8$

$$|x + 2| = 8$$

$x + 2 = 8$	$x + 2 = -8$
$x = 6$	$x = -10$

- Remove the absolute value sign and rewrite as two equations.
- Solve each equation.

Check: $|x + 2| = 8$

$|6 + 2|$	8
$|8|$	8
	$8 = 8$

$|x + 2| = 8$

$|-10 + 2|$	8
$|-8|$	8
	$8 = 8$

The solutions are 6 and -10.

➡ Solve: $|5 - 3x| - 8 = -4$

$$|5 - 3x| - 8 = -4$$
$$|5 - 3x| = 4$$

$5 - 3x = 4$	$5 - 3x = -4$
$-3x = -1$	$-3x = -9$
$x = \dfrac{1}{3}$	$x = 3$

- Solve for the absolute value.
- Remove the absolute value sign and rewrite as two equations.
- Solve each equation.

Check: $|5 - 3x| - 8 = -4$

$\left|5 - 3\left(\frac{1}{3}\right)\right| - 8$	-4
$|5 - 1| - 8$	-4
$4 - 8$	-4
	$-4 = -4$

$|5 - 3x| - 8 = -4$

$|5 - 3(3)| - 8$	-4
$|5 - 9| - 8$	-4
$4 - 8$	-4
	$-4 = -4$

The solutions are $\dfrac{1}{3}$ and 3.

Example 1
Solve: $|2 - x| = 12$

Solution
$$|2 - x| = 12$$

$2 - x = 12$	$2 - x = -12$
$-x = 10$	$-x = -14$
$x = -10$	$x = 14$

The solutions are -10 and 14.

Example 2
Solve: $|2x| = -4$

Solution
$$|2x| = -4$$

There is no solution to this equation because the absolute value of a number must be nonnegative.

You Try It 1
Solve: $|2x - 3| = 5$

Your solution

You Try It 2
Solve: $|x - 3| = -2$

Your solution

Solutions on p. S5

Example 3

Solve: $3 - |2x - 4| = -5$

Solution

$$3 - |2x - 4| = -5$$
$$-|2x - 4| = -8$$
$$|2x - 4| = 8$$

$2x - 4 = 8$ $2x - 4 = -8$

$2x = 12$ $2x = -4$

$x = 6$ $x = -2$

The solutions are 6 and -2.

You Try It 3

Solve: $5 - |3x + 5| = 3$

Your solution

Solution on p. S5

 Objective B   **To solve an absolute value inequality**   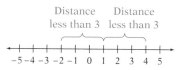

Recall that absolute value represents the distance between two points. For example, the solutions of the absolute value equation $|x - 1| = 3$ are the numbers whose distance from 1 is 3. Therefore the solutions are -2 and 4.

The solutions of the absolute value inequality $|x - 1| < 3$ are the numbers whose distance from 1 is less than 3. Therefore, the solutions are the numbers greater than -2 and less than 4. The solution set is $\{x | -2 < x < 4\}$.

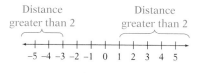

To solve an absolute value inequality of the form $|ax + b| < c$, solve the equivalent compound inequality $-c < ax + b < c$.

⇒ Solve: $|3x - 1| < 5$

$$|3x - 1| < 5$$
$$-5 < 3x - 1 < 5$$
$$-5 + 1 < 3x - 1 + 1 < 5 + 1$$
$$-4 < 3x < 6$$
$$\frac{-4}{3} < \frac{3x}{3} < \frac{6}{3}$$
$$-\frac{4}{3} < x < 2$$

• Solve the equivalent compound inequality.

The solution set is $\left\{x \left| -\frac{4}{3} < x < 2\right.\right\}$.

The solutions of the absolute value inequality $|x + 1| > 2$ are the numbers whose distance from -1 is greater than 2. Therefore, the solutions are the numbers that are less than -3 or greater than 1. The solution set of $|x + 1| > 2$ is $\{x | x < -3 \text{ or } x > 1\}$.

To solve an absolute value inequality of the form $|ax + b| > c$, solve the equivalent compound inequality $ax + b < -c$ or $ax + b > c$.

⇒ Solve: $|3 - 2x| > 1$

$$3 - 2x < -1 \quad \text{or} \quad 3 - 2x > 1$$
$$-2x < -4 \qquad\qquad -2x > -2 \qquad \bullet \text{ Solve each inequality.}$$
$$x > 2 \qquad\qquad\quad x < 1$$
$$\{x \,|\, x > 2\} \qquad\quad \{x \,|\, x < 1\}$$

The solution of a compound inequality with *or* is the union of the solution sets of the two inequalities.

$$\{x \,|\, x > 2\} \cup \{x \,|\, x < 1\} = \{x \,|\, x > 2 \text{ or } x < 1\}$$

The rules for solving these absolute value inequalities are summarized below.

Solutions of Absolute Value Inequalities

To solve an absolute value inequality of the form $|ax + b| < c$, solve the equivalent compound inequality $-c < ax + b < c$.

To solve an absolute value inequality of the form $|ax + b| > c$, solve the equivalent compound inequality $ax + b < -c$ or $ax + b > c$.

Example 4   Solve: $|4x - 3| < 5$

Solution   Solve the equivalent compound inequality.

$$-5 < 4x - 3 < 5$$
$$-5 + 3 < 4x - 3 + 3 < 5 + 3$$
$$-2 < 4x < 8$$
$$\frac{-2}{4} < \frac{4x}{4} < \frac{8}{4}$$
$$-\frac{1}{2} < x < 2$$
$$\left\{ x \,\middle|\, -\frac{1}{2} < x < 2 \right\}$$

You Try It 4   Solve: $|3x + 2| < 8$

Your solution

Example 5   Solve: $|x - 3| < 0$

Solution   The absolute value of a number is greater than or equal to zero, since it measures the number's distance from zero on the number line. Therefore, the solution set of $|x - 3| < 0$ is the empty set.

You Try It 5   **Solve:** $|3x - 7| < 0$

Your solution

Solutions on p. S6

Example 6   Solve: $|x + 4| > -2$

Solution   The absolute value of a number is greater than or equal to zero. Therefore, the solution set of $|x + 4| > -2$ is the set of real numbers.

You Try It 6   Solve: $|2x + 7| \geq -1$

Your solution

Example 7   Solve: $|2x - 1| > 7$

Solution   Solve the equivalent compound inequality.

$$2x - 1 < -7 \quad \text{or} \quad 2x - 1 > 7$$
$$2x < -6 \qquad\qquad 2x > 8$$
$$x < -3 \qquad\qquad x > 4$$
$$\{x \mid x < -3\} \qquad\qquad \{x \mid x > 4\}$$
$$\{x \mid x < -3\} \cup \{x \mid x > 4\}$$
$$= \{x \mid x < -3 \text{ or } x > 4\}$$

You Try It 7   **Solve:** $|5x + 3| > 8$

Your solution

Solutions on p. S6

Objective C   **To solve application problems**

The **tolerance** of a component, or part, is the acceptable amount by which the component may vary from a given measurement. For example, the diameter of a piston may vary from the given measurement of 9 cm by 0.001 cm. This is written 9 cm ± 0.001 cm and is read "9 centimeters plus or minus 0.001 centimeter." The maximum diameter, or **upper limit,** of the piston is 9 cm + 0.001 cm = 9.001 cm. The minimum diameter, or **lower limit,** is 9 cm − 0.001 cm = 8.999 cm.

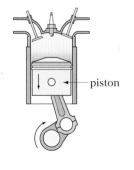
piston

The lower and upper limits of the diameter of the piston could also be found by solving the absolute value inequality $|d - 9| \leq 0.001$, where d is the diameter of the piston.

$$|d - 9| \leq 0.001$$
$$-0.001 \leq d - 9 \leq 0.001$$
$$-0.001 + 9 \leq d - 9 + 9 \leq 0.001 + 9$$
$$8.999 \leq d \leq 9.001$$

The lower and upper limits of the diameter of the piston are 8.999 cm and 9.001 cm.

Example 8

The diameter of a piston for an automobile is $3\frac{5}{16}$ in. with a tolerance of $\frac{1}{64}$ in. Find the lower and upper limits of the diameter of the piston.

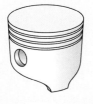

Strategy

To find the lower and upper limits of the diameter of the piston, let d represent the diameter of the piston, T the tolerance, and L the lower and upper limits of the diameter. Solve the absolute value inequality $|L - d| \le T$ for L.

Solution

$$|L - d| \le T$$

$$\left| L - 3\frac{5}{16} \right| \le \frac{1}{64}$$

$$-\frac{1}{64} \le L - 3\frac{5}{16} \le \frac{1}{64}$$

$$-\frac{1}{64} + 3\frac{5}{16} \le L - 3\frac{5}{16} + 3\frac{5}{16} \le \frac{1}{64} + 3\frac{5}{16}$$

$$3\frac{19}{64} \le L \le 3\frac{21}{64}$$

The lower and upper limits of the diameter of the piston are $3\frac{19}{64}$ in. and $3\frac{21}{64}$ in.

You Try It 8

A machinist must make a bushing that has a tolerance of 0.003 in. The diameter of the bushing is 2.55 in. Find the lower and upper limits of the diameter of the bushing.

Your strategy

Your solution

Solution on p. S6

2.5 Exercises

Objective A

1. Is 2 a solution of $|x - 8| = 6$?

2. Is -2 a solution of $|2x - 5| = 9$?

3. Is -1 a solution of $|3x - 4| = 7$?

4. Is 1 a solution of $|6x - 1| = -5$?

Solve.

5. $|x| = 7$

6. $|a| = 2$

7. $|b| = 4$

8. $|c| = 12$

9. $|-y| = 6$

10. $|-t| = 3$

11. $|-a| = 7$

12. $|-x| = 3$

13. $|x| = -4$

14. $|y| = -3$

15. $|-t| = -3$

16. $|-y| = -2$

17. $|x + 2| = 3$

18. $|x + 5| = 2$

19. $|y - 5| = 3$

20. $|y - 8| = 4$

21. $|a - 2| = 0$

22. $|a + 7| = 0$

23. $|x - 2| = -4$

24. $|x + 8| = -2$

25. $|3 - 4x| = 9$

26. $|2 - 5x| = 3$

27. $|2x - 3| = 0$

28. $|5x + 5| = 0$

29. $|3x - 2| = -4$

30. $|2x + 5| = -2$

31. $|x - 2| - 2 = 3$

32. $|x - 9| - 3 = 2$

33. $|3a + 2| - 4 = 4$

34. $|2a + 9| + 4 = 5$

35. $|2 - y| + 3 = 4$

36. $|8 - y| - 3 = 1$

37. $|2x - 3| + 3 = 3$

38. $|4x - 7| - 5 = -5$

39. $|2x - 3| + 4 = -4$

40. $|3x - 2| + 1 = -1$

41. $|6x - 5| - 2 = 4$

42. $|4b + 3| - 2 = 7$

43. $|3t + 2| + 3 = 4$

44. $|5x - 2| + 5 = 7$

45. $3 - |x - 4| = 5$

46. $2 - |x - 5| = 4$

47. $8 - |2x - 3| = 5$

48. $8 - |3x + 2| = 3$

49. $|2 - 3x| + 7 = 2$

50. $|1 - 5a| + 2 = 3$

51. $|8 - 3x| - 3 = 2$

52. $|6 - 5b| - 4 = 3$

53. $|2x - 8| + 12 = 2$

54. $|3x - 4| + 8 = 3$

55. $2 + |3x - 4| = 5$

56. $5 + |2x + 1| = 8$

57. $5 - |2x + 1| = 5$

58. $3 - |5x + 3| = 3$

59. $6 - |2x + 4| = 3$

60. $8 - |3x - 2| = 5$

61. $8 - |1 - 3x| = -1$

62. $3 - |3 - 5x| = -2$

63. $5 + |2 - x| = 3$

64. $6 + |3 - 2x| = 2$

Objective B

Solve.

65. $|x| > 3$

66. $|x| < 5$

67. $|x + 1| > 2$

68. $|x - 2| > 1$

69. $|x - 5| \le 1$

70. $|x - 4| \le 3$

71. $|2 - x| \ge 3$

72. $|3 - x| \ge 2$

73. $|2x + 1| < 5$

74. $|3x - 2| < 4$

75. $|5x + 2| > 12$

76. $|7x - 1| > 13$

77. $|4x - 3| \le -2$

78. $|5x + 1| \le -4$

79. $|2x + 7| > -5$

80. $|3x - 1| > -4$

81. $|4 - 3x| \ge 5$

82. $|7 - 2x| > 9$

83. $|5 - 4x| \le 13$

84. $|3 - 7x| < 17$

85. $|6 - 3x| \le 0$

86. $|10 - 5x| \ge 0$

87. $|2 - 9x| > 20$

88. $|5x - 1| < 16$

89. $|2x - 3| + 2 < 8$

90. $|3x - 5| + 1 < 7$

91. $|2 - 5x| - 4 > -2$

92. $|4 - 2x| - 9 > -3$

93. $8 - |2x - 5| < 3$

94. $12 - |3x - 4| > 7$

Objective C   *Application Problems*

95. The diameter of a bushing is 1.75 in. The bushing has a tolerance of 0.008 in. Find the lower and upper limits of the diameter of the bushing.

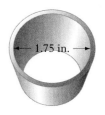
1.75 in.

96. A machinist must make a bushing that has a tolerance of 0.004 in. The diameter of the bushing is 3.48 in. Find the lower and upper limits of the diameter of the bushing.

97. An electric motor is designed to run on 220 volts plus or minus 25 volts. Find the lower and upper limits of voltage on which the motor will run.

98. A power strip is utilized on a computer to prevent the loss of program-ming by electrical surges. The power strip is designed to allow 110 volts plus or minus 16.5 volts. Find the lower and upper limits of voltage to the computer.

99. A piston rod for an automobile is $9\frac{5}{8}$ in. long with a tolerance of $\frac{1}{32}$ in. Find the lower and upper limits of the length of the piston rod.

100. A piston rod for an automobile is $9\frac{3}{8}$ in. long with a tolerance of $\frac{1}{64}$ in. Find the lower and upper limits of the length of the piston rod.

The tolerance of the resistors used in electronics is given as a percent. Use your calculator for the fol-lowing exercises.

101. Find the lower and upper limits of a 29,000-ohm resistor with a 2% tol-erance.

102. Find the lower and upper limits of a 15,000-ohm resistor with a 10% tol-erance.

103. Find the lower and upper limits of a 25,000-ohm resistor with a 5% tol-erance.

104. Find the lower and upper limits of a 56-ohm resistor with a 5% toler-ance.

APPLYING THE CONCEPTS

105. For what values of the variable is the equation true? Write the solution set in set-builder notation.
   a. $|x + 3| = x + 3$ **b.** $|a - 4| = 4 - a$

106. Write an absolute value inequality to represent all real numbers within 5 units of 2.

107. Replace the question mark with $\leq$, $\geq$, or $=$.
   a. $|x + y| \ ? \ |x| + |y|$ **b.** $|x - y| \ ? \ |x| - |y|$

   c. $||x| - |y|| \ ? \ |x| - |y|$ **d.** $\left|\frac{x}{y}\right| \ ? \ \frac{|x|}{|y|}, y \neq 0$

   e. $|xy| \ ? \ |x||y|$

Focus on Problem Solving

Understand the Problem

The first of the four steps that Polya advocated to solve problems is to *understand the problem*. This aspect of problem solving is frequently not given enough attention. There are various exercises that you can try to achieve a good understanding of a problem. Some of these are stated in the Focus on Problem Solving in the chapter entitled "Review of Real Numbers" and are reviewed here.

- Try to restate the problem in your own words.
- Determine what is known about this type of problem.
- Determine what information is given.
- Determine what information is unknown.
- Determine whether any of the information given is unnecessary.
- Determine the goal.

To illustrate this aspect of problem solving, consider the following famous ancient limerick.

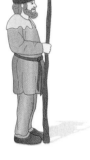

> As I was going to St. Ives,
> I met a man with seven wives;
> Each wife had seven sacks,
> Each sack had seven cats,
> Each cat had seven kits:
> Kits, cats, sacks, and wives,
> How many were going to St. Ives?

To answer the question in the limerick, we will ask and answer some of the questions listed above.

1. What is the goal?

The goal is to determine how many were going to St. Ives. (We know this from reading the last line of the limerick.)

Point of Interest

Try this brain teaser: You have two U.S. coins that add up to $.55. One is not a nickel. What are the two coins?

2. What information is necessary and what information is unnecessary?

The first line indicates that the poet was going to St. Ives. The next five lines describe a man the poet *met* on the way. This information is irrelevant.

The answer to the question, then, is 1. Only the poet was going to St. Ives.

There are many other examples of the importance, in problem solving, of recognizing *irrelevant information*. One that frequently makes the college circuit is posed in the form of a test. The first line of a 100-question test states, "Read the entire test before you begin." The last line of the test reads, "Choose any *one* question to answer." Many people ignore the information given in the first line and just begin the test, only to find out much later that they did a lot more work than was necessary.

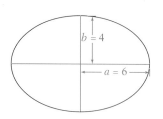

To illustrate another aspect of Polya's first step in the problem-solving process, consider the problem of finding the area of the oval-shaped region (called an *ellipse*) shown in the diagram at the left. This problem can be solved by doing some research to determine *what information is known about this type of problem*. Mathematicians have found a formula for the area of an ellipse. That formula is $A = \pi ab$, where a and b are as shown in the diagram. Therefore, $A = \pi(6)(4) = 24\pi \approx 75.40$ square units. Without the formula, this problem is difficult to solve. With the formula, it is fairly easy.

For each of the following problems, examine the problem in terms of the first step in Polya's problem-solving method. *Do not solve the problem.*

1. Johanna spent one-third of her allowance on a book. She then spent $5 for a sandwich and iced tea. The cost of the iced tea was one-fifth the cost of the sandwich. Find the cost of the iced tea.

2. A flight from Los Angeles to Boston took 6 h. What was the average speed of the plane?

3. A major league baseball is approximately 5 in. in diameter and is covered with cowhide. Approximately how much cowhide is used to cover 10 baseballs?

4. How many donuts are in seven baker's dozen?

5. The smallest prime number is 2. Twice the difference between the eighth and the seventh prime numbers is two more than the smallest prime number. How large is the smallest prime number?

Projects and Group Activities

Venn Diagrams   Diagrams can be very useful when working with sets. The set diagrams shown below are called **Venn diagrams.**

In the Venn diagram at the right, the rectangle represents the set U, and the circles A and B represent subsets of set U. The common area of the two circles represents the intersection of A and B.

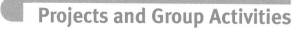
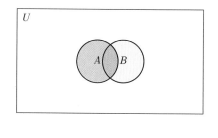

This Venn diagram represents the set $U = \{1, 2, 3, 4, 5, 6, 7, 8\}$. The sets $A = \{2, 3, 4, 5\}$ and $B = \{4, 5, 6, 7\}$ are subsets of U. Note that 1 and 8 are in set U but not in A or B. The numbers 4 and 5 are in both set A and set B.

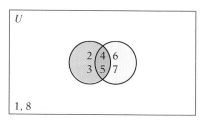

Sixty-five students at a small college enrolled in the following courses.

39 enrolled in English.
26 enrolled in mathematics.
35 enrolled in history.
19 enrolled in English and history.
11 enrolled in English and
   mathematics.
9 enrolled in mathematics and
   history.
2 enrolled in all three courses.

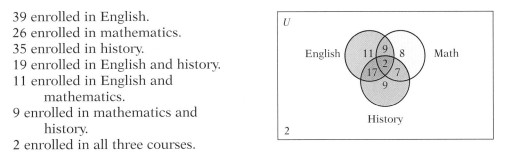

The Venn diagram is shown above. The diagram was drawn by placing the 2 students in the intersection of all three courses. Since 9 students took history and mathematics and 2 students are already in the intersection of history and mathematics, then 7 more students are placed in the intersection of history and mathematics. Continue in this manner until all information is used.

Complete the following.

1. Use the Venn diagram above that shows course enrollments.
   a. How many students enrolled only in English and mathematics?
   b. How many students enrolled in English but did not enroll in mathematics or history?
   c. How many students did not enroll in any of the three courses?
2. Sketch the Venn diagram for $U = \{1, 2, 3, 4, 5, 6, 7, 8\}$, $A = \{2, 3, 4, 5, 6\}$, $B = \{4, 5, 6, 7, 8\}$, and $C = \{4, 5\}$.
3. A busload of 50 scouts stopped at a fast-food restaurant and ordered hamburgers. The scouts could have pickles, tomatoes, or lettuce on their hamburgers.

   31 ordered pickles.
   36 ordered tomatoes.
   31 ordered lettuce.
   21 ordered pickles and tomatoes.
   24 ordered tomatoes and lettuce.
   22 ordered pickles and lettuce.
   17 ordered all three.

   a. How many scouts ordered a hamburger with pickles only?
   b. How many scouts ordered a hamburger with lettuce and tomatoes without the pickles?
   c. How many scouts ordered a hamburger without pickles, tomatoes, or lettuce?

Chapter Summary

Key Words   An *equation* expresses the equality of two mathematical expressions. A *solution,* or *root,* of an equation is a replacement value for the variable that will make the equation true. To *solve an equation* means to find its solutions. [p. 53]

A *contradiction* is an equation for which no value of the variable produces a true equation. A *conditional equation* is one that is true for at least one value of the variable but not all values of the variable. An *identity* is an equation that is true for all values of the variable. [p. 53]

An equation of the form $ax + b = c, a \neq 0$, is called a *first-degree equation*. [p. 53]

Equivalent equations are equations that have the same solution. [p. 53]

A *literal equation* is an equation that contains more than one variable. [p. 58]

A *value mixture problem* involves combining into a single blend two ingredients that have different prices. A *percent mixture problem* involves combining into a single blend two solutions that contain different amounts of a substance. [pp. 69, 71]

Uniform motion means that the speed of an object does not change. [p. 73]

The *solution set of an inequality* is a set of numbers, each element of which, when substituted for the variable, results in a true inequality. [p. 79]

A *compound inequality* is formed by joining two inequalities with a connective word such as "and" or "or." [p. 82]

The *absolute value* of a number is its distance from zero on the number line. An *absolute value equation* contains an absolute value symbol. [p. 91]

Essential Rules

Addition Property of Equations	If $a = b$, then $a + c = b + c$. [p. 53]
Multiplication Property of Equations	If $a = b$ and $c \neq 0$, then $ac = bc$. [p. 54]
Addition Property of Inequalities	If $a > b$, then $a + c > b + c$. If $a < b$, then $a + c < b + c$. [p. 54]
Multiplication Property of Inequalities	**Rule 1** If $a > b$ and $c > 0$, then $ac > bc$. If $a < b$ and $c > 0$, then $ac < bc$. **Rule 2** If $a > b$ and $c < 0$, then $ac < bc$. If $a < b$ and $c < 0$, then $ac > bc$. [p. 80]
Consecutive Integers	$n, n + 1, n + 2, \ldots$ [p. 63]
Consecutive Even Integers	$n, n + 2, n + 4, \ldots$ [p. 63]
Consecutive Odd Integers	$n, n + 2, n + 4, \ldots$ [p. 63]
Coin (or Stamp) Equation	Number of coins $\times$ value of coin in cents $=$ total value in cents [p. 65]
Value Mixture Equation	Amount $\times$ unit cost $=$ value $AC = V$ [p. 69]
Percent Mixture Equation	Amount $\times$ percent concentration $=$ quantity $Ar = Q$ [p. 71]
Uniform Motion Equation	Rate $\times$ time $=$ distance $rt = d$ [p. 73]
Solution of an Absolute Value Equation	If $a \geq 0$ and $|x| = a$, then $x = a$ or $x = -a$. [p. 91]

To solve an absolute value inequality of the form $|ax + b| < c$, solve the equivalent compound inequality $-c < ax + b < c$. [p. 93]

To solve an absolute value inequality of the form $|ax + b| > c$, solve the equivalent compound inequality $ax + b < -c$ or $ax + b > c$. [p. 94]

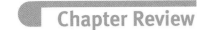

Chapter Review

1. Solve: $3t - 3 + 2t = 7t - 15$

2. Solve: $3x - 7 > -2$

3. Solve $P = 2L + 2W$ for L.

4. Solve: $x + 4 = -5$

5. Solve: $3x < 4$ and $x + 2 > -1$

6. Solve: $\frac{3}{5}x - 3 = 2x + 5$

7. Solve: $-\frac{2}{3}x = \frac{4}{9}$

8. Solve: $|x - 4| - 8 = -3$

9. Solve: $|2x - 5| < 3$

10. Solve: $\frac{2x - 3}{3} + 2 = \frac{2 - 3x}{5}$

11. Solve: $2(a - 3) = 5(4 - 3a)$

12. Solve: $5x - 2 > 8$ or $3x + 2 < -4$

13. Solve: $|4x - 5| \geq 3$

14. Solve $P = \frac{R - C}{n}$ for C.

15. Solve: $\frac{1}{2}x - \frac{5}{8} = \frac{3}{4}x + \frac{3}{2}$

16. Solve: $6 + |3x - 3| = 2$

17. Solve: $3x - 2 > x - 4$ or $7x - 5 < 3x + 3$

18. Solve: $2x - (3 - 2x) = 4 - 3(4 - 2x)$

19. A grocer mixed apple juice that costs $3.20 per gallon with 40 gal of cranberry juice that costs $5.50 per gallon. How much apple juice was used to make cranapple juice costing $4.20 per gallon?

20. A sales executive earns $800 per month plus 4% commission on the amount of sales. The executive's goal is to earn $3000 per month. What amount of sales will enable the executive to earn $3000 or more per month?

21. A coin collection contains thirty coins in nickels, dimes, and quarters. There are three more dimes than nickels. The value of the coins is $3.55. Find the number of quarters in the collection.

22. The diameter of a bushing is 2.75 in. The bushing has a tolerance of 0.003 in. Find the lower and upper limits of the diameter of the bushing.

23. The sum of two integers is twenty. Five times the smaller integer is two more than twice the larger integer. Find the two integers.

24. An average score of 80 to 90 in a psychology class receives a B grade. A student has scores of 92, 66, 72, and 88 on four tests. Find the range of scores on the fifth test that will give the student a B for the course.

25. Two planes are 1680 mi apart and are traveling toward each other. One plane is traveling 80 mph faster than the other plane. The planes meet in 1.75 h. Find the speed of each plane.

26. An alloy containing 30% tin is mixed with an alloy containing 70% tin. How many pounds of each were used to make 500 lb of an alloy containing 40% tin?

27. A piston rod for an automobile is $10\frac{3}{8}$ in. long with a tolerance of $\frac{1}{32}$ in. Find the lower and upper limits of the length of the piston rod.

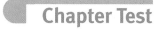

Chapter Test

1. Solve: $x - 2 = -4$

2. Solve: $b + \frac{3}{4} = \frac{5}{8}$

3. Solve: $-\frac{3}{4}y = -\frac{5}{8}$

4. Solve: $3x - 5 = 7$

5. Solve: $\frac{3}{4}y - 2 = 6$

6. Solve: $2x - 3 - 5x = 8 + 2x - 10$

7. Solve: $2[a - (2 - 3a) - 4] = a - 5$

8. Solve $E = IR + Ir$ for R.

9. Solve: $\frac{2x + 1}{3} - \frac{3x + 4}{6} = \frac{5x - 9}{9}$

10. Solve: $3x - 2 \geq 6x + 7$

11. Solve: $4 - 3(x + 2) < 2(2x + 3) - 1$

12. Solve: $4x - 1 > 5$ or $2 - 3x < 8$

13. Solve: $4 - 3x \geq 7$ and $2x + 3 \geq 7$

14. Solve: $|3 - 5x| = 12$

15. Solve: $2 - |2x - 5| = -7$

16. Solve: $|3x - 5| \leq 4$

17. Solve: $|4x - 3| > 5$

18. Gambelli Agency rents cars for $12 per day plus 10¢ for every mile driven. McDougal Rental rents cars for $24 per day with unlimited mileage. How many miles a day can you drive a Gambelli Agency car if it is to cost you less than a McDougal Rental car?

19. A machinist must make a bushing that has a tolerance of 0.002 in. The diameter of the bushing is 2.65 in. Find the lower and upper limits of the diameter of the bushing.

20. The sum of two integers is fifteen. Eight times the smaller integer is one less than three times the larger integer. Find the integers.

21. A stamp collection contains 11¢, 15¢, and 24¢ stamps. There are twice as many 11¢ stamps as 15¢ stamps. There are thirty stamps in all, with a value of $4.40. How many 24¢ stamps are in the collection?

22. A butcher combines 100 lb of hamburger that costs $1.60 per pound with 60 lb of hamburger that costs $3.20 per pound. Find the cost of the hamburger mixture.

23. A jogger runs a distance at a speed of 8 mph and returns the same distance running at a speed of 6 mph. Find the total distance that the jogger ran if the total time running was one hour and forty-five minutes.

24. Two trains are 250 mi apart and are traveling toward each other. One train is traveling 5 mph faster than the other train. The trains pass each other in 2 h. Find the speed of each train.

25. How many ounces of pure water must be added to 60 oz of an 8% salt solution to make a 3% salt solution?

Cumulative Review

1. Simplify: $-4 - (-3) - 8 + (-2)$

2. Simplify: $-2^2 \cdot 3^3$

3. Simplify: $4 - (2 - 5)^2 \div 3 + 2$

4. Simplify: $4 \div \dfrac{\frac{3}{8} - 1}{5} \cdot 2$

5. Evaluate $2a^2 - (b - c)^2$ when $a = 2$, $b = 3$, and $c = -1$.

6. Evaluate $\dfrac{a - b^2}{b - c}$ when $a = 2$, $b = -3$, and $c = 4$.

7. Identify the property that justifies the statement.
$(2x + 3y) + 2 = (3y + 2x) + 2$

8. Translate and simplify "the sum of three times a number and six added to the product of three and the number."

9. Solve $F = \dfrac{evB}{c}$ for B.

10. Simplify: $5[y - 2(3 - 2y) + 6]$

11. Find $A \cap B$, given $A = \{-4, -2, 0, 2\}$ and $B = \{-4, 0, 4, 8\}$.

12. Graph the solution set of $\{x \mid x \le 3\} \cap \{x \mid x > -1\}$.

$$\overset{\displaystyle \xleftarrow{\;\;|\;\;|\;\;|\;\;|\;\;|\;\;|\;\;|\;\;|\;\;|\;\;|\;\;|\;\;}\;\;}{-5\ -4\ -3\ -2\ -1\ \ 0\ \ 1\ \ 2\ \ 3\ \ 4\ \ 5}$$

13. Solve $Ax + By + C = 0$ for y.

14. Solve: $-\dfrac{5}{6}b = -\dfrac{5}{12}$

15. Solve: $2x + 5 = 5x + 2$

16. Solve: $\dfrac{5}{12}x - 3 = 7$

17. Solve: $2[3 - 2(3 - 2x)] = 2(3 + x)$

18. Solve: $3[2x - 3(4 - x)] = 2(1 - 2x)$

19. Solve: $\dfrac{1}{2}y - \dfrac{2}{3}y + \dfrac{5}{12} = \dfrac{3}{4}y - \dfrac{1}{2}$

20. Solve: $\dfrac{3x - 1}{4} - \dfrac{4x - 1}{12} = \dfrac{3 + 5x}{8}$

21. Solve: $3 - 2(2x - 1) \geq 3(2x - 2) + 1$

22. Solve: $3x + 2 \leq 5$ and $x + 5 > 1$

23. Solve: $|3 - 2x| = 5$

24. Solve: $3 - |2x - 3| = -8$

25. Solve: $|3x - 1| > 5$

26. Solve: $|2x - 4| < 8$

27. A bank offers two types of checking accounts. One account has a charge of $5 per month plus 4¢ for each check. The second account has a charge of $2 per month plus 10¢ for each check. How many checks can a customer who has the second type of account write if it is to cost the customer less than the first type of account?

28. Four times the sum of the first and third of three consecutive odd integers is one less than seven times the middle integer. Find the first integer.

29. A coin purse contains dimes and quarters. The number of dimes is five less than twice the number of quarters. The total value of the coins is $4.00. Find the number of dimes in the coin purse.

30. A silversmith combined pure silver that costs $8.50 per ounce with 100 oz of a silver alloy that costs $4.00 per ounce. How many ounces of pure silver were used to make an alloy of silver costing $6.00 per ounce?

31. Two planes are 1400 mi apart and are traveling toward each other. One plane is traveling 120 mph faster than the other plane. The planes meet in 2.5 h. Find the speed of the slower plane.

32. The diameter of a bushing is 2.45 in. The bushing has a tolerance of 0.001 in. Find the lower and upper limits of the diameter of the bushing.

33. How many liters of a 12% acid solution must be mixed with 4 L of a 5% acid solution to make an 8% acid solution?

Chapter 3

Linear Functions and Inequalities in Two Variables

Objectives

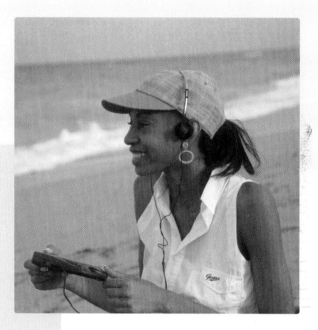

Companies spend time and money doing research in order to determine what price to charge for a product like this portable CD player. The price depends on what consumers are willing to pay and how many consumers are expected to purchase the product at that price. A higher price will not result in higher revenue for the company if consumers won't pay that price. Linear functions can be used to estimate sales of a product given different prices for that product, as shown in **Example 5 on page 160.**

Need help? For on-line student resources, such as section quizzes, visit this textbook's web site at **math.college.hmco.com/students**.

Prep Test

For Exercises 1 to 3, simplify.

1. $-4(x - 3)$

2. $\sqrt{(-6)^2 + (-8)^2}$

3. $\dfrac{3 - (-5)}{2 - 6}$

4. Evaluate $-2x + 5$ for $x = -3$.

5. Evaluate $\dfrac{2r}{r - 1}$ for $r = 5$.

6. Evaluate $2p^3 - 3p + 4$ for $p = -1$.

7. Evaluate $\dfrac{x_1 + x_2}{2}$ for $x_1 = 7$ and $x_2 = -5$.

8. Given $3x - 4y = 12$, find the value of x when $y = 0$.

9. Solve $2x - y = 7$ for y.

Go Figure

If $\boxed{5} = 4$ and $\textcircled{5} = 6$ and $y = x - 1$, which of the following has the largest value?

$\boxed{x}$ $\textcircled{x}$ $\boxed{y}$ $\textcircled{y}$

3.1 The Rectangular Coordinate System

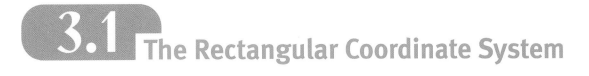

Objective A   To graph points in a rectangular coordinate system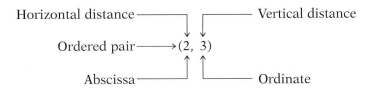

Point of Interest

A rectangular coordinate system is also called a **Cartesian coordinate system,** in honor of Descartes.

Before the 15th century, geometry and algebra were considered separate branches of mathematics. That all changed when René Descartes, a French mathematician who lived from 1596 to 1650, founded **analytic geometry.** In this geometry, a *coordinate system* is used to study relationships between variables.

A **rectangular coordinate system** is formed by two number lines, one horizontal and one vertical, that intersect at the zero point of each line. The point of intersection is called the **origin.** The two lines are called **coordinate axes,** or simply **axes.**

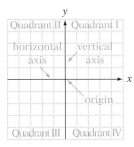

The axes determine a **plane,** which can be thought of as a large, flat sheet of paper. The two axes divide the plane into four regions called **quadrants.** The quadrants are numbered counterclockwise from I to IV.

Point of Interest

Gottfried Leibniz introduced the words *abscissa* and *ordinate. Abscissa* is from Latin meaning "to cut off." Originally, Leibnitz used the phrase *abscissa linea,* "cut off a line" (axis). The root of *ordinate* is also a Latin word used to suggest a sense of order.

Each point in the plane can be identified by a pair of numbers called an **ordered pair.** The first number of the pair measures a horizontal distance and is called the **abscissa.** The second number of the pair measures a vertical distance and is called the **ordinate.** The **coordinates** of the point are the numbers in the ordered pair associated with the point. The abscissa is also called the **first coordinate** of the ordered pair, and the ordinate is also called the **second coordinate** of the ordered pair.

Horizontal distance ———————┐ ┌——————— Vertical distance

Ordered pair ——————→ (2, 3)

Abscissa ——————————┘ └——————— Ordinate

To **graph** or **plot** an ordered pair in the plane, place a dot at the location given by the ordered pair. The **graph of an ordered pair** is the dot drawn at the coordinates of the point in the plane. The points whose coordinates are (3, 4) and (−2.5, −3) are graphed in the figure at the right.

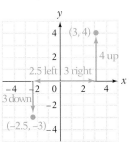

The points whose coordinates are (3, −1) and (−1, 3) are graphed at the right. Note that the graphs are in different locations. The *order* of the coordinates of an ordered pair is important.

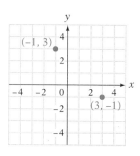

When drawing a rectangular coordinate system, we often label the horizontal axis x and the vertical axis y. In this case, the coordinate system is called an **xy-coordinate system.** The coordinates of the points are given by ordered pairs (x, y), where the abscissa is called the **x-coordinate** and the ordinate is called the **y-coordinate.**

The xy-coordinate system is used to graph **equations in two variables.** Examples of equations in two variables are shown at the right. A **solution of an equation in two variables** is an ordered pair (x, y) whose coordinates make the equation a true statement.

$$y = 3x + 7$$
$$y = x^2 - 4x + 3$$
$$x^2 + y^2 = 25$$
$$x = \frac{y}{y^2 + 4}$$

➡ Is the ordered pair (−3, 7) a solution of the equation $y = -2x + 1$?

$$y = -2x + 1$$

7	$-2(-3) + 1$
7	$6 + 1$

$7 = 7$

Yes, the ordered pair (−3, 7) is a solution of the equation.

- Replace x by −3 and y by 7.
- Simplify.
- Compare the results. If the resulting equation is a true statement, the ordered pair is a solution of the equation. If it is not a true statement, the ordered pair is not a solution of the equation.

Besides the ordered pair (−3, 7), there are many other ordered-pair solutions of the equation $y = -2x + 1$. For example, (−5, 11), (0, 1), $\left(-\frac{3}{2}, 4\right)$, and (4, −7) are also solutions of the equation.

In general, an equation in two variables has an infinite number of solutions. By choosing any value of x and substituting that value into the equation, we can calculate a corresponding value of y. The resulting ordered-pair solution (x, y) of the equation can be graphed in a rectangular coordinate system.

➡ Graph the solutions (x, y) of $y = x^2 - 1$ when x equals −2, −1, 0, 1, and 2.

Substitute each value of x into the equation and solve for y. It is convenient to record the ordered-pair solutions in a table similar to the one shown below. Then graph the ordered pairs, as shown at the left.

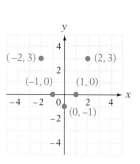

x	$y = x^2 - 1$	y	(x, y)
−2	$y = (-2)^2 - 1$	3	(−2, 3)
−1	$y = (-1)^2 - 1$	0	(−1, 0)
0	$y = 0^2 - 1$	−1	(0, −1)
1	$y = 1^2 - 1$	0	(1, 0)
2	$y = 2^2 - 1$	3	(2, 3)

Example 1

Determine the ordered-pair solution of

$y = \dfrac{x}{x - 2}$ corresponding to $x = 4$.

Solution

$y = \dfrac{4}{4 - 2} = \dfrac{4}{2} = 2$ • Replace x by 4 and solve for y.

The ordered-pair solution is (4, 2).

You Try It 1

Determine the ordered-pair solution of

$y = \dfrac{3x}{x + 1}$ corresponding to $x = -2$.

Your solution

Example 2

Graph the ordered-pair solutions of $y = x^2 - x$ when $x = -1, 0, 1,$ and 2.

Solution

x	y
-1	2
0	0
1	0
2	2

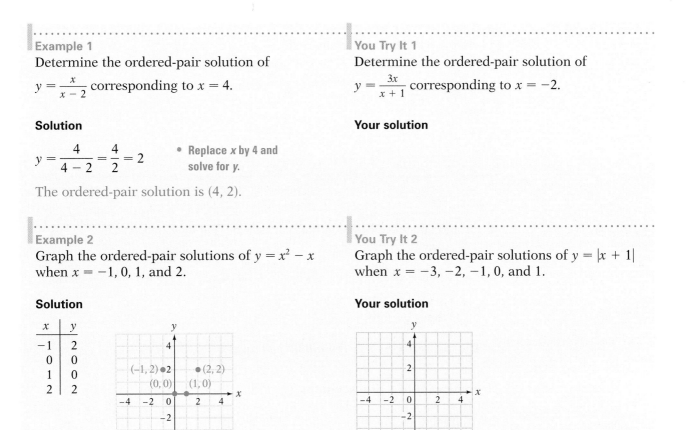

You Try It 2

Graph the ordered-pair solutions of $y = |x + 1|$ when $x = -3, -2, -1, 0,$ and 1.

Your solution

Solutions on p. S6

Objective B   **To find the length and midpoint of a line segment**   〈 3 〉

The distance between two points in an xy-coordinate system can be calculated by using the Pythagorean Theorem.

Point of Interest

Although Pythagoras (c. 550 B.C.) is given credit for discovering the Pythagorean Theorem, a clay tablet called Plimpton 322 shows that the Babylonians knew the theorem 1000 years before Pythagoras!

> **Pythagorean Theorem**
>
> If a and b are the lengths of the legs of a right triangle and c is the length of the hypotenuse, then $a^2 + b^2 = c^2$.

Consider the two points and the right triangle shown at the right. The vertical distance between $P_1(x_1, y_1)$ and $P_2(x_2, y_2)$ is $|y_2 - y_1|$.

The horizontal distance between the points $P_1(x_1, y_1)$ and $P_2(x_2, y_2)$ is $|x_2 - x_1|$.

The quantity d^2 is calculated by applying the Pythagorean Theorem to the right triangle.

The distance d is the square root of d^2.

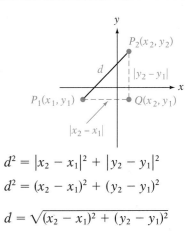

$d^2 = |x_2 - x_1|^2 + |y_2 - y_1|^2$

$d^2 = (x_2 - x_1)^2 + (y_2 - y_1)^2$

$d = \sqrt{(x_2 - x_1)^2 + (y_2 - y_1)^2}$

Because $(x_2 - x_1)^2 = (x_1 - x_2)^2$ and $(y_2 - y_1)^2 = (y_1 - y_2)^2$, the distance formula is usually written in the following form.

The Distance Formula

If $P_1(x_1, y_1)$ and $P_2(x_2, y_2)$ are two points in the plane, then the distance d between the two points is given by

$d = \sqrt{(x_1 - x_2)^2 + (y_1 - y_2)^2}.$

⇒ Find the distance between the points $(-6, 1)$ and $(-2, 4)$.

$d = \sqrt{(x_1 - x_2)^2 + (y_1 - y_2)^2}$ • Use the distance formula.

$ = \sqrt{[-6 - (-2)]^2 + (1 - 4)^2}$ • Let $(x_1, y_1) = (-6, 1)$ and $(x_2, y_2) = (-2, 4)$.

$ = \sqrt{(-4)^2 + (-3)^2} = \sqrt{16 + 9}$

$ = \sqrt{25} = 5$

The distance between the points is 5 units.

The **midpoint of a line segment** is equidistant from its endpoints. The coordinates of the midpoint of the line segment P_1P_2 are (x_m, y_m). The intersection of the horizontal line segment through P_1 and the vertical line segment through P_2 is Q, with coordinates (x_2, y_1).

The x-coordinate x_m of the midpoint of the line segment P_1P_2 is the same as the x-coordinate of the midpoint of the line segment P_1Q. It is the average of the x-coordinates of the points P_1 and P_2.

$$x_m = \frac{x_1 + x_2}{2}$$

Similarly, the y-coordinate y_m of the midpoint of the line segment P_1P_2 is the same as the y-coordinate of the midpoint of the line segment P_2Q. It is the average of the y-coordinates of the points P_1 and P_2.

$$y_m = \frac{y_1 + y_2}{2}$$

The Midpoint Formula

If $P_1(x_1, y_1)$ and $P_2(x_2, y_2)$ are the endpoints of a line segment, then the coordinates of the midpoint (x_m, y_m) of the line segment are given by

$$x_m = \frac{x_1 + x_2}{2} \quad \text{and} \quad y_m = \frac{y_1 + y_2}{2}$$

➡ Find the coordinates of the midpoint of the line segment with endpoints $(-3, 5)$ and $(1, -7)$.

$$x_m = \frac{x_1 + x_2}{2} \qquad y_m = \frac{y_1 + y_2}{2}$$ • Use the midpoint formula.

$$= \frac{-3 + 1}{2} \qquad = \frac{5 + (-7)}{2}$$ • Let $(x_1, y_1) = (-3, 5)$ and $(x_2, y_2) = (1, -7)$.

$$= -1 \qquad = -1$$

The coordinates of the midpoint are $(-1, -1)$.

Example 3

Find the distance, to the nearest hundredth, between the points whose coordinates are $(-3, 2)$ and $(4, -1)$.

Solution

$(x_1, y_1) = (-3, 2); (x_2, y_2) = (4, -1)$

$d = \sqrt{(x_1 - x_2)^2 + (y_1 - y_2)^2}$

$\quad = \sqrt{(-3 - 4)^2 + [2 - (-1)]^2}$

$\quad = \sqrt{(-7)^2 + 3^2} = \sqrt{49 + 9}$

$\quad = \sqrt{58} \approx 7.62$

You Try It 3

Find the distance, to the nearest hundredth, between the points whose coordinates are $(5, -2)$ and $(-4, 3)$.

Your solution

Example 4

Find the coordinates of the midpoint of the line segment with endpoints $(-5, 4)$ and $(-3, 7)$.

Solution

$(x_1, y_1) = (-5, 4); (x_2, y_2) = (-3, 7)$

$x_m = \frac{x_1 + x_2}{2} \qquad y_m = \frac{y_1 + y_2}{2}$

$\quad = \frac{-5 + (-3)}{2} \qquad = \frac{4 + 7}{2}$

$\quad = -4 \qquad\qquad = \frac{11}{2}$

The midpoint is $\left(-4, \frac{11}{2}\right)$.

You Try It 4

Find the coordinates of the midpoint of the line segment with endpoints $(-3, -5)$ and $(-2, 3)$.

Your solution

Solutions on p. S6

Objective C   **To graph a scatter diagram**

Discovering a relationship between two variables is an important task in the study of mathematics. These relationships occur in many forms and in a wide variety of applications. Here are some examples.

• A botanist wants to know the relationship between the number of bushels of wheat yielded per acre and the amount of watering per acre.

• An environmental scientist wants to know the relationship between the incidence of skin cancer and the amount of ozone in the atmosphere.

• A business analyst wants to know the relationship between the price of a product and the number of products that are sold at that price.

A researcher may investigate the relationship between two variables by means of *regression analysis,* which is a branch of statistics. The study of the relationship between the two variables may begin with a **scatter diagram,** which is a graph of the ordered pairs of the known data.

The following table shows randomly selected data from the participants 40 years old and older and their times (in minutes) for a recent Boston Marathon.

Age (x)	55	46	53	40	40	44	54	44	41	50
Time (y)	254	204	243	194	281	197	238	300	232	216

TAKE NOTE

The jagged portion of the horizontal axis in the figure at the right indicates that the numbers between 0 and 40 are missing.

The scatter diagram for these data is shown at the right. Each ordered pair represents the age and time for a participant. For instance, the ordered pair (53, 243) indicates that a 53-year-old participant ran the marathon in 243 min.

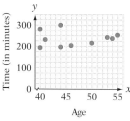

- -

Example 5

The grams of sugar and the grams of fiber in a 1-ounce serving of six breakfast cereals are shown in the table below.

	Sugar (x)	Fiber (y)
Wheaties	4	3
Rice Krispies	3	0
Total	5	3
Life	6	2
Kix	3	1
Grape-Nuts	7	5

Draw a scatter diagram of these data.

Strategy

To draw a scatter diagram:

- Draw a coordinate grid with the horizontal axis representing the grams of sugar and the vertical axis the grams of fiber.
- Graph the ordered pairs (4, 3), (3, 0), (5, 3), (6, 2), (3, 1), and (7, 5).

Solution

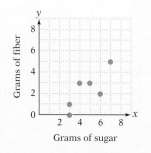

You Try It 5

According to Forester Research, the growth of the software rental market, in billions of dollars, is projected to be as shown in the table below.

Year	Market (in billions of dollars)
1999	1
2000	2
2001	4
2002	7
2003	11

Draw a scatter diagram of these data.

Your strategy

Your solution

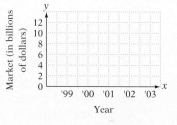

Solution on p. S7

3.1 Exercises

Objective A

1. Graph the ordered pairs $(1, -1)$, $(2, 0)$, $(3, 2)$, and $(-1, 4)$.

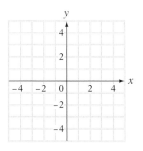

2. Graph the ordered pairs $(-1, -3)$, $(0, -4)$, $(0, 4)$, and $(3, -2)$.

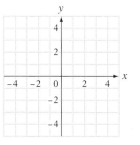

3. Find the coordinates of each of the points.

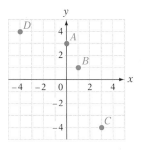

4. Find the coordinates of each of the points.

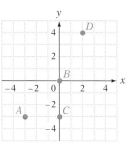

5. Draw a line through all points with an abscissa of 2.

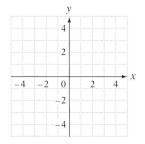

6. Draw a line through all points with an abscissa of -3.

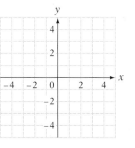

7. Draw a line through all points with an ordinate of -3.

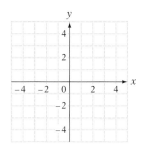

8. Draw a line through all points with an ordinate of 4.

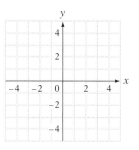

9. Graph the ordered-pair solutions of $y = x^2$ when $x = -2, -1, 0, 1,$ and 2.

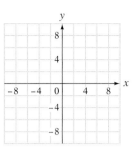

10. Graph the ordered-pair solutions of $y = -x^2 + 1$ when $x = -2, -1, 0, 1,$ and 2.

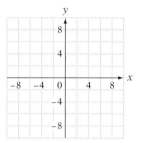

11. Graph the ordered-pair solutions of $y = |x + 1|$ when $x = -5, -3, 0, 3,$ and 5.

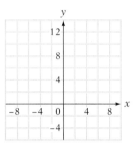

12. Graph the ordered-pair solutions of $y = -2|x|$ when $x = -3, -1, 0, 1,$ and 3.

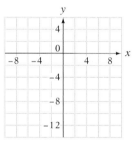

13. Graph the ordered-pair solutions of $y = -x^2 + 2$ when $x = -2, -1, 0, 1,$ and 2.

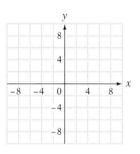

14. Graph the ordered-pair solutions of $y = -x^2 + 4$ when $x = -3, -1, 0, 1,$ and 3.

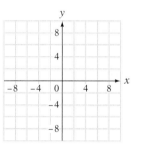

15. Graph the ordered-pair solutions of $y = x^3 - 2$ when $x = -1, 0, 1,$ and 2.

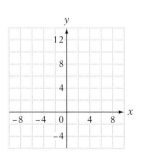

16. Graph the ordered-pair solutions of $y = -x^3 + 1$ when $x = -1, 0, 1,$ and $\frac{3}{2}$.

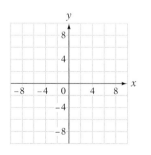

Objective B

Find the distance to the nearest hundredth between the given points. Then find the coordinates of the midpoint of the line segment connecting the points.

17. $P_1(3, 5)$ and $P_2(5, 1)$

18. $P_1(-2, 3)$ and $P_2(4, -1)$

19. $P_1(0, 3)$ and $P_2(-2, 4)$

20. $P_1(6, -1)$ and $P_2(-3, -2)$

21. $P_1(-3, -5)$ and $P_2(2, -4)$

22. $P_1(-7, -5)$ and $P_2(-2, -1)$

23. $P_1(5, -2)$ and $P_2(-2, 5)$

24. $P_1(3, -6)$ and $P_2(6, 0)$

25. $P_1(5, -5)$ and $P_2(2, -5)$

26. $P_1(-2, -3)$ and $P_2(-2, 5)$

27. $P_1\left(\dfrac{3}{2}, -\dfrac{4}{3}\right)$ and $P_2\left(-\dfrac{1}{2}, \dfrac{7}{3}\right)$

28. $P_1(4.5, -6.3)$ and $P_2(-1.7, -4.5)$

Objective C

Solve.

29. The temperature of a chemical reaction is measured at intervals of 10 min and recorded in the scatter diagram at the right.
   a. Find the temperature of the reaction after 20 min.
   b. After how many minutes is the temperature 160°F?

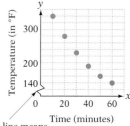

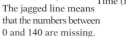
The jagged line means that the numbers between 0 and 140 are missing.

30. The amount of a substance that can be dissolved in a fixed amount of water usually increases as the temperature of the water increases. Cerium selenate, however, does not behave in this manner. The graph at the right shows the number of grams of cerium selenate that will dissolve in 100 mg of water for various temperatures, in degrees Celsius.
   a. Determine the temperature at which 25 g of cerium selenate will dissolve.

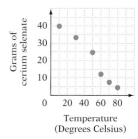

   b. Determine the number of grams of cerium selenate that will dissolve when the temperature is 80°C.

31. Past experience of executives of a car company shows that the profit of a dealership will depend on the total income of all the residents of the town in which the dealership is located. The table below shows the profits of several dealerships and the total incomes of the towns. Draw a scatter diagram for these data.

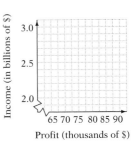

Profit (thousands of $)	65	85	81	77	89	69
Total Income (billions of $)	2.2	2.6	2.5	2.4	2.7	2.3

32. A power company suggests that a larger power plant can produce energy more efficiently and therefore at lower cost to consumers. The table below shows the output and average cost for power plants of various sizes. Draw a scatter diagram for these data.

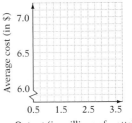

Output (millions of watts)	0.7	2.2	2.6	3.2	2.8	3.5
Average Cost (in dollars)	6.9	6.5	6.3	6.4	6.5	6.1

APPLYING THE CONCEPTS

33. Graph the ordered pairs (x, x^2), where $x \in \{-2, -1, 0, 1, 2\}$.

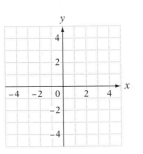

34. Graph the ordered pairs $\left(x, \dfrac{1}{x}\right)$, where

$$x \in \left\{-2, -1, -\frac{1}{2}, -\frac{1}{3}, \frac{1}{3}, \frac{1}{2}, 1, 2\right\}.$$

35. Describe the graph of all the ordered pairs (x, y) that are 5 units from the origin.

36. Consider two distinct fixed points in the plane. Describe the graph of all the points (x, y) that are equidistant from these fixed points.

37. Draw a line passing through every point whose abscissa equals its ordinate.

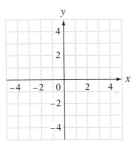

38. Draw a line passing through every point whose ordinate is the additive inverse of its abscissa.

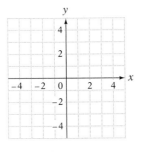

3.2 Introduction to Functions

Objective A   **To evaluate a function**

In mathematics and its applications, there are many times when it is necessary to investigate a relationship between two quantities. Here is a financial application: Consider a person who is planning to finance the purchase of a car. If the current interest rate for a five-year loan is 9%, the equation that describes the relationship between the amount that is borrowed B and the monthly payment P is $P = 0.020758B$.

For each amount the purchaser may borrow (B), there is a certain monthly payment (P). The relationship between the amount borrowed and the payment can be recorded as ordered pairs, where the first coordinate is the amount borrowed and the second coordinate is the monthly payment. Some of these ordered pairs are shown at the right.

$0.020758B = P$

(5000, 103.79)
(6000, 124.55)
(7000, 145.31)
(8000, 166.06)

A relationship between two quantities is not always given by an equation. The table at the right describes a grading scale that defines a relationship between a score on a test and a letter grade. For each score, the table assigns only one letter grade. The ordered pair (84, B) indicates that a score of 84 receives a letter grade of B.

Score	Grade
90–100	A
80–89	B
70–79	C
60–69	D
0–59	F

The graph at the right also shows a relationship between two quantities. It is a graph of the viscosity V of SAE 40 motor oil at various temperatures T. Ordered pairs can be approximated from the graph. The ordered pair (120, 250) indicates that the viscosity of the oil at 120°F is 250 units.

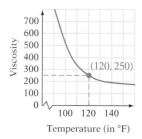

In each of these examples, there is a rule (an equation, a table, or a graph) that determines a certain set of ordered pairs.

Definition of Relation

A **relation** is a set of ordered pairs.

Here are some of the ordered pairs for the relations given above.

Relation	Some of the Ordered Pairs of the Relation
Car Payment	(2500, 51.90), (3750, 77.84), (4396, 91.25)
Grading Scale	(78, C), (98, A), (70, C), (81, B), (94, A)
Oil Viscosity	(100, 500), (120, 250), (130, 200), (150, 180)

Each of these three relations is actually a special type of relation called a *function*. Functions play an important role in mathematics and its applications.

Definition of Function

A **function** is a relation in which no two ordered pairs have the same first coordinate and different second coordinates.

The **domain** of a function is the set of the first coordinates of all the ordered pairs of the function. The **range** is the set of the second coordinates of all the ordered pairs of the function.

For the function defined by the ordered pairs

$$\{(2, 3), (4, 5), (6, 7), (8, 9)\}$$

the domain is $\{2, 4, 6, 8\}$ and the range is $\{3, 5, 7, 9\}$.

➡ Find the domain and range of the function $\{(2, 3), (4, 6), (6, 8), (10, 6)\}$.

The domain is $\{2, 4, 6, 10\}$.   • The domain of the function is the set of the first components of the ordered pairs.

The range is $\{3, 6, 8\}$.   • The range of the function is the set of the second components of the ordered pairs.

For each element of the domain of a function there is a corresponding element in the range of the function. A possible diagram for the function above is

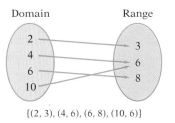

$$\{(2, 3), (4, 6), (6, 8), (10, 6)\}$$

Functions defined by tables or graphs, such as those described at the beginning of this section, have important applications. However, a major focus of this text is functions defined by equations in two variables.

The *square function*, which pairs each real number with its square, can be defined by the equation

$$y = x^2$$

This equation states that for a given value of x in the domain, the corresponding value of y in the range is the square of x. For instance, if $x = 6$, then $y = 36$ and if $x = -7$, then $y = 49$. Because the value of y *depends* on the value of x, y is called the **dependent variable** and x is called the **independent variable.**

A function can be thought of as a rule that pairs one number with another number. For instance, the **square function** pairs a number with its square. The ordered pairs for the values shown at the right are $(-5, 25)$, $\left(\frac{3}{5}, \frac{9}{25}\right)$, $(0, 0)$, and $(3, 9)$. For this function, the second coordinate is the square of the first coordinate. If we let x represent the first coordinate, then the second coordinate is x^2 and we have the ordered pair (x, x^2).

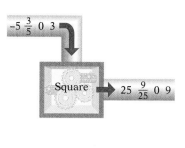

A function cannot have two ordered pairs with *different* second coordinates and the same first coordinate. However, a function may contain ordered pairs with the *same* second coordinate. For instance, the square function has the ordered pairs $(-3, 9)$ and $(3, 9)$; the second coordinates are the same but the first coordinates are different.

The **double function** pairs a number with twice that number. The ordered pairs for the values shown at the right are $(-5, -10)$, $\left(\frac{3}{5}, \frac{6}{5}\right)$, $(0, 0)$, and $(3, 6)$. For this function, the second coordinate is twice the first coordinate. If we let x represent the first coordinate, then the second coordinate is $2x$ and we have the ordered pair $(x, 2x)$.

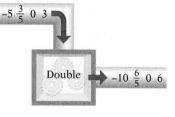

Not every equation in two variables defines a function. For instance, consider the equation

$$y^2 = x^2 + 9$$

Because

$$5^2 = 4^2 + 9 \quad \text{and} \quad (-5)^2 = 4^2 + 9$$

the ordered pairs $(4, 5)$ and $(4, -5)$ are both solutions of the equation. Consequently, there are two ordered pairs that have the same first coordinate (4) but *different* second coordinates (5 and -5). Therefore, the equation does not define a function. Other ordered pairs for this equation are $(0, -3)$, $(0, 3)$, $\left(\sqrt{7}, -4\right)$, and $\left(\sqrt{7}, 4\right)$. A graphical representation of these ordered pairs is shown below.

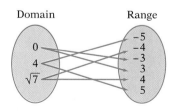

Note from this graphical representation that each element from the domain has two arrows pointing to two different elements in the range. Any time this occurs, the situation does not represent a function. However, this diagram *does* represent a relation. The relation for the values shown is $\left\{(0, -3), (0, 3), (4, -5), (4, 5), \left(\sqrt{7}, -4\right), \left(\sqrt{7}, 4\right)\right\}$.

The phrase "y is a function of x," or the same phrase with different variables, is used to describe an equation in two variables that defines a function. To emphasize that the equation represents a function, **functional notation** is used.

Just as the variable x is commonly used to represent a number, the letter f is commonly used to name a function. The square function is written in functional notation as follows.

This is the value of the function.
It is the number that is paired with x.

$$f(x) = x^2$$

The name of the function is f.   This is an algebraic expression that defines the relationship between the dependent and independent variables.

The symbol $f(x)$ is read "the value of f at x" or "f of x."

It is important to note that $f(x)$ does *not* mean f times x. The symbol $f(x)$ is the **value of the function** and represents the value of the dependent variable for a given value of the independent variable. We often write $y = f(x)$ to emphasize the relationship between the independent variable x and the dependent variable y. Remember that y and $f(x)$ are different symbols for the same number.

The letters used to represent a function are somewhat arbitrary. All of the following equations represent the same function.

$$\left. \begin{array}{l} f(x) = x^2 \\ s(t) = t^2 \\ P(v) = v^2 \end{array} \right\} \text{ Each equation represents the square function.}$$

The process of determining $f(x)$ for a given value of x is called **evaluating the function**. For instance, to evaluate $f(x) = x^2$ when $x = 4$, replace x by 4 and simplify.

$$f(x) = x^2$$
$$f(4) = 4^2 = 16$$

The *value* of the function is 16 when $x = 4$. An ordered pair of the function is (4, 16).

➡ Evaluate $g(t) = 3t^2 - 5t + 1$ when $t = -2$.

$$g(t) = 3t^2 - 5t + 1$$
$$g(-2) = 3(-2)^2 - 5(-2) + 1 \quad \bullet \textbf{ Replace } t \textbf{ by } -2 \textbf{ and then simplify.}$$
$$= 3(4) - 5(-2) + 1$$
$$= 12 + 10 + 1 = 23$$

When t is -2, the value of the function is 23.
Therefore, an ordered pair of the function is $(-2, 23)$.

It is possible to evaluate a function for a variable expression.

➡ Evaluate $P(z) = 3z - 7$ when $z = 3 + h$.

$$P(z) = 3z - 7$$
$$P(3 + h) = 3(3 + h) - 7 \quad \bullet \textbf{ Replace } z \textbf{ by } 3 + h \textbf{ and then simplify.}$$
$$= 9 + 3h - 7$$
$$= 3h + 2$$

When z is $3 + h$, the value of the function is $3h + 2$.
Therefore, an ordered pair of the function is $(3 + h, 3h + 2)$.

Recall that the range of a function is found by applying the function to each element of the domain. If the domain contains an infinite number of elements, it may be difficult to find the range. However, if the domain has a finite number of elements, then the range can be found by evaluating the function for each element in the domain.

⇒ Find the range of $f(x) = x^3 + x$ if the domain is $\{-2, -1, 0, 1, 2\}$.

$$f(x) = x^3 + x$$
$$f(-2) = (-2)^3 + (-2) = -10$$
$$f(-1) = (-1)^3 + (-1) = -2$$
$$f(0) = 0^3 + 0 = 0$$
$$f(1) = 1^3 + 1 = 2$$
$$f(2) = 2^3 + 2 = 10$$

• Replace *x* by each member of the domain. The range includes the values of $f(-2)$, $f(-1)$, $f(0)$, $f(1)$, and $f(2)$.

The range is $\{-10, -2, 0, 2, 10\}$.

When a function is represented by an equation, the domain of the function is all real numbers for which the value of the function is a real number. For instance:

e domain of $f(x) = x^2$ is all real numbers, because the square of every real
nber is a real number.

domain of $g(x) = \dfrac{1}{x - 2}$ is all real numbers except 2, because when $x = 2$,

$\dfrac{1}{2 - 2} = \dfrac{1}{0}$, which is not a real number.

ain of the grading-scale function is the set of
mbers from 0 to 100. In set-builder notation,
itten $\{x | 0 \le x \le 100, x \in \text{whole numbers}\}$. The
, B, C, D, F}.

Score	Grade
90–100	A
80–89	B
70–79	C
60–69	D
0–59	F

es, if any, are excluded from the domain of $f(x) = 2x^2 - 7x + 1$?

e value of $2x^2 - 7x + 1$ is a real number for any value of *x*, the
the function is all real numbers. No values are excluded from the
$(x) = 2x^2 - 7x + 1$.

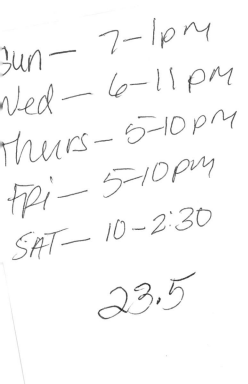

Sun — 7–1pm
Wed — 6–11 pm
Thurs — 5–10 pm
Fri — 5–10 pm
Sat — 10–2:30

23.5

···

 me function

,, (10, 6)}.

Solution
Domain: $\{2, 4, 6, 10\}$;
Range: $\{3, 6, 8\}$

You Try It 1
Find the domain and range of the function
$\{(-1, 5), (3, 5), (4, 5), (6, 5)\}$.

Your solution

Example 2

Given $p(r) = 5r^3 - 6r - 2$, find $p(-3)$.

Solution

$$p(r) = 5r^3 - 6r - 2$$
$$p(-3) = 5(-3)^3 - 6(-3) - 2$$
$$= 5(-27) + 18 - 2$$
$$= -135 + 18 - 2 = -119$$

You Try It 2

Evaluate $G(x) = \dfrac{3x}{x + 2}$ when $x = -4$.

Your solution

Example 3

Evaluate $Q(r) = 2r + 5$ when $r = h + 3$.

Solution

$$Q(r) = 2r + 5$$
$$Q(h + 3) = 2(h + 3) + 5$$
$$= 2h + 6 + 5$$
$$= 2h + 11$$

You Try It 3

Evaluate $f(x) = x^2 - 11$ when $x = 3h$.

Your solution

Example 4

Find the range of $f(x) = x^2 - 1$ if the domain is $\{-2, -1, 0, 1, 2\}$.

Solution

To find the range, evaluate the function at each element of the domain.

$$f(x) = x^2 - 1$$
$$f(-2) = (-2)^2 - 1 = 4 - 1 = 3$$
$$f(-1) = (-1)^2 - 1 = 1 - 1 = 0$$
$$f(0) = 0^2 - 1 = 0 - 1 = -1$$
$$f(1) = 1^2 - 1 = 1 - 1 = 0$$
$$f(2) = 2^2 - 1 = 4 - 1 = 3$$

The range is $\{-1, 0, 3\}$. Note that 0 and 3 are listed only once.

You Try It 4

Find the range of $h(z) = 3z + 1$ if the domain is $\left\{0, \dfrac{1}{3}, \dfrac{2}{3}, 1\right\}$.

Your solution

Example 5

What is the domain of $f(x) = 2x^2 - 7x + 1$?

Solution

Because $2x^2 - 7x + 1$ evaluates to a real number for any value of x, the domain of the function is all real numbers.

You Try It 5

What value is excluded from the domain of
$$f(x) = \dfrac{2}{x - 5}?$$

Your solution

Solutions on p. S7

3.2 Exercises

1. In your own words, explain what a function is.

2. What is the domain of a function? What is the range of a function?

3. Does the diagram below represent a function? Explain your answer.

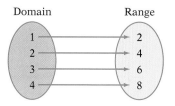

4. Does the diagram below represent a function? Explain your answer.

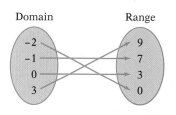

5. Does the diagram below represent a function? Explain your answer.

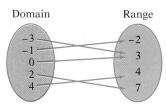

6. Does the diagram below represent a function? Explain your answer.

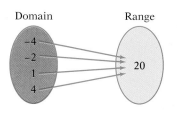

7. Does the diagram below represent a function? Explain your answer.

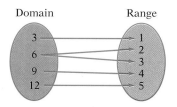

8. Does the diagram below represent a function? Explain your answer.

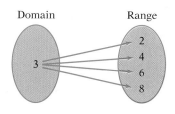

State whether each of the following relations is a function.

9. $\{(0, 0), (2, 4), (3, 6), (4, 8), (5, 10)\}$

10. $\{(1, 3), (3, 5), (5, 7), (7, 9)\}$

11. $\{(-2, -1), (-4, -5), (0, -1), (3, 5)\}$

12. $\{(-3, -1), (-1, -1), (0, 1), (2, 6)\}$

13. $\{(-2, 3), (-1, 3), (0, -3), (1, 3), (2, 3)\}$

14. $\{(0, 0), (1, 0), (2, 0), (3, 0), (4, 0)\}$

15. $\{(1, 1), (4, 2), (9, 3), (1, -1), (4, -2)\}$

16. $\{(3, 1), (3, 2), (3, 3), (3, 4)\}$

Solve.

17. The table at the right shows the cost to send an overnight package using United Parcel Service.
   a. Does this table define a function?
   b. Given $x = 2.75$ lb, find y.

Weight in pounds (x)	Cost (y)
$0 < x \leq 1$	$24.05
$1 < x \leq 2$	$26.83
$2 < x \leq 3$	$29.62
$3 < x \leq 4$	$32.40
$4 < x \leq 5$	$35.18

18. The table at the right shows the cost to send an "Express Mail" package using the U.S. Postal Service.
   a. Does this table define a function?
   b. Given $x = 0.5$ lb, find y.

Weight in pounds (x)	Cost (y)
$0 < x \leq 0.5$	$9.70
$0.5 < x \leq 1$	$11.75
$1 < x \leq 2$	$15.75
$2 < x \leq 3$	$18.75
$3 < x \leq 4$	$21.25

Given $f(x) = 5x - 4$, evaluate:

19. $f(3)$ **20.** $f(-2)$ **21.** $f(0)$ **22.** $f(-1)$

Given $G(t) = 4 - 3t$, evaluate:

23. $G(0)$ **24.** $G(-3)$ **25.** $G(-2)$ **26.** $G(4)$

Given $q(r) = r^2 - 4$, evaluate:

27. $q(3)$ **28.** $q(4)$ **29.** $q(-2)$ **30.** $q(-5)$

Given $F(x) = x^2 + 3x - 4$, evaluate:

31. $F(4)$ **32.** $F(-4)$ **33.** $F(-3)$ **34.** $F(-6)$

Given $H(p) = \dfrac{3p}{p + 2}$, evaluate:

35. $H(1)$ **36.** $H(-3)$ **37.** $H(t)$ **38.** $H(v)$

Given $s(t) = t^3 - 3t + 4$, evaluate:

39. $s(-1)$ **40.** $s(2)$ **41.** $s(a)$ **42.** $s(w)$

43. Given $P(x) = 4x + 7$, write $P(-2 + h) - P(-2)$ in simplest form.

44. Given $G(t) = 9 - 2t$, write $G(-3 + h) - G(-3)$ in simplest form.

45. Game Engineering has just completed the programming and testing for a new computer game. The cost to manufacture and package the game depends on the number of units Game Engineering plans to sell. The table at the right shows the cost per game for packaging various quantities. Evaluate this function when

a. $x = 7000$ **b.** $x = 20,000$

Number of Games Manufactured	Cost to Manufacture One Game
$0 < x \leq 2500$	$6.00
$2500 < x \leq 5000$	$5.50
$5000 < x \leq 10,000$	$4.75
$10,000 < x \leq 20,000$	$4.00
$20,000 < x \leq 40,000$	$3.00

46. Airport administrators have a tendency to price airport parking at a rate that discourages people from using the parking lot for long periods of time. The rate structure for an airport is given in the table at the right. Evaluate this function when

a. $t = 2.5$ h   **b.** $t = 7$ h

Hours Parked	Cost
$0 < t \leq 1$	$1.00
$1 < t \leq 2$	$3.00
$2 < t \leq 4$	$6.50
$4 < t \leq 7$	$10.00
$7 < t \leq 12$	$14.00

47. A real estate appraiser charges a fee that depends on the estimated value, V, of the property. A table giving the fees charged for various estimated values of the real estate appears at the right. Evaluate this function when

a. $V = \$5,000,000$ **b.** $V = \$767,000$

Value of Property	Appraisal Fee
$V < 100,000$	$350
$100.000 \leq V < 500,000$	$525
$500,000 \leq V < 1,000,000$	$950
$1,000,000 \leq V < 5,000,000$	$2500
$5,000,000 \leq V < 10,000,000$	$3000

48. The cost to mail a priority overnight package by Federal Express depends on the weight, w, of the package. A table of the costs for selected weights is given at the right. Evaluate this function when

a. $w = 2$ lb 3 oz   **b.** $w = 1.9$ lb

Weight (lb)	Cost
$0 < w < 1$	$24.00
$1 \leq w < 2$	$26.75
$2 \leq w < 3$	$29.25
$3 \leq w < 4$	$32.00
$4 \leq w < 5$	$34.00

Find the domain and range of the function.

49. $\{(1, 1), (2, 4), (3, 7), (4, 10), (5, 13)\}$

50. $\{(2, 6), (4, 18), (6, 38), (8, 66), (10, 102)\}$

51. $\{(0, 1), (2, 2), (4, 3), (6, 4)\}$

52. $\{(0, 1), (1, 2), (4, 3), (9, 4)\}$

53. $\{(1, 0), (3, 0), (5, 0), (7, 0), (9, 0)\}$

54. $\{(-2, -4), (2, 4), (-1, 1), (1, 1), (-3, 9), (3, 9)\}$

55. $\{(0, 0), (1, 1), (-1, 1), (2, 2), (-2, 2)\}$

56. $\{(0, -5), (5, 0), (10, 5), (15, 10)\}$

57. $\{(-2, -3), (-1, 6), (0, 7), (2, 3), (1, 9)\}$

58. $\{(-8, 0), (-4, 2), (-2, 4), (0, -4), (4, 4)\}$

What values are excluded from the domain of the function?

59. $f(x) = \dfrac{1}{x - 1}$

60. $g(x) = \dfrac{1}{x + 4}$

61. $h(x) = \dfrac{x + 3}{x + 8}$

62. $F(x) = \dfrac{2x - 5}{x - 4}$

63. $f(x) = 3x + 2$

64. $g(x) = 4 - 2x$

65. $G(x) = x^2 + 1$

66. $H(x) = \dfrac{1}{2}x^2$

67. $f(x) = \dfrac{x - 1}{x}$

68. $g(x) = \dfrac{2x + 5}{7}$

69. $H(x) = x^2 - x + 1$

70. $f(x) = 3x^2 + x + 4$

71. $f(x) = \dfrac{2x - 5}{3}$

72. $g(x) = \dfrac{3 - 5x}{5}$

73. $H(x) = \dfrac{x - 2}{x + 2}$

74. $h(x) = \dfrac{3 - x}{6 - x}$

75. $f(x) = \dfrac{x - 2}{2}$

76. $G(x) = \dfrac{2}{x - 2}$

Find the range of the function defined by each equation and the given domain.

77. $f(x) = 4x - 3$; domain = $\{0, 1, 2, 3\}$

78. $G(x) = 3 - 5x$; domain = $\{-2, -1, 0, 1, 2\}$

79. $g(x) = 5x - 8$; domain = $\{-3, -1, 0, 1, 3\}$

80. $h(x) = 3x - 7$; domain = $\{-4, -2, 0, 2, 4\}$

81. $h(x) = x^2$; domain = $\{-2, -1, 0, 1, 2\}$

82. $H(x) = 1 - x^2$; domain = $\{-2, -1, 0, 1, 2\}$

83. $f(x) = 2x^2 - 2x + 2$;
domain = $\{-4, -2, 0, 4\}$

84. $G(x) = -2x^2 + 5x - 2$;
domain = $\{-3, -1, 0, 1, 3\}$

85. $H(x) = \dfrac{5}{1-x}$; domain = $\{-2, 0, 2\}$

86. $g(x) = \dfrac{4}{4-x}$; domain = $\{-5, 0, 3\}$

87. $f(x) = \dfrac{2}{x-4}$; domain = $\{-2, 0, 2, 6\}$

88. $g(x) = \dfrac{x}{3-x}$; domain = $\{-2, -1, 0, 1, 2\}$

89. $H(x) = 2 - 3x - x^2$; domain = $\{-5, 0, 5\}$

90. $G(x) = 4 - 3x - x^3$; domain = $\{-3, 0, 3\}$

APPLYING THE CONCEPTS

91. Explain the words *relation* and *function*. Include in your explanation how the meanings of the two words differ.

92. Give a real-world example of a relation that is not a function. Is it possible to give an example of a function that is not a relation? If so, give one. If not, explain why it is not possible.

93. Find the set of ordered pairs (x, y) determined by the equation $y = x^3$, where $x \in \{-2, -1, 0, 1, 2\}$. Does the set of ordered pairs define a function? Why or why not?

94. Find the set of ordered pairs (x, y) determined by the equation $|y| = x$, where $x \in \{0, 1, 2, 3\}$. Does the set of ordered pairs define a function? Why or why not?

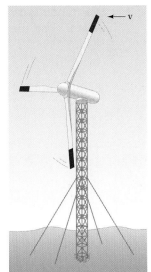

95. The power a windmill can generate is a function of the velocity of the wind. The function can be approximated by $P = f(v) = 0.015v^3$, where P is the power in watts and v is the velocity of the wind in meters per second. How much power will be produced by a windmill when the velocity of the wind is 15 m/s?

96. The distance, s (in feet), a car will skid on a certain road surface after the brakes are applied is a function of the car's velocity, v (in miles per hour). The function can be approximated by $s = f(v) = 0.017v^2$. How far will a car skid after its brakes are applied if it is traveling 60 mph?

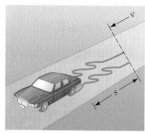

Each of the following graphs defines a function. Evaluate the function by estimating the ordinate (which is the value of the function) for the given value of t.

97. The graph at the right shows the speed, v, in feet per second that a parachutist is falling during the first 20 s after jumping out of a plane. Estimate the speed at which the parachutist is falling when

  a. $t = 5$ s   **b.** $t = 15$ s

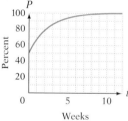

98. The graph at the right shows what an industrial psychologist has determined to be the average percent score, P, for an employee taking a performance test t weeks after training begins. Estimate the score an employee would receive on this test when

  a. $t = 4$ weeks   **b.** $t = 10$ weeks

99. The graph at the right shows the temperature, T, in degrees Fahrenheit of a can of cola t hours after it is placed in a refrigerator. Use the graph to estimate the temperature of the cola when

  a. $t = 5$ h   **b.** $t = 15$ h

100. The graph at the right shows the decrease in the heart rate, r, of a runner (in beats per minute) t minutes after the completion of a race. Use the graph to estimate the heart rate of a runner when

  a. $t = 5$ min   **b.** $t = 20$ min

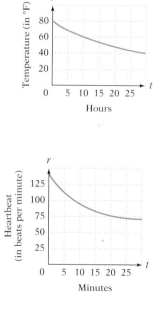

3.3 Linear Functions

Objective A   **To graph a linear function**   ‹ 3 ›

Recall that the ordered pairs of a function can be written as $(x, f(x))$ or (x, y). The **graph of a function** is a graph of the ordered pairs (x, y) that belong to the function. Certain functions have characteristic graphs. A function that can be written in the form $f(x) = mx + b$ (or $y = mx + b$) is called a **linear function** because its graph is a straight line.

Examples of linear functions are shown at the right. Note that the exponent on each variable is 1.

$$f(x) = 2x + 5 \qquad (m = 2, b = 5)$$
$$P(t) = 3t - 2 \qquad (m = 3, b = -2)$$
$$y = -2x \qquad (m = -2, b = 0)$$
$$y = -\frac{2}{3}x + 1 \qquad \left(m = -\frac{2}{3}, b = 1\right)$$
$$g(z) = z - 2 \qquad (m = 1, b = -2)$$

The equation $y = x^2 + 4x + 3$ is not a linear function because it includes a term with a variable squared. The equation $f(x) = \dfrac{3}{x - 2}$ is not a linear function because a variable occurs in the denominator. Another example of an equation that is not a linear function is $y = \sqrt{x} + 4$; this equation contains a variable within a radical and so is not a linear function.

Consider $f(x) = 2x + 1$. Evaluating the linear function when $x = -2, -1, 0, 1,$ and 2 produces some of the ordered pairs of the function. It is convenient to record the results in a table similar to the one at the right. The graph of the ordered pairs is shown in the leftmost figure below.

x	$f(x) = 2x + 1$	y	(x, y)
-2	$2(-2) + 1$	-3	$(-2, -3)$
-1	$2(-1) + 1$	-1	$(-1, -1)$
0	$2(0) + 1$	1	$(0, 1)$
1	$2(1) + 1$	3	$(1, 3)$
2	$2(2) + 1$	5	$(2, 5)$

Evaluating the function when x is not an integer produces more ordered pairs to graph, such as $\left(-\dfrac{5}{2}, -4\right)$ and $\left(\dfrac{3}{2}, 4\right)$, as shown in the middle figure below. Evaluating the function for still other values of x would result in more and more ordered pairs being graphed. The result would be so many dots that the graph would look like the straight line shown in the rightmost figure, which is the graph of $f(x) = 2x + 1$.

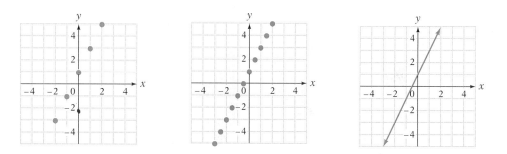

No matter what value of x is chosen, $2x + 1$ is a real number. This means the domain of $f(x) = 2x + 1$ is all real numbers. Therefore, we can use any real number when evaluating the function. Normally, however, values such as π or $\sqrt{5}$ are not used because it is difficult to graph the resulting ordered pairs.

Note from the graph of $f(x) = 2x + 1$ shown at the right that $(-1.5, -2)$ and $(3, 7)$ are the coordinates of points on the graph and that $f(-1.5) = -2$ and $f(3) = 7$. Note also that the point whose coordinates are $(2, 1)$ is not a point on the graph and that $f(2) \neq 1$. Every point on the graph is an ordered pair that belongs to the function, and every ordered pair that belongs to the function corresponds to a point on the graph.

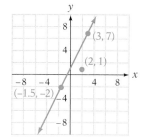

Whether an equation is written as $f(x) = mx + b$ or as $y = mx + b$, the equation represents a linear function, and the graph of the equation is a straight line.

Because the graph of a linear function is a straight line, and a straight line is determined by two points, the graph of a linear function can be drawn by finding only two of the ordered pairs of the function. However, it is recommended that you find at least *three* ordered pairs to ensure accuracy.

➡ Graph: $f(x) = -\dfrac{1}{2}x + 3$

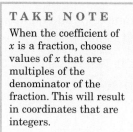

TAKE NOTE

When the coefficient of x is a fraction, choose values of x that are multiples of the denominator of the fraction. This will result in coordinates that are integers.

x	$y = f(x)$
-4	5
0	3
2	2

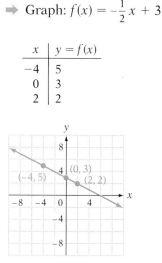

- Find at least three ordered pairs. When the coefficient of x is a fraction, choose values of x that will simplify the calculations. The ordered pairs can be displayed in a table.

- Graph the ordered pairs and draw a line through the points.

. . . .

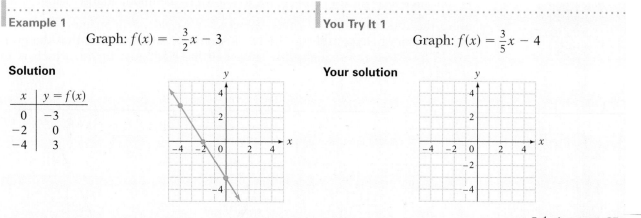

Example 1

Graph: $f(x) = -\dfrac{3}{2}x - 3$

Solution

x	$y = f(x)$
0	-3
-2	0
-4	3

You Try It 1

Graph: $f(x) = \dfrac{3}{5}x - 4$

Your solution

Solution on p. S7

Example 2

Graph: $y = \dfrac{2}{3}x$

Solution

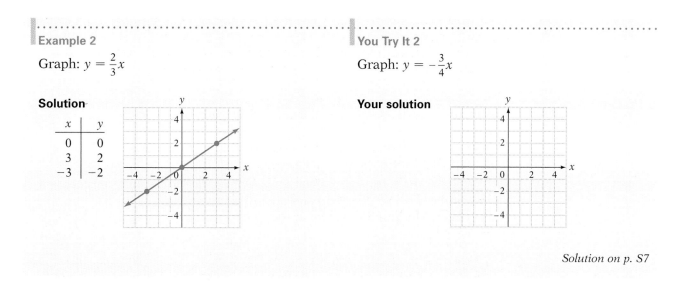

x	y
0	0
3	2
-3	-2

You Try It 2

Graph: $y = -\dfrac{3}{4}x$

Your solution

Solution on p. S7

Objective B   To graph an equation of the form $Ax + By = C$

An equation of the form $Ax + By = C$, where A and B are coefficients and C is a constant, is also a *linear equation in two variables.* This equation can be written in the form $y = mx + b$.

➡ Write $4x - 3y = 6$ in the form $y = mx + b$.

$$4x - 3y = 6$$
$$-3y = -4x + 6$$
$$y = \frac{4}{3}x - 2$$

• Subtract 4x from each side of the equation.

• Divide each side of the equation by -3. This is the form $y = mx + b.\ m = \dfrac{4}{3}, b = -2$

To graph an equation of the form $Ax + By = C$, first solve the equation for y. Then follow the same procedure used for graphing an equation of the form $y = mx + b$.

➡ Graph: $3x + 2y = 6$

$$3x + 2y = 6$$
$$2y = -3x + 6$$
$$y = -\frac{3}{2}x + 3$$

• Solve the equation for y.

x	y
0	3
2	0
4	-3

• Find at least three solutions.

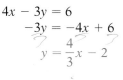

• Graph the ordered pairs in a rectangular coordinate system. Draw a straight line through the points.

An equation in which one of the variables is missing has a graph that is either a horizontal or a vertical line.

The equation $y = -2$ can be written

$$0 \cdot x + y = -2$$

Because $0 \cdot x = 0$ for all values of x, y is always -2 for every value of x.

Some of the possible ordered pairs are given in the table below. The graph is shown at the right.

x	y
-2	-2
0	-2
3	-2

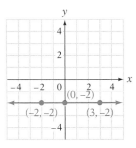

Graph of $y = b$

The graph of $y = b$ is a horizontal line passing through the point $(0, b)$.

The equation $y = -2$ represents a function. Some of the ordered pairs of this function are $(-2, -2)$, $(0, -2)$, and $(3, -2)$. In functional notation we would write $f(x) = -2$. This function is an example of a constant function. No matter what value of x is selected, $f(x) = -2$.

Definition of the Constant Function

A function given by $f(x) = b$, where b is a constant, is a **constant function**. The graph of the constant function is a horizontal line passing through $(0, b)$.

For each value in the domain of a constant function, the value of the function is the same (constant). For instance, if $f(x) = 4$, then $f(2) = 4$, $f(3) = 4$, $f(\sqrt{3}) = 4$, $f(\pi) = 4$, and so on. The value of $f(x)$ is 4 for all values of x.

➡ Graph: $y + 4 = 0$

Solve for y.
$y + 4 = 0$
  $y = -4$

The graph of $y = -4$ is a horizontal line passing through $(0, -4)$.

Evaluate $P(t) = -7$ when $t = 6$.

$$P(t) = -7$$
$$P(6) = -7$$

• The value of the constant function is the same for all values of the variable.

For the equation $y = -2$, the coefficient of x was zero. For the equation $x = 2$, the coefficient of y is zero. For instance, the equation $x = 2$ can be written

$$x + 0 \cdot y = 2$$

No matter what value of y is chosen, $0 \cdot y = 0$ and therefore x is always 2.

Some of the possible ordered pairs are given in the table below. The graph is shown at the right.

x	y
2	6
2	1
2	-4

Graph of $x = a$

The graph of $x = a$ is a vertical line passing through the point $(a, 0)$.

TAKE NOTE

The equation $y = b$ represents a function. The equation $x = a$ does *not* represent a function. Remember, not all equations represent functions.

Recall that a function is a set of ordered pairs in which no two ordered pairs have the same first coordinate and different second coordinates. Because $(2, 6)$, $(2, 1)$, and $(2, -4)$ are ordered pairs belonging to the equation $x = 2$, this equation does not represent a function, and the graph is not the graph of a function.

....

Example 3
Graph: $2x + 3y = 9$

Solution
Solve the equation for y.

$$2x + 3y = 9$$
$$3y = -2x + 9$$
$$y = -\frac{2}{3}x + 3$$

x	y
-3	5
0	3
3	1

You Try It 3
Graph: $-3x + 2y = 4$

Your solution

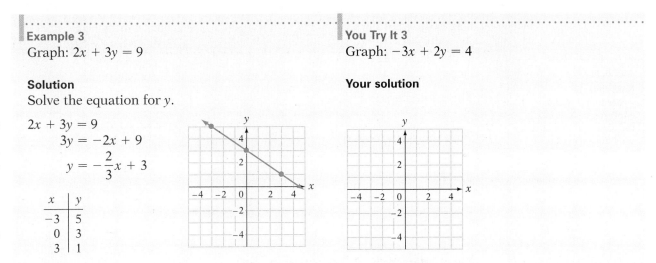

Solution on p. S7

Example 4

Graph: $x = -4$

Solution

The graph of an equation of the form $x = a$ is a vertical line passing through the point whose coordinates are $(a, 0)$.

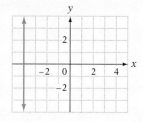

You Try It 4

Graph: $y - 3 = 0$

Your solution

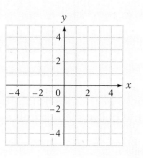

Solution on p. S7

Objective C   **To find the x- and the y-intercepts of a straight line**

The graph of the equation $x - 2y = 4$ is shown at the right. The graph crosses the x-axis at the point $(4, 0)$. This point is called the **x-intercept**. The graph also crosses the y-axis at the point $(0, -2)$. This point is called the **y-intercept.**

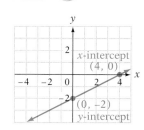

TAKE NOTE

The x-intercept occurs when $y = 0$.

The y-intercept occurs when $x = 0$.

➡ Find the x- and y-intercepts of the graph of the equation $3x + 4y = -12$.

To find the x-intercept, let $y = 0$. (Any point on the x-axis has y-coordinate 0.)

$$3x + 4y = -12$$
$$3x + 4(0) = -12$$
$$3x = -12$$
$$x = -4$$

The x-intercept is $(-4, 0)$.

To find the y-intercept, let $x = 0$. (Any point on the y-axis has x-coordinate 0.)

$$3x + 4y = -12$$
$$3(0) + 4y = -12$$
$$4y = -12$$
$$y = -3$$

The y-intercept is $(0, -3)$.

TAKE NOTE

Note that the y-coordinate of the y-intercept $(0, 7)$ has the same value as the constant term of $y = \frac{2}{3}x + 7$.

As shown below, this is always true:

$$y = mx + b$$
$$y = m(0) + b$$
$$y = b$$

Thus, the y-intercept is $(0, b)$.

➡ Find the y-intercept of $y = \frac{2}{3}x + 7$.

$$y = \frac{2}{3}x + 7$$

$$y = \frac{2}{3}(0) + 7 \qquad \bullet \text{ To find the } y\text{-intercept, let } x = 0.$$

$$y = 7$$

The y-intercept is $(0, 7)$.

For any equation of the form $y = mx + b$, the y-intercept is $(0, b)$.

A linear equation can be graphed by finding the *x*- and *y*-intercepts and then drawing a line through the two points.

 Graph $3x - 2y = 6$ by using the *x*- and *y*-intercepts.

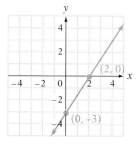

$$3x - 2y = 6$$
$$3x - 2(0) = 6$$
$$3x = 6$$
$$x = 2$$

- To find the *x*-intercept, let $y = 0$. Then solve for *x*.

$$3x - 2y = 6$$
$$3(0) - 2y = 6$$
$$-2y = 6$$
$$y = -3$$

- To find the *y*-intercept, let $x = 0$. Then solve for *y*.

The *x*-intercept is (2, 0). The *y*-intercept is (0, −3).

Example 5

Graph $4x - y = 4$ by using the *x*- and *y*-intercepts.

Solution

x-intercept:

$$4x - y = 4$$
$$4x - 0 = 4$$
$$4x = 4$$
$$x = 1$$
$$(1, 0)$$

y-intercept:

$$4x - y = 4$$
$$4(0) - y = 4$$
$$-y = 4$$
$$y = -4$$
$$(0, -4)$$

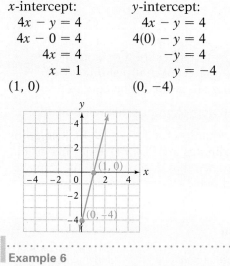

You Try It 5

Graph $3x - y = 2$ by using the *x*- and *y*-intercepts.

Your solution

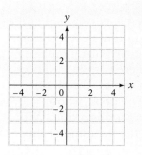

Example 6

Graph $y = \frac{2}{3}x - 2$ by using the *x*- and *y*-intercepts.

Solution

x-intercept:

$$y = \frac{2}{3}x - 2$$
$$0 = \frac{2}{3}x - 2$$
$$-\frac{2}{3}x = -2$$
$$x = 3$$
$$(3, 0)$$

y-intercept:

$$(0, b)$$
$$b = -2$$
$$(0, -2)$$

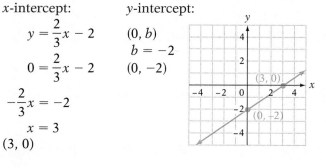

You Try It 6

Graph $y = \frac{1}{4}x + 1$ by using the *x*- and *y*-intercepts.

Your solution

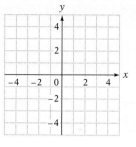

Solutions on pp. S7–S8

Objective D   **To solve application problems**

There are a variety of applications of linear functions.

⇒ The heart rate, R, after t minutes for a person taking a brisk walk can be approximated by the equation $R = 2t + 70$.
   a. Graph this equation for $0 \leq t \leq 10$.
   b. The point whose coordinates are (5, 82) is on the graph. Write a sentence that describes the meaning of this ordered pair.

a.

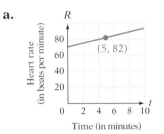

b. The ordered pair (5, 82) means that after 5 min the person's heart rate is 82 beats per minute.

Example 7

An electronics technician charges $45 plus $1 per minute to repair defective wiring in a home or apartment. The equation that describes the total cost, C, to have defective wiring repaired is given by $C = t + 45$, where t is the number of minutes the technician works. Graph this equation for $0 \leq t \leq 60$. The point whose coordinates are (15, 60) is on this graph. Write a sentence that describes the meaning of this ordered pair.

Solution

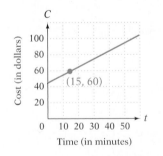

The ordered pair (15, 60) means that it costs $60 for the technician to work 15 min.

You Try It 7

The height h (in inches) of a person and the length L (in inches) of that person's stride while walking are related. The equation $h = \frac{3}{4}L + 50$ approximates this relationship. Graph this equation for $15 \leq L \leq 40$. The point whose coordinates are (32, 74) is on this graph. Write a sentence that describes the meaning of this ordered pair.

Your solution

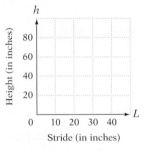

Solution on p. S8

3.3 Exercises

Objective A

Graph.

1. $y = 3x - 4$

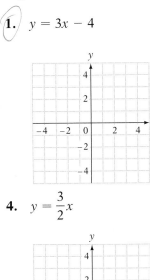

2. $y = -2x + 3$

3. $y = -\dfrac{2}{3}x$

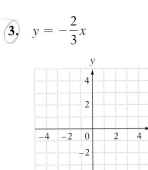

4. $y = \dfrac{3}{2}x$

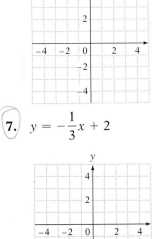

5. $y = \dfrac{2}{3}x - 4$

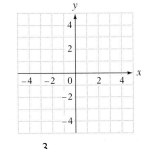

6. $y = \dfrac{3}{4}x + 2$

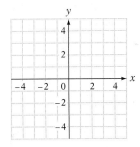

7. $y = -\dfrac{1}{3}x + 2$

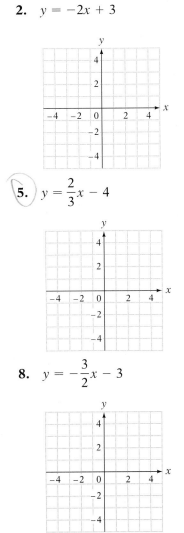

8. $y = -\dfrac{3}{2}x - 3$

9. $y = \dfrac{3}{5}x - 1$

Objective B

Graph.

10. $2x - y = 3$

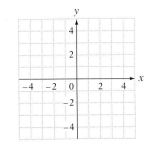

11. $2x + y = -3$

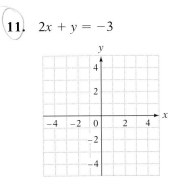

12. $2x + 5y = 10$

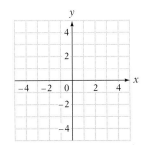

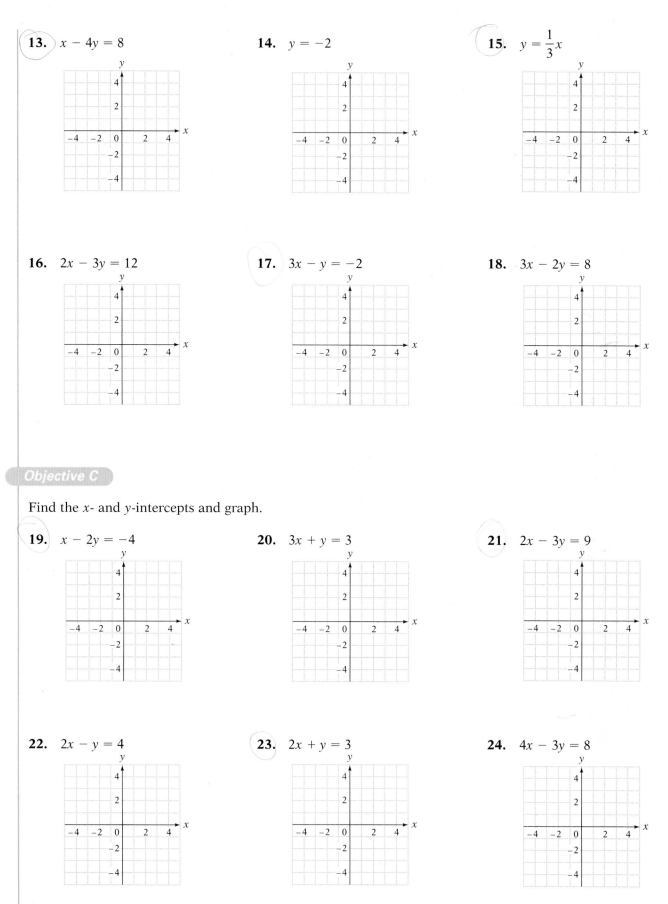

13. $x - 4y = 8$

14. $y = -2$

15. $y = \dfrac{1}{3}x$

16. $2x - 3y = 12$

17. $3x - y = -2$

18. $3x - 2y = 8$

Objective C

Find the *x*- and *y*-intercepts and graph.

19. $x - 2y = -4$

20. $3x + y = 3$

21. $2x - 3y = 9$

22. $2x - y = 4$

23. $2x + y = 3$

24. $4x - 3y = 8$

25. $3x + 2y = 4$

26. $2x - 3y = 4$

27. $4x - 3y = 6$

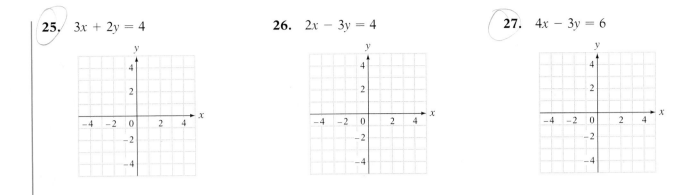

Objective D   *Application Problems*

28. A bee beats its wings approximately 100 times per second. The equation that describes the total number of times a bee beats its wings is given by $B = 100t$. Graph this equation for $0 \le t \le 60$. The ordered pair $(35, 3500)$ is on this graph. Write a sentence that describes the meaning of this ordered pair.

29. Marlys receives $11 per hour as a mathematics tutor. The equation that describes Marlys's wages is $W = 11t$, where t is the number of hours she spends tutoring. Graph this equation for $0 \le t \le 20$. The ordered pair $(15, 165)$ is on the graph. Write a sentence that describes the meaning of this ordered pair.

30. The monthly cost for receiving messages from a telephone answering service is $8.00 plus $.20 a message. The equation that describes the cost is $C = 0.20n + 8.00$, where n is the number of messages received. Graph this equation for $0 \le n \le 40$. The point $(32, 14.40)$ is on the graph. Write a sentence that describes the meaning of this ordered pair.

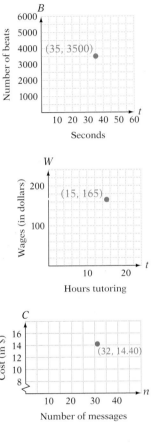

31. The Rattler roller coaster at Fiesta in San Antonio, Texas, has a maximum speed of approximately 110 ft/s. The equation that describes the total number of feet traveled by the roller coaster in t seconds at this speed is given by $D = 110t$. Graph this equation for $0 \le t \le 10$. The point $(5, 550)$ is on this graph. Write a sentence that describes the meaning of this ordered pair.

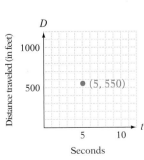

32. The cost of manufacturing skis is $5000 for startup and $80 per pair of skis manufactured. The equation that describes the cost of manufacturing n pairs of skis is $C = 80n + 5000$. Graph this equation for $0 \le n \le 2000$. The point $(50, 9000)$ is on the graph. Write a sentence that describes the meaning of this ordered pair.

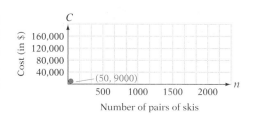

33. The cost of manufacturing compact disks is $3000 for startup and $5 per disk. The equation that describes the cost of manufacturing n compact disks is $C = 5n + 3000$. Graph this equation for $0 \le n \le 10,000$. The point $(6000, 33,000)$ is on the graph. Write a sentence that describes the meaning of this ordered pair.

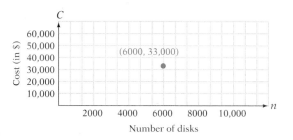

APPLYING THE CONCEPTS

34. Explain what the graph of an equation represents.

35. Explain how to graph the equation of a straight line by plotting points.

36. Explain how to graph the equation of a straight line by using its x- and y-intercepts.

37. Explain why you cannot graph the equation $4x + 3y = 0$ by using just its intercepts.

An equation of the form $\dfrac{x}{a} + \dfrac{y}{b} = 1$, where $a \ne 0$ and $b \ne 0$, is called the *intercept form of a straight line* because $(a, 0)$ and $(0, b)$ are the x- and y-intercepts of the graph of the equation. Graph each of the following.

38. $\dfrac{x}{3} + \dfrac{y}{5} = 1$ **39.** $\dfrac{x}{2} + \dfrac{y}{3} = 1$ **40.** $\dfrac{x}{3} - \dfrac{y}{2} = 1$

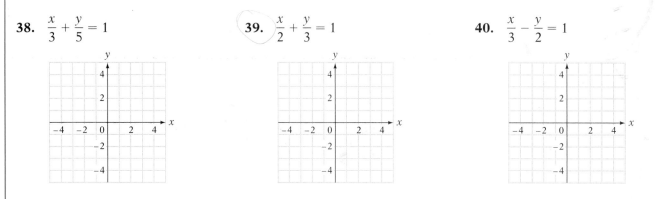

3.4 Slope of a Straight Line

Objective A   To find the slope of a line given two points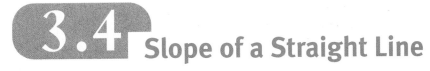

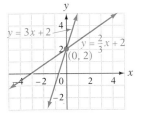

The graphs of $y = 3x + 2$ and $y = \frac{2}{3}x + 2$ are shown at the left. Each graph crosses the y-axis at the point $(0, 2)$, but the graphs have different slants. The **slope** of a line is a measure of the slant of the line. The symbol for slope is m.

The slope of a line containing two points is the ratio of the change in the y values between the two points to the change in the x values. The line containing the points whose coordinates are $(-1, -3)$ and $(5, 2)$ is shown below.

The change in the y values is the difference between the y-coordinates of the two points.

$$\text{Change in } y = 2 - (-3) = 5$$

The change in the x values is the difference between the x-coordinates of the two points.

$$\text{Change in } x = 5 - (-1) = 6$$

The slope of the line between the two points is the ratio of the change in y to the change in x.

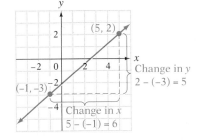

$$\text{Slope} = m = \frac{\text{change in } y}{\text{change in } x} = \frac{5}{6} \qquad m = \frac{2 - (-3)}{5 - (-1)} = \frac{5}{6}$$

In general, if $P_1(x_1, y_1)$ and $P_2(x_2, y_2)$ are two points on a line, then

$$\text{Change in } y = y_2 - y_1 \qquad \text{Change in } x = x_2 - x_1$$

Using these ideas, we can state a formula for slope.

Slope Formula

The slope of the line containing the two points $P_1(x_1, y_1)$ and $P_2(x_2, y_2)$ is given by

$$m = \frac{y_2 - y_1}{x_2 - x_1}, \, x_1 \neq x_2$$

Frequently, the Greek letter Δ is used to designate the change in a variable. Using this notation, we can write the equations for the change in y and the change in x as follows:

$$\text{Change in } y = \Delta y = y_2 - y_1 \qquad \text{Change in } x = \Delta x = x_2 - x_1$$

Using this notation, the slope formula is written $m = \frac{\Delta y}{\Delta x}$.

➡ Find the slope of the line containing the points whose coordinates are $(-2, 0)$ and $(4, 5)$.

Let $P_1 = (-2, 0)$ and $P_2 = (4, 5)$. (It does not matter which point is named P_1 or P_2; the slope will be the same.)

$$m = \frac{y_2 - y_1}{x_2 - x_1} = \frac{5 - 0}{4 - (-2)} = \frac{5}{6}$$

A line that slants upward to the right always has a **positive slope.**

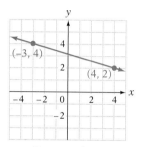

Positive slope

➡ Find the slope of the line containing the points whose coordinates are $(-3, 4)$ and $(4, 2)$.

Let $P_1 = (-3, 4)$ and $P_2 = (4, 2)$.

$$m = \frac{y_2 - y_1}{x_2 - x_1} = \frac{2 - 4}{4 - (-3)} = \frac{-2}{7} = -\frac{2}{7}$$

A line that slants downward to the right always has a **negative slope.**

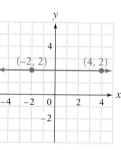

Negative slope

➡ Find the slope of the line containing the points whose coordinates are $(-2, 2)$ and $(4, 2)$.

Let $P_1 = (-2, 2)$ and $P_2 = (4, 2)$.

$$m = \frac{y_2 - y_1}{x_2 - x_1} = \frac{2 - 2}{4 - (-2)} = \frac{0}{6} = 0$$

A horizontal line has **zero slope.**

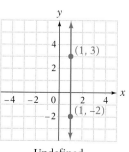

Zero slope

➡ Find the slope of the line containing the points whose coordinates are $(1, -2)$ and $(1, 3)$.

Let $P_1 = (1, -2)$ and $P_2 = (1, 3)$.

$$m = \frac{y_2 - y_1}{x_2 - x_1} = \frac{3 - (-2)}{1 - 1} = \frac{5}{0} \quad \text{Not a real number}$$

The slope of a vertical line is **undefined.**

Undefined

Point of Interest

One of the motivations for the discovery of calculus was to solve a more complicated version of the distance-rate problem at the right.

You may be familiar with twirling a ball on the end of a string. If you release the string, the ball flies off in a path as shown below.

The question that mathematicians tried to answer was essentially, "What is the slope of the line represented by the arrow?"

Answering questions similar to this led to the development of one aspect of calculus.

There are many applications of slope. Here are two examples.

The first record for the one-mile run was recorded in 1865 in England. Richard Webster ran the mile in 4 min 36.5 s. His average speed was approximately 19 feet per second.

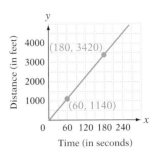

The graph at the right shows the distance Webster ran during that run. From the graph, note that after 60 s (1 min) he had traveled 1140 ft and that after 180 s (3 min) he had traveled 3420 ft.

Let the point (60, 1140) be (x_1, y_1) and the point (180, 3420) be (x_2, y_2). The slope of the line between these two points is

$$m = \frac{y_2 - y_1}{x_2 - x_1}$$

$$= \frac{3420 - 1140}{180 - 60} = \frac{2280}{120} = 19$$

Note that the slope of the line is the same as Webster's average speed, 19 feet per second.

Average speed is related to slope.

The following example is related to economics.

As a result of cheaper digital networks and more competition, the cost of wireless phone use is decreasing. The average cost per minute in 1999 was 28¢. The projected average cost in 2003 is 20¢. (Source: The Strategic Group)

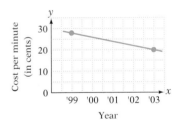

Let the point (1999, 28) be (x_1, y_1) and the point (2003, 20) be (x_2, y_2). The slope of the line between these two points is

$$m = \frac{y_2 - y_1}{x_2 - y_1}$$

$$= \frac{20 - 28}{2003 - 1999} = \frac{-8}{4} = -2$$

Note that if we interpret negative slope as decreasing cost, then the slope of the line represents the rate at which the per-minute cost of phone use is decreasing, 2¢ per year.

In general, any quantity that is expressed by using the word *per* is represented mathematically as slope. In the first example, the slope was 19 feet per second. In the second example, the slope was −2¢ per year.

Example 1

Find the slope of the line containing the points $(2, -5)$ and $(-4, 2)$.

Solution

Let $P_1 = (2, -5)$ and $P_2 = (-4, 2)$.

$$m = \frac{y_2 - y_1}{x_2 - x_1} = \frac{2 - (-5)}{-4 - 2} = \frac{7}{-6}$$

The slope is $-\dfrac{7}{6}$.

You Try It 1

Find the slope of the line containing the points $(4, -3)$ and $(2, 7)$.

Your solution

Example 2

Find the slope of the line containing the points $(-3, 4)$ and $(5, 4)$.

Solution

Let $P_1 = (-3, 4)$ and $P_2 = (5, 4)$.

$$m = \frac{y_2 - y_1}{x_2 - x_1}$$

$$= \frac{4 - 4}{5 - (-3)}$$

$$= \frac{0}{8} = 0$$

The slope of the line is zero.

You Try It 2

Find the slope of the line containing the points $(6, -1)$ and $(6, 7)$.

Your solution

Example 3

The graph below shows the relationship between the cost of an item and the sales tax. Find the slope of the line between the two points shown on the graph. Write a sentence that states the meaning of the slope.

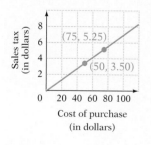

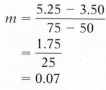

Solution

$$m = \frac{5.25 - 3.50}{75 - 50}$$

$$= \frac{1.75}{25}$$

$$= 0.07$$

A slope of 0.07 means that the sales tax is $.07 per dollar.

You Try It 3

The graph below shows the decrease in the value of a printing press for a period of 6 years. Find the slope of the line between the two points shown on the graph. Write a sentence that states the meaning of the slope.

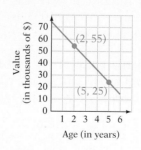

Your solution

Solutions on p. S8

Objective B   **To graph a line given a point and the slope**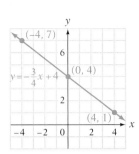

The graph of the equation $y = -\frac{3}{4}x + 4$ is shown at the right. The points $(-4, 7)$ and $(4, 1)$ are on the graph. The slope of the line is

$$m = \frac{7 - 1}{-4 - 4} = \frac{6}{-8} = -\frac{3}{4}$$

Note that the slope of the line has the same value as the coefficient of x. The y-intercept is $(0, 4)$.

Slope-Intercept Formula

The equation $y = mx + b$ is called the **slope-intercept form** of a straight line. The slope of the line is m, the coefficient of x. The y-intercept is $(0, b)$.

When the equation of a straight line is in the form $y = mx + b$, the graph can be drawn by using the slope and y-intercept. First locate the y-intercept. Use the slope to find a second point on the line. Then draw a line through the two points.

➡ Graph $y = \frac{5}{3}x - 4$ by using the slope and y-intercept.

The slope is the coefficient of x:

$$m = \frac{5}{3} = \frac{\text{change in } y}{\text{change in } x}$$

The y-intercept is $(0, -4)$.

TAKE NOTE

When graphing a line by using its slope and y-intercept, *always* start at the y-intercept.

Beginning at the y-intercept $(0, -4)$, move right 3 units (change in x) and then up 5 units (change in y).

The point whose coordinates are $(3, 1)$ is a second point on the graph. Draw a line through the points $(0, -4)$ and $(3, 1)$.

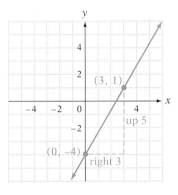

➡ Graph $x + 2y = 4$ by using the slope and y-intercept.

Solve the equation for y.

$$x + 2y = 4$$
$$2y = -x + 4$$
$$y = -\frac{1}{2}x + 2 \qquad \bullet\; m = -\frac{1}{2},$$
$$\text{y-intercept} = (0, 2)$$

Beginning at the y-intercept $(0, 2)$, move right 2 units (change in x) and then down 1 unit (change in y).

The point whose coordinates are $(2, 1)$ is a second point on the graph. Draw a line through the points $(0, 2)$ and $(2, 1)$.

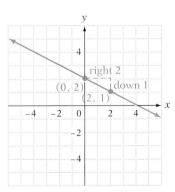

The graph of a line can be drawn when any point on the line and the slope of the line are given.

➡ Graph the line that passes through the point $(-4, -4)$ and has slope 2.

When the slope is an integer, write it as a fraction with denominator 1.

$$m = 2 = \frac{2}{1} = \frac{\text{change in } y}{\text{change in } x}$$

Locate $(-4, -4)$ in the coordinate plane. Beginning at that point, move right 1 unit (change in x) and then up 2 units (change in y).

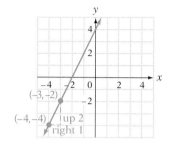

The point whose coordinates are $(-3, -2)$ is a second point on the graph. Draw a line through the points $(-4, -4)$ and $(-3, -2)$.

> **TAKE NOTE**
>
> This example differs from the preceding two in that a point other than the y-intercept is used. In this case, start at the given point.

Example 4

Graph $y = -\frac{3}{2}x + 4$ by using the slope and y-intercept.

Solution

$m = -\frac{3}{2} = \frac{-3}{2}$

y-intercept $= (0, 4)$

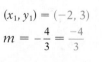

You Try It 4

Graph $2x + 3y = 6$ by using the slope and y-intercept.

Your solution

Example 5

Graph the line that passes through the point $(-2, 3)$ and has slope $-\frac{4}{3}$.

Solution

$(x_1, y_1) = (-2, 3)$

$m = -\frac{4}{3} = \frac{-4}{3}$

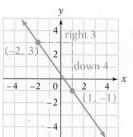

You Try It 5

Graph the line that passes through the point $(-3, -2)$ and has slope 3.

Your solution

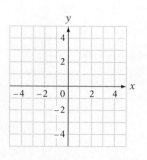

Solutions on p. S8

3.4 Exercises

Objective A

Find the slope of the line containing the points.

1. $P_1(1, 3), P_2(3, 1)$

2. $P_1(2, 3), P_2(5, 1)$

3. $P_1(-1, 4), P_2(2, 5)$

4. $P_1(3, -2), P_2(1, 4)$

5. $P_1(-1, 3), P_2(-4, 5)$

6. $P_1(-1, -2), P_2(-3, 2)$

7. $P_1(0, 3), P_2(4, 0)$

8. $P_1(-2, 0), P_2(0, 3)$

9. $P_1(2, 4), P_2(2, -2)$

10. $P_1(4, 1), P_2(4, -3)$

11. $P_1(2, 5), P_2(-3, -2)$

12. $P_1(4, 1), P_2(-1, -2)$

13. $P_1(2, 3), P_2(-1, 3)$

14. $P_1(3, 4), P_2(0, 4)$

15. $P_1(0, 4), P_2(-2, 5)$

16. $P_1(-2, 3), P_2(-2, 5)$

17. $P_1(-3, -1), P_2(-3, 4)$

18. $P_1(-2, -5), P_2(-4, -1)$

19. The graph below shows the relationship between the distance traveled by a motorist and the time of travel. Find the slope of the line between the two points shown on the graph. Write a sentence that states the meaning of the slope.

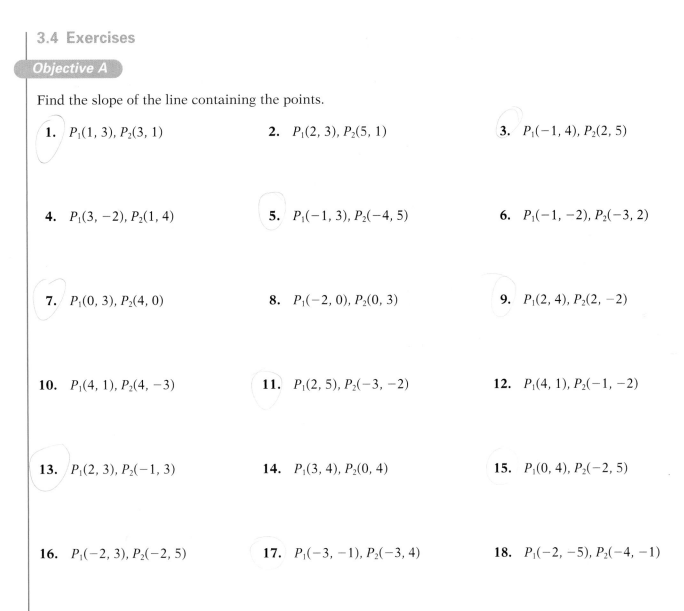

20. The graph below shows the number of people subscribing to a sports magazine of increasing popularity. Find the slope of the line between the two points shown on the graph. What is the meaning of the slope?

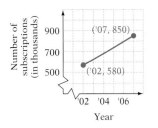

21. The graph below shows how the amount of gas in the tank of a car decreases as the car is driven. Find the slope of the line. Write a sentence that states the meaning of the slope.

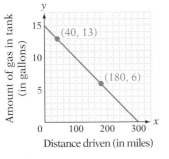

22. The troposphere extends from Earth's surface to an elevation of about 11 km. The graph below shows the decrease in the temperature of the troposphere as altitude increases. Find the slope of the line. Write a sentence that states the meaning of the slope.

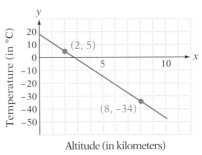

Altitude (in kilometers)

23. The graph below shows the relationship between distance and time for the 5000-meter run for the world record by Said Aouita in 1987. Find the slope of the line between the two points shown on the graph. Write a sentence that states the meaning of the slope.

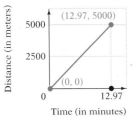

24. The graph below shows the relationship between distance and time for the 10,000-meter run for the world record by Arturo Barrios in 1989. Find the slope of the line between the two points shown on the graph. Write a sentence that states the meaning of the slope.

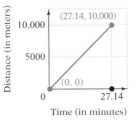

The American National Standards Institute (ANSI) states that the slope for a wheelchair ramp must not exceed $\frac{1}{12}$.

25. Does a ramp that is 6 in. high and 5 ft long meet the requirements of ANSI?

26. Does a ramp that is 12 in. high and 170 in. long meet the requirements of ANSI?

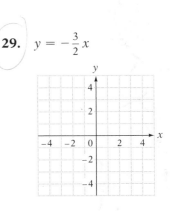

Objective B

Graph by using the slope and the y-intercept.

27. $y = \frac{1}{2}x + 2$

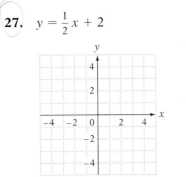

28. $y = \frac{2}{3}x - 3$

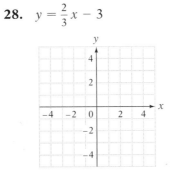

29. $y = -\frac{3}{2}x$

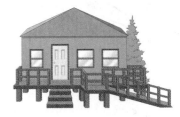

30. $y = \frac{3}{4}x$

31. $y = -\frac{1}{2}x + 2$

32. $y = \frac{2}{3}x - 1$

33. $x - 3y = 3$

34. $3x + 2y = 8$

35. $4x + y = 2$

36. Graph the line that passes through the point $(-1, -3)$ and has slope $\frac{4}{3}$.

37. Graph the line that passes through the point $(-2, -3)$ and has slope $\frac{5}{4}$.

38. Graph the line that passes through the point $(-3, 0)$ and has slope -3.

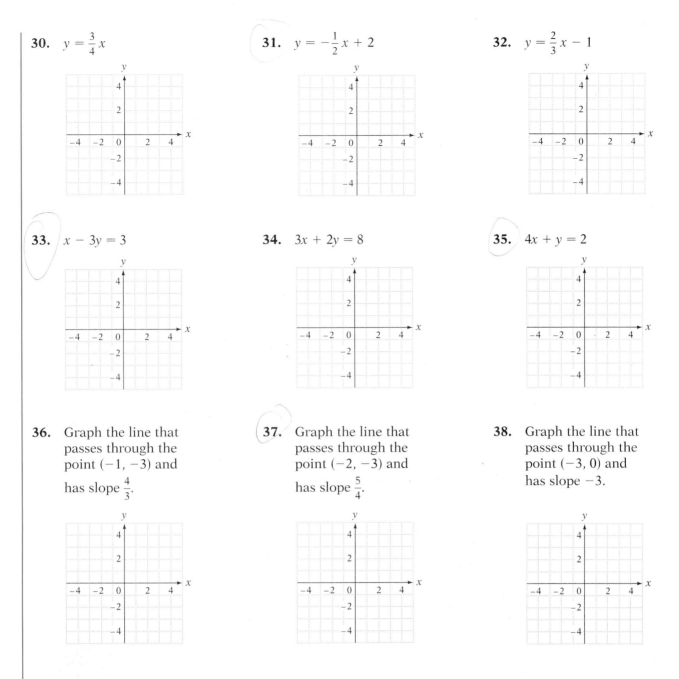

APPLYING THE CONCEPTS

Complete the following sentences.

39. If a line has a slope of 2, then the value of y <u>increases/decreases</u> by _____ as the value of x increases by 1.

40. If a line has a slope of -3, then the value of y <u>increases/decreases</u> by _____ as the value of x increases by 1.

41. If a line has a slope of -2, then the value of y <u>increases/decreases</u> by _____ as the value of x decreases by 1.

42. If a line has a slope of 3, then the value of y <u>increases/decreases</u> by _____ as the value of x decreases by 1.

43. If a line has a slope of $-\frac{2}{3}$, then the value of y <u>increases/decreases</u> by _____ as the value of x increases by 1.

44. If a line has a slope of $\frac{1}{2}$, then the value of y <u>increases/decreases</u> by _____ as the value of x increases by 1.

45. Match each equation with its graph.

   i. $y = -2x + 4$

   ii. $y = 2x - 4$

   iii. $y = 2$

   iv. $2x + 4y = 0$

   v. $y = \frac{1}{2}x + 4$

   vi. $y = -\frac{1}{4}x - 2$

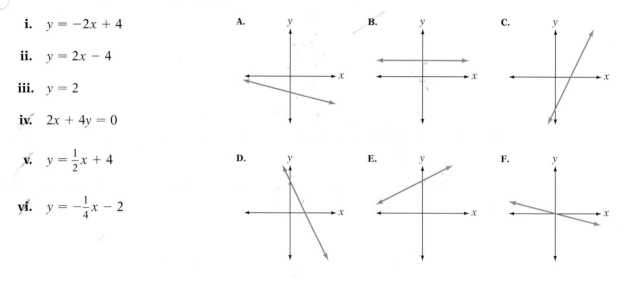

46. Explain how to graph the equation of a straight line by using its slope and a point on the line.

47. Explain why $y = mx + b$ is called the slope-intercept form of the equation of a straight line.

48. Explain how you can use the slope of a line to determine whether three given points lie on the same line. Then use your procedure to determine whether each of the following sets of points lie on the same line.
   a. $(2, 5), (-1, -1), (3, 7)$ **b.** $(-1, 5), (0, 3), (-3, 4)$

Determine the value of k such that the points whose coordinates are given below lie on the same line.

49. $(3, 2), (4, 6), (5, k)$ **50.** $(-2, 3), (1, 0), (k, 2)$

51. $(k, 1), (0, -1), (2, -2)$ **52.** $(4, -1), (3, -4), (k, k)$

3.5 Finding Equations of Lines

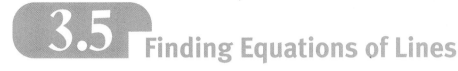

Objective A   **To find the equation of a line given a point and the slope**

When the slope of a line and a point on the line are known, the equation of the line can be determined. If the particular point is the y-intercept, use the slope-intercept form, $y = mx + b$, to find the equation.

➡ Find the equation of the line that contains the point $(0, 3)$ and has slope $\frac{1}{2}$.

The known point is the y-intercept, $(0, 3)$.

$y = mx + b$ • Use the slope-intercept form.

$y = \dfrac{1}{2}x + 3$ • Replace m with $\frac{1}{2}$, the given slope.

Replace b with 3, the y-coordinate of the y-intercept.

The equation of the line is $y = \dfrac{1}{2}x + 3$.

One method of finding the equation of a line when the slope and *any* point on the line are known involves using the *point-slope* formula. This formula is derived from the formula for the slope of a line as follows.

Let (x_1, y_1) be the given point on the line, and let (x, y) be any other point on the line. See the graph at the left.

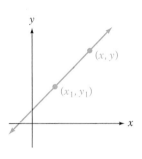

$\dfrac{y - y_1}{x - x_1} = m$ • Use the formula for the slope of a line.

$\dfrac{y - y_1}{x - x_1}(x - x_1) = m(x - x_1)$ • Multiply each side by $(x - x_1)$.

$y - y_1 = m(x - x_1)$ • Simplify.

Point-Slope Formula

Let m be the slope of a line, and let (x_1, y_1) be the coordinates of a point on the line. The equation of the line can be found from the **point-slope formula:**

$$y - y_1 = m(x - x_1)$$

➡ Find the equation of the line that contains the point whose coordinates are $(4, -1)$ and has slope $-\dfrac{3}{4}$.

$y - y_1 = m(x - x_1)$ • Use the point-slope formula.

$y - (-1) = \left(\dfrac{3}{4}\right)(x - 4)$ • $m = -\dfrac{3}{4}$, $(x_1, y_1) = (4, -1)$

$y + 1 = \dfrac{3}{4}x + 3$ • Simplify.

$y = \dfrac{3}{4}x + 2$ • Write the equation in the form $y = mx + b$.

The equation of the line is $y = -\dfrac{3}{4}x + 2$.

⇒ Find the equation of the line that passes through the point whose coordinates are (4, 3) and whose slope is undefined.

Because the slope is undefined, the point-slope formula cannot be used to find the equation. Instead, recall that when the slope is undefined, the line is vertical and that the equation of a vertical line is $x = a$, where a is the x-coordinate of the x-intercept. Because the line is vertical and passes through (4, 3), the x-intercept is (4, 0). The equation of the line is $x = 4$.

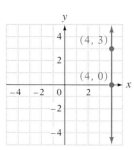

Example 1

Find the equation of the line that contains the point (3, 0) and has slope −4.

Solution

$m = -4 \qquad (x_1, y_1) = (3, 0)$

$y - y_1 = m(x - x_1)$
$y - 0 = -4(x - 3)$
$\quad\ y = -4x + 12$

The equation of the line is $y = -4x + 12$.

You Try It 1

Find the equation of the line that contains the point (−3, −2) and has slope $-\frac{1}{3}$.

Your solution

Example 2

Find the equation of the line that contains the point (−2, 4) and has slope 2.

Solution

$m = 2 \qquad (x_1, y_1) = (-2, 4)$

$y - y_1 = m(x - x_1)$
$y - 4 = 2[x - (-2)]$
$y - 4 = 2(x + 2)$
$y - 4 = 2x + 4$
$\quad\ y = 2x + 8$

The equation of the line is $y = 2x + 8$.

You Try It 2

Find the equation of the line that contains the point (4, −3) and has slope −3.

Your solution

Solutions on p. S8

Objective B   **To find the equation of a line given two points**

The point-slope formula and the formula for slope are used to find the equation of a line when two points are known.

⇒ Find the equation of the line containing the points $(3, 2)$ and $(-5, 6)$.

To use the point-slope formula, we must know the slope. Use the formula for slope to determine the slope of the line between the two given points. Let $(x_1, y_1) = (3, 2)$ and $(x_2, y_2) = (-5, 6)$.

$$m = \frac{y_2 - y_1}{x_2 - x_1} = \frac{6 - 2}{-5 - 3} = \frac{4}{-8} = -\frac{1}{2}$$

Now use the point-slope formula with $m = -\dfrac{1}{2}$ and $(x_1, y_1) = (3, 2)$.

$$y - y_1 = m(x - x_1)$$ • Use the point-slope formula.

$$y - 2 = \left(-\frac{1}{2}\right)(x - 3)$$ • $m = -\dfrac{1}{2}, (x_1, y_1) = (3, 2)$

$$y - 2 = -\frac{1}{2}x + \frac{3}{2}$$ • Simplify.

$$y = -\frac{1}{2}x + \frac{7}{2}$$

The equation of the line is $y = -\dfrac{1}{2}x + \dfrac{7}{2}$.

- -

Example 3

Find the equation of the line containing the points $(2, 3)$ and $(4, 1)$.

Solution

Let $(x_1, y_1) = (2, 3)$ and $(x_2, y_2) = (4, 1)$.

$$m = \frac{y_2 - y_1}{x_2 - x_1} = \frac{1 - 3}{4 - 2} = \frac{-2}{2} = -1$$

$$y - y_1 = m(x - x_1)$$
$$y - 3 = -1(x - 2)$$
$$y - 3 = -x + 2$$
$$y = -x + 5$$

The equation of the line is $y = -x + 5$.

You Try It 3

Find the equation of the line containing the points $(2, 0)$ and $(5, 3)$.

Your solution

- -

Example 4

Find the equation of the line containing the points $(2, -3)$ and $(2, 5)$.

Solution

Let $(x_1, y_1) = (2, -3)$ and $(x_2, y_2) = (2, 5)$.

$$m = \frac{y_2 - y_1}{x_2 - x_1} = \frac{5 - (-3)}{2 - 2} = \frac{8}{0}$$

The slope is undefined, so the graph of the line is vertical.

The equation of the line is $x = 2$.

You Try It 4

Find the equation of the line containing the points $(2, 3)$ and $(-5, 3)$.

Your solution

Solutions on p. S9

Objective C   **To solve application problems**

Linear functions can be used to model a variety of applications in science and business. For each application, data are collected and the independent and dependent variables are selected. Then a linear function is determined that models the data.

Example 5

Suppose a manufacturer has determined that at a price of $115, consumers will purchase 1 million portable CD players and that at a price of $90, consumers will purchase 1.25 million portable CD players. Describe this situation with a linear function. Use this function to predict how many portable CD players consumers will purchase if the price is $80.

Strategy

• Select the independent and dependent variables. Because you are trying to determine the number of CD players, that quantity is the *dependent* variable, y. The price of the CD players is the *independent* variable, x.

• From the given data, two ordered pairs are (115, 1) and (90, 1.25). (The ordinates are in millions of units.) Use these ordered pairs to determine the linear function.

• Evaluate this function when $x = 80$ to predict how many CD players consumers will purchase if the price is $80.

Solution

Let $(x_1, y_1) = (115, 1)$ and $(x_2, y_2) = (90, 1.25)$.

$$m = \frac{y_2 - y_1}{x_2 - x_1} = \frac{1.25 - 1}{90 - 115} = -\frac{0.25}{25} = -0.01$$

$$y - y_1 = m(x - x_1)$$

$$y - 1 = -0.01(x - 115)$$

$$y = -0.01x + 2.15$$

The linear function is $f(x) = -0.01x + 2.15$.

$$f(80) = -0.01(80) + 2.15 = 1.35$$

Consumers will purchase 1.35 million CD players at a price of $80.

You Try It 5

Gabriel Daniel Fahrenheit invented the mercury thermometer in 1717. In terms of readings on this thermometer, water freezes at 32°F and boils at 212°F. In 1742 Anders Celsius invented the Celsius temperature scale. On this scale, water freezes at 0°C and boils at 100°C. Determine a linear function that can be used to predict the Celsius temperature when the Fahrenheit temperature is known.

Your strategy

Your solution

Solution on p. S9

3.5 Exercises

Objective A

1. Explain how to find the equation of a line given its slope and its *y*-intercept.

2. What is the point-slope formula and how is it used?

Find the equation of the line that contains the given point and has the given slope.

3. Point (0, 5), $m = 2$

4. Point (0, 3), $m = 1$

5. Point (2, 3), $m = \frac{1}{2}$

6. Point (5, 1), $m = \frac{2}{3}$

7. Point (−1, 4), $m = \frac{5}{4}$

8. Point (−2, 1), $m = \frac{3}{2}$

9. Point (3, 0), $m = -\frac{5}{3}$

10. Point (−2, 0), $m = \frac{3}{2}$

11. Point (2, 3), $m = -3$

12. Point (1, 5), $m = -\frac{4}{5}$

13. Point (−1, 7), $m = -3$

14. Point (−2, 4), $m = -4$

15. Point (−1, −3), $m = \frac{2}{3}$

16. Point (−2, −4), $m = \frac{1}{4}$

17. Point (0, 0), $m = \frac{1}{2}$

18. Point (0, 0), $m = \frac{3}{4}$

19. Point (2, −3), $m = 3$

20. Point (4, −5), $m = 2$

21. Point (3, 5), $m = -\frac{2}{3}$

22. Point (5, 1), $m = -\frac{4}{5}$

23. Point (0, −3), $m = -1$

24. Point (2, 0), $m = \frac{5}{6}$

25. Point (1, −4), $m = \frac{7}{5}$

26. Point (3, 5), $m = -\frac{3}{7}$

27. Point $(4, -1)$, $m = -\frac{2}{5}$

28. Point $(-3, 5)$, $m = -\frac{1}{4}$

29. Point $(3, -4)$, slope is undefined

30. Point $(-2, 5)$, slope is undefined

31. Point $(-2, -5)$, $m = -\frac{5}{4}$

32. Point $(-3, -2)$, $m = -\frac{2}{3}$

33. Point $(-2, -3)$, $m = 0$

34. Point $(-3, -2)$, $m = 0$

35. Point $(4, -5)$, $m = -2$

36. Point $(-3, 5)$, $m = 3$

37. Point $(-5, -1)$, slope is undefined

38. Point $(0, 4)$, slope is undefined

Objective B

Find the equation of the line that contains the given points.

39. $P_1(0, 2)$, $P_2(3, 5)$

40. $P_1(0, 4)$, $P_2(1, 5)$

41. $P_1(0, -3)$, $P_2(-4, 5)$

42. $P_1(0, -2)$, $P_2(-3, 4)$

43. $P_1(2, 3)$, $P_2(5, 5)$

44. $P_1(4, 1)$, $P_2(6, 3)$

45. $P_1(-1, 3)$, $P_2(2, 4)$

46. $P_1(-1, 1)$, $P_2(4, 4)$

47. $P_1(-1, -2)$, $P_2(3, 4)$

48. $P_1(-3, -1)$, $P_2(2, 4)$

49. $P_1(0, 3)$, $P_2(2, 0)$

50. $P_1(0, 4)$, $P_2(2, 0)$

51. $P_1(-3, -1)$, $P_2(2, -1)$

52. $P_1(-3, -5)$, $P_2(4, -5)$

53. $P_1(-2, -3)$, $P_2(-1, -2)$

54. $P_1(4, 1), P_2(3, -2)$ **55.** $P_1(-2, 3), P_2(2, -1)$ **56.** $P_1(3, 1), P_2(-3, -2)$

57. $P_1(2, 3), P_2(5, -5)$ **58.** $P_1(7, 2), P_2(4, 4)$ **59.** $P_1(2, 0), P_2(0, -1)$

60. $P_1(0, 4), P_2(-2, 0)$ **61.** $P_1(3, -4), P_2(-2, -4)$ **62.** $P_1(-3, 3), P_2(-2, 3)$

63. $P_1(0, 0), P_2(4, 3)$ **64.** $P_1(2, -5), P_2(0, 0)$ **65.** $P_1(2, -1), P_2(-1, 3)$

66. $P_1(3, -5), P_2(-2, 1)$ **67.** $P_1(-2, 5), P_2(-2, -5)$ **68.** $P_1(3, 2), P_2(3, -4)$

69. $P_1(2, 1), P_2(-2, -3)$ **70.** $P_1(-3, -2), P_2(1, -4)$ **71.** $P_1(-4, -3), P_2(2, 5)$

72. $P_1(4, 5), P_2(-4, 3)$ **73.** $P_1(0, 3), P_2(3, 0)$ **74.** $P_1(1, -3), P_2(-2, 4)$

Objective C   *Application Problems*

75. The pilot of a Boeing 747 jet takes off from Boston's Logan Airport, which is at sea level, and climbs to a cruising altitude of 32,000 ft at a constant rate of 1200 ft/min. Write a linear equation for the height of the plane in terms of the time after takeoff. Use your equation to find the height of the plane 11 min after takeoff.

76. A manufacturer of graphing calculators determined that 10,000 calculators per week would be sold at a price of $95. At a price of $90, it was estimated that 12,000 calculators would be sold. Determine a linear function to predict the number of calculators that would be sold at a given price. Use this model to predict the number of calculators per week that would be sold at a price of $75.

77. The gas tank of a certain car contains 16 gal of gas when the driver of the car begins a trip. Each mile driven decreases the amount of gas in the tank by 0.032 gal. Write a linear equation for the number of gallons of gas in the tank in terms of the number of miles driven. Use your equation to find the number of gallons in the tank after driving 150 mi.

78. A jogger running at 9 mph burns approximately 14 calories per minute. Write a linear equation for the number of calories burned by the jogger in terms of the number of minutes run. Use your equation to find the number of calories burned after jogging for 32 min.

APPLYING THE CONCEPTS

79. Explain the similarities and differences between the point-slope formula and the slope-intercept form of a straight line.

80. Explain why the point-slope formula cannot be used to find the equation of a line that is parallel to the y-axis.

81. Refer to Example 5 in this section for each of the following.
   a. Explain the meaning of the slope of the graph of the linear function given in the example.
   b. Explain the meaning of the y-intercept.
   c. Explain the meaning of the x-intercept.

82. For an equation of the form $y = mx + b$, how does the graph of this equation change if the value of b changes and the value of m remains constant?

83. A line contains the points $(-3, 6)$ and $(6, 0)$. Find the coordinates of three other points that are on this line.

84. A line contains the points $(4, -1)$ and $(2, 1)$. Find the coordinates of three other points that are on this line.

85. Given that f is a linear function for which $f(1) = 3$ and $f(-1) = 5$, determine $f(4)$.

86. Given that f is a linear function for which $f(-3) = 2$ and $f(2) = 7$, determine $f(0)$.

87. Find the equation of the line that passes through the midpoint of the line segment between $P_1(2, 5)$ and $P_2(-4, 1)$ and has slope -2.

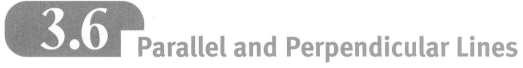

Parallel and Perpendicular Lines

Objective A   **To find parallel and perpendicular lines**

Two lines that have the same slope do not intersect and are called **parallel lines.**

The slope of each of the lines at the right is $\frac{2}{3}$.

The lines are parallel.

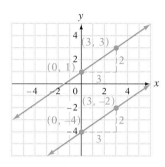

Slopes of Parallel Lines

Two nonvertical lines with slopes of m_1 and m_2 are parallel if and only if $m_1 = m_2$. Any two vertical lines are parallel.

⇒ Is the line containing the points $(-2, 1)$ and $(-5, -1)$ parallel to the line that contains the points $(1, 0)$ and $(4, 2)$?

$$m_1 = \frac{-1 - 1}{-5 - (-2)} = \frac{-2}{-3} = \frac{2}{3}$$

- Find the slope of the line through $(-2, 1)$ and $(-5, -1)$.

$$m_2 = \frac{2 - 0}{4 - 1} = \frac{2}{3}$$

- Find the slope of the line through $(1, 0)$ and $(4, 2)$.

Because $m_1 = m_2$, the lines are parallel.

⇒ Find the equation of the line that contains the point $(2, 3)$ and is parallel to the line $y = \frac{1}{2}x - 4$.

The slope of the given line is $\frac{1}{2}$. Because parallel lines have the same slope, the slope of the unknown line is also $\frac{1}{2}$.

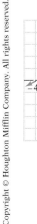

$$y - y_1 = m(x - x_1)$$

- Use the point-slope formula.

$$y - 3 = \frac{1}{2}(x - 2)$$

- $m = \frac{1}{2}$, $(x_1, y_1) = (2, 3)$

$$y - 3 = \frac{1}{2}x - 1$$

- Simplify.

$$y = \frac{1}{2}x + 2$$

- Write the equation in the form $y = mx + b$.

The equation of the line is $y = \frac{1}{2}x + 2$.

➡ Find the equation of the line that contains the point $(-1, 4)$ and is parallel to the line $2x - 3y = 5$.

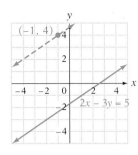

Because the lines are parallel, the slope of the unknown line is the same as the slope of the given line. Solve $2x - 3y = 5$ for y and determine its slope.

$$2x - 3y = 5$$
$$-3y = -2x + 5$$
$$y = \frac{2}{3}x - \frac{5}{3}$$

The slope of the given line is $\frac{2}{3}$. Because the lines are parallel, this is the slope of the unknown line. Use the point-slope formula to determine the equation.

$y - y_1 = m(x - x_1)$ • Use the point-slope formula.

$y - 4 = \frac{2}{3}[x - (-1)]$ • $m = \frac{2}{3}, (x_1, y_1) = (-1, 4)$

$y - 4 = \frac{2}{3}x + \frac{2}{3}$ • Simplify.

$y = \frac{2}{3}x + \frac{14}{3}$ • Write the equation in the form $y = mx + b$.

The equation of the line is $y = \frac{2}{3}x + \frac{14}{3}$.

Two lines that intersect at right angles are **perpendicular lines.**

Any horizontal line is perpendicular to any vertical line. For example, $x = 3$ is perpendicular to $y = -2$.

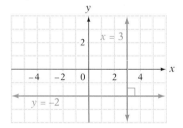

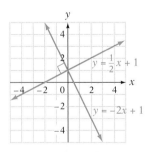

Slopes of Perpendicular Lines

If m_1 and m_2 are the slopes of two lines, neither of which is vertical, then the lines are perpendicular if and only if $m_1 \cdot m_2 = -1$.

A vertical line is perpendicular to a horizontal line.

Solving $m_1 \cdot m_2 = -1$ for m_1 gives $m_1 = -\dfrac{1}{m_2}$. This last equation states that the slopes of perpendicular lines are **negative reciprocals** of each other.

➡ Is the line that contains the points $(4, 2)$ and $(-2, 5)$ perpendicular to the line that contains the points $(-4, 3)$ and $(-3, 5)$?

$m_1 = \dfrac{5 - 2}{-2 - 4} = \dfrac{3}{-6} = -\dfrac{1}{2}$ • Find the slope of the line through $(4, 2)$ and $(-2, 5)$.

$m_2 = \dfrac{5 - 3}{-3 - (-4)} = \dfrac{2}{1} = 2$ • Find the slope of the line through $(-4, 3)$ and $(-3, 5)$.

$m_1 \cdot m_2 = -\dfrac{1}{2}(2) = -1$ • Find the product of the two slopes.

Because $m_1 \cdot m_2 = -1$, the lines are perpendicular.

➡ Are the graphs of the lines whose equations are $3x + 4y = 8$ and $8x + 6y = 5$ perpendicular?

To determine whether the lines are perpendicular, solve each equation for y and find the slope of each line. Then use the equation $m_1 \cdot m_2 = -1$.

$$3x + 4y = 8 \qquad\qquad\qquad 8x + 6y = 5$$
$$4y = -3x + 8 \qquad\qquad\quad 6y = -8x + 5$$
$$y = -\frac{3}{4}x + 2 \qquad\qquad\quad y = -\frac{4}{3}x + \frac{5}{6}$$

$$m_1 = -\frac{3}{4} \qquad\qquad\qquad\quad m_2 = -\frac{4}{3}$$

$$m_1 \cdot m_2 = \left(-\frac{3}{4}\right)\left(-\frac{4}{3}\right) = 1$$

Because $m_1 \cdot m_2 = 1 \neq -1$, the lines are not perpendicular.

➡ Find the equation of the line that contains the point $(-2, 1)$ and is perpendicular to the line $y = -\frac{2}{3}x + 2$.

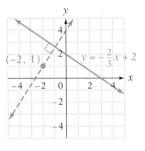

The slope of the given line is $-\frac{2}{3}$. The slope of the line perpendicular to the given line is the negative reciprocal of $-\frac{2}{3}$, which is $\frac{3}{2}$. Substitute this slope and the coordinates of the given point, $(-2, 1)$, into the point-slope formula.

$$y - y_1 = m(x - x_1)$$ • The point-slope formula

$$y - 1 = \frac{3}{2}[x - (-2)]$$ • $m = \frac{3}{2}$, $(x_1, y_1) = (-2, 1)$

$$y - 1 = \frac{3}{2}x + 3$$ • Simplify.

$$y = \frac{3}{2}x + 4$$ • Write the equation in the form $y = mx + b$.

The equation of the perpendicular line is $y = \frac{3}{2}x + 4$.

➡ Find the equation of the line that contains the point $(3, -4)$ and is perpendicular to the line $2x - y = -3$.

$$2x - y = -3$$ • Determine the slope of the given line by solving the equation for y.
$$-y = -2x - 3$$
$$y = 2x + 3$$ • The slope is 2.

The slope of the line perpendicular to the given line is $-\frac{1}{2}$, the negative reciprocal of 2. Now use the point-slope formula to find the equation of the line.

$$y - y_1 = m(x - x_1)$$ • The point-slope formula

$$y - (-4) = -\frac{1}{2}(x - 3)$$ • $m = -\frac{1}{2}$, $(x_1, y_1) = (3, -4)$

$$y + 4 = -\frac{1}{2}x + \frac{3}{2}$$ • Simplify.

$$y = -\frac{1}{2}x - \frac{5}{2}$$ • Write the equation in the form $y = mx + b$.

The equation of the perpendicular line is $y = -\frac{1}{2}x - \frac{5}{2}$.

Example 1

Is the line that contains the points $(-4, 2)$ and $(1, 6)$ parallel to the line that contains the points $(2, -4)$ and $(7, 0)$?

Solution

$$m_1 = \frac{6 - 2}{1 - (-4)} = \frac{4}{5}$$

$$m_2 = \frac{0 - (-4)}{7 - 2} = \frac{4}{5}$$

$$m_1 = m_2 = \frac{4}{5}$$

The lines are parallel.

You Try It 1

Is the line that contains the points $(-2, -3)$ and $(7, 1)$ perpendicular to the line that contains the points $(4, 1)$ and $(6, -5)$?

Your solution

Example 2

Are the lines $4x - y = -2$ and $x + 4y = -12$ perpendicular?

Solution

$$4x - y = -2 \qquad\qquad x + 4y = -12$$
$$-y = -4x - 2 \qquad\qquad 4y = -x - 12$$
$$y = 4x + 2 \qquad\qquad y = -\frac{1}{4}x - 3$$
$$m_1 = 4 \qquad\qquad\qquad m_2 = -\frac{1}{4}$$

$$m_1 \cdot m_2 = 4\left(-\frac{1}{4}\right) = -1$$

The lines are perpendicular.

You Try It 2

Are the lines $5x + 2y = 2$ and $5x + 2y = -6$ parallel?

Your solution

Example 3

Find the equation of the line that contains the point $(3, -1)$ and is parallel to the line $3x - 2y = 4$.

Solution

$$3x - 2y = 4$$
$$-2y = -3x + 4$$
$$y = \frac{3}{2}x - 2 \qquad \bullet \; m = \frac{3}{2}$$

$$y - y_1 = m(x - x_1)$$
$$y - (-1) = \frac{3}{2}(x - 3)$$
$$y + 1 = \frac{3}{2}x - \frac{9}{2}$$
$$y = \frac{3}{2}x - \frac{11}{2}$$

The equation of the line is $y = \frac{3}{2}x - \frac{11}{2}$.

You Try It 3

Find the equation of the line that contains the point $(-2, 2)$ and is perpendicular to the line $x - 4y = 3$.

Your solution

Solutions on p. S9

3.6 Exercises

Objective A

1. Explain how to determine whether the graphs of two lines are parallel.

2. Explain how to determine whether the graphs of two lines are perpendicular.

3. The slope of a line is -5. What is the slope of any line parallel to this line?

4. The slope of a line is $\frac{3}{2}$. What is the slope of any line parallel to this line?

5. The slope of a line is 4. What is the slope of any line perpendicular to this line?

6. The slope of a line is $-\frac{4}{5}$. What is the slope of any line perpendicular to this line?

7. Is the line $x = -2$ perpendicular to the line $y = 3$?

8. Is the line $y = \frac{1}{2}$ perpendicular to the line $y = -4$?

9. Is the line $x = -3$ parallel to the line $y = \frac{1}{3}$?

10. Is the line $x = 4$ parallel to the line $x = -4$?

11. Is the line $y = \frac{2}{3}x - 4$ parallel to the line $y = -\frac{3}{2}x - 4$?

12. Is the line $y = -2x + \frac{2}{3}$ parallel to the line $y = -2x + 3$?

13. Is the line $y = \frac{4}{3}x - 2$ perpendicular to the line $y = -\frac{3}{4}x + 2$?

14. Is the line $y = \frac{1}{2}x + \frac{3}{2}$ perpendicular to the line $y = -\frac{1}{2}x + \frac{3}{2}$?

15. Are the lines $2x + 3y = 2$ and $2x + 3y = -4$ parallel?

16. Are the lines $2x - 4y = 3$ and $2x + 4y = -3$ parallel?

17. Are the lines $x - 4y = 2$ and $4x + y = 8$ perpendicular?

18. Are the lines $4x - 3y = 2$ and $4x + 3y = -7$ perpendicular?

19. Is the line that contains the points $(3, 2)$ and $(1, 6)$ parallel to the line that contains the points $(-1, 3)$ and $(-1, -1)$?

20. Is the line that contains the points $(4, -3)$ and $(2, 5)$ parallel to the line that contains the points $(-2, -3)$ and $(-4, 1)$?

21. Is the line that contains the points $(-3, 2)$ and $(4, -1)$ perpendicular to the line that contains the points $(1, 3)$ and $(-2, -4)$?

22. Is the line that contains the points $(-1, 2)$ and $(3, 4)$ perpendicular to the line that contains the points $(-1, 3)$ and $(-4, 1)$?

23. Find the equation of the line containing the point $(-2, -4)$ and parallel to the line $2x - 3y = 2$.

24. Find the equation of the line containing the point $(3, 2)$ and parallel to the line $3x + y = -3$.

25. Find the equation of the line containing the point $(4, 1)$ and perpendicular to the line $y = -3x + 4$.

26. Find the equation of the line containing the point $(2, -5)$ and perpendicular to the line $y = \frac{5}{2}x - 4$.

27. Find the equation of the line containing the point $(-1, -3)$ and perpendicular to the line $3x - 5y = 2$.

28. Find the equation of the line containing the point $(-1, 3)$ and perpendicular to the line $2x + 4y = -1$.

APPLYING THE CONCEPTS

For Exercises 29 and 30, suppose a ball is being twirled at the end of a string and the center of rotation is the origin of a coordinate system. If the string breaks, the initial path of the ball is on a line that is perpendicular to the radius of the circle.

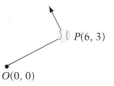

29. Suppose the string breaks when the ball is at the point whose coordinates are $P(6, 3)$. Find the equation of the line on which the initial path lies.

30. Suppose the string breaks when the ball is at the point whose coordinates are $P(2, 8)$. Find the equation of the line on which the initial path lies.

31. If the graphs of $A_1x + B_1y = C_1$ and $A_2x + B_2y = C_2$ are parallel, express $\frac{A_1}{B_1}$ in terms of A_2 and B_2.

32. If the graphs of $A_1x + B_1y = C_1$ and $A_2x + B_2y = C_2$ are perpendicular, express $\frac{A_1}{B_1}$ in terms of A_2 and B_2.

33. The graphs of $y = -\frac{1}{2}x + 2$ and $y = \frac{2}{3}x - 5$ intersect at the point whose coordinates are $(6, -1)$. Find the equation of a line whose graph intersects the graphs of the given lines to form a right triangle. (*Hint:* There is more than one answer to this question.)

34. A theorem from geometry states that a line passing through the center of a circle and through a point P on the circle is perpendicular to the tangent line at P. (See the figure at the right.)
   a. If the coordinates of P are $(5, 4)$ and the coordinates of C are $(3, 2)$, what is the equation of the tangent line?
   b. What are the x- and y-intercepts of the tangent line?

3.7 Inequalities in Two Variables

Objective A To graph the solution set of an inequality in two variables

The graph of the linear equation $y = x - 1$ separates the plane into three sets: the set of points on the line, the set of points above the line, and the set of points below the line.

The point whose coordinates are $(2, 1)$ is a solution of $y = x - 1$ and is a point on the line.

The point whose coordinates are $(2, 4)$ is a solution of $y > x - 1$ and is a point above the line.

The point whose coordinates are $(2, -2)$ is a solution of $y < x - 1$ and is a point below the line.

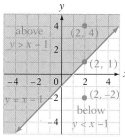

The set of points on the line are the solutions of the equation $y = x - 1$. The set of points above the line are the solutions of the inequality $y > x - 1$. These points form a **half-plane.** The set of points below the line are solutions of the inequality $y < x - 1$. These points also form a half-plane.

An inequality of the form $y > mx + b$ or $Ax + By > C$ is a **linear inequality in two variables.** (The inequality symbol could be replaced by $\geq$, $<$, or $\leq$.) The solution set of a linear inequality in two variables is a half-plane.

The following illustrates the procedure for graphing the solution set of a linear inequality in two variables.

⇒ Graph the solution set of $3x - 4y < 12$.

$3x - 4y < 12$ • Solve the inequality for *y*.
$\quad -4y < -3x + 12$
$\quad\quad y > \dfrac{3}{4}x - 3$

Change the inequality $y > \dfrac{3}{4}x - 3$ to the equality $y = \dfrac{3}{4}x - 3$, and graph the line.

If the inequality contains $\leq$ or $\geq$, the line belongs to the solution set and is shown by a *solid line*. If the inequality contains $<$ or $>$, the line is not part of the solution set and is shown by a *dotted line*.

If the inequality contains $>$ or $\geq$, shade the upper half-plane. If the inequality contains $<$ or $\leq$, shade the lower half-plane.

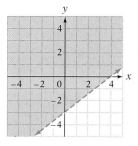

As a check, use the ordered pair $(0, 0)$ to determine whether the correct region of the plane has been shaded. If $(0, 0)$ is a solution of the inequality, then $(0, 0)$ should be in the shaded region. If $(0, 0)$ is not a solution of the inequality, then $(0, 0)$ should not be in the shaded region.

If the line passes through the point (0, 0), another point, such as (0, 1), must be used as a check.

From the graph of $y > \frac{3}{4}x - 3$, note that for a given value of x, more than one value of y can be paired with that value of x. For instance, (4, 1), (4, 3), (5, 1), and $\left(5, \frac{9}{4}\right)$ are all ordered pairs that belong to the graph. Because there are ordered pairs with the same first coordinate and different second coordinates, the inequality does not represent a function. The inequality is a relation but not a function.

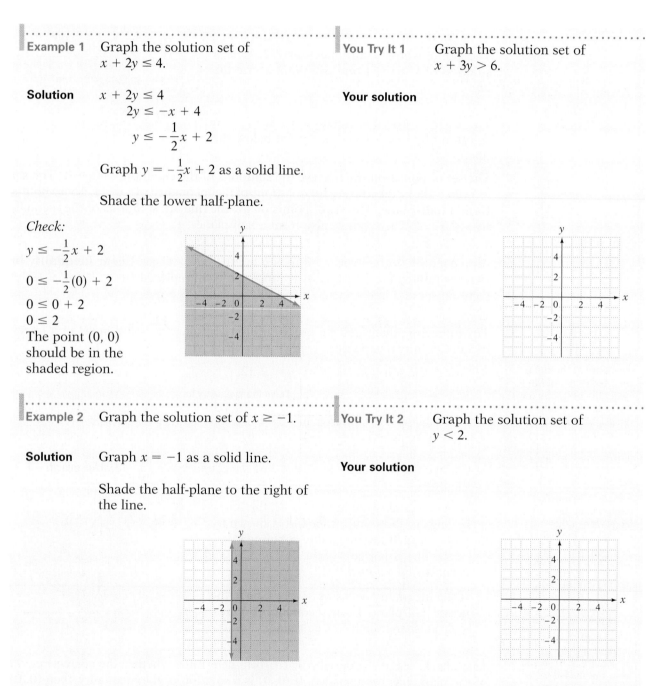

Example 1   Graph the solution set of $x + 2y \leq 4$.

Solution $x + 2y \leq 4$

$$2y \leq -x + 4$$

$$y \leq -\frac{1}{2}x + 2$$

Graph $y = -\frac{1}{2}x + 2$ as a solid line.

Shade the lower half-plane.

Check:

$$y \leq -\frac{1}{2}x + 2$$

$$0 \leq -\frac{1}{2}(0) + 2$$

$$0 \leq 0 + 2$$

$$0 \leq 2$$

The point (0, 0) should be in the shaded region.

You Try It 1   Graph the solution set of $x + 3y > 6$.

Your solution

Example 2   Graph the solution set of $x \geq -1$.

Solution   Graph $x = -1$ as a solid line.

Shade the half-plane to the right of the line.

You Try It 2   Graph the solution set of $y < 2$.

Your solution

Solutions on p. S9

3.7 Exercises

1. What is a half-plane?

2. Explain a method you can use to check that the graph of a linear inequality in two variables has been shaded correctly.

3. Is (0, 0) a solution of $y > 2x - 7$?

4. Is (0, 0) a solution of $y < 5x + 3$?

5. Is (0, 0) a solution of $y \leq -\frac{2}{3}x - 8$?

6. Is (0, 0) a solution of $y \geq -\frac{3}{4}x + 9$?

Graph the solution set.

7. $3x - 2y \geq 6$

8. $4x - 3y \leq 12$

9. $x + 3y < 4$

10. $3x - 5y > 15$

11. $4x - 5y > 10$

12. $4x + 3y < 9$

13. $x + 3y < 6$

14. $2x - 5y \leq 10$

15. $2x + 3y \geq 6$

16. $3x + 2y < 4$

17. $-x + 2y > -8$

18. $-3x + 2y > 2$

19. $y - 4 < 0$

20. $x + 2 \geq 0$

21. $6x + 5y < 15$

22. $3x - 5y < 10$

23. $-5x + 3y \geq -12$

24. $3x + 4y \geq 12$

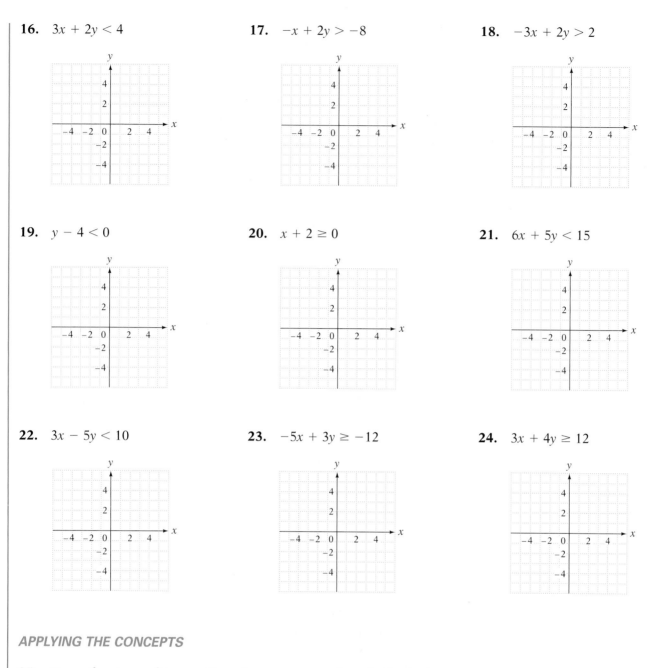

APPLYING THE CONCEPTS

25. Does the inequality $y < 3x - 1$ represent a function? Explain your answer.

26. Are there any points whose coordinates satisfy both $y \leq x + 3$ and $y \geq -\frac{1}{2}x + 1$? If so, give the coordinates of three such points. If not, explain why not.

27. Are there any points whose coordinates satisfy both $y \leq x - 1$ and $y \geq x + 2$? If so, give the coordinates of three such points. If not, explain why not.

Focus on Problem Solving

Find a Pattern

Polya's four recommended problem-solving steps are stated below.

1. Understand the problem.
2. Devise a plan.
3. Carry out the plan.
4. Review your solution.

One of the several ways of devising a plan is first to try to find a pattern. Karl Friedrich Gauss supposedly used this method to solve a problem that was given to his math class when he was in elementary school. As the story goes, his teacher wanted to grade some papers while the class worked on a math problem. The problem given to the class was to find the sum

$$1 + 2 + 3 + 4 + \cdots + 100$$

Gauss quickly solved the problem by seeing a pattern. Here is what he saw.

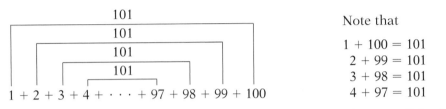

Note that
$$1 + 100 = 101$$
$$2 + 99 = 101$$
$$3 + 98 = 101$$
$$4 + 97 = 101$$

Gauss noted that there were 50 sums of 101. Therefore, the sum of the first 100 natural numbers is

$$1 + 2 + 3 + 4 + \cdots + 97 + 98 + 99 + 100 = 50(101) = 5050$$

Try to solve the following problems by finding a pattern.

1. Find the sum $2 + 4 + 6 + \cdots + 96 + 98 + 100$.

2. Find the sum $1 + 3 + 5 + \cdots + 97 + 99 + 101$.

3. Find another method of finding the sum $1 + 3 + 5 + \cdots + 97 + 99 + 101$ given in the preceding exercise.

4. Find the sum $\dfrac{1}{1 \cdot 2} + \dfrac{1}{2 \cdot 3} + \dfrac{1}{3 \cdot 4} + \cdots + \dfrac{1}{49 \cdot 50}$.

   Hint: $\dfrac{1}{1 \cdot 2} = \dfrac{1}{2}, \dfrac{1}{1 \cdot 2} + \dfrac{1}{2 \cdot 3} = \dfrac{2}{3}, \dfrac{1}{1 \cdot 2} + \dfrac{1}{2 \cdot 3} + \dfrac{1}{3 \cdot 4} = \dfrac{3}{4}$

2 points, 2 regions

3 points, 4 regions

4 points, 8 regions   5 points, ? regions

5. A *figurate number* is a number that can be represented by arranging that number of dots in rows to form a geometric figure such as a triangle, square, pentagon, or hexagon. For instance, the first four *triangular* numbers, 3, 6, 10, and 15, are shown below. What are the next two triangular numbers?

6. The following problem shows that checking a few cases does not always result in a conjecture that is true for *all* cases. Select any two points on a circle (see drawing in the left margin) and draw a *chord*, a line connecting the points. The chord divides the circle into 2 regions. Now select 3 different points and draw chords connecting each of the three points with every other point. The chords divide the circle into 4 regions. Now select 4 points and connect each of the points with every other point. Make a conjecture as to the relationship between the number of regions and the number of points on the circle. Does your conjecture work for 5 points? 6 points?

Projects and Group Activities

Evaluating a Function with a Graphing Calculator

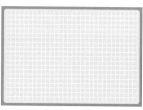

You can use a graphing calculator to evaluate some functions. Shown below are the keystrokes needed to evaluate $f(x) = 3x^2 + 2x - 1$ for $x = -2$ on a TI-83. (There are other methods of evaluating functions. This is just one of them.) Try these keystrokes. The calculator should display 7 as the value of the function.

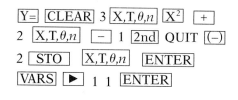

Introduction to Graphing Calculators

There are a variety of computer programs and calculators that can graph an equation. A computer or graphing calculator screen is divided into pixels. Depending on the computer or calculator, there are approximately 6000 to 790,000 pixels available on the screen. The greater the number of pixels, the smoother the graph will appear. A portion of a screen is shown at the left. Each little rectangle represents one pixel.

A graphing calculator draws a graph in a manner similar to the method we have used in this chapter. Values of x are chosen and ordered pairs calculated. Then a graph is drawn through those points by illuminating pixels (an abbreviation for "picture element") on the screen.

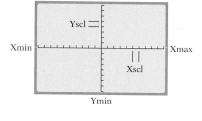

Graphing utilities can display only a portion of the xy-plane, called a window. The window [Xmin, Xmax] by [Ymin, Ymax] consists of those points (x, y) that satisfy both of the following inequalities:

$$Xmin \le x \le Xmax \text{ and } Ymin \le y \le Ymax$$

The user sets these values before a graph is drawn.

The numbers Xscl and Yscl are the distances between the tick marks that are drawn on the x- and y-axes. If you do not want tick marks on the axes, set Xscl = 0 and Yscl = 0.

The graph at the right is a portion of the graph of $y = \frac{1}{2}x + 1$, as it was drawn with a graphing calculator. The window is

Xmin = −4.7, Xmax = 4.7
Ymin = −3.1, Ymax = 3.1
Xscl = 1 and Yscl = 1

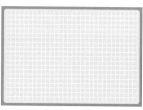

Calculator Screen

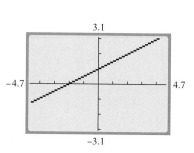

Using interval notation, this is written $[-4.7, 4.7]$ by $[-3.1, 3.1]$. The window $[-4.7, 4.7]$ by $[-3.1, 3.1]$ gives "nice" coordinates in the sense that each time the ▶ or the ◀ is pressed, the change in x is 0.1. The reason for this is that the horizontal distance from the middle of the first pixel to the middle of the last pixel is 94 units. By using Xmin $= -4.7$ and Xmax $= 4.7$, we have[1]

$$\text{Change in } x = \frac{\text{Xmax} - \text{Xmin}}{94} = \frac{4.7 - (-4.7)}{94} = \frac{9.4}{94} = 0.1$$

Similarly, the vertical distance from the middle of the first pixel to the middle of the last pixel is 62 units. Therefore, using Ymin $= -3.1$ and Ymax $= 3.1$ will give nice coordinates in the vertical direction.

Graph each of the following by using a graphing calculator.

> **TAKE NOTE**
>
> The Appendix "Guidelines for Using Graphing Calculators" contains keystroking suggestions to use with this project.

1. $y = 2x + 1$ For $2x$, you may enter $2 \times x$ or just $2x$. The times sign $\times$ is not necessary on many graphing calculators.

2. $y = -x + 2$ Many calculators use the $(-)$ key to enter a negative sign.

3. $3x + 2y = 6$ Solve for y. Then enter the equation.

4. $y = 50x$ You must adjust the viewing window. Try the window $[-4.7, 4.7]$ by $[-250, 250]$ with Yscl $= 50$.

5. $y = \dfrac{2}{3}x - 3$ You may enter $\frac{2}{3}x$ as $2x/3$ or $(2/3)x$. Although entering $2/3x$ works on some calculators, it is not recommended.

6. $4x + 3y = 75$ You must adjust the viewing window.

Chapter Summary

Key Words   A *function* is a set of ordered pairs in which no two ordered pairs that have the same first coordinate have different second coordinates. [p. 124]

Functional notation is used for those equations that represent functions. For the equation $y = f(x)$, x is the *independent variable* and y is the *dependent variable*. [pp. 125–126]

The *domain* of a function is the set of first coordinates of all the ordered pairs of the function. [p. 124]

The *range* of a function is the set of second coordinates of all the ordered pairs of the function. [p. 124]

A *linear function* is one that can be expressed in the form $f(x) = mx + b$. [p. 135]

The *graph of a function* is the graph of the ordered pairs that belong to the function. [p. 135]

[1]Some calculators have screen widths of 126 pixels. For those calculators, use Xmin $= -6.3$ and Xmax $= 6.3$ to obtain nice coordinates.

The point at which a graph crosses the *x*-axis is called the *x-intercept*. [p. 140]

The point at which a graph crosses the *y*-axis is called the *y-intercept*. [p. 140]

An equation of the form $Ax + By = C$ is a *linear equation in two variables*. [p. 137]

The *graph of a linear function* is a straight line. [p. 135]

The *slope* of a line is a measure of the slant, or tilt, of the line. The symbol for slope is *m*. [p. 147]

A line that slants upward to the right has a *positive slope*. [p. 148]

A line that slants downward to the right has a *negative slope*. [p. 148]

A horizontal line has *zero slope*. [p. 148]

The slope of a vertical line is *undefined*. [p. 148]

Two lines that have the same slope do not intersect and are called *parallel lines*. [p. 165]

Two lines that intersect at right angles are called *perpendicular lines*. [p. 166]

An inequality of the form $y > mx + b$ or $Ax + By > C$ is a *linear inequality in two variables*. (The symbol $>$ could be replaced by $\geq$, $<$, or $\leq$.) [p. 171]

The solution set of an inequality in two variables is a *half-plane*. [p. 171]

Essential Rules

Pythagorean Theorem
$a^2 + b^2 = c^2$ [p. 115]

Distance Formula
$d = \sqrt{(x_1 - x_2)^2 + (y_1 - y_2)^2}$ [p. 116]

Midpoint Formula
$\left(\dfrac{x_1 + x_2}{2}, \dfrac{y_1 + y_2}{2} \right)$ [p. 116]

Slope of a Straight Line
$m = \dfrac{y_2 - y_1}{x_2 - x_1}, x_1 \neq x_2$ [p. 147]

Slope-Intercept Formula
$y = mx + b$ [p. 151]

Point-Slope Formula
$y - y_1 = m(x - x_1)$ [p. 157]

Slopes of Parallel Lines
$m_1 = m_2$ [p. 165]

Slopes of Perpendicular Lines
$m_1 \cdot m_2 = -1$ [p. 166]

Chapter Review

1. Determine the ordered-pair solution of $y = \dfrac{x}{x-2}$ that corresponds to $x = 4$.

2. Given $P(x) = 3x + 4$, evaluate $P(-2)$ and $P(a)$.

3. Graph the ordered-pair solutions of $y = 2x^2 - 5$ when $x = -2, -1, 0, 1,$ and 2.

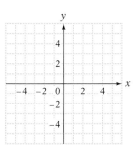

4. Draw a line through all the points with an abscissa of -3.

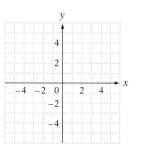

5. Find the range of $f(x) = x^2 + x - 1$ if the domain is $\{-2, -1, 0, 1, 2\}$.

6. Find the domain and range of the function $\{(-1, 0), (0, 2), (1, 2), (5, 4)\}$.

7. Find the midpoint and the length (to the nearest hundredth) of the line segment with endpoints $(-2, 4)$ and $(3, 5)$.

8. What value of x is excluded from the domain of $f(x) = \dfrac{x}{x+4}$?

9. Find the x- and y-intercepts and graph $y = -\dfrac{2}{3}x - 2$.

10. Graph $3x + 2y = -6$ by using the x- and y-intercepts.

11. Graph: $y = -2x + 2$

12. Graph: $4x - 3y = 12$

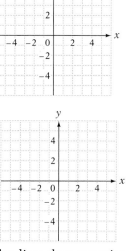

13. Find the slope of the line that contains the ordered pairs $(3, -2)$ and $(-1, 2)$.

14. Find the equation of the line that contains the ordered pair $(-3, 4)$ and has slope $\dfrac{5}{2}$.

15. Graph the solution set of $y \geq 2x - 3$.

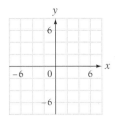

16. Graph the solution set of $3x - 2y < 6$.

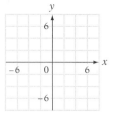

17. Find the equation of the line that contains the ordered pairs $(-2, 4)$ and $(4, -3)$.

18. Find the equation of the line that contains the ordered pair $(-2, -4)$ and is parallel to the graph of $4x - 2y = 7$.

19. Find the equation of the line that contains the ordered pair $(3, -2)$ and is parallel to the graph of $y = -3x + 4$.

20. Find the equation of the line that contains the ordered pair $(2, 5)$ and is perpendicular to the graph of the line $y = -\frac{2}{3}x + 6$.

21. Graph the line that passes through the point whose coordinates are $(-1, 4)$ and has slope $-\frac{1}{3}$.

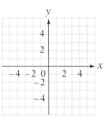

22. A car is traveling at 55 mph. The equation that describes the distance traveled is $d = 55t$. Graph this equation for $0 \leq t \leq 6$. The point whose coordinates are $(4, 220)$ is on the graph. Write a sentence that explains the meaning of this ordered pair.

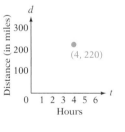

23. The graph at the right shows the relationship between the cost of manufacturing calculators and the number of calculators manufactured. Find the slope of the line between the two points shown on the graph. Write a sentence that explains the meaning of the slope of this line.

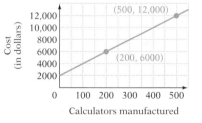

24. A building contractor estimates that the cost to build a new home is $25,000 plus $80 for each square foot of floor space. Determine a linear function that will give the cost to build a house that contains a given number of square feet. Use the model to determine the cost to build a house that contains 2000 ft².

Chapter Test

1. Graph the ordered-pair solutions of $P(x) = 2 - x^2$ when $x = -2, -1, 0, 1,$ and 2.

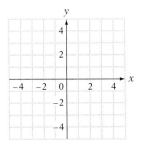

2. Find the ordered-pair solution of $y = 2x + 6$ that corresponds to $x = -3$.

3. Graph: $y = \dfrac{2}{3}x - 4$

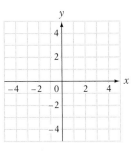

4. Graph: $2x + 3y = -3$

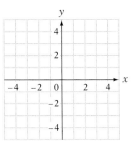

5. Find the equation of the vertical line that contains the point $(-2, 3)$.

6. Find the length, to the nearest hundredth, and the midpoint of the line segment with endpoints $(4, 2)$ and $(-5, 8)$.

7. Find the slope of the line that contains the points $(-2, 3)$ and $(4, 2)$.

8. Given $P(x) = 3x^2 - 2x + 1$, evaluate $P(2)$.

9. Graph $2x - 3y = 6$ by using the x- and the y-intercepts.

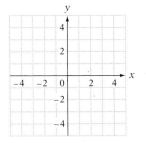

10. Graph the line that passes through the point $(-2, 3)$ and has slope $-\dfrac{3}{2}$.

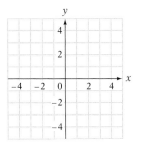

11. Find the equation of the line that contains the point $(-5, 2)$ and has slope $\frac{2}{5}$.

12. What value of x is excluded from the domain of $f(x) = \frac{2x + 1}{x}$?

13. Find the equation of the line that contains the points $(3, -4)$ and $(-2, 3)$.

14. Find the equation of the horizontal line that contains the point $(4, -3)$.

15. Find the domain and range of the function $\{(-4, 2), (-2, 2), (0, 0), (3, 5)\}$.

16. Find the equation of the line that contains the point $(1, 2)$ and is parallel to the line $y = -\frac{3}{2}x - 6$.

17. Find the equation of the line that contains the point $(-2, -3)$ and is perpendicular to the line $y = -\frac{1}{2}x - 3$.

18. Graph the solution set of $3x - 4y > 8$.

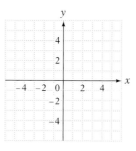

19. The graph below shows the relationship between the cost of a rental house and the depreciation allowed for income tax purposes. Find the slope between the two points shown on the graph. Write a sentence that states the meaning of the slope.

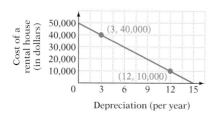

20. The director of a baseball school estimates that 100 students will enroll if the tuition is $250. For each $20 increase in tuition, six fewer students will enroll. Determine a linear function that will predict the number of students who will enroll at a given tuition. Use this model to predict enrollment when the tuition is $300.

Cumulative Review

1. Identify the property that justifies the statement $(x + y) \cdot 2 = 2 \cdot (x + y)$.

2. Solve: $3 - \dfrac{x}{2} = \dfrac{3}{4}$

3. Solve:
   $2[y - 2(3 - y) + 4] = 4 - 3y$

4. Solve:
   $\dfrac{1 - 3x}{2} + \dfrac{7x - 2}{6} = \dfrac{4x + 2}{9}$

5. Solve:
   $x - 3 < -4 \quad \text{or} \quad 2x + 2 > 3$

6. Solve:
   $8 - |2x - 1| = 4$

7. Solve: $|3x - 5| < 5$

8. Simplify: $4 - 2(4 - 5)^3 + 2$

9. Evaluate $(a - b)^2 \div (ab)$ when $a = 4$ and $b = -2$.

10. Graph: $\{x \mid x < -2\} \cup \{x \mid x > 0\}$

    $\overset{\displaystyle \longleftarrow \! + \! + \! + \! + \! + \! + \! + \! + \! + \! + \! + \! \longrightarrow}{-5 \ -4 \ -3 \ -2 \ -1 \ \ 0 \ \ 1 \ \ 2 \ \ 3 \ \ 4 \ \ 5}$

11. Solve $P = \dfrac{R - C}{n}$ for C.

12. Solve $2x + 3y = 6$ for x.

13. Solve: $3x - 1 < 4$ and $x - 2 > 2$

14. Given $P(x) = x^2 + 5$, evaluate $P(-3)$.

15. Find the ordered-pair solution of
    $y = -\dfrac{5}{4}x + 3$ that corresponds to $x = -8$.

16. Find the slope of the line that contains the points $(-1, 3)$ and $(3, -4)$.

17. Find the equation of the line that contains the point $(-1, 5)$ and has slope $\dfrac{3}{2}$.

18. Find the equation of the line that contains the points $(4, -2)$ and $(0, 3)$.

19. Find the equation of the line that contains the point (2, 4) and is parallel to the line $y = -\frac{3}{2}x + 2$.

20. Find the equation of the line that contains the point (4, 0) and is perpendicular to the line $3x - 2y = 5$.

21. A coin purse contains 17 coins with a value of $1.60. The purse contains nickels, dimes, and quarters. There are four times as many nickels as quarters. Find the number of dimes in the purse.

22. Two planes are 1800 mi apart and are traveling toward each other. One plane is traveling twice as fast as the other plane. The planes meet in 3 h. Find the speed of each plane.

23. A grocer combines coffee costing $9 per pound with coffee costing $6 per pound. How many pounds of each should be used to make 60 pounds of a blend costing $8 per pound?

24. Graph $3x - 5y = 15$ by using the x- and y-intercepts.

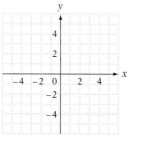

25. Graph the line that passes through the point (-3, 1) and has slope $-\frac{3}{2}$.

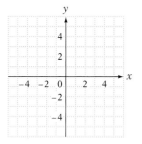

26. Graph the inequality $3x - 2y \geq 6$.

27. The relationship between the cost of a truck and the depreciation allowed for income tax purposes is shown in the graph at the right. Write the equation for the line that represents the depreciated value of the truck. Write a sentence that explains the meaning of the slope in the context of this exercise.

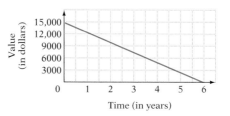

4

Systems of Linear Equations and Inequalities

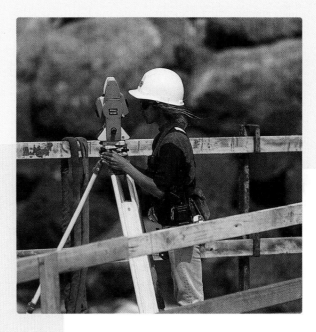

This surveyor is working on a construction site, using the tools necessary to find the area of a plot of land. When a plot is in the shape of a polygon, its vertices can be represented by rectangular coordinates. These coordinates are used in the surveyor's area formula to determine the area of the plot. **Exercise 35 on page 218**, which involves use of the surveyor's area formula, is an application of determinants.

Need help? For on-line student resources, such as section quizzes, visit this textbook's web site at **math.college.hmco.com/students**.

1. Simplify: $10\left(\dfrac{3}{5}x + \dfrac{1}{2}y\right)$

2. Evaluate $3x + 2y - z$ for $x = -1$, $y = 4$, and $z = -2$.

3. Given $3x - 2z = 4$, find the value of x when $z = -2$

4. Solve: $3x + 4(-2x - 5) = -5$

5. Solve: $0.45x + 0.06(-x + 4000) = 630$

6. Graph: $y = \dfrac{1}{2}x - 4$

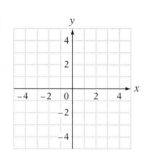

7. Graph: $3x - 2y = 6$

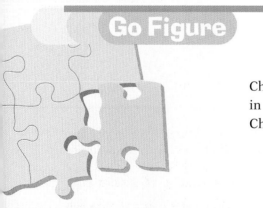

8. Graph: $y > -\dfrac{3}{5}x + 1$

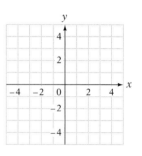

Copyright © Houghton Mifflin Company. All rights reserved.

Go Figure

Chris can beat Pat by 100 m in a 1000-meter race. Pat can beat Leslie by 10 m in a 100-meter race. If both run at this given rate, by how many meters will Chris beat Leslie in a 1000-meter race?

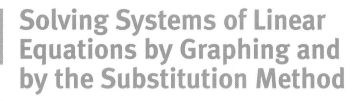

4.1 Solving Systems of Linear Equations by Graphing and by the Substitution Method

Objective A   **To solve a system of linear equations by graphing**

A **system of equations** is two or more equations considered together. The system at the right is a system of two linear equations in two variables. The graphs of the equations are straight lines.

$$3x + 4y = 7$$
$$2x - 3y = 6$$

A **solution of a system of equations in two variables** is an ordered pair that is a solution of each equation of the system.

➡ Is $(3, -2)$ a solution of the system
$$2x - 3y = 12$$
$$5x + 2y = 11?$$

$2x - 3y$	12
$2(3) - 3(-2)$	12
$6 - (-6)$	12
	$12 = 12$

$5x + 2y$	11
$5(3) + 2(-2)$	11
$15 + (-4)$	11
	$11 = 11$

• Replace x by 3 and y by -2.

Yes, because $(3, -2)$ is a solution of each equation, it is a solution of the system of equations.

A solution of a system of linear equations can be found by graphing the lines of the system on the same set of coordinate axes. Three examples of linear equations in two variables are shown below, along with the graphs of the equations of the systems.

System I
$x + 2y = 4$
$2x + y = -1$

System II
$2x + 3y = 6$
$4x + 6y = -12$

System III
$x - 2y = 4$
$2x - 4y = 8$

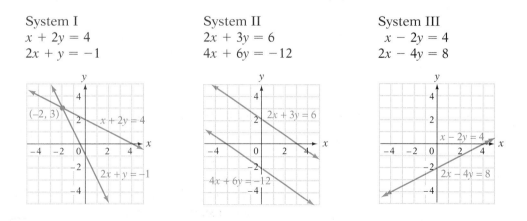

Check:

$x + 2y$	4
$-2 + 2(3)$	4
$-2 + 6$	4
	$4 = 4$

$2x + y$	-1
$2(-2) + 3$	-1
$-4 + 3$	-1
	$-1 = -1$

In System I, the two lines intersect at a single point whose coordinates are $(-2, 3)$. Because this point lies on both lines, it is a solution of each equation of the system of equations. We can check this by replacing x by -2 and y by 3. The check is shown at the left. The ordered pair $(-2, 3)$ is a solution of System I.

When the graphs of a system of equations intersect at only one point, the system is called an **independent system of equations.** System I is an independent system of equations.

System II from the preceding page and the graph of the equations of that system are shown again at the right. Note in this case that the graphs of the lines are parallel and do not intersect. Since the graphs do not intersect, there is no point that is on both lines. Therefore, the system of equations has no solution.

When a system of equations has no solution, it is called an **inconsistent system of equations**. System II is an inconsistent system of equations.

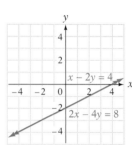

$2x + 3y = 6$
$4x + 6y = -12$

System III from the preceding page and the graph of the equations of that system are shown again at the right. Note that the graph of $x - 2y = 4$ lies directly on top of the graph of $2x - 4y = 8$. Thus the two lines intersect at an infinite number of points. Because the graphs intersect at an infinite number of points, there are an infinite number of solutions of this system of equations. Since each equation represents the same set of points, the solutions of the system of equations can be stated by using the ordered pairs of either one of the equations. Therefore, we can say "The solutions are the ordered pairs that satisfy $x - 2y = 4$," or we can solve the equation for y and say "The solutions are the ordered pairs that satisfy $y = \frac{1}{2}x - 2$."

We normally state this solution using ordered pairs—"The solutions are the ordered pairs $\left(x, \frac{1}{2}x - 2\right)$."

$x - 2y = 4$
$2x - 4y = 8$

When the two equations in a system of equations represent the same line, the system is called a **dependent system of equations**. System III is a dependent system of equations.

The above systems illustrate the three possibilities for a system of linear equations in two variables.

TAKE NOTE

Keep in mind the difference between independent, dependent, and inconsistent systems of equations. You should be able to express your understanding of these terms by using graphs.

1. The graphs intersect at one point.
   The solution of the system of equations is the ordered pair (x, y) whose coordinates are the point of intersection.
   The system of equations is independent.

2. The lines are parallel and never intersect.
   There is no solution of the system of equations.
   The system of equations is inconsistent.

3. The graphs are the same line, and they intersect at infinitely many points.
   There are an infinite number of solutions of the system of equations.
   The system of equations is dependent.

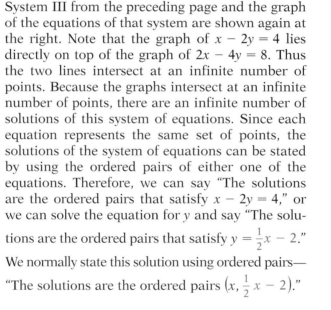

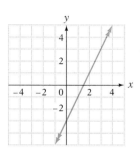

The Projects and Group
Activities feature at the end of
this chapter shows how to use
a graphing calculator to solve a
system of equations.

➡ Solve by graphing: $2x - y = 3$
$4x - 2y = 6$

Graph each line.
The system of equations is dependent.
Solve one of the equations for y.

$$2x - y = 3$$
$$-y = -2x + 3$$
$$y = 2x - 3$$

The solutions are the ordered pairs $(x, 2x - 3)$.

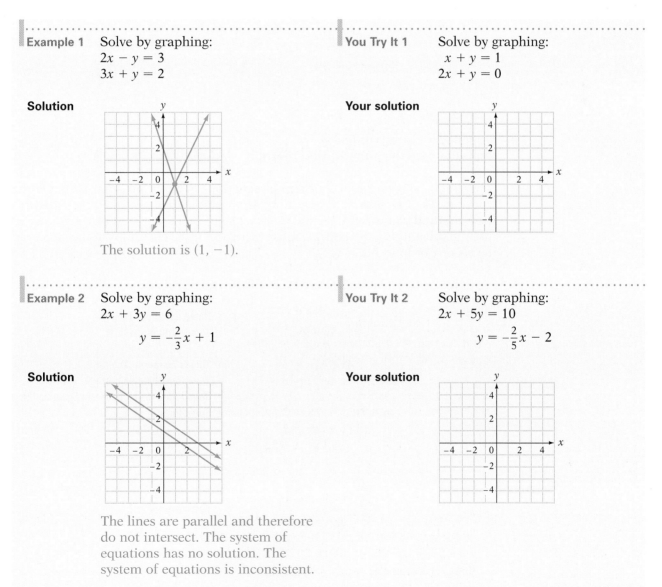

Example 1   Solve by graphing:
$2x - y = 3$
$3x + y = 2$

Solution

The solution is $(1, -1)$.

You Try It 1   Solve by graphing:
$x + y = 1$
$2x + y = 0$

Your solution

Example 2   Solve by graphing:
$2x + 3y = 6$

$y = -\dfrac{2}{3}x + 1$

Solution

The lines are parallel and therefore
do not intersect. The system of
equations has no solution. The
system of equations is inconsistent.

You Try It 2   Solve by graphing:
$2x + 5y = 10$

$y = -\dfrac{2}{5}x - 2$

Your solution

Solutions on p. S10

Example 3 Solve by graphing:
$$x - 2y = 6$$
$$y = \frac{1}{2}x - 3$$

Solution

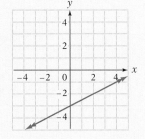

The system of equations is dependent. The solutions are the ordered pairs $\left(x, \frac{1}{2}x - 3\right)$.

You Try It 3 Solve by graphing:
$$3x - 4y = 12$$
$$y = \frac{3}{4}x - 3$$

Your solution

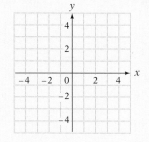

Solution on p. S10

Objective B | **To solve a system of linear equations by the substitution method**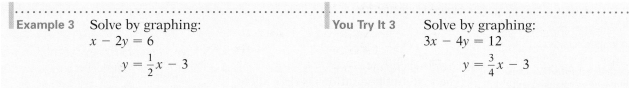

The graphical solution of a system of equations is based on approximating the coordinates of a point of intersection. An algebraic method called the **substitution method** can be used to find an exact solution of a system of equations. To use the substitution method, we must write one of the equations of the system in terms of x or in terms of y.

➡ Solve by the substitution method: (1) $3x + y = 5$
                                      (2) $4x + 5y = 3$

$$3x + y = 5$$
(3) $$y = -3x + 5$$

- **Solve equation (1) for y.**
  This is Equation (3).

(2) $$4x + 5y = 3$$
$$4x + 5(-3x + 5) = 3$$

- **This is Equation (2).**
- **Equation (3) states that $y = -3x + 5$. Substitute $-3x + 5$ for y in Equation (2).**

$$4x - 15x + 25 = 3$$
$$-11x + 25 = 3$$
$$-11x = -22$$
$$x = 2$$

- **Solve for x.**

(3) $$y = -3x + 5$$
$$y = -3(2) + 5$$
$$y = -6 + 5$$
$$y = -1$$

- **Substitute the value of x into Equation (3) and solve for y.**

The solution is the ordered pair $(2, -1)$.

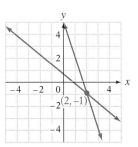

The graph of the system of equations is shown at the left. Note that the graphs intersect at the point whose coordinates are $(2, -1)$, the solution of the system of equations.

➡ Solve by the substitution method: (1) $6x + 2y = 8$
(2) $3x + y = 2$

(3)
$$3x + y = 2$$
$$y = -3x + 2$$

- We will solve equation (2) for y.
- This is Equation (3).

(1)
$$6x + 2y = 8$$
$$6x + 2(-3x + 2) = 8$$

$$6x - 6x + 4 = 8$$
$$0x + 4 = 8$$
$$4 = 8$$

- This is Equation (1).
- Equation (3) states that $y = -3x + 2$. Substitute $-3x + 2$ for y in Equation (1).
- Solve for x.

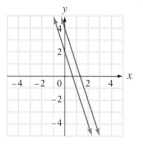

This is not a true equation.
The system of equations has no solution.
The system of equations is inconsistent.

The graph of the system of equations is shown at the left. Note that the lines are parallel.

....

Example 4   Solve by substitution:
(1) $3x - 2y = 4$
(2) $-x + 4y = -3$

You Try It 4   Solve by substitution:
$$3x - y = 3$$
$$6x + 3y = -4$$

Solution   Solve Equation (2) for x.
$$-x + 4y = -3$$
$$-x = -4y - 3$$
$$x = 4y + 3 \quad \text{• Equation (3)}$$

Your solution

Substitute $4y + 3$ for x in
Equation (1).
$$3x - 2y = 4 \quad \text{• Equation (1)}$$
$$3(4y + 3) - 2y = 4 \quad \text{• } x = 4y + 3$$
$$12y + 9 - 2y = 4$$
$$10y + 9 = 4$$
$$10y = -5$$
$$y = -\frac{5}{10} = -\frac{1}{2}$$

Substitute the value of y into
Equation (3).
$$x = 4y + 3 \quad \text{• Equation (3)}$$
$$= 4\left(-\frac{1}{2}\right) + 3 \quad \text{• } y = -\frac{1}{2}$$
$$= -2 + 3 = 1$$

The solution is $\left(1, -\frac{1}{2}\right)$.

Solution on p. S10

Example 5 Solve by substitution and graph:
$$3x - 3y = 2$$
$$y = x + 2$$

You Try It 5 Solve by substitution and graph:
$$y = 2x - 3$$
$$2x - 4y = 5$$

Solution
$$3x - 3y = 2$$
$$3x - 3(x + 2) = 2$$
$$3x - 3x - 6 = 2$$
$$-6 = 2$$

Your solution

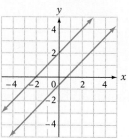

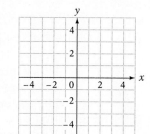

This is not a true equation. The lines are parallel, so the system is inconsistent. The system does not have a solution.

Example 6 Solve by substitution and graph:
$$9x + 3y = 12$$
$$y = -3x + 4$$

You Try It 6 Solve by substitution and graph:
$$6x - 3y = 6$$
$$2x - y = 2$$

Solution
$$9x + 3y = 12$$
$$9x + 3(-3x + 4) = 12$$
$$9x - 9x + 12 = 12$$
$$12 = 12$$
This is a true equation. The system is dependent.

Your solution

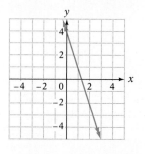

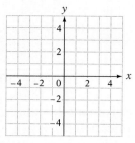

The solutions are the ordered pairs $(x, -3x + 4)$.

Solutions on pp. S10–S11

Objective C   **To solve investment problems**

The annual simple interest that an investment earns is given by the equation $Pr = I$, where P is the principal, or the amount invested, r is the simple interest rate, and I is the simple interest.

For instance, if you invest $500 at a simple interest rate of 5%, then the interest earned after one year is calculated as follows.

$$Pr = I$$
$$500(0.05) = I \qquad \bullet \text{ Replace } P \text{ by 500 and } r \text{ by 0.05 (5%).}$$
$$25 = I \qquad \bullet \text{ Simplify.}$$

The amount of interest earned is $25.

➡ You have a total of $5000 to invest in two simple interest accounts. On one account, the money market fund, the annual simple interest rate is 3.5%. On the second account, the bond fund, the annual simple interest rate is 7.5%. If you earn $245 per year from these two investments, how much do you have invested in each account?

> **Strategy for Solving Simple-Interest Investment Problems**
>
> **1.** For each amount invested, use the equation $Pr = I$. Write a numerical or variable expression for the principal, the interest rate, and the interest earned.

Amount invested at 3.5%: x
Amount invested at 7.5%: y

	Principal, P	$\cdot$	Interest rate, r	$=$	Interest earned, I
Amount at 3.5%	x	$\cdot$	0.035	$=$	$0.035x$
Amount at 7.5%	y	$\cdot$	0.075	$=$	$0.075y$

> **2.** Write a system of equations. One equation will express the relationship between the amounts invested. The second equation will express the relationship between the amounts of interest earned by the investments.

The total amount invested is $5000: $x + y = 5000$
The total annual interest earned is $245: $0.035x + 0.075y = 245$

Solve the system of equations.   (1) $\qquad\qquad x + y = 5000$
(2) $0.035x + 0.075y = 245$

Solve equation (1) for y: (3) $y = -x + 5000$
Substitute into Equation (2):   (2) $0.035x + 0.075(-x + 5000) = 245$
$$0.035x - 0.075x + 375 = 245$$
$$-0.04x = -130$$
$$x = 3250$$

Substitute the value of x into Equation (3) and solve for y.

$$y = -x + 5000$$
$$y = -3250 + 5000 = 1750$$

The amount invested at 3.5% is $3250.
The amount invested at 7.5% is $1750.

Example 7

An investment of $4000 is made at an annual simple interest rate of 4.9%. How much additional money must be invested at an annual simple interest rate of 7.4% so that the total interest earned is 6.4% of the total investment?

You Try It 7

An investment club invested $13,600 into two simple interest accounts. On one account, the annual simple interest rate is 4.2%. On the other, the annual simple interest rate is 6%. How much should be invested in each account so that both accounts earn the same annual interest?

Strategy

• Amount invested at 4.9%: 4000
  Amount invested at 7.4%: x
  Amount invested at 6.4%: y

	Principal	Rate	Interest
Amount at 4.9%	4000	0.049	0.049(4000)
Amount at 7.4%	x	0.074	0.074x
Amount at 6.4%	y	0.064	0.064y

Your strategy

• The amount invested at 6.4% (y) is $4000 more than the amount invested at 7.4% (x):
  $y = x + 4000$
• The sum of the interest earned at 4.9% and the interest earned at 7.4% equals the interest earned at 6.4%:
  $0.049(4000) + 0.074x = 0.064y$

Solution

$$y = x + 4000 \quad (1)$$
$$0.049(4000) + 0.074x = 0.064y \quad (2)$$

Replace y in Equation (2) by $x + 4000$ from Equation (1). Then solve for x.

$$0.049(4000) + 0.074x = 0.064(x + 4000)$$
$$196 + 0.074x = 0.064x + 256$$
$$0.01x = 60$$
$$x = 6000$$

$6000 must be invested at an annual simple interest rate of 7.4%.

Your solution

Solution on p. S11

4.1 Exercises

Objective A

Determine whether the ordered pair is a solution of the system of equations.

1. $(0, -1)$
$3x - 2y = 2$
$x + 2y = 6$

2. $(2, 1)$
$x + y = 3$
$2x - 3y = 1$

3. $(-3, -5)$
$x + y = -8$
$2x + 5y = -31$

4. $(1, -1)$
$3x - y = 4$
$7x + 2y = -5$

State whether the system of equations is independent, inconsistent, or dependent.

5.   **6.**   **7.**   **8.**

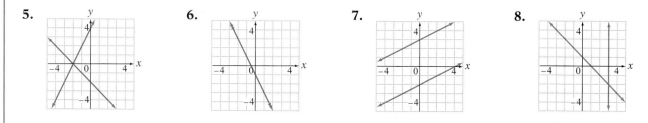

Solve by graphing.

9. $x + y = 2$
$x - y = 4$

10. $x + y = 1$
$3x - y = -5$

11. $x - y = -2$
$x + 2y = 10$

12. $2x - y = 5$
$3x + y = 5$

13. $3x - 2y = 6$
$y = 3$

14. $x = 4$
$3x - 2y = 4$

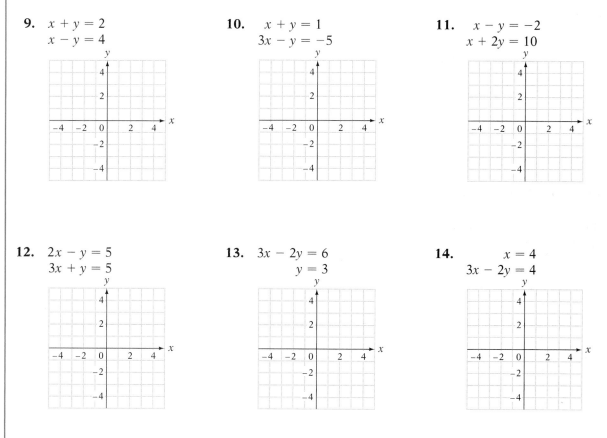

15. $x = 4$
$y = -1$

16. $x + 2 = 0$
$y - 1 = 0$

17. $2x + y = 3$
$x - 2 = 0$

18. $x - 3y = 6$
$y + 3 = 0$

19. $x - y = 6$
$x + y = 2$

20. $2x + y = 2$
$-x + y = 5$

21. $y = x - 5$
$2x + y = 4$

22. $2x - 5y = 4$
$y = x + 1$

23. $y = \dfrac{1}{2}x - 2$
$x - 2y = 8$

24. $2x + 3y = 6$
$y = -\dfrac{2}{3}x + 1$

25. $2x - 5y = 10$
$y = \dfrac{2}{5}x - 2$

26. $3x - 2y = 6$
$y = \dfrac{3}{2}x - 3$

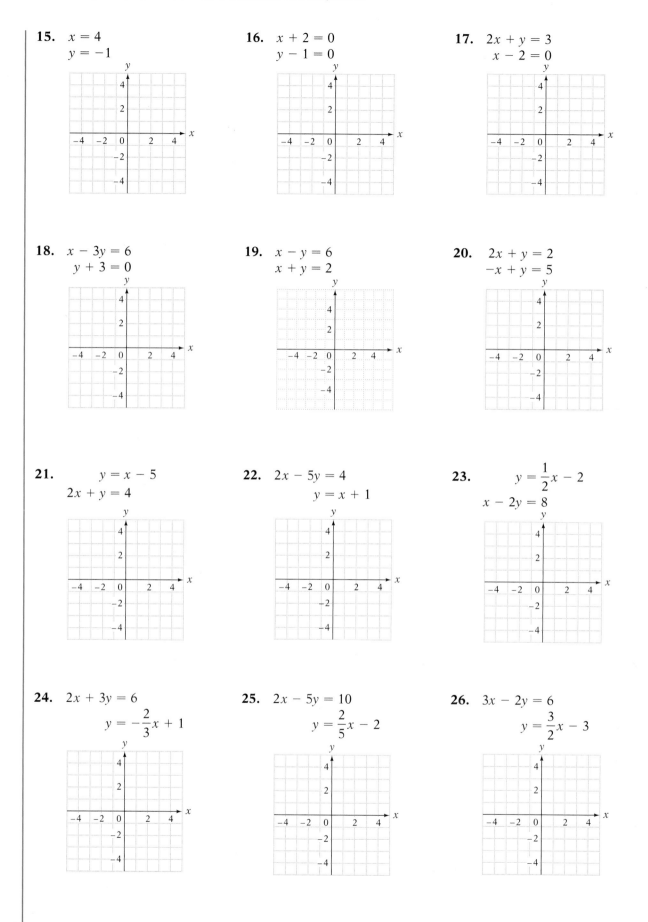

Objective B

Solve by the substitution method.

27. $y = -x + 1$
$2x - y = 5$

28. $x = 3y + 1$
$x - 2y = 6$

29. $x = 2y - 3$
$3x + y = 5$

30. $4x - 3y = 5$
$y = 2x - 3$

31. $3x + 5y = -1$
$y = 2x - 8$

32. $5x - 2y = 9$
$y = 3x - 4$

33. $4x - 3y = 2$
$y = 2x + 1$

34. $x = 2y + 4$
$4x + 3y = -17$

35. $3x - 2y = -11$
$x = 2y - 9$

36. $5x + 4y = -1$
$y = 2 - 2x$

37. $3x + 2y = 4$
$y = 1 - 2x$

38. $2x - 5y = -9$
$y = 9 - 2x$

39. $5x + 2y = 15$
$x = 6 - y$

40. $7x - 3y = 3$
$x = 2y + 2$

41. $3x - 4y = 6$
$x = 3y + 2$

42. $2x + 2y = 7$
$y = 4x + 1$

43. $3x + 7y = -5$
$y = 6x - 5$

44. $3x + y = 5$
$2x + 3y = 8$

45. $3x - y = 10$
$6x - 2y = 5$

46. $6x - 4y = 3$
$3x - 2y = 9$

47. $3x + 4y = 14$
$2x + y = 1$

48. $5x + 3y = 8$
$3x + y = 8$

49. $3x + 5y = 0$
$x - 4y = 0$

50. $2x - 7y = 0$
$3x + y = 0$

51. $2x - 4y = 16$
$-x + 2y = -8$

52. $3x - 12y = -24$
$-x + 4y = 8$

53. $y = 3x + 2$
$y = 2x + 3$

54. $y = 3x - 7$
$y = 2x - 5$

55. $y = 3x + 1$
$y = 6x - 1$

56. $y = 2x - 3$
$y = 4x - 4$

Objective C   *Application Problems*

57. The Community Relief Charity Group is earning 3.5% simple interest on the $2800 it invested in a savings account. It also earns 4.2% simple interest on an insured bond fund. The annual interest earned from both accounts is $329. How much is invested in the insured bond fund?

58. Two investments earn an annual income of $575. One investment earns an annual simple interest rate of 8.5%, and the other investment earns an annual simple interest rate of 6.4%. The total amount invested is $8000. How much is invested in each account?

59. An investment club invested $6000 at an annual simple interest rate of 4.0%. How much additional money must be invested at an annual simple interest rate of 6.5% so that the total annual interest earned will be 5% of the total investment?

60. A small company invested $30,000 by putting part of it into a savings account that earned 3.2% annual simple interest and the remainder in a risky stock fund that earned 12.6% annual simple interest. If the company earned $1665 annually from the investments, how much was in each account?

61. An account executive divided $42,000 between two simple interest accounts. On the tax-free account the annual simple interest rate is 3.5%, and on the money market fund the annual simple interest rate is 4.5%. How much should be invested in each account so that both accounts earn the same annual interest?

62. The Ridge Investment Club placed $33,000 into two simple interest accounts. On one account, the annual simple interest rate is 6.5%. On the other, the annual simple interest rate is 4.5%. How much should be invested in each account so that both accounts earn the same annual interest?

63. The Cross Creek Investment Club has $20,000 to invest. The members of the club decided to invest $16,000 of their money in two bond funds. The first, a mutual bond fund, earns annual simple interest of 4.5%. The second account, a corporate bond fund, earns 8% annual simple interest. If the members earned $1070 from these two accounts, how much was invested in the mutual bond fund?

64. Cabin Financial Service Group recommends that a client purchase for $10,000 a corporate bond that earns 5% annual simple interest. How much additional money must be placed in U.S. government securities that earn a simple interest rate of 3.5% so that the total annual interest earned from the two investments is 4% of the total investment?

APPLYING THE CONCEPTS

Use a graphing calculator to estimate solutions to the following systems of equations. See the Projects and Group Activities at the end of this chapter for assistance.

65. $y = -\dfrac{1}{2}x + 2$

   $y = 2x - 1$

66. $y = \sqrt{2}x - 1$

   $y = -\sqrt{3}x + 1$

67. $y = \pi x - \dfrac{2}{3}$

   $y = -x + \dfrac{\pi}{2}$

4.2 Solving Systems of Linear Equations by the Addition Method

Objective A **To solve a system of two linear equations in two variables by the addition method** ⟨ 4 ⟩

The **addition method** is an alternative method for solving a system of equations. This method is based on the Addition Property of Equations. Use the addition method when it is not convenient to solve one equation for one variable in terms of another variable.

Note for the system of equations at the right the effect of adding Equation (2) to Equation (1). Because $-3y$ and $3y$ are additive inverses, adding the equations results in an equation with only one variable.

$$(1)\ 5x - 3y = 14$$
$$(2)\ 2x + 3y = -7$$
$$7x + 0y = 7$$
$$7x = 7$$

The solution of the resulting equation is the first component of the ordered-pair solution of the system.

$$7x = 7$$
$$x = 1$$

The second component is found by substituting the value of x into Equation (1) or (2) and then solving for y. Equation (1) is used here.

$$(1)\ \ 5x - 3y = 14$$
$$5(1) - 3y = 14$$
$$5 - 3y = 14$$
$$-3y = 9$$
$$y = -3$$

The solution is $(1, -3)$.

Sometimes each equation of the system of equations must be multiplied by a constant so that the coefficients of one of the variable terms are opposites.

➡ Solve by the addition method:
$$(1)\ 3x + 4y = 2$$
$$(2)\ 2x + 5y = -1$$

To eliminate x, multiply Equation (1) by 2 and Equation (2) by -3. Note at the right how the constants are chosen.

$$2(3x + 4y) = 2 \cdot 2$$
$$-3(2x + 5y) = -3(-1)$$

- The negative is used so that the coefficients will be opposites.

$$\begin{array}{r} 6x + 8y = 4 \\ -6x - 15y = 3 \\ \hline -7y = 7 \\ y = -1 \end{array}$$

- 2 times Equation (1).
- -3 times Equation (2).
- Add the equations.
- Solve for y.

Substitute the value of y into Equation (1) or Equation (2) and solve for x. Equation (1) will be used here.

$$(1)\ \ \ \ \ 3x + 4y = 2$$
$$3x + 4(-1) = 2$$
$$3x - 4 = 2$$
$$3x = 6$$
$$x = 2$$

- Substitute -1 for y.
- Solve for x.

The solution is $(2, -1)$.

Point of Interest

There are records of Babylonian mathematicians solving systems of equations 3600 years ago. Here is a system of equations from that time (in our modern notation):

$$\frac{2}{3}x = \frac{1}{2}y + 500$$
$$x + y = 1800$$

We say *modern notation* for many reasons. Foremost is the fact that using variables did not become widespread until the 17th century. There are many other reasons, however. The equals sign had not been invented, 2 and 3 did not look like they do today, and zero had not even been considered as a possible number.

⇒ Solve by the addition method: (1) $\dfrac{2}{3}x + \dfrac{1}{2}y = 4$

(2) $\dfrac{1}{4}x - \dfrac{3}{8}y = -\dfrac{3}{4}$

Clear fractions. Multiply each equation by the LCM of the denominators.

$$6\left(\dfrac{2}{3}x + \dfrac{1}{2}y\right) = 6(4)$$

$$8\left(\dfrac{1}{4}x - \dfrac{3}{8}y\right) = 8\left(-\dfrac{3}{4}\right)$$

$$\begin{aligned}
4x + 3y &= 24 \\
\underline{2x - 3y} &= \underline{-6} \\
6x &= 18 \\
x &= 3
\end{aligned}$$

• Eliminate *y*. Add the equations. Then solve for *x*.

$$\dfrac{2}{3}x + \dfrac{1}{2}y = 4$$

• This is Equation (1).

$$\dfrac{2}{3}(3) + \dfrac{1}{2}y = 4$$

• Substitute *x* = 3 into Equation (1) and solve for *y*.

$$2 + \dfrac{1}{2}y = 4$$

$$\dfrac{1}{2}y = 2$$

$$y = 4$$

The solution is (3, 4).

⇒ Solve by the addition method: (1) $2x - y = 3$
(2) $4x - 2y = 6$

Eliminate *y*. Multiply Equation (1) by −2.

(1) $-2(2x - y) = -2(3)$ • −2 times Equation (1).
(3) $-4x + 2y = -6$ • This is Equation (3).

Add Equation (3) to Equation (2).

(2) $4x - 2y = 6$
(3) $\underline{-4x + 2y = -6}$
         $0 = 0$

The equation $0 = 0$ indicates that the system of equations is dependent. This means that the graphs of the two lines are the same. Therefore, the solutions of the system of equations are the ordered-pair solutions of the equation of the line. Solve Equation (1) for *y*.

$$\begin{aligned}
2x - y &= 3 \\
-y &= -2x + 3 \\
y &= 2x - 3
\end{aligned}$$

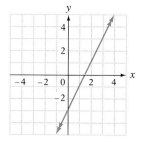

The ordered-pair solutions are (x, y), where $y = 2x - 3$. These ordered pairs are usually written $(x, 2x - 3)$, where $2x - 3$ is substituted for *y*.

Example 1 Solve by the addition method:
(1) $3x - 2y = 2x + 5$
(2) $2x + 3y = -4$

Solution Write Equation (1) in the form
$Ax + By = C$.

$$3x - 2y = 2x + 5$$
$$x - 2y = 5$$

Solve the system:

$$x - 2y = 5$$
$$2x + 3y = -4$$

Eliminate x.

$$-2(x - 2y) = -2(5)$$
$$2x + 3y = -4$$

$$-2x + 4y = -10$$
$$2x + 3y = -4$$

Add the equations.

$$7y = -14$$
$$y = -2$$

Replace y in Equation (2).

$$2x + 3y = -4$$
$$2x + 3(-2) = -4$$
$$2x - 6 = -4$$
$$2x = 2$$
$$x = 1$$

The solution is $(1, -2)$.

You Try It 1 Solve by the addition method:
$2x + 5y = 6$
$3x - 2y = 6x + 2$

Your solution

Example 2 Solve by the addition method:
(1) $4x - 8y = 36$
(2) $3x - 6y = 27$

Solution Eliminate x.

$$3(4x - 8y) = 3(36)$$
$$-4(3x - 6y) = -4(27)$$

$$12x - 24y = 108$$
$$-12x + 24y = -108$$

Add the equations.

$$0 = 0$$

The system of equations is
dependent. The solutions are the
ordered pairs
$\left(x, \frac{1}{2}x - \frac{9}{2}\right)$.

You Try It 2 Solve by the addition method:
$2x + y = 5$
$4x + 2y = 6$

Your solution

Solutions on p. S11

Objective B **To solve a system of three linear equations in three variables by the addition method**

An equation of the form $Ax + By + Cz = D$, where A, B, and C are the coefficients of the variables and D is a constant, is a **linear equation in three variables.** Examples of this type of equation are shown at the right.

$$2x + 4y - 3z = 7$$

$$x - 6y + z = -3$$

Graphing an equation in three variables requires a third coordinate axis perpendicular to the xy-plane. The third axis is commonly called the z-axis. The result is a three-dimensional coordinate system called the **xyz-coordinate system.** To help visualize a three-dimensional coordinate system, think of a corner of a room: the floor is the xy-plane, one wall is the yz-plane, and the other wall is the xz-plane. A three-dimensional coordinate system is shown at the right.

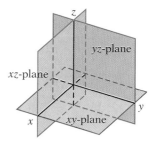

The graph of a point in an xyz-coordinate system is an **ordered triple** (x, y, z). Graphing an ordered triple requires three moves, the first in the direction of the x-axis, the second in the direction of the y-axis, and the third in the direction of the z-axis. The graph of the points $(-4, 2, 3)$ and $(3, 4, -2)$ is shown at the right.

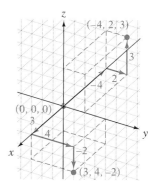

The graph of a linear equation in three variables is a plane. That is, if all the solutions of a linear equation in three variables were plotted in an xyz-coordinate system, the graph would look like a large piece of paper extending infinitely. The graph of $x + y + z = 3$ is shown at the right.

There are different ways in which three planes can be oriented in an *xyz*-coordinate system. The systems of equations represented by the planes below are inconsistent.

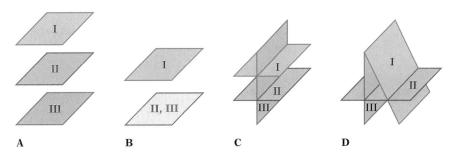

Graphs of Inconsistent Systems of Equations

For a system of three equations in three variables to have a solution, the graphs of the planes must intersect at a single point, they must intersect along a common line, or all equations must have a graph that is the same plane. These situations are shown in the figures below.

The three planes shown in Figure E intersect at a point. A system of equations represented by planes that intersect at a point is independent.

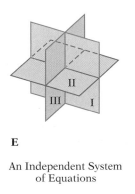

An Independent System
of Equations

The planes shown in Figures F and G intersect along a common line. The system of equations represented by the planes in Figure H has a graph that is the same plane. The systems of equations represented by the graphs below are dependent.

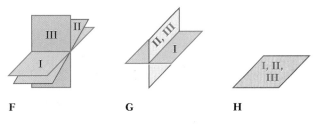

Dependent Systems of Equations

Point of Interest

In the early 1980s, Stephen Hoppe became interested in winning Monopoly strategies. Finding these strategies required solving a system that contained 123 equations in 123 variables!

Just as a solution of an equation in two variables is an ordered pair (x, y), a **solution of an equation in three variables** is an ordered triple (x, y, z). For example, $(2, 1, -3)$ is a solution of the equation $2x - y - 2z = 9$. The ordered triple $(1, 3, 2)$ is not a solution.

A **system of linear equations in three variables** is shown at the right. A **solution of a system of equations in three variables** is an ordered triple that is a solution of each equation of the system.

$$x - 2y + z = 6$$
$$3x + y - 2z = 2$$
$$2x - 3y + 5z = 1$$

A system of linear equations in three variables can be solved by using the addition method. First, eliminate one variable from any two of the given equations. Then eliminate the same variable from any other two equations. The result will be a system of two equations in two variables. Solve this system by the addition method.

⇒ Solve: (1) $\quad x + 4y - z = 10$
(2) $\quad 3x + 2y + z = 4$
(3) $\quad 2x - 3y + 2z = -7$

Eliminate z from Equations (1) and (2) by adding the two equations.

$$x + 4y - z = 10$$
$$3x + 2y + z = 4$$
(4) $\quad\quad 4x + 6y = 14$ • **Add the equations.**

Eliminate z from Equations (1) and (3). Multiply Equation (1) by 2 and add to Equation (3).

$$2x + 8y - 2z = 20 \quad • \textbf{ 2 times Equation (1).}$$
$$2x - 3y + 2z = -7 \quad • \textbf{ This is Equation (3).}$$
(5) $\quad\quad 4x + 5y = 13 \quad • \textbf{ Add the equations.}$

Using Equations (4) and (5), solve the system of two equations in two variables.

(4) $\quad 4x + 6y = 14$
(5) $\quad 4x + 5y = 13$

Eliminate x. Multiply Equation (5) by -1 and add to Equation (4).

$$4x + 6y = 14 \quad\quad • \textbf{ This is Equation (4).}$$
$$-4x - 5y = -13 \quad • \textbf{ −1 times Equation (5).}$$
$$y = 1 \quad\quad\quad • \textbf{ Add the equations.}$$

Substitute the value of y into Equation (4) or Equation (5) and solve for x. Equation (4) is used here.

$$4x + 6y = 14 \quad • \textbf{ This is Equation (4).}$$
$$4x + 6(1) = 14 \quad • \textbf{ \textit{y} = 1}$$
$$4x = 8 \quad\quad • \textbf{ Solve for \textit{x}.}$$
$$x = 2$$

Substitute the value of y and the value of x into one of the equations in the original system. Equation (2) is used here.

$$3x + 2y + z = 4$$
$$3(2) + 2(1) + z = 4 \quad • \textbf{ \textit{x} = 2, \textit{y} = 1}$$
$$6 + 2 + z = 4$$
$$8 + z = 4$$
$$z = -4$$

The solution is $(2, 1, -4)$.

⇒ Solve: (1) $2x - 3y - z = 1$
          (2) $x + 4y + 3z = 2$
          (3) $4x - 6y - 2z = 5$

Eliminate x from Equations (1) and (2).

$2x - 3y - z = 1$ • This is Equation (1).
$-2x - 8y - 6z = -4$ • -2 times Equation (2).
$\overline{-11y - 7z = -3}$ • Add the equations.

Eliminate x from Equations (1) and (3).

$-4x + 6y + 2z = -2$ • -2 times Equation (1).
$4x - 6y - 2z = 5$ • This is Equation (3).
$\overline{0 = 3}$ • Add the equations.

The equation $0 = 3$ is not a true equation. The system of equations is inconsistent and therefore has no solution.

⇒ Solve: (1) $3x - z = -1$
          (2) $2y - 3z = 10$
          (3) $x + 3y - z = 7$

Eliminate x from Equations (1) and (3). Multiply Equation (3) by -3 and add to Equation (1).

          $3x - z = -1$ • This is Equation (1).
      $-3x - 9y + 3z = -21$ • -3 times Equation (3).
(4) $-9y + 2z = -22$ • Add the equations.

Use Equations (2) and (4) to form a system of equations in two variables.

(2) $2y - 3z = 10$
(4) $-9y + 2z = -22$

Eliminate z. Multiply Equation (2) by 2 and Equation (4) by 3.

$4y - 6z = 20$ • 2 times Equation (2).
$-27y + 6z = -66$ • 3 times Equation (4).
$\overline{-23y = -46}$ • Add the equations.
$y = 2$ • Solve for y.

Substitute the value of y into Equation (2) or Equation (4) and solve for z. Equation (2) is used here.

(2) $2y - 3z = 10$ • This is Equation (2).
      $2(2) - 3z = 10$ • $y = 2$
        $4 - 3z = 10$ • Solve for z.
          $-3z = 6$
            $z = -2$

Substitute the value of z into Equation (1) and solve for x.

(1) $3x - z = -1$ • This is Equation (1).
      $3x - (-2) = -1$ • $z = -2$
        $3x + 2 = -1$ • Solve for x.
          $3x = -3$
            $x = -1$

The solution is $(-1, 2, -2)$.

Example 3   Solve: (1) $3x - y + 2z = 1$
(2) $2x + 3y + 3z = 4$
(3) $x + y - 4z = -9$

You Try It 3   Solve: $x - y + z = 6$
$2x + 3y - z = 1$
$x + 2y + 2z = 5$

Solution   Eliminate y. Add
Equations (1) and (3).

$3x - y + 2z = 1$
$x + y - 4z = -9$
$\overline{ 4x - 2z = -8}$

Multiply each side of the equation
by $\frac{1}{2}$.

(4) $2x - z = -4$

Multiply Equation (1) by 3 and
add to Equation (2).

$9x - 3y + 6z = 3$
$2x + 3y + 3z = 4$
(5) $\overline{ 11x + 9z = 7}$

Solve the system of two
equations.

(4) $2x - z = -4$
(5) $11x + 9z = 7$

Multiply Equation (4) by 9 and
add to Equation (5).

$18x - 9z = -36$
$11x + 9z = 7$
$\overline{ 29x = -29}$
$x = -1$

Replace x by -1 in
Equation (4).

$2x - z = -4$
$2(-1) - z = -4$
$-2 - z = -4$
$-z = -2$
$z = 2$

Replace x by -1 and z by 2 in
Equation (3).

$x + y - 4z = -9$
$-1 + y - 4(2) = -9$
$-9 + y = -9$
$y = 0$

The solution is $(-1, 0, 2)$.

Your solution

Solution on pp. S11–S12

4.2 Exercises

Objective A

Solve by the addition method.

1. $x - y = 5$
$x + y = 7$

2. $x + y = 1$
$2x - y = 5$

3. $3x + y = 4$
$x + y = 2$

4. $x - 3y = 4$
$x + 5y = -4$

5. $3x + y = 7$
$x + 2y = 4$

6. $x - 2y = 7$
$3x - 2y = 9$

7. $2x + 3y = -1$
$x + 5y = 3$

8. $x + 5y = 7$
$2x + 7y = 8$

9. $3x - y = 4$
$6x - 2y = 8$

10. $x - 2y = -3$
$-2x + 4y = 6$

11. $2x + 5y = 9$
$4x - 7y = -16$

12. $8x - 3y = 21$
$4x + 5y = -9$

13. $4x - 6y = 5$
$2x - 3y = 7$

14. $3x + 6y = 7$
$2x + 4y = 5$

15. $3x - 5y = 7$
$x - 2y = 3$

16. $3x + 4y = 25$
$2x + y = 10$

17. $x + 3y = 7$
$-2x + 3y = 22$

18. $2x - 3y = 14$
$5x - 6y = 32$

19. $3x + 2y = 16$
$2x - 3y = -11$

20. $2x - 5y = 13$
$5x + 3y = 17$

21. $4x + 4y = 5$
$2x - 8y = -5$

22. $3x + 7y = 16$
$4x - 3y = 9$

23. $5x + 4y = 0$
$3x + 7y = 0$

24. $3x - 4y = 0$
$4x - 7y = 0$

25. $5x + 2y = 1$
$2x + 3y = 7$

26. $3x + 5y = 16$
$5x - 7y = -4$

27. $3x - 6y = 6$
$9x - 3y = 8$

28. $\dfrac{2}{3}x - \dfrac{1}{2}y = 3$
$\dfrac{1}{3}x - \dfrac{1}{4}y = \dfrac{3}{2}$

29. $\dfrac{3}{4}x + \dfrac{1}{3}y = -\dfrac{1}{2}$
$\dfrac{1}{2}x - \dfrac{5}{6}y = -\dfrac{7}{2}$

30. $\dfrac{2}{5}x - \dfrac{1}{3}y = 1$
$\dfrac{3}{5}x + \dfrac{2}{3}y = 5$

31. $\dfrac{5x}{6} + \dfrac{y}{3} = \dfrac{4}{3}$
$\dfrac{2x}{3} - \dfrac{y}{2} = \dfrac{11}{6}$

32. $\dfrac{3x}{4} + \dfrac{2y}{5} = -\dfrac{3}{20}$
$\dfrac{3x}{2} - \dfrac{y}{4} = \dfrac{3}{4}$

33. $\dfrac{2x}{5} - \dfrac{y}{2} = \dfrac{13}{2}$
$\dfrac{3x}{4} - \dfrac{y}{5} = \dfrac{17}{2}$

34. $\dfrac{x}{2} + \dfrac{y}{3} = \dfrac{5}{12}$
$\dfrac{x}{2} - \dfrac{y}{3} = \dfrac{1}{12}$

35. $\dfrac{3x}{2} - \dfrac{y}{4} = -\dfrac{11}{12}$
$\dfrac{x}{3} - y = -\dfrac{5}{6}$

36. $\dfrac{3x}{4} - \dfrac{2y}{3} = 0$
$\dfrac{5x}{4} - \dfrac{y}{3} = \dfrac{7}{12}$

37. $4x - 5y = 3y + 4$
$2x + 3y = 2x + 1$

38. $5x - 2y = 8x - 1$
$2x + 7y = 4y + 9$

39. $2x + 5y = 5x + 1$
$3x - 2y = 3y + 3$

40. $4x - 8y = 5$
$8x + 2y = 1$

41. $5x + 2y = 2x + 1$
$2x - 3y = 3x + 2$

42. $3x + 3y = y + 1$
$x + 3y = 9 - x$

Objective B

Solve by the addition method.

43. $x + 2y - z = 1$
$2x - y + z = 6$
$x + 3y - z = 2$

44. $x + 3y + z = 6$
$3x + y - z = -2$
$2x + 2y - z = 1$

45. $2x - y + 2z = 7$
$x + y + z = 2$
$3x - y + z = 6$

46. $x - 2y + z = 6$
$x + 3y + z = 16$
$3x - y - z = 12$

47. $3x + y = 5$
$3y - z = 2$
$x + z = 5$

48. $2y + z = 7$
$2x - z = 3$
$x - y = 3$

49. $x - y + z = 1$
$2x + 3y - z = 3$
$-x + 2y - 4z = 4$

50. $2x + y - 3z = 7$
$x - 2y + 3z = 1$
$3x + 4y - 3z = 13$

51. $2x + 3z = 5$
$3y + 2z = 3$
$3x + 4y = -10$

52. $3x + 4z = 5$
$2y + 3z = 2$
$2x - 5y = 8$

53. $2x + 4y - 2z = 3$
$x + 3y + 4z = 1$
$x + 2y - z = 4$

54. $x - 3y + 2z = 1$
$x - 2y + 3z = 5$
$2x - 6y + 4z = 3$

55. $2x + y - z = 5$
$x + 3y + z = 14$
$3x - y + 2z = 1$

56. $3x - y - 2z = 11$
$2x + y - 2z = 11$
$x + 3y - z = 8$

57. $3x + y - 2z = 2$
$x + 2y + 3z = 13$
$2x - 2y + 5z = 6$

58. $4x + 5y + z = 6$
$2x - y + 2z = 11$
$x + 2y + 2z = 6$

59. $2x - y + z = 6$
$3x + 2y + z = 4$
$x - 2y + 3z = 12$

60. $3x + 2y - 3z = 8$
$2x + 3y + 2z = 10$
$x + y - z = 2$

61. $3x - 2y + 3z = -4$
$2x + y - 3z = 2$
$3x + 4y + 5z = 8$

62. $3x - 3y + 4z = 6$
$4x - 5y + 2z = 10$
$x - 2y + 3z = 4$

63. $3x - y + 2z = 2$
$4x + 2y - 7z = 0$
$2x + 3y - 5z = 7$

64. $2x + 2y + 3z = 13$
$-3x + 4y - z = 5$
$5x - 3y + z = 2$

65. $2x - 3y + 7z = 0$
$x + 4y - 4z = -2$
$3x + 2y + 5z = 1$

66. $5x + 3y - z = 5$
$3x - 2y + 4z = 13$
$4x + 3y + 5z = 22$

APPLYING THE CONCEPTS

67. Explain, graphically, the following situations when they are related to a system of three linear equations in three variables.
a. The system of equations has no solution.
b. The system of equations has exactly one solution.
c. The system of equations has infinitely many solutions.

68. Describe the graph of each of the following equations in an *xyz*-coordinate system.
a. $x = 3$ **b.** $y = 4$ **c.** $z = 2$ **d.** $y = x$

The following systems are not linear systems of equations. However, they can be solved by using a modification of the addition method. Solve each system of equations.

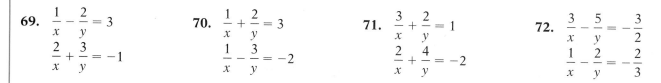

69. $\dfrac{1}{x} - \dfrac{2}{y} = 3$
$\dfrac{2}{x} + \dfrac{3}{y} = -1$

70. $\dfrac{1}{x} + \dfrac{2}{y} = 3$
$\dfrac{1}{x} - \dfrac{3}{y} = -2$

71. $\dfrac{3}{x} + \dfrac{2}{y} = 1$
$\dfrac{2}{x} + \dfrac{4}{y} = -2$

72. $\dfrac{3}{x} - \dfrac{5}{y} = -\dfrac{3}{2}$
$\dfrac{1}{x} - \dfrac{2}{y} = -\dfrac{2}{3}$

73. For Exercise 69, solve each equation of the system for y. Then use a graphing calculator to verify the solution you determined algebraically. Note that the graphs are not straight lines.

74. For Exercise 70, solve each equation of the system for y. Then use a graphing calculator to verify the solution you determined algebraically. Note that the graphs are not straight lines.

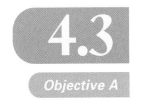

Solving Systems of Equations by Using Determinants

Objective A To evaluate a determinant 『4』

Point of Interest

The word *matrix* was first used in a mathematical context in 1850. The root of this word is the Latin *mater*, meaning *mother*. A matrix was thought of as an object from which something else originates. The idea was that a determinant, discussed below, originated (was born) from a matrix. Today, matrices are one of the most widely used tools of applied mathematics.

A **matrix** is a rectangular array of numbers. Each number in the matrix is called an **element** of the matrix. The matrix at the right, with three rows and four columns, is called a 3×4 (read "3 by 4") matrix.

$$A = \begin{pmatrix} 1 & -3 & 2 & 4 \\ 0 & 4 & -3 & 2 \\ 6 & -5 & 4 & -1 \end{pmatrix}$$

A matrix of m rows and n columns is said to be of **order $m \times n$.** The matrix above has order 3×4. The notation a_{ij} refers to the element of a matrix in the ith row and the jth column. For matrix A, $a_{23} = -3$, $a_{31} = 6$, and $a_{13} = 2$.

A **square matrix** is one that has the same number of rows as columns. A 2×2 matrix and a 3×3 matrix are shown at the right.

$$\begin{pmatrix} -1 & 3 \\ 5 & 2 \end{pmatrix} \qquad \begin{pmatrix} 4 & 0 & 1 \\ 5 & -3 & 7 \\ 2 & 1 & 4 \end{pmatrix}$$

Associated with every square matrix is a number called its **determinant.**

TAKE NOTE

Note that vertical bars are used to represent the determinant and that parentheses are used to represent the matrix.

Determinant of a 2 × 2 Matrix

The **determinant** of a 2×2 matrix $\begin{pmatrix} a_{11} & a_{12} \\ a_{21} & a_{22} \end{pmatrix}$ is written $\begin{vmatrix} a_{11} & a_{12} \\ a_{21} & a_{22} \end{vmatrix}$. The value of this determinant is given by the formula

$$\begin{vmatrix} a_{11} & a_{12} \\ a_{21} & a_{22} \end{vmatrix} = a_{11}a_{22} - a_{12}a_{21}$$

➡ Find the value of the determinant $\begin{vmatrix} 3 & 4 \\ -1 & 2 \end{vmatrix}$.

$$\begin{vmatrix} 3 & 4 \\ -1 & 2 \end{vmatrix} = 3 \cdot 2 - 4(-1) = 6 - (-4) = 10$$

The value of the determinant is 10.

For a square matrix whose order is 3×3 or greater, the value of the determinant is found by using 2×2 determinants.

The **minor of an element** in a 3×3 determinant is the 2×2 determinant that is obtained by eliminating the row and column that contain that element.

➡ Find the minor of -3 for the determinant $\begin{vmatrix} 2 & -3 & 4 \\ 0 & 4 & 8 \\ -1 & 3 & 6 \end{vmatrix}$.

The minor of -3 is the 2×2 determinant created by eliminating the row and column that contain -3.

Eliminate the row and column as shown: $\begin{vmatrix} 2 & -3 & 4 \\ 0 & 4 & 8 \\ -1 & 3 & 6 \end{vmatrix}$

The minor of -3 is $\begin{vmatrix} 0 & 8 \\ -1 & 6 \end{vmatrix}$.

Cofactor of an Element of a Matrix

The **cofactor** of an element of a matrix is $(-1)^{i+j}$ times the minor of that element, where i is the row number of the element and j is the column number of the element.

For the determinant $\begin{vmatrix} 3 & -2 & 1 \\ 2 & -5 & -4 \\ 0 & 3 & 1 \end{vmatrix}$, find the cofactor of -2 and of -5.

Because -2 is in the first row and the second column, $i = 1$ and $j = 2$. Thus $(-1)^{i+j} = (-1)^{1+2} = (-1)^3 = -1$. The cofactor of -2 is $(-1)\begin{vmatrix} 2 & -4 \\ 0 & 1 \end{vmatrix}$.

Because -5 is in the second row and the second column, $i = 2$ and $j = 2$. Thus $(-1)^{i+j} = (-1)^{2+2} = (-1)^4 = 1$. The cofactor of -5 is $1 \cdot \begin{vmatrix} 3 & 1 \\ 0 & 1 \end{vmatrix}$.

Note from this example that the cofactor of an element is -1 times the minor of that element or 1 times the minor of that element, depending on whether the sum $i + j$ is an odd or even integer.

The value of a 3×3 or larger determinant can be found by **expanding by cofactors** of *any* row or *any* column. The result of expanding by cofactors using the first row of a 3×3 matrix is shown below.

$$\begin{vmatrix} a_{11} & a_{12} & a_{13} \\ a_{21} & a_{22} & a_{23} \\ a_{31} & a_{32} & a_{33} \end{vmatrix} = a_{11}(-1)^{1+1}\begin{vmatrix} a_{22} & a_{23} \\ a_{32} & a_{33} \end{vmatrix} + a_{12}(-1)^{1+2}\begin{vmatrix} a_{21} & a_{23} \\ a_{31} & a_{33} \end{vmatrix} + a_{13}(-1)^{1+3}\begin{vmatrix} a_{21} & a_{22} \\ a_{31} & a_{32} \end{vmatrix}$$

$$= a_{11}\begin{vmatrix} a_{22} & a_{23} \\ a_{32} & a_{33} \end{vmatrix} - a_{12}\begin{vmatrix} a_{21} & a_{23} \\ a_{31} & a_{33} \end{vmatrix} + a_{13}\begin{vmatrix} a_{21} & a_{22} \\ a_{31} & a_{32} \end{vmatrix}$$

Find the value of the determinant $\begin{vmatrix} 2 & -3 & 2 \\ 1 & 3 & -1 \\ 0 & -2 & 2 \end{vmatrix}$.

Expand by cofactors of the first row.

$$\begin{vmatrix} 2 & -3 & 2 \\ 1 & 3 & -1 \\ 0 & -2 & 2 \end{vmatrix} = 2\begin{vmatrix} 3 & -1 \\ -2 & 2 \end{vmatrix} - (-3)\begin{vmatrix} 1 & -1 \\ 0 & 2 \end{vmatrix} + 2\begin{vmatrix} 1 & 3 \\ 0 & -2 \end{vmatrix}$$

$$= 2(6 - 2) - (-3)(2 - 0) + 2(-2 - 0)$$
$$= 2(4) - (-3)(2) + 2(-2) = 8 - (-6) + (-4) = 10$$

To illustrate a statement made earlier, the value of this determinant will now be found by expanding by cofactors using the second column.

$$\begin{vmatrix} 2 & -3 & 2 \\ 1 & 3 & -1 \\ 0 & -2 & 2 \end{vmatrix} = -3(-1)^{1+2}\begin{vmatrix} 1 & -1 \\ 0 & 2 \end{vmatrix} + 3(-1)^{2+2}\begin{vmatrix} 2 & 2 \\ 0 & 2 \end{vmatrix} + (-2)(-1)^{3+2}\begin{vmatrix} 2 & 2 \\ 1 & -1 \end{vmatrix}$$

$$= -3(-1)\begin{vmatrix} 1 & -1 \\ 0 & 2 \end{vmatrix} + 3(1)\begin{vmatrix} 2 & 2 \\ 0 & 2 \end{vmatrix} + (-2)(-1)\begin{vmatrix} 2 & 2 \\ 1 & -1 \end{vmatrix}$$

$$= 3(2 - 0) + 3(4 - 0) + 2(-2 - 2)$$
$$= 3(2) + 3(4) + 2(-4) = 6 + 12 + (-8) = 10$$

Note that the value of the determinant is the same whether the first row or the second column is used to expand by cofactors. *Any row or column* can be used to evaluate a determinant by expanding by cofactors.

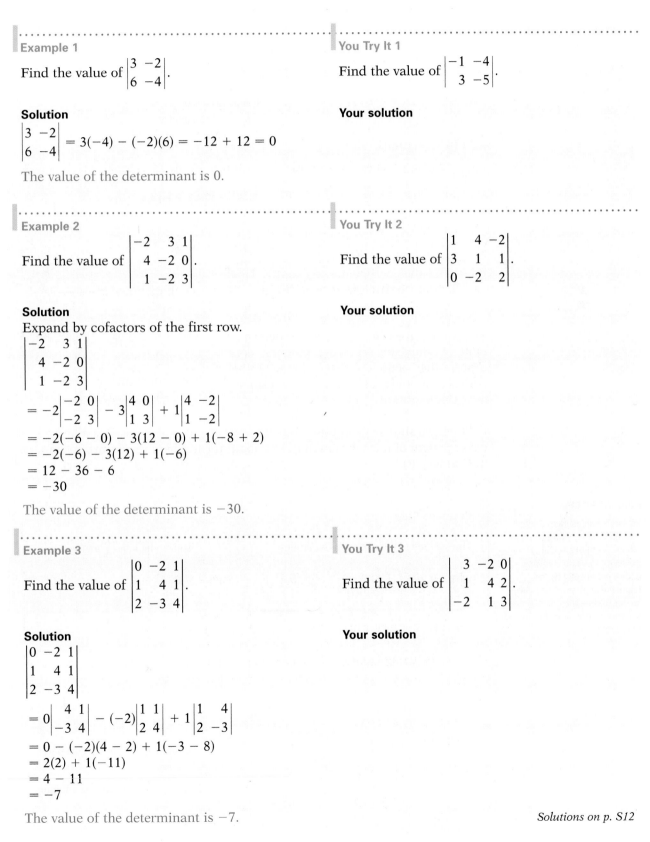

Example 1

Find the value of $\begin{vmatrix} 3 & -2 \\ 6 & -4 \end{vmatrix}$.

Solution

$\begin{vmatrix} 3 & -2 \\ 6 & -4 \end{vmatrix} = 3(-4) - (-2)(6) = -12 + 12 = 0$

The value of the determinant is 0.

You Try It 1

Find the value of $\begin{vmatrix} -1 & -4 \\ 3 & -5 \end{vmatrix}$.

Your solution

Example 2

Find the value of $\begin{vmatrix} -2 & 3 & 1 \\ 4 & -2 & 0 \\ 1 & -2 & 3 \end{vmatrix}$.

Solution
Expand by cofactors of the first row.

$\begin{vmatrix} -2 & 3 & 1 \\ 4 & -2 & 0 \\ 1 & -2 & 3 \end{vmatrix}$

$= -2\begin{vmatrix} -2 & 0 \\ -2 & 3 \end{vmatrix} - 3\begin{vmatrix} 4 & 0 \\ 1 & 3 \end{vmatrix} + 1\begin{vmatrix} 4 & -2 \\ 1 & -2 \end{vmatrix}$

$= -2(-6 - 0) - 3(12 - 0) + 1(-8 + 2)$

$= -2(-6) - 3(12) + 1(-6)$

$= 12 - 36 - 6$

$= -30$

The value of the determinant is -30.

You Try It 2

Find the value of $\begin{vmatrix} 1 & 4 & -2 \\ 3 & 1 & 1 \\ 0 & -2 & 2 \end{vmatrix}$.

Your solution

Example 3

Find the value of $\begin{vmatrix} 0 & -2 & 1 \\ 1 & 4 & 1 \\ 2 & -3 & 4 \end{vmatrix}$.

Solution

$\begin{vmatrix} 0 & -2 & 1 \\ 1 & 4 & 1 \\ 2 & -3 & 4 \end{vmatrix}$

$= 0\begin{vmatrix} 4 & 1 \\ -3 & 4 \end{vmatrix} - (-2)\begin{vmatrix} 1 & 1 \\ 2 & 4 \end{vmatrix} + 1\begin{vmatrix} 1 & 4 \\ 2 & -3 \end{vmatrix}$

$= 0 - (-2)(4 - 2) + 1(-3 - 8)$

$= 2(2) + 1(-11)$

$= 4 - 11$

$= -7$

The value of the determinant is -7.

You Try It 3

Find the value of $\begin{vmatrix} 3 & -2 & 0 \\ 1 & 4 & 2 \\ -2 & 1 & 3 \end{vmatrix}$.

Your solution

Solutions on p. S12

Objective B **To solve a system of equations by using Cramer's Rule**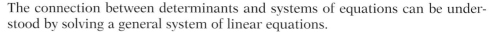

The connection between determinants and systems of equations can be understood by solving a general system of linear equations.

Solve: (1) $a_1x + b_1y = c_1$
(2) $a_2x + b_2y = c_2$

Eliminate y. Multiply Equation (1) by b_2 and Equation (2) by $-b_1$.

$a_1b_2x + b_1b_2y = c_1b_2$ • b_2 times Equation (1).
$-a_2b_1x - b_1b_2y = -c_2b_1$ • $-b_1$ times Equation (2).

Add the equations.

$a_1b_2x - a_2b_1x = c_1b_2 - c_2b_1$

$(a_1b_2 - a_2b_1)x = c_1b_2 - c_2b_1$ • Solve for x, assuming $a_1b_2 - a_2b_1 \neq 0$.

$$x = \frac{c_1b_2 - c_2b_1}{a_1b_2 - a_2b_1}$$

The denominator $a_1b_2 - a_2b_1$ is the determinant of the coefficients of x and y. This is called the **coefficient determinant.**

$$a_1b_2 - a_2b_1 = \begin{vmatrix} a_1 & b_1 \\ a_2 & b_2 \end{vmatrix}$$

coefficients of x ⟶
coefficients of y ⟶

The numerator $c_1b_2 - c_2b_1$ is the determinant obtained by replacing the first column in the coefficient determinant by the constants c_1 and c_2. This is called a **numerator determinant.**

$$c_1b_2 - c_2b_1 = \begin{vmatrix} c_1 & b_1 \\ c_2 & b_2 \end{vmatrix}$$

constants of ⟶
the equations

Following a similar procedure and eliminating x, it is also possible to express the y-component of the solution in determinant form. These results are summarized in Cramer's Rule.

Point of Interest

Cramer's Rule is named after Gabriel Cramer, who used it in a book he published in 1750. However, this rule was also published in 1683 by the Japanese mathematician Seki Kown. That publication occurred seven years before Cramer's birth.

Cramer's Rule

The solution of the system of equations $\begin{matrix} a_1x + b_1y = c_1 \\ a_2x + b_2y = c_2 \end{matrix}$ is given by $x = \dfrac{D_x}{D}$ and $y = \dfrac{D_y}{D}$,

where $D = \begin{vmatrix} a_1 & b_1 \\ a_2 & b_2 \end{vmatrix}$, $D_x = \begin{vmatrix} c_1 & b_1 \\ c_2 & b_2 \end{vmatrix}$, $D_y = \begin{vmatrix} a_1 & c_1 \\ a_2 & c_2 \end{vmatrix}$, and $D \neq 0$.

⟹ Solve by using Cramer's Rule: $3x - 2y = 1$
$2x + 5y = 3$

$$D = \begin{vmatrix} 3 & -2 \\ 2 & 5 \end{vmatrix} = 19$$ • Find the value of the coefficient determinant.

$$D_x = \begin{vmatrix} 1 & -2 \\ 3 & 5 \end{vmatrix} = 11, \quad D_y = \begin{vmatrix} 3 & 1 \\ 2 & 3 \end{vmatrix} = 7$$ • Find the value of each of the numerator determinants.

$$x = \frac{D_x}{D} = \frac{11}{19}, \quad y = \frac{D_y}{D} = \frac{7}{19}$$ • Use Cramer's Rule to write the solution.

The solution is $\left(\dfrac{11}{19}, \dfrac{7}{19}\right)$.

A procedure similar to that followed for two equations in two variables can be used to extend Cramer's Rule to three equations in three variables.

Cramer's Rule for a System of Three Equations in Three Variables

$$a_1x + b_1y + c_1z = d_1$$

The solution of the system of equations
$$a_2x + b_2y + c_2z = d_2$$
$$a_3x + b_3y + c_3z = d_3$$

is given by $x = \dfrac{D_x}{D}$, $y = \dfrac{D_y}{D}$, and $z = \dfrac{D_z}{D}$, where

$$D = \begin{vmatrix} a_1 & b_1 & c_1 \\ a_2 & b_2 & c_2 \\ a_3 & b_3 & c_3 \end{vmatrix},\ D_x = \begin{vmatrix} d_1 & b_1 & c_1 \\ d_2 & b_2 & c_2 \\ d_3 & b_3 & c_3 \end{vmatrix},\ D_y = \begin{vmatrix} a_1 & d_1 & c_1 \\ a_2 & d_2 & c_2 \\ a_3 & d_3 & c_3 \end{vmatrix},\ D_z = \begin{vmatrix} a_1 & b_1 & d_1 \\ a_2 & b_2 & d_2 \\ a_3 & b_3 & d_3 \end{vmatrix}, \text{ and } D \neq 0.$$

⇒ Solve by using Cramer's Rule:
$$2x - y + z = 1$$
$$x + 3y - 2z = -2$$
$$3x + y + 3z = 4$$

Find the value of the coefficient determinant.

$$D = \begin{vmatrix} 2 & -1 & 1 \\ 1 & 3 & -2 \\ 3 & 1 & 3 \end{vmatrix} = 2\begin{vmatrix} 3 & -2 \\ 1 & 3 \end{vmatrix} - (-1)\begin{vmatrix} 1 & -2 \\ 3 & 3 \end{vmatrix} + 1\begin{vmatrix} 1 & 3 \\ 3 & 1 \end{vmatrix}$$
$$= 2(11) + 1(9) + 1(-8)$$
$$= 23$$

Find the value of each of the numerator determinants.

$$D_x = \begin{vmatrix} 1 & -1 & 1 \\ -2 & 3 & -2 \\ 4 & 1 & 3 \end{vmatrix} = 1\begin{vmatrix} 3 & -2 \\ 1 & 3 \end{vmatrix} - (-1)\begin{vmatrix} -2 & -2 \\ 4 & 3 \end{vmatrix} + 1\begin{vmatrix} -2 & 3 \\ 4 & 1 \end{vmatrix}$$
$$= 1(11) + 1(2) + 1(-14)$$
$$= -1$$

$$D_y = \begin{vmatrix} 2 & 1 & 1 \\ 1 & -2 & -2 \\ 3 & 4 & 3 \end{vmatrix} = 2\begin{vmatrix} -2 & -2 \\ 4 & 3 \end{vmatrix} - 1\begin{vmatrix} 1 & -2 \\ 3 & 3 \end{vmatrix} + 1\begin{vmatrix} 1 & -2 \\ 3 & 4 \end{vmatrix}$$
$$= 2(2) - 1(9) + 1(10)$$
$$= 5$$

$$D_z = \begin{vmatrix} 2 & -1 & 1 \\ 1 & 3 & -2 \\ 3 & 1 & 4 \end{vmatrix} = 2\begin{vmatrix} 3 & -2 \\ 1 & 4 \end{vmatrix} - (-1)\begin{vmatrix} 1 & -2 \\ 3 & 4 \end{vmatrix} + 1\begin{vmatrix} 1 & 3 \\ 3 & 1 \end{vmatrix}$$
$$= 2(14) + 1(10) + 1(-8)$$
$$= 30$$

Use Cramer's Rule to write the solution.

$$x = \frac{D_x}{D} = \frac{-1}{23}, \qquad y = \frac{D_y}{D} = \frac{5}{23}, \qquad z = \frac{D_z}{D} = \frac{30}{23}$$

The solution is $\left(-\dfrac{1}{23}, \dfrac{5}{23}, \dfrac{30}{23}\right)$.

Example 4

Solve by using Cramer's Rule.

$6x - 9y = 5$
$4x - 6y = 4$

Solution

$$D = \begin{vmatrix} 6 & -9 \\ 4 & -6 \end{vmatrix} = 0$$

Because $D = 0$, $\frac{D_x}{D}$ is undefined. Therefore, the system is dependent or inconsistent.

You Try It 4

Solve by using Cramer's Rule.

$3x - y = 4$
$6x - 2y = 5$

Your solution

Example 5

Solve by using Cramer's Rule.

$3x - y + z = 5$
$x + 2y - 2z = -3$
$2x + 3y + z = 4$

Solution

$$D = \begin{vmatrix} 3 & -1 & 1 \\ 1 & 2 & -2 \\ 2 & 3 & 1 \end{vmatrix} = 28,$$

$$D_x = \begin{vmatrix} 5 & -1 & 1 \\ -3 & 2 & -2 \\ 4 & 3 & 1 \end{vmatrix} = 28,$$

$$D_y = \begin{vmatrix} 3 & 5 & 1 \\ 1 & -3 & -2 \\ 2 & 4 & 1 \end{vmatrix} = 0,$$

$$D_z = \begin{vmatrix} 3 & -1 & 5 \\ 1 & 2 & -3 \\ 2 & 3 & 4 \end{vmatrix} = 56$$

$$x = \frac{D_x}{D} = \frac{28}{28} = 1, \ y = \frac{D_y}{D} = \frac{0}{28} = 0,$$

$$z = \frac{D_z}{D} = \frac{56}{28} = 2$$

The solution is $(1, 0, 2)$.

You Try It 5

Solve by using Cramer's Rule.

$2x - y + z = -1$
$3x + 2y - z = 3$
$x + 3y + z = -2$

Your solution

Solutions on p. S12

4.3 Exercises

Objective A

1. How do you find the value of the determinant associated with a 2×2 matrix?

2. What is the cofactor of a given element in a matrix?

Evaluate the determinant.

3. $\begin{vmatrix} 2 & -1 \\ 3 & 4 \end{vmatrix}$

4. $\begin{vmatrix} 5 & 1 \\ -1 & 2 \end{vmatrix}$

5. $\begin{vmatrix} 6 & -2 \\ -3 & 4 \end{vmatrix}$

6. $\begin{vmatrix} -3 & 5 \\ 1 & 7 \end{vmatrix}$

7. $\begin{vmatrix} 3 & 6 \\ 2 & 4 \end{vmatrix}$

8. $\begin{vmatrix} 5 & -10 \\ 1 & -2 \end{vmatrix}$

9. $\begin{vmatrix} 1 & -1 & 2 \\ 3 & 2 & 1 \\ 1 & 0 & 4 \end{vmatrix}$

10. $\begin{vmatrix} 4 & 1 & 3 \\ 2 & -2 & 1 \\ 3 & 1 & 2 \end{vmatrix}$

11. $\begin{vmatrix} 3 & -1 & 2 \\ 0 & 1 & 2 \\ 3 & 2 & -2 \end{vmatrix}$

12. $\begin{vmatrix} 4 & 5 & -2 \\ 3 & -1 & 5 \\ 2 & 1 & 4 \end{vmatrix}$

13. $\begin{vmatrix} 4 & 2 & 6 \\ -2 & 1 & 1 \\ 2 & 1 & 3 \end{vmatrix}$

14. $\begin{vmatrix} 3 & 6 & -3 \\ 4 & -1 & 6 \\ -1 & -2 & 3 \end{vmatrix}$

Objective B

Solve by using Cramer's Rule.

15. $2x - 5y = 26$
$5x + 3y = 3$

16. $3x + 7y = 15$
$2x + 5y = 11$

17. $x - 4y = 8$
$3x + 7y = 5$

18. $5x + 2y = -5$
$3x + 4y = 11$

19. $2x + 3y = 4$
$6x - 12y = -5$

20. $5x + 4y = 3$
$15x - 8y = -21$

21. $2x + 5y = 6$
$6x - 2y = 1$

22. $7x + 3y = 4$
$5x - 4y = 9$

23. $-2x + 3y = 7$
$4x - 6y = 9$

24. $9x + 6y = 7$
$3x + 2y = 4$

25. $2x - 5y = -2$
$3x - 7y = -3$

26. $8x + 7y = -3$
$2x + 2y = 5$

27. $2x - y + 3z = 9$
$x + 4y + 4z = 5$
$3x + 2y + 2z = 5$

28. $3x - 2y + z = 2$
$2x + 3y + 2z = -6$
$3x - y + z = 0$

29. $3x - y + z = 11$
$x + 4y - 2z = -12$
$2x + 2y - z = -3$

30. $x + 2y + 3z = 8$
$2x - 3y + z = 5$
$3x - 4y + 2z = 9$

31. $4x - 2y + 6z = 1$
$3x + 4y + 2z = 1$
$2x - y + 3z = 2$

32. $x - 3y + 2z = 1$
$2x + y - 2z = 3$
$3x - 9y + 6z = -3$

APPLYING THE CONCEPTS

33. Determine whether the following statements are always true, sometimes true, or never true.
a. The determinant of a matrix is a positive number.
b. A determinant can be evaluated by expanding about any row or column of the matrix.
c. Cramer's Rule can be used to solve a system of linear equations in three variables.

34. Show that $\begin{vmatrix} a & b \\ c & d \end{vmatrix} = -\begin{vmatrix} c & d \\ a & b \end{vmatrix}$.

35. Surveyors use a formula to find the area of a plot of land. The *surveyor's area formula* states that if the vertices (x_1, y_1), (x_2, y_2), . . .,(x_n, y_n) of a simple polygon are listed counterclockwise around the perimeter, then the area of the polygon is

$$A = \frac{1}{2}\left\{ \begin{vmatrix} x_1 & x_2 \\ y_1 & y_2 \end{vmatrix} + \begin{vmatrix} x_2 & x_3 \\ y_2 & y_3 \end{vmatrix} + \begin{vmatrix} x_3 & x_4 \\ y_3 & y_4 \end{vmatrix} + \cdots + \begin{vmatrix} x_n & x_1 \\ y_n & y_1 \end{vmatrix} \right\}$$

Use the surveyor's area formula to find the area of the polygon with vertices $(9, -3)$, $(26, 6)$, $(18, 21)$, $(16, 10)$, and $(1, 11)$. Measurements are given in feet.

Solve the following systems of equations by using Cramer's Rule and a calculator.

36. $3x - 7y + z = 2$
$x + y - z = -1$
$-x + 2y + 3z = 0$

37. $3x + 2y - z = 1$
$x - 4y - 5z = 2$
$3x + 4y + z = 1$

38. $x + 2y + 3z = 1$
$4x + 5y + 6z = 2$
$7x + 8y + 9z = 3$

4.4 Application Problems

Objective A **To solve rate-of-wind or rate-of-current problems**

Motion problems that involve an object moving with or against a wind or current normally require two variables to solve.

A motorboat traveling with the current can go 24 mi in 2 h. Against the current it takes 3 h to go the same distance. Find the rate of the motorboat in calm water and the rate of the current.

> **Strategy for Solving Rate-of-Wind or Rate-of-Current Problems**
>
> **1.** Choose one variable to represent the rate of the object in calm conditions and a second variable to represent the rate of the wind or current. Using these variables, express the rate of the object with and against the wind or current. Use the equation $rt = d$ to write expressions for the distance traveled by the object. The results can be recorded in a table.

Rate of the boat in calm water: x
Rate of the current: y

	Rate	·	Time	=	Distance
With the current	$x + y$	·	2	=	$2(x + y)$
Against the current	$x - y$	·	3	=	$3(x - y)$

> **2.** Determine how the expressions for distance are related.

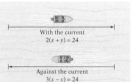

With the current
$2(x + y) = 24$

Against the current
$3(x - y) = 24$

The distance traveled with the current is 24 mi: $2(x + y) = 24$
The distance traveled against the current is 24 mi: $3(x - y) = 24$

Solve the system of equations.

$$2(x + y) = 24 \xrightarrow{\text{Multiply by } \frac{1}{2}} \frac{1}{2} \cdot 2(x + y) = \frac{1}{2} \cdot 24 \longrightarrow x + y = 12$$

$$3(x - y) = 24 \xrightarrow{\text{Multiply by } \frac{1}{3}} \frac{1}{3} \cdot 3(x - y) = \frac{1}{3} \cdot 24 \longrightarrow x - y = 8$$

$$2x = 20$$
$$x = 10$$
• Add the equations.

Replace x by 10 in the equation $x + y = 12$.
Solve for y.

$$x + y = 12$$
$$10 + y = 12$$
$$y = 2$$

The rate of the boat in calm water is 10 mph.
The rate of the current is 2 mph.

Example 1

Flying with the wind, a plane flew 1000 mi in 5 h. Flying against the wind, the plane could fly only 500 mi in the same amount of time. Find the rate of the plane in calm air and the rate of the wind.

You Try It 1

A rowing team rowing with the current traveled 18 mi in 2 h. Against the current, the team rowed 10 mi in 2 h. Find the rate of the rowing team in calm water and the rate of the current.

Strategy

• Rate of the plane in still air: p
  Rate of the wind: w

	Rate	Time	Distance
With wind	$p + w$	5	$5(p + w)$
Against wind	$p - w$	5	$5(p - w)$

• The distance traveled with the wind is 1000 mi.
  The distance traveled against the wind is 500 mi.

$$5(p + w) = 1000$$
$$5(p - w) = 500$$

Your strategy

Solution

$$5(p + w) = 1000 \qquad \frac{1}{5} \cdot 5(p + w) = \frac{1}{5} \cdot 1000$$

$$5(p - w) = 500 \qquad \frac{1}{5} \cdot 5(p - w) = \frac{1}{5} \cdot 500$$

$$p + w = 200$$
$$p - w = 100$$
$$2p = 300$$
$$p = 150$$

$$p + w = 200$$
$$150 + w = 200$$
$$w = 50$$

The rate of the plane in calm air is 150 mph.
The rate of the wind is 50 mph.

Your solution

Solution on p. S12

Objective B   **To solve application problems**

The application problems in this section are varieties of problems solved earlier in the text. Each of the strategies for the problems in this section will result in a system of equations.

A store owner purchased twenty 60-watt light bulbs and 30 fluorescent bulbs for a total cost of $80. A second purchase, at the same prices, included thirty 60-watt light bulbs and 10 fluorescent bulbs for a total cost of $50. Find the cost of a 60-watt bulb and that of a fluorescent bulb.

> **Strategy for Solving Application Problems**
>
> **1.** Choose a variable to represent each of the unknown quantities. Write numerical or variable expressions for all the remaining quantities. These results may be recorded in tables, one for each condition.

Cost of 60-watt bulb: b
Cost of fluorescent bulb: f

First Purchase

	Amount	·	Unit Cost	=	Value
60-watt	20	·	b	=	$20b$
Fluorescent	30	·	f	=	$30f$

Second Purchase

	Amount	·	Unit Cost	=	Value
60-watt	30	·	b	=	$30b$
Fluorescent	10	·	f	=	$10f$

> **2.** Determine a system of equations. The strategies presented in the chapter on First-Degree Equations and Inequalities can be used to determine the relationships among the expressions in the tables. Each table will give one equation of the system of equations.

The total of the first purchase was \$80: $20b + 30f = 80$
The total of the second purchase was \$50: $30b + 10f = 50$

Solve the system of equations: (1) $20b + 30f = 80$
(2) $30b + 10f = 50$

$$\begin{aligned} 60b + 90f &= 240 \\ -60b - 20f &= -100 \\ 70f &= 140 \\ f &= 2 \end{aligned}$$

- 3 times Equation (1).
- −2 times Equation (2).

Replace f by 2 in Equation (1) and solve for b.

$$\begin{aligned} 20b + 30f &= 80 \\ 20b + 30(2) &= 80 \\ 20b + 60 &= 80 \\ 20b &= 20 \\ b &= 1 \end{aligned}$$

The cost of a 60-watt bulb was \$1.00.
The cost of a fluorescent bulb was \$2.00.

Some application problems may require more than two variables, as shown in Example 2 and You Try It 2 on the next page.

Example 2

An investor has a total of $20,000 deposited in three different accounts, which earn annual interest rates of 9%, 7%, and 5%. The amount deposited in the 9% account is twice the amount in the 7% account. If the total annual interest earned for the three accounts is $1300, how much is invested in each account?

Strategy

• Amount invested at 9%: x
  Amount invested at 7%: y
  Amount invested at 5%: z

	Principal	Rate	Interest
Amount at 9%	x	0.09	$0.09x$
Amount at 7%	y	0.07	$0.07y$
Amount at 5%	z	0.05	$0.05z$

• The amount invested at 9% (x) is twice the amount invested at 7% (y): $x = 2y$
  The sum of the interest earned for all three accounts is $1300:
  $0.09x + 0.07y + 0.05z = 1300$
  The total amount invested is $20,000:
  $x + y + z = 20,000$

Solution

(1) $x = 2y$
(2) $0.09x + 0.07y + 0.05z = 1300$
(3) $x + y + z = 20,000$
Solve the system of equations. Substitute $2y$ for x in Equation (2) and Equation (3).
$0.09(2y) + 0.07y + 0.05z = 1300$
               $2y + y + z = 20,000$
(4) $0.25y + 0.05z = 1300$ • $0.09(2y) + 0.07y = 0.25y$
(5) $3y + z = 20,000$ • $2y + y = 3y$
Solve the system of equations in two variables by multiplying Equation (5) by -0.05 and adding to Equation (4).
  $0.25y + 0.05z = 1300$
$\underline{-0.15y - 0.05z = -1000}$
        $0.10y = 300$
           $y = 3000$
Substituting the value of y into Equation (1),
$x = 6000$.
Substituting the values of x and y into Equation (3), $z = 11,000$.
The investor placed $6000 in the 9% account, $3000 in the 7% account, and $11,000 in the 5% account.

You Try It 2

A coin bank contains only nickels, dimes, and quarters. The value of the 19 coins in the bank is $2. If there are twice as many nickels as dimes, find the number of each type of coin in the bank.

Your strategy

Your solution

Solution on p. S13

4.4 Exercises

Objective A   *Rate-of-Wind or Rate-of-Current Problems*

1. A motorboat traveling with the current went 36 mi in 2 h. Against the current, it took 3 h to travel the same distance. Find the rate of the boat in calm water and the rate of the current.

2. A cabin cruiser traveling with the current went 45 mi in 3 h. Against the current, it took 5 h to travel the same distance. Find the rate of the cabin cruiser in calm water and the rate of the current.

3. A jet plane flying with the wind went 2200 mi in 4 h. Against the wind, the plane could fly only 1820 mi in the same amount of time. Find the rate of the plane in calm air and the rate of the wind.

4. Flying with the wind, a small plane flew 300 mi in 2 h. Against the wind, the plane could fly only 270 mi in the same amount of time. Find the rate of the plane in calm air and the rate of the wind.

5. A rowing team rowing with the current traveled 20 km in 2 h. Rowing against the current, the team rowed 12 km in the same amount of time. Find the rate of the team in calm water and the rate of the current.

6. A motorboat traveling with the current went 72 km in 3 h. Against the current, the boat could go only 48 km in the same amount of time. Find the rate of the boat in calm water and the rate of the current.

7. A turboprop plane flying with the wind flew 800 mi in 4 h. Flying against the wind, the plane required 5 h to travel the same distance. Find the rate of the wind and the rate of the plane in calm air.

8. Flying with the wind, a pilot flew 600 mi between two cities in 4 h. The return trip against the wind took 5 h. Find the rate of the plane in calm air and the rate of the wind.

9. A plane flying with a tailwind flew 600 mi in 5 h. Against the wind, the plane required 6 h to fly the same distance. Find the rate of the plane in calm air and the rate of the wind.

10. Flying with the wind, a plane flew 720 mi in 3 h. Against the wind, the plane required 4 h to fly the same distance. Find the rate of the plane in calm air and the rate of the wind.

Objective B *Application Problems*

11. A coin bank contains only nickels and dimes. The total value of the coins in the bank is $2.50. If the nickels were dimes and the dimes were nickels, the total value of the coins would be $3.50. Find the number of nickels in the bank.

12. A merchant mixed 10 lb of cinnamon tea with 5 lb of spice tea. The 15-pound mixture cost $40. A second mixture included 12 lb of the cinnamon tea and 8 lb of the spice tea. The 20-pound mixture cost $54. Find the cost per pound of the cinnamon tea and the spice tea.

13. A carpenter purchased 60 ft of redwood and 80 ft of pine for a total cost of $286. A second purchase, at the same prices, included 100 ft of redwood and 60 ft of pine for a total cost of $396. Find the cost per foot of redwood and pine.

14. The total value of the quarters and dimes in a coin bank is $5.75. If the quarters were dimes and the dimes were quarters, the total value of the coins would be $6.50. Find the number of quarters in the bank.

15. A contractor buys 16 yd of nylon carpet and 20 yd of wool carpet for $1840. A second purchase, at the same prices, includes 18 yd of nylon carpet and 25 yd of wool carpet for $2200. Find the cost per yard of the wool carpet.

16. During one month, a homeowner used 500 units of electricity and 100 units of gas for a total cost of $352. The next month, 400 units of electricity and 150 units of gas were used for a total cost of $304. Find the cost per unit of gas.

17. A company manufactures both mountain bikes and trail bikes. The cost of materials for a mountain bike is $70, and the cost of materials for a trail bike is $50. The cost of labor to manufacture a mountain bike is $80 and the cost of labor to manufacture a trail bike is $40. During a week in which the company has budgeted $2500 for materials and $2600 for labor, how many mountain bikes does the company plan to manufacture?

18. A company manufactures both color and black-and-white monitors. The cost of materials for a black-and-white monitor is $40, and the cost of materials for a color monitor is $160. The cost of labor to manufacture a black-and-white monitor is $60, and the cost of labor to manufacture a color monitor is $100. During a week in which the company has budgeted $8400 for materials and $5600 for labor, how many color monitors does the company plan to manufacture?

19. A chemist has two alloys, one of which is 10% gold and 15% lead, and the other of which is 30% gold and 40% lead. How many grams of each of the two alloys should be used to make an alloy that contains 60 g of gold and 88 g of lead?

20. A pharmacist has two vitamin-supplement powders. The first powder is 20% vitamin B1 and 10% vitamin B2. The second is 15% vitamin B1 and 20% vitamin B2. How many milligrams of each of the two powders should the pharmacist use to make a mixture that contains 130 mg of vitamin B1 and 80 mg of vitamin B2?

21. On Monday, a computer manufacturing company sent out three shipments. The first order, which contained a bill for $114,000, was for 4 Model II, 6 Model VI, and 10 Model IX computers. The second shipment, which contained a bill for $72,000, was for 8 Model II, 3 Model VI, and 5 Model IX computers. The third shipment, which contained a bill for $81,000, was for 2 Model II, 9 Model VI, and 5 Model IX computers. What does the manufacturer charge for each Model VI computer?

22. A relief organization supplies blankets, cots, and lanterns to victims of fires, floods, and other natural disasters. One week the organization purchased 15 blankets, 5 cots, and 10 lanterns for a total cost of $1250. The next week, at the same prices, the organization purchased 20 blankets, 10 cots, and 15 lanterns for a total cost of $2000. The next week, at the same prices, the organization purchased 10 blankets, 15 cots, and 5 lanterns for a total cost of $1625. Find the cost of one blanket, the cost of one cot, and the cost of one lantern.

23. A science museum charges $10 for a regular admission ticket, but members receive a discount of $3 and students are admitted for $5. Last Saturday, 750 tickets were sold for a total of $5400. If 20 more student tickets than regular tickets were sold, how many of each type of ticket were sold?

24. The total value of the nickels, dimes, and quarters in a coin bank is $4. If the nickels were dimes and the dimes were nickels, the total value of the coins would be $3.75. If the quarters were dimes and the dimes were quarters, the total value would be $6.25. Find the number of nickels, dimes, and quarters in the bank.

25. An investor has a total of $25,000 deposited in three different accounts, which earn annual interest rates of 8%, 6%, and 4%. The amount deposited in the 8% account is twice the amount in the 6% account. If the three accounts earn total annual interest of $1520, how much money is deposited in each account?

APPLYING THE CONCEPTS

26. Two angles are complementary. The larger angle is 9° more than eight times the measure of the smaller angle. Find the measures of the two angles. (Complementary angles are two angles whose sum is 90°.)

27. Two angles are supplementary. The larger angle is 40° more than three times the measure of the smaller angle. Find the measures of the two angles. (Supplementary angles are two angles whose sum is 180°.)

28. The sum of the ages of a gold coin and a silver coin is 75 years. The age of the gold coin 10 years from now will be 5 years less than the age of the silver coin 10 years ago. Find the present ages of the two coins.

29. The difference between the ages of an oil painting and a watercolor is 35 years. The age of the oil painting 5 years from now will be twice the age that the watercolor was five years ago. Find the present ages of each.

30. A coin bank contains only dimes and quarters. The total value of all the coins is $1. How many of each type of coin are in the bank? (*Hint:* There is more than one answer to this problem.)

31. An investor has a total of $25,000 deposited in three different accounts, which earn annual interest rates of 8%, 6%, and 4%. The amount deposited in the 8% account is $1000 more than the amount in the 4% account. If the investor wishes to earn $1520 from the three accounts, show that there is more than one way the investor can allocate money to each of the accounts to reach the goal. Explain why this happens here while in Exercise 25 there is only one way to allocate the $25,000. *Suggestion:* Suppose the amount invested at 4% is $8000 and then find the other amounts. Now suppose $10,000 is invested at 4% and then find the other amounts.

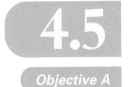

4.5 Solving Systems of Linear Inequalities

Objective A   **To graph the solution set of a system of linear inequalities**

Point of Interest

Large systems of linear inequalities containing over 100 inequalities have been used to solve application problems in such diverse areas as providing health care and hardening a nuclear missile silo.

Two or more inequalities considered together are called a **system of inequalities**. The **solution set of a system of inequalities** is the intersection of the solution sets of the individual inequalities. To graph the solution set of a system of inequalities, first graph the solution set of each inequality. The solution set of the system of inequalities is the region of the plane represented by the intersection of the shaded areas.

➡ Graph the solution set: $2x - y \leq 3$
$ 3x + 2y > 8$

Solve each inequality for y.

$$2x - y \leq 3 \qquad\qquad 3x + 2y > 8$$
$$-y \leq -2x + 3 \qquad\quad 2y > -3x + 8$$
$$y \geq 2x - 3 \qquad\qquad y > -\frac{3}{2}x + 4$$

Graph $y = 2x - 3$ as a solid line. Because the inequality is $\geq$, shade above the line.

Graph $y = -\frac{3}{2}x + 4$ as a dotted line. Because the inequality is $>$, shade above the line.

The solution set of the system is the region of the plane represented by the intersection of the solution sets of the individual inequalities.

TAKE NOTE

You can use a test point to check that the correct region has been denoted as the solution set. We can see from the graph that the point $(2, 4)$ is in the solution set and, as shown below, it is a solution of each inequality in the system. This indicates that the solution set as graphed is correct.

$$2x - y \leq 3$$
$$2(2) - (4) \leq 3$$
$$0 \leq 3 \quad \text{True}$$

$$3x + 2y > 8$$
$$3(2) + 2(4) > 8$$
$$14 > 8 \quad \text{True}$$

➡ Graph the solution set: $-x + 2y \geq 4$
$ x - 2y \geq 6$

Solve each inequality for y.

$$-x + 2y \geq 4 \qquad\qquad x - 2y \geq 6$$
$$2y \geq x + 4 \qquad\qquad -2y \geq -x + 6$$
$$y \geq \frac{1}{2}x + 2 \qquad\qquad y \leq \frac{1}{2}x - 3$$

Shade above the solid line $y = \frac{1}{2}x + 2$.

Shade below the solid line $y = \frac{1}{2}x - 3$.

Because the solution sets of the two inequalities do not intersect, the solution of the system is the empty set.

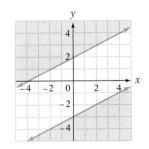

Example 1

Graph the solution set: $y \geq x - 1$
$y < -2x$

Solution

Shade above the solid line $y = x - 1$.
Shade below the dotted line $y = -2x$.

The solution of the system is the intersection of the solution sets of the individual inequalities.

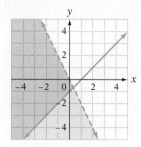

You Try It 1

Graph the solution set: $y \geq 2x - 3$
$y > -3x$

Your solution

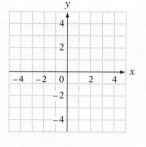

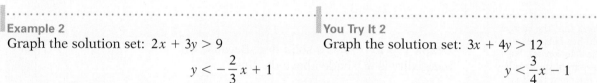

Example 2

Graph the solution set: $2x + 3y > 9$
$$y < -\frac{2}{3}x + 1$$

Solution

$2x + 3y > 9$
$3y > -2x + 9$
$$y > -\frac{2}{3}x + 3$$

Graph above the dotted line $y = -\frac{2}{3}x + 3$.

Graph below the dotted line $y = -\frac{2}{3}x + 1$.

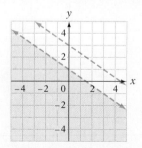

The intersection of the system is the empty set, because the solution sets of the two inequalities do not intersect.

You Try It 2

Graph the solution set: $3x + 4y > 12$
$$y < \frac{3}{4}x - 1$$

Your solution

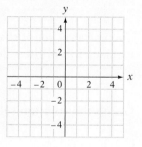

Solutions on p. S13

4.5 Exercises

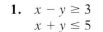

Graph the solution set.

1. $x - y \geq 3$
$x + y \leq 5$

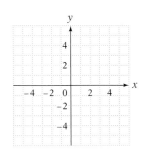

2. $2x - y < 4$
$x + y < 5$

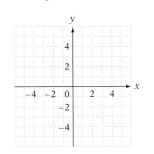

3. $3x - y < 3$
$2x + y \geq 2$

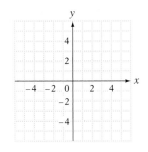

4. $x + 2y \leq 6$
$x - y \leq 3$

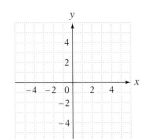

5. $2x + y \geq -2$
$6x + 3y \leq 6$

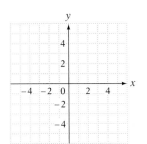

6. $x + y \geq 5$
$3x + 3y \leq 6$

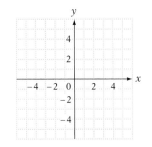

7. $3x - 2y < 6$
$y \leq 3$

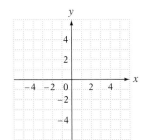

8. $x \leq 2$
$3x + 2y > 4$

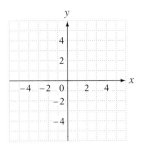

9. $y > 2x - 6$
$x + y < 0$

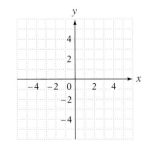

10. $x < 3$
$y < -2$

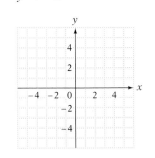

11. $x + 1 \geq 0$
$y - 3 \leq 0$

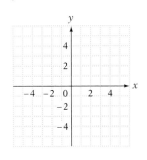

12. $5x - 2y \geq 10$
$3x + 2y \geq 6$

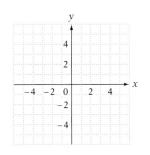

13. $2x + y \geq 4$
$3x - 2y < 6$

14. $3x - 4y < 12$
$x + 2y < 6$

15. $x - 2y \leq 6$
$2x + 3y \leq 6$

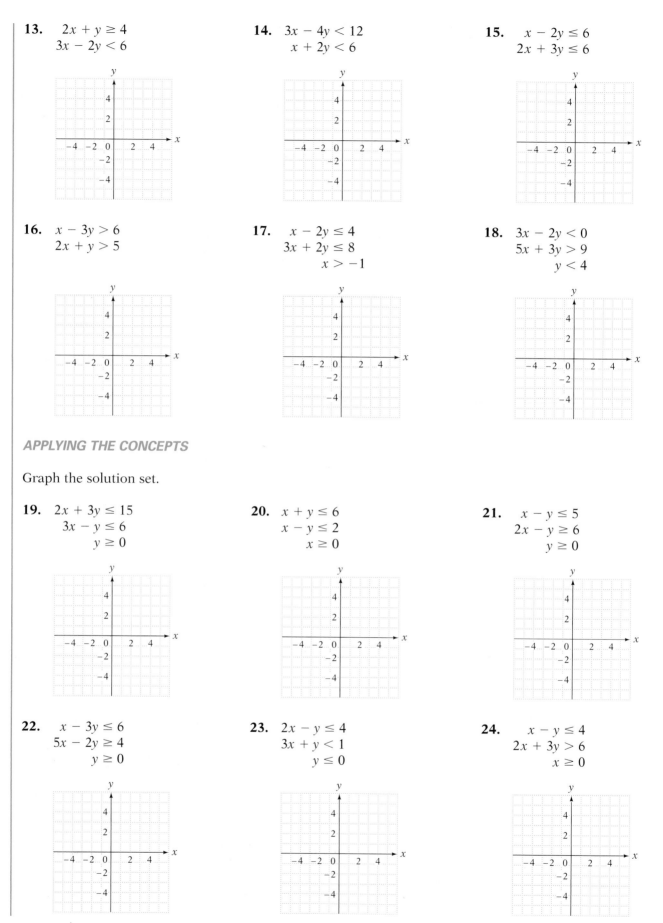

16. $x - 3y > 6$
$2x + y > 5$

17. $x - 2y \leq 4$
$3x + 2y \leq 8$
$x > -1$

18. $3x - 2y < 0$
$5x + 3y > 9$
$y < 4$

APPLYING THE CONCEPTS

Graph the solution set.

19. $2x + 3y \leq 15$
$3x - y \leq 6$
$y \geq 0$

20. $x + y \leq 6$
$x - y \leq 2$
$x \geq 0$

21. $x - y \leq 5$
$2x - y \geq 6$
$y \geq 0$

22. $x - 3y \leq 6$
$5x - 2y \geq 4$
$y \geq 0$

23. $2x - y \leq 4$
$3x + y < 1$
$y \leq 0$

24. $x - y \leq 4$
$2x + 3y > 6$
$x \geq 0$

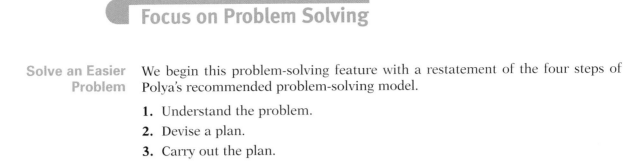

Focus on Problem Solving

Solve an Easier Problem

We begin this problem-solving feature with a restatement of the four steps of Polya's recommended problem-solving model.

1. Understand the problem.

2. Devise a plan.

3. Carry out the plan.

4. Review your solution.

One of the several methods of devising a plan is to try to solve an easier problem. Suppose you are in charge of your softball league, which consists of 15 teams. You must devise a schedule in which each team plays every other team once. How many games must be scheduled?

To solve this problem, we will attempt an easier problem first. Suppose that your league contains only a small number of teams. For instance, if there were only 1 team, you would schedule 0 games. If there were 2 teams, you would schedule 1 game. If there were 3 teams, you would schedule 3 games. The diagram at the left shows that 6 games must be scheduled when there are 4 teams in the league.

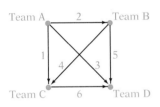

Here is a table of our results so far. (Remember that making a table is another strategy to be used in problem solving.)

Number of Teams	Number of Games	Possible Pattern
1	0	0
2	1	1
3	3	1 + 2
4	6	1 + 2 + 3

1. Draw a diagram with 5 dots to represent the teams. Draw lines from each dot to a second dot, and determine the number of games required.

2. What is the apparent pattern for the number of games required?

3. Assuming that the pattern continues, how many games must be scheduled for the 15 teams of the original problem?

After solving a problem, good problem solvers ask whether it is possible to solve the problem in a different manner. Here is a possible alternative method of solving the scheduling problem.

Begin with one of the 15 teams (say team A) and ask, "How many games must this team play?" Because there are 14 teams left to play, you must schedule 14 games. Now move to team B. It is already scheduled to play team A, and it does not play itself, so there are 13 teams left for it to play. Consequently, you must schedule 14 + 13 games.

4. Continue this reasoning for the remaining teams and determine the number of games that must be scheduled. Does this answer correspond to the answer you obtained using the first method?

5. Making connections to other problems you have solved is an important step toward becoming an excellent problem solver. How does the answer to the scheduling of teams relate to the triangular numbers discussed in the Focus on Problem Solving in the chapter titled Linear Functions and Inequalities in Two Variables?

Projects and Group Activities

Using a Graphing Calculator to Solve a System of Equations

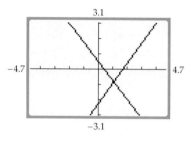

A graphing calculator can be used to solve a system of equations. For this procedure to work on most calculators, it is necessary that the point of intersection be on the screen. This means that you may have to experiment with Xmin, Xmax, Ymin, and Ymax values until the graphs intersect on the screen.

To solve a system of equations graphically, solve each equation for y. Then graph the equations of the system. Their point of intersection is the solution.

For instance, to solve the system of equations

$$4x - 3y = 7$$
$$5x + 4y = 2$$

first solve each equation for y.

$$4x - 3y = 7 \Rightarrow y = \frac{4}{3}x - \frac{7}{3}$$

$$5x + 4y = 2 \Rightarrow y = -\frac{5}{4}x + \frac{1}{2}$$

The keystrokes needed to solve this system using a TI-83 are given below. We are using a viewing window of $[-4.7, 4.7]$ by $[-3.1, 3.1]$. The approximate solution is $(1.096774, -0.870968)$.

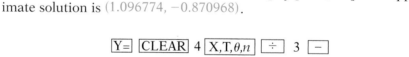

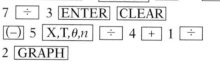

Once the calculator has drawn the graphs, use the TRACE feature and move the cursor to the approximate point of intersection. This will give you an approximate solution of the system of equations. A more accurate solution can be found by using the following keystrokes.

2nd CALC 5 ENTER ENTER ENTER

Some of the exercises in the first section of this chapter asked you to solve a system of equations by graphing. Try those exercises again, this time using your graphing calculator.

Current models of calculators do not allow you to solve graphically a system of equations in three variables. However, these calculators do have matrix and determinant operations that can be used to solve these systems.

Solving a First-Degree Equation with a Graphing Calculator

A first-degree equation in one variable can be solved graphically[1] by using a graphing calculator. The idea is to rewrite the equation as a system of equations and then solve the system of equations as illustrated on the preceding page. For instance, to solve the equation

$$2x + 1 = 5x - 5$$

write the equation as the following system of equations.

$$y = 2x + 1$$
$$y = 5x - 5$$

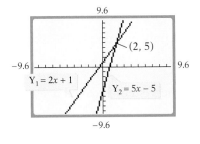

Notice that we have used the left and right sides of the original equation to form the system of equations.

Now graph the equations and find the point of intersection. This is shown at the left using a viewing window of $[-9.6, 9.6]$ by $[-9.6, 9.6]$. The solution of the original equation, $2x + 1 = 5x - 5$, is the x-coordinate of the point of intersection. Thus the solution of the equation is 2.

Recall that not all equations have a solution. Consider the equation

$$x - 3(x + 2) = 2x - 2(2x - 2)$$

The graphs of the left and right sides of the equation are shown at the left. Note that the lines appear to be parallel and therefore do not intersect. Since the lines do not intersect, there is no solution of the system of equations and thus no solution of the original equation. We algebraically verify this result below.

$$x - 3(x + 2) = 2x - 2(2x - 2)$$
$$x - 3x - 6 = 2x - 4x + 4 \qquad \bullet \text{ Use the Distributive Property.}$$
$$-2x - 6 = -2x + 4 \qquad \bullet \text{ Simplify.}$$
$$-6 = 4 \qquad \bullet \text{ Add } 2x \text{ to each side of the equation.}$$

Because $-6 = 4$ is not a true equation, there is no solution.

Note in the third line of the solution that we obtained the expressions $-2x - 6$ and $-2x + 4$. Consider the graphs of the equations $y = -2x - 6$ and $y = -2x + 4$; the slopes of the lines are equal (both are -2) and therefore the graphs of the lines are parallel.

Before we leave graphical solutions, a few words of caution. Graphs can be deceiving and appear not to intersect when they do, and appear to intersect when they do not. For instance, an attempt to graphically solve

$$2x - 10 = 1.8x + 3.6$$

is shown at the left.

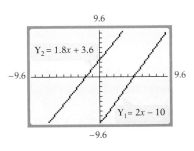

If you use the viewing window $[-9.6, 9.6]$ by $[-9.6, 9.6]$, the graphs will appear to be parallel. However, the graphs of $y = 2x - 10$ and $y = 1.8x + 3.6$ do not have the same slope and therefore must intersect. For this equation, you need a larger viewing window to see the point of intersection.

Solve the following equations by using a graphing calculator.

1. $3x - 1 = 5x + 1$

2. $3x + 2 = 4$

3. $3 + 2(2x - 4) = 5(x - 3)$

4. $2x - 4 = 5x - 3(x + 2) + 2$

5. Explain how Problem 4 relates to an *identity* as explained in Section 1 in the chapter on First-Degree Equations and Inequalities.

6. Find an appropriate viewing window so that you can determine the solution of the equation $2x - 10 = 1.8x + 3.6$ given above. What is the solution?

[1]Some graphing calculators offer nongraphical techniques for solving an equation.

Chapter Summary

Key Words

Equations considered together are called a *system of equations*. [p. 187]

A *solution of a system of equations* in two variables is an ordered pair that is a solution of each equation of the system. [p. 187]

When the graphs of a system of equations intersect at only one point, the system is called an *independent system of equations*. [p. 187]

When the graphs of a system of equations coincide, the system is called a *dependent system of equations*. [p. 188]

When a system of equations has no solution, it is called an *inconsistent system of equations*. [p. 188]

An equation of the form $Ax + By + Cz = D$, where A, B, and C are coefficients of the variables and D is a constant, is called a *linear equation in three variables*. [p. 202]

A *solution of a system of equations in three variables* is an ordered triple that is a solution of each equation of the system. [p. 204]

A *matrix* is a rectangular array of numbers. [p. 211]

A *square matrix* has the same number of rows as columns. [p. 211]

A *determinant* is a number associated with a square matrix. [p. 211]

The *minor of an element* in a 3×3 determinant is the 2×2 determinant obtained by eliminating the row and column that contain that element. [p. 211]

The *cofactor* of an element of a matrix is $(-1)^{i+j}$ times the minor of that element, where i is the row number of the element and j is its column number. [p. 212]

The evaluation of the determinant of a 3×3 or larger matrix is accomplished by *expanding by cofactors*. [p. 212]

Inequalities considered together are called a *system of inequalities*. [p. 227]

The *solution set of a system of inequalities* is the intersection of the solution sets of the individual inequalities. [p. 227]

Essential Rules

A system of equations can be solved by the graphing method, the substitution method, or the addition method. [pp. 187, 190, 199]

Cramer's Rule is a method of solving a system of equations by using determinants. [pp. 214–215]

Chapter Review

1. Solve by substitution: $2x - 6y = 15$
   $x = 4y + 8$

2. Solve by the addition method: $3x + 2y = 2$
   $x + y = 3$

3. Solve by graphing: $x + y = 3$
   $3x - 2y = -6$

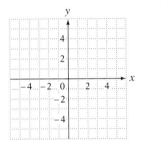

4. Solve by graphing: $2x - y = 4$
   $y = 2x - 4$

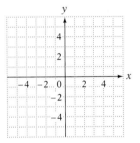

5. Solve by substitution: $3x + 12y = 18$
   $x + 4y = 6$

6. Solve by the addition method: $5x - 15y = 30$
   $x - 3y = 6$

7. Solve by the addition method.
   $3x - 4y - 2z = 17$
   $4x - 3y + 5z = 5$
   $5x - 5y + 3z = 14$

8. Solve by the addition method.
   $3x + y = 13$
   $2y + 3z = 5$
   $x + 2z = 11$

9. Evaluate the determinant.
   $$\begin{vmatrix} 6 & 1 \\ 2 & 5 \end{vmatrix}$$

10. Evaluate the determinant.
    $$\begin{vmatrix} 1 & 5 & -2 \\ -2 & 1 & 4 \\ 4 & 3 & -8 \end{vmatrix}$$

11. Solve by using Cramer's Rule:
$$2x - y = 7$$
$$3x + 2y = 7$$

12. Solve by using Cramer's Rule:
$$3x - 4y = 10$$
$$2x + 5y = 15$$

13. Solve by using Cramer's Rule:
$$x + y + z = 0$$
$$x + 2y + 3z = 5$$
$$2x + y + 2z = 3$$

14. Solve by using Cramer's Rule:
$$x + 3y + z = 6$$
$$2x + y - z = 12$$
$$x + 2y - z = 13$$

15. Graph the solution set:
$$x + 3y \le 6$$
$$2x - y \ge 4$$

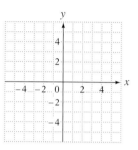

16. Graph the solution set:
$$2x + 4y \ge 8$$
$$x + y \le 3$$

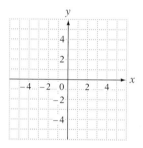

17. A cabin cruiser traveling with the current went 60 mi in 3 h. Against the current, it took 5 h to travel the same distance. Find the rate of the cabin cruiser in calm water and the rate of the current.

18. A pilot flying with the wind flew 600 mi in 3 h. Flying against the wind, the pilot required 4 h to travel the same distance. Find the rate of the plane in calm air and the rate of the wind.

19. At a movie theater, admission tickets are $5 for children and $8 for adults. The receipts for one Friday evening were $2500. The next day there were three times as many children as the preceding evening and only half the number of adults as the night before, yet the receipts were still $2500. Find the number of children who attended on Friday evening.

20. A trust administrator divides $20,000 between two accounts. One account earns an annual simple interest rate of 3%, and the second account earns an annual simple interest rate of 7%. The total annual income from the two accounts is $1200. How much is invested in each account?

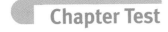

Chapter Test

1. Solve by graphing: $2x - 3y = -6$
$\qquad\qquad\qquad\quad 2x - y = 2$

2. Solve by graphing: $x - 2y = -5$
$\qquad\qquad\qquad\quad 3x + 4y = -15$

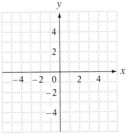

3. Graph the solution set: $2x - y < 3$
$\qquad\qquad\qquad\qquad\quad 4x + 3y < 11$

4. Graph the solution set: $x + y > 2$
$\qquad\qquad\qquad\qquad\quad 2x - y < -1$

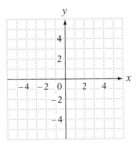

5. Solve by substitution: $3x + 2y = 4$
$\qquad\qquad\qquad\qquad\quad x = 2y - 1$

6. Solve by substitution: $5x + 2y = -23$
$\qquad\qquad\qquad\qquad\quad 2x + y = -10$

7. Solve by substitution: $y = 3x - 7$
$\qquad\qquad\qquad\qquad\quad y = -2x + 3$

8. Solve by the addition method:
$3x + 4y = -2$
$2x + 5y = 1$

9. Solve by the addition method:
$4x - 6y = 5$
$6x - 9y = 4$

10. Solve by the addition method:
$3x - y = 2x + y - 1$
$5x + 2y = y + 6$

11. Solve by the addition method:
$2x + 4y - z = 3$
$x + 2y + z = 5$
$4x + 8y - 2z = 7$

12. Solve by the addition method:
$x - y - z = 5$
$2x + z = 2$
$3y - 2z = 1$

13. Evaluate the determinant: $\begin{vmatrix} 3 & -1 \\ -2 & 4 \end{vmatrix}$

14. Evaluate the determinant: $\begin{vmatrix} 1 & -2 & 3 \\ 3 & 1 & 1 \\ 2 & -1 & -2 \end{vmatrix}$

15. Solve by using Cramer's Rule:
$$x - y = 3$$
$$2x + y = -4$$

16. Solve by using Cramer's Rule:
$$5x + 2y = 9$$
$$3x + 5y = -7$$

17. Solve by using Cramer's Rule:
$$x - y + z = 2$$
$$2x - y - z = 1$$
$$x + 2y - 3z = -4$$

18. A plane flying with the wind went 350 mi in 2 h. The return trip, flying against the wind, took 2.8 h. Find the rate of the plane in calm air and the rate of the wind.

19. A clothing manufacturer purchased 60 yd of cotton and 90 yd of wool for a total cost of $1800. Another purchase, at the same prices, included 80 yd of cotton and 20 yd of wool for a total cost of $1000. Find the cost per yard of the cotton and of the wool.

20. The annual interest earned on two investments is $549. One investment is in a 2.7% tax-free annual simple interest account, and the other investment is in a 5.1% annual simple interest CD. The total amount invested is $15,000. How much is invested in each account?

Cumulative Review

1. Solve: $\frac{3}{2}x - \frac{3}{8} + \frac{1}{4}x = \frac{7}{12}x - \frac{5}{6}$

2. Find the equation of the line that contains the points $(2, -1)$ and $(3, 4)$.

3. Simplify: $3[x - 2(5 - 2x) - 4x] + 6$

4. Evaluate $a + bc \div 2$ when $a = 4$, $b = 8$, and $c = -2$.

5. Solve: $2x - 3 < 9$ or $5x - 1 < 4$

6. Solve: $|x - 2| - 4 < 2$

7. Solve: $|2x - 3| > 5$

8. Given $f(x) = 3x^3 - 2x^2 + 1$, evaluate $f(-3)$.

9. Find the range of $f(x) = 3x^2 - 2x$ if the domain is $\{-2, -1, 0, 1, 2\}$.

10. Given $F(x) = x^2 - 3$, find $F(2)$.

11. Given $f(x) = 3x - 4$, write $f(2 + h) - f(2)$ in simplest form.

12. Graph the solution set of $\{x \mid x \le 2\} \cap \{x \mid x > -3\}$.

    ```
    ←—+—+—+—+—+—+—+—+—+—+—+—→
      -5 -4 -3 -2 -1  0  1  2  3  4  5
    ```

13. Find the equation of the line that contains the point $(-2, 3)$ and has slope $-\frac{2}{3}$.

14. Find the equation of the line that contains the point $(-1, 2)$ and is perpendicular to the line $2x - 3y = 7$.

15. Find the distance, to the nearest hundredth, between the points $(-4, 2)$ and $(2, 0)$.

16. Find the midpoint of the line segment connecting the points $(-4, 3)$ and $(3, 5)$.

17. Graph $2x - 5y = 10$ by using the slope and y-intercept.

    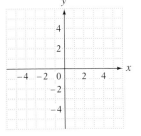

18. Graph the solution set of the inequality $3x - 4y \ge 8$.

    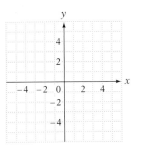

19. Solve by graphing.
$$5x - 2y = 10$$
$$3x + 2y = 6$$

20. Solve by graphing.
$$3x - 2y \geq 4$$
$$x + y < 3$$

21. Solve by the addition method.
$$3x + 2z = 1$$
$$2y - z = 1$$
$$x + 2y = 1$$

22. Evaluate the determinant.
$$\begin{vmatrix} 2 & -5 & 1 \\ 3 & 1 & 2 \\ 6 & -1 & 4 \end{vmatrix}$$

23. Solve by using Cramer's Rule.
$$4x - 3y = 17$$
$$3x - 2y = 12$$

24. Solve by substitution.
$$3x - 2y = 7$$
$$y = 2x - 1$$

25. A coin purse contains 40 coins in nickels, dimes, and quarters. There are three times as many dimes as quarters. The total value of the coins is $4.10. How many nickels are in the coin purse?

26. How many milliliters of pure water must be added to 100 ml of a 4% salt solution to make a 2.5% salt solution?

27. Flying with the wind, a small plane required 2 h to fly 150 mi. Against the wind, it took 3 h to fly the same distance. Find the rate of the wind.

28. A restaurant manager buys 100 lb of hamburger and 50 lb of steak for a total cost of $540. A second purchase, at the same prices, includes 150 lb of hamburger and 100 lb of steak. The total cost is $960. Find the price of one pound of steak.

29. Find the lower and upper limits of a 12,000-ohm resistor with a 15% tolerance.

30. The graph shows the relationship between the monthly income and the sales of an account executive. Find the slope of the line between the two points shown on the graph. Write a sentence that states the meaning of the slope.

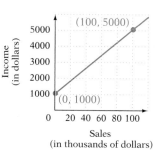

Chapter **5** **Polynomials**

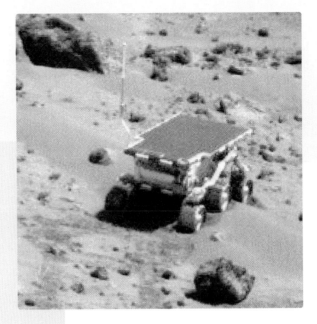

Objectives

Section 5.1

A To multiply monomials
B To divide monomials and simplify expressions with negative exponents
C To write a number using scientific notation
D To solve application problems

Section 5.2

A To evaluate polynomial functions
B To add or subtract polynomials

Section 5.3

A To multiply a polynomial by a monomial
B To multiply two polynomials
C To multiply polynomials that have special products
D To solve application problems

Section 5.4

A To divide polynomials
B To divide polynomials by using synthetic division
C To evaluate a polynomial using synthetic division

Section 5.5

A To factor a monomial from a polynomial
B To factor by grouping
C To factor a trinomial of the form $x^2 + bx + c$
D To factor $ax^2 + bx + c$

Section 5.6

A To factor the difference of two perfect squares or a perfect-square trinomial
B To factor the sum or the difference of two perfect cubes
C To factor a trinomial that is quadratic in form
D To factor completely

Section 5.7

A To solve an equation by factoring
B To solve application problems

The Pathfinder mission produced the closest look to date at the surface of Mars. This photo of the Pathfinder Sojourner rover was taken at Mermaid Dune on Mars on August 8, 1997. Images such as this, transmitted from Mars to Earth, traveled a distance of over 100 million miles. Distances this great, as well as very small measurements, are generally expressed in scientific notation, as illustrated in **Exercises 117–125 on page 254**.

 Need help? For on-line student resources, such as section quizzes, visit this textbook's web site at **math.college.hmco.com/students**.

For Exercises 1 to 5, simplify.

1. $-4(3y)$

2. $(-2)^3$

3. $-4a - 8b + 7a$

4. $3x - 2[y - 4(x + 1) + 5]$

5. $-(x - y)$

6. Write 40 as a product of prime numbers.

7. Find the GCF of 16, 20, and 24.

8. Evaluate $x^3 - 2x^2 + x + 5$ for $x = -2$.

9. Solve: $3x + 1 = 0$

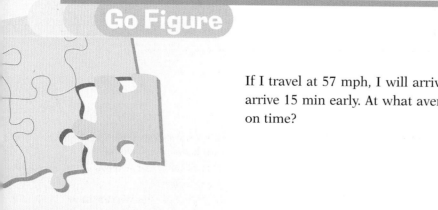

If I travel at 57 mph, I will arrive 20 min late. If I travel at 64 mph, I will arrive 15 min early. At what average speed should I travel to arrive exactly on time?

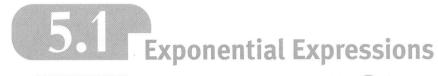

5.1 Exponential Expressions

Objective A   **To multiply monomials**   « 5 »

A **monomial** is a number, a variable, or a product of a number and variables.

The examples at the right are monomials. The **degree of a monomial** is the sum of the exponents of the variables.

x	degree 1 $(x = x^1)$
$3x^2$	degree 2
$4x^2y$	degree 3
$6x^3y^4z^2$	degree 9

In this chapter, the variable n is considered a positive integer when used as an exponent.

x^n degree n

The degree of a nonzero constant term is zero.

6   degree 0

The expression $5\sqrt{x}$ is not a monomial because $\sqrt{x}$ cannot be written as a product of variables. The expression $\dfrac{x}{y}$ is not a monomial because it is a quotient of variables.

The expression x^4 is an exponential expression. The exponent, 4, indicates the number of times the base, x, occurs as a factor.

The product of exponential expressions with the *same* base can be simplified by writing each expression in factored form and writing the result with an exponent.

$$x^3 \cdot x^4 = \underbrace{(x \cdot x \cdot x)}_{\text{3 factors}} \cdot \underbrace{(x \cdot x \cdot x \cdot x)}_{\text{4 factors}}$$
$$\underbrace{\hphantom{x^3 \cdot x^4 = (x \cdot x \cdot x) \cdot (x \cdot x \cdot x \cdot x)}}_{\text{7 factors}}$$
$$= x^7$$

$$x^3 \cdot x^4 = x^{3+4} = x^7$$

Note that adding the exponents results in the same product.

Rule for Multiplying Exponential Expressions

If m and n are positive integers, then $x^m \cdot x^n = x^{m+n}$.

⇨ Simplify: $(-4x^5y^3)(3xy^2)$

$(-4x^5y^3)(3xy^2) = (-4 \cdot 3)(x^5 \cdot x)(y^3 \cdot y^2)$

- Use the Commutative and Associative Properties of Multiplication to rearrange and group factors.

$= -12(x^{5+1})(y^{3+2})$

- Multiply variables with the same base by adding their exponents.

$= -12x^6y^5$

- Simplify.

As shown below, the power of a monomial can be simplified by writing the power in factored form and then using the Rule for Multiplying Exponential Expressions. It can also be simplified by multiplying each exponent inside the parentheses by the exponent outside the parentheses.

$$(a^2)^3 = a^2 \cdot a^2 \cdot a^2$$
$$= a^{2+2+2}$$
$$= a^6$$

$$(x^3y^4)^2 = (x^3y^4)(x^3y^4)$$
$$= x^{3+3}y^{4+4}$$
$$= x^6y^8$$

- Write in factored form. Then use the Rule for Multiplying Exponential Expressions.

$$(a^2)^3 = a^{2\cdot3} = a^6$$

$$(x^3y^4)^2 = x^{3\cdot2}y^{4\cdot2} = x^6y^8$$

- Multiply each exponent inside the parentheses by the exponent outside the parentheses.

Rule for Simplifying the Power of an Exponential Expression

If m and n are positive integers, then $(x^m)^n = x^{mn}$.

Rule for Simplifying Powers of Products

If m, n, and p are positive integers, then $(x^m y^n)^p = x^{mp} y^{np}$.

➡ Simplify: $(x^4)^5$

$$(x^4)^5 = x^{4\cdot5}$$
$$= x^{20}$$

- Use the Rule for Simplifying Powers of Exponential Expressions to multiply the exponents.

➡ Simplify: $(2a^3b^4)^3$

$$(2a^3b^4)^3 = 2^{1\cdot3}a^{3\cdot3}b^{4\cdot3}$$
$$= 2^3a^9b^{12}$$
$$= 8a^9b^{12}$$

- Use the Rule for Simplifying Powers of Products to multiply each exponent inside the parentheses by the exponent outside the parentheses.

Example 1   Simplify: $(2xy^2)(-3xy^4)^3$

Solution
$$(2xy^2)(-3xy^4)^3 = (2xy^2)[(-3)^3x^3y^{12}]$$
$$= (2xy^2)(-27x^3y^{12})$$
$$= -54x^4y^{14}$$

You Try It 1   Simplify: $(-3a^2b^4)(-2ab^3)^4$

Your solution

Example 2   Simplify: $(x^{n+2})^5$

Solution $(x^{n+2})^5 = x^{5n+10}$

You Try It 2   Simplify: $(y^{n-3})^2$

Your solution

Example 3   Simplify: $[(2xy^2)^2]^3$

Solution
$$[(2xy^2)^2]^3 = [2^2x^2y^4]^3 = [4x^2y^4]^3$$
$$= 4^3x^6y^{12} = 64x^6y^{12}$$

You Try It 3   Simplify: $[(ab^3)^3]^4$

Your solution

Solutions on p. S13

Objective B

To divide monomials and simplify expressions with negative exponents

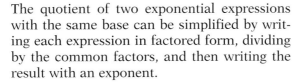

The quotient of two exponential expressions with the same base can be simplified by writing each expression in factored form, dividing by the common factors, and then writing the result with an exponent.

$$\frac{x^5}{x^2} = \frac{\overset{1}{\cancel{x}} \cdot \overset{1}{\cancel{x}} \cdot x \cdot x \cdot x}{\underset{1}{\cancel{x}} \cdot \underset{1}{\cancel{x}}} = x^3$$

Note that subtracting the exponents gives the same result.

$$\frac{x^5}{x^2} = x^{5-2} = x^3$$

To divide two monomials with the same base, subtract the exponents of the like bases.

➡ Simplify: $\dfrac{z^8}{z^2}$

$\dfrac{z^8}{z^2} = z^{8-2}$ • The bases are the same. Subtract the exponents.

$\quad = z^6$

➡ Simplify: $\dfrac{a^5 b^9}{a^4 b}$

$\dfrac{a^5 b^9}{a^4 b} = a^{5-4} b^{9-1}$ • Subtract the exponents of the like bases.

$\quad = ab^8$

Consider the expression $\dfrac{x^4}{x^4}$, $x \neq 0$. This expression can be simplified, as shown below, by subtracting exponents or by dividing by common factors.

$$\frac{x^4}{x^4} = x^{4-4} = x^0 \qquad\qquad \frac{x^4}{x^4} = \frac{\overset{1}{\cancel{x}} \cdot \overset{1}{\cancel{x}} \cdot \overset{1}{\cancel{x}} \cdot \overset{1}{\cancel{x}}}{\underset{1}{\cancel{x}} \cdot \underset{1}{\cancel{x}} \cdot \underset{1}{\cancel{x}} \cdot \underset{1}{\cancel{x}}} = 1$$

The equations $\dfrac{x^4}{x^4} = x^0$ and $\dfrac{x^4}{x^4} = 1$ suggest the following definition of x^0.

TAKE NOTE

In the example at the right, we indicated that $z \neq 0$. If we try to evaluate $(16z^5)^0$ when $z = 0$, we have $[16(0)^5]^0 = [16(0)]^0 = 0^0$. However, 0^0 is not defined and so we must assume that $z \neq 0$. To avoid stating this for every example or exercise, we will assume that variables cannot take on values that result in the expression 0^0.

Definition of Zero as an Exponent
If $x \neq 0$, then $x^0 = 1$. The expression 0^0 is not defined.

➡ Simplify: $(16z^5)^0$, $z \neq 0$

$(16z^5)^0 = 1$ • Any nonzero expression to the zero power is 1.

➡ Simplify: $-(7x^4 y^3)^0$

$-(7x^4 y^3)^0 = -(1) = -1$ • The negative outside the parentheses is not affected by the exponent.

Point of Interest

In the 15th century, the expression 12^{2m} was used to mean $12x^{-2}$. The use of $\overline{m}$ reflects an Italian influence, where m was used for minus and p was used for plus. It was understood that $2\overline{m}$ referred to an unnamed variable. Isaac Newton, in the 17th century, advocated the use of a negative exponent, the symbol we use today.

Consider the expression $\dfrac{x^4}{x^6}$, $x \neq 0$. This expression can be simplified, as shown below, by subtracting exponents or by dividing by common factors.

$$\dfrac{x^4}{x^6} = x^{4-6} = x^{-2} \qquad\qquad \dfrac{x^4}{x^6} = \dfrac{\overset{1}{\cancel{x}} \cdot \overset{1}{\cancel{x}} \cdot \overset{1}{\cancel{x}} \cdot \overset{1}{\cancel{x}}}{\underset{1}{\cancel{x}} \cdot \underset{1}{\cancel{x}} \cdot \underset{1}{\cancel{x}} \cdot \underset{1}{\cancel{x}} \cdot x \cdot x} = \dfrac{1}{x^2}$$

The equations $\dfrac{x^4}{x^6} = x^{-2}$ and $\dfrac{x^4}{x^6} = \dfrac{1}{x^2}$ suggest that $x^{-2} = \dfrac{1}{x^2}$.

Definition of a Negative Exponent

If $x \neq 0$ and n is a positive integer, then

$$x^{-n} = \dfrac{1}{x^n} \quad \text{and} \quad \dfrac{1}{x^{-n}} = x^n$$

TAKE NOTE

Note from the example at the right that 2^{-4} is a *positive* number. A negative exponent does not change the sign of a number.

➡ Evaluate: 2^{-4}

$2^{-4} = \dfrac{1}{2^4}$ • Use the Definition of a Negative Exponent.

$\qquad = \dfrac{1}{16}$ • Evaluate the expression.

The expression $\left(\dfrac{x^3}{y^4}\right)^2$, $y \neq 0$, can be simplified by squaring $\dfrac{x^3}{y^4}$ or by multiplying each exponent in the quotient by the exponent outside the parentheses.

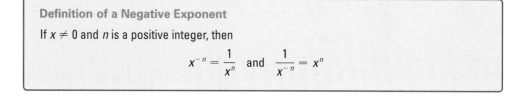

Rule for Simplifying Powers of Quotients

If m, n, and p are integers and $y \neq 0$, then $\left(\dfrac{x^m}{y^n}\right)^p = \dfrac{x^{mp}}{y^{np}}$.

➡ Simplify: $\left(\dfrac{a^2}{b^3}\right)^{-2}$

$\left(\dfrac{a^2}{b^3}\right)^{-2} = \dfrac{a^{2(-2)}}{b^{3(-2)}}$ • Use the Rule for Simplifying Powers of Quotients.

$\qquad = \dfrac{a^{-4}}{b^{-6}} = \dfrac{b^6}{a^4}$ • Use the Definition of a Negative Exponent to write the expression with positive exponents.

The preceding example suggests the following rule.

> **Rule for Negative Exponents on Fractional Expressions**
>
> If $a \neq 0$, $b \neq 0$, and n is a positive integer, then $\left(\dfrac{a}{b}\right)^{-n} = \left(\dfrac{b}{a}\right)^{n}$.

An exponential expression is in simplest form when it is written with only positive exponents.

➡ Simplify: $7d^{-5}$, $d \neq 0$

$$7d^{-5} = 7 \cdot \frac{1}{d^5} = \frac{7}{d^5}$$

- Use the Definition of a Negative Exponent to rewrite the expression with a positive exponent.

➡ Simplify: $\dfrac{2}{5a^{-4}}$

$$\frac{2}{5a^{-4}} = \frac{2}{5} \cdot \frac{1}{a^{-4}} = \frac{2}{5} \cdot a^4 = \frac{2a^4}{5}$$

- Use the Definition of a Negative Exponent to rewrite the expression with a positive exponent.

Now that zero and negative exponents have been defined, a rule for dividing exponential expressions can be stated.

> **Rule for Dividing Exponential Expressions**
>
> If m and n are integers and $x \neq 0$, then $\dfrac{x^m}{x^n} = x^{m-n}$.

➡ Simplify: $\dfrac{x^4}{x^9}$

$$\frac{x^4}{x^9} = x^{4-9}$$
$$= x^{-5}$$
$$= \frac{1}{x^5}$$

- Use the Rule for Dividing Exponential Expressions.

- Subtract the exponents.

- Use the Definition of a Negative Exponent to rewrite the expression with a positive exponent.

The rules for simplifying exponential expressions and powers of exponential expressions are true for all integers. These rules are restated here for convenience.

> **Rules of Exponents**
>
> If m, n, and p are integers, then
>
> $x^m \cdot x^n = x^{m+n}$ $(x^m)^n = x^{mn}$ $(x^m y^n)^p = x^{mp} y^{np}$
>
> $\dfrac{x^m}{x^n} = x^{m-n}, x \neq 0$ $\left(\dfrac{x^m}{y^n}\right)^p = \dfrac{x^{mp}}{y^{np}}, y \neq 0$ $x^{-n} = \dfrac{1}{x^n}, x \neq 0$
>
> $x^0 = 1, x \neq 0$

➡ Simplify: $(3ab^{-4})(-2a^{-3}b^7)$

$$(3ab^{-4})(-2a^{-3}b^7) = [3 \cdot (-2)](a^{1+(-3)}b^{-4+7})$$
$$= -6a^{-2}b^3$$
$$= -\frac{6b^3}{a^2}$$

- When multiplying expressions, add the exponents on like bases.

➡ Simplify: $\dfrac{4a^{-2}b^5}{6a^5b^2}$

$$\frac{4a^{-2}b^5}{6a^5b^2} = \frac{\cancel{2} \cdot 2a^{-2}b^5}{\cancel{2} \cdot 3a^5b^2} = \frac{2a^{-2}b^5}{3a^5b^2}$$
$$= \frac{2a^{-2-5}b^{5-2}}{3}$$
$$= \frac{2a^{-7}b^3}{3} = \frac{2b^3}{3a^7}$$

- Divide the coefficients by their common factor.

- Use the Rule for Dividing Exponential Expressions.

- Use the Definition of a Negative Exponent to rewrite the expression with a positive exponent.

➡ Simplify: $\left[\dfrac{6m^2n^3}{8m^7n^2}\right]^{-3}$

$$\left[\frac{6m^2n^3}{8m^7n^2}\right]^{-3} = \left[\frac{3m^{2-7}n^{3-2}}{4}\right]^{-3}$$
$$= \left[\frac{3m^{-5}n}{4}\right]^{-3}$$
$$= \frac{3^{-3}m^{15}n^{-3}}{4^{-3}}$$
$$= \frac{4^3m^{15}}{3^3n^3} = \frac{64m^{15}}{27n^3}$$

- Simplify inside the brackets.

- Subtract the exponents.

- Use the Rule for Simplifying Powers of Quotients.

- Use the Definition of a Negative Exponent to rewrite the expression with positive exponents. Then simplify.

Example 4

Simplify: $\dfrac{-28x^6z^{-3}}{42x^{-1}z^4}$

Solution

$$\frac{-28x^6z^{-3}}{42x^{-1}z^4} = -\frac{14 \cdot 2x^{6-(-1)}z^{-3-4}}{14 \cdot 3}$$
$$= -\frac{2x^7z^{-7}}{3} = -\frac{2x^7}{3z^7}$$

You Try It 4

Simplify: $\dfrac{20r^{-2}t^{-5}}{-16r^{-3}s^{-2}}$

Your solution

Example 5

Simplify: $\dfrac{(3a^{-1}b^4)^{-3}}{(6^{-1}a^{-3}b^{-4})^3}$

Solution

$$\frac{(3a^{-1}b^4)^{-3}}{(6^{-1}a^{-3}b^{-4})^3} = \frac{3^{-3}a^3b^{-12}}{6^{-3}a^{-9}b^{-12}} = 3^{-3} \cdot 6^3a^{12}b^0$$
$$= \frac{6^3a^{12}}{3^3} = \frac{216a^{12}}{27} = 8a^{12}$$

You Try It 5

Simplify: $\dfrac{(9u^{-6}v^4)^{-1}}{(6u^{-3}v^{-2})^{-2}}$

Your solution

Solutions on p. S13

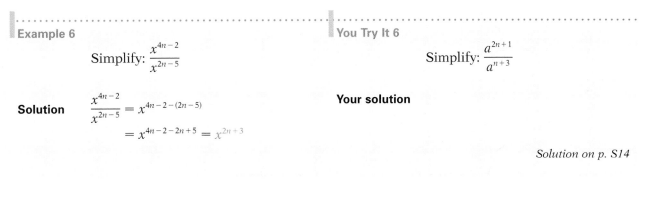

Example 6

Simplify: $\dfrac{x^{4n-2}}{x^{2n-5}}$

Solution

$$\dfrac{x^{4n-2}}{x^{2n-5}} = x^{4n-2-(2n-5)}$$
$$= x^{4n-2-2n+5} = x^{2n+3}$$

You Try It 6

Simplify: $\dfrac{a^{2n+1}}{a^{n+3}}$

Your solution

Solution on p. S14

Objective C   **To write a number using scientific notation**

Point of Interest

Astronomers measure the distance of some stars by using the parsec. One parsec is approximately 1.91×10^{13} mi.

TAKE NOTE

There are two steps involved in writing a number in scientific notation: (1) determine the number between 1 and 10, and (2) determine the exponent on 10.

Integer exponents are used to represent the very large and very small numbers encountered in the fields of science and engineering. For example, the mass of the electron is 0.00000000000000000000000000009 g. Numbers such as this are difficult to read and write, so a more convenient system for writing such numbers has been developed. It is called **scientific notation.**

To express a number in scientific notation, write the number as the product of a number between 1 and 10 and a power of 10. The form for scientific notation is **$a \times 10^n$**, where $1 \le a < 10$.

For numbers greater than 10, move the decimal point to the right of the first digit. The exponent n is positive and equal to the number of places the decimal point has been moved.

$$965{,}000 = 9.65 \times 10^5$$
$$3{,}600{,}000 = 3.6 \times 10^6$$
$$92{,}000{,}000{,}000 = 9.2 \times 10^{10}$$

For numbers less than 1, move the decimal point to the right of the first nonzero digit. The exponent n is negative. The absolute value of the exponent is equal to the number of places the decimal point has been moved.

$$0.0002 = 2 \times 10^{-4}$$
$$0.0000000974 = 9.74 \times 10^{-8}$$
$$0.000000000086 = 8.6 \times 10^{-11}$$

Converting a number written in scientific notation to decimal notation requires moving the decimal point.

When the exponent is positive, move the decimal point to the right the same number of places as the exponent.

$$1.32 \times 10^4 = 13{,}200$$
$$1.4 \times 10^8 = 140{,}000{,}000$$

When the exponent is negative, move the decimal point to the left the same number of places as the absolute value of the exponent.

$$1.32 \times 10^{-2} = 0.0132$$
$$1.4 \times 10^{-4} = 0.00014$$

Numerical calculations involving numbers that have more digits than a hand-held calculator is able to handle can be performed using scientific notation.

⇒ Simplify: $\dfrac{220{,}000 \times 0.000000092}{0.0000011}$

$$\dfrac{220{,}000 \times 0.000000092}{0.0000011} = \dfrac{2.2 \times 10^5 \times 9.2 \times 10^{-8}}{1.1 \times 10^{-6}}$$

• Write the numbers in scientific notation.

$$= \dfrac{(2.2)(9.2) \times 10^{5+(-8)-(-6)}}{1.1}$$

• Simplify.

$$= 18.4 \times 10^3 = 18{,}400$$

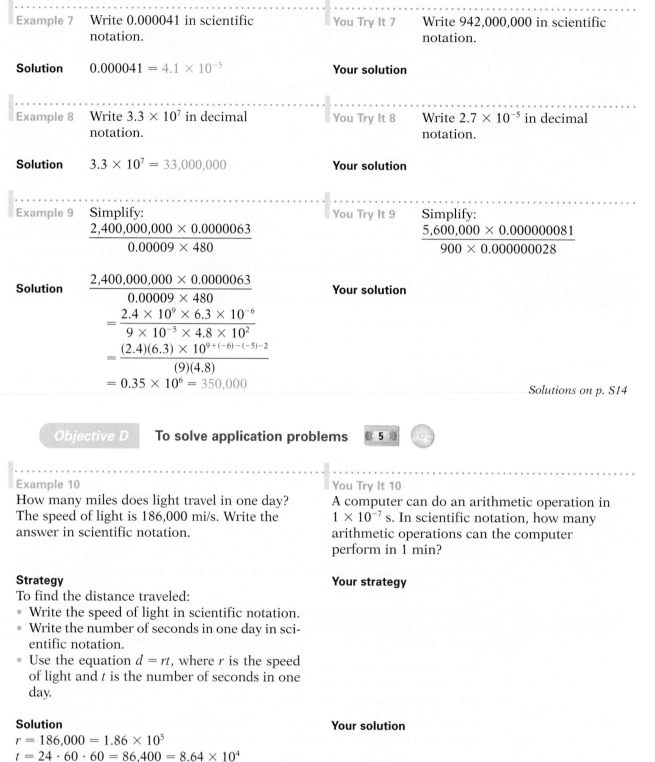

Example 7 Write 0.000041 in scientific notation.

Solution $0.000041 = 4.1 \times 10^{-5}$

You Try It 7 Write 942,000,000 in scientific notation.

Your solution

Example 8 Write 3.3×10^7 in decimal notation.

Solution $3.3 \times 10^7 = 33,000,000$

You Try It 8 Write 2.7×10^{-5} in decimal notation.

Your solution

Example 9 Simplify:
$$\frac{2,400,000,000 \times 0.0000063}{0.00009 \times 480}$$

Solution
$$\frac{2,400,000,000 \times 0.0000063}{0.00009 \times 480}$$
$$= \frac{2.4 \times 10^9 \times 6.3 \times 10^{-6}}{9 \times 10^{-5} \times 4.8 \times 10^2}$$
$$= \frac{(2.4)(6.3) \times 10^{9+(-6)-(-5)-2}}{(9)(4.8)}$$
$$= 0.35 \times 10^6 = 350,000$$

You Try It 9 Simplify:
$$\frac{5,600,000 \times 0.000000081}{900 \times 0.000000028}$$

Your solution

Solutions on p. S14

Objective D **To solve application problems** ▸ 5 ▸

Example 10

How many miles does light travel in one day? The speed of light is 186,000 mi/s. Write the answer in scientific notation.

You Try It 10

A computer can do an arithmetic operation in 1×10^{-7} s. In scientific notation, how many arithmetic operations can the computer perform in 1 min?

Strategy
To find the distance traveled:
• Write the speed of light in scientific notation.
• Write the number of seconds in one day in scientific notation.
• Use the equation $d = rt$, where r is the speed of light and t is the number of seconds in one day.

Your strategy

Solution
$r = 186,000 = 1.86 \times 10^5$
$t = 24 \cdot 60 \cdot 60 = 86,400 = 8.64 \times 10^4$
$d = rt$
$d = (1.86 \times 10^5)(8.64 \times 10^4)$
$= 1.86 \times 8.64 \times 10^9$
$= 16.0704 \times 10^9$
$= 1.60704 \times 10^{10}$

Light travels 1.60704×10^{10} mi in one day.

Your solution

Solution on p. S14

5.1 Exercises

Objective A

Simplify.

1. $(ab^3)(a^3b)$

2. $(-2ab^4)(-3a^2b^4)$

3. $(9xy^2)(-2x^2y^2)$

4. $(x^2y)^2$

5. $(x^2y^4)^4$

6. $(-2ab^2)^3$

7. $(-3x^2y^3)^4$

8. $(2^2a^2b^3)^3$

9. $(3^3a^5b^3)^2$

10. $(xy)(x^2y)^4$

11. $(x^2y^2)(xy^3)^3$

12. $[(2x)^4]^2$

13. $[(3x)^3]^2$

14. $[(x^2y)^4]^5$

15. $[(ab)^3]^6$

16. $[(2ab)^3]^2$

17. $[(2xy)^3]^4$

18. $[(3x^2y^3)^2]^2$

19. $[(2a^4b^3)^3]^2$

20. $y^n \cdot y^{2n}$

21. $x^n \cdot x^{n+1}$

22. $y^{2n} \cdot y^{4n+1}$

23. $y^{3n} \cdot y^{3n-2}$

24. $(a^n)^{2n}$

25. $(a^{n-3})^{2n}$

26. $(y^{2n-1})^3$

27. $(x^{3n+2})^5$

28. $(b^{2n-1})^n$

29. $(2xy)(-3x^2yz)(x^2y^3z^3)$

30. $(x^2z^4)(2xyz^4)(-3x^3y^2)$

31. $(3b^5)(2ab^2)(-2ab^2c^2)$

32. $(-c^3)(-2a^2bc)(3a^2b)$

33. $(-2x^2y^3z)(3x^2yz^4)$

34. $(2a^2b)^3(-3ab^4)^2$

35. $(-3ab^3)^3(-2^2a^2b)^2$

36. $(4ab)^2(-2ab^2c^3)^3$

37. $(-2ab^2)(-3a^4b^5)^3$

Objective B

Simplify.

38. 2^{-3}

39. $\dfrac{1}{3^{-5}}$

40. $\dfrac{1}{x^{-4}}$

41. $\dfrac{1}{y^{-3}}$

42. $\dfrac{2x^{-2}}{y^4}$ **43.** $\dfrac{a^3}{4b^{-2}}$ **44.** $x^{-3}y$ **45.** xy^{-4}

46. $-5x^0$ **47.** $\dfrac{1}{2x^0}$ **48.** $\dfrac{(2x)^0}{-2^3}$ **49.** $\dfrac{-3^{-2}}{(2y)^0}$

50. $\dfrac{y^{-7}}{y^{-8}}$ **51.** $\dfrac{y^{-2}}{y^6}$ **52.** $(x^2y^{-4})^2$ **53.** $(x^3y^5)^{-2}$

54. $\dfrac{x^{-2}y^{-11}}{xy^{-2}}$ **55.** $\dfrac{x^4y^3}{x^{-1}y^{-2}}$ **56.** $\dfrac{a^{-1}b^{-3}}{a^4b^{-5}}$ **57.** $\dfrac{a^6b^{-4}}{a^{-2}b^5}$

58. $(2a^{-1})^{-2}(2a^{-1})^4$ **59.** $(3a)^{-3}(9a^{-1})^{-2}$ **60.** $(x^{-2}y)^2(xy)^{-2}$ **61.** $(x^{-1}y^2)^{-3}(x^2y^{-4})^{-3}$

62. $\dfrac{50b^{10}}{70b^5}$ **63.** $\dfrac{x^3y^6}{x^6y^2}$ **64.** $\dfrac{x^{17}y^5}{-x^7y^{10}}$ **65.** $\dfrac{-6x^2y}{12x^4y}$

66. $\dfrac{2x^2y^4}{(3xy^2)^3}$ **67.** $\dfrac{-3ab^2}{(9a^2b^4)^3}$ **68.** $\left(\dfrac{-12a^2b^3}{9a^5b^9}\right)^3$ **69.** $\left(\dfrac{12x^3y^2z}{18xy^3z^4}\right)^4$

70. $\dfrac{(4x^2y)^2}{(2xy^3)^3}$ **71.** $\dfrac{(3a^2b)^3}{(-6ab^3)^2}$ **72.** $\dfrac{(-4x^2y^3)^2}{(2xy^2)^3}$ **73.** $\dfrac{(-3a^2b^3)^2}{(-2ab^4)^3}$

74. $\dfrac{(-4xy^3)^3}{(-2x^7y)^4}$ **75.** $\dfrac{(-8x^2y^2)^4}{(16x^3y^7)^2}$ **76.** $\dfrac{a^{5n}}{a^{3n}}$ **77.** $\dfrac{b^{6n}}{b^{10n}}$

78. $\dfrac{-x^{5n}}{x^{2n}}$ **79.** $\dfrac{y^{2n}}{-y^{8n}}$ **80.** $\dfrac{x^{2n-1}}{x^{n-3}}$ **81.** $\dfrac{y^{3n+2}}{y^{2n+4}}$

82. $\dfrac{a^{3n}b^n}{a^nb^{2n}}$ **83.** $\dfrac{x^ny^{3n}}{x^ny^{5n}}$ **84.** $\dfrac{a^{3n-2}b^{n+1}}{a^{2n+1}b^{2n+2}}$ **85.** $\dfrac{x^{2n-1}y^{n-3}}{x^{n+4}y^{n+3}}$

86. $\left(\dfrac{4^{-2}xy^{-3}}{x^{-3}y}\right)^3\left(\dfrac{8^{-1}x^{-2}y}{x^4y^{-1}}\right)^{-2}$ **87.** $\left(\dfrac{9ab^{-2}}{8a^{-2}b}\right)^{-2}\left(\dfrac{3a^{-2}b}{2a^2b^{-2}}\right)^3$ **88.** $\left(\dfrac{2ab^{-1}}{ab}\right)^{-1}\left(\dfrac{3a^{-2}b}{a^2b^2}\right)^{-2}$

Objective C

Write in scientific notation.

89. 0.00000467

90. 0.00000005

91. 0.00000000017

92. 4,300,000

93. 200,000,000,000

94. 9,800,000,000

Write in decimal notation.

95. 1.23×10^{-7}

96. 6.2×10^{-12}

97. 8.2×10^{15}

98. 6.34×10^{5}

99. 3.9×10^{-2}

100. 4.35×10^{9}

Simplify. Write the answer in decimal notation.

101. $(3 \times 10^{-12})(5 \times 10^{16})$

102. $(8.9 \times 10^{-5})(3.2 \times 10^{-6})$

103. $(0.0000065)(3,200,000,000,000)$

104. $(480,000)(0.0000000096)$

105. $\dfrac{9 \times 10^{-3}}{6 \times 10^{5}}$

106. $\dfrac{2.7 \times 10^{4}}{3 \times 10^{-6}}$

107. $\dfrac{0.0089}{500,000,000}$

108. $\dfrac{4800}{0.00000024}$

109. $\dfrac{0.00056}{0.000000000004}$

110. $\dfrac{0.000000346}{0.0000005}$

111. $\dfrac{(3.2 \times 10^{-11})(2.9 \times 10^{15})}{8.1 \times 10^{-3}}$

112. $\dfrac{(6.9 \times 10^{27})(8.2 \times 10^{-13})}{4.1 \times 10^{15}}$

113. $\dfrac{(0.00000004)(84,000)}{(0.0003)(1,400,000)}$

114. $\dfrac{(720)(0.0000000039)}{(26,000,000,000)(0.018)}$

Objective D   *Application Problems*

Solve. Write the answer in scientific notation.

115. How many kilometers does light travel in one day? The speed of light is 300,000 km/s.

116. How many meters does light travel in 8 h? The speed of light is 300,000,000 m/s.

117. It took 11 min for the commands from a computer on Earth to travel to the rover Sojourner on Mars, a distance of 119 million miles. How fast did the signals from Earth to Mars travel? Write the answer in scientific notation.

118. In 2000, the national debt was 6.7×10^{12} dollars. How much would each American citizen have to pay in order to pay off the debt? Use 280,000,000 as the number of citizens.

119. A high-speed centrifuge makes 9×10^7 revolutions each minute. Find the time in seconds for the centrifuge to make one revolution.

120. How long does it take light to travel to Earth from the sun? The sun is 9.3×10^7 mi from Earth, and light travels 1.86×10^5 mi/s.

121. The mass of Earth is 5.9×10^{27} g. The mass of the sun is 2×10^{33} g. How many times heavier is the sun than Earth?

122. One astronomical unit (A.U.) is 9.3×10^7 mi. The star Pollux in the constellation Gemini is 1.8228×10^{12} mi from Earth. Find the distance from Pollux to Earth in astronomical units.

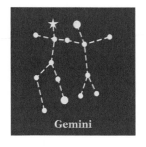

Gemini

123. The weight of 31 million orchid seeds is one ounce. Find the weight of one orchid seed.

124. One light year, an astronomical unit of distance, is the distance that light will travel in one year. Light travels 1.86×10^5 mi/s. Find the measure of one light year in miles. Use a 365-day year.

125. The Coma cluster of galaxies is approximately 2.8×10^8 light years from Earth. Find the distance, in miles, from the Coma cluster to Earth. (See Exercise 124 for the definition of one light year. Use a 365-day year.)

APPLYING THE CONCEPTS

126. Correct the error in the following expressions. Explain which rule or property was used incorrectly.
a. $x^0 = 0$
b. $(x^4)^5 = x^9$
c. $x^2 \cdot x^3 = x^6$

127. Simplify.   **a.** $1 + [1 + (1 + 2^{-1})^{-1}]^{-1}$
          b. $2 - [2 - (2 - 2^{-1})^{-1}]^{-1}$

5.2   Introduction to Polynomials

Objective A   **To evaluate polynomial functions**   〔 5 〕 ◉

A **polynomial** is a variable expression in which the terms are monomials.

A polynomial of one term is a **monomial.** $5x$

A polynomial of two terms is a **binomial.** $5x^2y + 6x$

A polynomial of three terms is a **trinomial.** $3x^2 + 9xy - 5y$

Polynomials with more than three terms do not have special names.

The **degree of a polynomial** is the greatest of the degrees of any of its terms.

$$3x + 2 \qquad\qquad \text{degree} \quad 1$$
$$3x^2 + 2x - 4 \qquad \text{degree} \quad 2$$
$$4x^3y^2 + 6x^4 \qquad \text{degree} \quad 5$$
$$3x^{2n} - 5x^n - 2 \qquad \text{degree} \ 2n$$

The terms of a polynomial in one variable are usually arranged so that the exponents of the variable decrease from left to right. This is called **descending order.**

$$2x^2 - x + 8$$

$$3y^3 - 3y^2 + y - 12$$

For a polynomial in more than one variable, descending order may refer to any one of the variables.

The polynomial at the right is shown first in descending order of the x variable and then in descending order of the y variable.

$$2x^2 + 3xy + 5y^2$$

$$5y^2 + 3xy + 2x^2$$

Polynomial functions have many applications in mathematics. In general, a **polynomial function** is an expression whose terms are monomials. The **linear function** given by $f(x) = mx + b$ is an example of a polynomial function. It is a polynomial function of degree one. A second-degree polynomial function, called a **quadratic function,** is given by the equation $f(x) = ax^2 + bx + c$, $a \neq 0$. A third-degree polynomial function is called a **cubic function.**

To evaluate a polynomial function, replace the variable by its value and simplify.

⟹ Given $P(x) = x^3 - 3x^2 + 4$, evaluate $P(-3)$.

$$P(x) = x^3 - 3x^2 + 4$$
$$P(-3) = (-3)^3 - 3(-3)^2 + 4 \qquad \bullet \ \textbf{Substitute } -3 \textbf{ for } x \textbf{ and simplify.}$$
$$= -27 - 27 + 4$$
$$= -50$$

The **leading coefficient** of a polynomial function is the coefficient of the variable with the largest exponent. The **constant term** is the term without a variable.

⇒ Find the leading coefficient, the constant term, and the degree of the polynomial function $P(x) = 7x^4 - 3x^2 + 2x - 4$.

The leading coefficient is 7, the constant term is -4, and the degree is 4.

The three equations below do not represent polynomial functions.

$f(x) = 3x^2 + 2x^{-1}$ A polynomial function does not have a variable raised to a negative power.

$g(x) = 2\sqrt{x} - 3$ A polynomial function does not have a variable expression within a radical.

$h(x) = \dfrac{x}{x - 1}$ A polynomial function does not have a variable in the denominator of a fraction.

The graph of a linear function is a straight line and can be found by plotting just two points. The graph of a polynomial function of degree greater than one is a curve. Consequently, many points may have to be found before an accurate graph can be drawn.

Evaluating the quadratic function given by the equation $f(x) = x^2 - x - 6$ when $x = -3, -2, -1, 0, 1, 2, 3$, and 4 gives the points shown in Figure 1 below. For instance, $f(-3) = 6$, so $(-3, 6)$ is graphed; $f(2) = -4$, so $(2, -4)$ is graphed; and $f(4) = 6$, so $(4, 6)$ is graphed. Evaluating the function when x is not an integer, such as when $x = -\dfrac{3}{2}$ and $x = \dfrac{5}{2}$, produces more points to graph, as shown in Figure 2. Connecting the points with a smooth curve results in Figure 3, which is the graph of f.

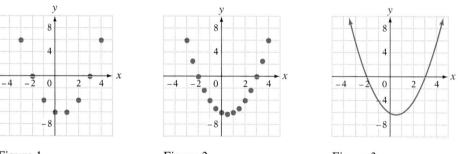

Figure 1   Figure 2   Figure 3

Here is an example of graphing a cubic function, $P(x) = x^3 - 2x^2 - 5x + 6$. Evaluating the function when $x = -2, -1, 0, 1, 2, 3$, and 4 gives the graph in Figure 4 below. Evaluating at some noninteger values gives the graph in Figure 5. Finally, connecting the dots with a smooth curve gives the graph in Figure 6.

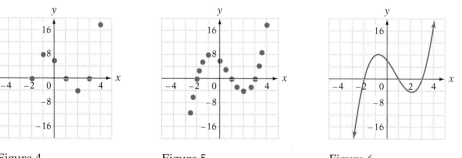

Figure 4   Figure 5   Figure 6

Example 1

Given $P(x) = x^3 + 3x^2 - 2x + 8$, evaluate $P(-2)$.

Solution

$$P(x) = x^3 + 3x^2 - 2x + 8$$
$$P(-2) = (-2)^3 + 3(-2)^2 - 2(-2) + 8$$
$$= (-8) + 3(4) + 4 + 8$$
$$= -8 + 12 + 4 + 8$$
$$= 16 \quad \text{4}$$

You Try It 1

Given $R(x) = -2x^4 - 5x^3 + 2x - 8$, evaluate $R(2)$.

Your solution

Example 2

Find the leading coefficient, the constant term, and the degree of the polynomial.
$P(x) = 5x^6 - 4x^5 - 3x^2 + 7$

Solution

The leading coefficient is 5, the constant term is 7, and the degree is 6.

You Try It 2

Find the leading coefficient, the constant term, and the degree of the polynomial.
$r(x) = -3x^4 + 3x^3 + 3x^2 - 2x - 12$

Your solution

Example 3

Which of the following is a polynomial function?
  a. $P(x) = 3x^{\frac{1}{2}} + 2x^2 - 3$
  b. $T(x) = 3\sqrt{x} - 2x^2 - 3x + 2$
  c. $R(x) = 14x^3 - \pi x^2 + 3x + 2$

Solution

  a. This is not a polynomial function. A polynomial function does not have a variable raised to a fractional power.
  b. This is not a polynomial function. A polynomial function does not have a variable expression within a radical.
  c. This is a polynomial function.

You Try It 3

Which of the following is a polynomial function?
  a. $R(x) = 5x^{14} - 5$
  b. $V(x) = -x^{-1} + 2x - 7$
  c. $P(x) = 2x^4 - 3\sqrt{x} - 3$

Your solution

Example 4

Graph $f(x) = x^2 - 2$.

Solution

x	$y = f(x)$
-3	7
-2	2
-1	-1
0	-2
1	-1
2	2
3	7

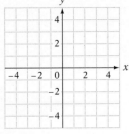

You Try It 4

Graph $f(x) = x^2 + 2x - 3$.

Your solution

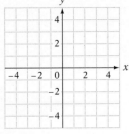

Solutions on p. S14

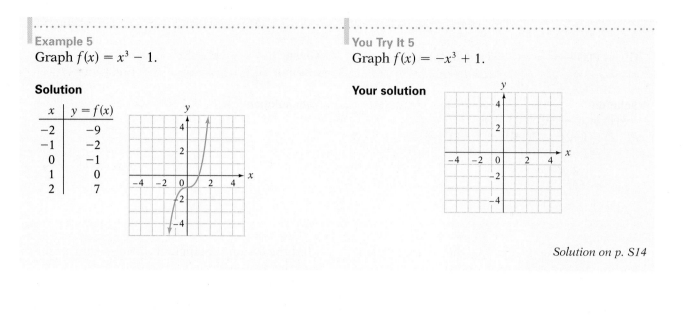

Example 5

Graph $f(x) = x^3 - 1$.

Solution

x	$y = f(x)$
-2	-9
-1	-2
0	-1
1	0
2	7

You Try It 5

Graph $f(x) = -x^3 + 1$.

Your solution

Solution on p. S14

Objective B | **To add or subtract polynomials** ⟨ 5 ⟩

Polynomials can be added by combining like terms. Either a vertical or a horizontal format can be used.

➡ Simplify $(3x^2 + 2x - 7) + (7x^3 - 3 + 4x^2)$. Use a horizontal format.

Use the Commutative and Associative Properties of Addition to rearrange and group like terms.

$(3x^2 + 2x - 7) + (7x^3 - 3 + 4x^2)$
$= 7x^3 + (3x^2 + 4x^2) + 2x + (-7 - 3)$ • **Combine like terms.**
$= 7x^3 + 7x^2 + 2x - 10$

➡ Simplify $(4x^2 + 5x - 3) + (7x^3 - 7x + 1) + (2x - 3x^2 + 4x^3 + 1)$. Use a vertical format.

Arrange the terms of each polynomial in descending order with like terms in the same column.

$$
\begin{array}{r}
4x^2 + 5x - 3 \\
7x^3 \qquad - 7x + 1 \\
4x^3 - 3x^2 + 2x + 1 \\
\hline
11x^3 + x^2 \qquad - 1
\end{array}
$$

• **Add the terms in each column.**

The **additive inverse of the polynomial** $x^2 + 5x - 4$ is $-(x^2 + 5x - 4)$.

To simplify the additive inverse of a polynomial, change the sign of every term inside the parentheses. $-(x^2 + 5x - 4) = -x^2 - 5x + 4$

To subtract two polynomials, add the additive inverse of the second polynomial to the first.

⇒ Simplify $(3x^2 - 7xy + y^2) - (-4x^2 + 7xy - 3y^2)$. Use a horizontal format.

Rewrite the subtraction as the addition of the additive inverse.

$$(3x^2 - 7xy + y^2) - (-4x^4 + 7xy - 3y^2)$$
$$= (3x^2 - 7xy + y^2) + (4x^2 - 7xy + 3y^2)$$
$$= 7x^2 - 14xy + 4y^2 \qquad \bullet \text{ Combine like terms.}$$

⇒ Simplify $(6x^3 - 3x + 7) - (3x^2 - 5x + 12)$. Use a vertical format.

Rewrite subtraction as the addition of the additive inverse.

$$(6x^3 - 3x + 7) - (3x^2 - 5x + 12) = (6x^3 - 3x + 7) + (-3x^2 + 5x - 12)$$

Arrange the terms of each polynomial in descending order with like terms in the same column.

$$
\begin{array}{r}
6x^3 \qquad - 3x + 7 \\
- 3x^2 + 5x - 12 \\
\hline
6x^3 - 3x^2 + 2x - 5
\end{array}
$$
• Combine the terms in each column.

Function notation can be used when adding or subtracting polynomials.

⇒ Given $P(x) = 3x^2 - 2x + 4$ and $R(x) = -5x^3 + 4x + 7$, find $P(x) + R(x)$.

$$P(x) + R(x) = (3x^2 - 2x + 4) + (-5x^3 + 4x + 7)$$
$$= -5x^3 + 3x^2 + 2x + 11$$

⇒ Given $P(x) = -5x^2 + 8x - 4$ and $R(x) = -3x^2 - 5x + 9$, find $P(x) - R(x)$.

$$P(x) - R(x) = (-5x^2 + 8x - 4) - (-3x^2 - 5x + 9)$$
$$= (-5x^2 + 8x - 4) + (3x^2 + 5x - 9)$$
$$= -2x^2 + 13x - 13$$

⇒ Given $P(x) = 3x^2 - 5x + 6$ and $R(x) = 2x^2 - 5x - 7$, find $S(x)$, the sum of the two polynomials.

$$S(x) = P(x) + R(x) = (3x^2 - 5x + 6) + (2x^2 - 5x - 7)$$
$$= 5x^2 - 10x - 1$$

Note from the preceding example that evaluating $P(x) = 3x^2 - 5x + 6$ and $R(x) = 2x^2 - 5x - 7$ at, for example, $x = 3$ and then adding the values is the same as evaluating $S(x) = 5x^2 - 10x - 1$ at 3.

$$P(3) = 3(3)^2 - 5(3) + 6 = 27 - 15 + 6 = 18$$
$$R(3) = 2(3)^2 - 5(3) - 7 = 18 - 15 - 7 = -4$$

$$P(3) + R(3) = 18 + (-4) = 14$$

$$S(3) = 5(3)^2 - 10(3) - 1 = 5(9) - 30 - 1 = 45 - 30 - 1 = 14$$

Example 6

Simplify:

$(4x^2 - 3xy + 7y^2) + (-3x^2 + 7xy + y^2)$

Use a vertical format.

Solution

$$\begin{array}{r} 4x^2 - 3xy + 7y^2 \\ -3x^2 + 7xy + y^2 \\ \hline x^2 + 4xy + 8y^2 \end{array}$$

You Try It 6

Simplify:

$(-3x^2 - 4x + 9) + (-5x^2 - 7x + 1)$

Use a vertical format.

Your solution

Example 7

Simplify:

$(3x^2 - 2x + 4) - (7x^2 + 3x - 12)$

Use a vertical format.

Solution

Add the additive inverse of $7x^2 + 3x - 12$ to $3x^2 - 2x + 4$.

$$\begin{array}{r} 3x^2 - 2x + 4 \\ -7x^2 - 3x + 12 \\ \hline -4x^2 - 5x + 16 \end{array}$$

You Try It 7

Simplify:

$(-5x^2 + 2x - 3) - (6x^2 + 3x - 7)$

Use a vertical format.

Your solution

Example 8

Given $P(x) = -3x^2 + 2x - 6$ and $R(x) = 4x^3 - 3x + 4$, find $S(x) = P(x) + R(x)$. Evaluate $S(-2)$.

Solution

$$\begin{aligned} S(x) &= (-3x^2 + 2x - 6) + (4x^3 - 3x + 4) \\ &= 4x^3 - 3x^2 - x - 2 \end{aligned}$$

$$\begin{aligned} S(-2) &= 4(-2)^3 - 3(-2)^2 - (-2) - 2 \\ &= 4(-8) - 3(4) + 2 - 2 \\ &= -44 \end{aligned}$$

You Try It 8

Given $P(x) = 4x^3 - 3x^2 + 2$ and $R(x) = -2x^2 + 2x - 3$, find $S(x) = P(x) + R(x)$. Evaluate $S(-1)$.

Your solution

Example 9

Given $P(x) = (2x^{2n} - 3x^n + 7)$ and $R(x) = (3x^{2n} + 3x^n + 5)$, find $D(x) = P(x) - R(x)$.

Solution

$$\begin{aligned} D(x) &= (2x^{2n} - 3x^n + 7) - (3x^{2n} + 3x^n + 5) \\ &= (2x^{2n} - 3x^n + 7) + (-3x^{2n} - 3x^n - 5) \\ &= -x^{2n} - 6x^n + 2 \end{aligned}$$

You Try It 9

Given $P(x) = (5x^{2n} - 3x^n - 7)$ and $R(x) = (-2x^{2n} - 5x^n + 8)$, find $D(x) = P(x) - R(x)$.

Your solution

Solutions on p. S14

5.2 Exercises

Objective A

1. Given $P(x) = 3x^2 - 2x - 8$, evaluate $P(3)$.

2. Given $P(x) = -3x^2 - 5x + 8$, evaluate $P(-5)$.

3. Given $R(x) = 2x^3 - 3x^2 + 4x - 2$, evaluate $R(2)$.

4. Given $R(x) = -x^3 + 2x^2 - 3x + 4$, evaluate $R(-1)$.

5. Given $f(x) = x^4 - 2x^2 - 10$, evaluate $f(-1)$.

6. Given $f(x) = x^5 - 2x^3 + 4x$, evaluate $f(2)$.

Which of the following define a polynomial function? For those that are polynomial functions, identify **a.** the leading coefficient, **b.** the constant term, and **c.** the degree.

7. $P(x) = -x^2 + 3x + 8$

8. $P(x) = 3x^4 - 3x - 7$

9. $R(x) = \dfrac{x}{x + 1}$

10. $R(x) = \dfrac{3x^2 - 2x + 1}{x}$

11. $f(x) = \sqrt{x} - x^2 + 2$

12. $f(x) = x^2 - \sqrt{x + 2} - 8$

13. $g(x) = 3x^5 - 2x^2 + \pi$

14. $g(x) = -4x^5 - 3x^2 + x - \sqrt{7}$

15. $P(x) = 3x^2 - 5x^3 + 2$

16. $P(x) = x^2 - 5x^4 - x^6$

17. $R(x) = 14$

18. $R(x) = \dfrac{1}{x} + 2$

Graph.

19. $P(x) = x^2 - 1$

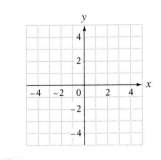

20. $P(x) = 2x^2 + 3$

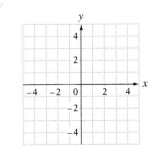

21. $R(x) = x^3 + 2$

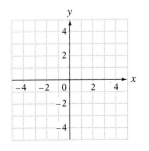

22. $R(x) = x^4 + 1$

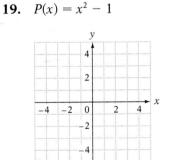

23. $f(x) = x^3 - 2x$

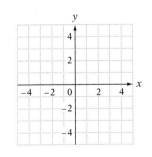

24. $f(x) = x^2 - x - 2$

Objective B

Simplify. Use a vertical format.

25. $(5x^2 + 2x - 7) + (x^2 - 8x + 12)$

26. $(3x^2 - 2x + 7) + (-3x^2 + 2x - 12)$

27. $(x^2 - 3x + 8) - (2x^2 - 3x + 7)$

28. $(2x^2 + 3x - 7) - (5x^2 - 8x - 1)$

Simplify. Use a horizontal format for Exercises 29–32.

29. $(3y^2 - 7y) + (2y^2 - 8y + 2)$

30. $(-2y^2 - 4y - 12) + (5y^2 - 5y)$

31. $(2a^2 - 3a - 7) - (-5a^2 - 2a - 9)$

32. $(3a^2 - 9a) - (-5a^2 + 7a - 6)$

33. Given $P(x) = x^2 - 3xy + y^2$ and $R(x) = 2x^2 - 3y^2$, find $P(x) + R(x)$.

34. Given $P(x) = x^{2n} + 7x^n - 3$ and $R(x) = -x^{2n} + 2x^n + 8$, find $P(x) + R(x)$.

35. Given $P(x) = 3x^2 + 2y^2$ and $R(x) = -5x^2 + 2xy - 3y^2$, find $P(x) - R(x)$.

36. Given $P(x) = 2x^{2n} - x^n - 1$ and $R(x) = 5x^{2n} + 7x^n + 1$, find $P(x) - R(x)$.

37. Given $P(x) = 3x^4 - 3x^3 - x^2$ and $R(x) = 3x^3 - 7x^2 + 2x$, find $S(x) = P(x) + R(x)$. Evaluate $S(2)$.

38. Given $P(x) = 3x^4 - 2x + 1$ and $R(x) = 3x^5 - 5x - 8$, find $S(x) = P(x) + R(x)$. Evaluate $S(-1)$.

APPLYING THE CONCEPTS

39. For what value of k is the given equation an identity?
a. $(2x^3 + 3x^2 + kx + 5) - (x^3 + 2x^2 + 3x + 7) = x^3 + x^2 + 5x - 2$
b. $(6x^3 + kx^2 - 2x - 1) - (4x^3 - 3x^2 + 1) = 2x^3 - x^2 - 2x - 2$

40. If $P(x)$ is a third-degree polynomial and $Q(x)$ is a fourth-degree polynomial, what can be said about the degree of $P(x) + Q(x)$? Give some examples of polynomials that support your answer.

41. If $P(x)$ is a fifth-degree polynomial and $Q(x)$ is a fourth-degree polynomial, what can be said about the degree of $P(x) - Q(x)$? Give some examples of polynomials that support your answer.

42. The deflection D (in inches) of a beam that is uniformly loaded is given by the polynomial function $D(x) = 0.005x^4 - 0.1x^3 + 0.5x^2$, where x is the distance from one end of the beam. See the figure at the right. The maximum deflection occurs when x is the midpoint of the beam. Determine the maximum deflection for the beam in the diagram.

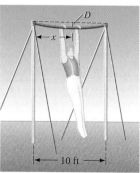

5.3 Multiplication of Polynomials

Objective A   **To multiply a polynomial by a monomial**

To multiply a polynomial by a monomial, use the Distributive Property and the Rule for Multiplying Exponential Expressions.

➡ Multiply: $-3x^2(2x^2 - 5x + 3)$

$-3x^2(2x^2 - 5x + 3)$

$= -3x^2(2x^2) - (-3x^2)(5x) + (-3x^2)(3)$ • Use the Distributive Property.

$= -6x^4 + 15x^3 - 9x^2$ • Use the Rule for Multiplying Exponential Expressions.

➡ Simplify: $2x^2 - 3x[2 - x(4x + 1) + 2]$

$2x^2 - 3x[2 - x(4x + 1) + 2]$

$= 2x^2 - 3x[2 - 4x^2 - x + 2]$ • Use the Distributive Property to remove the parentheses.

$= 2x^2 - 3x[-4x^2 - x + 4]$ • Simplify.

$= 2x^2 + 12x^3 + 3x^2 - 12x$ • Use the Distributive Property to remove the brackets.

$= 12x^3 + 5x^2 - 12x$ • Simplify.

➡ Multiply: $x^n(x^n - x^2 + 3)$

$x^n(x^n - x^2 + 3)$

$= x^n(x^n) - x^n(x^2) + x^n(3)$ • Use the Distributive Property.

$= x^{2n} - x^{n+2} + 3x^n$

Example 1
Multiply: $(3a^2 - 2a + 4)(-3a)$

Solution
$(3a^2 - 2a + 4)(-3a)$
$= 3a^2(-3a) - 2a(-3a) + 4(-3a)$
$= -9a^3 + 6a^2 - 12a$

You Try It 1
Multiply: $(2b^2 - 7b - 8)(-5b)$

Your solution

Solution on p. S14

Example 2

Simplify: $y - 3y[y - 2(3y - 6) + 2]$

Solution

$y - 3y[y - 2(3y - 6) + 2]$
$= y - 3y[y - 6y + 12 + 2]$
$= y - 3y[-5y + 14]$
$= y + 15y^2 - 42y$
$= 15y^2 - 41y$

You Try It 2

Simplify: $x^2 - 2x[x - x(4x - 5) + x^2]$

Your solution

Example 3

Multiply: $x^{n+2}(x^{n-1} + 2x - 1)$

Solution

$x^{n+2}(x^{n-1} + 2x - 1)$
$= x^{n+2}(x^{n-1}) + (x^{n+2})(2x) - (x^{n+2})(1)$
$= x^{n+2+(n-1)} + 2x^{n+2+1} - x^{n+2}$
$= x^{2n+1} + 2x^{n+3} - x^{n+2}$

You Try It 3

Multiply: $y^{n+3}(y^{n-2} - 3y^2 + 2)$

Your solution

Solutions on p. S15

Objective B **To multiply two polynomials** ⟦ 5 ⟧ ◉

The product of two polynomials is the polynomial obtained by multiplying each term of one polynomial by each term of the other polynomial and then combining like terms.

➡ Multiply: $(2x^2 - 2x + 1)(3x + 2)$

Use the Distributive Property to multiply the trinomial by each term of the binomial.

$(2x^2 - 2x + 1)(3x + 2) = (2x^2 - 2x + 1)(3x) + (2x^2 - 2x + 1)(2)$

$= (6x^3 - 6x^2 + 3x) + (4x^2 - 4x + 2)$

$= 6x^3 - 2x^2 - x + 2$

A convenient method of multiplying two polynomials is to use a vertical format similar to that used for multiplication of whole numbers.

➡ Multiply: $(3x^2 - 4x + 8)(2x - 7)$

$$
\begin{array}{r}
3x^2 - 4x + 8 \\
2x - 7 \\
\hline
-21x^2 + 28x - 56 \\
6x^3 - 8x^2 + 16x \\
\hline
6x^3 - 29x^2 + 44x - 56
\end{array}
$$

$-21x^2 + 28x - 56 = -7(3x^2 - 4x + 8)$

$6x^3 - 8x^2 + 16x = 2x(3x^2 - 4x + 8)$

• Like terms are in the same column.

• Combine like terms.

It is frequently necessary to find the product of two binomials. The product can be found by using a method called **FOIL,** which is based on the Distributive Property. The letters of FOIL stand for **F**irst, **O**uter, **I**nner, and **L**ast.

Multiply: $(3x - 2)(2x + 5)$

Multiply the **F**irst terms.	$(3x - 2)(2x + 5)$	$3x \cdot 2x = 6x^2$
Multiply the **O**uter terms.	$(3x - 2)(2x + 5)$	$3x \cdot 5 = 15x$
Multiply the **I**nner terms.	$(3x - 2)(2x + 5)$	$-2 \cdot 2x = -4x$
Multiply the **L**ast terms.	$(3x - 2)(2x + 5)$	$-2 \cdot 5 = -10$

$$\text{F} \qquad \text{O} \qquad \text{I} \qquad \text{L}$$

Add the products. $(3x - 2)(2x + 5) \quad = \quad 6x^2 + 15x - 4x - 10$

Combine like terms. $= \quad 6x^2 + 11x - 10$

⇒ Multiply: $(6x - 5)(3x - 4)$

$$(6x - 5)(3x - 4) = 6x(3x) + 6x(-4) + (-5)(3x) + (-5)(-4)$$
$$= 18x^2 - 24x - 15x + 20$$
$$= 18x^2 - 39x + 20$$

Example 4

Multiply: $(4a^3 - 3a + 7)(a - 5)$

Solution

$$
\begin{array}{r}
4a^3 \qquad - 3a + 7 \\
a - 5 \\
\hline
-20a^3 \qquad + 15a - 35 \\
4a^4 \qquad - 3a^2 + 7a \qquad\quad \\
\hline
4a^4 - 20a^3 - 3a^2 + 22a - 35
\end{array}
$$

You Try It 4

Multiply: $(-2b^2 + 5b - 4)(-3b + 2)$

Your solution

Example 5

Multiply: $(5a - 3b)(2a + 7b)$

Solution

$(5a - 3b)(2a + 7b)$
$= 10a^2 + 35ab - 6ab - 21b^2$
$= 10a^2 + 29ab - 21b^2$

You Try It 5

Multiply: $(3x - 4)(2x - 3)$

Your solution

Example 6

Multiply: $(a^n - 2b^n)(3a^n - b^n)$

Solution

$(a^n - 2b^n)(3a^n - b^n)$
$= 3a^{2n} - a^n b^n - 6a^n b^n + 2b^{2n}$
$= 3a^{2n} - 7a^n b^n + 2b^{2n}$

You Try It 6

Multiply: $(2x^n + y^n)(x^n - 4y^n)$

Your solution

Solutions on p. S15

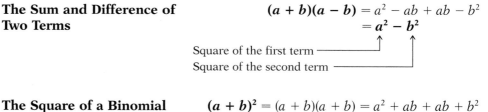

Objective C   **To multiply polynomials that have special products**

Using FOIL, a pattern can be found for the product of the sum and difference of two terms and for the square of a binomial.

The Sum and Difference of Two Terms

$$(a + b)(a - b) = a^2 - ab + ab - b^2$$
$$= a^2 - b^2$$

Square of the first term ———————
Square of the second term ———————

The Square of a Binomial

$$(a + b)^2 = (a + b)(a + b) = a^2 + ab + ab + b^2$$
$$= a^2 + 2ab + b^2$$

Square of the first term ———————
Twice the product of the two terms ———————
Square of the second term ———————

⟹ Multiply: $(4x + 3)(4x - 3)$

$(4x + 3)(4x - 3)$ is the sum and difference of the same two terms. The product is the difference of the squares of the terms.

$$(4x + 3)(4x - 3) = (4x)^2 - 3^2$$
$$= 16x^2 - 9$$

⟹ Simplify: $(2x - 3y)^2$

$(2x - 3y)^2$ is the square of a binomial.

$$(2x - 3y)^2 = (2x)^2 + 2(2x)(-3y) + (-3y)^2$$
$$= 4x^2 - 12xy + 9y^2$$

Example 7

Multiply: $(2a - 3)(2a + 3)$

Solution

$(2a - 3)(2a + 3) = 4a^2 - 9$

You Try It 7

Multiply: $(3x - 7)(3x + 7)$

Your solution

Example 8

Multiply: $(x^n + 5)(x^n - 5)$

Solution

$(x^n + 5)(x^n - 5) = x^{2n} - 25$

You Try It 8

Multiply: $(2x^n + 3)(2x^n - 3)$

Your solution

Solutions on p. S15

Example 9

Simplify: $(2x + 7y)^2$

Solution

$(2x + 7y)^2 = 4x^2 + 28xy + 49y^2$

You Try It 9

Simplify: $(3x - 4y)^2$

Your solution

Example 10

Simplify: $(x^{2n} - 2)^2$

Solution

$(x^{2n} - 2)^2 = x^{4n} - 4x^{2n} + 4$

You Try It 10

Simplify: $(2x^n - 8)^2$

Your solution

Solutions on p. S15

Objective D   **To solve application problems**   ‹ 5 ›

Example 11

The length of a rectangle is $(2x + 3)$ ft. The width is $(x - 5)$ ft. Find the area of the rectangle in terms of the variable x.

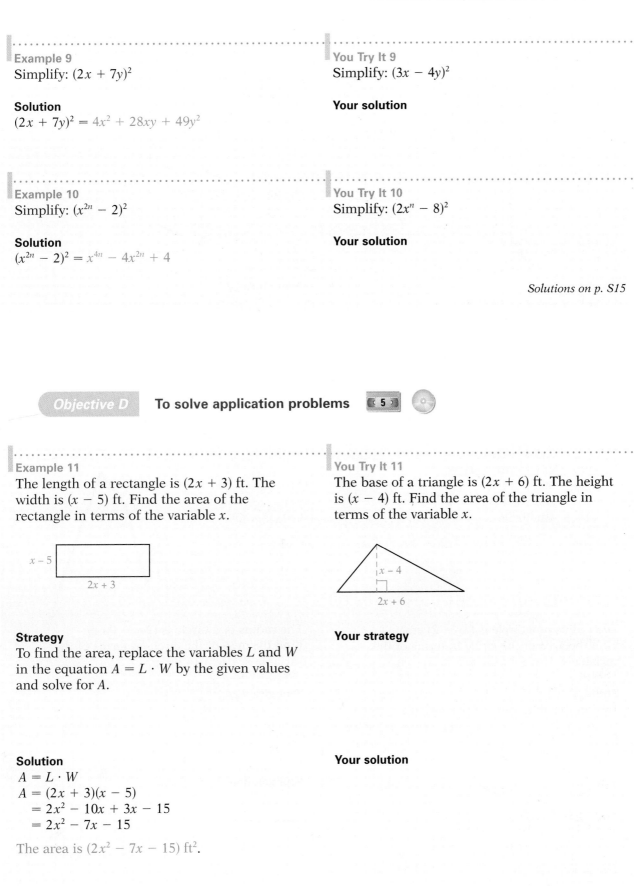

$x - 5$

$2x + 3$

Strategy

To find the area, replace the variables L and W in the equation $A = L \cdot W$ by the given values and solve for A.

Solution

$A = L \cdot W$

$A = (2x + 3)(x - 5)$

$ = 2x^2 - 10x + 3x - 15$

$ = 2x^2 - 7x - 15$

The area is $(2x^2 - 7x - 15)$ ft^2.

You Try It 11

The base of a triangle is $(2x + 6)$ ft. The height is $(x - 4)$ ft. Find the area of the triangle in terms of the variable x.

$x - 4$

$2x + 6$

Your strategy

Your solution

Solution on p. S15

Example 12

The corners are cut from a rectangular piece of cardboard measuring 8 in. by 12 in. The sides are folded up to make a box. Find the volume of the box in terms of the variable x, where x is the length of the side of the square cut from each corner of the rectangle.

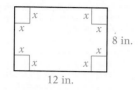

Strategy

Length of the box: $12 - 2x$
Width of the box: $8 - 2x$
Height of the box: x
To find the volume, replace the variables L, W, and H in the equation $V = L \cdot W \cdot H$ and solve for V.

Solution
$$V = L \cdot W \cdot H$$
$$\begin{aligned} V &= (12 - 2x)(8 - 2x)x \\ &= (96 - 24x - 16x + 4x^2)x \\ &= (96 - 40x + 4x^2)x \\ &= 96x - 40x^2 + 4x^3 \\ &= 4x^3 - 40x^2 + 96x \end{aligned}$$

The volume is $(4x^3 - 40x^2 + 96x)$ in^3.

You Try It 12

Find the volume of the rectangular solid shown in the diagram below. All dimensions are in feet.

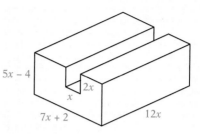

Your strategy

Your solution

Example 13

The radius of a circle is $(3x - 2)$ cm. Find the area of the circle in terms of the variable x. Use 3.14 for π.

Strategy

To find the area, replace the variable r in the equation $A = \pi r^2$ by the given value and solve for A.

Solution
$$A = \pi r^2$$
$$\begin{aligned} A &= 3.14(3x - 2)^2 \\ &= 3.14(9x^2 - 12x + 4) \\ &= 28.26x^2 - 37.68x + 12.56 \end{aligned}$$

The area is $(28.26x^2 - 37.68x + 12.56)$ cm^2.

You Try It 13

The radius of a circle is $(2x + 3)$ cm. Find the area of the circle in terms of the variable x. Use 3.14 for π.

Your strategy

Your solution

Solutions on p. S15

5.3 Exercises

Objective A

Simplify.

1. $2x(x - 3)$

2. $2a(2a + 4)$

3. $3x^2(2x^2 - x)$

4. $-4y^2(4y - 6y^2)$

5. $3xy(2x - 3y)$

6. $-4ab(5a - 3b)$

7. $x^n(x + 1)$

8. $y^n(y^{2n} - 3)$

9. $x^n(x^n + y^n)$

10. $x - 2x(x - 2)$

11. $2b + 4b(2 - b)$

12. $-2y(3 - y) + 2y^2$

13. $-2a^2(3a^2 - 2a + 3)$

14. $4b(3b^3 - 12b^2 - 6)$

15. $3b(3b^4 - 3b^2 + 8)$

16. $(2x^2 - 3x - 7)(-2x^2)$

17. $(-3y^2 - 4y + 2)(y^2)$

18. $(6b^4 - 5b^2 - 3)(-2b^3)$

19. $-5x^2(4 - 3x + 3x^2 + 4x^3)$

20. $-2y^2(3 - 2y - 3y^2 + 2y^3)$

21. $-2x^2y(x^2 - 3xy + 2y^2)$

22. $3ab^2(3a^2 - 2ab + 4b^2)$

23. $x^n(x^{2n} + x^n + x)$

24. $x^{2n}(x^{2n-2} + x^{2n} + x)$

25. $a^{n+1}(a^n - 3a + 2)$

26. $a^{n+4}(a^{n-2} + 5a^2 - 3)$

27. $2y^2 - y[3 - 2(y - 4) - y]$

28. $3x^2 - x[x - 2(3x - 4)]$

29. $2y - 3[y - 2y(y - 3) + 4y]$

30. $4a^2 - 2a[3 - a(2 - a + a^2)]$

Multiply.

31. $(x - 2)(x + 7)$

32. $(y + 8)(y + 3)$

33. $(2y - 3)(4y + 7)$

34. $(5x - 7)(3x - 8)$

35. $2(2x - 3y)(2x + 5y)$

36. $-3(7x - 3y)(2x - 9y)$

37. $(xy + 4)(xy - 3)$

38. $(xy - 5)(2xy + 7)$

39. $(2x^2 - 5)(x^2 - 5)$

40. $(x^2 - 4)(x^2 - 6)$

41. $(5x^2 - 5y)(2x^2 - y)$

42. $(x^2 - 2y^2)(x^2 + 4y^2)$

43. $(x^n + 2)(x^n - 3)$

44. $(x^n - 4)(x^n - 5)$

45. $(2a^n - 3)(3a^n + 5)$

46. $(5b^n - 1)(2b^n + 4)$

47. $(2a^n - b^n)(3a^n + 2b^n)$

48. $(3x^n + b^n)(x^n + 2b^n)$

49. $(x + 5)(x^3 - 3x + 4)$

50. $(a + 2)(a^3 - 3a^2 + 7)$

51. $(2a - 3b)(5a^2 - 6ab + 4b^2)$

52. $(3a + b)(2a^2 - 5ab - 3b^2)$

53. $(2y^2 - 1)(y^3 - 5y^2 - 3)$

54. $(2b^2 - 3)(3b^2 - 3b + 6)$

55. $(2x - 5)(2x^4 - 3x^3 - 2x + 9)$

56. $(2a - 5)(3a^4 - 3a^2 + 2a - 5)$

57. $(x^2 + 2x - 3)(x^2 - 5x + 7)$

58. $(x^2 - 3x + 1)(x^2 - 2x + 7)$

59. $(a - 2)(2a - 3)(a + 7)$

60. $(b - 3)(3b - 2)(b - 1)$

61. $(x^n + 1)(x^{2n} + x^n + 1)$

62. $(a^{2n} - 3)(a^{5n} - a^{2n} + a^n)$

63. $(x^n + y^n)(x^n - 2x^ny^n + 3y^n)$

64. $(x^n - y^n)(x^{2n} - 3x^ny^n - y^{2n})$

Objective C

Simplify.

65. $(3x - 2)(3x + 2)$

66. $(4y + 1)(4y - 1)$

67. $(6 - x)(6 + x)$

68. $(10 + b)(10 - b)$

69. $(2a - 3b)(2a + 3b)$

70. $(5x - 7y)(5x + 7y)$

71. $(x^2 + 1)(x^2 - 1)$

72. $(x^2 + y^2)(x^2 - y^2)$

73. $(x^n + 3)(x^n - 3)$

74. $(x^n + y^n)(x^n - y^n)$

75. $(x - 5)^2$

76. $(y + 2)^2$

77. $(3a + 5b)^2$

78. $(5x - 4y)^2$

79. $(x^2 - 3)^2$

80. $(x^2 + y^2)^2$

81. $(2x^2 - 3y^2)^2$

82. $(x^n - 1)^2$

83. $(a^n - b^n)^2$

84. $(2x^n + 5y^n)^2$

85. $y^2 - (x - y)^2$

86. $a^2 + (a + b)^2$

87. $(x - y)^2 - (x + y)^2$

88. $(a + b)^2 + (a - b)^2$

Objective D   *Application Problems*

89. The length of a rectangle is $(3x - 2)$ ft. The width is $(x + 4)$ ft. Find the area of the rectangle in terms of the variable x.

90. The base of a triangle is $(x - 4)$ ft. The height is $(3x + 2)$ ft. Find the area of the triangle in terms of the variable x.

91. Find the area of the figure shown below. All dimensions given are in meters.

92. Find the area of the figure shown below. All dimensions given are in feet.

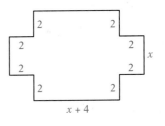

93. The length of the side of a cube is $(x + 3)$ cm. Find the volume of the cube in terms of the variable x.

94. The length of a box is $(3x + 2)$ cm, the width is $(x - 4)$ cm, and the height is x cm. Find the volume of the box in terms of the variable x.

95. Find the volume of the figure shown below. All dimensions given are in inches.

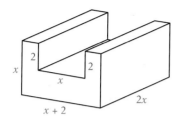

96. Find the volume of the figure shown below. All dimensions given are in centimeters.

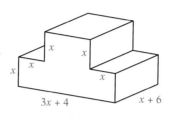

97. The radius of a circle is $(5x + 4)$ in. Find the area of the circle in terms of the variable x. Use 3.14 for π.

98. The radius of a circle is $(x - 2)$ in. Find the area of the circle in terms of the variable x. Use 3.14 for π.

APPLYING THE CONCEPTS

99. Find the product.
   a. $(a - b)(a^2 + ab + b^2)$
   b. $(x + y)(x^2 - xy + y^2)$

100. Correct the error in each of the following.
   a. $(x + 3)^2 = x^2 + 9$
   b. $(a - b)^2 = a^2 - b^2$

101. For what value of k is the given equation an identity?
   a. $(3x - k)(2x + k) = 6x^2 + 5x - k^2$
   b. $(4x + k)^2 = 16x^2 + 8x + k^2$

102. Complete.
   a. If $m = n + 1$, then $\dfrac{a^m}{a^n} = $ _____ .
   b. If $m = n + 2$, then $\dfrac{a^m}{a^n} = $ _____ .

103. Subtract the product of $4a + b$ and $2a - b$ from $9a^2 - 2ab$.

104. Subtract the product of $5x - y$ and $x + 3y$ from $6x^2 + 12xy - 2y^2$.

5.4 Division of Polynomials

Objective A   **To divide polynomials**

Some rational expressions cannot be simplified by factoring and dividing by the common factors. In these cases, long division of polynomials is used.

To divide two polynomials, use a method similar to that used for division of whole numbers. To check division of polynomials, use

Dividend = (quotient × divisor) + remainder

➡ Simplify: $(x^2 + 5x - 7) \div (x + 3)$

Step 1

$$\begin{array}{r} x \\ x + 3\overline{)x^2 + 5x - 7} \\ \underline{x^2 + 3x} \downarrow \\ 2x - 7 \end{array}$$

Think: $x\overline{)x^2} = \dfrac{x^2}{x} = x$

Multiply: $x(x + 3) = x^2 + 3x$

Subtract: $(x^2 + 5x) - (x^2 + 3x) = 2x$

Step 2

$$\begin{array}{r} x + 2 \\ x + 3\overline{)x^2 + 5x - 7} \\ \underline{x^2 + 3x} \\ 2x - 7 \\ \underline{2x + 6} \\ -13 \end{array}$$

Think: $x\overline{)2x} = \dfrac{2x}{x} = 2$

Multiply: $2(x + 3) = 2x + 6$

Subtract: $(2x - 7) - (2x + 6) = -13$

The remainder is -13.

Check: $(x + 2)(x + 3) + (-13) = x^2 + 3x + 2x + 6 - 13 = x^2 + 5x - 7$

$(x^2 + 5x - 7) \div (x + 3) = x + 2 - \dfrac{13}{x + 3}$

➡ Simplify: $\dfrac{6 - 6x^2 + 4x^3}{2x + 3}$

Arrange the terms in descending order. Note that there is no term of x in $4x^3 - 6x^2 + 6$. Insert a zero for the missing term so that like terms will be in the same columns.

$$\begin{array}{r} 2x^2 - 6x + 9 \\ 2x + 3\overline{)4x^3 - 6x^2 + 0x + 6} \\ \underline{4x^3 + 6x^2} \\ -12x^2 + 0x \\ \underline{-12x^2 - 18x} \\ 18x + 6 \\ \underline{18x + 27} \\ -21 \end{array}$$

$\dfrac{4x^3 - 6x^2 + 6}{2x + 3} = 2x^2 - 6x + 9 - \dfrac{21}{2x + 3}$

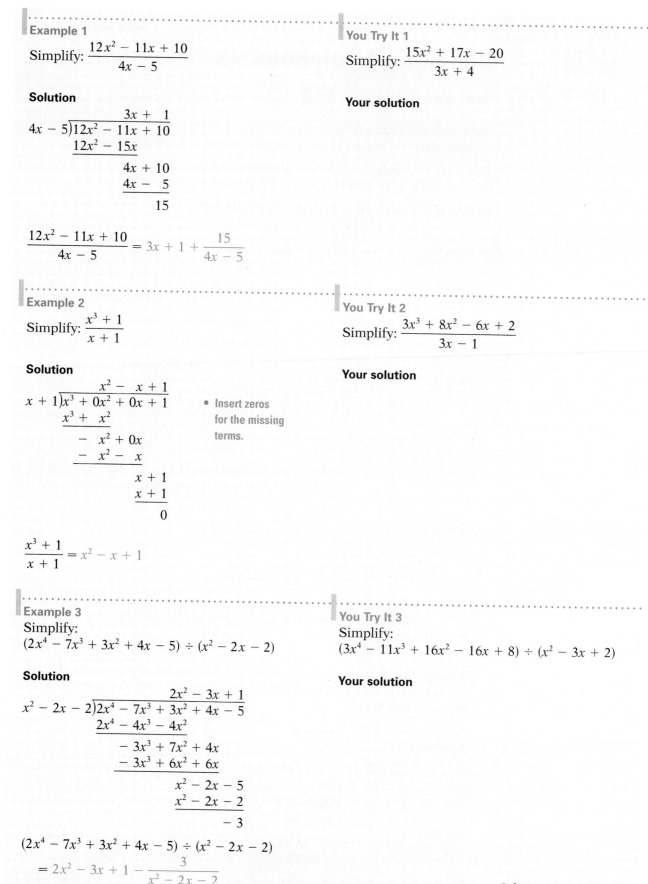

Example 1

Simplify: $\dfrac{12x^2 - 11x + 10}{4x - 5}$

Solution

$$
\begin{array}{r}
3x + 1 \\
4x - 5 \overline{\smash{)}\ 12x^2 - 11x + 10} \\
\underline{12x^2 - 15x} \\
4x + 10 \\
\underline{4x - 5} \\
15
\end{array}
$$

$$\frac{12x^2 - 11x + 10}{4x - 5} = 3x + 1 + \frac{15}{4x - 5}$$

You Try It 1

Simplify: $\dfrac{15x^2 + 17x - 20}{3x + 4}$

Your solution

Example 2

Simplify: $\dfrac{x^3 + 1}{x + 1}$

Solution

$$
\begin{array}{r}
x^2 - x + 1 \\
x + 1 \overline{\smash{)}\ x^3 + 0x^2 + 0x + 1} \\
\underline{x^3 + x^2} \\
-x^2 + 0x \\
\underline{-x^2 - x} \\
x + 1 \\
\underline{x + 1} \\
0
\end{array}
$$

• Insert zeros for the missing terms.

$$\frac{x^3 + 1}{x + 1} = x^2 - x + 1$$

You Try It 2

Simplify: $\dfrac{3x^3 + 8x^2 - 6x + 2}{3x - 1}$

Your solution

Example 3

Simplify:
$(2x^4 - 7x^3 + 3x^2 + 4x - 5) \div (x^2 - 2x - 2)$

Solution

$$
\begin{array}{r}
2x^2 - 3x + 1 \\
x^2 - 2x - 2 \overline{\smash{)}\ 2x^4 - 7x^3 + 3x^2 + 4x - 5} \\
\underline{2x^4 - 4x^3 - 4x^2} \\
-3x^3 + 7x^2 + 4x \\
\underline{-3x^3 + 6x^2 + 6x} \\
x^2 - 2x - 5 \\
\underline{x^2 - 2x - 2} \\
-3
\end{array}
$$

$(2x^4 - 7x^3 + 3x^2 + 4x - 5) \div (x^2 - 2x - 2)$

$= 2x^2 - 3x + 1 - \dfrac{3}{x^2 - 2x - 2}$

You Try It 3

Simplify:
$(3x^4 - 11x^3 + 16x^2 - 16x + 8) \div (x^2 - 3x + 2)$

Your solution

Solutions on pp. S15–S16

Objective B **To divide polynomials by using synthetic division**

Synthetic division is a shorter method of dividing a polynomial by a binomial of the form $x - a$.

→ Simplify $(3x^2 - 4x + 6) \div (x - 2)$ by using long division.

$$
\begin{array}{r}
3x + 2 \\
x - 2\overline{)3x^2 - 4x + 6} \\
\underline{3x^2 - 6x} \\
2x + 6 \\
\underline{2x - 4} \\
10
\end{array}
$$

$$(3x^2 - 4x + 6) \div (x - 2) = 3x + 2 + \frac{10}{x - 2}$$

The variables can be omitted because the position of a term indicates the power of the term.

$$
\begin{array}{r}
3 \quad 2 \\
-2\overline{)3 \;\; -4 \quad 6} \\
\underline{3 \;\; -6} \\
2 \quad 6 \\
\underline{2 \;\; -4} \\
10
\end{array}
$$

Each number shown in color is exactly the same as the number above it. Removing the colored numbers condenses the vertical spacing.

$$
\begin{array}{r}
3 \quad 2 \\
-2\overline{)3 \;\; -4 \quad 6} \\
\underline{-6 \;\; -4} \\
2 \quad 10
\end{array}
$$

The number in color in the top row is the same as the one in the bottom row. Writing the 3 from the top row in the bottom row allows the spacing to be condensed even further.

$$
\begin{array}{c|ccc}
-2 & 3 & -4 & 6 \\
   &   & -6 & -4 \\
\hline
   & 3 & 2 & 10
\end{array}
$$

Terms of Remainder
the quotient

Because the degree of the dividend $(3x^2 - 4x + 6)$ is 2 and the degree of the divisor $(x - 2)$ is 1, the degree of the quotient is $2 - 1 = 1$. This means that, using the terms of the quotient given above, that quotient is $3x + 2$. The remainder is 10.

In general, the degree of the quotient of two polynomials is the difference between the degree of the dividend and the degree of the divisor.

By replacing the constant term in the divisor by its additive inverse, we may add rather than subtract terms. This is illustrated in the following example.

➡ Simplify: $(3x^3 + 6x^2 - x - 2) \div (x + 3)$

The additive inverse of the binomial constant

$$
\begin{array}{c|cccc}
 & \text{Coefficients of the polynomial} \\
\downarrow & \overbrace{} \\
-3 & 3 & 6 & -1 & -2 \\
 & \downarrow \\
\hline
 & 3
\end{array}
$$

• Bring down the 3.

$$
\begin{array}{c|cccc}
-3 & 3 & 6 & -1 & -2 \\
 & & -9 \\
\hline
 & 3 & -3
\end{array}
$$

• Multiply $-3(3)$ and add the product to 6.

$$
\begin{array}{c|cccc}
-3 & 3 & 6 & -1 & -2 \\
 & & -9 & 9 \\
\hline
 & 3 & -3 & 8
\end{array}
$$

• Multiply $-3(-3)$ and add the product to -1.

$$
\begin{array}{c|cccc}
-3 & 3 & 6 & -1 & -2 \\
 & & -9 & 9 & -24 \\
\hline
 & 3 & -3 & 8 & -26
\end{array}
$$

• Multiply $-3(8)$ and add the product to -2.

$\underbrace{\qquad\qquad}_{\text{Terms of the quotient}}$ $\underbrace{\qquad}_{\text{Remainder}}$

The degree of the dividend is 3 and the degree of the divisor is 1. Therefore, the degree of the quotient is $3 - 1 = 2$.

$$(3x^3 + 6x^2 - x - 2) \div (x + 3) = 3x^2 - 3x + 8 - \frac{26}{x + 3}$$

➡ Simplify: $(2x^3 - x + 2) \div (x - 2)$

The additive inverse of the binomial constant

$$
\begin{array}{c|cccc}
 & \text{Coefficients of the polynomial} \\
\downarrow & \overbrace{} \\
2 & 2 & 0 & -1 & 2 \\
 & \downarrow \\
\hline
 & 2
\end{array}
$$

• Insert a 0 for the missing term and bring down the 2.

$$
\begin{array}{c|cccc}
2 & 2 & 0 & -1 & 2 \\
 & & 4 \\
\hline
 & 2 & 4
\end{array}
$$

• Multiply $2(2)$ and add the product to 0.

$$
\begin{array}{c|cccc}
2 & 2 & 0 & -1 & 2 \\
 & & 4 & 8 \\
\hline
 & 2 & 4 & 7
\end{array}
$$

• Multiply $2(4)$ and add the product to -1.

$$
\begin{array}{c|cccc}
2 & 2 & 0 & -1 & 2 \\
 & & 4 & 8 & 14 \\
\hline
 & 2 & 4 & 7 & 16
\end{array}
$$

• Multiply $2(7)$ and add the product to 2.

$\underbrace{\qquad\qquad}_{\text{Terms of the quotient}}$ $\underbrace{\qquad}_{\text{Remainder}}$

$$(2x^3 - x + 2) \div (x - 2) = 2x^2 + 4x + 7 + \frac{16}{x - 2}$$

Example 4

Simplify: $(7 - 3x + 5x^2) \div (x - 1)$

Solution

Arrange the coefficients in decreasing powers of x.

$$
\begin{array}{r|rrr}
1 & 5 & -3 & 7 \\
  &   & 5 & 2 \\
\hline
  & 5 & 2 & 9
\end{array}
$$

$(5x^2 - 3x + 7) \div (x - 1) = 5x + 2 + \dfrac{9}{x - 1}$

You Try It 4

Simplify: $(6x^2 + 8x - 5) \div (x + 2)$

Your solution

Example 5

Simplify: $(2x^3 + 4x^2 - 3x + 12) \div (x + 4)$

Solution

$$
\begin{array}{r|rrrr}
-4 & 2 & 4 & -3 & 12 \\
   &   & -8 & 16 & -52 \\
\hline
   & 2 & -4 & 13 & -40
\end{array}
$$

$(2x^3 + 4x^2 - 3x + 12) \div (x + 4)$
$\quad = 2x^2 - 4x + 13 - \dfrac{40}{x + 4}$

You Try It 5

Simplify: $(5x^3 - 12x^2 - 8x + 16) \div (x - 2)$

Your solution

Example 6

Simplify: $(3x^4 - 8x^2 + 2x + 1) \div (x + 2)$

Solution

Insert a zero for the missing term.

$$
\begin{array}{r|rrrrr}
-2 & 3 & 0 & -8 & 2 & 1 \\
   &   & -6 & 12 & -8 & 12 \\
\hline
   & 3 & -6 & 4 & -6 & 13
\end{array}
$$

$(3x^4 - 8x^2 + 2x + 1) \div (x + 2)$
$\quad = 3x^3 - 6x^2 + 4x - 6 + \dfrac{13}{x + 2}$

You Try It 6

Simplify: $(2x^4 - 3x^3 - 8x^2 - 2) \div (x - 3)$

Your solution

Solutions on p. S16

Objective C   **To evaluate a polynomial using synthetic division**   5

A polynomial can be evaluated by using synthetic division. Consider the polynomial $P(x) = 2x^4 - 3x^3 + 4x^2 - 5x + 1$. One way to evaluate the polynomial when $x = 2$ is to replace x by 2 and then simplify the numerical expression.

$$P(x) = 2x^4 - 3x^3 + 4x^2 - 5x + 1$$
$$
\begin{aligned}
P(2) &= 2(2)^4 - 3(2)^3 + 4(2)^2 - 5(2) + 1 \\
     &= 2(16) - 3(8) + 4(4) - 5(2) + 1 \\
     &= 32 - 24 + 16 - 10 + 1 \\
     &= 15
\end{aligned}
$$

Now use synthetic division to divide $(2x^4 - 3x^3 + 4x^2 - 5x + 1) \div (x - 2)$.

$$
\begin{array}{r|rrrrr}
2 & 2 & -3 & 4 & -5 & 1 \\
  &   & 4 & 2 & 12 & 14 \\
\hline
  & 2 & 1 & 6 & 7 & 15 \\
\end{array}
$$

$$\underbrace{}_{\text{Terms of the quotient}} \qquad \underbrace{}_{\text{Remainder}}$$

Note that the remainder is 15, which is the same value as $P(2)$. This is not a coincidence. The following theorem states that this situation is always true.

Remainder Theorem

If the polynomial $P(x)$ is divided by $x - a$, the remainder is $P(a)$.

➡ Evaluate $P(x) = x^4 - 3x^2 + 4x - 5$ when $x = -2$ by using the Remainder Theorem.

The value at which the polynomial is evaluated

$$
\begin{array}{r|rrrrr}
-2 & 1 & 0 & -3 & 4 & -5 \\
   &   & -2 & 4 & -2 & -4 \\
\hline
   & 1 & -2 & 1 & 2 & -9 \\
\end{array}
$$

• A 0 is inserted for the x^3 term.

⟵ The remainder

$P(-2) = -9$

Example 7

Use synthetic division to evaluate $P(x) = x^2 - 6x + 4$ when $x = 3$.

Solution

$$
\begin{array}{r|rrr}
3 & 1 & -6 & 4 \\
  &   & 3 & -9 \\
\hline
  & 1 & -3 & -5 \\
\end{array}
$$

$P(3) = -5$

You Try It 7

Use synthetic division to evaluate $P(x) = 2x^2 - 3x - 5$ when $x = 2$.

Your solution

Example 8

Use synthetic division to evaluate $P(x) = -x^4 + 3x^3 + 2x^2 - x - 5$ when $x = -2$.

Solution

$$
\begin{array}{r|rrrrr}
-2 & -1 & 3   & 2 & -1 & -5 \\
   &   & 2   & -10 & 16 & -30 \\
\hline
   & -1 & 5   & -8 & 15 & -35 \\
\end{array}
$$

$P(-2) = -35$

You Try It 8

Use synthetic division to evaluate $P(x) = 2x^3 - 5x^2 + 7$ when $x = -3$.

Your solution

Solutions on p. S16

5.4 Exercises

Divide by using long division.

1. $(x^2 + 3x - 40) \div (x - 5)$

2. $(x^2 - 14x + 24) \div (x - 2)$

3. $(x^3 - 3x + 2) \div (x - 3)$

4. $(x^3 + 4x^2 - 8) \div (x + 4)$

5. $(6x^2 + 13x + 8) \div (2x + 1)$

6. $(12x^2 + 13x - 14) \div (3x - 2)$

7. $(10x^2 + 9x - 5) \div (2x - 1)$

8. $(18x^2 - 3x + 2) \div (3x + 2)$

9. $(8x^3 - 9) \div (2x - 3)$

10. $(64x^3 + 4) \div (4x + 2)$

11. $(6x^4 - 13x^2 - 4) \div (2x^2 - 5)$

12. $(12x^4 - 11x^2 + 10) \div (3x^2 + 1)$

13. $\dfrac{-10 - 33x + 3x^3 - 8x^2}{3x + 1}$

14. $\dfrac{10 - 49x + 38x^2 - 8x^3}{1 - 4x}$

15. $\dfrac{x^3 - 5x^2 + 7x - 4}{x - 3}$

16. $\dfrac{2x^3 - 3x^2 + 6x + 4}{2x + 1}$

17. $\dfrac{16x^2 - 13x^3 + 2x^4 + 20 - 9x}{x - 5}$

18. $\dfrac{x - x^2 + 5x^3 + 3x^4 - 2}{x + 2}$

19. $\dfrac{2x^3 + 4x^2 - x + 2}{x^2 + 2x - 1}$

20. $\dfrac{3x^3 - 2x^2 + 5x - 4}{x^2 - x + 3}$

21. $\dfrac{x^4 + 2x^3 - 3x^2 - 6x + 2}{x^2 - 2x - 1}$

22. $\dfrac{x^4 - 3x^3 + 4x^2 - x + 1}{x^2 + x - 3}$

23. $\dfrac{x^4 + 3x^2 - 4x + 5}{x^2 + 2x + 3}$

24. $\dfrac{x^4 + 2x^3 - x + 2}{x^2 - x - 1}$

Objective B

Divide by using synthetic division.

25. $(2x^2 - 6x - 8) \div (x + 1)$

26. $(3x^2 + 19x + 20) \div (x + 5)$

27. $(3x^2 - 14x + 16) \div (x - 2)$

28. $(4x^2 - 23x + 28) \div (x - 4)$

29. $(3x^2 - 4) \div (x - 1)$

30. $(4x^2 - 8) \div (x - 2)$

31. $(2x^3 - x^2 + 6x + 9) \div (x + 1)$

32. $(3x^3 + 10x^2 + 6x - 4) \div (x + 2)$

33. $(18 + x - 4x^3) \div (2 - x)$

34. $(12 - 3x^2 + x^3) \div (x + 3)$

35. $(2x^3 + 5x^2 - 5x + 20) \div (x + 4)$

36. $(5x^3 + 3x^2 - 17x + 6) \div (x + 2)$

37. $\dfrac{5 + 5x - 8x^2 + 4x^3 - 3x^4}{2 - x}$

38. $\dfrac{3 - 13x - 5x^2 + 9x^3 - 2x^4}{3 - x}$

39. $\dfrac{3x^4 + 3x^3 - x^2 + 3x + 2}{x + 1}$

40. $\dfrac{4x^4 + 12x^3 - x^2 - x + 2}{x + 3}$

41. $\dfrac{2x^4 - x^2 + 2}{x - 3}$

42. $\dfrac{x^4 - 3x^3 - 30}{x + 2}$

Objective C

Use the Remainder Theorem to evaluate each of the polynomials.

43. $P(x) = 2x^2 - 3x - 1; P(3)$

44. $Q(x) = 3x^2 - 5x - 1; Q(2)$

45. $R(x) = x^3 - 2x^2 + 3x - 1; R(4)$

46. $F(x) = x^3 + 4x^2 - 3x + 2; F(3)$

47. $P(z) = 2z^3 - 4z^2 + 3z - 1; P(-2)$

48. $R(t) = 3t^3 + t^2 - 4t + 2; R(-3)$

49. $Z(p) = 2p^3 - p^2 + 3; Z(-3)$

50. $P(y) = 3y^3 + 2y^2 - 5; P(-2)$

51. $Q(x) = x^4 + 3x^3 - 2x^2 + 4x - 9; Q(2)$

52. $Y(z) = z^4 - 2z^3 - 3z^2 - z + 7; Z(3)$

53. $F(x) = 2x^4 - x^3 + 2x - 5; F(-3)$

54. $Q(x) = x^4 - 2x^3 + 4x - 2; Q(-2)$

55. $P(x) = x^3 - 3; P(5)$

56. $S(t) = 4t^3 + 5; S(-4)$

57. $R(t) = 4t^4 - 3t^2 + 5; R(-3)$

58. $P(z) = 2z^4 + z^2 - 3; Z(-4)$

59. $Q(x) = x^5 - 4x^3 - 2x^2 + 5x - 2; Q(2)$

60. $T(x) = 2x^5 + 4x^4 - x^2 + 4; T(3)$

61. $R(x) = 2x^5 - x^3 + 4x - 1; R(-2)$

62. $P(x) = x^5 - x^3 + 4x + 1; P(-3)$

APPLYING THE CONCEPTS

63. Divide by using long division.

    a. $\dfrac{a^3 + b^3}{a + b}$
    b. $\dfrac{x^5 + y^5}{x + y}$
    c. $\dfrac{x^6 - y^6}{x + y}$

64. For what value of k will the remainder be zero?

    a. $(x^3 - x^2 - 3x + k) \div (x + 3)$
    b. $(2x^3 - x + k) \div (x - 1)$

65. Divide.

    a. $(2x^3 + 7x^2 + 2x - 8) \div (4x + 8)$
    b. $(4x^3 + 13x^2 - 22x + 24) \div (6x - 12)$

    c. $(2x^4 - 3x^3 + 4x^2 + x - 10) \div (x^2 - x + 1)$
    d. $(x^4 + 4x^3 + 2x^2 - x + 5) \div (x^2 - 2x - 3)$

66. Show how synthetic division can be modified so that the divisor can be of the form $ax + b$.

5.5 Factoring Polynomials

Objective A   **To factor a monomial from a polynomial**

The greatest common factor (GCF) of two or more monomials is the product of the common factors with the smallest exponents.

$$16a^4b = 2^4a^4b$$
$$40a^2b^5 = 2^3 \cdot 5a^2b^5$$

$$\text{GCF} = 2^3a^2b = 8a^2b$$

Note that the exponent of each variable in the GCF is the same as the smallest exponent of the variable in either of the monomials.

To **factor a polynomial** means to write the polynomial as a product of other polynomials.

In the example at the right, $3x$ is the GCF of the terms $3x^2$ and $6x$. $3x$ is a **common monomial factor** of the terms of the binomial. $x - 2$ is a **binomial factor** of $3x^2 - 6x$.

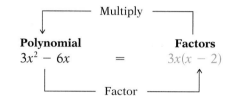

➡ Factor: $4x^3y^2 + 12x^3y + 20xy^3$

The GCF of $4x^3y^2$, $12x^3y$, and $20xy^3$ is $4xy$.

$4x^3y^2 + 12x^3y + 20xy^3$

$$= 4xy\left(\frac{4x^3y^2}{4xy} + \frac{12x^3y}{4xy} + \frac{20xy^3}{4xy}\right)$$

$$= 4xy(x^2y + 3x^2 + 5y^2)$$

• Find the GCF of the terms of the polynomial.

• Factor the GCF from each term of the polynomial. Think of this as dividing each term of the polynomial by the GCF.

Example 1
Factor: $4x^2y^2 - 6xy^2 + 12xy^3$

Solution
The GCF of $4x^2y^2$, $6xy^2$, and $12xy^3$ is $2xy^2$.

$4x^2y^2 - 6xy^2 + 12xy^3 = 2xy^2(2x - 3 + 6y)$

You Try It 1
Factor: $3x^3y - 6x^2y^2 - 3xy^3$

Your solution

Solution on p. S16

Example 2

Factor: $x^{2n} + x^{n+1} + x^n$

Solution

The GCF of x^{2n}, x^{n+1}, and x^n is x^n.

$x^{2n} + x^{n+1} + x^n = x^n(x^n + x + 1)$

You Try It 2

Factor: $6t^{2n} - 9t^n$

Your solution

Solution on p. S16

Objective B   **To factor by grouping**   〈 5 〉

In the examples at the right, the binomials in parentheses are binomial factors.

$$4x^4(2x - 3)$$
$$-2r^2s(5r + 2s)$$

The Distributive Property is used to factor a common binomial factor from an expression.

⇒ Factor: $4a(2b + 3) - 5(2b + 3)$

The common binomial factor is $(2b + 3)$. Use the Distributive Property to write the expression as a product of factors.

$4a(2b + 3) - 5(2b + 3) = (2b + 3)(4a - 5)$

Consider the binomial $y - x$. Factoring -1 from this binomial gives

$$y - x = -(x - y)$$

This equation is used to factor a common binomial from an expression.

TAKE NOTE

For the simplification at the right,

$6r(r - s) - 7(s - r)$
$= 6r(r - s) - 7[(-1)(r - s)]$
$= 6r(r - s) + 7(r - s)$

⇒ Factor: $6r(r - s) - 7(s - r)$

$6r(r - s) - 7(s - r) = 6r(r - s) + 7(r - s)$ • $s - r = -(r - s)$
$\qquad\qquad\qquad\qquad = (r - s)(6r + 7)$

Some polynomials can be factored by grouping terms so that a common binomial factor is found.

⇒ Factor: $3xz - 4yz - 3xa + 4ya$

$3xz - 4yz - 3xa + 4ya$
$\quad = (3xz - 4yz) - (3xa - 4ya)$ • Group the first two terms and the last two terms. Note that $-3xa + 4ya = -(3xa - 4ya)$.

$\quad = z(3x - 4y) - a(3x - 4y)$ • Factor the GCF from each group.
$\quad = (3x - 4y)(z - a)$ • Write the expression as the product of factors.

➡ Factor: $8y^2 + 4y - 6ay - 3a$

$8y^2 + 4y - 6ay - 3a$
$= (8y^2 + 4y) - (6ay + 3a)$

* Group the first two terms and the last two terms. Note that $-6ay - 3a = -(6ay + 3a)$.

$= 4y(2y + 1) - 3a(2y + 1)$
$= (2y + 1)(4y - 3a)$

* Factor the GCF from each group.
* Write the expression as the product of factors.

Example 3

Factor: $x^2(5y - 2) - 7(5y - 2)$

Solution

$x^2(5y - 2) - 7(5y - 2)$
$= (5y - 2)(x^2 - 7)$

You Try It 3

Factor: $3(6x - 7y) - 2x^2(6x - 7y)$

Your solution

Example 4

Factor: $15x^2 + 6x - 5xz - 2z$

Solution

$15x^2 + 6x - 5xz - 2z$
$= (15x^2 + 6x) - (5xz + 2z)$
$= 3x(5x + 2) - z(5x + 2)$
$= (5x + 2)(3x - z)$

You Try It 4

Factor: $4a^2 - 6a - 6ax + 9x$

Your solution

Solutions on p. S16

Objective C   **To factor a trinomial of the form $x^2 + bx + c$**

A **quadratic trinomial** is a trinomial of the form $ax^2 + bx + c$, where a, b, and c are nonzero integers. The degree of a quadratic trinomial is 2. Here are examples of quadratic trinomials:

$$4x^2 - 3x - 7 \qquad\qquad z^2 + z + 10 \qquad\qquad 2y^2 + 4y - 9$$
$$(a = 4, b = -3, c = -7) \qquad (a = 1, b = 1, c = 10) \qquad (a = 2, b = 4, c = -9)$$

To **factor a quadratic trinomial** means to express the trinomial as the product of two binomials. For example,

Trinomial		**Factored Form**
$2x^2 - x - 1$	$=$	$(2x + 1)(x - 1)$
$y^2 - 3y + 2$	$=$	$(y - 1)(y - 2)$

In this objective, trinomials of the form $x^2 + bx + c$ ($a = 1$) will be factored. The next objective deals with trinomials where $a \neq 1$.

The method by which factors of a trinomial are found is based on FOIL. Consider the following binomial products, noting the relationship between the constant terms of the binomials and the terms of the trinomial.

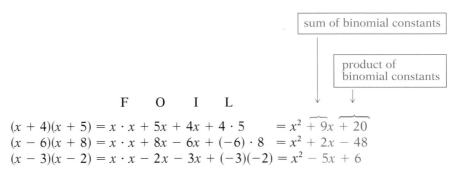

$$(x + 4)(x + 5) = x \cdot x + 5x + 4x + 4 \cdot 5 \quad = x^2 + 9x + 20$$
$$(x - 6)(x + 8) = x \cdot x + 8x - 6x + (-6) \cdot 8 \; = x^2 + 2x - 48$$
$$(x - 3)(x - 2) = x \cdot x - 2x - 3x + (-3)(-2) = x^2 - 5x + 6$$

Observe two important points from these examples.

1. The constant term of the trinomial is the *product* of the constant terms of the binomials. The coefficient of x in the trinomial is the *sum* of the constant terms of the binomials.

2. When the constant term of the trinomial is positive, the constant terms of the binomials have the *same* sign. When the constant term of the trinomial is negative, the constant terms of the binomials have *opposite* signs.

➡ Factor: $x^2 - 7x + 12$

The constant term is positive. The signs of the binomial constants will be the same.

Find two negative factors of 12 whose sum is -7.

Negative Factors of 12	Sum
$-1, -12$	-13
$-2, -6$	-8
$-3, -4$	-7

Write the trinomial in factored form. $\quad x^2 - 7x + 12 = (x - 3)(x - 4)$

Check: $(x - 3)(x - 4) = x^2 - 4x - 3x + 12 = x^2 - 7x + 12$

➡ Factor: $y^2 + 10y - 24$

The constant term is negative. The signs of the binomial constants will be opposite.

Find two factors of -24 that have opposite signs and whose sum is 10. All of the possible factors are shown at the right. In practice, once the correct pair is found, the remaining choices need not be checked.

Factors of -24	Sum
$-1, \quad 24$	23
$1, -24$	-23
$-2, \quad 12$	10
$2, -12$	-10
$-3, \quad 8$	5
$3, -8$	-5
$-4, \quad 6$	2
$4, -6$	-2

Write the trinomial in factored form. $\quad y^2 + 10y - 24 = (y - 2)(y + 12)$

Check: $(y - 2)(y + 12) = y^2 + 12y - 2y - 24 = y^2 + 10y - 24$

When only integers are used, some trinomials do not factor. For example, to factor $x^2 + 11x + 5$, it would be necessary to find two positive integers whose product is 5 and whose sum is 11. This is not possible, because the only positive factors of 5 are 1 and 5, and the sum of 1 and 5 is 6. The polynomial $x^2 + 11x + 5$ is a *prime polynomial*. Such a polynomial is said to be **nonfactorable over the integers.** Binomials of the form $x + a$ or $x - a$ are also prime polynomials.

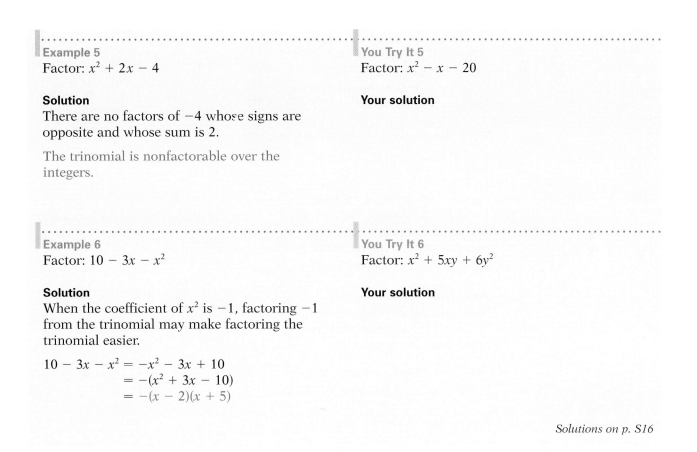

Example 5

Factor: $x^2 + 2x - 4$

Solution

There are no factors of -4 whose signs are opposite and whose sum is 2.

The trinomial is nonfactorable over the integers.

You Try It 5

Factor: $x^2 - x - 20$

Your solution

Example 6

Factor: $10 - 3x - x^2$

Solution

When the coefficient of x^2 is -1, factoring -1 from the trinomial may make factoring the trinomial easier.

$$10 - 3x - x^2 = -x^2 - 3x + 10$$
$$= -(x^2 + 3x - 10)$$
$$= -(x - 2)(x + 5)$$

You Try It 6

Factor: $x^2 + 5xy + 6y^2$

Your solution

Solutions on p. S16

Objective D   **To factor $ax^2 + bx + c$**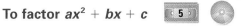

There are various methods of factoring trinomials of the form $ax^2 + bx + c$, where $a \neq 1$. Factoring by using trial factors and factoring by grouping will be discussed in this objective. Factoring by using trial factors is illustrated first.

To use the trial factor method, use the factors of a and the factors of c to write all of the possible binomial factors of the trinomial. Then use FOIL to determine the correct factorization. To reduce the number of trial factors that must be considered, remember the following.

1. Use the signs of the constant term and the coefficient of x in the trinomial to determine the signs of the binomial factors. If the constant term positive, the signs of the binomial factors will be the same as the sign of the efficient of x in the trinomial. If the sign of the constant term is negative, t onstant terms in the binomials will have opposite signs.

2. If the terms of the trinomial do not have a common factor, then rms in either one of the binomial factors will not have a common factor

➡ Factor: $3x^2 - 8x + 4$

The terms have no common factor.
The constant term is positive.
The coefficient of x is negative.
The binomial constants will be negative.

Positive Factors of 3 (coefficient of x^2)	Negative Factors of 4 (constant term)
1, 3	−1, −4
	−2, −2

Write trial factors. Use the *Outer* and *Inner* products of FOIL to determine the middle term of the trinomial.

Trial Factors	Middle Term
$(x - 1)(3x - 4)$	$-4x - 3x = -7x$
$(x - 4)(3x - 1)$	$-x - 12x = -13x$
$(x - 2)(3x - 2)$	$-2x - 6x = -8x$

Write the trinomial in factored form.

$$3x^2 - 8x + 4 = (x - 2)(3x - 2)$$

➡ Factor: $5 - 3x - 2x^2$

Factor -1 from the trinomial.
$5 - 3x - 2x^2 = -(2x^2 + 3x - 5)$
The constant term, -5, is negative; the signs of the binomial constants will be opposites.

Positive Factors of 2 (coefficient of x^2)	Factors of −5 (constant term)
1, 2	1, −5
	−1, 5

Write trial factors. Use the *Outer* and *Inner* products of FOIL to determine the middle term of the trinomial.

Trial Factors	Middle Term
$(x + 1)(2x - 5)$	$-5x + 2x = -3x$
$(x - 5)(2x + 1)$	$x - 10x = -9x$
$(x - 1)(2x + 5)$	$5x - 2x = 3x$
$(x + 5)(2x - 1)$	$-x + 10x = 9x$

Write the trinomial in factored form.

$$5 - 3x - 2x^2 = -(x - 1)(2x + 5)$$

➡ Factor: $10y^3 + 44y^2 - 30y$

The GCF is $2y$. Factor the GCF from the terms.

$$10y^3 + 44y^2 - 30y = 2y(5y^2 + 22y - 15)$$

Factor the trinomial.
The constant term is negative.
The binomial constants will have opposite signs.

Positive Factors of 5 (coefficient of y^2)	Factors of −15 (constant term)
1, 5	−1, 15
	1, −15
	−3, 5
	3, −5

Write trial factors. Use the *Outer* and *Inner* products of FOIL to determine the middle term of the trinomial.

It is not necessary to test trial factors that have a common factor.

Trial Factors	Middle Term
$(y - 1)(5y + 15)$	common factor
$(y + 15)(5y - 1)$	$-y + 75y = 74y$
$(y + 1)(5y - 15)$	common factor
$(y - 15)(5y + 1)$	$y - 75y = -74y$
$(y - 3)(5y + 5)$	common factor
$(y + 5)(5y - 3)$	$-3y + 25y = 22y$
$(y + 3)(5y - 5)$	common factor
$(y - 5)(5y + 3)$	$3y - 25y = -22y$

Write the trinomial in factored form.

$$10y^3 + 44y^2 - 30y = 2y(y + 5)(5y - 3)$$

For previous examples, all the trial factors were listed. Once the correct factors have been found, however, the remaining trial factors can be omitted.

Trinomials of the form $ax^2 + bx + c$ can also be factored by grouping. This method is an extension of the method discussed in the preceding objective.

To factor $ax^2 + bx + c$, first find two factors of $a \cdot c$ whose sum is b. Then use factoring by grouping to write the factorization of the trinomial.

For the trinomial $3x^2 - 8x + 4$, $a = 3$, $b = -8$, and $c = 4$. To find two factors of $a \cdot c$ whose sum is b, first find the product $a \cdot c$ ($a \cdot c = 3 \cdot 4 = 12$). Then find two factors of 12 whose sum is -8. (-2 and -6 are two factors of 12 whose sum is -8.)

⇒ Factor: $3x^2 + 11x + 8$

Find two positive factors of 24 ($ac = 3 \cdot 8$) whose sum is 11, the coefficient of x.

Positive Factors of 24	Sum
1, 24	25
2, 12	14
3, 8	11

The required sum has been found. The remaining factors need not be checked.

Use the factors of 24 whose sum is 11 to write $11x$ as $3x + 8x$.
Factor by grouping.

$$3x^2 + 11x + 8 = 3x^2 + 3x + 8x + 8$$
$$= (3x^2 + 3x) + (8x + 8)$$
$$= 3x(x + 1) + 8(x + 1)$$
$$= (x + 1)(3x + 8)$$

Check: $(x + 1)(3x + 8) = 3x^2 + 8x + 3x + 8 = 3x^2 + 11x + 8$

⇒ Factor: $4z^2 - 17z - 21$

Find two factors of -84 [$ac = 4 \cdot (-21)$] whose sum is -17, the coefficient of z.

Once the required sum is found, the remaining factors need not be checked.

Factors of -84	Sum
1, -84	-83
-1, 84	83
2, -42	-40
-2, 42	40
3, -28	-25
-3, 28	25
4, -21	-17

Use the factors of -84 whose sum is -17 to write $-17z$ as $4z - 21z$.

$$4z^2 - 17z - 21 = 4z^2 + 4z - 21z - 21$$

Factor by grouping. Recall that $-21z - 21 = -(21z + 21)$.

$$= (4z^2 + 4z) - (21z + 21)$$
$$= 4z(z + 1) - 21(z + 1)$$
$$= (z + 1)(4z - 21)$$

Check: $(z + 1)(4z - 21) = 4z^2 - 21z + 4z - 21 = 4z^2 - 17z - 21$

⇒ Factor: $3x^2 - 11x + 4$

Find two negative factors of 12 (3 · 4) whose sum is -11.

Negative Factors of 12	Sum
$-1, -12$	-13
$-2, -6$	-8
$-3, -4$	-7

Because no integer factors of 12 have a sum of -11, $3x^2 - 11x + 4$ is nonfactorable over the integers. $3x^2 - 11x + 4$ is a prime polynomial over the integers.

Either method of factoring discussed in this objective will always lead to a correct factorization of trinomials of the form $ax^2 + bx + c$ that are not prime polynomials.

....

Example 7

Factor: $6x^2 + 11x - 10$

Solution

$6x^2 + 11x - 10 = (2x + 5)(3x - 2)$

You Try It 7

Factor: $4x^2 + 15x - 4$

Your solution

....

Example 8

Factor: $12x^2 - 32x + 5$

Solution

$12x^2 - 32x + 5 = (6x - 1)(2x - 5)$

You Try It 8

Factor: $10x^2 + 39x + 14$

Your solution

....

Example 9

Factor: $30y + 2xy - 4x^2y$

Solution

The GCF of $30y$, $2xy$, and $4x^2y$ is $2y$.

$30y + 2xy - 4x^2y = 2y(15 + x - 2x^2)$
$= -2y(2x^2 - x - 15)$
$= -2y(2x + 5)(x - 3)$

You Try It 9

Factor: $3a^3b^3 + 3a^2b^2 - 60ab$

Your solution

Solutions on p. S16

5.5 Exercises

Objective A

Factor.

1. $6a^2 - 15a$

2. $32b^2 + 12b$

3. $4x^3 - 3x^2$

4. $12a^5b^2 + 16a^4b$

5. $3a^2 - 10b^3$

6. $9x^2 + 14y^4$

7. $x^5 - x^3 - x$

8. $y^4 - 3y^2 - 2y$

9. $16x^2 - 12x + 24$

10. $2x^5 + 3x^4 - 4x^2$

11. $5b^2 - 10b^3 + 25b^4$

12. $x^2y^4 - x^2y - 4x^2$

13. $x^{2n} - x^n$

14. $2a^{5n} + a^{2n}$

15. $x^{3n} - x^{2n}$

16. $y^{4n} + y^{2n}$

17. $a^{2n+2} + a^2$

18. $b^{n+5} - b^5$

19. $12x^2y^2 - 18x^3y + 24x^2y$

20. $14a^4b^4 - 42a^3b^3 + 28a^3b^2$

21. $24a^3b^2 - 4a^2b^2 - 16a^2b^4$

22. $10x^2y + 20x^2y^2 + 30x^2y^3$

23. $y^{2n+2} + y^{n+2} - y^2$

24. $a^{2n+2} + a^{2n+1} + a^n$

Objective B

Factor.

25. $x(a + 2) - 2(a + 2)$

26. $3(x + y) + a(x + y)$

27. $a(x - 2) - b(2 - x)$

28. $3(a - 7) - b(7 - a)$ **29.** $x(a - 2b) + y(2b - a)$ **30.** $b(3 - 2c) - 5(2c - 3)$

31. $xy + 4y - 2x - 8$ **32.** $ab + 7b - 3a - 21$ **33.** $ax + bx - ay - by$

34. $2ax - 3ay - 2bx + 3by$ **35.** $x^2y - 3x^2 - 2y + 6$ **36.** $a^2b + 3a^2 + 2b + 6$

37. $6 + 2y + 3x^2 + x^2y$ **38.** $15 + 3b - 5a^2 - a^2b$ **39.** $2ax^2 + bx^2 - 4ay - 2by$

40. $4a^2x + 2a^2y - 6bx - 3by$ **41.** $6xb + 3ax - 4by - 2ay$ **42.** $a^2x - 3a^2y + 2x - 6y$

43. $x^ny - 5x^n + y - 5$ **44.** $a^nx^n + 2a^n + x^n + 2$ **45.** $2x^3 - x^2 + 4x - 2$

Objective C

Factor.

46. $x^2 - 8x + 15$ **47.** $x^2 + 12x + 20$ **48.** $a^2 + 12a + 11$

49. $a^2 + a - 72$ **50.** $b^2 + 2b - 35$ **51.** $a^2 + 7a + 6$

52. $y^2 - 16y + 39$ **53.** $y^2 - 18y + 72$ **54.** $b^2 + 4b - 32$

55. $x^2 + x - 132$ **56.** $a^2 - 15a + 56$ **57.** $x^2 + 15x + 50$

58. $y^2 + 13y + 12$ **59.** $b^2 - 6b - 16$ **60.** $x^2 + 4x - 5$

61. $a^2 - 3ab + 2b^2$ **62.** $a^2 + 11ab + 30b^2$ **63.** $a^2 + 8ab - 33b^2$

64. $x^2 - 14xy + 24y^2$

65. $x^2 + 5xy + 6y^2$

66. $y^2 + 2xy - 63x^2$

67. $2 + x - x^2$

68. $21 - 4x - x^2$

69. $5 + 4x - x^2$

70. $50 + 5a - a^2$

71. $x^2 - 5x + 6$

72. $x^2 - 7x - 12$

Objective D

Factor.

73. $2x^2 + 7x + 3$

74. $2x^2 - 11x - 40$

75. $6y^2 + 5y - 6$

76. $4y^2 - 15y + 9$

77. $6b^2 - b - 35$

78. $2a^2 + 13a + 6$

79. $3y^2 - 22y + 39$

80. $12y^2 - 13y - 72$

81. $6a^2 - 26a + 15$

82. $5x^2 + 26x + 5$

83. $4a^2 - a - 5$

84. $11x^2 - 122x + 11$

85. $10x^2 - 29x + 10$

86. $2x^2 + 5x + 12$

87. $4x^2 - 6x + 1$

88. $6x^2 + 5xy - 21y^2$

89. $6x^2 + 41xy - 7y^2$

90. $4a^2 + 43ab + 63b^2$

91. $7a^2 + 46ab - 21b^2$

92. $10x^2 - 23xy + 12y^2$

93. $18x^2 + 27xy + 10y^2$

94. $24 + 13x - 2x^2$

95. $6 - 7x - 5x^2$

96. $8 - 13x + 6x^2$

97. $30 + 17a - 20a^2$

98. $15 - 14a - 8a^2$

99. $35 - 6b - 8b^2$

100. $12y^3 + 22y^2 - 70y$

101. $5y^4 - 29y^3 + 20y^2$

102. $30a^2 + 85ab + 60b^2$

103. $20x^2 - 38x^3 - 30x^4$

104. $4x^2y^2 - 32xy + 60$

105. $a^4b^4 - 3a^3b^3 - 10a^2b^2$

106. $2a^2b^4 + 9ab^3 - 18b^2$

107. $90a^2b^2 + 45ab + 10$

108. $3x^3y^2 + 12x^2y - 96x$

109. $4x^4 - 45x^2 + 80$

110. $x^4 + 2x^2 + 15$

111. $2a^5 + 14a^3 + 20a$

112. $3b^6 - 9b^4 - 30b^2$

113. $3x^4y^2 - 39x^2y^2 + 120y^2$

114. $2a^3b^3 - 10a^2b^2 + 12ab$

115. $3x^3y^3 + 6x^2y^2 - 24xy$

116. $2x^4 + 14x^2 + 24$

117. $y^5 - 8y^3 + 15y$

118. $12x + x^2 - 6x^3$

119. $3y - 16y^2 + 16y^3$

120. $12x^{2n} - 30x^n + 12$

121. $12y^{2n} - 51y^n + 45$

122. $2x^{3n} + 4x^{2n} - 30x^n$

123. $x^{3n} + 10x^{2n} + 16x^n$

APPLYING THE CONCEPTS

124. Write the areas of the shaded regions in factored form.

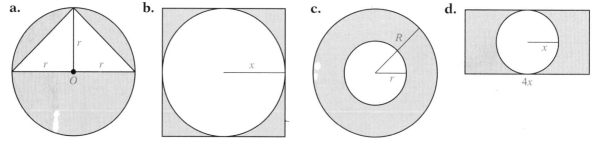

125. Find all integers k such that the trinomial can be factored.

a. $x^2 + kx + 8$

b. $x^2 + kx - 6$

c. $2x^2 + kx + 3$

d. $2x^2 + kx - 5$

e. $3x^2 + kx + 5$

f. $2x^2 + kx - 3$

5.6 Special Factoring

Objective A

To factor the difference of two perfect squares or a perfect-square trinomial

The product of a term and itself is called a **perfect square.** The exponents on variables of perfect squares are always even numbers.

Term		Perfect Square
5	$5 \cdot 5 =$	25
x	$x \cdot x =$	x^2
$3y^4$	$3y^4 \cdot 3y^4 =$	$9y^8$
x^n	$x^n \cdot x^n =$	x^{2n}

The **square root** of a perfect square is one of the two equal factors of the perfect square. "$\sqrt{}$" is the symbol for square root. To find the exponent of the square root of a variable term, multiply the exponent by $\frac{1}{2}$.

$$\sqrt{25} = 5$$
$$\sqrt{x^2} = x$$
$$\sqrt{9y^8} = 3y^4$$
$$\sqrt{x^{2n}} = x^n$$

The **difference of two perfect squares** is the product of the sum and difference of two terms. The factors of the difference of two perfect squares are the sum and difference of the square roots of the perfect squares.

> **Factors of the Difference of Two Perfect Squares**
>
> $a^2 - b^2 = (a + b)(a - b)$

The **sum of two perfect squares,** $a^2 + b^2$, is nonfactorable over the integers.

⇒ Factor: $4x^2 - 81y^2$

Write the binomial as the difference of two perfect squares.

$$4x^2 - 81y^2 = (2x)^2 - (9y)^2$$

The factors are the sum and difference of the square roots of the perfect squares.

$$= (2x + 9y)(2x - 9y)$$

A **perfect-square trinomial** is the square of a binomial.

> **Factors of a Perfect-Square Trinomial**
>
> $a^2 + 2ab + b^2 = (a + b)^2$
> $a^2 - 2ab + b^2 = (a - b)^2$

In factoring a perfect-square trinomial, remember that the terms of the binomial are the square roots of the perfect squares of the trinomial. The sign in the binomial is the sign of the middle term of the trinomial.

➡ Factor: $4x^2 + 12x + 9$

Because $4x^2$ is a perfect square $[4x^2 = (2x)^2]$ and 9 is a perfect square $(9 = 3^2)$, try factoring $4x^2 + 12x + 9$ as the square of a binomial.

$4x^2 + 12x + 9 \stackrel{?}{=} (2x + 3)^2$

Check: $(2x + 3)^2 = (2x + 3)(2x + 3) = 4x^2 + 6x + 6x + 9 = 4x^2 + 12x + 9$

The check verifies that $4x^2 + 12x + 9 = (2x + 3)^2$.

It is important to check a proposed factorization as we did above. The next example illustrates the importance of this check.

➡ Factor: $x^2 + 13x + 36$

Because x^2 is a perfect square and 36 is a perfect square, try factoring $x^2 + 13x + 36$ as the square of a binomial.

$x^2 + 13x + 36 \stackrel{?}{=} (x + 6)^2$

Check: $(x + 6)^2 = (x + 6)(x + 6) = x^2 + 6x + 6x + 36 = x^2 + 12x + 36$

In this case, the proposed factorization of $x^2 + 13x + 36$ does *not* check. Try another factorization. The numbers 4 and 9 are factors of 36 whose sum is 13.

$x^2 + 13x + 36 = (x + 4)(x + 9)$

Example 1
Factor: $25x^2 - 1$

Solution
$25x^2 - 1 = (5x)^2 - (1)^2$
$\qquad\quad = (5x + 1)(5x - 1)$

You Try It 1
Factor: $x^2 - 36y^4$

Your solution

Example 2
Factor: $4x^2 - 20x + 25$

Solution
$4x^2 - 20x + 25 = (2x - 5)^2$

You Try It 2
Factor: $9x^2 + 12x + 4$

Your solution

Example 3
Factor: $(x + y)^2 - 4$

Solution
$(x + y)^2 - 4 = (x + y)^2 - (2)^2$
$\qquad\qquad\quad = (x + y + 2)(x + y - 2)$

You Try It 3
Factor: $(a + b)^2 - (a - b)^2$

Your solution

Solutions on p. S16

Objective B   **To factor the sum or the difference of two perfect cubes**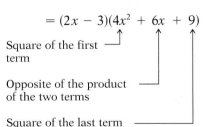

The product of the same three factors is called a **perfect cube.** The exponents on variables of perfect cubes are always divisible by 3.

Term		Perfect Cube
2	$2 \cdot 2 \cdot 2 = 2^3 =$	8
$3y$	$3y \cdot 3y \cdot 3y = (3y)^3 =$	$27y^3$
y^2	$y^2 \cdot y^2 \cdot y^2 = (y^2)^3 =$	y^6

The **cube root** of a perfect cube is one of the three equal factors of the perfect cube. "$\sqrt[3]{}$" is the symbol for cube root. To find the exponent of the cube root of a variable term, multiply the exponent by $\frac{1}{3}$.

$\sqrt[3]{8} = 2$
$\sqrt[3]{27y^3} = 3y$
$\sqrt[3]{y^6} = y^2$

The following rules are used to factor the sum or difference of two perfect cubes.

Factoring the Sum or Difference of Two Perfect Cubes

$a^3 + b^3 = (a + b)(a^2 - ab + b^2)$
$a^3 - b^3 = (a - b)(a^2 + ab + b^2)$

To factor $8x^3 - 27$:

Write the binomial as the difference of two perfect cubes.

$$8x^3 - 27 = (2x)^3 - 3^3$$

The terms of the binomial factor are the cube roots of the perfect cubes. The sign of the binomial factor is the same sign as in the given binomial. The trinomial factor is obtained from the binomial factor.

$$= (2x - 3)(4x^2 + 6x + 9)$$

Square of the first term ⟶

Opposite of the product of the two terms ⟶

Square of the last term ⟶

⟹ Factor: $a^3 + 64y^3$

$a^3 + 64y^3 = a^3 + (4y)^3$

• Write the binomial as the sum of two perfect cubes.

$= (a + 4y)(a^2 - 4ay + 16y^2)$

• Factor.

⟹ Factor: $64y^4 - 125y$

$64y^4 - 125y = y(64y^3 - 125)$

• Factor out y, the GCF.

$= y[(4y)^3 - 5^3]$

• Write the binomial as the difference of two perfect cubes.

$= y(4y - 5)(16y^2 + 20y + 25)$

• Factor.

Example 4
Factor: $x^3y^3 - 1$

Solution
$x^3y^3 - 1 = (xy)^3 - 1^3$
$= (xy - 1)(x^2y^2 + xy + 1)$

You Try It 4
Factor: $a^3b^3 - 27$

Your solution

Example 5
Factor: $64c^3 + 8d^3$

Solution
$64c^3 + 8d^3 = 8(8c^3 + d^3)$
$= 8[(2c)^3 + d^3]$
$= 8(2c + d)(4c^2 - 2cd + d^2)$

You Try It 5
Factor: $8x^3 + y^3z^3$

Your solution

Example 6
Factor: $(x + y)^3 - x^3$

Solution
$(x + y)^3 - x^3$
$= [(x + y) - x][(x + y)^2 + x(x + y) + x^2]$
$= y(x^2 + 2xy + y^2 + x^2 + xy + x^2)$
$= y(3x^2 + 3xy + y^2)$

You Try It 6
Factor: $(x - y)^3 + (x + y)^3$

Your solution

Solutions on p. S16

Objective C **To factor a trinomial that is quadratic in form**

Certain trinomials that are not quadratic can be expressed as quadratic trinomials by making suitable variable substitutions. A trinomial is **quadratic in form** if it can be written as $au^2 + bu + c$.

As shown below, the trinomials $x^4 + 5x^2 + 6$ and $2x^2y^2 + 3xy - 9$ are quadratic in form.

$$x^4 + 5x^2 + 6 \qquad\qquad 2x^2y^2 + 3xy - 9$$

$$(x^2)^2 + 5(x^2) + 6 \qquad\qquad 2(xy)^2 + 3(xy) - 9$$

Let $u = x^2$. $\quad u^2 + 5u + 6 \qquad\qquad$ Let $u = xy$. $\quad 2u^2 + 3u - 9$

When we use this method to factor a trinomial that is quadratic in form, the variable part of the first term in each binomial will be u.

TAKE NOTE
The trinomial $x^4 + 5x^2 + 6$ was shown to be quadratic in form on the preceding page.

⇒ Factor: $x^4 + 5x^2 + 6$

$$x^4 + 5x^2 + 6 = u^2 + 5u + 6$$ • Let $u = x^2$.
$$= (u + 3)(u + 2)$$ • Factor.
$$= (x^2 + 3)(x^2 + 2)$$ • Replace u by x^2.

Here is an example in which $u = \sqrt{x}$.

⇒ Factor: $x - 2\sqrt{x} - 15$

$$x - 2\sqrt{x} - 15 = u^2 - 2u - 15$$ • Let $u = \sqrt{x}$. Then $u^2 = x$.
$$= (u - 5)(u + 3)$$ • Factor.
$$= (\sqrt{x} - 5)(\sqrt{x} + 3)$$ • Replace u by $\sqrt{x}$.

Example 7
Factor: $6x^2y^2 - xy - 12$

Solution
Let $u = xy$.

$$6x^2y^2 - xy - 12 = 6u^2 - u - 12$$
$$= (3u + 4)(2u - 3)$$
$$= (3xy + 4)(2xy - 3)$$

You Try It 7
Factor: $3x^4 + 4x^2 - 4$

Your solution

Solution on p. S17

Objective D　**To factor completely**　❮ 5 ❯

When factoring a polynomial completely, ask the following questions about the polynomial.

1. Is there a common factor? If so, factor out the GCF.

2. If the polynomial is a binomial, is it the difference of two perfect squares, the sum of two perfect cubes, or the difference of two perfect cubes? If so, factor.

3. If the polynomial is a trinomial, is it a perfect-square trinomial or the product of two binomials? If so, factor.

4. Can the polynomial be factored by grouping? If so, factor.

5. Is each factor nonfactorable over the integers? If not, factor.

TAKE NOTE
Remember that you may have to factor more than once in order to write the polynomial as a product of *prime* factors.

Example 8

Factor: $6a^3 + 15a^2 - 36a$

Solution

$$6a^3 + 15a^2 - 36a = 3a(2a^2 + 5a - 12)$$
$$= 3a(2a - 3)(a + 4)$$

You Try It 8

Factor: $18x^3 - 6x^2 - 60x$

Your solution

Example 9

Factor: $x^2y + 2x^2 - y - 2$

Solution

$$x^2y + 2x^2 - y - 2 = (x^2y + 2x^2) - (y + 2)$$
$$= x^2(y + 2) - (y + 2)$$
$$= (y + 2)(x^2 - 1)$$
$$= (y + 2)(x + 1)(x - 1)$$

You Try It 9

Factor: $4x - 4y - x^3 + x^2y$

Your solution

Example 10

Factor: $x^{4n} - y^{4n}$

Solution

$$x^{4n} - y^{4n} = (x^{2n})^2 - (y^{2n})^2$$
$$= (x^{2n} + y^{2n})(x^{2n} - y^{2n})$$
$$= (x^{2n} + y^{2n})[(x^n)^2 - (y^n)^2]$$
$$= (x^{2n} + y^{2n})(x^n + y^n)(x^n - y^n)$$

You Try It 10

Factor: $x^{4n} - x^{2n}y^{2n}$

Your solution

Example 11

Factor: $x^{n+3} + x^n y^3$

Solution

$$x^{n+3} + x^n y^3 = x^n(x^3 + y^3)$$
$$= x^n(x + y)(x^2 - xy + y^2)$$

You Try It 11

Factor: $ax^5 - ax^2y^6$

Your solution

Solutions on p. S17

5.6 Exercises

Objective A

Determine which of the following expressions are perfect squares.

1. $4; 8; 25x^6; 12y^{10}; 100x^4y^4$

2. $9; 18; 15a^8; 49b^{12}; 64a^{16}b^2$

Find the square root of the expression.

3. $16z^8$

4. $36d^{10}$

5. $81a^4b^6$

6. $25m^2n^{12}$

Factor.

7. $x^2 - 16$

8. $y^2 - 49$

9. $4x^2 - 1$

10. $81x^2 - 4$

11. $16x^2 - 121$

12. $49y^2 - 36$

13. $1 - 9a^2$

14. $16 - 81y^2$

15. $x^2y^2 - 100$

16. $a^2b^2 - 25$

17. $x^2 + 4$

18. $a^2 + 16$

19. $25 - a^2b^2$

20. $64 - x^2y^2$

21. $a^{2n} - 1$

22. $b^{2n} - 16$

23. $x^2 - 12x + 36$

24. $y^2 - 6y + 9$

25. $b^2 - 2b + 1$

26. $a^2 + 14a + 49$

27. $16x^2 - 40x + 25$

28. $49x^2 + 28x + 4$

29. $4a^2 + 4a - 1$

30. $9x^2 + 12x - 4$

31. $b^2 + 7b + 14$

32. $y^2 - 5y + 25$

33. $x^2 + 6xy + 9y^2$

34. $4x^2y^2 + 12xy + 9$

35. $25a^2 - 40ab + 16b^2$

36. $4a^2 - 36ab + 81b^2$

37. $x^{2n} + 6x^n + 9$ **38.** $y^{2n} - 16y^n + 64$ **39.** $(x - 4)^2 - 9$

40. $16 - (a - 3)^2$ **41.** $(x - y)^2 - (a + b)^2$ **42.** $(x - 2y)^2 - (x + y)^2$

Objective B

Determine which of the following expressions are perfect cubes.

43. $4; 8; x^9; a^8b^8; 27c^{15}d^{18}$ **44.** $9; 27; y^{12}; m^3n^6; 64mn^9$

Find the cube root of the expression.

45. $8x^9$ **46.** $27y^{15}$ **47.** $64a^6b^{18}$ **48.** $125c^{12}d^3$

Factor.

49. $x^3 - 27$ **50.** $y^3 + 125$ **51.** $8x^3 - 1$

52. $64a^3 + 27$ **53.** $x^3 - y^3$ **54.** $x^3 - 8y^3$

55. $m^3 + n^3$ **56.** $27a^3 + b^3$ **57.** $64x^3 + 1$

58. $1 - 125b^3$ **59.** $27x^3 - 8y^3$ **60.** $64x^3 + 27y^3$

61. $x^3y^3 + 64$ **62.** $8x^3y^3 + 27$ **63.** $16x^3 - y^3$

64. $27x^3 - 8y^2$ **65.** $8x^3 - 9y^3$ **66.** $27a^3 - 16$

67. $(a - b)^3 - b^3$ **68.** $a^3 + (a + b)^3$ **69.** $x^{6n} + y^{3n}$

70. $x^{3n} + y^{3n}$ **71.** $x^{3n} + 8$ **72.** $a^{3n} + 64$

Objective C

Factor.

73. $x^2y^2 - 8xy + 15$

74. $x^2y^2 - 8xy - 33$

75. $x^2y^2 - 17xy + 60$

76. $a^2b^2 + 10ab + 24$

77. $x^4 - 9x^2 + 18$

78. $y^4 - 6y^2 - 16$

79. $b^4 - 13b^2 - 90$

80. $a^4 + 14a^2 + 45$

81. $x^4y^4 - 8x^2y^2 + 12$

82. $a^4b^4 + 11a^2b^2 - 26$

83. $x^{2n} + 3x^n + 2$

84. $a^{2n} - a^n - 12$

85. $3x^2y^2 - 14xy + 15$

86. $5x^2y^2 - 59xy + 44$

87. $6a^2b^2 - 23ab + 21$

88. $10a^2b^2 + 3ab - 7$

89. $2x^4 - 13x^2 - 15$

90. $3x^4 + 20x^2 + 32$

91. $2x^{2n} - 7x^n + 3$

92. $4x^{2n} + 8x^n - 5$

93. $6a^{2n} + 19a^n + 10$

Objective D

Factor.

94. $5x^2 + 10x + 5$

95. $12x^2 - 36x + 27$

96. $3x^4 - 81x$

97. $27a^4 - a$

98. $7x^2 - 28$

99. $20x^2 - 5$

100. $y^4 - 10y^3 + 21y^2$

101. $y^5 + 6y^4 - 55y^3$

102. $x^4 - 16$

103. $16x^4 - 81$

104. $8x^5 - 98x^3$

105. $16a - 2a^4$

106. $x^3y^3 - x^3$

107. $a^3b^6 - b^3$

108. $x^6y^6 - x^3y^3$

109. $8x^4 - 40x^3 + 50x^2$

110. $6x^5 + 74x^4 + 24x^3$

111. $x^4 - y^4$

112. $16a^4 - b^4$

113. $x^6 + y^6$

114. $x^4 - 5x^2 - 4$

115. $a^4 - 25a^2 - 144$

116. $3b^5 - 24b^2$

117. $16a^4 - 2a$

118. $x^4y^2 - 5x^3y^3 + 6x^2y^4$

119. $a^4b^2 - 8a^3b^3 - 48a^2b^4$

120. $16x^3y + 4x^2y^2 - 42xy^3$

121. $24a^2b^2 - 14ab^3 - 90b^4$

122. $x^3 - 2x^2 - x + 2$

123. $x^3 - 2x^2 - 4x + 8$

124. $4x^2y^2 - 4x^2 - 9y^2 + 9$

125. $4x^4 - x^2 - 4x^2y^2 + y^2$

126. $a^{2n+2} - 6a^{n+2} + 9a^2$

127. $x^{2n+1} + 2x^{n+1} + x$

128. $2x^{n+2} - 7x^{n+1} + 3x^n$

129. $3b^{n+2} + 4b^{n+1} - 4b^n$

APPLYING THE CONCEPTS

130. Factor: $x^2(x - 3) - 3x(x - 3) + 2(x - 3)$

131. Given that $(x - 3)$ and $(x + 4)$ are factors of $x^3 + 6x^2 - 7x - 60$, explain how you can find a third *first-degree* factor of $x^3 + 6x^2 - 7x - 60$. Then find the factor.

5.7 Solving Equations by Factoring

Objective A   **To solve an equation by factoring**   〈5〉

Consider the equation $ab = 0$. If a is not zero, then b must be zero. Conversely, if b is not zero, then a must be zero. This is summarized in the **Principle of Zero Products.**

Principle of Zero Products

If the product of two factors is zero, then at least one of the factors must be zero.

If $ab = 0$, then $a = 0$ or $b = 0$.

The Principle of Zero Products is used to solve equations.

➡ Solve: $(x - 4)(x + 2) = 0$

By the Principle of Zero Products, if $(x - 4)(x + 2) = 0$, then $x - 4 = 0$ or $x + 2 = 0$.

$$x - 4 = 0 \qquad\qquad x + 2 = 0$$
$$x = 4 \qquad\qquad x = -2$$

Check:

$$\begin{array}{c|c} (x - 4)(x + 2) = 0 \\ \hline (4 - 4)(4 + 2) & 0 \\ 0 \cdot 6 & 0 \\ & 0 = 0 \end{array} \qquad \begin{array}{c|c} (x - 4)(x + 2) = 0 \\ \hline (-2 - 4)(-2 + 2) & 0 \\ -6 \cdot 0 & 0 \\ & 0 = 0 \end{array}$$

-2 and 4 check as solutions. The solutions are -2 and 4.

An equation of the form $ax^2 + bx + c = 0$, $a \neq 0$, is a **quadratic equation.** A quadratic equation is in **standard form** when the polynomial is written in descending order and equal to zero.

Some quadratic equations can be solved by factoring and then using the Principle of Zero Products.

➡ Solve: $2x^2 - x = 1$

TAKE NOTE

Note the steps involved in solving a quadratic equation by factoring:
1. Write in standard form.
2. Factor.
3. Set each factor equal to 0.
4. Solve each equation.
5. Check the solutions.

$$2x^2 - x = 1$$
$$2x^2 - x - 1 = 0 \qquad \bullet \text{ Write the equation in standard form.}$$
$$(2x + 1)(x - 1) = 0 \qquad \bullet \text{ Factor.}$$
$$2x + 1 = 0 \qquad x - 1 = 0 \qquad \bullet \text{ Use the Principle of Zero Products.}$$
$$2x = -1 \qquad\quad x = 1 \qquad \bullet \text{ Solve each equation.}$$
$$x = -\frac{1}{2}$$

The solutions are $-\frac{1}{2}$ and 1. **You should check these solutions.**

⇒ Solve: $t^4 - 2t^2 + 1 = 0$

$$t^4 - 2t^2 + 1 = 0$$
$$(t^2 - 1)(t^2 - 1) = 0$$
$$(t - 1)(t + 1)(t - 1)(t + 1) = 0$$

$t - 1 = 0, \quad t + 1 = 0, \quad t - 1 = 0, \quad t + 1 = 0$
$t = 1 \qquad t = -1 \qquad t = 1 \qquad t = -1$

- Factor the trinomial.
- Factor completely.

- Use the Principle of Zero Products.

The solutions are -1 and 1. These solutions are called **repeated solutions.**

⋯⋯⋯⋯⋯⋯⋯⋯⋯⋯⋯⋯⋯⋯⋯⋯⋯⋯⋯⋯⋯⋯⋯⋯⋯⋯⋯⋯⋯⋯⋯⋯⋯⋯⋯

Example 1

Solve: $x^2 + (x + 2)^2 = 100$

Solution
$$x^2 + (x + 2)^2 = 100$$
$$x^2 + x^2 + 4x + 4 = 100$$
$$2x^2 + 4x - 96 = 0$$
$$2(x^2 + 2x - 48) = 0$$
$$2(x - 6)(x + 8) = 0$$
$x - 6 = 0 \qquad\qquad x + 8 = 0$
$\quad x = 6 \qquad\qquad\quad x = -8$

The solutions are -8 and 6.

You Try It 1

Solve: $(x + 4)(x - 1) = 14$

Your solution

Solution on p. S17

Objective B **To solve application problems** ‹ 5 ›

⋯⋯⋯⋯⋯⋯⋯⋯⋯⋯⋯⋯⋯⋯⋯⋯⋯⋯⋯⋯⋯⋯⋯⋯⋯⋯⋯⋯⋯⋯⋯⋯⋯⋯⋯

Example 2

The length of a rectangle is 8 in. more than the width. The area of the rectangle is 240 in². Find the width of the rectangle.

Strategy
Draw a diagram.
Then use the
formula for the
area of a rectangle.

Solution
$$A = LW$$
$$240 = (x + 8)x$$
$$240 = x^2 + 8x$$
$$0 = x^2 + 8x - 240$$
$$0 = (x + 20)(x - 12)$$
$x + 20 = 0 \qquad x - 12 = 0$
$\quad x = -20 \qquad\quad x = 12$

The width cannot be negative.
The width is 12 in.

You Try It 2

The height of a triangle is 3 cm more than the length of the base of the triangle. The area of the triangle is 54 cm². Find the height of the triangle and the length of the base.

Your strategy

Your solution

Solution on p. S17

5.7 Exercises

Objective A

Solve.

1. $(x - 5)(x + 3) = 0$

2. $(x - 2)(x + 6) = 0$

3. $(x + 7)(x - 8) = 0$

4. $x(2x + 5)(x - 6) = 0$

5. $2x(3x - 2)(x + 4) = 0$

6. $6x(3x - 7)(x + 7) = 0$

7. $x^2 + 2x - 15 = 0$

8. $t^2 + 3t - 10 = 0$

9. $z^2 - 4z + 3 = 0$

10. $6x^2 - 9x = 0$

11. $r^2 - 10 = 3r$

12. $t^2 - 12 = 4t$

13. $4t^2 = 4t + 3$

14. $5y^2 + 11y = 12$

15. $4v^2 - 4v + 1 = 0$

16. $9s^2 - 6s + 1 = 0$

17. $x^2 - 9 = 0$

18. $t^2 - 16 = 0$

19. $4y^2 - 1 = 0$

20. $9z^2 - 4 = 0$

21. $x + 15 = x(x - 1)$

22. $x^2 - x - 2 = (2x - 1)(x - 3)$

23. $v^2 + v + 5 = (3v + 2)(v - 4)$

24. $z^2 + 5z - 4 = (2z + 1)(z - 4)$

25. $4x^2 + x - 10 = (x - 2)(x + 1)$

26. $x^3 + 2x^2 - 15x = 0$

27. $c^3 + 3c^2 - 10c = 0$

28. $z^4 - 18z^2 + 81 = 0$

29. $y^4 - 8y^2 + 16 = 0$

30. $x^3 + x^2 - 4x - 4 = 0$

31. $a^3 + a^2 - 9a - 9 = 0$

32. $2x^3 - x^2 - 2x + 1 = 0$

33. $3x^3 + 2x^2 - 12x - 8 = 0$

34. $2x^3 + 3x^2 - 18x - 27 = 0$

35. $5x^3 + 2x^2 - 20x - 8 = 0$

Objective B   Application Problems

36. The sum of a number and its square is 72. Find the number.

37. The sum of a number and its square is 210. Find the number.

38. The length of a rectangle is 2 ft more than twice the width. The area of the rectangle is 84 ft^2. Find the length and width of the rectangle.

w

$2w + 2$

39. The length of a rectangle is 8 cm more than three times the width. The area of the rectangle is 380 cm^2. Find the length and width of the rectangle.

40. The height of a triangle is 8 cm more than the length of the base. The area of the triangle is 64 cm^2. Find the base and height of the triangle.

$b + 8$

b

41. One leg of a right triangle is 2 ft more than twice the other leg. The hypotenuse is 1 ft more than the longer leg. Find the length of the hypotenuse of the right triangle.

42. An object is thrown downward, with an initial speed of 16 ft/s, from the top of a building 480 ft high. How many seconds later will the object hit the ground? Use the equation $d = vt + 16t^2$, where d is the distance in feet, v is the initial speed, and t is the time in seconds.

43. A stone is thrown into a well with an initial speed of 8 ft/s. The well is 80 ft deep. How many seconds later will the stone hit the bottom of the well? Use the equation $d = vt + 16t^2$, where d is the distance in feet, v is the initial speed, and t is the time in seconds.

APPLYING THE CONCEPTS

44. Write an equation whose solutions are 3, 2, and -1.

45. A rectangular piece of cardboard is 10 in. longer than it is wide. Squares 2 in. on a side are to be cut from each corner, and then the sides are to be folded up to make an open box with a volume of 112 in^3. Find the length and width of the piece of cardboard.

46. The following seems to show that $1 = 2$. Explain the error.

$$a = b$$
$$a^2 = ab$$ • Multiply each side of the equation by a.
$$a^2 - b^2 = ab - b^2$$ • Subtract b^2 from each side of the equation.
$$(a - b)(a + b) = b(a - b)$$ • Factor.
$$a + b = b$$ • Divide each side by $a - b$.
$$b + b = b$$ • Because $a = b$, substitute b for a.
$$2b = b$$
$$2 = 1$$ • Divide both sides by b.

Focus on Problem Solving

Find a Counterexample

When you are faced with an assertion, it may be that the assertion is false. For instance, consider the statement "Every prime number is an odd number." This assertion is false because the prime number 2 is an even number.

Finding an example that illustrates that an assertion is false is called finding a *counterexample*. The number 2 is a counterexample to the assertion that every prime number is an odd number.

If you are given an unfamiliar problem, one strategy to consider as a means of solving the problem is to try to find a counterexample. For each of the following problems, answer *true* if the assertion is always true. If the assertion is not true, answer *false* and give a counterexample. If there are terms used that you do not understand, consult a reference to find the meaning of the term.

1. If x is a real number, then x^2 is always positive.

2. The product of an odd integer and an even integer is an even integer.

3. If m is a positive integer, then $2m + 1$ is always a positive odd integer.

4. If $x < y$, then $x^2 < y^2$.

5. Given any three positive numbers a, b, and c, it is possible to construct a triangle whose sides have lengths a, b, and c.

6. The product of two irrational numbers is an irrational number.

7. If n is a positive integer greater than 2, then $1 \cdot 2 \cdot 3 \cdot 4 \cdot \cdots \cdot n + 1$ is a prime number.

8. Draw a polygon with more than three sides. Select two different points inside the polygon and join the points with a line segment. The line segment always lies completely inside the polygon.

9. Let A, B, and C be three points in the plane that are not collinear. Let d_1 be the distance from A to B, and let d_2 be the distance from A to C. Then the distance between B and C is less than $d_1 + d_2$.

10. Consider the line segment AB shown at the left. Two points, C and D, are randomly selected on the line segment and three new segments are formed: AC, CD, and DB. The three new line segments can always be connected to form a triangle.

It may not be easy to establish that an assertion is true or to find a counterexample to the assertion. For instance, consider the assertion that every positive integer greater than 3 can be written as the sum of two primes. For example, $6 = 3 + 3$, $8 = 3 + 5$, $9 = 2 + 7$. Is this assertion always true? (*Note:* This assertion, called Goldbach's conjecture, has never been proved, nor has a counterexample been found!)

Projects and Group Activities

Reverse Polish Notation

Following the Order of Operations Agreement is sometimes referred to as algebraic logic. It is a system used by many calculators. Algebraic logic is not the only system used by calculators. Another system is called RPN logic. RPN stands for Reverse Polish Notation.

During the 1950s, Jan Lukasiewicz, a Polish logician, developed a parenthesis-free notational system for writing mathematical expressions. In this system, the operator (such as addition, multiplication, or division) follows the operands (the numbers to be added, multiplied, or divided). For RPN calculators, an [ENTER] key is used to store a number temporarily until another number and the operation are entered. Here are some examples, along with the algebraic logic equivalent.

To Calculate	RPN Logic	Algebraic Logic
1. $3 + 4$	3 [ENTER] 4 [+]	3 [+] 4 [=]
2. $5 \times 6 \times 7$	5 [ENTER] 6 [×] 7 [×]	5 [×] 6 [×] 7 [=]
3. $4 \times (7 + 3)$	4 [ENTER] 7 [ENTER] 3 [+] [×]	4 [×] [(] 7 [+] 3 [)] [=]
4. $(3 + 4) \times (5 + 2)$	3 [ENTER] 4 [+] 5 [ENTER] 2 [+] [×]	[(] 3 [+] 4 [)] [×] [(] 5 [+] 2 [)] [=]

Examples 3 and 4 above illustrate the concept of being "parenthesis-free." Note that the examples under RPN logic do not require parentheses, whereas those under algebraic logic do. If the parentheses keys are not used for Examples 3 and 4, a calculator, following algebraic logic, would use the Order of Operations Agreement. The result in Example 3 would be

$$4 \times 7 + 3 = 28 + 3 = 31 \qquad \text{instead of} \qquad 4 \times (7 + 3) = 4 \times 10 = 40$$

RPN logic may seem strange, but it is actually very efficient. A glimpse of its efficiency can be seen from Example 4. For the RPN operation, nine keys were pushed, whereas 12 keys were pushed for the algebraic operation. Note also that RPN logic does not use the "equals" key. The result of an operation is displayed after the operation key is pressed.

Try the following exercises, which ask you to change from algebraic logic to RPN logic or to evaluate an RPN logic expression.

Change to RPN logic.

1. $12 \div 6$ **2.** $7 \times 5 + 6$ **3.** $(9 + 3) \div 4$

4. $(5 + 7) \div (2 + 4)$ **5.** $6 \times 7 \times 10$ **6.** $(1 + 4 \times 5) \div 7$

Evaluate each of the following expressions by using RPN logic.

7. 18 [ENTER] 2 [÷]

8. 6 [ENTER] 5 [ENTER] 3 [×] [+]

9. 3 [ENTER] 5 [×] 4 [+]

10. 7 [ENTER] 4 [ENTER] 5 [×] [+] 3 [÷]

11. 228 [ENTER] 6 [ENTER] 9 [ENTER] 12 [×] [+] [÷]

12. 1 [ENTER] 1 [ENTER] 1 [ENTER] 3 [+] [÷] [+]

Pythagorean Triples

Recall that the Pythagorean Theorem states that if a and b are the lengths of the legs of a right triangle and c is the length of the hypotenuse, then $a^2 + b^2 = c^2$.

$$c^2 = a^2 + b^2$$

For instance, the triangle with legs 3 and 4 and hypotenuse 5 is a right triangle because $3^2 + 4^2 = 5^2$. The numbers 3, 4, and 5 are called a **Pythagorean triple** because they are natural numbers that satisfy the equation of the Pythagorean Theorem.

1. Determine if the numbers are a Pythagorean triple.
  a. 5, 7, and 9
  b. 8, 15, and 17
  c. 11, 60, and 61
  d. 28, 45, and 53

Mathematicians have investigated Pythagorean triples and have found formulas that will generate these triples. One such set of formulas is

$$a = m^2 - n^2 \qquad b = 2mn \qquad c = m^2 + n^2, \text{ where } m > n$$

For instance, let $m = 2$ and $n = 1$. Then $a = 2^2 - 1^2 = 3$, $b = 2(2)(1) = 4$, and $c = 2^2 + 1^2 = 5$. This is the Pythagorean triple given above.

2. Find the Pythagorean triple produced by each of the following.
  a. $m = 3$ and $n = 1$
  b. $m = 5$ and $n = 2$
  c. $m = 4$ and $n = 2$
  d. $m = 6$ and $n = 1$

3. Find values of m and n that yield the Pythagorean triple 11, 60, 61.

4. Verify that $a^2 + b^2 = c^2$ when $a = m^2 - n^2$, $b = 2mn$, and $c = m^2 + n^2$.

5. The early Greek builders used a rope with 12 equally spaced knots to make right-angle corners for buildings. Explain how they used the rope.

6. Find three odd integers a, b, and c such that $a^2 + b^2 = c^2$.

Chapter Summary

Key Words

A *monomial* is a number, a variable, or a product of a number and variables. [p. 243]

A *polynomial* is a variable expression in which the terms are monomials. A *binomial* is a polynomial of two terms. A *trinomial* is a polynomial of three terms. [p. 255]

The *degree of a polynomial* in one variable is the greatest of the degrees of any of its terms. [p. 255]

A number written in *scientific notation* is a number written in the form $a \times 10^n$, where $1 \le a < 10$. [p. 249]

Synthetic division is a shorter method of dividing a polynomial by a binomial of the form $x - a$. This method uses only the coefficients of the variable terms. [p. 275]

To *factor a polynomial* means to write the polynomial as the product of other polynomials. A polynomial is *nonfactorable over the integers* if it does not factor using only integers. Such a polynomial is called a *prime polynomial.* [pp. 283, 287]

A *quadratic trinomial* is a polynomial of the form $ax^2 + bx + c$, where a, b, and c are nonzero integers. To *factor a quadratic trinomial* means to express the trinomial as the product of two binomials. [p. 285]

The product of a term and itself is a *perfect square.* The *square root* of a perfect square is one of the two equal factors of the perfect square. [p. 245]

The product of the same three factors is called a *perfect cube.* The *cube root* of a perfect cube is one of the three equal factors of the perfect cube. [p. 297]

An equation of the form $ax^2 + bx + c = 0$, $a \ne 0$, is a *quadratic equation.* [p. 312]

Essential Rules	**Definition of a Negative Exponent**	For $x \ne 0$, $x^{-n} = \dfrac{1}{x^n}$ and $\dfrac{1}{x^{-n}} = x^n$. [p. 246]
	Rule for Negative Exponents on Fractional Expressions	For $a \ne 0$, $b \ne 0$, $\left(\dfrac{a}{b}\right)^{-n} = \left(\dfrac{b}{a}\right)^n$. [p. 247]
	Rule for Multiplying Exponential Expressions	$x^m \cdot x^n = x^{m+n}$ [p. 243]
	Rule for Simplifying the Power of an Exponential Expression	$(x^m)^n = x^{mn}$ [p. 244]
	Rule for Simplifying Powers of Products	$(x^m y^n)^p = x^{mp} y^{np}$ [p. 244]
	Rule for Dividing Exponential Expressions	For $x \ne 0$, $\dfrac{x^m}{x^n} = x^{m-n}$. [p. 247]
	Zero as an Exponent	For $x \ne 0$, $x^0 = 1$. [p. 245]
	Rule for Simplifying Powers of Quotients	For $y \ne 0$, $\left(\dfrac{x^m}{y^n}\right)^p = \dfrac{x^{mp}}{y^{np}}$. [p. 246]
	Remainder Theorem	If the polynomial $P(x)$ is divided by $x - a$, the remainder is $P(a)$. [p. 278]
	The Product of the Sum and Difference of Two Terms, and the Difference of Two Squares	$(a + b)(a - b) = a^2 - b^2$ [pp. 266, 295]
	The Square of a Binomial, and a Perfect Square Trinomial	$(a + b)^2 = a^2 + 2ab + b^2$ $(a - b)^2 = a^2 - 2ab + b^2$ [pp. 266, 295]
	The Sum or Difference of Two Cubes	$a^3 + b^3 = (a + b)(a^2 - ab + b^2)$ $a^3 - b^3 = (a - b)(a^2 + ab + b^2)$ [p. 297]
	Principle of Zero Products	If $ab = 0$, then $a = 0$ or $b = 0$. [p. 305]

Chapter Review

1. Factor: $18a^5b^2 - 12a^3b^3 + 30a^2b$

2. Simplify: $\dfrac{15x^2 + 2x - 2}{3x - 2}$

3. Simplify: $(2x^{-1}y^2z^5)^4(-3x^3yz^{-3})^2$

4. Factor: $2ax + 4bx - 3ay - 6by$

5. Factor: $12 + x - x^2$

6. Use the Remainder Theorem to evaluate $P(x) = x^3 - 2x^2 + 3x - 5$ when $x = 2$.

7. Simplify: $(5x^2 - 8xy + 2y^2) - (x^2 - 3y^2)$

8. Factor: $24x^2 + 38x + 15$

9. Factor: $4x^2 + 12xy + 9y^2$

10. Simplify: $(-2a^2b^4)(3ab^2)$

11. Factor: $64a^3 - 27b^3$

12. Simplify: $\dfrac{4x^3 + 27x^2 + 10x + 2}{x + 6}$

13. Given $P(x) = 2x^3 - x + 7$, evaluate $P(-2)$.

14. Factor: $x^2 - 3x - 40$

15. Factor: $x^2y^2 - 9$

16. Multiply: $4x^2y(3x^3y^2 + 2xy - 7y^3)$

17. Factor: $x^{2n} - 12x^n + 36$

18. Solve: $6x^2 + 60 = 39x$

19. Simplify: $5x^2 - 4x[x - 3(3x + 2) + x]$

20. Factor: $3a^6 - 15a^4 - 18a^2$

21. Simplify: $(4x - 3y)^2$

22. Simplify: $\dfrac{x^4 - 4}{x - 4}$

23. Factor: $15x^4 + x^2 - 6$

24. Simplify: $\dfrac{(2a^4b^{-3}c^2)^3}{(2a^3b^2c^{-1})^4}$

25. Multiply: $(x - 4)(3x + 2)(2x - 3)$

26. Factor: $21x^4y^4 + 23x^2y^2 + 6$

27. Solve: $x^3 + 16 = x(x + 16)$

28. Multiply: $(5a + 2b)(5a - 2b)$

29. Write 2.54×10^{-3} in decimal notation.

30. Factor: $6x^2 - 31x + 18$

31. Graph $y = x^2 + 1$.

32. Identify **a.** the leading coefficient, **b.** the constant term, and **c.** the degree of the polynomial $P(x) = 3x^5 - 6x^2 + 7x + 8$.

33. The mass of the moon is 3.7×10^{-8} times the mass of the sun. The mass of the sun is 2.19×10^{27} tons. Find the mass of the moon. Write the answer in scientific notation.

34. The sum of a number and its square is 56. Find the number.

35. The most distant object visible from Earth without the aid of a telescope is the Great Galaxy of Andromeda. It takes light from the Great Galaxy of Andromeda 2.2×10^6 years to travel to Earth. Light travels about 6.7×10^8 mph. How far from Earth is the Great Galaxy of Andromeda? Use a 365-day year.

36. The length of a rectangle is $(5x + 3)$ cm. The width is $(2x - 7)$ cm. Find the area of the rectangle in terms of the variable x.

Chapter Test

1. Factor: $16t^2 + 24t + 9$

2. Multiply: $-6rs^2(3r - 2s - 3)$

3. Given $P(x) = 3x^2 - 8x + 1$, evaluate $P(2)$.

4. Factor: $27x^3 - 8$

5. Factor: $16x^2 - 25$

6. Multiply: $(3t^3 - 4t^2 + 1)(2t^2 - 5)$

7. Simplify: $-5x[3 - 2(2x - 4) - 3x]$

8. Factor: $12x^3 + 12x^2 - 45x$

9. Solve: $6x^3 + x^2 - 6x - 1 = 0$

10. Simplify:
$(6x^3 - 7x^2 + 6x - 7) - (4x^3 - 3x^2 + 7)$

11. Write the number 0.000000501 in scientific notation.

12. Simplify: $\dfrac{14x^2 + x + 1}{7x - 3}$

13. Multiply: $(7 - 5x)(7 + 5x)$

14. Factor: $6a^4 - 13a^2 - 5$

15. Multiply: $(3a + 4b)(2a - 7b)$

16. Factor: $3x^4 - 23x^2 - 36$

17. Simplify: $(-4a^2b)^3(-ab^4)$

18. Solve: $6x^2 = x + 1$

19. Use the Remainder Theorem to evaluate $P(x) = -x^3 + 4x - 8$ when $x = -2$.

20. Simplify: $\dfrac{(2a^{-4}b^2)^3}{4a^{-2}b^{-1}}$

21. Simplify: $\dfrac{x^3 - 2x^2 - 5x + 7}{x + 3}$

22. Factor: $12 - 17x + 6x^2$

23. Factor: $6x^2 - 4x - 3xa + 2a$

24. Write the number of seconds in one week in scientific notation.

25. An arrow is shot into the air with an upward velocity of 48 ft/s from a hill 32 ft high. How many seconds later will the arrow be 64 ft above the ground? Use the equation $h = 32 + 48t - 16t^2$, where h is the height in feet and t is the time in seconds.

26. The length of a rectangle is $(5x + 1)$ ft. The width is $(2x - 1)$ ft. Find the area of the rectangle in terms of the variable x.

Cumulative Review

1. Simplify: $8 - 2[-3 - (-1)]^2 + 4$

2. Evaluate $\dfrac{2a - b}{b - c}$ when $a = 4$, $b = -2$, and $c = 6$.

3. Identify the property that justifies the statement $2x + (-2x) = 0$.

4. Simplify: $2x - 4[x - 2(3 - 2x) + 4]$

5. Solve: $\dfrac{2}{3} - y = \dfrac{5}{6}$

6. Solve: $8x - 3 - x = -6 + 3x - 8$

7. Divide: $\dfrac{x^3 - 3}{x - 3}$

8. Solve: $3 - |2 - 3x| = -2$

9. Given $P(x) = 3x^2 - 2x + 2$, evaluate $P(-2)$.

10. What values of x are excluded from the domain of the function $f(x) = \dfrac{x + 1}{x + 2}$?

11. Find the range of the function given by $F(x) = 3x^2 - 4$ if the domain is $\{-2, -1, 0, 1, 2\}$.

12. Find the slope of the line containing the points $(-2, 3)$ and $(4, 2)$.

13. Find the equation of the line that contains the point $(-1, 2)$ and has slope $-\dfrac{3}{2}$.

14. Find the equation of the line that contains the point $(-2, 4)$ and is perpendicular to the line $3x + 2y = 4$.

15. Solve by using Cramer's Rule:
$2x - 3y = 2$
$x + y = -3$

16. Solve by the addition method:
$x - y + z = 0$
$2x + y - 3z = -7$
$-x + 2y + 2z = 5$

17. Graph $3x - 4y = 12$ by using the x- and y-intercepts.

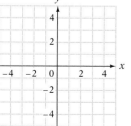

18. Graph the solution set: $-3x + 2y < 6$

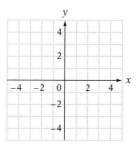

19. Solve by graphing:
$$x - 2y = 3$$
$$-2x + y = -3$$

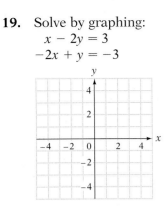

20. Graph the solution set:
$$2x + y < 3$$
$$-6x + 3y \geq 4$$

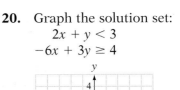

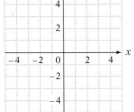

21. Simplify: $(4a^{-2}b^3)(2a^3b^{-1})^{-2}$

22. Simplify: $\dfrac{(5x^3y^{-3}z)^{-2}}{y^4z^{-2}}$

23. Simplify: $3 - (3 - 3^{-1})^{-1}$

24. Multiply: $(2x + 3)(2x^2 - 3x + 1)$

25. Factor: $-4x^3 + 14x^2 - 12x$

26. Factor: $a(x - y) - b(y - x)$

27. Factor: $x^4 - 16$

28. Factor: $2x^3 - 16$

29. The sum of two integers is twenty-four. The difference between four times the smaller integer and nine is three less than twice the larger integer. Find the integers.

30. How many ounces of pure gold that costs \$360 per ounce must be mixed with 80 oz of an alloy that costs \$120 per ounce to make a mixture that costs \$200 per ounce?

31. Two bicycles are 25 mi apart and are traveling toward each other. One cyclist is traveling at $\dfrac{2}{3}$ the rate of the other cyclist. They pass in 2 h. Find the rate of each cyclist.

32. A space vehicle travels 2.4×10^5 mi from Earth to the moon at an average velocity of 2×10^4 mph. How long does it take the vehicle to reach the moon?

33. The graph shows the relationship between the distance traveled and the time of travel. Find the slope of the line between the two points on the graph. Write a sentence that states the meaning of the slope.

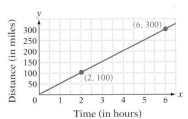

Chapter 6

Rational Expressions

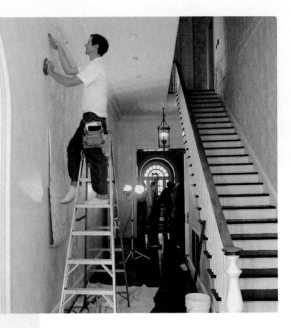

Objectives

Section 6.1

A To simplify a rational expression
B To multiply rational expressions
C To divide rational expressions

Section 6.2

A To rewrite rational expressions in terms of a common denominator
B To add or subtract rational expressions

Section 6.3

A To simplify a complex fraction

Section 6.4

A To solve a proportion
B To solve application problems

Section 6.5

A To solve a fractional equation
B To solve work problems
C To solve uniform motion problems

Section 6.6

A To solve variation problems

New wallpaper is applied as part of this mansion's renovation process. Interior decorators use rates and proportions to estimate the amount of material needed on a job. A decorator, knowing the total wall area of a room and how much wall area one roll of wallpaper will cover, can use a proportion to calculate how many rolls of wallpaper need to be purchased. **Exercise 20 on page 370** is an example of the type of question a decorator would need to answer before purchasing wallpaper.

Need help? For on-line student resources, such as section quizzes, visit this textbook's web site at **math.college.hmco.com/students**.

1. Find the LCM of 10 and 25.

For Exercises 2 to 5, add, subtract, multiply, or divide.

2. $-\dfrac{3}{8} \cdot \dfrac{4}{9}$

3. $-\dfrac{4}{5} \div \dfrac{8}{15}$

4. $-\dfrac{5}{6} + \dfrac{7}{8}$

5. $-\dfrac{3}{8} - \left(-\dfrac{7}{12}\right)$

6. Simplify: $\dfrac{\frac{2}{3} - \frac{1}{4}}{\frac{1}{8} - 2}$

7. Evaluate $\dfrac{2x - 3}{x^2 - x + 1}$ for $x = 2$.

8. Solve: $4(2x + 1) = 3(x - 2)$

9. Solve: $10\left(\dfrac{t}{2} + \dfrac{t}{5}\right) = 10(1)$

10. Two planes start from the same point and fly in opposite directions. The first plane is flying 20 mph slower than the second plane. In 2 h, the planes are 480 mi apart. Find the rate of each plane.

Go Figure

If 6 machines can fill 12 boxes of cereal in 7 minutes, how many boxes of cereal can be filled by 14 machines in 12 minutes?

6.1 Multiplication and Division of Rational Expressions

Objective A   **To simplify a rational expression**

An expression in which the numerator or denominator is a polynomial is called a **rational expression.** Examples of rational expressions are shown at the right.

$$\frac{9}{z} \qquad \frac{3x + 4}{2x^2 + 1} \qquad \frac{x^3 - x + 1}{x^2 - 3x - 5}$$

The expression $\dfrac{\sqrt{x} + 3}{x}$ is not a rational expression because $\sqrt{x} + 3$ is not a polynomial.

A function that is written in terms of a rational expression is a **rational function.** Each of the following equations represents a rational function.

$$f(x) = \frac{x^2 + 3}{2x - 1} \qquad g(t) = \frac{3}{t^2 - 4} \qquad R(z) = \frac{z^2 + 3z - 1}{z^2 + z - 12}$$

To evaluate a rational function, replace the variable by its value. Then simplify.

➡ Evaluate $f(-2)$ given $f(x) = \dfrac{x^2}{3x^2 - x - 9}$.

$$f(x) = \frac{x^2}{3x^2 - x - 9}$$

$$f(-2) = \frac{(-2)^2}{3(-2)^2 - (-2) - 9} = \frac{4}{12 + 2 - 9} = \frac{4}{5}$$

Because division by zero is not defined, the domain of a rational function must exclude those numbers for which the value of the polynomial in the denominator is zero.

The graph of $f(x) = \dfrac{1}{x - 2}$ is shown at the right. Note that the graph never intersects the line $x = 2$ (shown as a dotted line). This value of x is excluded from the domain of $f(x) = \dfrac{1}{x - 2}$.

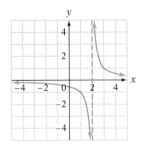

➡ Determine the domain of $g(x) = \dfrac{x^2 + 4}{3x - 6}$.

The domain of g must exclude values of x for which the denominator is zero. To find these values, set the denominator equal to zero and solve for x.

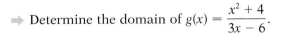

$3x - 6 = 0$ • Set the denominator equal to zero.

$3x = 6$ • Solve for x.

$x = 2$ • This value must be *excluded* from the domain.

The domain of g is $\{x \mid x \neq 2\}$.

A rational expression is in simplest form when the numerator and denominator have no common factors. The Multiplication Property of One is used to write a rational expression in simplest form.

➡ Simplify: $\dfrac{x^2 - 25}{x^2 + 13x + 40}$

$$\frac{x^2 - 25}{x^2 + 13x + 40} = \frac{(x - 5)}{(x + 8)} \cdot \boxed{\frac{(x + 5)}{(x + 5)}} = \frac{x - 5}{x + 8} \cdot 1 = \frac{x - 5}{x + 8} \qquad x \neq -8, -5$$

The requirement $x \neq -8, -5$ is necessary because division by 0 is not allowed.

The simplification above is usually shown with slashes to indicate that a common factor has been removed:

$$\frac{x^2 - 25}{x^2 + 13x + 40} = \frac{(x - 5)\cancel{(x + 5)}^{1}}{(x + 8)\cancel{(x + 5)}_{1}} = \frac{x - 5}{x + 8} \qquad x \neq -8, -5$$

We will show a simplification with slashes. We will not show the restrictions that prevent division by zero. Nonetheless, those restrictions *always* are implied.

➡ Simplify: $\dfrac{12 + 5x - 2x^2}{2x^2 - 3x - 20}$

$$\frac{12 + 5x - 2x^2}{2x^2 - 3x - 20} = \frac{(4 - x)(3 + 2x)}{(x - 4)(2x + 5)}$$

• Factor the numerator and denominator.

$$= \frac{\cancel{(4 - x)}^{-1}(3 + 2x)}{\cancel{(x - 4)}_{1}(2x + 5)}$$

• $\dfrac{4 - x}{x - 4} = \dfrac{-\cancel{(x - 4)}^{1}}{\cancel{x - 4}_{1}} = \dfrac{-1}{1} = -1$

$$= -\frac{2x + 3}{2x + 5}$$

• Write the answer in simplest form.

TAKE NOTE

Recall that
$b - a = -(a - b)$.

Therefore,
$4 - x = -(x - 4)$.

In general,
$\dfrac{b - a}{a - b} = \dfrac{-\cancel{(a - b)}^{1}}{\cancel{a - b}_{1}} = \dfrac{-1}{1}$
$= -1.$

....

Example 1

Given $f(x) = \dfrac{3x - 4}{x^2 - 2x + 1}$, find $f(-2)$.

Solution

$f(x) = \dfrac{3x - 4}{x^2 - 2x + 1}$

$f(-2) = \dfrac{3(-2) - 4}{(-2)^2 - 2(-2) + 1} = \dfrac{-6 - 4}{4 + 4 + 1}$

$= \dfrac{-10}{9} = -\dfrac{10}{9}$

You Try It 1

Given $f(x) = \dfrac{3 - 5x}{x^2 + 5x + 6}$, find $f(2)$.

Your solution

Solution on p. S17

Example 2

Find the domain of $f(x) = \dfrac{x^2 - 1}{x^2 - 2x - 15}$.

Solution

Set the denominator equal to zero. Then solve for x.

$x^2 - 2x - 15 = 0$
$(x - 5)(x + 3) = 0$

$x - 5 = 0 \quad x + 3 = 0$
$\quad\quad x = 5 \quad\quad\quad x = -3$

The domain is $\{x \,|\, x \neq -3, 5\}$.

You Try It 2

Find the domain of $f(x) = \dfrac{2x + 1}{2x^2 - 7x + 3}$.

Your solution

Example 3

Simplify: $\dfrac{6x^3 - 9x^2}{12x^2 - 18x}$

Solution

$\dfrac{6x^3 - 9x^2}{12x^2 - 18x} = \dfrac{3x^2(2x - 3)}{6x(2x - 3)}$

$= \dfrac{3x^2\cancel{(2x - 3)}}{6x\cancel{(2x - 3)}} = \dfrac{x}{2}$

You Try It 3

Simplify: $\dfrac{6x^4 - 24x^3}{12x^3 - 48x^2}$

Your solution

Example 4

Simplify: $\dfrac{12x^3y^2 + 6x^3y^3}{6x^2y^2}$

Solution

$\dfrac{12x^3y^2 + 6x^3y^3}{6x^2y^2} = \dfrac{6x^3y^2(2 + y)}{6x^2y^2}$

$= x(2 + y)$

You Try It 4

Simplify: $\dfrac{21a^3b - 14a^3b^2}{7a^2b}$

Your solution

Example 5

Simplify: $\dfrac{2x - 8x^2}{16x^3 - 28x^2 + 6x}$

Solution

$\dfrac{2x - 8x^2}{16x^3 - 28x^2 + 6x} = \dfrac{2x(1 - 4x)}{2x(8x^2 - 14x + 3)}$

$= \dfrac{2x(1 - 4x)}{2x(4x - 1)(2x - 3)}$

$= \dfrac{2x\overset{-1}{\cancel{(1 - 4x)}}}{2x\underset{1}{\cancel{(4x - 1)}}(2x - 3)}$

$= -\dfrac{1}{2x - 3}$

You Try It 5

Simplify: $\dfrac{20x - 15x^2}{15x^3 - 5x^2 - 20x}$

Your solution

Solutions on pp. S17–S18

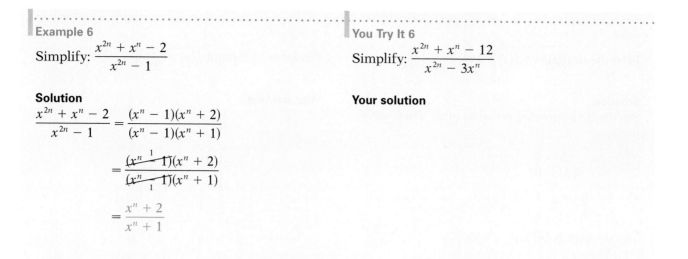

Example 6

Simplify: $\dfrac{x^{2n} + x^n - 2}{x^{2n} - 1}$

Solution

$\dfrac{x^{2n} + x^n - 2}{x^{2n} - 1} = \dfrac{(x^n - 1)(x^n + 2)}{(x^n - 1)(x^n + 1)}$

$\qquad = \dfrac{\overset{1}{\cancel{(x^n - 1)}}(x^n + 2)}{\underset{1}{\cancel{(x^n - 1)}}(x^n + 1)}$

$\qquad = \dfrac{x^n + 2}{x^n + 1}$

You Try It 6

Simplify: $\dfrac{x^{2n} + x^n - 12}{x^{2n} - 3x^n}$

Your solution

Solution on p. S18

Objective B **To multiply rational expressions**

The product of two fractions is a fraction whose numerator is the product of the numerators of the two fractions and whose denominator is the product of the denominators of the two fractions.

$$\frac{a}{b} \cdot \frac{c}{d} = \frac{ac}{bd} \qquad \frac{5}{a + 2} \cdot \frac{b - 3}{3} = \frac{5(b - 3)}{(a + 2)3} = \frac{5b - 15}{3a + 6}$$

The product of two rational expressions can often be simplified by factoring the numerator and the denominator.

➡ Simplify: $\dfrac{x^2 - 2x}{2x^2 + x - 15} \cdot \dfrac{2x^2 - x - 10}{x^2 - 4}$

$\dfrac{x^2 - 2x}{2x^2 + x - 15} \cdot \dfrac{2x^2 - x - 10}{x^2 - 4}$

$\quad = \dfrac{x(x - 2)}{(x + 3)(2x - 5)} \cdot \dfrac{(x + 2)(2x - 5)}{(x + 2)(x - 2)}$ • Factor the numerator and the denominator of each fraction.

$\quad = \dfrac{x(x - 2)(x + 2)(2x - 5)}{(x + 3)(2x - 5)(x + 2)(x - 2)}$ • Multiply.

$\quad = \dfrac{x\overset{1}{\cancel{(x - 2)}}\overset{1}{\cancel{(x + 2)}}\overset{1}{\cancel{(2x - 5)}}}{(x + 3)\underset{1}{\cancel{(2x - 5)}}\underset{1}{\cancel{(x + 2)}}\underset{1}{\cancel{(x - 2)}}}$ • Simplify.

$\quad = \dfrac{x}{x + 3}$ • Write the answer in simplest form.

Example 7

Simplify: $\dfrac{2x^2 - 6x}{3x - 6} \cdot \dfrac{6x - 12}{8x^3 - 12x^2}$

Solution

$$\dfrac{2x^2 - 6x}{3x - 6} \cdot \dfrac{6x - 12}{8x^3 - 12x^2} = \dfrac{2x(x - 3)}{3(x - 2)} \cdot \dfrac{6(x - 2)}{4x^2(2x - 3)}$$

$$= \dfrac{2x(x - 3) \cdot 6(x - 2)}{3(x - 2) \cdot 4x^2(2x - 3)}$$

$$= \dfrac{2x(x - 3) \cdot 6\overset{1}{\cancel{(x - 2)}}}{3\underset{1}{\cancel{(x - 2)}} \cdot 4x^2(2x - 3)}$$

$$= \dfrac{x - 3}{x(2x - 3)}$$

You Try It 7

Simplify: $\dfrac{12 + 5x - 3x^2}{x^2 + 2x - 15} \cdot \dfrac{2x^2 + x - 45}{3x^2 + 4x}$

Your solution

Example 8

Simplify: $\dfrac{6x^2 + x - 2}{6x^2 + 7x + 2} \cdot \dfrac{2x^2 + 9x + 4}{4 - 7x - 2x^2}$

Solution

$$\dfrac{6x^2 + x - 2}{6x^2 + 7x + 2} \cdot \dfrac{2x^2 + 9x + 4}{4 - 7x - 2x^2}$$

$$= \dfrac{(2x - 1)(3x + 2)}{(3x + 2)(2x + 1)} \cdot \dfrac{(2x + 1)(x + 4)}{(1 - 2x)(4 + x)}$$

$$= \dfrac{(2x - 1)(3x + 2) \cdot (2x + 1)(x + 4)}{(3x + 2)(2x + 1) \cdot (1 - 2x)(4 + x)}$$

$$= \dfrac{\overset{-1}{\cancel{(2x - 1)}}\overset{1}{\cancel{(3x + 2)}}\overset{1}{\cancel{(2x + 1)}}\overset{1}{\cancel{(x + 4)}}}{\underset{1}{\cancel{(3x + 2)}}\underset{1}{\cancel{(2x + 1)}}\underset{1}{\cancel{(1 - 2x)}}\underset{1}{\cancel{(x + 4)}}}$$

$$= -1$$

You Try It 8

Simplify: $\dfrac{2x^2 - 13x + 20}{x^2 - 16} \cdot \dfrac{2x^2 + 9x + 4}{6x^2 - 7x - 5}$

Your solution

Solutions on p. S18

Objective C   **To divide rational expressions**

The **reciprocal** of a rational expression is the rational expression with the numerator and denominator interchanged.

Rational
Expression $\left\{ \begin{array}{cc} \dfrac{a}{b} & \dfrac{b}{a} \\ \dfrac{a^2 - 2y}{4} & \dfrac{4}{a^2 - 2y} \end{array} \right\}$ Reciprocal

To divide two rational expressions, multiply by the reciprocal of the divisor.

$$\dfrac{a}{b} \div \dfrac{c}{d} = \dfrac{a}{b} \cdot \dfrac{d}{c} = \dfrac{ad}{bc}$$

$$\dfrac{2}{a} \div \dfrac{5}{b} = \dfrac{2}{a} \cdot \dfrac{b}{5} = \dfrac{2b}{5a}$$

$$\dfrac{x + y}{2} \div \dfrac{x - y}{5} = \dfrac{x + y}{2} \cdot \dfrac{5}{x - y} = \dfrac{(x + y)5}{2(x - y)} = \dfrac{5x + 5y}{2x - 2y}$$

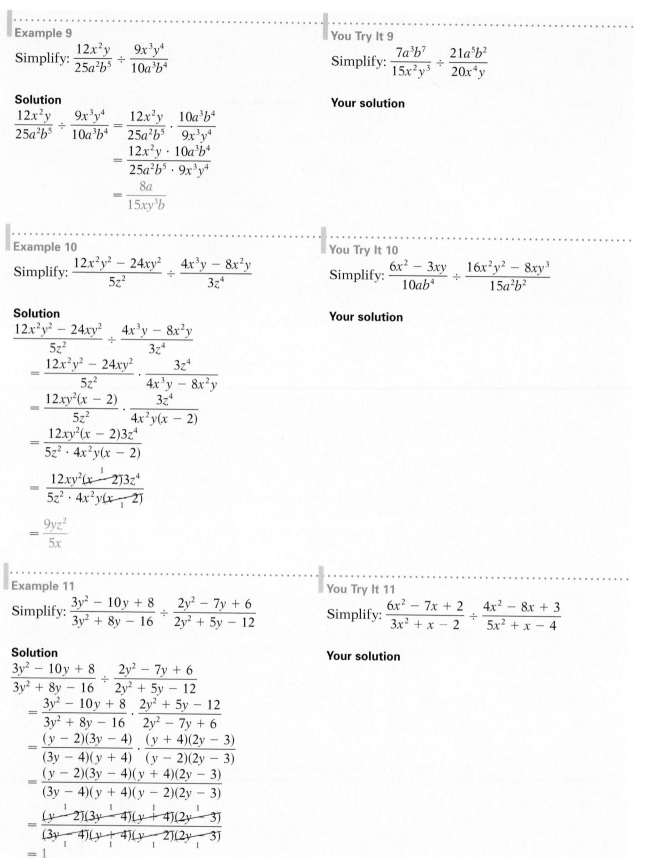

Example 9

Simplify: $\dfrac{12x^2y}{25a^2b^5} \div \dfrac{9x^3y^4}{10a^3b^4}$

Solution

$$\dfrac{12x^2y}{25a^2b^5} \div \dfrac{9x^3y^4}{10a^3b^4} = \dfrac{12x^2y}{25a^2b^5} \cdot \dfrac{10a^3b^4}{9x^3y^4}$$

$$= \dfrac{12x^2y \cdot 10a^3b^4}{25a^2b^5 \cdot 9x^3y^4}$$

$$= \dfrac{8a}{15xy^3b}$$

You Try It 9

Simplify: $\dfrac{7a^3b^7}{15x^2y^3} \div \dfrac{21a^5b^2}{20x^4y}$

Your solution

Example 10

Simplify: $\dfrac{12x^2y^2 - 24xy^2}{5z^2} \div \dfrac{4x^3y - 8x^2y}{3z^4}$

Solution

$$\dfrac{12x^2y^2 - 24xy^2}{5z^2} \div \dfrac{4x^3y - 8x^2y}{3z^4}$$

$$= \dfrac{12x^2y^2 - 24xy^2}{5z^2} \cdot \dfrac{3z^4}{4x^3y - 8x^2y}$$

$$= \dfrac{12xy^2(x-2)}{5z^2} \cdot \dfrac{3z^4}{4x^2y(x-2)}$$

$$= \dfrac{12xy^2(x-2)3z^4}{5z^2 \cdot 4x^2y(x-2)}$$

$$= \dfrac{12xy^2\overset{1}{\cancel{(x-2)}}3z^4}{5z^2 \cdot 4x^2y\underset{1}{\cancel{(x-2)}}}$$

$$= \dfrac{9yz^2}{5x}$$

You Try It 10

Simplify: $\dfrac{6x^2 - 3xy}{10ab^4} \div \dfrac{16x^2y^2 - 8xy^3}{15a^2b^2}$

Your solution

Example 11

Simplify: $\dfrac{3y^2 - 10y + 8}{3y^2 + 8y - 16} \div \dfrac{2y^2 - 7y + 6}{2y^2 + 5y - 12}$

Solution

$$\dfrac{3y^2 - 10y + 8}{3y^2 + 8y - 16} \div \dfrac{2y^2 - 7y + 6}{2y^2 + 5y - 12}$$

$$= \dfrac{3y^2 - 10y + 8}{3y^2 + 8y - 16} \cdot \dfrac{2y^2 + 5y - 12}{2y^2 - 7y + 6}$$

$$= \dfrac{(y-2)(3y-4)}{(3y-4)(y+4)} \cdot \dfrac{(y+4)(2y-3)}{(y-2)(2y-3)}$$

$$= \dfrac{(y-2)(3y-4)(y+4)(2y-3)}{(3y-4)(y+4)(y-2)(2y-3)}$$

$$= \dfrac{\overset{1}{\cancel{(y-2)}}\overset{1}{\cancel{(3y-4)}}\overset{1}{\cancel{(y+4)}}\overset{1}{\cancel{(2y-3)}}}{\underset{1}{\cancel{(3y-4)}}\underset{1}{\cancel{(y+4)}}\underset{1}{\cancel{(y-2)}}\underset{1}{\cancel{(2y-3)}}}$$

$$= 1$$

You Try It 11

Simplify: $\dfrac{6x^2 - 7x + 2}{3x^2 + x - 2} \div \dfrac{4x^2 - 8x + 3}{5x^2 + x - 4}$

Your solution

Solutions on p. S18

6.1 Exercises

Objective A

1. What is a rational function? Give an example of a rational function.

2. What values are excluded from the domain of a rational function?

3. Given $f(x) = \dfrac{2}{x-3}$, find $f(4)$.

4. Given $f(x) = \dfrac{-7}{5-x}$, find $f(-2)$.

5. Given $f(x) = \dfrac{x-2}{x+4}$, find $f(-2)$.

6. Given $f(x) = \dfrac{x-3}{2x-1}$, find $f(3)$.

7. Given $f(x) = \dfrac{1}{x^2 - 2x + 1}$, find $f(-2)$.

8. Given $f(x) = \dfrac{-3}{x^2 - 4x + 2}$, find $f(-1)$.

9. Given $f(x) = \dfrac{x-2}{2x^2 + 3x + 8}$, find $f(3)$.

10. Given $f(x) = \dfrac{x^2}{3x^2 - 3x + 5}$, find $f(4)$.

11. Given $f(x) = \dfrac{x^2 - 2x}{x^3 - x + 4}$, find $f(-1)$.

12. Given $f(x) = \dfrac{8 - x^2}{x^3 - x^2 + 4}$, find $f(-3)$.

Find the domain of the function.

13. $f(x) = \dfrac{4}{x-3}$

14. $G(x) = \dfrac{-2}{x+2}$

15. $H(x) = \dfrac{x}{x+4}$

16. $F(x) = \dfrac{3x}{x-5}$

17. $h(x) = \dfrac{5x}{3x+9}$

18. $f(x) = \dfrac{-2x}{6-2x}$

19. $q(x) = \dfrac{4-x}{(x-4)(3x-2)}$

20. $p(x) = \dfrac{2x+1}{(2x+5)(3x-6)}$

21. $f(x) = \dfrac{2x-1}{x^2 + x - 6}$

22. $G(x) = \dfrac{3 - 4x}{x^2 + 4x - 5}$

23. $f(x) = \dfrac{x+1}{x^2 + 1}$

24. $g(x) = \dfrac{x^2 + 1}{x^2}$

25. When is a rational expression in simplest form?

26. Are the rational expressions $\dfrac{x(x-2)}{2(x-2)}$ and $\dfrac{x}{2}$ equal for all values of x? Why or why not?

Simplify.

27. $\dfrac{4-8x}{4}$

28. $\dfrac{8y+2}{2}$

29. $\dfrac{6x^2-2x}{2x}$

30. $\dfrac{3y-12y^2}{3y}$

31. $\dfrac{8x^2(x-3)}{4x(x-3)}$

32. $\dfrac{16y^4(y+8)}{12y^3(y+8)}$

33. $\dfrac{-36a^2-48a}{18a^3+24a^2}$

34. $\dfrac{a^2+4a}{4a-16}$

35. $\dfrac{3x-6}{x^2+2x}$

36. $\dfrac{16x^3-8x^2+12x}{4x}$

37. $\dfrac{3x^3y^3-12x^2y^2+15xy}{3xy}$

38. $\dfrac{-10a^4-20a^3+30a^2}{-10a^2}$

39. $\dfrac{x^{2n}+x^n y^n}{x^{2n}-y^{2n}}$

40. $\dfrac{a^{2n}-b^{2n}}{5a^{3n}+5a^{2n}b^n}$

41. $\dfrac{x^2-7x+12}{x^2-9x+20}$

42. $\dfrac{6-x-x^2}{3x^2-10x+8}$

43. $\dfrac{3x^2+10x-8}{8-14x+3x^2}$

44. $\dfrac{14-19x-3x^2}{3x^2-23x+14}$

45. $\dfrac{a^2-b^2}{a^3+b^3}$

46. $\dfrac{x^4-y^4}{x^2+y^2}$

47. $\dfrac{8x^3-y^3}{4x^2-y^2}$

48. $\dfrac{x^2-4}{a(x+2)-b(x+2)}$

49. $\dfrac{x^2(a-2)-a+2}{ax^2-ax}$

50. $\dfrac{x^4+3x^2+2}{x^4-1}$

51. $\dfrac{x^4-2x^2-3}{x^4+2x^2+1}$

52. $\dfrac{x^2y^2+4xy-21}{x^2y^2-10xy+21}$

53. $\dfrac{6x^2y^2+11xy+4}{9x^2y^2+9xy-4}$

54. $\dfrac{a^{2n}-a^n-2}{a^{2n}+3a^n+2}$

55. $\dfrac{a^{2n}+a^n-12}{a^{2n}-2a^n-3}$

56. $\dfrac{a^{2n}-1}{a^{2n}-2a^n+1}$

57. $\dfrac{a^{2n} + 2a^{n}b^{n} + b^{2n}}{a^{2n} - b^{2n}}$

58. $\dfrac{(x - 3) - b(x - 3)}{b(x + 3) - x - 3}$

59. $\dfrac{x^2(a + b) + a + b}{x^4 - 1}$

Simplify.

60. $\dfrac{27a^2b^5}{16xy^2} \cdot \dfrac{20x^2y^3}{9a^2b}$

61. $\dfrac{15x^2y^4}{24ab^3} \cdot \dfrac{28a^2b^4}{35xy^4}$

62. $\dfrac{3x - 15}{4x^2 - 2x} \cdot \dfrac{20x^2 - 10x}{15x - 75}$

63. $\dfrac{2x^2 + 4x}{8x^2 - 40x} \cdot \dfrac{6x^3 - 30x^2}{3x^2 + 6x}$

64. $\dfrac{x^2y^3}{x^2 - 4x - 5} \cdot \dfrac{2x^2 - 13x + 15}{x^4y^3}$

65. $\dfrac{2x^2 - 5x + 3}{x^6y^3} \cdot \dfrac{x^4y^4}{2x^2 - x - 3}$

66. $\dfrac{x^2 - 3x + 2}{x^2 - 8x + 15} \cdot \dfrac{x^2 + x - 12}{8 - 2x - x^2}$

67. $\dfrac{x^2 + x - 6}{12 + x - x^2} \cdot \dfrac{x^2 + x - 20}{x^2 - 4x + 4}$

68. $\dfrac{x^{n+1} + 2x^n}{4x^2 - 6x} \cdot \dfrac{8x^2 - 12x}{x^{n+1} - x^n}$

69. $\dfrac{x^{2n} + 2x^n}{x^{n+1} + 2x} \cdot \dfrac{x^2 - 3x}{x^{n+1} - 3x^n}$

70. $\dfrac{x^{2n} - x^n - 6}{x^{2n} + x^n - 2} \cdot \dfrac{x^{2n} - 5x^n - 6}{x^{2n} - 2x^n - 3}$

71. $\dfrac{x^{2n} + 3x^n + 2}{x^{2n} - x^n - 6} \cdot \dfrac{x^{2n} + x^n - 12}{x^{2n} - 1}$

72. $\dfrac{x^3 - y^3}{2x^2 + xy - 3y^2} \cdot \dfrac{2x^2 + 5xy + 3y^2}{x^2 + xy + y^2}$

73. $\dfrac{x^4 - 5x^2 + 4}{3x^2 - 4x - 4} \cdot \dfrac{3x^2 - 10x - 8}{x^2 - 4}$

74. $\dfrac{x^2 + x - 6}{3x^2 + 5x - 12} \cdot \dfrac{2x^2 - 14x}{16x^2 - 4x} \cdot \dfrac{12x^2 - 19x + 4}{x^2 + 5x - 14}$

75. $\dfrac{x^2 - y^2}{x^2 + xy + y^2} \cdot \dfrac{x^2 - xy}{3x^2 - 3xy} \cdot \dfrac{x^3 - y^3}{x^2 - 2xy + y^2}$

Simplify.

76. $\dfrac{6x^2y^4}{35a^2b^5} \div \dfrac{12x^3y^3}{7a^4b^5}$

77. $\dfrac{12a^4b^7}{13x^2y^2} \div \dfrac{18a^5b^6}{26xy^3}$

78. $\dfrac{2x - 6}{6x^2 - 15x} \div \dfrac{4x^2 - 12x}{18x^3 - 45x^2}$

79. $\dfrac{4x^2 - 4y^2}{6x^2y^2} \div \dfrac{3x^2 + 3xy}{2x^2y - 2xy^2}$

80. $\dfrac{2x^2 - 2y^2}{14x^2y^4} \div \dfrac{x^2 + 2xy + y^2}{35xy^3}$

81. $\dfrac{8x^3 + 12x^2y}{4x^2 - 9y^2} \div \dfrac{16x^2y^2}{4x^2 - 12xy + 9y^2}$

82. $\dfrac{x^2 - 8x + 15}{x^2 + 2x - 35} \div \dfrac{15 - 2x - x^2}{x^2 + 9x + 14}$

83. $\dfrac{2x^2 + 13x + 20}{8 - 10x - 3x^2} \div \dfrac{6x^2 - 13x - 5}{9x^2 - 3x - 2}$

84. $\dfrac{x^{2n} + x^n}{2x - 2} \div \dfrac{4x^n + 4}{x^{n+1} - x^n}$

85. $\dfrac{x^{2n} - 4}{4x^n + 8} \div \dfrac{x^{n+1} - 2x}{4x^3 - 12x^2}$

86. $\dfrac{14 + 17x - 6x^2}{3x^2 + 14x + 8} \div \dfrac{4x^2 - 49}{2x^2 + 15x + 28}$

87. $\dfrac{16x^2 - 9}{6 - 5x - 4x^2} \div \dfrac{16x^2 + 24x + 9}{4x^2 + 11x + 6}$

88. $\dfrac{2x^{2n} - x^n - 6}{x^{2n} - x^n - 2} \div \dfrac{2x^{2n} + x^n - 3}{x^{2n} - 1}$

89. $\dfrac{x^{4n} - 1}{x^{2n} + x^n - 2} \div \dfrac{x^{2n} + 1}{x^{2n} + 3x^n + 2}$

90. $\dfrac{6x^2 + 6x}{3x + 6x^2 + 3x^3} \div \dfrac{x^2 - 1}{1 - x^3}$

91. $\dfrac{x^3 + y^3}{2x^3 + 2x^2y} \div \dfrac{3x^3 - 3x^2y + 3xy^2}{6x^2 - 6y^2}$

92. $\dfrac{x^2 - 9}{x^2 - x - 6} \div \dfrac{2x^2 + x - 15}{x^2 + 7x + 10} \cdot \dfrac{2x^2 - x - 10}{x^2 - 25}$

93. $\dfrac{3x^2 + 10x - 8}{x^2 + 4x + 3} \cdot \dfrac{x^2 + 6x + 9}{3x^2 - 5x + 2} \div \dfrac{x^2 + x - 12}{x^2 - 1}$

APPLYING THE CONCEPTS

Simplify.

94. $\dfrac{3x^2 + 6x}{4x^2 - 16} \cdot \dfrac{2x + 8}{x^2 + 2x} \div \dfrac{3x - 9}{5x - 20}$

95. $\dfrac{5y^2 - 20}{3y^2 - 12y} \cdot \dfrac{9y^3 + 6y}{2y^2 - 4y} \div \dfrac{y^3 + 2y^2}{2y^2 - 8y}$

6.2 Addition and Subtraction of Rational Expressions

Objective A   **To rewrite rational expressions in terms of a common denominator**

When adding or subtracting rational expressions, it is frequently necessary to express the rational expressions in terms of a common denominator. This common denominator is the least common multiple (LCM) of the denominators.

The LCM of two or more polynomials is the simplest polynomial that contains the factors of each polynomial. To find the LCM, first factor each polynomial completely. The LCM is the product of each factor the greatest number of times it occurs in any one factorization.

⟹ Find the LCM of $3x^2 + 15x$ and $6x^4 + 24x^3 - 30x^2$.

Factor each polynomial.

$3x^2 + 15x = 3x(x + 5)$
$6x^4 + 24x^3 - 30x^2 = 6x^2(x^2 + 4x - 5)$
$\qquad\qquad\qquad\qquad = 6x^2(x - 1)(x + 5)$

The LCM is the product of the LCM of the numerical coefficients and each variable factor the greatest number of times it occurs in any one factorization.

$\textbf{LCM} = 6x^2(x - 1)(x + 5)$

TAKE NOTE

This is the only section in this text in which the numerator in the final answer is in unfactored form. In building fractions, the fraction is not in simplest form, and leaving the fractions in unfactored form removes the temptation to cancel.

⟹ Write the fractions $\dfrac{5}{6x^2y}$ and $\dfrac{a}{9xy^3}$ in terms of the LCM of the denominators.

The LCM is $18x^2y^3$.

$\dfrac{5}{6x^2y} = \dfrac{5}{6x^2y} \cdot \dfrac{3y^2}{3y^2} = \dfrac{15y^2}{18x^2y^3}$

$\dfrac{a}{9xy^3} = \dfrac{a}{9xy^3} \cdot \dfrac{2x}{2x} = \dfrac{2ax}{18x^2y^3}$

- Find the LCM of the denominators.

- For each fraction, multiply the numerator and denominator by the factor whose product with the denominator is the LCM.

⟹ Write the fractions $\dfrac{x + 2}{x^2 - 2x}$ and $\dfrac{5x}{3x - 6}$ in terms of the LCM of the denominators.

TAKE NOTE

$x^2 - 2x = x(x - 2)$
$3x - 6 = 3(x - 2)$
The LCM of $x(x - 2)$ and $3(x - 2)$ is $3x(x - 2)$.

The LCM is $3x(x - 2)$.

$\dfrac{x + 2}{x^2 - 2x} = \dfrac{x + 2}{x(x - 2)} \cdot \dfrac{3}{3} = \dfrac{3x + 6}{3x(x - 2)}$

$\dfrac{5x}{3x - 6} = \dfrac{5x}{3(x - 2)} \cdot \dfrac{x}{x} = \dfrac{5x^2}{3x(x - 2)}$

- Find the LCM of the denominators.

- For each fraction, multiply the numerator and denominator by the factor whose product with the denominator is the LCM.

Example 1

Write the fractions $\dfrac{3x}{x-1}$ and $\dfrac{4}{2x+5}$ in terms of the LCM of the denominators.

Solution

The LCM of $x-1$ and $2x+5$ is $(x-1)(2x+5)$.

$\dfrac{3x}{x-1} = \dfrac{3x}{x-1} \cdot \dfrac{2x+5}{2x+5} = \dfrac{6x^2+15x}{(x-1)(2x+5)}$

$\dfrac{4}{2x+5} = \dfrac{4}{2x+5} \cdot \dfrac{x-1}{x-1} = \dfrac{4x-4}{(x-1)(2x+5)}$

You Try It 1

Write the fractions $\dfrac{2x}{2x-5}$ and $\dfrac{3}{x+4}$ in terms of the LCM of the denominators.

Your solution

Example 2

Write the fractions $\dfrac{2a-3}{a^2-2a}$ and $\dfrac{a+1}{2a^2-a-6}$ in terms of the LCM of the denominators.

Solution

$a^2 - 2a = a(a-2)$;
$2a^2 - a - 6 = (2a+3)(a-2)$

The LCM is $a(a-2)(2a+3)$.

$\dfrac{2a-3}{a^2-2a} = \dfrac{2a-3}{a(a-2)} \cdot \dfrac{2a+3}{2a+3}$

$\qquad = \dfrac{4a^2-9}{a(a-2)(2a+3)}$

$\dfrac{a+1}{2a^2-a-6} = \dfrac{a+1}{(a-2)(2a+3)} \cdot \dfrac{a}{a}$

$\qquad = \dfrac{a^2+a}{a(a-2)(2a+3)}$

You Try It 2

Write the fractions $\dfrac{3x}{2x^2-11x+15}$ and $\dfrac{x-2}{x^2-3x}$ in terms of the LCM of the denominators.

Your solution

Example 3

Write the fractions $\dfrac{2x-3}{3x-x^2}$ and $\dfrac{3x}{x^2-4x+3}$ in terms of the LCM of the denominators.

Solution

$3x - x^2 = x(3-x) = -x(x-3)$;
$x^2 - 4x + 3 = (x-3)(x-1)$

The LCM is $x(x-3)(x-1)$.

$\dfrac{2x-3}{3x-x^2} = -\dfrac{2x-3}{x(x-3)} \cdot \dfrac{x-1}{x-1}$

$\qquad = -\dfrac{2x^2-5x+3}{x(x-3)(x-1)}$

$\dfrac{3x}{x^2-4x+3} = \dfrac{3x}{(x-3)(x-1)} \cdot \dfrac{x}{x}$

$\qquad = \dfrac{3x^2}{x(x-3)(x-1)}$

You Try It 3

Write the fractions $\dfrac{2x-7}{2x-x^2}$ and $\dfrac{3x-2}{3x^2-5x-2}$ in terms of the LCM of the denominators.

Your solution

Solutions on p. S18

Objective B   **To add or subtract rational expressions**

When adding rational expressions in which the denominators are the same, add the numerators. The denominator of the sum is the common denominator.

$$\frac{a}{c} + \frac{b}{c} = \frac{a+b}{c}$$

$$\frac{4x}{15} + \frac{8x}{15} = \frac{4x+8x}{15} = \frac{12x}{15} = \frac{4x}{5}$$

- Note that the sum is written in simplest form.

$$\frac{a}{a^2 - b^2} + \frac{b}{a^2 - b^2} = \frac{a+b}{a^2 - b^2} = \frac{a+b}{(a-b)(a+b)} = \frac{\overset{1}{(a+b)}}{(a-b)\underset{1}{(a+b)}} = \frac{1}{a-b}$$

When subtracting rational expressions in which the denominators are the same, subtract the numerators. The denominator of the difference is the common denominator. Write the answer in simplest form.

$$\frac{7x-12}{2x^2+5x-12} - \frac{3x-6}{2x^2+5x-12} = \frac{(7x-12)-(3x-6)}{2x^2+5x-12} = \frac{4x-6}{2x^2+5x-12}$$

$$= \frac{2\overset{1}{(2x-3)}}{\underset{1}{(2x-3)}(x+4)} = \frac{2}{x+4}$$

Before two rational expressions with *different* denominators can be added or subtracted, each rational expression must be expressed in terms of a common denominator. This common denominator is the LCM of the denominators of the rational expressions.

➡ Simplify: $\dfrac{x}{x-3} - \dfrac{x+1}{x-2}$

The LCM is $(x-3)(x-2)$.

- Find the LCM of the denominators.

$$\frac{x}{x-3} - \frac{x+1}{x-2} = \frac{x}{x-3} \cdot \frac{x-2}{x-2} - \frac{x+1}{x-2} \cdot \frac{x-3}{x-3}$$

- Express each fraction in terms of the LCM.

$$= \frac{x(x-2) - (x+1)(x-3)}{(x-3)(x-2)}$$

- Subtract the fractions.

$$= \frac{(x^2 - 2x) - (x^2 - 2x - 3)}{(x-3)(x-2)}$$

$$= \frac{3}{(x-3)(x-2)}$$

- Simplify.

➡ Simplify: $\dfrac{3x}{2x-3} + \dfrac{3x+6}{2x^2+x-6}$

The LCM of $2x-3$ and $2x^2+x-6$ is $(2x-3)(x+2)$.

- Find the LCM of the denominators.

$$\frac{3x}{2x-3} + \frac{3x+6}{2x^2+x-6} = \frac{3x}{2x-3} \cdot \frac{x+2}{x+2} + \frac{3x+6}{(2x-3)(x+2)}$$

- Express each fraction in terms of the LCM.

$$= \frac{3x(x+2) + (3x+6)}{(2x-3)(x+2)}$$

- Add the fractions.

$$= \frac{(3x^2 + 6x) + (3x+6)}{(2x-3)(x+2)}$$

$$= \frac{3x^2 + 9x + 6}{(2x-3)(x+2)} = \frac{3\overset{}{(x+2)}(x+1)}{(2x-3)\underset{}{(x+2)}}$$

- Simplify.

$$= \frac{3(x+1)}{2x-3}$$

Example 4

Simplify: $\dfrac{2}{x} - \dfrac{3}{x^2} + \dfrac{1}{xy}$

Solution

The LCM is x^2y.

$$\dfrac{2}{x} - \dfrac{3}{x^2} + \dfrac{1}{xy} = \dfrac{2}{x} \cdot \dfrac{xy}{xy} - \dfrac{3}{x^2} \cdot \dfrac{y}{y} + \dfrac{1}{xy} \cdot \dfrac{x}{x}$$

$$= \dfrac{2xy}{x^2y} - \dfrac{3y}{x^2y} + \dfrac{x}{x^2y} = \dfrac{2xy - 3y + x}{x^2y}$$

You Try It 4

Simplify: $\dfrac{2}{b} - \dfrac{1}{a} + \dfrac{4}{ab}$

Your solution

Example 5

Simplify: $\dfrac{x}{2x - 4} - \dfrac{4 - x}{x^2 - 2x}$

Solution

$2x - 4 = 2(x - 2);\ x^2 - 2x = x(x - 2)$

The LCM is $2x(x - 2)$.

$$\dfrac{x}{2x - 4} - \dfrac{4 - x}{x^2 - 2x} = \dfrac{x}{2(x - 2)} \cdot \dfrac{x}{x} - \dfrac{4 - x}{x(x - 2)} \cdot \dfrac{2}{2}$$

$$= \dfrac{x^2 - (4 - x)2}{2x(x - 2)}$$

$$= \dfrac{x^2 - (8 - 2x)}{2x(x - 2)} = \dfrac{x^2 + 2x - 8}{2x(x - 2)}$$

$$= \dfrac{(x + 4)(x - 2)}{2x(x - 2)} = \dfrac{(x + 4)\overset{1}{\cancel{(x - 2)}}}{2x\underset{1}{\cancel{(x - 2)}}}$$

$$= \dfrac{x + 4}{2x}$$

You Try It 5

Simplify: $\dfrac{a - 3}{a^2 - 5a} + \dfrac{a - 9}{a^2 - 25}$

Your solution

Example 6

Simplify: $\dfrac{x}{x + 1} - \dfrac{2}{x - 2} - \dfrac{3}{x^2 - x - 2}$

Solution

The LCM is $(x + 1)(x - 2)$.

$$\dfrac{x}{x + 1} - \dfrac{2}{x - 2} - \dfrac{3}{x^2 - x - 2}$$

$$= \dfrac{x}{x + 1} \cdot \dfrac{x - 2}{x - 2} - \dfrac{2}{x - 2} \cdot \dfrac{x + 1}{x + 1}$$

$$- \dfrac{3}{(x + 1)(x - 2)}$$

$$= \dfrac{x(x - 2) - 2(x + 1) - 3}{(x + 1)(x - 2)}$$

$$= \dfrac{x^2 - 2x - 2x - 2 - 3}{(x + 1)(x - 2)}$$

$$= \dfrac{x^2 - 4x - 5}{(x + 1)(x - 2)} = \dfrac{\overset{1}{\cancel{(x + 1)}}(x - 5)}{\underset{1}{\cancel{(x + 1)}}(x - 2)} = \dfrac{x - 5}{x - 2}$$

You Try It 6

Simplify: $\dfrac{2x}{x - 4} - \dfrac{x - 1}{x + 1} + \dfrac{2}{x^2 - 3x - 4}$

Your solution

Solutions on p. S19

6.2 Exercises

Objective A

Write each fraction in terms of the LCM of the denominators.

1. $\dfrac{3}{4x^2y}, \dfrac{17}{12xy^4}$

2. $\dfrac{5}{16a^3b^3}, \dfrac{7}{30a^5b}$

3. $\dfrac{x-2}{3x(x-2)}, \dfrac{3}{6x^2}$

4. $\dfrac{5x-1}{4x(2x+1)}, \dfrac{2}{5x^3}$

5. $\dfrac{3x-1}{2x^2-10x}, -3x$

6. $\dfrac{4x-3}{3x(x-2)}, 2x$

7. $\dfrac{3x}{2x-3}, \dfrac{5x}{2x+3}$

8. $\dfrac{2}{7y-3}, \dfrac{-3}{7y+3}$

9. $\dfrac{2x}{x^2-9}, \dfrac{x+1}{x-3}$

10. $\dfrac{3x}{16-x^2}, \dfrac{2x}{16-4x}$

11. $\dfrac{3}{3x^2-12y^2}, \dfrac{5}{6x-12y}$

12. $\dfrac{2x}{x^2-36}, \dfrac{x-1}{6x-36}$

13. $\dfrac{3x}{x^2-1}, \dfrac{5x}{x^2-2x+1}$

14. $\dfrac{x^2+2}{x^3-1}, \dfrac{3}{x^2+x+1}$

15. $\dfrac{x-3}{8-x^3}, \dfrac{2}{4+2x+x^2}$

16. $\dfrac{2x}{x^2+x-6}, \dfrac{-4x}{x^2+5x+6}$

17. $\dfrac{2x}{x^2+2x-3}, \dfrac{-x}{x^2+6x+9}$

18. $\dfrac{3x}{2x^2-x-3}, \dfrac{-2x}{2x^2-11x+12}$

19. $\dfrac{-4x}{4x^2-16x+15}, \dfrac{3x}{6x^2-19x+10}$

20. $\dfrac{3}{2x^2+5x-12}, \dfrac{2x}{3-2x}, \dfrac{3x-1}{x+4}$

21. $\dfrac{5}{6x^2-17x+12}, \dfrac{2x}{4-3x}, \dfrac{x+1}{2x-3}$

22. $\dfrac{3x}{x-4}, \dfrac{4}{x+5}, \dfrac{x+2}{20-x-x^2}$

23. $\dfrac{2x}{x-3}, \dfrac{-2}{x+5}, \dfrac{x-1}{15-2x-x^2}$

24. $\dfrac{2}{x^{2n}-1}, \dfrac{5}{x^{2n}+2x^n+1}$

25. $\dfrac{x-5}{x^{2n}+3x^n+2}, \dfrac{2x}{x^n+2}$

Objective B

Simplify.

26. $\dfrac{3}{2xy} - \dfrac{7}{2xy} - \dfrac{9}{2xy}$

27. $-\dfrac{3}{4x^2} + \dfrac{8}{4x^2} - \dfrac{3}{4x^2}$

28. $\dfrac{x}{x^2-3x+2} - \dfrac{2}{x^2-3x+2}$

29. $\dfrac{3x}{3x^2+x-10} - \dfrac{5}{3x^2+x-10}$

30. $\dfrac{3}{2x^2y} - \dfrac{8}{5x} - \dfrac{9}{10xy}$

31. $\dfrac{2}{5ab} - \dfrac{3}{10a^2b} + \dfrac{4}{15ab^2}$

32. $\dfrac{2}{3x} - \dfrac{3}{2xy} + \dfrac{4}{5xy} - \dfrac{5}{6x}$

33. $\dfrac{3}{4ab} - \dfrac{2}{5a} + \dfrac{3}{10b} - \dfrac{5}{8ab}$

34. $\dfrac{2x-1}{12x} - \dfrac{3x+4}{9x}$

35. $\dfrac{3x-4}{6x} - \dfrac{2x-5}{4x}$

36. $\dfrac{3x+2}{4x^2y} - \dfrac{y-5}{6xy^2}$

37. $\dfrac{2y-4}{5xy^2} + \dfrac{3-2x}{10x^2y}$

38. $\dfrac{2x}{x-3} - \dfrac{3x}{x-5}$

39. $\dfrac{3a}{a-2} - \dfrac{5a}{a+1}$

40. $\dfrac{3}{2a-3} + \dfrac{2a}{3-2a}$

41. $\dfrac{x}{2x-5} - \dfrac{2}{5x-2}$

42. $\dfrac{1}{x+h} - \dfrac{1}{h}$

43. $\dfrac{1}{a-b} + \dfrac{1}{b}$

44. $\dfrac{2}{x} - 3 - \dfrac{10}{x-4}$

45. $\dfrac{6a}{a-3} - 5 + \dfrac{3}{a}$

46. $\dfrac{1}{2x-3} - \dfrac{5}{2x} + 1$

47. $\dfrac{5}{x} - \dfrac{5x}{5 - 6x} + 2$

48. $\dfrac{3}{x^2 - 1} + \dfrac{2x}{x^2 + 2x + 1}$

49. $\dfrac{1}{x^2 - 6x + 9} - \dfrac{1}{x^2 - 9}$

50. $\dfrac{x}{x + 3} - \dfrac{3 - x}{x^2 - 9}$

51. $\dfrac{1}{x + 2} - \dfrac{3x}{x^2 + 4x + 4}$

52. $\dfrac{2x - 3}{x + 5} - \dfrac{x^2 - 4x - 19}{x^2 + 8x + 15}$

53. $\dfrac{-3x^2 + 8x + 2}{x^2 + 2x - 8} - \dfrac{2x - 5}{x + 4}$

54. $\dfrac{x^n}{x^{2n} - 1} - \dfrac{2}{x^n + 1}$

55. $\dfrac{2}{x^n - 1} + \dfrac{x^n}{x^{2n} - 1}$

56. $\dfrac{2}{x^n - 1} - \dfrac{6}{x^{2n} + x^n - 2}$

57. $\dfrac{2x^n - 6}{x^{2n} - x^n - 6} + \dfrac{x^n}{x^n + 2}$

58. $\dfrac{2x - 2}{4x^2 - 9} - \dfrac{5}{3 - 2x}$

59. $\dfrac{x^2 + 4}{4x^2 - 36} - \dfrac{13}{x + 3}$

60. $\dfrac{x - 2}{x + 1} - \dfrac{3 - 12x}{2x^2 - x - 3}$

61. $\dfrac{3x - 4}{4x + 1} + \dfrac{3x + 6}{4x^2 + 9x + 2}$

62. $\dfrac{x + 1}{x^2 + x - 6} - \dfrac{x + 2}{x^2 + 4x + 3}$

63. $\dfrac{x + 1}{x^2 + x - 12} - \dfrac{x - 3}{x^2 + 7x + 12}$

64. $\dfrac{x^2 + 6x}{x^2 + 3x - 18} - \dfrac{2x - 1}{x + 6} + \dfrac{x - 2}{3 - x}$

65. $\dfrac{2x^2 - 2x}{x^2 - 2x - 15} - \dfrac{2}{x + 3} + \dfrac{x}{5 - x}$

66. $\dfrac{7 - 4x}{2x^2 - 9x + 10} + \dfrac{x - 3}{x - 2} - \dfrac{x + 1}{2x - 5}$

67. $\dfrac{x}{3x + 4} + \dfrac{3x + 2}{x - 5} - \dfrac{7x^2 + 24x + 28}{3x^2 - 11x - 20}$

68. $\dfrac{32x - 9}{2x^2 + 7x - 15} + \dfrac{x - 2}{3 - 2x} + \dfrac{3x + 2}{x + 5}$

69. $\dfrac{x + 1}{1 - 2x} - \dfrac{x + 3}{4x - 3} + \dfrac{10x^2 + 7x - 9}{8x^2 - 10x + 3}$

70. $\dfrac{x^2}{x^3 - 8} - \dfrac{x + 2}{x^2 + 2x + 4}$

71. $\dfrac{2x}{4x^2 + 2x + 1} + \dfrac{4x + 1}{8x^3 - 1}$

72. $\dfrac{2x^2}{x^4 - 1} - \dfrac{1}{x^2 - 1} + \dfrac{1}{x^2 + 1}$

73. $\dfrac{x^2 - 12}{x^4 - 16} + \dfrac{1}{x^2 - 4} - \dfrac{1}{x^2 + 4}$

74. $\left(\dfrac{x + 8}{4} + \dfrac{4}{x} \right) \div \dfrac{x + 4}{16x^2}$

75. $\left(\dfrac{a - 3}{a^2} - \dfrac{a - 3}{9} \right) \div \dfrac{a^2 - 9}{3a}$

76. $\dfrac{3}{x - 2} - \dfrac{x^2 + x}{2x^3 + 3x^2} \cdot \dfrac{2x^2 + x - 3}{x^2 + 3x + 2}$

77. $\dfrac{x^2 - 4x + 4}{2x + 1} \cdot \dfrac{2x^2 + x}{x^3 - 4x} - \dfrac{3x - 2}{x + 1}$

78. $\left(\dfrac{x - y}{x^2} - \dfrac{x - y}{y^2} \right) \div \dfrac{x^2 - y^2}{xy}$

79. $\left(\dfrac{a - 2b}{b} + \dfrac{b}{a} \right) \div \left(\dfrac{b + a}{a} - \dfrac{2a}{b} \right)$

80. $\dfrac{2}{x - 3} - \dfrac{x}{x^2 - x - 6} \cdot \dfrac{x^2 - 2x - 3}{x^2 - x}$

81. $\dfrac{2x}{x^2 - x - 6} - \dfrac{6x - 6}{2x^2 - 9x + 9} \div \dfrac{x^2 + x - 2}{2x - 3}$

APPLYING THE CONCEPTS

82. Correct the following expressions.

a. $\dfrac{3}{4} + \dfrac{x}{5} = \dfrac{3 + x}{4 + 5}$

b. $\dfrac{4x + 5}{4} = x + 5$

c. $\dfrac{1}{x} + \dfrac{1}{y} = \dfrac{1}{x + y}$

83. Simplify.

a. $\left(\dfrac{b}{6} - \dfrac{6}{b} \right) \div \left(\dfrac{6}{b} - 4 + \dfrac{b}{2} \right)$

b. $\left(\dfrac{x + 1}{2x - 1} - \dfrac{x - 1}{2x + 1} \right) \cdot \left(\dfrac{2x - 1}{x} - \dfrac{2x - 1}{x^2} \right)$

84. Rewrite each fraction as the sum of two fractions in simplest form.

a. $\dfrac{3x + 6y}{xy}$

b. $\dfrac{4a^2 + 3ab}{a^2 b^2}$

c. $\dfrac{3m^2 n + 2mn^2}{12m^3 n^2}$

85. Let $f(x) = \dfrac{x}{x + 2}$, $g(x) = \dfrac{4}{x - 3}$, and $S(x) = \dfrac{x^2 + x + 8}{x^2 - x - 6}$. Evaluate $f(4)$, $g(4)$, and $S(4)$. Does $f(4) + g(4) = S(4)$? Let a be a real number ($a \neq -2, a \neq 3$). Express $S(a)$ in terms of $f(a)$ and $g(a)$.

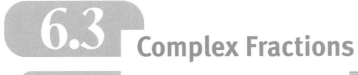

Complex Fractions

Objective A **To simplify a complex fraction**

A **complex fraction** is a fraction whose numerator or denominator contains one or more fractions. Examples of complex fractions are shown below.

$$\frac{5}{2 + \dfrac{1}{2}}, \qquad \frac{5 + \dfrac{1}{y}}{5 - \dfrac{1}{y}}, \qquad \frac{x + 4 + \dfrac{1}{x + 2}}{x - 2 + \dfrac{1}{x + 2}}$$

➡ Simplify: $\dfrac{\dfrac{1}{x} + \dfrac{1}{y}}{\dfrac{1}{x} - \dfrac{1}{y}}$

The LCM of x and y is xy.

- Find the LCM of the denominators of the fractions in the numerator and denominator.
- Multiply the numerator and denominator of the complex fraction by the LCM.

$$\frac{\dfrac{1}{x} + \dfrac{1}{y}}{\dfrac{1}{x} - \dfrac{1}{y}} = \frac{\dfrac{1}{x} + \dfrac{1}{y}}{\dfrac{1}{x} - \dfrac{1}{y}} \cdot \frac{xy}{xy}$$

$$= \frac{\dfrac{1}{x} \cdot xy + \dfrac{1}{y} \cdot xy}{\dfrac{1}{x} \cdot xy - \dfrac{1}{y} \cdot xy} = \frac{y + x}{y - x}$$

Example 1

Simplify: $\dfrac{1 + \dfrac{1}{y - 2}}{1 - \dfrac{2}{y + 1}}$

Solution
The LCM of $y - 2$ and $y + 1$ is $(y - 2)(y + 1)$.

$$\frac{1 + \dfrac{1}{y - 2}}{1 - \dfrac{2}{y + 1}} = \frac{1 + \dfrac{1}{y - 2}}{1 - \dfrac{2}{y + 1}} \cdot \frac{(y - 2)(y + 1)}{(y - 2)(y + 1)}$$

$$= \frac{(y - 2)(y + 1) + \dfrac{1}{y - 2}(y - 2)(y + 1)}{(y - 2)(y + 1) - \dfrac{2}{y + 1}(y - 2)(y + 1)}$$

$$= \frac{y^2 - y - 2 + y + 1}{y^2 - y - 2 - 2y + 4} = \frac{y^2 - 1}{y^2 - 3y + 2}$$

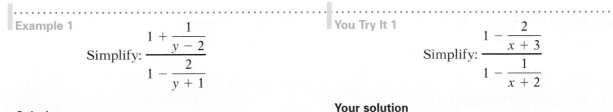

$$= \frac{(y - 1)(y + 1)}{(y - 1)(y - 2)} = \frac{y + 1}{y - 2}$$

You Try It 1

Simplify: $\dfrac{1 - \dfrac{2}{x + 3}}{1 - \dfrac{1}{x + 2}}$

Your solution

Solution on p. S19

Example 2

Simplify: $\dfrac{2x - 1 + \dfrac{7}{x + 4}}{3x - 8 + \dfrac{17}{x + 4}}$

Solution

The LCM of $x + 4$ and $x + 4$ is $x + 4$.

$$\dfrac{2x - 1 + \dfrac{7}{x + 4}}{3x - 8 + \dfrac{17}{x + 4}} = \dfrac{2x - 1 + \dfrac{7}{x + 4}}{3x - 8 + \dfrac{17}{x + 4}} \cdot \dfrac{x + 4}{x + 4}$$

$$= \dfrac{(2x - 1)(x + 4) + \dfrac{7}{x + 4}(x + 4)}{(3x - 8)(x + 4) + \dfrac{17}{x + 4}(x + 4)}$$

$$= \dfrac{2x^2 + 7x - 4 + 7}{3x^2 + 4x - 32 + 17}$$

$$= \dfrac{2x^2 + 7x + 3}{3x^2 + 4x - 15}$$

$$= \dfrac{(2x + 1)(x + 3)}{(3x - 5)(x + 3)}$$

$$= \dfrac{(2x + 1)\cancel{(x + 3)}}{(3x - 5)\cancel{(x + 3)}} = \dfrac{2x + 1}{3x - 5}$$

You Try It 2

Simplify: $\dfrac{2x + 5 + \dfrac{14}{x - 3}}{4x + 16 + \dfrac{49}{x - 3}}$

Your solution

Example 3

Simplify: $1 + \dfrac{a}{2 + \dfrac{1}{a}}$

Solution

The LCM of the denominators is a.

$$1 + \dfrac{a}{2 + \dfrac{1}{a}} = 1 + \dfrac{a}{2 + \dfrac{1}{a}} \cdot \dfrac{a}{a}$$

$$= 1 + \dfrac{a \cdot a}{2 \cdot a + \dfrac{1}{a} \cdot a} = 1 + \dfrac{a^2}{2a + 1}$$

The LCM of 1 and $2a + 1$ is $2a + 1$.

$$1 + \dfrac{a^2}{2a + 1} = 1 \cdot \dfrac{2a + 1}{2a + 1} + \dfrac{a^2}{2a + 1}$$

$$= \dfrac{2a + 1}{2a + 1} + \dfrac{a^2}{2a + 1} = \dfrac{2a + 1 + a^2}{2a + 1}$$

$$= \dfrac{a^2 + 2a + 1}{2a + 1} = \dfrac{(a + 1)^2}{2a + 1}$$

You Try It 3

Simplify: $2 - \dfrac{1}{2 - \dfrac{1}{x}}$

Your solution

Solutions on p. S19

6.3 Exercises

Objective A

1. What is a complex fraction?

2. What is the general goal of simplifying a complex fraction?

Simplify.

3. $\dfrac{2 - \dfrac{1}{3}}{4 + \dfrac{11}{3}}$

4. $\dfrac{3 + \dfrac{5}{2}}{8 - \dfrac{3}{2}}$

5. $\dfrac{3 - \dfrac{2}{3}}{5 + \dfrac{5}{6}}$

6. $\dfrac{5 - \dfrac{3}{4}}{2 + \dfrac{1}{2}}$

7. $\dfrac{1 + \dfrac{1}{x}}{1 - \dfrac{1}{x^2}}$

8. $\dfrac{\dfrac{1}{y^2} - 1}{1 + \dfrac{1}{y}}$

9. $\dfrac{a - 2}{\dfrac{4}{a} - a}$

10. $\dfrac{\dfrac{25}{a} - a}{5 + a}$

11. $\dfrac{\dfrac{1}{a^2} - \dfrac{1}{a}}{\dfrac{1}{a^2} + \dfrac{1}{a}}$

12. $\dfrac{\dfrac{1}{b} + \dfrac{1}{2}}{\dfrac{4}{b^2} - 1}$

13. $\dfrac{2 - \dfrac{4}{x + 2}}{5 - \dfrac{10}{x + 2}}$

14. $\dfrac{4 + \dfrac{12}{2x - 3}}{5 + \dfrac{15}{2x - 3}}$

15. $\dfrac{\dfrac{3}{2a - 3} + 2}{\dfrac{-6}{2a - 3} - 4}$

16. $\dfrac{\dfrac{-5}{b - 5} - 3}{\dfrac{10}{b - 5} + 6}$

17. $\dfrac{\dfrac{x}{x + 1} - \dfrac{1}{x}}{\dfrac{x}{x + 1} + \dfrac{1}{x}}$

18. $\dfrac{\dfrac{2a}{a - 1} - \dfrac{3}{a}}{\dfrac{1}{a - 1} + \dfrac{2}{a}}$

19. $\dfrac{1 - \dfrac{1}{x} - \dfrac{6}{x^2}}{1 - \dfrac{4}{x} + \dfrac{3}{x^2}}$

20. $\dfrac{1 - \dfrac{3}{x} - \dfrac{10}{x^2}}{1 + \dfrac{11}{x} + \dfrac{18}{x^2}}$

21. $\dfrac{1 + \dfrac{1}{x} - \dfrac{12}{x^2}}{\dfrac{9}{x^2} + \dfrac{3}{x} - 2}$

22. $\dfrac{\dfrac{15}{x^2} - \dfrac{2}{x} - 1}{\dfrac{4}{x^2} - \dfrac{5}{x} + 4}$

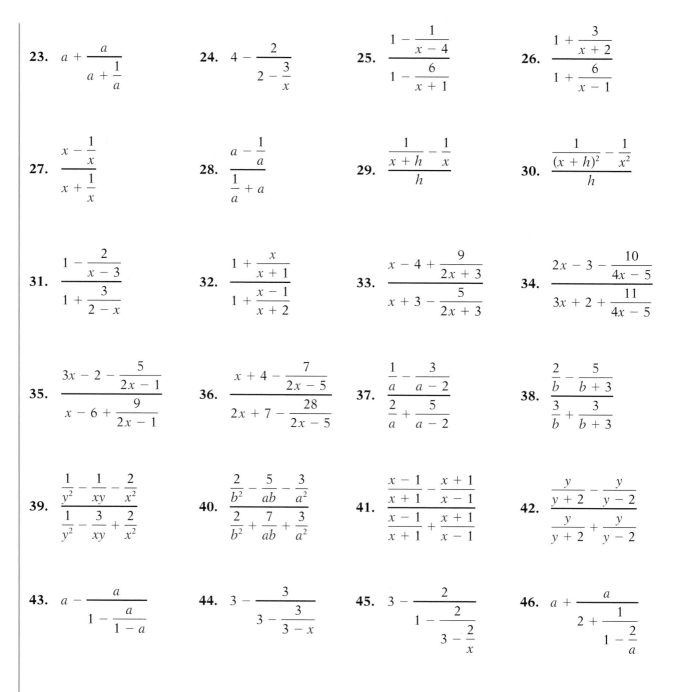

23. $a + \dfrac{a}{a + \dfrac{1}{a}}$

24. $4 - \dfrac{2}{2 - \dfrac{3}{x}}$

25. $\dfrac{1 - \dfrac{1}{x - 4}}{1 - \dfrac{6}{x + 1}}$

26. $\dfrac{1 + \dfrac{3}{x + 2}}{1 + \dfrac{6}{x - 1}}$

27. $\dfrac{x - \dfrac{1}{x}}{x + \dfrac{1}{x}}$

28. $\dfrac{a - \dfrac{1}{a}}{\dfrac{1}{a} + a}$

29. $\dfrac{\dfrac{1}{x + h} - \dfrac{1}{x}}{h}$

30. $\dfrac{\dfrac{1}{(x + h)^2} - \dfrac{1}{x^2}}{h}$

31. $\dfrac{1 - \dfrac{2}{x - 3}}{1 + \dfrac{3}{2 - x}}$

32. $\dfrac{1 + \dfrac{x}{x + 1}}{1 + \dfrac{x - 1}{x + 2}}$

33. $\dfrac{x - 4 + \dfrac{9}{2x + 3}}{x + 3 - \dfrac{5}{2x + 3}}$

34. $\dfrac{2x - 3 - \dfrac{10}{4x - 5}}{3x + 2 + \dfrac{11}{4x - 5}}$

35. $\dfrac{3x - 2 - \dfrac{5}{2x - 1}}{x - 6 + \dfrac{9}{2x - 1}}$

36. $\dfrac{x + 4 - \dfrac{7}{2x - 5}}{2x + 7 - \dfrac{28}{2x - 5}}$

37. $\dfrac{\dfrac{1}{a} - \dfrac{3}{a - 2}}{\dfrac{2}{a} + \dfrac{5}{a - 2}}$

38. $\dfrac{\dfrac{2}{b} - \dfrac{5}{b + 3}}{\dfrac{3}{b} + \dfrac{3}{b + 3}}$

39. $\dfrac{\dfrac{1}{y^2} - \dfrac{1}{xy} - \dfrac{2}{x^2}}{\dfrac{1}{y^2} - \dfrac{3}{xy} + \dfrac{2}{x^2}}$

40. $\dfrac{\dfrac{2}{b^2} - \dfrac{5}{ab} - \dfrac{3}{a^2}}{\dfrac{2}{b^2} + \dfrac{7}{ab} + \dfrac{3}{a^2}}$

41. $\dfrac{\dfrac{x - 1}{x + 1} - \dfrac{x + 1}{x - 1}}{\dfrac{x - 1}{x + 1} + \dfrac{x + 1}{x - 1}}$

42. $\dfrac{\dfrac{y}{y + 2} - \dfrac{y}{y - 2}}{\dfrac{y}{y + 2} + \dfrac{y}{y - 2}}$

43. $a - \dfrac{a}{1 - \dfrac{a}{1 - a}}$

44. $3 - \dfrac{3}{3 - \dfrac{3}{3 - x}}$

45. $3 - \dfrac{2}{1 - \dfrac{2}{3 - \dfrac{2}{x}}}$

46. $a + \dfrac{a}{2 + \dfrac{1}{1 - \dfrac{2}{a}}}$

APPLYING THE CONCEPTS

47. Find the sum of the reciprocals of three consecutive even integers.

48. If $a = \dfrac{b^2 + 4b + 4}{b^2 - 4}$ and $b = \dfrac{1}{c}$, express a in terms of c.

49. Simplify.

a. $\dfrac{x^{-1}}{x^{-1} + 2^{-1}}$

b. $\dfrac{-x^{-1} + x}{x^{-1} - x}$

c. $\dfrac{x^{-1} - x^{-2} - 6x^{-3}}{x^{-1} - 4x^{-3}}$

6.4 Ratio and Proportion

Objective A   **To solve a proportion**

Quantities such as 3 feet, 5 liters, and 2 miles are number quantities written with units. In these examples the units are feet, liters, and miles.

A **ratio** is the quotient of two quantities that have the same units.

The weekly wages of a painter are $800. The painter spends $150 a week for food. The ratio of the wages spent for food to the total weekly wages is written:

$$\frac{\$150}{\$800} = \frac{150}{800} = \frac{3}{16}$$ A ratio is in simplest form when the two numbers do not have a common factor. Note that the units are not written.

A **rate** is the quotient of two quantities that have different units.

A car travels 180 mi on 3 gal of gas. The miles-to-gallons rate is:

$$\frac{180 \text{ mi}}{3 \text{ gal}} = \frac{60 \text{ mi}}{1 \text{ gal}}$$ A rate is in simplest form when the two numbers do not have a common factor. The units are written as part of the rate.

A **proportion** is an equation that states the equality of two ratios or rates. For example, $\frac{90 \text{ km}}{4 \text{ L}} = \frac{45 \text{ km}}{2 \text{ L}}$ and $\frac{3}{4} = \frac{x + 2}{16}$ are proportions.

➡ Solve the proportion: $\frac{2}{7} = \frac{x}{5}$

$$\frac{2}{7} = \frac{x}{5}$$

$$35 \cdot \frac{2}{7} = 35 \cdot \frac{x}{5}$$

$$10 = 7x$$

$$\frac{10}{7} = x$$

- Multiply each side of the proportion by the LCM of the denominators.
- Solve the equation.

The solution is $\frac{10}{7}$.

....

Example 1

Solve the proportion: $\frac{3}{12} = \frac{5}{x + 5}$

Solution

$$\frac{3}{12} = \frac{5}{x + 5}$$

$$\frac{3}{12} \cdot 12(x + 5) = \frac{5}{x + 5} \cdot 12(x + 5)$$

$$3(x + 5) = 5 \cdot 12$$

$$3x + 15 = 60$$

$$3x = 45$$

$$x = 15$$

The solution is 15.

You Try It 1

Solve the proportion: $\frac{5}{x - 2} = \frac{3}{4}$

Your solution

Solution on p. S19

Example 2

Solve the proportion: $\dfrac{3}{x-2} = \dfrac{4}{2x+1}$

Solution

$$\frac{3}{x-2} = \frac{4}{2x+1}$$

$$\frac{3}{x-2}(2x+1)(x-2) = \frac{4}{2x+1}(2x+1)(x-2)$$

$$3(2x+1) = 4(x-2)$$

$$6x+3 = 4x-8$$

$$2x+3 = -8$$

$$2x = -11$$

$$x = -\frac{11}{2}$$

The solution is $-\dfrac{11}{2}$.

You Try It 2

Solve the proportion: $\dfrac{5}{2x-3} = \dfrac{-2}{x+1}$

Your solution

Solution on p. S20

Objective B   **To solve application problems**

Example 3

A stock investment of 50 shares pays a dividend of $106. At this rate, how many additional shares are required to earn a dividend of $424?

Strategy

To find the additional number of shares that are required, write and solve a proportion using x to represent the additional number of shares. Then $50 + x$ is the total number of shares of stock.

Solution

$$\frac{106}{50} = \frac{424}{50+x}$$

$$\frac{53}{25} = \frac{424}{50+x}$$

$$\frac{53}{25}(25)(50+x) = \frac{424}{50+x}(25)(50+x)$$

$$53(50+x) = 424(25)$$

$$2650 + 53x = 10{,}600$$

$$53x = 7950$$

$$x = 150$$

An additional 150 shares of stock are required.

You Try It 3

Two pounds of cashews cost $5.80. At this rate, how much would 15 lb of cashews cost?

Your strategy

Your solution

Solution on p. S20

6.4 Exercises

Objective A

1. How does a ratio differ from a rate?

2. What is a proportion?

Solve the proportion.

3. $\dfrac{x}{30} = \dfrac{3}{10}$

4. $\dfrac{5}{15} = \dfrac{x}{75}$

5. $\dfrac{2}{x} = \dfrac{8}{30}$

6. $\dfrac{80}{16} = \dfrac{15}{x}$

7. $\dfrac{x+1}{10} = \dfrac{2}{5}$

8. $\dfrac{5-x}{10} = \dfrac{3}{2}$

9. $\dfrac{4}{x+2} = \dfrac{3}{4}$

10. $\dfrac{8}{3} = \dfrac{24}{x+3}$

11. $\dfrac{x}{4} = \dfrac{x-2}{8}$

12. $\dfrac{8}{x-5} = \dfrac{3}{x}$

13. $\dfrac{16}{2-x} = \dfrac{4}{x}$

14. $\dfrac{6}{x-5} = \dfrac{1}{x}$

15. $\dfrac{8}{x-2} = \dfrac{4}{x+1}$

16. $\dfrac{4}{x-4} = \dfrac{2}{x-2}$

17. $\dfrac{x}{3} = \dfrac{x+1}{7}$

18. $\dfrac{x-3}{2} = \dfrac{3+x}{5}$

19. $\dfrac{8}{3x-2} = \dfrac{2}{2x+1}$

20. $\dfrac{3}{2x-4} = \dfrac{-5}{x+2}$

21. $\dfrac{3x+1}{3x-4} = \dfrac{x}{x-2}$

22. $\dfrac{x-2}{x-5} = \dfrac{2x}{2x+5}$

Objective B   *Application Problems*

23. If a 56-gram serving of pasta contains 7 g of protein, how many grams of protein are in a 454-gram box of the pasta?

24. A movie theatre that can seat 1500 people has a parking lot with 600 spaces. At the same rate, how many parking spaces should a new theatre with 2200 seats create?

25. A team of biologists captured and tagged approximately 2400 northern squawfish at the Bonneville Dam on the Colorado River. Later, 225 squawfish were recaptured and nine of them had tags. How many squawfish would you estimate are in the area?

26. The United States Postal Service delivers 200 billion pieces of mail each year. How many pieces of mail does the Postal Service deliver per week? Assume the Postal Service delivers mail 300 days during the year and 6 days per week.

27. Of 250 people who purchased home computers from a national company, 45 received machines with defective CD-ROM drives. At this rate, how many of the 2800 computers the company sold nationwide would you expect to have bad CD-ROM drives?

28. During a recent year, 3293 cars were stolen in Montgomery County, Maryland. Police were able to recover 2305 of these. At this rate, if 19,000 cars were stolen throughout the state of Maryland, how many of these would you expect to be recovered?

29. On an architectural drawing, $\frac{1}{4}$ in. represents one foot. Using this scale, find the dimensions of a room that measures $4\frac{1}{2}$ in. by 6 in. on the drawing.

30. One hundred forty-four ceramic tiles are required to tile a 25-square-foot area. At this rate, how many tiles are required to tile 275 ft^2?

31. One and one-half ounces of a medication are required for a 140-pound adult. At the same rate, how many additional ounces of medication are required for 210-pound adult?

32. After working for a company for 9 months, an employee accrued 6 vacation days. At this rate, how much longer will it be before she can take a 3-week vacation (15 work days)?

33. Walking 4 mi in 2 h will use up 650 calories. Walking at the same rate, how many miles would a person need to walk to lose one pound? (The burning of 3500 calories is equivalent to the loss of one pound.) Round to the nearest hundredth.

34. To raise money, a school is sponsoring a magazine subscription drive. By visiting 90 homes in the neighborhood, one student was able to sell $375 worth of subscriptions. At this rate, how many additional homes will the student have to visit in order to meet a $1000 goal?

APPLYING THE CONCEPTS

35. If $\frac{a}{b} = \frac{c}{d}$, show that $\frac{a+b}{b} = \frac{c+d}{d}$.

36. Three people put their money together to buy lottery tickets. The first person put in $20, the second person put in $25, and the third person put in $30. One of the tickets was a winning ticket. If they won $4.5 million, what was the first person's share of the winnings?

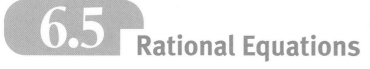

Rational Equations

Objective A   **To solve a fractional equation**   ◄ 6 ►

To solve an equation containing fractions, **clear denominators** by multiplying each side of the equation by the LCM of the denominators. Then solve for the variable.

After each side of a fractional equation has been multiplied by the LCM of the denominators, the resulting equation is sometimes a quadratic equation. The solutions to the resulting equation must be checked, because multiplying each side of an equation by a variable expression may produce an equation that has a solution that is not a solution of the original equation.

> **TAKE NOTE**
>
> If each side of an equation is multiplied by a variable expression, it is essential that the solutions be checked. As Example 1 on the next page shows, a proposed solution to an equation may not check when it is substituted into the original equation.

⇒ Solve: $\dfrac{3x}{x-5} = x + \dfrac{12}{x-5}$

$$\frac{3x}{x-5} = x + \frac{12}{x-5}$$

$$(x-5)\left(\frac{3x}{x-5}\right) = (x-5)\left(x + \frac{12}{x-5}\right)$$ • Multiply each side of the equation by the LCM of the denominators.

$$3x = (x-5)x + (x-5)\left(\frac{12}{x-5}\right)$$

$$3x = x^2 - 5x + 12$$

$$0 = x^2 - 8x + 12$$ • Solve the quadratic equation by factoring.

$$0 = (x-2)(x-6)$$

$$x = 2 \qquad x = 6$$

2 and 6 check as solutions.

The solutions are 2 and 6.

⇒ Solve the literal equation $\dfrac{1}{R_1} + \dfrac{1}{R_2} = \dfrac{1}{R}$ for R.

$$\frac{1}{R_1} + \frac{1}{R_2} = \frac{1}{R}$$

$$RR_1R_2\left(\frac{1}{R_1} + \frac{1}{R_2}\right) = RR_1R_2\left(\frac{1}{R}\right)$$ • Multiply each side of the equation by the LCM of the denominators.

$$\frac{RR_1R_2}{R_1} + \frac{RR_1R_2}{R_2} = \frac{RR_1R_2}{R}$$

$$RR_2 + RR_1 = R_1R_2$$

$$R(R_2 + R_1) = R_1R_2$$ • Factor.

$$\frac{R(R_1 + R_2)}{R_1 + R_2} = \frac{R_1R_2}{R_1 + R_2}$$

$$R = \frac{R_1R_2}{R_1 + R_2}$$

This literal equation is true for all real numbers, $R_1 + R_2 \neq 0$.

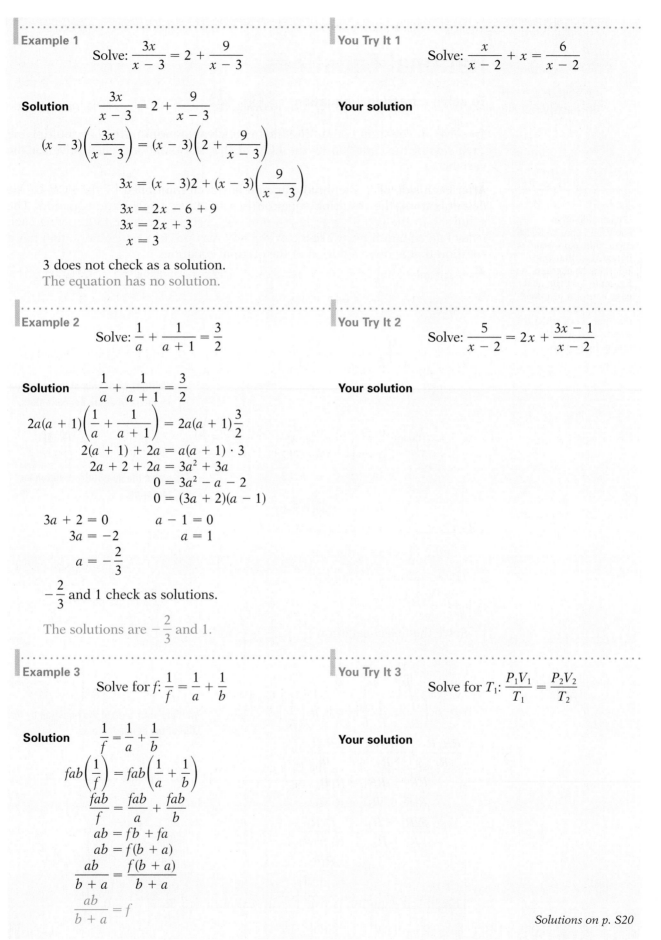

Example 1

Solve: $\dfrac{3x}{x-3} = 2 + \dfrac{9}{x-3}$

Solution

$$\dfrac{3x}{x-3} = 2 + \dfrac{9}{x-3}$$

$$(x-3)\left(\dfrac{3x}{x-3}\right) = (x-3)\left(2 + \dfrac{9}{x-3}\right)$$

$$3x = (x-3)2 + (x-3)\left(\dfrac{9}{x-3}\right)$$

$$3x = 2x - 6 + 9$$
$$3x = 2x + 3$$
$$x = 3$$

3 does not check as a solution.
The equation has no solution.

You Try It 1

Solve: $\dfrac{x}{x-2} + x = \dfrac{6}{x-2}$

Your solution

Example 2

Solve: $\dfrac{1}{a} + \dfrac{1}{a+1} = \dfrac{3}{2}$

Solution

$$\dfrac{1}{a} + \dfrac{1}{a+1} = \dfrac{3}{2}$$

$$2a(a+1)\left(\dfrac{1}{a} + \dfrac{1}{a+1}\right) = 2a(a+1)\dfrac{3}{2}$$

$$2(a+1) + 2a = a(a+1) \cdot 3$$
$$2a + 2 + 2a = 3a^2 + 3a$$
$$0 = 3a^2 - a - 2$$
$$0 = (3a+2)(a-1)$$

$$3a + 2 = 0 \qquad a - 1 = 0$$
$$3a = -2 \qquad a = 1$$
$$a = -\dfrac{2}{3}$$

$-\dfrac{2}{3}$ and 1 check as solutions.

The solutions are $-\dfrac{2}{3}$ and 1.

You Try It 2

Solve: $\dfrac{5}{x-2} = 2x + \dfrac{3x-1}{x-2}$

Your solution

Example 3

Solve for f: $\dfrac{1}{f} = \dfrac{1}{a} + \dfrac{1}{b}$

Solution

$$\dfrac{1}{f} = \dfrac{1}{a} + \dfrac{1}{b}$$

$$fab\left(\dfrac{1}{f}\right) = fab\left(\dfrac{1}{a} + \dfrac{1}{b}\right)$$

$$\dfrac{fab}{f} = \dfrac{fab}{a} + \dfrac{fab}{b}$$

$$ab = fb + fa$$
$$ab = f(b+a)$$

$$\dfrac{ab}{b+a} = \dfrac{f(b+a)}{b+a}$$

$$\dfrac{ab}{b+a} = f$$

You Try It 3

Solve for T_1: $\dfrac{P_1V_1}{T_1} = \dfrac{P_2V_2}{T_2}$

Your solution

Solutions on p. S20

Objective B   **To solve work problems**

Point of Interest

The following problem was recorded in the *Jiuzhang*, a Chinese text that dates to the Han dynasty (about 200 B.C. to A.D. 200). "A reservoir has five channels bringing water to it. The first can fill the reservoir in $\frac{1}{3}$ day, the second in 1 day, the third in $2\frac{1}{2}$ days, the fourth in 3 days, and the fifth in 5 days. If all channels are open, how long does it take to fill the reservoir?" This problem is the earliest known work problem.

Rate of work is that part of a task that is completed in one unit of time. If a mason can build a retaining wall in 12 h, then in 1 h the mason can build $\frac{1}{12}$ of the wall. The mason's rate of work is $\frac{1}{12}$ of the wall each hour. If an apprentice can build the wall in x hours, the rate of work for the apprentice is $\frac{1}{x}$ of the wall each hour.

In solving a work problem, the goal is to determine the time it takes to complete a task. The basic equation that is used to solve work problems is

Rate of work × time worked = part of task completed

For example, if a pipe can fill a tank in 5 h, then in 2 h the pipe will fill $\frac{1}{5} \times 2 = \frac{2}{5}$ of the tank. In t hours, the pipe will fill $\frac{1}{5} \times t = \frac{t}{5}$ of the tank.

➡ A mason can build a wall in 10 h. An apprentice can build a wall in 15 h. How long will it take to build the wall when they work together?

> **Strategy for Solving a Work Problem**
>
> 1. For each person or machine, write a numerical or variable expression for the rate of work, the time worked, and the part of the task completed. The results can be recorded in a table.

Unknown time to build the wall working together: t

	Rate of Work	·	Time Worked	=	Part of Task Completed
Mason	$\frac{1}{10}$	·	t	=	$\frac{t}{10}$
Apprentice	$\frac{1}{15}$	·	t	=	$\frac{t}{15}$

> 2. Determine how the parts of the task completed are related. Use the fact that the sum of the parts of the task completed must equal 1, the complete task.

The sum of the part of the task completed by the mason and the part of the task completed by the apprentice is 1.

$$\frac{t}{10} + \frac{t}{15} = 1$$
$$30\left(\frac{t}{10} + \frac{t}{15}\right) = 30(1)$$
$$3t + 2t = 30$$
$$5t = 30$$
$$t = 6$$

Working together, they will build the wall in 6 h.

Example 4

An electrician requires 12 h to wire a house. The electrician's apprentice can wire a house in 16 h. After working alone on a job for 4 h, the electrician quits, and the apprentice completes the task. How long does it take the apprentice to finish wiring the house?

Strategy

• Time required for the apprentice to finish wiring the house: t

	Rate	Time	Part
Electrician	$\dfrac{1}{12}$	4	$\dfrac{4}{12}$
Apprentice	$\dfrac{1}{16}$	t	$\dfrac{t}{16}$

• The sum of the part of the task completed by the electrician and the part of the task completed by the apprentice is 1.

Solution

$$\frac{4}{12} + \frac{t}{16} = 1$$

$$\frac{1}{3} + \frac{t}{16} = 1$$

$$48\left(\frac{1}{3} + \frac{t}{16}\right) = 48(1)$$

$$16 + 3t = 48$$

$$3t = 32$$

$$t = \frac{32}{3}$$

It will take the apprentice $10\frac{2}{3}$ h to finish wiring the house.

You Try It 4

Two water pipes can fill a tank with water in 6 h. The larger pipe working alone can fill the tank in 9 h. How long will it take the smaller pipe working alone to fill the tank?

Your strategy

Your solution

Solution on p. S20

Objective C   **To solve uniform motion problems**

A car that travels constantly in a straight line at 55 mph is in uniform motion. **Uniform motion** means that the speed of an object does not change.

The basic equation used to solve uniform motion problems is

Distance = rate × time

An alternative form of this equation can be written by solving the equation for time.

$$\frac{\textbf{Distance}}{\textbf{Rate}} = \textbf{time}$$

This form of the equation is useful when the total time of travel for two objects or the time of travel between two points is known.

➡ A motorist drove 150 mi on country roads before driving 50 mi on mountain roads. The rate of speed on the country roads was three times the rate on the mountain roads. The time spent traveling the 200 mi was 5 h. Find the rate of the motorist on the country roads.

Strategy for Solving a Uniform Motion Problem

1. For each object, write a numerical or variable expression for the distance, rate, and time. The results can be recorded in a table.

Unknown rate of speed on the mountain roads: r
Rate of speed on the country roads: $3r$

	Distance	÷	Rate	=	Time
Country roads	150	÷	$3r$	=	$\frac{150}{3r}$
Mountain roads	50	÷	r	=	$\frac{50}{r}$

2. Determine how the times traveled by each object are related. For example, it may be known that the times are equal, or the total time may be known.

The total time of the trip is 5 h.

$$\frac{150}{3r} + \frac{50}{r} = 5$$

$$3r\left(\frac{150}{3r} + \frac{50}{r}\right) = 3r(5)$$

$$150 + 150 = 15r$$

$$300 = 15r$$

$$20 = r$$

The rate of speed on the country roads was $3r$.
Replace r with 20 and evaluate.

$$3r = 3(20) = 60$$

The rate of speed on the country roads was 60 mph.

Example 5

A marketing executive traveled 810 mi on a corporate jet in the same amount of time as it took to travel an additional 162 mi by helicopter. The rate of the jet was 360 mph faster than the rate of the helicopter. Find the rate of the jet.

Strategy

* Rate of the helicopter: r
  Rate of the jet: $r + 360$

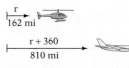

	Distance	Rate	Time
Jet	810	$r + 360$	$\dfrac{810}{r + 360}$
Helicopter	162	r	$\dfrac{162}{r}$

* The time traveled by jet is equal to the time traveled by helicopter.

Solution

$$\frac{810}{r + 360} = \frac{162}{r}$$

$$r(r + 360)\left(\frac{810}{r + 360}\right) = r(r + 360)\left(\frac{162}{r}\right)$$

$$810r = (r + 360)162$$

$$810r = 162r + 58{,}320$$

$$648r = 58{,}320$$

$$r = 90$$

$$r + 360 = 90 + 360 = 450$$

The rate of the jet was 450 mph.

You Try It 5

A plane can fly at a rate of 150 mph in calm air. Traveling with the wind, the plane flew 700 mi in the same amount of time as it flew 500 mi against the wind. Find the rate of the wind.

Your strategy

Your solution

Solution on p. S21

6.5 Exercises

Objective A

Solve.

1. $\dfrac{x}{2} + \dfrac{5}{6} = \dfrac{x}{3}$

2. $\dfrac{x}{5} - \dfrac{2}{9} = \dfrac{x}{15}$

3. $\dfrac{8}{2x - 1} = 2$

4. $3 = \dfrac{18}{3x - 4}$

5. $1 - \dfrac{3}{y} = 4$

6. $7 + \dfrac{6}{y} = 5$

7. $\dfrac{3}{x - 2} = \dfrac{4}{x}$

8. $\dfrac{5}{x} = \dfrac{2}{x + 3}$

9. $\dfrac{6}{2y + 3} = \dfrac{6}{y}$

10. $\dfrac{3}{x - 4} + 2 = \dfrac{5}{x - 4}$

11. $\dfrac{5}{y + 3} - 2 = \dfrac{7}{y + 3}$

12. $5 + \dfrac{8}{a - 2} = \dfrac{4a}{a - 2}$

13. $\dfrac{-4}{a - 4} = 3 - \dfrac{a}{a - 4}$

14. $\dfrac{x}{x + 1} - x = \dfrac{-4}{x + 1}$

15. $\dfrac{2x}{x + 2} + 3x = \dfrac{-5}{x + 2}$

16. $\dfrac{x}{x - 3} + x = \dfrac{3x - 4}{x - 3}$

17. $\dfrac{x}{2x - 9} - 3x = \dfrac{10}{9 - 2x}$

18. $\dfrac{2}{4y^2 - 9} + \dfrac{1}{2y - 3} = \dfrac{3}{2y + 3}$

19. $\dfrac{5}{x - 2} - \dfrac{2}{x + 2} = \dfrac{3}{x^2 - 4}$

20. $\dfrac{5}{x^2 - 7x + 12} = \dfrac{2}{x - 3} + \dfrac{5}{x - 4}$

21. $\dfrac{9}{x^2 + 7x + 10} = \dfrac{5}{x + 2} - \dfrac{3}{x + 5}$

Solve the formula for the given variable.

22. $F = \dfrac{Gm_1m_2}{r^2}$; m_2 (Physics)

23. $\dfrac{P_1V_1}{T_1} = \dfrac{P_2V_2}{T_2}$; P_2 (Chemistry)

24. $\dfrac{1}{R} = \dfrac{1}{R_1} + \dfrac{1}{R_2}$; R_2 (Physics)

25. $\dfrac{1}{f} = \dfrac{1}{a} + \dfrac{1}{b}$; b (Physics)

Objective B   *Application Problems*

26. A ski resort can manufacture enough man-made snow to open its steepest run in 12 h, whereas naturally falling snow would take 36 h to provide enough snow. If the resort makes snow at the same time that it is snowing naturally, how long will it take until the run can open?

27. An experienced bricklayer can work twice as fast as an apprentice bricklayer. After they work together on a job for 6 h, the experienced bricklayer quits. The apprentice requires 10 more hours to finish the job. How long would it take the experienced bricklayer working alone to do the job?

28. A roofer requires 8 h to shingle a roof. After the roofer and an apprentice work on a roof for 2 h, the roofer moves on to another job. The apprentice requires 10 more hours to finish the job. How long would it take the apprentice working alone to do the job?

29. An employee is faxing a long document that will take 40 min to send on her fastest machine. After 15 min, she realizes the fax won't make her deadline and she starts sending some of the pages with a second slower fax machine. Twenty minutes later the document finishes faxing. How long would it have taken the slower machine alone to send the fax?

30. The larger of two printers being used to print the payroll for a major corporation requires 30 min to print the payroll. After both printers have been operating for 10 min, the larger printer malfunctions. The smaller printer requires 40 more minutes to complete the payroll. How long would it take the smaller printer working alone to print the payroll?

31. Three machines are filling soda bottles. The machines can fill the daily quota of soda bottles in 10 h, 12 h, and 15 h, respectively. How long would it take to fill the daily quota of soda bottles with all three machines working?

32. A goat can eat all the grass in a farmer's field in 12 days, whereas a cow can finish it in 15 days and a horse in 20 days. How long before all the grass is eaten if all three animals are allowed to graze in the field?

33. The inlet pipe can fill a water tank in 45 min. The outlet pipe can empty the tank in 30 min. How long would it take to empty a full tank with both pipes open?

34. Three computers can print out a task in 20 min, 30 min, and 60 min, respectively. How long would it take to complete the task with all three computers working?

35. Two circus clowns are blowing up balloons, but some of the balloons are popping before they can be sold. The first clown can blow up a balloon every 2 min, the second clown requires 3 min for each balloon, and every 5 min one balloon pops. How long will it take the clowns, working together, to have 76 balloons?

36. An oil tank has two inlet pipes and one outlet pipe. One inlet pipe can fill the tank in 12 h, and the other inlet pipe can fill the tank in 20 h. The outlet pipe can empty the tank in 10 h. How long would it take to fill the tank with all three pipes open?

37. Two clerks are addressing advertising envelopes for a company. One clerk can address one envelope every 30 s, whereas it takes 40 s for the second clerk to address one envelope. How long will it take them, working together, to address 140 envelopes?

38. A single-engine airplane carries enough fuel for an 8-hour flight. After the airplane has been flying for 1 h, the fuel tank begins to leak at a rate that would empty the tank in 12 h. How long after the leak begins does the plane have until it will run out of fuel?

Objective C   *Application Problems*

39. The rate of a bicyclist is 7 mph faster than the rate of a long-distance runner. The bicyclist travels 30 mi in the same amount of time as it takes the runner to travel 16 mi. Find the rate of the runner.

40. A passenger train travels 240 mi in the same amount of time as it takes a freight train to travel 168 mi. The rate of the passenger train is 18 mph faster than the rate of the freight train. Find the rate of each train.

41. A tortoise and a hare have joined a 360-foot race. Since the hare can run 180 times faster than the tortoise, it reaches the finish line 14 min and 55 s before the tortoise. How fast was each animal running?

42. A cabin cruiser travels 20 mi in the same amount of time as it takes a power boat to travel 45 mi. The rate of the cabin cruiser is 10 mph less than the rate of the power boat. Find the rate of the cabin cruiser.

43. A cyclist and a jogger start from a town at the same time and head for a destination 30 mi away. The rate of the cyclist is twice the rate of the jogger. The cyclist arrives 3 h ahead of the jogger. Find the rate of the cyclist.

44. A Porsche 911 Turbo has a top speed that is 20 mph faster than a Dodge Viper's top speed. At top speed, the Porsche can travel 630 mi in the same amount of time that the Viper can travel 560 mi. What is the speed of each car?

45. A canoe can travel 8 mph in still water. Rowing with the current of a river, the canoe can travel 15 mi in the same amount of time as it takes to travel 9 mi against the current. Find the rate of the current.

46. A tour boat used for river excursions can travel 7 mph in calm water. The amount of time it takes to travel 20 mi with the current is the same as the amount of time it takes to travel 8 mi against the current. Find the rate of the current.

47. Some military fighter planes are capable of flying at a rate of 950 mph. One of these planes flew 6150 mi with the wind in the same amount of time it took to fly 5250 mi against the wind. Find the rate of the wind.

48. A plane can fly at a rate of 175 mph in calm air. Traveling with the wind, the plane flew 615 mi in the same amount of time as it took to fly 435 mi against the wind. Find the rate of the wind.

49. A jet can travel 550 mph in calm air. Flying with the wind, the jet can travel 3059 mi in the same amount of time as it takes to fly 2450 mi against the wind. Find the rate of the wind. Round to the nearest hundredth.

50. A jet-ski can comfortably travel across calm water at 35 mph. If a rider traveled 8 mi down a river in the same amount of time it took to travel 6 mi back up the river, find the rate of the river's current.

51. A river excursion motorboat can travel 6 mph in calm water. A recent trip, going 16 mi down a river and then returning, took 6 h. Find the rate of the river's current.

52. A pilot can fly a plane at 125 mph in calm air. A recent trip of 300 mi flying with the wind and 300 mi returning against the wind took 5 h. Find the rate of the wind.

APPLYING THE CONCEPTS

53. Solve for x.

   a. $\dfrac{x-2}{y} = \dfrac{x+2}{5y}$ **b.** $\dfrac{x}{x+y} = \dfrac{2x}{4y}$ **c.** $\dfrac{x-y}{x} = \dfrac{2x}{9y}$

54. The denominator of a fraction is 4 more than the numerator. If the numerator and denominator of the fraction are increased by 3, the new fraction is $\dfrac{5}{6}$. Find the original fraction.

55. Because of bad weather conditions, a bus driver reduced the usual speed along a 165-mile bus route by 5 mph. The bus arrived only 15 min later than its usual arrival time. How fast does the bus usually travel?

56. Explain the difference between dealing with the denominator in simplifying a rational expression and dealing with the denominator in solving a rational equation.

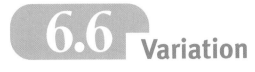

Variation

Objective A   **To solve variation problems**

Direct variation is a special function that can be expressed as the equation $y = kx$, where k is a constant. The equation $y = kx$ is read "y varies directly as x" or "y is directly proportional to x." The constant k is called the **constant of variation** or the **constant of proportionality.**

The circumference (C) of a circle varies directly as the diameter (d). The direct variation equation is written $C = \pi d$. The constant of variation is π.

A nurse makes \$20 per hour. The total wage (w) of the nurse is directly proportional to the number of hours (h) worked. The equation of variation is $w = 20h$. The constant of proportionality is 20.

A direct variation equation can be written in the form $y = kx^n$, where n is a positive number. For example, the equation $y = kx^2$ is read "y varies directly as the square of x."

The area (A) of a circle varies directly as the square of the radius (r) of the circle. The direct variation equation is $A = \pi r^2$. The constant of variation is π.

➡ Given that V varies directly as r and that $V = 20$ when $r = 4$, find the constant of variation and the variation equation.

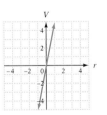

The graph of $V = 5r$

$V = kr$	• Write the basic direct variation equation.
$20 = k \cdot 4$	• Replace V and r by the given values. Then solve for k.
$5 = k$	• This is the constant of variation.
$V = 5r$	• Write the direct variation equation by substituting the value of k into the basic direct variation equation.

➡ The tension (T) in a spring varies directly as the distance (x) it is stretched. If $T = 8$ lb when $x = 2$ in., find T when $x = 4$ in.

The graph of $T = 4x$

$T = kx$	• Write the basic direct variation equation.
$8 = k \cdot 2$	• Replace T and x by the given values.
$4 = k$	• Solve for the constant of variation.
$T = 4x$	• Write the direct variation equation.

To find T when $x = 4$ in., substitute 4 for x in the equation and solve for T.

$T = 4x$
$T = 4 \cdot 4 = 16$

The tension is 16 lb.

Inverse variation is a function that can be expressed as the equation $y = \dfrac{k}{x}$, where k is a constant. The equation $y = \dfrac{k}{x}$ is read "y varies inversely as x" or "y is inversely proportional to x."

In general, an inverse variation equation can be written $y = \dfrac{k}{x^n}$, where n is a positive number. For example, the equation $y = \dfrac{k}{x^2}$ is read "y varies inversely as the square of x."

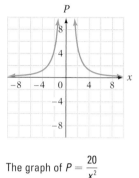

➡ Given that P varies inversely as the square of x and that $P = 5$ when $x = 2$, find the constant of variation and the variation equation.

$P = \dfrac{k}{x^2}$ • Write the basic inverse variation equation.

$5 = \dfrac{k}{2^2}$ • Replace P and x by the given values. Then solve for k.

$5 = \dfrac{k}{4}$

$20 = k$ • This is the constant of variation.

$P = \dfrac{20}{x^2}$ • Write the inverse variation equation by substituting the value of k into the basic inverse variation equation.

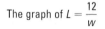

The graph of $P = \dfrac{20}{x^2}$

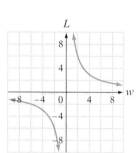

The graph of $L = \dfrac{12}{w}$

➡ The length (L) of a rectangle with fixed area is inversely proportional to the width (w). If $L = 6$ ft when $w = 2$ ft, find L when $w = 3$ ft.

$L = \dfrac{k}{w}$ • Write the basic inverse variation equation.

$6 = \dfrac{k}{2}$ • Replace L and w by the given values.

$12 = k$ • Solve for the constant of variation.

$L = \dfrac{12}{w}$ • Write the inverse variation equation.

To find L when $w = 3$ ft, substitute 3 for w in the equation and solve for L.

$L = \dfrac{12}{w}$

$L = \dfrac{12}{3} = 4$

The length is 4 ft.

Joint variation is a variation in which a variable varies directly as the product of two or more other variables. A joint variation can be expressed as the equation $z = kxy$, where k is a constant. The equation $z = kxy$ is read "z varies jointly as x and y."

The area (A) of a triangle varies jointly as the base (b) and the height (h). The joint variation equation is written $A = \dfrac{1}{2}bh$. The constant of variation is $\dfrac{1}{2}$.

A **combined variation** is a variation in which two or more types of variation occur at the same time. For example, in physics, the volume (V) of a gas varies directly as the temperature (T) and inversely as the pressure (P). This combined variation is written $V = \dfrac{kT}{P}$.

⇒ A ball is being twirled on the end of a string. The tension (T) in the string is directly proportional to the square of the speed (v) of the ball and inversely proportional to the length (r) of the string. If the tension is 96 lb when the length of the string is 0.5 ft and the speed is 4 ft/s, find the tension when the length of the string is 1 ft and the speed is 5 ft/s.

$$T = \frac{kv^2}{r}$$ • Write the basic combined variation equation.

$$96 = \frac{k \cdot 4^2}{0.5}$$ • Replace T, v, and r by the given values.

$$96 = \frac{k \cdot 16}{0.5}$$

$$96 = k \cdot 32$$ • Solve for the constant of variation.

$$3 = k$$

$$T = \frac{3v^2}{r}$$ • Write the combined variation equation.

To find T when $r = 1$ ft and $v = 5$ ft/s, substitute 1 for r and 5 for v and solve for T.

$$T = \frac{3v^2}{r} = \frac{3 \cdot 5^2}{1} = 3 \cdot 25 = 75$$

The tension is 75 lb.

- -

Example 1

The amount (A) of medication prescribed for a person is directly related to the person's weight (W). For a 50-kilogram person, 2 ml of medication are prescribed. How many milliliters of medication are required for a person who weighs 75 kg?

Strategy

To find the required amount of medication:
• Write the basic direct variation equation, replace the variables by the given values, and solve for k.
• Write the direct variation equation, replacing k by its value. Substitute 75 for W and solve for A.

Solution

$$A = kW$$
$$2 = k \cdot 50$$
$$\frac{1}{25} = k$$

$$A = \frac{1}{25}W = \frac{1}{25} \cdot 75 = 3$$

The required amount of medication is 3 ml.

You Try It 1

The distance (s) a body falls from rest varies directly as the square of the time (t) of the fall. An object falls 64 ft in 2 s. How far will it fall in 5 s?

Your strategy

Your solution

Solution on p. S21

Example 2

A company that produces personal computers has determined that the number of computers it can sell (s) is inversely proportional to the price (P) of the computer. Two thousand computers can be sold when the price is $2500. How many computers can be sold if the price of a computer is $2000?

Strategy

To find the number of computers:
- Write the basic inverse variation equation, replace the variables by the given values, and solve for k.
- Write the inverse variation equation, replacing k by its value. Substitute 2000 for P and solve for s.

Solution

$$s = \frac{k}{P}$$

$$2000 = \frac{k}{2500}$$

$$5{,}000{,}000 = k$$

$$s = \frac{5{,}000{,}000}{P} = \frac{5{,}000{,}000}{2000} = 2500$$

At $2000 each, 2500 computers can be sold.

You Try It 2

The resistance (R) to the flow of electric current in a wire of fixed length is inversely proportional to the square of the diameter (d) of the wire. If a wire of diameter 0.01 cm has a resistance of 0.5 ohm, what is the resistance in a wire that is 0.02 cm in diameter?

Your strategy

Your solution

Example 3

The pressure (P) of a gas varies directly as the temperature (T) and inversely as the volume (V). When $T = 50°$ and $V = 275$ in³, $P = 20$ lb/in². Find the pressure of a gas when $T = 60°$ and $V = 250$ in³.

Strategy

To find the pressure:
- Write the basic combined variation equation, replace the variables by the given values, and solve for k.
- Write the combined variation equation, replacing k by its value. Substitute 60 for T and 250 for V, and solve for P.

Solution

$$P = \frac{kT}{V}$$

$$20 = \frac{k \cdot 50}{275}$$

$$110 = k$$

$$P = \frac{110T}{V} = \frac{110 \cdot 60}{250} = 26.4$$

The pressure is 26.4 lb/in².

You Try It 3

The strength (s) of a rectangular beam varies jointly as its width (W) and as the square of its depth (d) and inversely as its length (L). If the strength of a beam 2 in. wide, 12 in. deep, and 12 ft long is 1200 lb, find the strength of a beam 4 in. wide, 8 in. deep, and 16 ft long.

Your strategy

Your solution

Solutions on p. S21

6.6 Exercises

Objective A *Application Problems*

1. The profit (*P*) realized by a company varies directly as the number of products it sells (*s*). If a company makes a profit of $4000 on the sale of 250 products, what is the profit when the company sells 5000 products?

2. The income (*I*) of a computer analyst varies directly as the number of hours (*h*) worked. If the analyst earns $256 for working 8 h, how much will the analyst earn by working 36 h?

3. The pressure (*p*) on a diver in the water varies directly as the depth (*d*). If the pressure is 4.5 lb/in² when the depth is 10 ft, what is the pressure when the depth is 15 ft?

4. The distance (*d*) a spring will stretch varies directly as the force (*f*) applied to the spring. If a force of 6 lb is required to stretch a spring 3 in., what force is required to stretch the spring 4 in.?

5. The distance (*d*) an object will fall is directly proportional to the square of the time (*t*) of the fall. If an object falls 144 ft in 3 s, how far will the object fall in 10 s?

6. The period (*p*) of a pendulum, or the time it takes the pendulum to make one complete swing, varies directly as the square root of the length (*L*) of the pendulum. If the period of a pendulum is 1.5 s when the length is 2 ft, find the period when the length is 5 ft. Round to the nearest hundredth.

7. The distance (*s*) a ball will roll down an inclined plane is directly proportional to the square of the time (*t*). If the ball rolls 6 ft in 1 s, how far will it roll in 3 s?

8. The stopping distance (*s*) of a car varies directly as the square of its speed (*v*). If a car traveling 30 mph requires 63 ft to stop, find the stopping distance for a car traveling 55 mph.

9. The length (*L*) of a rectangle of fixed area varies inversely as the width (*w*). If the length of a rectangle is 8 ft when the width is 5 ft, find the length of the rectangle when the width is 4 ft.

10. The time (t) for a car to travel between two cities is inversely proportional to the rate (r) of travel. If it takes 5 h to travel between the cities at a rate of 55 mph, find the time to travel between the two cities at a rate of 65 mph. Round to the nearest tenth.

11. The speed (v) of a gear varies inversely as the number of teeth (t). If a gear that has 45 teeth makes 24 revolutions per minute, how many revolutions per minute will a gear that has 36 teeth make?

12. The pressure (p) of a liquid varies directly as the product of the depth (d) and the density (D) of the liquid. If the pressure is 150 lb/in^2 when the depth is 100 in. and the density is 1.2, find the pressure when the density remains the same and the depth is 75 in.

13. The current (I) in a wire varies directly as the voltage (v) and inversely as the resistance (r). If the current is 10 amps when the voltage is 110 volts and the resistance is 11 ohms, find the current when the voltage is 180 volts and the resistance is 24 ohms.

14. The repulsive force (f) between the north poles of two magnets is inversely proportional to the square of the distance (d) between them. If the repulsive force is 20 lb when the distance is 4 in., find the repulsive force when the distance is 2 in.

15. The intensity (I) of a light source is inversely proportional to the square of the distance (d) from the source. If the intensity is 12 foot-candles at a distance of 10 ft, what is the intensity when the distance is 5 ft?

APPLYING THE CONCEPTS

16. In the inverse variation equation $y = \dfrac{k}{x}$, what is the effect on x if y doubles?

17. In the direct variation equation $y = kx$, what is the effect on y when x doubles?

Complete using the word *directly* or *inversely*.

18. If a varies directly as b and inversely as c, then c varies _____ as b and _____ as a.

19. If a varies _____ as b and c, then abc is constant.

20. If the length of a rectangle is held constant, the area of the rectangle varies _____ as the width.

21. If the area of a rectangle is held constant, the length of the rectangle varies _____ as the width.

Focus on Problem Solving

Implication Sentences that are constructed using "If..., then..." occur frequently in problem-solving situations. These sentences are called **implications.** The sentence "If it rains, then I will stay home" is an implication. The phrase *it rains* is called the **antecedent** of the implication. The phrase *I will stay home* is the **consequent** of the implication.

The sentence "If $x = 4$, then $x^2 = 16$" is a true sentence. The **contrapositive** of an implication is formed by switching the antecedent and the consequent and then negating each one. The contrapositive of "If $x = 4$, then $x^2 = 16$" is "If $x^2 \neq 16$, then $x \neq 4$." This sentence is also true. It is a principle of logic that an implication and its contrapositive are either both true or both false.

The **converse** of an implication is formed by switching the antecedent and the consequent. The converse of "If $x = 4$, then $x^2 = 16$" is "If $x^2 = 16$, then $x = 4$." Note that the converse is not a true statement, because if $x^2 = 16$, then x could equal -4. The converse of an implication may or may not be true.

Those statements for which the implication and the converse are both true are very important. They can be stated in the form "x if and only if y." For instance, a number is divisible by 5 if and only if the last digit of the number is 0 or 5. The implication "If a number is divisible by 5, then it ends in 0 or 5" and the converse "If a number ends in 0 or 5, then it is divisible by 5" are both true.

For each of the following, state the contrapositive and the converse of the implication. If the converse and the implication are both true statements, write a sentence using the phrase "if and only if."

1. If I live in Chicago, then I live in Illinois.

2. If today is June 1, then yesterday was May 31.

3. If today is not Thursday, then tomorrow is not Friday.

4. If a number is divisible by 8, then it is divisible by 4.

5. If a number is an even number, then it is divisible by 2.

6. If a number is a multiple of 6, then it is a multiple of 3.

7. If $4z = 20$, then $z = 5$.

8. If an angle measures 90°, then it is a right angle.

9. If p is a prime number greater than 2, then p is an odd number.

10. If the equation of a graph is $y = mx + b$, then the graph of the equation is a straight line.

11. If $a = 0$ or $b = 0$, then $ab = 0$.

12. If the coordinates of a point are $(5, 0)$, then the point is on the x-axis.

13. If a quadrilateral is a square, then the quadrilateral has four sides of equal length.

14. If $x = y$, then $x^2 = y^2$.

Projects and Group Activities

Continued Fractions   The following complex fraction is called a **continued fraction.**

$$1 + \cfrac{1}{1 + \cfrac{1}{1 + \cfrac{1}{1 + \cfrac{1}{1 + \cdots}}}}$$

The dots indicate that the pattern continues to repeat forever.

A **convergent** of a continued fraction is an approximation of the repeated pattern. For instance,

$$c_2 = 1 + \cfrac{1}{1 + \cfrac{1}{1 + 1}} \qquad c_3 = 1 + \cfrac{1}{1 + \cfrac{1}{1 + \cfrac{1}{1 + 1}}} \qquad c_4 = 1 + \cfrac{1}{1 + \cfrac{1}{1 + \cfrac{1}{1 + \cfrac{1}{1 + 1}}}}$$

1. Calculate c_5 for the continued fraction above.

This particular continued fraction is related to the golden rectangle, which has been used in architectural designs as diverse as the Parthenon in Athens, built around 440 B.C., and the United Nations building. A golden rectangle is one for which

$$\frac{\text{length}}{\text{width}} = \frac{\text{length} + \text{width}}{\text{length}}$$

An example of a golden rectangle is shown at the right.

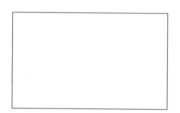

Here is another continued fraction that was discovered by Leonhard Euler (1707–1793). Calculating the convergents of this continued fraction yields approximations that are closer and closer to π.

$$\pi = 3 + \cfrac{1^2}{6 + \cfrac{3^2}{6 + \cfrac{5^2}{6 + \cfrac{7^2}{6 + \cdots}}}}$$

2. Calculate $c_5 = 3 + \cfrac{1^2}{6 + \cfrac{3^2}{6 + \cfrac{5^2}{6 + \cfrac{7^2}{6 + \cfrac{9^2}{6 + 11^2}}}}}$.

Graphing Rational Functions

The domain of a function is the set of the first coordinates of all the ordered pairs of the function. When a function is given by an equation, the domain of the function is all real numbers for which the function evaluates to a real number. For rational functions, we must exclude from the domain all those values of the variable for which the denominator of the rational function is zero. The graphing calculator is a useful tool to show the graphs of functions with excluded values in the domain of the function.

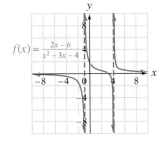

$f(x) = \dfrac{2x - 6}{x^2 - 3x - 4}$

The domain is
$\{x \mid x \neq -1, x \neq 4\}$.

The graph of the function $f(x) = \dfrac{2x - 6}{x^2 - 3x - 4}$ is shown at the left. Note that the graph never intersects the lines $x = -1$ and $x = 4$ (shown as dashed lines). These are the two values excluded from the domain of f. The graph of the function f gets closer to the dashed line $x = 4$ as x gets closer to 4. The graph of the function f also gets closer to $x = -1$ as x gets closer to -1. The lines that are "approached" by the function are called **asymptotes.**

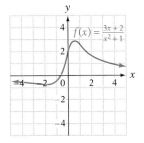

$f(x) = \dfrac{3x + 2}{x^2 + 1}$

The domain is
$\{x \mid x \in \text{real numbers}\}$.

The graph of the function $f(x) = \dfrac{3x + 2}{x^2 + 1}$ is shown at the left. The domain must exclude values of x for which $x^2 + 1 = 0$. It is not possible that $x^2 + 1 = 0$, because $x^2 \geq 0$, and a positive number added to a number equal to or greater than zero cannot equal zero. Therefore, there are no real numbers that must be excluded from the domain of f.

Exercises 13–24 in Section 6.1 ask you to find the domains of rational functions by algebraic means. Try these exercises again, this time using a graphing calculator.

Graphing Variation Equations

1. Graph $y = kx$ when $k = 2$.
   What kind of function does the graph represent?

2. Graph $y = kx$ when $k = \dfrac{1}{2}$.
   What kind of function does the graph represent?

3. Graph $y = \dfrac{k}{x}$ when $k = 2$ and $x > 0$.
   Is this the graph of a function?

Chapter Summary

Key Words

A fraction in which the numerator and denominator are polynomials is called a *rational expression*. [p. 321]

A rational expression is in *simplest form* when the numerator and denominator have no common factors. [p. 322]

The *least common multiple* (LCM) of two or more polynomials is the simplest polynomial that contains the factors of each polynomial. [p. 331]

The *reciprocal* of a rational expression is the rational expression with the numerator and denominator interchanged. [p. 325]

A *complex fraction* is a fraction whose numerator or denominator contains one or more fractions. [p. 339]

A *ratio* is the quotient of two quantities that have the same units. [p. 343]

A *rate* is the quotient of two quantities that have different units. [p. 343]

A *proportion* is an equation that states the equality of two ratios or rates. [p. 343]

Direct variation is a special function that can be expressed as the equation $y = kx$, where k is a constant called the *constant of variation* or the *constant of proportionality*. [p. 357]

Inverse variation is a function that can be expressed as the equation $y = \dfrac{k}{x}$, where k is a constant. [p. 358]

Joint variation is a variation in which a variable varies directly as the product of two or more variables. A joint variation can be expressed as the equation $z = kxy$, where k is a constant. [p. 358]

Combined variation is a variation in which two or more types of variation occur at the same time. [p. 358]

Essential Rules

To Multiply Fractions: $\dfrac{a}{b} \cdot \dfrac{c}{d} = \dfrac{ac}{bd}$ [p. 324]

To Divide Fractions: $\dfrac{a}{b} \div \dfrac{c}{d} = \dfrac{a}{b} \cdot \dfrac{d}{c}$ [p. 325]

To Add Fractions: $\dfrac{a}{c} + \dfrac{b}{c} = \dfrac{a + b}{c}$ [p. 333]

To Subtract Fractions: $\dfrac{a}{c} - \dfrac{b}{c} = \dfrac{a - b}{c}$ [p. 333]

Equation for Work Problems: $\dfrac{\text{Rate of}}{\text{work}} \times \text{time worked} = \dfrac{\text{part of task}}{\text{completed}}$ [p. 349]

Uniform Motion Equation: Distance = rate × time [p. 351]

Chapter Review

1. Simplify: $\dfrac{a^6b^4 + a^4b^6}{a^5b^4 - a^4b^4} \cdot \dfrac{a^2 - b^2}{a^4 - b^4}$

2. Simplify: $\dfrac{x}{x - 3} - 4 - \dfrac{2x - 5}{x + 2}$

3. Given $P(x) = \dfrac{x}{x - 3}$, find $P(4)$.

4. Solve: $\dfrac{3x - 2}{x + 6} = \dfrac{3x + 1}{x + 9}$

5. Simplify: $\dfrac{\dfrac{3x + 4}{3x - 4} + \dfrac{3x - 4}{3x + 4}}{\dfrac{3x - 4}{3x + 4} - \dfrac{3x + 4}{3x - 4}}$

6. Write each fraction in terms of the LCM of the denominators.

$\dfrac{4x}{4x - 1}, \dfrac{3x - 1}{4x + 1}$

7. Solve $S = \dfrac{a}{1 - r}$ for r.

8. Given $P(x) = \dfrac{x^2 - 2}{3x^2 - 2x + 5}$, find $P(-2)$.

9. Solve: $\dfrac{10}{5x + 3} = \dfrac{2}{10x - 3}$

10. Determine the domain of
$f(x) = \dfrac{2x - 7}{3x^2 + 3x - 18}$.

11. Simplify: $\dfrac{3x^4 + 11x^2 - 4}{3x^4 + 13x^2 + 4}$

12. Determine the domain of $g(x) = \dfrac{2x}{x - 3}$.

13. Simplify: $\dfrac{x^3 - 8}{x^3 + 2x^2 + 4x} \cdot \dfrac{x^3 + 2x^2}{x^2 - 4}$

14. Simplify: $\dfrac{3x^2 + 2}{x^2 - 4} - \dfrac{9x - x^2}{x^2 - 4}$

15. Solve $Q = \dfrac{N - S}{N}$ for N.

16. Solve: $\dfrac{30}{x^2 + 5x + 4} + \dfrac{10}{x + 4} = \dfrac{4}{x + 1}$

17. Simplify: $x + \dfrac{\dfrac{4}{x} - 1}{\dfrac{1}{x} - \dfrac{3}{x^2}}$

18. Write each fraction in terms of the LCM of the denominators.

$$\dfrac{x - 3}{x - 5}, \dfrac{x}{x^2 - 9x + 20}, \dfrac{1}{4 - x}$$

19. Solve: $\dfrac{6}{2x - 3} = \dfrac{5}{x + 5} + \dfrac{5}{2x^2 + 7x - 15}$

20. Simplify: $\dfrac{x^{n+1} + x}{x^{2n} - 1} \div \dfrac{x^{n+2} - x^2}{x^{2n} - 2x^n + 1}$

21. Simplify: $\dfrac{27x^3 - 8}{9x^3 + 6x^2 + 4x} \div \dfrac{9x^2 - 12x + 4}{9x^2 - 4}$

22. Simplify: $\dfrac{6x}{3x^2 - 7x + 2} - \dfrac{2}{3x - 1} + \dfrac{3x}{x - 2}$

23. Simplify: $\dfrac{x^3 - 27}{x^2 - 9}$

24. Simplify: $\dfrac{5}{3a^2b^3} + \dfrac{7}{8ab^4}$

25. A helicopter travels 9 mi in the same amount of time that it takes an airplane to travel 10 mi. The rate of the airplane is 20 mph faster than the rate of the helicopter. Find the rate of the helicopter.

26. A student reads 2 pages of text in 5 min. At the same rate, how long will it take the student to read 150 pages?

27. The current (I) in an electric circuit varies inversely as the resistance (R). If the current in the circuit is 4 amps when the resistance is 50 ohms, find the current in the circuit when the resistance is 100 ohms.

28. On a certain map, 2.5 in. represents 10 mi. How many miles would be represented by 12 in.?

29. The stopping distance (s) of a car varies directly as the square of the speed (v) of the car. For a car traveling at 50 mph, the stopping distance is 170 ft. Find the stopping distance of a car that is traveling at 30 mph.

30. An electrician requires 65 min to install a ceiling fan. The electrician and an apprentice working together take 40 min to install the fan. How long would it take the apprentice working alone to install the ceiling fan?

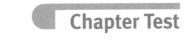

Chapter Test

1. Solve the proportion: $\dfrac{3}{x+1} = \dfrac{2}{x}$

2. Simplify: $\dfrac{x^2+x-6}{x^2+7x+12} \div \dfrac{x^2-3x+2}{x^2+6x+8}$

3. Simplify: $\dfrac{2x-1}{x+2} - \dfrac{x}{x-3}$

4. Write each fraction in terms of the LCM of the denominators.

$\dfrac{x+1}{x^2+x-6}, \dfrac{2x}{x^2-9}$

5. Solve: $\dfrac{4x}{2x-1} = 2 - \dfrac{1}{2x-1}$

6. Simplify: $\dfrac{v^3-4v}{2v^2-5v+2}$

7. Simplify: $\dfrac{3x^2-12}{5x-15} \cdot \dfrac{2x^2-18}{x^2+5x+6}$

8. Determine the domain of $f(x) = \dfrac{3x^2-x+1}{x^2-9}$.

9. Simplify: $\dfrac{1 - \dfrac{1}{x} - \dfrac{12}{x^2}}{1 + \dfrac{6}{x} + \dfrac{9}{x^2}}$

10. Simplify: $\dfrac{1 - \dfrac{1}{x+2}}{1 - \dfrac{3}{x+4}}$

11. Simplify: $\dfrac{2x^2-x-3}{2x^2-5x+3} \div \dfrac{3x^2-x-4}{x^2-1}$

12. Solve: $\dfrac{4x}{x+1} - x = \dfrac{2}{x+1}$

13. Simplify: $\dfrac{2a^2-8a+8}{4+4a-3a^2}$

14. Solve $\dfrac{1}{r} = \dfrac{1}{2} - \dfrac{2}{t}$ for t.

15. Given $f(x) = \dfrac{3 - x^2}{x^3 - 2x^2 + 4}$, find $f(-1)$.

16. Simplify: $\dfrac{x + 2}{x^2 + 3x - 4} - \dfrac{2x}{x^2 - 1}$

17. Solve the proportion: $\dfrac{x + 1}{2x + 5} = \dfrac{x - 3}{x}$

18. A cyclist travels 20 mi in the same amount of time as it takes a hiker to walk 6 mi. The rate of the cyclist is 7 mph faster than the rate of the hiker. Find the rate of the cyclist.

19. The electrical resistance (r) of a cable varies directly as its length (l) and inversely as the square of its diameter (d). If a cable 16,000 ft long and $\dfrac{1}{4}$ in. in diameter has a resistance of 3.2 ohms, what is the resistance of a cable that is 8000 ft long and $\dfrac{1}{2}$ in. in diameter?

20. An interior designer uses 2 rolls of wallpaper for every 45 ft² of wall space in an office. At this rate, how many rolls of wallpaper are needed for an office that has 315 ft² of wall space?

21. One landscaper can till the soil for a lawn in 30 min, whereas it takes a second landscaper 15 min to do the same job. How long would it take to till the soil for the lawn with both landscapers working together?

22. The current (I) in an electric circuit is inversely proportional to the resistance (R). If the current is 0.25 amps when the resistance is 8 ohms, find the resistance when the current is 1.25 amps.

Cumulative Review

1. Simplify: $8 - 4[-3 - (-2)]^2 \div 5$

2. Solve: $\dfrac{2x - 3}{6} - \dfrac{x}{9} = \dfrac{x - 4}{3}$

3. Solve: $5 - |x - 4| = 2$

4. Find the domain of $\dfrac{x}{x - 3}$.

5. Given $P(x) = \dfrac{x - 1}{2x - 3}$, find $P(-2)$.

6. Write 0.000000035 in scientific notation.

7. Simplify: $\dfrac{(2a^{-2}b^3)^{-2}}{(4a)^{-1}}$

8. Solve: $x - 3(1 - 2x) \geq 1 - 4(2 - 2x)$

9. Simplify: $(2a^2 - 3a + 1)(-2a^2)$

10. Factor: $2x^{2n} + 3x^n - 2$

11. Factor: $x^3y^3 - 27$

12. Simplify: $\dfrac{x^4 + x^3y - 6x^2y^2}{x^3 - 2x^2y}$

13. Find the equation of the line that contains the point $(-2, -1)$ and is parallel to the line $3x - 2y = 6$.

14. Solve: $8x^2 - 6x - 9 = 0$

15. Simplify: $\dfrac{4x^3 + 2x^2 - 10x + 1}{x - 2}$

16. Simplify: $\dfrac{16x^2 - 9y^2}{16x^2y - 12xy^2} \div \dfrac{4x^2 - xy - 3y^2}{12x^2y^2}$

17. Write each fraction in terms of the LCM of the denominators. $\dfrac{xy}{2x^2 + 2x}, \dfrac{2}{2x^4 - 2x^3 - 4x^2}$

18. Simplify: $\dfrac{5x}{3x^2 - x - 2} - \dfrac{2x}{x^2 - 1}$

19. Graph $-3x + 5y = -15$ by using the x- and y-intercepts.

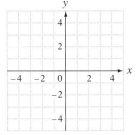

20. Graph the solution set: $\begin{array}{l} x + y \leq 3 \\ -2x + y > 4 \end{array}$

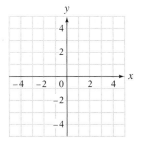

21. Evaluate the determinant:

$$\begin{vmatrix} 6 & 5 \\ 2 & -3 \end{vmatrix}$$

22. Simplify: $\dfrac{x - 4 + \dfrac{5}{x + 2}}{x + 2 - \dfrac{1}{x + 2}}$

23. Solve: $\begin{aligned} x + y + z &= 3 \\ -2x + y + 3z &= 2 \\ 2x - 4y + z &= -1 \end{aligned}$

24. Solve: $|3x - 2| > 4$

25. Solve the proportion: $\dfrac{2}{x - 3} = \dfrac{5}{2x - 3}$

26. Solve: $\dfrac{3}{x^2 - 36} = \dfrac{2}{x - 6} - \dfrac{5}{x + 6}$

27. Solve $I = \dfrac{E}{R + r}$ for r.

28. Simplify: $(1 - x^{-1})^{-1}$

29. The sum of two integers is fifteen. Five times the smaller integer is five more than twice the larger integer. Find the integers.

30. How many pounds of almonds that cost $5.40 per pound must be mixed with 50 lb of peanuts that cost $2.60 per pound to make a mixture that costs $4.00 per pound?

31. A pre-election survey showed that three out of five voters would vote in an election. At this rate, how many people would be expected to vote in a city of 125,000?

32. A new computer can work six times faster than an older computer. Working together, the computers can complete a job in 12 min. How long would it take the new computer working alone to complete the job?

33. A plane can fly at a rate of 300 mph in calm air. Traveling with the wind, the plane flew 900 mi in the same amount of time as it took to fly 600 mi against the wind. Find the rate of the wind.

34. Two people start from the same point on a circular exercise track that is 0.25 mi in circumference. The first person walks at 3 mph, and the second person jogs at 5 mph. After 1 h, how far apart (along the track) are the two people?

Chapter 7

Exponents and Radicals

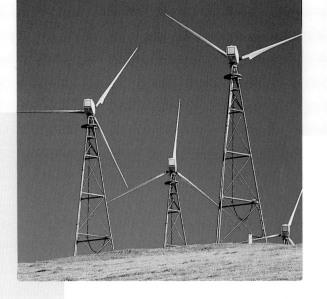

Alternate sources of energy are important today given the rising cost of natural gas and oil, as well as the concern for depletion of Earth's resources. Windmills provide an alternate source by harnessing energy in the form of wind. Windmills like the ones pictured here, which are used to generate electricity, are known as wind turbine generators. The amount of power a windmill produces is dependent on the velocity of the wind and can be modeled by a radical equation, such as the one in **Exercise 28 on page 414**.

Need help? For on-line student resources, such as section quizzes, visit this textbook's web site at **math.college.hmco.com/students**.

1. Complete: $48 = ? \cdot 3$

For Exercises 2 to 7, simplify.

2. 2^5

3. $6\left(\dfrac{3}{2}\right)$

4. $\dfrac{1}{2} - \dfrac{2}{3} + \dfrac{1}{4}$

5. $(3 - 7x) - (4 - 2x)$

6. $\dfrac{3x^5y^6}{12x^4y}$

7. $(3x - 2)^2$

For Exercises 8 and 9, multiply.

8. $(2 + 4x)(5 - 3x)$

9. $(6x - 1)(6x + 1)$

10. Solve: $x^2 - 14x - 5 = 10$

Go Figure

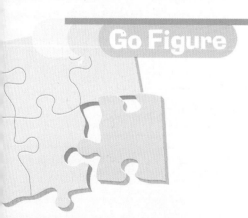

You are planning a large dinner party. If you seat 5 people at each table, you end up with only 2 people at the last table. If you seat 3 people at each table, you have 9 people left over with no place to sit. There are less than 10 tables. How many guests are coming to the dinner party?

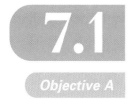

Rational Exponents and Radical Expressions

Objective A   **To simplify expressions with rational exponents**

Point of Interest

Nicolas Chuquet (c. 1475), a French physician, wrote an algebra text in which he used a notation for expressions with fractional exponents. He wrote $R^2 6$ to mean $6^{1/2}$ and $R^3 15$ to mean $15^{1/3}$. This was an improvement over earlier notations that used words for these expressions.

In this section, the definition of an exponent is extended beyond integers so that any rational number can be used as an exponent. The definition is expressed in such a way that the Rules of Exponents hold true for rational exponents.

Consider the expression $(a^{1/n})^n$ for $a > 0$ and n a positive integer. Now simplify, assuming that the Rule for Simplifying the Power of an Exponential Expression is true.

$$(a^{1/n})^n = a^{\frac{1}{n} \cdot n} = a^1 = a$$

Because $(a^{1/n})^n = a$, the number $a^{1/n}$ is the number whose nth power is a.

If $a > 0$ and n is a positive number, then $a^{1/n}$ is called the nth root of a.

$25^{1/2} = 5$ because $(5)^2 = 25$.

$8^{1/3} = 2$ because $(2)^3 = 8$.

In the expression $a^{1/n}$, if a is a negative number and n is a positive even integer, then $a^{1/n}$ is not a real number.

$(-4)^{1/2}$ is not a real number, because there is no real number whose second power is -4.

When n is a positive odd integer, a can be a positive or a negative number.

$(-27)^{1/3} = -3$ because $(-3)^3 = -27$.

Using the definition of $a^{1/n}$ and the Rules of Exponents, it is possible to define any exponential expression that contains a rational exponent.

> **Rule for Rational Exponents**
>
> If m and n are positive integers and $a^{1/n}$ is a real number, then
> $$a^{m/n} = (a^{1/n})^m$$

The expression $a^{m/n}$ can also be written $a^{m/n} = a^{m \cdot \frac{1}{n}} = (a^m)^{1/n}$. However, rewriting $a^{m/n}$ as $(a^m)^{1/n}$ is not as useful as rewriting it as $(a^{1/n})^m$. See the Take Note at the top of the next page.

As shown above, expressions that contain rational exponents do not always represent real numbers when the base of the exponential expression is a negative number. For this reason, **all variables in this chapter represent positive numbers unless otherwise stated.**

TAKE NOTE

Although we can simplify an expression by rewriting $a^{m/n}$ in the form $(a^m)^{1/n}$, it is usually easier to simplify the form $(a^{1/n})^m$. For instance, simplifying $(27^{1/3})^2$ is easier than simplifying $(27^2)^{1/3}$.

➡ Simplify: $27^{2/3}$

$$27^{2/3} = (3^3)^{2/3}$$
• Rewrite 27 as 3^3.

$$= 3^{3(2/3)}$$
• Multiply the exponents.

$$= 3^2$$
• Simplify.

$$= 9$$

TAKE NOTE

Note that $32^{-2/5} = \dfrac{1}{4}$, a positive number. The negative exponent does not affect the sign of a number.

➡ Simplify: $32^{-2/5}$

$$32^{-2/5} = (2^5)^{-2/5}$$
• Rewrite 32 as 2^5.

$$= 2^{-2}$$
• Multiply the exponents.

$$= \frac{1}{2^2}$$
• Use the Rule of Negative Exponents.

$$= \frac{1}{4}$$
• Simplify.

➡ Simplify: $a^{1/2} \cdot a^{2/3} \cdot a^{-1/4}$

$$a^{1/2} \cdot a^{2/3} \cdot a^{-1/4} = a^{1/2 + 2/3 - 1/4}$$
• Use the Rule for Multiplying Exponential Expressions.

$$= a^{6/12 + 8/12 - 3/12}$$

$$= a^{11/12}$$
• Simplify.

➡ Simplify: $(x^6 y^4)^{3/2}$

$$(x^6 y^4)^{3/2} = x^{6(3/2)} y^{4(3/2)}$$
• Use the Rule for Simplifying Powers of Products.

$$= x^9 y^6$$
• Simplify.

➡ Simplify: $\left(\dfrac{8a^3 b^{-4}}{64a^{-9} b^2}\right)^{2/3}$

$$\left(\frac{8a^3 b^{-4}}{64a^{-9} b^2}\right)^{2/3} = \left(\frac{2^3 a^3 b^{-4}}{2^6 a^{-9} b^2}\right)^{2/3}$$
• Rewrite 8 as 2^3 and 64 as 2^6.

$$= (2^{-3} a^{12} b^{-6})^{2/3}$$
• Use the Rule for Dividing Exponential Expressions.

$$= 2^{-2} a^8 b^{-4}$$
• Use the Rule for Simplifying Powers of Products.

$$= \frac{a^8}{2^2 b^4} = \frac{a^8}{4b^4}$$
• Use the Rule of Negative Exponents and simplify.

Example 1   Simplify: $64^{-2/3}$

Solution $64^{-2/3} = (2^6)^{-2/3} = 2^{-4}$
$$= \frac{1}{2^4} = \frac{1}{16}$$

You Try It 1   Simplify: $16^{-3/4}$

Your solution

Example 2   Simplify: $(-49)^{3/2}$

Solution   The base of the exponential expression is a negative number, while the denominator of the exponent is a positive even number.

Therefore, $(-49)^{3/2}$ is not a real number.

You Try It 2   Simplify: $(-81)^{3/4}$

Your solution

Example 3   Simplify: $(x^{1/2}y^{-3/2}z^{1/4})^{-3/2}$

Solution $(x^{1/2}y^{-3/2}z^{1/4})^{-3/2}$
$$= x^{-3/4}y^{9/4}z^{-3/8}$$
$$= \frac{y^{9/4}}{x^{3/4}z^{3/8}}$$

You Try It 3   Simplify: $(x^{3/4}y^{1/2}z^{-2/3})^{-4/3}$

Your solution

Example 4
Simplify: $\dfrac{x^{1/2}y^{-5/4}}{x^{-4/3}y^{1/3}}$

Solution $\dfrac{x^{1/2}y^{-5/4}}{x^{-4/3}y^{1/3}}$
$$= x^{3/6-(-8/6)}y^{-15/12-4/12}$$
$$= x^{11/6}y^{-19/12} = \frac{x^{11/6}}{y^{19/12}}$$

You Try It 4
Simplify: $\left(\dfrac{16a^{-2}b^{4/3}}{9a^4b^{-2/3}}\right)^{-1/2}$

Your solution

Solutions on p. S22

Objective B   **To write exponential expressions as radical expressions and to write radical expressions as exponential expressions**

Point of Interest

The radical sign was introduced in 1525 in a book by Christoff Rudolff called *Coss*. He modified the symbol to indicate square roots, cube roots, and fourth roots. The idea of using an index, as we do in our modern notation, did not occur until some years later.

Recall that $a^{1/n}$ is the nth root of a. The expression $\sqrt[n]{a}$ is another symbol for the nth root of a.

If a is a real number, then $a^{1/n} = \sqrt[n]{a}$.

In the expression $\sqrt[n]{a}$, the symbol $\sqrt{}$ is called a **radical,** n is the **index** of the radical, and a is the **radicand.** When $n = 2$, the radical expression represents a square root and the index 2 is usually not written.

An exponential expression with a rational exponent can be written as a radical expression.

> **Rule for Writing Exponential Expressions as Radical Expressions**
>
> If $a^{1/n}$ is a real number, then $a^{m/n} = a^{m \cdot 1/n} = (a^m)^{1/n} = \sqrt[n]{a^m}$.

The expression $a^{m/n}$ can also be written $a^{m/n} = (a^{1/n})^m = (\sqrt[n]{a})^m$.

The exponential expression at the right has been written as a radical expression.

$$y^{2/3} = (y^2)^{1/3}$$
$$= \sqrt[3]{y^2}$$

The radical expressions at the right have been written as exponential expressions.

$$\sqrt[5]{x^6} = (x^6)^{1/5} = x^{6/5}$$
$$\sqrt{17} = (17)^{1/2} = 17^{1/2}$$

⇒ Write $(5x)^{2/5}$ as a radical expression.

$$(5x)^{2/5} = \sqrt[5]{(5x)^2}$$

• The denominator of the rational exponent is the index of the radical. The numerator is the power of the radicand.

$$= \sqrt[5]{25x^2}$$

• Simplify.

⇒ Write $\sqrt[3]{x^4}$ as an exponential expression with a rational exponent.

$$\sqrt[3]{x^4} = (x^4)^{1/3}$$

• The index of the radical is the denominator of the rational exponent. The power of the radicand is the numerator of the rational exponent.

$$= x^{4/3}$$

• Simplify.

⇒ Write $\sqrt[3]{a^3 + b^3}$ as an exponential expression with a rational exponent.

$$\sqrt[3]{a^3 + b^3} = (a^3 + b^3)^{1/3}$$

Note that $(a^3 + b^3)^{1/3} \neq a + b$.

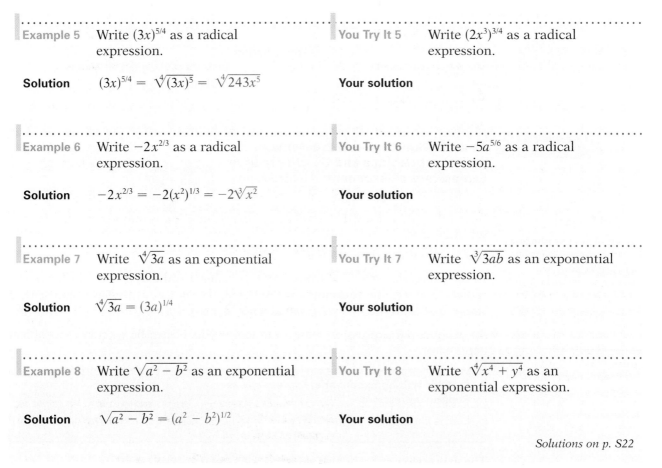

Example 5 Write $(3x)^{5/4}$ as a radical expression.

Solution $(3x)^{5/4} = \sqrt[4]{(3x)^5} = \sqrt[4]{243x^5}$

You Try It 5 Write $(2x^3)^{3/4}$ as a radical expression.

Your solution

Example 6 Write $-2x^{2/3}$ as a radical expression.

Solution $-2x^{2/3} = -2(x^2)^{1/3} = -2\sqrt[3]{x^2}$

You Try It 6 Write $-5a^{5/6}$ as a radical expression.

Your solution

Example 7 Write $\sqrt[4]{3a}$ as an exponential expression.

Solution $\sqrt[4]{3a} = (3a)^{1/4}$

You Try It 7 Write $\sqrt[3]{3ab}$ as an exponential expression.

Your solution

Example 8 Write $\sqrt{a^2 - b^2}$ as an exponential expression.

Solution $\sqrt{a^2 - b^2} = (a^2 - b^2)^{1/2}$

You Try It 8 Write $\sqrt[4]{x^4 + y^4}$ as an exponential expression.

Your solution

Solutions on p. S22

Objective C   **To simplify expressions of the form** $\sqrt[n]{a^n}$ 〖 7 〗 ◎

Every positive number has two square roots, one a positive number and one a negative number. For example, because $(5)^2 = 25$ and $(-5)^2 = 25$, there are two square roots of 25: 5 and -5.

The symbol $\sqrt{}$ is used to indicate the positive or **principal square root.** To indicate the negative square root of a number, a negative sign is placed in front of the radical.

$$\sqrt{25} = 5$$

$$-\sqrt{25} = -5$$

The square root of zero is zero.

$$\sqrt{0} = 0$$

The square root of a negative number is not a real number, because the square of a real number must be positive.

$\sqrt{-25}$ is not a real number.

Note that

$$\sqrt{(-5)^2} = \sqrt{25} = 5 \quad \text{and} \quad \sqrt{5^2} = \sqrt{25} = 5$$

This is true for all real numbers and is stated as the following result.

For any real number a, $\sqrt{a^2} = |a|$ and $-\sqrt{a^2} = -|a|$. If a is a positive real number, then $\sqrt{a^2} = a$ and $(\sqrt{a})^2 = a$.

Besides square roots, we can also determine cube roots, fourth roots, and so on.

$\sqrt[3]{8} = 2$, because $2^3 = 8$.

- The cube root of a positive number is positive.

$\sqrt[3]{-8} = -2$, because $(-2)^3 = -8$.

- The cube root of a negative number is negative.

$\sqrt[4]{625} = 5$, because $5^4 = 625$.

$\sqrt[5]{243} = 3$, because $3^5 = 243$.

The following properties hold true for finding the nth root of a real number.

If n is an even integer, then $\sqrt[n]{a^n} = |a|$ and $-\sqrt[n]{a^n} = -|a|$. If n is an odd integer, then $\sqrt[n]{a^n} = a$.

For example,

$$\sqrt[6]{y^6} = |y| \qquad -\sqrt[12]{x^{12}} = -|x| \qquad \sqrt[5]{b^5} = b$$

For the remainder of this chapter, we will assume that variable expressions inside a radical represent positive numbers. Therefore, it is not necessary to use the absolute value signs.

TAKE NOTE

Note that when the index is an even natural number, the nth root requires absolute value symbols.

$\sqrt[6]{y^6} = |y|$ but $\sqrt[5]{y^5} = y$

Because we stated that variables within radicals represent *positive* numbers, we will omit the absolute value symbols when writing an answer.

⟹ Simplify: $\sqrt[4]{x^4y^8}$

$\sqrt[4]{x^4y^8} = (x^4y^8)^{1/4}$

- The radicand is a perfect fourth power because the exponents on the variables are divisible by 4. Write the radical expressions as an exponential expression.

$= xy^2$

- Use the Rule for Simplifying Powers of Products.

➡ Simplify: $\sqrt[3]{125c^9d^6}$

$\sqrt[3]{125c^9d^6} = (5^3c^9d^6)^{1/3}$

- The radicand is a perfect cube because 125 is a perfect cube ($125 = 5^3$) and all the exponents on the variables are divisible by 3.

$= 5c^3d^2$

- Use the Rule for Simplifying Powers of Products.

Note that **a variable expression is a perfect power if the exponents on the factors are evenly divisible by the index of the radical.**

The chart below shows roots of perfect powers. Knowledge of these roots is very helpful when simplifying radical expressions.

Square Roots		Cube Roots	Fourth Roots	Fifth Roots
$\sqrt{1} = 1$	$\sqrt{36} = 6$	$\sqrt[3]{1} = 1$	$\sqrt[4]{1} = 1$	$\sqrt[5]{1} = 1$
$\sqrt{4} = 2$	$\sqrt{49} = 7$	$\sqrt[3]{8} = 2$	$\sqrt[4]{16} = 2$	$\sqrt[5]{32} = 2$
$\sqrt{9} = 3$	$\sqrt{64} = 8$	$\sqrt[3]{27} = 3$	$\sqrt[4]{81} = 3$	$\sqrt[5]{243} = 3$
$\sqrt{16} = 4$	$\sqrt{81} = 9$	$\sqrt[3]{64} = 4$	$\sqrt[4]{256} = 4$	
$\sqrt{25} = 5$	$\sqrt{100} = 10$	$\sqrt[3]{125} = 5$	$\sqrt[4]{625} = 5$	

TAKE NOTE

From the chart, $\sqrt[5]{243} = 3$, which means that $3^5 = 243$. From this we know that $(-3)^5 = -243$, which means $\sqrt[5]{-243} = -3$.

➡ Simplify: $\sqrt[5]{-243x^5y^{15}}$

$\sqrt[5]{-243x^5y^{15}} = -3xy^3$

- From the chart, 243 is a perfect fifth power, and each exponent is divisible by 5. Therefore, the radicand is a perfect fifth power.

Example 9 Simplify: $\sqrt[3]{-125a^6b^9}$

Solution The radicand is a perfect cube.

$\sqrt[3]{-125a^6b^9} = -5a^2b^3$ • Divide each exponent by 3.

You Try It 9 Simplify: $\sqrt[3]{-8x^{12}y^3}$

Your solution

Example 10 Simplify: $-\sqrt[4]{16a^4b^8}$

Solution The radicand is a perfect fourth power.

$-\sqrt[4]{16a^4b^8} = -2ab^2$ • Divide each exponent by 4.

You Try It 10 Simplify: $-\sqrt[4]{81x^{12}y^8}$

Your solution

Solutions on p. S22

7.1 Exercises

Simplify.

1. $8^{1/3}$

2. $16^{1/2}$

3. $9^{3/2}$

4. $25^{3/2}$

5. $27^{-2/3}$

6. $64^{-1/3}$

7. $32^{2/5}$

8. $16^{3/4}$

9. $(-25)^{5/2}$

10. $(-36)^{1/4}$

11. $\left(\dfrac{25}{49}\right)^{-3/2}$

12. $\left(\dfrac{8}{27}\right)^{-2/3}$

13. $x^{1/2}x^{1/2}$

14. $a^{1/3}a^{5/3}$

15. $y^{-1/4}y^{3/4}$

16. $x^{2/5}\cdot x^{-4/5}$

17. $x^{-2/3}\cdot x^{3/4}$

18. $x\cdot x^{-1/2}$

19. $a^{1/3}\cdot a^{3/4}\cdot a^{-1/2}$

20. $y^{-1/6}\cdot y^{2/3}\cdot y^{1/2}$

21. $\dfrac{a^{1/2}}{a^{3/2}}$

22. $\dfrac{b^{1/3}}{b^{4/3}}$

23. $\dfrac{y^{-3/4}}{y^{1/4}}$

24. $\dfrac{x^{-3/5}}{x^{1/5}}$

25. $\dfrac{y^{2/3}}{y^{-5/6}}$

26. $\dfrac{b^{3/4}}{b^{-3/2}}$

27. $(x^2)^{-1/2}$

28. $(a^8)^{-3/4}$

29. $(x^{-2/3})^6$

30. $(y^{-5/6})^{12}$

31. $(a^{-1/2})^{-2}$

32. $(b^{-2/3})^{-6}$

33. $(x^{-3/8})^{-4/5}$

34. $(y^{-3/2})^{-2/9}$

35. $(a^{1/2}\cdot a)^2$

36. $(b^{2/3}\cdot b^{1/6})^6$

37. $(x^{-1/2}\cdot x^{3/4})^{-2}$

38. $(a^{1/2}\cdot a^{-2})^3$

39. $(y^{-1/2}\cdot y^{2/3})^{2/3}$

40. $(b^{-2/3}\cdot b^{1/4})^{-4/3}$

41. $(x^8y^2)^{1/2}$

42. $(a^3b^9)^{2/3}$

43. $(x^4y^2z^6)^{3/2}$ **44.** $(a^8b^4c^4)^{3/4}$ **45.** $(x^{-3}y^6)^{-1/3}$ **46.** $(a^2b^{-6})^{-1/2}$

47. $(x^{-2}y^{1/3})^{-3/4}$ **48.** $(a^{-2/3}b^{2/3})^{3/2}$ **49.** $\left(\dfrac{x^{1/2}}{y^2}\right)^4$ **50.** $\left(\dfrac{b^{-3/4}}{a^{-1/2}}\right)^8$

51. $\dfrac{x^{1/4}\cdot x^{-1/2}}{x^{2/3}}$ **52.** $\dfrac{b^{1/2}\cdot b^{-3/4}}{b^{1/4}}$ **53.** $\left(\dfrac{y^{2/3}\cdot y^{-5/6}}{y^{1/9}}\right)^9$ **54.** $\left(\dfrac{a^{1/3}\cdot a^{-2/3}}{a^{1/2}}\right)^4$

55. $\left(\dfrac{b^2\cdot b^{-3/4}}{b^{-1/2}}\right)^{-1/2}$ **56.** $\dfrac{(x^{-5/6}\cdot x^3)^{-2/3}}{x^{4/3}}$ **57.** $(a^{2/3}b^2)^6(a^3b^3)^{1/3}$ **58.** $(x^3y^{-1/2})^{-2}(x^{-3}y^2)^{1/6}$

59. $(16m^{-2}n^4)^{-1/2}(mn^{1/2})$ **60.** $(27m^3n^{-6})^{1/3}(m^{-1/3}n^{5/6})^6$

61. $\left(\dfrac{x^{1/2}y^{-3/4}}{y^{2/3}}\right)^{-6}$ **62.** $\left(\dfrac{x^{1/2}y^{-5/4}}{y^{-3/4}}\right)^{-4}$ **63.** $\left(\dfrac{2^{-6}b^{-3}}{a^{-1/2}}\right)^{-2/3}$ **64.** $\left(\dfrac{49c^{5/3}}{a^{-1/4}b^{5/6}}\right)^{-3/2}$

65. $y^{3/2}(y^{1/2}-y^{-1/2})$ **66.** $y^{3/5}(y^{2/5}+y^{-3/5})$ **67.** $a^{-1/4}(a^{5/4}-a^{9/4})$ **68.** $x^{4/3}(x^{2/3}+x^{-1/3})$

69. $x^n\cdot x^{3n}$ **70.** $a^{2n}\cdot a^{-5n}$ **71.** $x^n\cdot x^{n/2}$ **72.** $a^{n/2}\cdot a^{-n/3}$

73. $\dfrac{y^{n/2}}{y^{-n}}$ **74.** $\dfrac{b^{m/3}}{b^m}$ **75.** $(x^{2n})^n$ **76.** $(x^{5n})^{2n}$

77. $(x^{n/4}y^{n/8})^8$ **78.** $(x^{n/2}y^{n/3})^6$ **79.** $(x^{n/5}y^{n/10})^{20}$ **80.** $(x^{n/2}y^{n/5})^{10}$

Objective B

Rewrite the exponential expression as a radical expression.

81. $3^{1/4}$

82. $5^{1/2}$

83. $a^{3/2}$

84. $b^{4/3}$

85. $(2t)^{5/2}$

86. $(3x)^{2/3}$

87. $-2x^{2/3}$

88. $-3a^{2/5}$

89. $(a^2b)^{2/3}$

90. $(x^2y^3)^{3/4}$

91. $(a^2b^4)^{3/5}$

92. $(a^3b^7)^{3/2}$

93. $(4x - 3)^{3/4}$

94. $(3x - 2)^{1/3}$

95. $x^{-2/3}$

96. $b^{-3/4}$

Rewrite the radical expression as an exponential expression.

97. $\sqrt{14}$

98. $\sqrt{7}$

99. $\sqrt[3]{x}$

100. $\sqrt[4]{y}$

101. $\sqrt[3]{x^4}$

102. $\sqrt[4]{a^3}$

103. $\sqrt[5]{b^3}$

104. $\sqrt[4]{b^5}$

105. $\sqrt[3]{2x^2}$

106. $\sqrt[5]{4y^7}$

107. $-\sqrt{3x^5}$

108. $-\sqrt[4]{4x^5}$

109. $3x\sqrt[3]{y^2}$

110. $2y\sqrt{x^3}$

111. $\sqrt{a^2 - 2}$

112. $\sqrt{3 - y^2}$

Objective C

Simplify.

113. $\sqrt{x^{16}}$

114. $\sqrt{y^{14}}$

115. $-\sqrt{x^8}$

116. $-\sqrt{a^6}$

117. $\sqrt[3]{x^3y^9}$

118. $\sqrt[3]{a^6b^{12}}$

119. $-\sqrt[3]{x^{15}y^3}$

120. $-\sqrt[3]{a^9b^9}$

121. $\sqrt{16a^4b^{12}}$

122. $\sqrt{25x^8y^2}$

123. $\sqrt{-16x^4y^2}$

124. $\sqrt{-9a^4b^8}$

125. $\sqrt[3]{27x^9}$

126. $\sqrt[3]{8a^{21}b^6}$

127. $\sqrt[3]{-64x^9y^{12}}$

128. $\sqrt[3]{-27a^3b^{15}}$

129. $-\sqrt[4]{x^8y^{12}}$

130. $-\sqrt[4]{a^{16}b^4}$

131. $\sqrt[5]{x^{20}y^{10}}$

132. $\sqrt[5]{a^5b^{25}}$

133. $\sqrt[4]{81x^4y^{20}}$

134. $\sqrt[4]{16a^8b^{20}}$

135. $\sqrt[5]{32a^5b^{10}}$

136. $\sqrt[5]{-32x^{15}y^{20}}$

APPLYING THE CONCEPTS

137. Determine whether the following statements are true or false. If the statement is false, correct the right-hand side of the equation.
   a. $\sqrt{(-2)^2} = -2$ **b.** $\sqrt[3]{(-3)^3} = -3$ **c.** $\sqrt[n]{a} = a^{1/n}$

   d. $\sqrt[n]{a^n + b^n} = a + b$ **e.** $(a^{1/2} + b^{1/2})^2 = a + b$ **f.** $\sqrt[m]{a^n} = a^{mn}$

138. Simplify.
   a. $\sqrt[3]{\sqrt{x^6}}$ **b.** $\sqrt[4]{\sqrt{a^8}}$ **c.** $\sqrt{\sqrt{81y^8}}$

   d. $\sqrt{\sqrt[n]{a^{4n}}}$ **e.** $\sqrt[n]{\sqrt{b^{6n}}}$ **f.** $\sqrt{\sqrt[3]{x^{12}y^{24}}}$

139. If x is any real number, is $\sqrt{x^2} = x$ always true? Show why or why not.

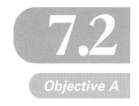

Operations on Radical Expressions

Objective A   **To simplify radical expressions**

If a number is not a perfect power, its root can only be approximated; examples include $\sqrt{5}$ and $\sqrt[3]{3}$. These numbers are **irrational numbers.** Their decimal representations never terminate or repeat.

$$\sqrt{5} = 2.2360679\ldots \qquad \sqrt[3]{3} = 1.4422495\ldots$$

A radical expression is in simplest form when the radicand contains no factor that is a perfect power. The Product Property of Radicals is used to simplify radical expressions whose radicands are not perfect powers.

> **The Product Property of Radicals**
>
> If $\sqrt[n]{a}$ and $\sqrt[n]{b}$ are positive real numbers, then $\sqrt[n]{ab} = \sqrt[n]{a} \cdot \sqrt[n]{b}$ and $\sqrt[n]{a} \cdot \sqrt[n]{b} = \sqrt[n]{ab}$.

➡ Simplify: $\sqrt{48}$

$\sqrt{48} = \sqrt{16 \cdot 3}$ • Write the radicand as the product of a perfect square and a factor that does not contain a perfect square.

$\quad = \sqrt{16}\,\sqrt{3}$ • Use the Product Property of Radicals to write the expression as a product.

$\quad = 4\sqrt{3}$ • Simplify $\sqrt{16}$.

Note that 48 must be written as the product of a perfect square and *a factor that does not contain a perfect square.* Therefore, it would not be correct to rewrite $\sqrt{48}$ as $\sqrt{4 \cdot 12}$ and simplify the expression as shown at the right. Although 4 is a perfect square factor of 48, 12 contains a perfect square $(12 = 4 \cdot 3)$ and thus $\sqrt{12}$ can be simplified further. Remember to find the largest perfect power that is a factor of the radicand.

$\sqrt{48} = \sqrt{4 \cdot 12}$
$\quad = \sqrt{4}\,\sqrt{12}$
$\quad = 2\sqrt{12}$

Not in simplest form

➡ Simplify: $\sqrt{18x^2y^3}$

$\sqrt{18x^2y^3} = \sqrt{9x^2y^2 \cdot 2y}$ • Write the radicand as the product of a perfect square and factors that do not contain a perfect square.

$\quad = \sqrt{9x^2y^2}\,\sqrt{2y}$ • Use the Product Property of Radicals to write the expression as a product.

$\quad = 3xy\sqrt{2y}$ • Simplify.

⇨ Simplify: $\sqrt[3]{x^7}$

$\sqrt[3]{x^7} = \sqrt[3]{x^6 \cdot x}$ • Write the radicand as the product of a perfect cube and a factor that does not contain a perfect cube.

$= \sqrt[3]{x^6} \sqrt[3]{x}$ • Use the Product Property of Radicals to write the expression as a product.

$= x^2 \sqrt[3]{x}$ • Simplify.

⇨ Simplify: $\sqrt[4]{32x^7}$

$\sqrt[4]{32x^7} = \sqrt[4]{16x^4(2x^3)}$ • Write the radicand as the product of a perfect fourth power and factors that do not contain a perfect fourth power.

$= \sqrt[4]{16x^4} \sqrt[4]{2x^3}$ • Use the Product Property of Radicals to write the expression as a product.

$= 2x \sqrt[4]{2x^3}$ • Simplify.

Example 1 Simplify: $\sqrt[4]{x^9}$

Solution $\sqrt[4]{x^9} = \sqrt[4]{x^8 \cdot x} = \sqrt[4]{x^8} \sqrt[4]{x}$
$= x^2 \sqrt[4]{x}$

You Try It 1 Simplify: $\sqrt[5]{x^7}$

Your solution

Example 2 Simplify: $\sqrt[3]{-27a^5b^{12}}$

Solution $\sqrt[3]{-27a^5b^{12}}$
$= \sqrt[3]{-27a^3b^{12}(a^2)}$
$= \sqrt[3]{-27a^3b^{12}} \sqrt[3]{a^2}$
$= -3ab^4 \sqrt[3]{a^2}$

You Try It 2 Simplify: $\sqrt[3]{-64x^8y^{18}}$

Your solution

Solutions on p. S22

Objective B   **To add or subtract radical expressions**   ⟨ 7 ⟩   ◎

The Distributive Property is used to simplify the sum or difference of radical expressions that have the same radicand and the same index. For example,

$$3\sqrt{5} + 8\sqrt{5} = (3 + 8)\sqrt{5} = 11\sqrt{5}$$
$$2\sqrt[3]{3x} - 9\sqrt[3]{3x} = (2 - 9)\sqrt[3]{3x} = -7\sqrt[3]{3x}$$

Radical expressions that are in simplest form and have unlike radicands or different indices cannot be simplified by the Distributive Property. The expressions below cannot be simplified by the Distributive Property.

$$3\sqrt[3]{2} - 6\sqrt[4]{3} \qquad\qquad 2\sqrt[4]{4x} + 3\sqrt[3]{4x}$$

⇒ Simplify: $3\sqrt{32x^2} - 2x\sqrt{2} + \sqrt{128x^2}$

$3\sqrt{32x^2} - 2x\sqrt{2} + \sqrt{128x^2}$

$= 3\sqrt{16x^2}\sqrt{2} - 2x\sqrt{2} + \sqrt{64x^2}\sqrt{2}$

$= 3 \cdot 4x\sqrt{2} - 2x\sqrt{2} + 8x\sqrt{2}$

$= 12x\sqrt{2} - 2x\sqrt{2} + 8x\sqrt{2}$

$= 18x\sqrt{2}$

- First simplify each term. Then combine like terms by using the Distributive Property.

Example 3
Simplify: $5b\sqrt[4]{32a^7b^5} - 2a\sqrt[4]{162a^3b^9}$

Solution
$5b\sqrt[4]{32a^7b^5} - 2a\sqrt[4]{162a^3b^9}$

$= 5b\sqrt[4]{16a^4b^4 \cdot 2a^3b} - 2a\sqrt[4]{81b^8 \cdot 2a^3b}$

$= 5b \cdot 2ab\sqrt[4]{2a^3b} - 2a \cdot 3b^2\sqrt[4]{2a^3b}$

$= 10ab^2\sqrt[4]{2a^3b} - 6ab^2\sqrt[4]{2a^3b}$

$= 4ab^2\sqrt[4]{2a^3b}$

You Try It 3
Simplify: $3xy\sqrt[3]{81x^5y} - \sqrt[3]{192x^8y^4}$

Your solution

Solution on p. S22

Objective C   **To multiply radical expressions**

The Product Property of Radicals is used to multiply radical expressions with the same index.

$$\sqrt{3x} \cdot \sqrt{5y} = \sqrt{3x \cdot 5y} = \sqrt{15xy}$$

⇒ Simplify: $\sqrt[3]{2a^5b}\sqrt[3]{16a^2b^2}$

$\sqrt[3]{2a^5b}\sqrt[3]{16a^2b^2} = \sqrt[3]{32a^7b^3}$

$= \sqrt[3]{8a^6b^3 \cdot 4a}$

$= 2a^2b\sqrt[3]{4a}$

- Use the Product Property of Radicals to multiply the radicands.

- Simplify.

⇒ Simplify: $\sqrt{2x}(\sqrt{8x} - \sqrt{3})$

$\sqrt{2x}(\sqrt{8x} - \sqrt{3}) = \sqrt{2x}(\sqrt{8x}) - \sqrt{2x}(\sqrt{3})$

$= \sqrt{16x^2} - \sqrt{6x}$

$= 4x - \sqrt{6x}$

- Use the Distributive Property.

- Simplify.

⇒ Simplify: $(2\sqrt{5} - 3)(3\sqrt{5} + 4)$

$(2\sqrt{5} - 3)(3\sqrt{5} + 4) = 6(\sqrt{5})^2 + 8\sqrt{5} - 9\sqrt{5} - 12$

$= 30 + 8\sqrt{5} - 9\sqrt{5} - 12$

$= 18 - \sqrt{5}$

- Use the FOIL method to multiply the numbers.

- Combine like terms.

➡ Simplify: $(4\sqrt{a} - \sqrt{b})(2\sqrt{a} + 5\sqrt{b})$

$(4\sqrt{a} - \sqrt{b})(2\sqrt{a} + 5\sqrt{b})$
$= 8(\sqrt{a})^2 + 20\sqrt{ab} - 2\sqrt{ab} - 5(\sqrt{b})^2$ • Use the FOIL method.
$= 8a + 18\sqrt{ab} - 5b$

The expressions $a + b$ and $a - b$ are **conjugates** of each other. Recall that $(a + b)(a - b) = a^2 - b^2$. This identity is used to simplify conjugate radical expressions.

➡ Simplify: $(\sqrt{11} - 3)(\sqrt{11} + 3)$

$(\sqrt{11} - 3)(\sqrt{11} + 3) = (\sqrt{11})^2 - 3^2 = 11 - 9$ • The radical expressions are conjugates.
$= 2$

Example 4
Simplify: $\sqrt{3x}(\sqrt{27x^2} - \sqrt{3x})$

Solution
$\sqrt{3x}(\sqrt{27x^2} - \sqrt{3x}) = \sqrt{81x^3} - \sqrt{9x^2}$
$= \sqrt{81x^2 \cdot x} - \sqrt{9x^2}$
$= 9x\sqrt{x} - 3x$

You Try It 4
Simplify: $\sqrt{5b}(\sqrt{3b} - \sqrt{10})$

Your solution

Example 5
Simplify: $(2\sqrt[3]{x} - 3)(3\sqrt[3]{x} - 4)$

Solution
$(2\sqrt[3]{x} - 3)(3\sqrt[3]{x} - 4)$
$= 6\sqrt[3]{x^2} - 8\sqrt[3]{x} - 9\sqrt[3]{x} + 12$
$= 6\sqrt[3]{x^2} - 17\sqrt[3]{x} + 12$

You Try It 5
Simplify: $(2\sqrt[3]{2x} - 3)(\sqrt[3]{2x} - 5)$

Your solution

Example 6
Simplify: $(2\sqrt{x} - \sqrt{2y})(2\sqrt{x} + \sqrt{2y})$

Solution
$(2\sqrt{x} - \sqrt{2y})(2\sqrt{x} + \sqrt{2y})$
$= (2\sqrt{x})^2 - (\sqrt{2y})^2$
$= 4x - 2y$

You Try It 6
Simplify: $(\sqrt{a} - 3\sqrt{y})(\sqrt{a} + 3\sqrt{y})$

Your solution

Solutions on p. S22

Objective D   **To divide radical expressions**

The Quotient Property of Radicals is used to divide radical expressions with the same index.

> **The Quotient Property of Radicals**
>
> If $\sqrt[n]{a}$ and $\sqrt[n]{b}$ are real numbers, and $b \neq 0$, then
>
> $$\sqrt[n]{\frac{a}{b}} = \frac{\sqrt[n]{a}}{\sqrt[n]{b}} \quad \text{and} \quad \frac{\sqrt[n]{a}}{\sqrt[n]{b}} = \sqrt[n]{\frac{a}{b}}$$

➡ Simplify: $\sqrt[3]{\dfrac{81x^5}{y^6}}$

$$\sqrt[3]{\frac{81x^5}{y^6}} = \frac{\sqrt[3]{81x^5}}{\sqrt[3]{y^6}}$$

- Use the Quotient Property of Radicals.

$$= \frac{\sqrt[3]{27x^3 \cdot 3x^2}}{\sqrt[3]{y^6}} = \frac{3x\sqrt[3]{3x^2}}{y^2}$$

- Simplify each radical expression.

➡ Simplify: $\dfrac{\sqrt{5a^4b^7c^2}}{\sqrt{ab^3c}}$

$$\frac{\sqrt{5a^4b^7c^2}}{\sqrt{ab^3c}} = \sqrt{\frac{5a^4b^7c^2}{ab^3c}}$$

- Use the Quotient Property of Radicals.

$$= \sqrt{5a^3b^4c}$$

- Simplify the radicand.

$$= \sqrt{a^2b^4 \cdot 5ac} = ab^2\sqrt{5ac}$$

A radical expression is in simplest form when no radical remains in the denominator of the radical expression. The procedure used to remove a radical from the denominator is called **rationalizing the denominator.**

➡ Simplify: $\dfrac{5}{\sqrt{2}}$

$$\frac{5}{\sqrt{2}} = \frac{5}{\sqrt{2}} \cdot 1 = \frac{5}{\sqrt{2}} \cdot \frac{\sqrt{2}}{\sqrt{2}}$$

- Multiply by $\dfrac{\sqrt{2}}{\sqrt{2}}$, which equals 1.

$$= \frac{5\sqrt{2}}{2}$$

- $\sqrt{2} \cdot \sqrt{2} = (\sqrt{2})^2 = 2$

➡ Simplify: $\dfrac{3x}{\sqrt[3]{4x}}$

$$\frac{3x}{\sqrt[3]{4x}} = \frac{3x}{\sqrt[3]{4x}} \cdot \frac{\sqrt[3]{2x^2}}{\sqrt[3]{2x^2}}$$

- Because $4x \cdot 2x^2 = 8x^3$, a perfect cube, multiply the expression by $\dfrac{\sqrt[3]{2x^2}}{\sqrt[3]{2x^2}}$, which equals 1.

$$= \frac{3\sqrt[3]{2x^2}}{\sqrt[3]{8x^3}} = \frac{3\sqrt[3]{2x^2}}{2x}$$

- Simplify.

TAKE NOTE

Multiplying by $\dfrac{\sqrt[3]{4x}}{\sqrt[3]{4x}}$ will not rationalize the denominator of $\dfrac{3}{\sqrt[3]{4x}}$.

$$\frac{3}{\sqrt[3]{4x}} \cdot \frac{\sqrt[3]{4x}}{\sqrt[3]{4x}} = \frac{3\sqrt[3]{4x}}{\sqrt[3]{16x^2}}$$

Because $16x^2$ is not a perfect cube, the denominator still contains a radical expression.

To simplify a fraction that has a square-root radical expression with two terms in the denominator, multiply the numerator and denominator by the conjugate of the denominator. Then simplify.

➡ Simplify: $\dfrac{\sqrt{x} - \sqrt{y}}{\sqrt{x} + \sqrt{y}}$

$$\dfrac{\sqrt{x} - \sqrt{y}}{\sqrt{x} + \sqrt{y}} = \dfrac{\sqrt{x} - \sqrt{y}}{\sqrt{x} + \sqrt{y}} \cdot \dfrac{\sqrt{x} - \sqrt{y}}{\sqrt{x} - \sqrt{y}}$$

$$= \dfrac{(\sqrt{x})^2 - \sqrt{xy} - \sqrt{xy} + (\sqrt{y})^2}{(\sqrt{x})^2 - (\sqrt{y})^2} = \dfrac{x - 2\sqrt{xy} + y}{x - y}$$

Example 7

Simplify: $\dfrac{5}{\sqrt{5x}}$

Solution

$\dfrac{5}{\sqrt{5x}} = \dfrac{5}{\sqrt{5x}} \cdot \dfrac{\sqrt{5x}}{\sqrt{5x}} = \dfrac{5\sqrt{5x}}{(\sqrt{5x})^2}$

$= \dfrac{5\sqrt{5x}}{5x} = \dfrac{\sqrt{5x}}{x}$

You Try It 7

Simplify: $\dfrac{y}{\sqrt{3y}}$

Your solution

Example 8

Simplify: $\dfrac{3x}{\sqrt[4]{2x}}$

Solution

$\dfrac{3x}{\sqrt[4]{2x}} = \dfrac{3x}{\sqrt[4]{2x}} \cdot \dfrac{\sqrt[4]{8x^3}}{\sqrt[4]{8x^3}}$

$= \dfrac{3x\sqrt[4]{8x^3}}{\sqrt[4]{16x^4}} = \dfrac{3x\sqrt[4]{8x^3}}{2x}$

$= \dfrac{3\sqrt[4]{8x^3}}{2}$

You Try It 8

Simplify: $\dfrac{3x}{\sqrt[3]{3x^2}}$

Your solution

Example 9

Simplify: $\dfrac{3}{5 - 2\sqrt{3}}$

Solution

$\dfrac{3}{5 - 2\sqrt{3}} = \dfrac{3}{5 - 2\sqrt{3}} \cdot \dfrac{5 + 2\sqrt{3}}{5 + 2\sqrt{3}} = \dfrac{15 + 6\sqrt{3}}{5^2 - (2\sqrt{3})^2}$

$= \dfrac{15 + 6\sqrt{3}}{25 - 12} = \dfrac{15 + 6\sqrt{3}}{13}$

You Try It 9

Simplify: $\dfrac{3 + \sqrt{6}}{2 - \sqrt{6}}$

Your solution

Solutions on p. S22

7.2 Exercises

Objective A

Simplify.

1. $\sqrt{x^4y^3z^5}$

2. $\sqrt{x^3y^6z^9}$

3. $\sqrt{8a^3b^8}$

4. $\sqrt{24a^9b^6}$

5. $\sqrt{45x^2y^3z^5}$

6. $\sqrt{60xy^7z^{12}}$

7. $\sqrt{-9x^3}$

8. $\sqrt{-x^2y^5}$

9. $\sqrt[3]{a^{16}b^8}$

10. $\sqrt[3]{a^5b^8}$

11. $\sqrt[3]{-125x^2y^4}$

12. $\sqrt[3]{-216x^5y^9}$

13. $\sqrt[3]{a^4b^5c^6}$

14. $\sqrt[3]{a^8b^{11}c^{15}}$

15. $\sqrt[4]{16x^9y^5}$

16. $\sqrt[4]{64x^8y^{10}}$

Objective B

Simplify.

17. $2\sqrt{x} - 8\sqrt{x}$

18. $3\sqrt{y} + 12\sqrt{y}$

19. $\sqrt{8} - \sqrt{32}$

20. $\sqrt{27a} - \sqrt{8a}$

21. $\sqrt{18b} + \sqrt{75b}$

22. $2\sqrt{2x^3} + 4x\sqrt{8x}$

23. $3\sqrt{8x^2y^3} - 2x\sqrt{32y^3}$

24. $2\sqrt{32x^2y^3} - xy\sqrt{98y}$

25. $2a\sqrt{27ab^5} + 3b\sqrt{3a^3b}$

26. $\sqrt[3]{128} + \sqrt[3]{250}$

27. $\sqrt[3]{16} - \sqrt[3]{54}$

28. $2\sqrt[3]{3a^4} - 3a\sqrt[3]{81a}$

29. $2b\sqrt[3]{16b^2} + \sqrt[3]{128b^5}$

30. $3\sqrt[3]{x^5y^7} - 8xy\sqrt[3]{x^2y^4}$

31. $3\sqrt[4]{32a^5} - a\sqrt[4]{162a}$

32. $2a\sqrt[4]{16ab^5} + 3b\sqrt[4]{256a^5b}$

33. $2\sqrt{50} - 3\sqrt{125} + \sqrt{98}$

34. $3\sqrt{108} - 2\sqrt{18} - 3\sqrt{48}$

35. $\sqrt{9b^3} - \sqrt{25b^3} + \sqrt{49b^3}$

36. $\sqrt{4x^7y^5} + 9x^2\sqrt{x^3y^5} - 5xy\sqrt{x^5y^3}$

37. $2x\sqrt{8xy^2} - 3y\sqrt{32x^3} + \sqrt{4x^3y^3}$

38. $5a\sqrt{3a^3b} + 2a^2\sqrt{27ab} - 4\sqrt{75a^5b}$

39. $\sqrt[3]{54xy^3} - 5\sqrt[3]{2xy^3} + y\sqrt[3]{128x}$

40. $2\sqrt[3]{24x^3y^4} + 4x\sqrt[3]{81y^4} - 3y\sqrt[3]{24x^3y}$

41. $2a\sqrt[4]{32b^5} - 3b\sqrt[4]{162a^4b} + \sqrt[4]{2a^4b^5}$

42. $6y\sqrt[4]{48x^5} - 2x\sqrt[4]{243xy^4} - 4\sqrt[4]{3x^5y^4}$

Objective C

Simplify.

43. $\sqrt{8}\,\sqrt{32}$

44. $\sqrt{14}\,\sqrt{35}$

45. $\sqrt[3]{4}\,\sqrt[3]{8}$

46. $\sqrt[3]{6}\,\sqrt[3]{36}$

47. $\sqrt{x^2y^5}\,\sqrt{xy}$

48. $\sqrt{a^3b}\,\sqrt{ab^4}$

49. $\sqrt{2x^2y}\,\sqrt{32xy}$

50. $\sqrt{5x^3y}\,\sqrt{10x^3y^4}$

51. $\sqrt[3]{x^2y}\,\sqrt[3]{16x^4y^2}$

52. $\sqrt[3]{4a^2b^3}\,\sqrt[3]{8ab^5}$

53. $\sqrt[4]{12ab^3}\,\sqrt[4]{4a^5b^2}$

54. $\sqrt[4]{36a^2b^4}\,\sqrt[4]{12a^5b^3}$

55. $\sqrt{3}\,(\sqrt{27} - \sqrt{3})$

56. $\sqrt{10}\,(\sqrt{10} - \sqrt{5})$

57. $\sqrt{x}\,(\sqrt{x} - \sqrt{2})$

58. $\sqrt{y}\,(\sqrt{y} - \sqrt{5})$

59. $\sqrt{2x}\,(\sqrt{8x} - \sqrt{32})$

60. $\sqrt{3a}\,(\sqrt{27a^2} - \sqrt{a})$

61. $(\sqrt{x} - 3)^2$

62. $(\sqrt{2x} + 4)^2$

63. $(4\sqrt{5} + 2)^2$

64. $2\sqrt{3x^2} \cdot 3\sqrt{12xy^3} \cdot \sqrt{6x^3y}$

65. $2\sqrt{14xy} \cdot 4\sqrt{7x^2y} \cdot 3\sqrt{8xy^2}$

66. $\sqrt[3]{8ab} \ \sqrt[3]{4a^2b^3} \ \sqrt[3]{9ab^4}$

67. $\sqrt[3]{2a^2b} \ \sqrt[3]{4a^3b^2} \ \sqrt[3]{8a^5b^6}$

68. $(\sqrt{2} - 3)(\sqrt{2} + 4)$

69. $(\sqrt{5} - 5)(2\sqrt{5} + 2)$

70. $(\sqrt{y} - 2)(\sqrt{y} + 2)$

71. $(\sqrt{x} - y)(\sqrt{x} + y)$

72. $(\sqrt{2x} - 3\sqrt{y})(\sqrt{2x} + 3\sqrt{y})$

73. $(2\sqrt{3x} - \sqrt{y})(2\sqrt{3x} + \sqrt{y})$

Objective D

74. When is a radical expression in simplest form?

75. Explain what it means to rationalize the denominator of a radical expression and how to do so.

Simplify.

76. $\dfrac{\sqrt{32x^2}}{\sqrt{2x}}$

77. $\dfrac{\sqrt{60y^4}}{\sqrt{12y}}$

78. $\dfrac{\sqrt{42a^3b^5}}{\sqrt{14a^2b}}$

79. $\dfrac{\sqrt{65ab^4}}{\sqrt{5ab}}$

80. $\dfrac{1}{\sqrt{5}}$

81. $\dfrac{1}{\sqrt{2}}$

82. $\dfrac{1}{\sqrt{2x}}$

83. $\dfrac{2}{\sqrt{3y}}$

84. $\dfrac{5}{\sqrt{5x}}$

85. $\dfrac{9}{\sqrt{3a}}$

86. $\sqrt{\dfrac{x}{5}}$

87. $\sqrt{\dfrac{y}{2}}$

88. $\dfrac{3}{\sqrt[3]{2}}$

89. $\dfrac{5}{\sqrt[3]{9}}$

90. $\dfrac{3}{\sqrt[3]{4x^2}}$

91. $\dfrac{5}{\sqrt[3]{3y}}$

92. $\dfrac{\sqrt{40x^3y^2}}{\sqrt{80x^2y^3}}$ **93.** $\dfrac{\sqrt{15a^2b^5}}{\sqrt{30a^5b^3}}$ **94.** $\dfrac{\sqrt{24a^2b}}{\sqrt{18ab^4}}$ **95.** $\dfrac{\sqrt{12x^3y}}{\sqrt{20x^4y}}$

96. $\dfrac{5}{\sqrt{3}-2}$ **97.** $\dfrac{-2}{1-\sqrt{2}}$ **98.** $\dfrac{-3}{2-\sqrt{3}}$ **99.** $\dfrac{-4}{3-\sqrt{2}}$

100. $\dfrac{2}{\sqrt{5}+2}$ **101.** $\dfrac{5}{2-\sqrt{7}}$ **102.** $\dfrac{3}{\sqrt{y}-2}$ **103.** $\dfrac{-7}{\sqrt{x}-3}$

104. $\dfrac{\sqrt{2}-\sqrt{3}}{\sqrt{2}+\sqrt{3}}$ **105.** $\dfrac{\sqrt{3}+\sqrt{4}}{\sqrt{2}+\sqrt{3}}$ **106.** $\dfrac{2+3\sqrt{7}}{5-2\sqrt{7}}$

107. $\dfrac{2+3\sqrt{5}}{1-\sqrt{5}}$ **108.** $\dfrac{2\sqrt{3}-1}{3\sqrt{3}+2}$ **109.** $\dfrac{2\sqrt{a}-\sqrt{b}}{4\sqrt{a}+3\sqrt{b}}$

110. $\dfrac{2\sqrt{x}-4}{\sqrt{x}+2}$ **111.** $\dfrac{3\sqrt{y}-y}{\sqrt{y}+2y}$ **112.** $\dfrac{3\sqrt{x}-4\sqrt{y}}{3\sqrt{x}-2\sqrt{y}}$

APPLYING THE CONCEPTS

113. Determine whether the following statements are true or false. If the statement is false, correct the right-hand side of the equation.

a. $\sqrt[2]{3}\cdot\sqrt[3]{4}=\sqrt[5]{12}$ **b.** $\sqrt{3}\cdot\sqrt{3}=3$ **c.** $\sqrt[3]{x}\cdot\sqrt[3]{x}=x$

d. $\sqrt{x}+\sqrt{y}=\sqrt{x+y}$ **e.** $\sqrt[2]{2}+\sqrt[3]{3}=\sqrt[5]{2+3}$ **f.** $8\sqrt[5]{a}-2\sqrt[5]{a}=6\sqrt[5]{a}$

114. Multiply: $(\sqrt[3]{a}+\sqrt[3]{b})(\sqrt[3]{a^2}-\sqrt[3]{ab}+\sqrt[3]{b^2})$

115. Rewrite $\dfrac{\sqrt[4]{(a+b)^3}}{\sqrt{a+b}}$ as an expression with a single radical.

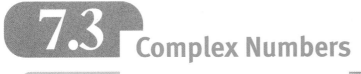

Complex Numbers

Objective A   **To simplify a complex number**

The radical expression $\sqrt{-4}$ is not a real number, because there is no real number whose square is -4. However, the solution of an algebraic equation is sometimes the square root of a negative number.

For example, the equation $x^2 + 1 = 0$ does not have a real number solution, because there is no real number whose square is a negative number.

$$x^2 + 1 = 0$$
$$x^2 = -1$$

Around the 17th century, a new number, called an **imaginary number,** was defined so that a negative number would have a square root. The letter i was chosen to represent the number whose square is -1.

$$i^2 = -1$$

An imaginary number is defined in terms of i.

> **Definition of $\sqrt{-a}$**
>
> If a is a positive real number, then the principal square root of negative a is the imaginary number $i\sqrt{a}$.
>
> $$\sqrt{-a} = i\sqrt{a}$$

Point of Interest

The first written occurrence of an imaginary number was in a book published in 1545 by Hieronimo Cardan, where he wrote (in our modern notation) $5 + \sqrt{-15}$. He went on to say that the number "is as refined as it is useless." It was not until the 20th century that applications of complex numbers were found.

Here are some examples.

$$\sqrt{-16} = i\sqrt{16} = 4i$$
$$\sqrt{-12} = i\sqrt{12} = 2i\sqrt{3}$$
$$\sqrt{-21} = i\sqrt{21}$$
$$\sqrt{-1} = i\sqrt{1} = i$$

It is customary to write i in front of a radical to avoid confusing $\sqrt{a}\,i$ with $\sqrt{ai}$.

The real numbers and imaginary numbers make up the complex numbers.

> **Complex Number**
>
> A **complex number** is a number of the form $a + bi$, where a and b are real numbers and $i = \sqrt{-1}$. The number a is the **real part** of $a + bi$, and the number b is the **imaginary part.**

TAKE NOTE

The *imaginary part* of a complex number is a real number. As another example, the imaginary part of $6 - 8i$ is -8.

Examples of complex numbers are shown at the right.

Real Part	Imaginary Part
a +	bi
3 +	$2i$
8 −	$10i$

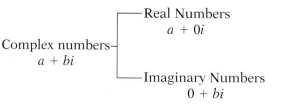

Complex numbers—
$a + bi$
┌─ Real Numbers
│ $a + 0i$
│
└─ Imaginary Numbers
     $0 + bi$

A *real number* is a complex number in which $b = 0$.

An *imaginary number* is a complex number in which $a = 0$.

⇒ Simplify: $\sqrt{20} - \sqrt{-50}$

$$\sqrt{20} - \sqrt{-50} = \sqrt{20} - i\sqrt{50}$$
$$= \sqrt{4^2 \cdot 5} - i\sqrt{25^2 \cdot 2}$$
$$= 2\sqrt{5} - 5i\sqrt{2}$$

• Write the complex number in the form $a + bi$.
• Use the Product Property of Radicals to simplify each radical.

....

Example 1
Simplify: $\sqrt{-80}$

Solution
$\sqrt{-80} = i\sqrt{80} = i\sqrt{16 \cdot 5} = 4i\sqrt{5}$

You Try It 1
Simplify: $\sqrt{-45}$

Your solution

....

Example 2
Simplify: $\sqrt{25} + \sqrt{-40}$

Solution
$$\sqrt{25} + \sqrt{-40} = \sqrt{25} + i\sqrt{40}$$
$$= \sqrt{25} + i\sqrt{4 \cdot 10}$$
$$= 5 + 2i\sqrt{10}$$

You Try It 2
Simplify: $\sqrt{98} - \sqrt{-60}$

Your solution

Solutions on p. S22

Objective B   **To add or subtract complex numbers**   ◀ 7 ▶

Addition and Subtraction of Complex Numbers

To add two complex numbers, add the real parts and add the imaginary parts. To subtract two complex numbers, subtract the real parts and subtract the imaginary parts.

$$(a + bi) + (c + di) = (a + c) + (b + d)i$$
$$(a + bi) - (c + di) = (a - c) + (b - d)i$$

⇒ Simplify: $(3 - 7i) - (4 - 2i)$

$$(3 - 7i) - (4 - 2i) = (3 - 4) + [-7 - (-2)]i$$

• Subtract the real parts and subtract the imaginary parts of the complex numbers.

$$= -1 - 5i$$

⇒ Simplify: $(3 + \sqrt{-12}) + (7 - \sqrt{-27})$

$(3 + \sqrt{-12}) + (7 - \sqrt{-27})$

$= (3 + i\sqrt{12}) + (7 - i\sqrt{27})$

$= (3 + i\sqrt{4 \cdot 3}) + (7 - i\sqrt{9 \cdot 3})$

$= (3 + 2i\sqrt{3}) + (7 - 3i\sqrt{3})$

$= 10 - i\sqrt{3}$

- Write each complex number in the form $a + bi$.
- Use the Product Property of Radicals to simplify each radical.

- Add the complex numbers.

Example 3
Simplify: $(3 + 2i) + (6 - 5i)$

Solution
$(3 + 2i) + (6 - 5i) = 9 - 3i$

You Try It 3
Simplify: $(-4 + 2i) - (6 - 8i)$

Your solution

Example 4
Simplify: $(9 - \sqrt{-8}) - (5 + \sqrt{-32})$

Solution
$(9 - \sqrt{-8}) - (5 + \sqrt{-32})$
$= (9 - i\sqrt{8}) - (5 + i\sqrt{32})$
$= (9 - i\sqrt{4 \cdot 2}) - (5 + i\sqrt{16 \cdot 2})$
$= (9 - 2i\sqrt{2}) - (5 + 4i\sqrt{2})$
$= 4 - 6i\sqrt{2}$

You Try It 4
Simplify: $(16 - \sqrt{-45}) - (3 + \sqrt{-20})$

Your solution

Example 5
Simplify: $(6 + 4i) + (-6 - 4i)$

Solution
$(6 + 4i) + (-6 - 4i) = 0 + 0i = 0$

This illustrates that the additive inverse of $a + bi$ is $-a - bi$.

You Try It 5
Simplify: $(3 - 2i) + (-3 + 2i)$

Your solution

Solutions on pp. S22–S23

Objective C   **To multiply complex numbers**

When multiplying complex numbers, we often find that the term i^2 is a part of the product. Recall that $i^2 = -1$.

⇒ Simplify: $2i \cdot 3i$

$2i \cdot 3i = 6i^2$ • Multiply the imaginary numbers.

$= 6(-1)$ • Replace i^2 by -1.

$= -6$ • Simplify.

⇒ Simplify: $\sqrt{-6} \cdot \sqrt{-24}$

$\sqrt{-6} \cdot \sqrt{-24} = i\sqrt{6} \cdot i\sqrt{24}$ • Write each radical as the product of a real number and i.

$= i^2\sqrt{144}$ • Multiply the imaginary numbers.

$= -\sqrt{144}$ • Replace i^2 by -1.

$= -12$ • Simplify the radical expression.

Note from the last example that it would have been incorrect to multiply the radicands of the two radical expressions. To illustrate,

$$\sqrt{-6} \cdot \sqrt{-24} = \sqrt{(-6)(-24)} = \sqrt{144} = 12, \, not \, -12$$

The Product Property of Radicals does not hold true when both radicands are negative and the index is an even number.

⇒ Simplify: $4i(3 - 2i)$

$4i(3 - 2i) = 12i - 8i^2$ • Use the Distributive Property to remove parentheses.

$= 12i - 8(-1)$ • Replace i^2 by -1.

$= 8 + 12i$ • Write the answer in the form $a + bi$.

The product of two complex numbers is defined as follows.

The Product of Two Complex Numbers

$(a + bi)(c + di) = (ac - bd) + (ad + bc)i$

One way to remember this rule is to use the FOIL method.

⇒ Simplify: $(2 + 4i)(3 - 5i)$

$(2 + 4i)(3 - 5i) = 6 - 10i + 12i - 20i^2$ • Use the FOIL method to find the product.

$= 6 + 2i - 20i^2$

$= 6 + 2i - 20(-1)$ • Replace i^2 by -1.

$= 26 + 2i$ • Write the answer in the form $a + bi$.

The conjugate of $a + bi$ is $a - bi$.
The product of conjugates, $(a + bi)(a - bi)$, is the real number $a^2 + b^2$.

$(a + bi)(a - bi) = a^2 - b^2i^2$

$= a^2 - b^2(-1)$

$= a^2 + b^2$

⇒ Simplify: $(2 + 3i)(2 - 3i)$

$(2 + 3i)(2 - 3i) = 2^2 + 3^2$ • The product of the conjugates is $2^2 + 3^2$.

$= 4 + 9$

$= 13$

Note that the product of a complex number and its conjugate is a real number.

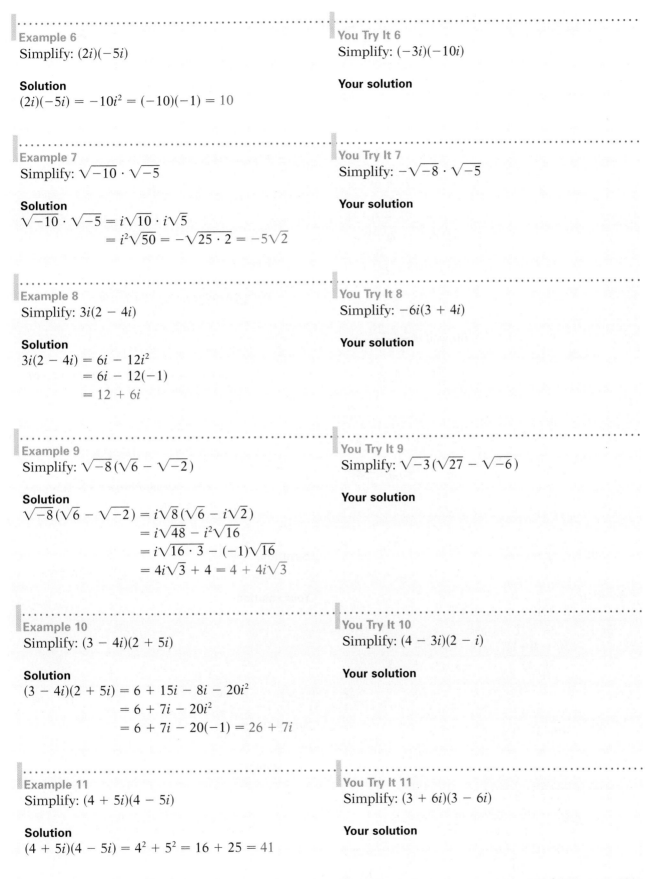

Example 6

Simplify: $(2i)(-5i)$

Solution

$(2i)(-5i) = -10i^2 = (-10)(-1) = 10$

You Try It 6

Simplify: $(-3i)(-10i)$

Your solution

Example 7

Simplify: $\sqrt{-10} \cdot \sqrt{-5}$

Solution

$\sqrt{-10} \cdot \sqrt{-5} = i\sqrt{10} \cdot i\sqrt{5}$
$= i^2\sqrt{50} = -\sqrt{25 \cdot 2} = -5\sqrt{2}$

You Try It 7

Simplify: $-\sqrt{-8} \cdot \sqrt{-5}$

Your solution

Example 8

Simplify: $3i(2 - 4i)$

Solution

$3i(2 - 4i) = 6i - 12i^2$
$= 6i - 12(-1)$
$= 12 + 6i$

You Try It 8

Simplify: $-6i(3 + 4i)$

Your solution

Example 9

Simplify: $\sqrt{-8}(\sqrt{6} - \sqrt{-2})$

Solution

$\sqrt{-8}(\sqrt{6} - \sqrt{-2}) = i\sqrt{8}(\sqrt{6} - i\sqrt{2})$
$= i\sqrt{48} - i^2\sqrt{16}$
$= i\sqrt{16 \cdot 3} - (-1)\sqrt{16}$
$= 4i\sqrt{3} + 4 = 4 + 4i\sqrt{3}$

You Try It 9

Simplify: $\sqrt{-3}(\sqrt{27} - \sqrt{-6})$

Your solution

Example 10

Simplify: $(3 - 4i)(2 + 5i)$

Solution

$(3 - 4i)(2 + 5i) = 6 + 15i - 8i - 20i^2$
$= 6 + 7i - 20i^2$
$= 6 + 7i - 20(-1) = 26 + 7i$

You Try It 10

Simplify: $(4 - 3i)(2 - i)$

Your solution

Example 11

Simplify: $(4 + 5i)(4 - 5i)$

Solution

$(4 + 5i)(4 - 5i) = 4^2 + 5^2 = 16 + 25 = 41$

You Try It 11

Simplify: $(3 + 6i)(3 - 6i)$

Your solution

Solutions on p. S23

Objective D **To divide complex numbers** ▸ 7 ◂

A rational expression containing one or more complex numbers is in simplest form when no imaginary number remains in the denominator.

➡ Simplify: $\dfrac{2 - 3i}{2i}$

$$\dfrac{2 - 3i}{2i} = \dfrac{2 - 3i}{2i} \cdot \dfrac{i}{i}$$

• Multiply the numerator and denominator by $\dfrac{i}{i}$.

$$= \dfrac{2i - 3i^2}{2i^2}$$

$$= \dfrac{2i - 3(-1)}{2(-1)}$$

• Replace i^2 by -1.

$$= \dfrac{3 + 2i}{-2}$$

• Simplify.

$$= -\dfrac{3}{2} - i$$

• Write the answer in the form $a + bi$.

➡ Simplify: $\dfrac{3 + 2i}{1 + i}$

$$\dfrac{3 + 2i}{1 + i} = \dfrac{3 + 2i}{1 + i} \cdot \dfrac{1 - i}{1 - i}$$

• Multiply the numerator and denominator by the conjugate of $1 + i$.

$$= \dfrac{3 - 3i + 2i - 2i^2}{1^2 + 1^2}$$

$$= \dfrac{3 - i - 2(-1)}{1 + 1}$$

• Replace i^2 by -1 and simplify.

$$= \dfrac{5 - i}{2} = \dfrac{5}{2} - \dfrac{1}{2}i$$

• Write the answer in the form $a + bi$.

Example 12

Simplify: $\dfrac{5 + 4i}{3i}$

Solution

$$\dfrac{5 + 4i}{3i} = \dfrac{5 + 4i}{3i} \cdot \dfrac{i}{i} = \dfrac{5i + 4i^2}{3i^2}$$

$$= \dfrac{5i + 4(-1)}{3(-1)} = \dfrac{-4 + 5i}{-3} = \dfrac{4}{3} - \dfrac{5}{3}i$$

You Try It 12

Simplify: $\dfrac{2 - 3i}{4i}$

Your solution

Example 13

Simplify: $\dfrac{5 - 3i}{4 + 2i}$

Solution

$$\dfrac{5 - 3i}{4 + 2i} = \dfrac{5 - 3i}{4 + 2i} \cdot \dfrac{4 - 2i}{4 - 2i}$$

$$= \dfrac{20 - 10i - 12i + 6i^2}{4^2 + 2^2}$$

$$= \dfrac{20 - 22i + 6(-1)}{20}$$

$$= \dfrac{14 - 22i}{20} = \dfrac{7 - 11i}{10} = \dfrac{7}{10} - \dfrac{11}{10}i$$

You Try It 13

Simplify: $\dfrac{2 + 5i}{3 - 2i}$

Your solution

Solutions on p. S23

7.3 Exercises

Objective A

1. What is an imaginary number? What is a complex number?

2. Are all real numbers also complex numbers? Are all complex numbers also real numbers?

Simplify.

3. $\sqrt{-4}$

4. $\sqrt{-64}$

5. $\sqrt{-98}$

6. $\sqrt{-72}$

7. $\sqrt{-27}$

8. $\sqrt{-75}$

9. $\sqrt{16} + \sqrt{-4}$

10. $\sqrt{25} + \sqrt{-9}$

11. $\sqrt{12} - \sqrt{-18}$

12. $\sqrt{60} - \sqrt{-48}$

13. $\sqrt{160} - \sqrt{-147}$

14. $\sqrt{96} - \sqrt{-125}$

Objective B

Simplify.

15. $(2 + 4i) + (6 - 5i)$

16. $(6 - 9i) + (4 + 2i)$

17. $(-2 - 4i) - (6 - 8i)$

18. $(3 - 5i) + (8 - 2i)$

19. $(8 - \sqrt{-4}) - (2 + \sqrt{-16})$

20. $(5 - \sqrt{-25}) - (11 - \sqrt{-36})$

21. $(12 - \sqrt{-50}) + (7 - \sqrt{-8})$

22. $(5 - \sqrt{-12}) - (9 + \sqrt{-108})$

23. $(\sqrt{8} + \sqrt{-18}) + (\sqrt{32} - \sqrt{-72})$

24. $(\sqrt{40} - \sqrt{-98}) - (\sqrt{90} + \sqrt{-32})$

Objective C

Simplify.

25. $(7i)(-9i)$

26. $(-6i)(-4i)$

27. $\sqrt{-2}\,\sqrt{-8}$

28. $\sqrt{-5}\sqrt{-45}$

29. $\sqrt{-3}\sqrt{-6}$

30. $\sqrt{-5}\sqrt{-10}$

31. $2i(6 + 2i)$

32. $-3i(4 - 5i)$

33. $\sqrt{-2}(\sqrt{8} + \sqrt{-2})$

34. $\sqrt{-3}(\sqrt{12} - \sqrt{-6})$

35. $(5 - 2i)(3 + i)$

36. $(2 - 4i)(2 - i)$

37. $(6 + 5i)(3 + 2i)$

38. $(4 - 7i)(2 + 3i)$

39. $(1 - i)\left(\dfrac{1}{2} + \dfrac{1}{2}i\right)$

40. $\left(\dfrac{4}{5} - \dfrac{2}{5}i\right)\left(1 + \dfrac{1}{2}i\right)$

41. $\left(\dfrac{6}{5} + \dfrac{3}{5}i\right)\left(\dfrac{2}{3} - \dfrac{1}{3}i\right)$

42. $(2 - i)\left(\dfrac{2}{5} + \dfrac{1}{5}i\right)$

Objective D

Simplify.

43. $\dfrac{3}{i}$

44. $\dfrac{4}{5i}$

45. $\dfrac{2 - 3i}{-4i}$

46. $\dfrac{16 + 5i}{-3i}$

47. $\dfrac{4}{5 + i}$

48. $\dfrac{6}{5 + 2i}$

49. $\dfrac{2}{2 - i}$

50. $\dfrac{5}{4 - i}$

51. $\dfrac{1 - 3i}{3 + i}$

52. $\dfrac{2 + 12i}{5 + i}$

53. $\dfrac{\sqrt{-10}}{\sqrt{8} - \sqrt{-2}}$

54. $\dfrac{\sqrt{-2}}{\sqrt{12} - \sqrt{-8}}$

55. $\dfrac{2 - 3i}{3 + i}$

56. $\dfrac{3 + 5i}{1 - i}$

57. $\dfrac{5 + 3i}{3 - i}$

58. $\dfrac{3 - 2i}{2i + 3}$

APPLYING THE CONCEPTS

59. **a.** Is $3i$ a solution of $2x^2 + 18 = 0$?
b. Is $3 + i$ a solution of $x^2 - 6x + 10 = 0$?

60. Evaluate i^n for $n = 0, 1, 2, 3, 4, 5, 6,$ and 7. Make a conjecture about the value of i^n for any natural number. Using your conjecture, evaluate i^{76}.

7.4 Solving Equations Containing Radical Expressions

Objective A To solve a radical equation

An equation that contains a variable expression in a radicand is a **radical equation.**

$$\left.\begin{array}{r} \sqrt[3]{2x - 5} + x = 7 \\ \sqrt{x + 1} - \sqrt{x} = 4 \end{array}\right\} \begin{array}{l} \text{Radical} \\ \text{equations} \end{array}$$

The following property is used to solve a radical equation.

The Property of Raising Each Side of an Equation to a Power

If two numbers are equal, then the same powers of the numbers are equal.

If $a = b$, then $a^n = b^n$.

➡ Solve: $\sqrt{x - 2} - 6 = 0$

$$\sqrt{x - 2} - 6 = 0$$
$$\sqrt{x - 2} = 6 \qquad \bullet \text{ Isolate the radical by adding 6 to each side of the equation.}$$
$$(\sqrt{x - 2})^2 = 6^2 \qquad \bullet \text{ Square each side of the equation.}$$
$$x - 2 = 36 \qquad \bullet \text{ Simplify and solve for } x.$$
$$x = 38$$

$$\begin{array}{r} Check: \quad \sqrt{x - 2} - 6 = 0 \\ \hline \sqrt{38 - 2} - 6 \;\big|\; 0 \\ \sqrt{36} - 6 \;\big|\; 0 \\ 6 - 6 \;\big|\; 0 \\ 0 = 0 \end{array}$$

38 checks as a solution. The solution is 38.

➡ Solve: $\sqrt[3]{x + 2} = -3$

$$\sqrt[3]{x + 2} = -3$$
$$(\sqrt[3]{x + 2})^3 = (-3)^3 \qquad \bullet \text{ Cube each side of the equation.}$$
$$x + 2 = -27 \qquad \bullet \text{ Solve the resulting equation.}$$
$$x = -29$$

$$\begin{array}{r} Check: \quad \sqrt[3]{x + 2} = -3 \\ \hline \sqrt[3]{-29 + 2} \;\big|\; -3 \\ \sqrt[3]{-27} \;\big|\; -3 \\ -3 = -3 \end{array}$$

-29 checks as a solution. The solution is -29.

Raising each side of an equation to an even power may result in an equation that has a solution that is not a solution of the original equation. This is called an **extraneous solution.** Here is an example:

➡️ Solve: $\sqrt{2x - 1} + \sqrt{x} = 2$

$$\sqrt{2x - 1} + \sqrt{x} = 2$$
$$\sqrt{2x - 1} = 2 - \sqrt{x}$$
$$(\sqrt{2x - 1})^2 = (2 - \sqrt{x})^2$$
$$2x - 1 = 4 - 4\sqrt{x} + x$$
$$x - 5 = -4\sqrt{x}$$
$$(x - 5)^2 = (-4\sqrt{x})^2$$

- Solve for one of the radical expressions.
- Square each side. Recall that $(a - b)^2 = a^2 - 2ab + b^2$.

$$x^2 - 10x + 25 = 16x$$
$$x^2 - 26x + 25 = 0$$
$$(x - 25)(x - 1) = 0$$
$$x = 25 \text{ or } x = 1$$

- Solve the quadratic equation by factoring.

Check:

$$\frac{\sqrt{2x - 1} + \sqrt{x} = 2}{\sqrt{2(25) - 1} + \sqrt{25} \,\big|\, 2}$$
$$7 + 5 \,\big|\, 2$$
$$12 \neq 2$$

$$\frac{\sqrt{2x - 1} + \sqrt{x} = 2}{\sqrt{2(1) - 1} + \sqrt{1} \,\big|\, 2}$$
$$1 + 1 \,\big|\, 2$$
$$2 = 2$$

25 does not check as a solution. 1 checks as a solution. The solution is 1. Here 25 is an extraneous solution.

Example 1

Solve: $\sqrt{x - 1} + \sqrt{x + 4} = 5$

Solution
$$\sqrt{x - 1} + \sqrt{x + 4} = 5$$
$$\sqrt{x + 4} = 5 - \sqrt{x - 1}$$
$$(\sqrt{x + 4})^2 = (5 - \sqrt{x - 1})^2$$
$$x + 4 = 25 - 10\sqrt{x - 1} + x - 1$$
$$2 = \sqrt{x - 1}$$
$$2^2 = (\sqrt{x - 1})^2$$
$$4 = x - 1$$
$$5 = x$$

The solution checks. The solution is 5.

You Try It 1

Solve: $\sqrt{x} - \sqrt{x + 5} = 1$

Your solution

Example 2

Solve: $\sqrt[3]{3x - 1} = -4$

Solution
$$\sqrt[3]{3x - 1} = -4$$
$$(\sqrt[3]{3x - 1})^3 = (-4)^3$$
$$3x - 1 = -64$$
$$3x = -63$$
$$x = -21$$

The solution checks. The solution is -21.

You Try It 2

Solve: $\sqrt[4]{x - 8} = 3$

Your solution

Solutions on p. S23

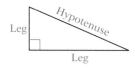

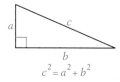

Objective B   **To solve application problems**

A right triangle contains one 90° angle. The side opposite the 90° angle is called the **hypotenuse.** The other two sides are called **legs.**

Pythagoras, a Greek mathematician, discovered that the square of the hypotenuse of a right triangle is equal to the sum of the squares of the two legs. Recall that this is called the **Pythagorean Theorem.**

. . . .

Example 3

A ladder 20 ft long is leaning against a building. How high on the building will the ladder reach when the bottom of the ladder is 8 ft from the building? Round to the nearest tenth.

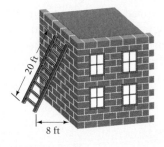

Strategy

To find the distance, use the Pythagorean Theorem. The hypotenuse is the length of the ladder. One leg is the distance from the bottom of the ladder to the base of the building. The distance along the building from the ground to the top of the ladder is the unknown leg.

Solution

$$c^2 = a^2 + b^2$$
$$20^2 = 8^2 + b^2$$
$$400 = 64 + b^2$$
$$336 = b^2$$
$$(336)^{1/2} = (b^2)^{1/2}$$
$$\sqrt{336} = b$$
$$18.3 \approx b$$

The distance is 18.3 ft.

You Try It 3

Find the diagonal of a rectangle that is 6 cm long and 3 cm wide. Round to the nearest tenth.

Your strategy

Your solution

Solution on p. S23

Example 4

An object is dropped from a high building. Find the distance the object has fallen when its speed reaches 96 ft/s. Use the equation $v = \sqrt{64d}$, where v is the speed of the object in feet per second and d is the distance in feet.

Strategy

To find the distance the object has fallen, replace v in the equation with the given value and solve for d.

Solution

$$v = \sqrt{64d}$$
$$96 = \sqrt{64d}$$
$$(96)^2 = (\sqrt{64d})^2$$
$$9216 = 64d$$
$$144 = d$$

The object has fallen 144 ft.

You Try It 4

How far would a submarine periscope have to be above the water to locate a ship 5.5 mi away? The equation for the distance in miles that the lookout can see is $d = \sqrt{1.5h}$, where h is the height in feet above the surface of the water. Round to the nearest hundredth.

Your strategy

Your solution

Example 5

Find the length of a pendulum that makes one swing in 1.5 s. The equation for the time of one swing is given by $T = 2\pi\sqrt{\dfrac{L}{32}}$, where T is the time in seconds and L is the length in feet. Use 3.14 for π. Round to the nearest hundredth.

Strategy

To find the length of the pendulum, replace T in the equation with the given value and solve for L.

Solution

$$T = 2\pi\sqrt{\dfrac{L}{32}}$$

$$1.5 = 2(3.14)\sqrt{\dfrac{L}{32}}$$

$$\dfrac{1.5}{2(3.14)} = \sqrt{\dfrac{L}{32}}$$

$$\left[\dfrac{1.5}{2(3.14)}\right]^2 = \left(\sqrt{\dfrac{L}{32}}\right)^2$$

$$\left(\dfrac{1.5}{6.28}\right)^2 = \dfrac{L}{32}$$

$$1.83 \approx L$$

The length of the pendulum is 1.83 ft.

You Try It 5

Find the distance required for a car to reach a velocity of 88 ft/s when the acceleration is 22 ft/s². Use the equation $v = \sqrt{2as}$, where v is the velocity in feet per second, a is the acceleration, and s is the distance in feet.

Your strategy

Your solution

Solutions on pp. S23–S24

7.4 Exercises

Objective A

Solve.

1. $\sqrt[3]{4x} = -2$

2. $\sqrt[3]{6x} = -3$

3. $\sqrt{3x - 2} = 5$

4. $\sqrt{3x + 9} - 12 = 0$

5. $\sqrt{4x - 3} - 5 = 0$

6. $\sqrt{4x - 2} = \sqrt{3x + 9}$

7. $\sqrt{2x + 4} = \sqrt{5x - 9}$

8. $\sqrt[3]{x - 2} = 3$

9. $\sqrt[3]{2x - 6} = 4$

10. $\sqrt[3]{3x - 9} = \sqrt[3]{2x + 12}$

11. $\sqrt[3]{x - 12} = \sqrt[3]{5x + 16}$

12. $\sqrt[4]{4x + 1} = 2$

13. $\sqrt[3]{2x - 3} + 5 = 2$

14. $\sqrt[3]{x - 4} + 7 = 5$

15. $\sqrt{x} + \sqrt{x - 5} = 5$

16. $\sqrt{x + 3} + \sqrt{x - 1} = 2$

17. $\sqrt{2x + 5} - \sqrt{2x} = 1$

18. $\sqrt{3x} - \sqrt{3x - 5} = 1$

19. $\sqrt{2x} - \sqrt{x - 1} = 1$

20. $\sqrt{2x - 5} + \sqrt{x + 1} = 3$

21. $\sqrt{2x + 2} + \sqrt{x} = 3$

Objective B   *Application Problems*

22. An object is dropped from a bridge. Find the distance the object has fallen when its speed reaches 100 ft/s. Use the equation $v = \sqrt{64d}$, where v is the speed in feet per second and d is the distance in feet.

23. The time it takes for an object to fall a certain distance is given by the equation $t = \sqrt{\dfrac{2d}{g}}$, where t is the time in seconds, d is the distance in feet, and g is the acceleration due to gravity. The acceleration due to gravity on Earth is 32 feet per second per second. If an object is dropped from atop a tall building, how far will it fall in 6 s?

24. The time it takes for an object to fall a certain distance is given by the equation $t = \sqrt{\dfrac{2d}{g}}$, where t is the time in seconds, d is the distance in feet, and g is the acceleration due to gravity. If an astronaut above the moon's surface drops an object, how far will it have fallen in 3 s? The acceleration on the moon's surface is 5.5 feet per second per second.

25. High definition television (HDTV) gives consumers a wider viewing area, more like a film screen in a theater. A regular television with a 27-inch diagonal measurement has a screen 16.2 in. tall. An HDTV screen with the same 16.2-inch height would have a diagonal measuring 33 in. How many inches wider is the HDTV screen? Round to the nearest hundredth.

26. At what height above Earth's surface would a satellite be in orbit if it is traveling at a speed of 7500 m/s? Use the equation $v = \sqrt{\dfrac{4 \times 10^{14}}{h + 6.4 \times 10^{6}}}$, where v is the speed of the satellite in meters per second and h is the height above Earth's surface in meters. Round to the nearest thousand.

27. Find the length of a pendulum that makes one swing in 3 s. The equation for the time of one swing of a pendulum is given by $T = 2\pi \sqrt{\dfrac{L}{32}}$, where T is the time in seconds and L is the length in feet. Round to the nearest hundredth.

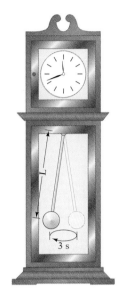

APPLYING THE CONCEPTS

28. Solve the following equations. Describe the solution by using the following terms: integer, rational number, irrational number, real number, and imaginary number. Note that more than one term may be used to describe the answer.
  a. $x^2 + 3 = 7$ **b.** $x^2 + 1 = 0$ **c.** $x^{3/4} = 8$

29. Solve: $\sqrt{3x - 2} = \sqrt{2x - 3} + \sqrt{x - 1}$

30. Solve $a^2 + b^2 = c^2$ for a.

31. Beginning with a square with sides of length s, draw a line segment from the midpoint of the base of the square to a vertex, as shown. Call this distance a. Make a rectangle whose width is that of the square and whose length is $\dfrac{s}{2} + a$. Find:

  a. The area of the rectangle in terms of s.
  b. The ratio of the length to the width.

The rectangle formed in this way is called a golden rectangle. It was described at the beginning of this chapter.

32. Solve $V = \dfrac{4}{3}\pi r^3$ for r.

33. Find the length of the side labeled x.

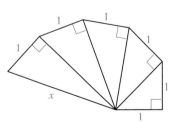

Focus on Problem Solving

Another Look at Polya's Four-Step Process

Polya's four general steps to follow when attempting to solve a problem are to understand the problem, devise a plan, carry out the plan, and review the solution. (See the Focus on Problem Solving in the chapter entitled "Review of Real Numbers.") In the process of devising a plan (Step 2), it may be appropriate to write a mathematical expression or an equation. We will illustrate this with the following problem.

Number the letters of the alphabet in sequence from 1 to 26. (See the list at the left.) Find a word for which the product of the numerical values of the letters of the word equals 1,000,000. We will agree that a "word" is any sequence of letters that contains at least one vowel but is not necessarily in the dictionary.

A = 1
B = 2
C = 3
D = 4
E = 5
F = 6
G = 7
H = 8
I = 9
J = 10
K = 11
L = 12
M = 13
N = 14
O = 15
P = 16
Q = 17
R = 18
S = 19
T = 20
U = 21
V = 22
W = 23
X = 24
Y = 25
Z = 26

Understand the Problem

Consider REZB. The product of the values of the letters is $18 \cdot 5 \cdot 26 \cdot 2 = 4680$. This "word" is a sequence of letters with at least one vowel. However, the product of the numerical values of the letters is not 1,000,000. Thus this word does not solve our problem.

Devise a Plan

Actually, we should have known that the product of the values of the letters in REZB could not equal 1,000,000. The letter R has a factor of 9, and the letter Z has a factor of 13. Neither of these two numbers is a factor of 1,000,000. Consequently, R and Z cannot be letters in the word we are trying to find. This observation leads to an important observation: each of the letters that make up our word must be a factor of 1,000,000. To find these letters, consider the prime factorization of 1,000,000.

$$1,000,000 = 2^6 \cdot 5^6$$

Looking at the prime factorization, we note that only letters that contain 2 or 5 as factors are possible candidates. These letters are B, D, E, H, J, P, T, and Y. One additional point: Because 1 times any number is the number, the letter A can be part of any word we construct.

Our task is now to construct a word from these letters such that the product is 1,000,000. From the prime factorization above, we must have 2 as a factor six times and 5 as a factor six times.

Carry Out the Plan

We must construct a word with the characteristics described in our plan. Here is a possibility:

THEBEYE

Review the Solution

You should multiply the values of all the letters and verify that the product is 1,000,000. To ensure that you have an understanding of the problem, try to find other "words" that satisfy the conditions of the problem.

Projects and Group Activities

Complex Numbers   The idea that a negative number might be the solution of an equation was not readily accepted by mathematicians until the middle of the 16th century. Hieronimo Cardano was the first to take notice of these numbers in his book *Ars Magna* (*The Great Art*), which was published in 1545. However, he called these solutions fictitious. In that same book, in one case, he also considered a solution of an equation of the form $5 + \sqrt{-15}$, a number we now call a complex number.

Complex numbers were thought to have no real or geometric interpretation and therefore to be "imaginary." One mathematician called them "ghosts of real numbers." Then, in 1806, Jean-Robert Argand published a book that included a geometric description of complex numbers. (Actually, Caspar Wessel first proposed this idea in 1797. However, his work went unnoticed.)

> **TAKE NOTE**
>
> 4 is the complex number $4 + 0i$, and $2i$ is the complex number $0 + 2i$.

An Argand diagram consists of a horizontal and a vertical axis. The horizontal axis is the *real axis;* the vertical axis is the *imaginary axis.* A complex number of the form $a + bi$ is shown as the point in the plane whose coordinates are (a, b). The graphs of $2 + 4i$, $-4 - 3i$, 4, and $2i$ are shown at the right.

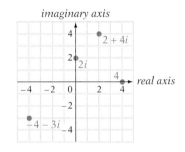

By using the Argand diagram, it is possible to give various interpretations of complex numbers and the operations on them. We will focus on only one small aspect: a geometric explanation of multiplying by i, the imaginary unit.

Before doing this, recall that real number multiplication is repeated addition. For instance, $2 \cdot 3 = 2 + 2 + 2$. Geometrically, this can be shown on a number line.

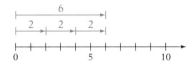

Thus, multiplying by 3 triples the size of the arrow that represents 2. In general, multiplying a real number by a number greater than 1 increases the magnitude of the number.

Consider the complex number $3 + 4i$. Multiply this number by i.

$$i(3 + 4i) = -4 + 3i$$

The two complex numbers are shown in the graph at the right.

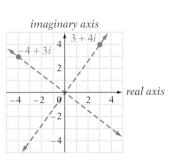

1. Find the slope of the line that passes through the origin and the point $(3, 4)$.

2. Find the slope of the line that passes through the origin and the point $(-4, 3)$.

3. Find the product of the slopes of the two lines. What does this product tell you?

4. Find the distance of the points (3, 4) and (−4, 3) from the origin.

A conclusion that can be drawn from the answers to the foregoing questions is that multiplying a number by i rotates the line on which the number lies by 90°. This is shown in the figure at the right.

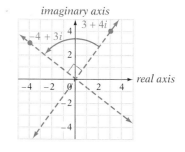

This property of complex numbers is important in the study of electricity and magnetism. Complex numbers are used extensively by electrical engineers to study electrical circuits.

Solving Radical Equations with a Graphing Calculator

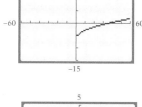

The radical equation $\sqrt{x-2}-6=0$ is solved algebraically on page 403. To solve $\sqrt{x-2}-6=0$ with a graphing calculator, use the left side of the equation to write an equation in two variables.

$$y = \sqrt{x-2} - 6$$

The graph of $y = \sqrt{x-2} - 6$ is shown at the left. The solution set of the equation $y = \sqrt{x-2} - 6$ is the set of ordered pairs (x, y) whose coordinates make the equation a true statement. The x-coordinate at which $y = 0$ is the solution of the equation $\sqrt{x-2} - 6 = 0$. The solution is the x-intercept of the curve given by the equation $y = \sqrt{x-2} - 6$. The solution is 38.

A graphing calculator can be used to find any rational or irrational solution to a predetermined degree of accuracy. The graph of $y = \sqrt[3]{x-2} + 0.8$ is shown at the left. The graph intersects the x-axis at $x \approx 1.488$. The solution of the radical equation $\sqrt[3]{x-2} = -0.8$ is 1.488.

To graph the equation $y = \sqrt[3]{x-2} + 0.8$ on a TI-83, enter

$\boxed{Y=}$ $\boxed{\text{MATH}}$ 4 $\boxed{X, T, \theta, n}$ $\boxed{-}$ 2 $\boxed{)}$ $\boxed{+}$ .8 $\boxed{\text{ZOOM}}$ 6

Enter $\boxed{\text{2nd}}$ $\boxed{\text{CALC}}$ 2 to find the x-intercept of the equation.

You solved Exercises 1–21 on page 407 by algebraic means. Try some of those exercises again, this time using a graphing calculator.

Use a graphing calculator to find the solutions of the equation's below. Round to the nearest thousandth.

1. $\sqrt{x + 0.3} = 1.3$ **2.** $\sqrt[3]{x + 1.2} = -1.1$ **3.** $\sqrt[3]{3x - 1.5} = 1.4$

Chapter Summary

Key Words
The *nth root of a* is $a^{1/n}$. The expression $\sqrt[n]{a}$ is another symbol for the *n*th root of *a*. [pp. 375, 377]

In the expression $\sqrt[n]{a}$, the symbol $\sqrt{}$ is called a *radical, n* is the *index* of the radical, and *a* is the *radicand*. [p. 375]

The symbol $\sqrt{}$ is used to indicate the positive or *principal square root* of a number. [p. 379]

Conjugates are binomial expressions that differ only in the sign of a term. The expressions $a - b$ and $a + b$ are conjugates. [p. 388]

The procedure used to remove a radical from the denominator is called *rationalizing the denominator*. [p. 389]

A *complex number* is a number of the form $a + bi$, where *a* and *b* are real numbers and $i = \sqrt{-1}$. [p. 395]

For the complex number $a + bi$, *a* is called the *real part* of the complex number and *b* is called the *imaginary part* of the complex number. [p. 395]

Essential Rules

Rational Exponents
If *m* and *n* are positive integers and $a^{1/n}$ is a real number, then $a^{m/n} = (a^{1/n})^m$. [p. 375]

The Product Property of Radicals
If $\sqrt[n]{a}$ and $\sqrt[n]{b}$ are positive real numbers, then $\sqrt[n]{a}\,\sqrt[n]{b} = \sqrt[n]{ab}$ and $\sqrt[n]{ab} = \sqrt[n]{a}\,\sqrt[n]{b}$. [p. 385]

The Quotient Property of Radicals
If $\sqrt[n]{a}$ and $\sqrt[n]{b}$ are positive real numbers and $b \neq 0$, $\dfrac{\sqrt[n]{a}}{\sqrt[n]{b}} = \sqrt[n]{\dfrac{a}{b}}$ and $\sqrt[n]{\dfrac{a}{b}} = \dfrac{\sqrt[n]{a}}{\sqrt[n]{b}}$. [p. 389]

Addition and Subtraction of Complex Numbers
$(a + bi) + (c + di) = (a + c) + (b + d)i$
$(a + bi) - (c + di) = (a - c) + (b - d)i$
[p. 396]

Multiplication of Complex Numbers
$(a + bi)(c + di) = (ac - bd) + (ad + bc)i$
[p. 398]

Property of Raising Each Side of an Equation to a Power
If *a* and *b* are real numbers and $a = b$, then $a^n = b^n$. [p. 403]

Pythagorean Theorem
$c^2 = a^2 + b^2$ [p. 405]

Chapter Review

1. Simplify: $(16x^{-4}y^{12})^{1/4}(100x^6y^{-2})^{1/2}$

2. Solve: $\sqrt[4]{3x - 5} = 2$

3. Simplify: $(6 - 5i)(4 + 3i)$

4. Rewrite $7y\sqrt[3]{x^2}$ as an exponential expression.

5. Simplify: $(\sqrt{3} + 8)(\sqrt{3} - 2)$

6. Solve: $\sqrt{4x + 9} + 10 = 11$

7. Simplify: $\dfrac{x^{-3/2}}{x^{7/2}}$

8. Simplify: $\dfrac{8}{\sqrt{3y}}$

9. Simplify: $\sqrt[3]{-8a^6b^{12}}$

10. Simplify: $\sqrt{50a^4b^3} - ab\sqrt{18a^2b}$

11. Simplify: $\dfrac{x + 2}{\sqrt{x} + \sqrt{2}}$

12. Simplify: $\dfrac{5 + 2i}{3i}$

13. Simplify: $\sqrt{18a^3b^6}$

14. Simplify:
$(\sqrt{50} + \sqrt{-72}) - (\sqrt{162} - \sqrt{-8})$

15. Simplify: $3x\sqrt[3]{54x^8y^{10}} - 2x^2y\sqrt[3]{16x^5y^7}$

16. Simplify: $\sqrt[3]{16x^4y}\,\sqrt[3]{4xy^5}$

17. Simplify: $i(3 - 7i)$

18. Rewrite $3x^{3/4}$ as a radical expression.

19. Simplify: $\sqrt[5]{-64a^8b^{12}}$

20. Simplify: $\dfrac{5+9i}{1-i}$

21. Simplify: $\sqrt{-12}\,\sqrt{-6}$

22. Solve: $\sqrt{x-5}+\sqrt{x+6}=11$

23. Simplify: $\sqrt[4]{81a^8b^{12}}$

24. Simplify: $\sqrt{-50}$

25. Simplify: $(-8+3i)-(4-7i)$

26. Simplify: $(5-\sqrt{6})^2$

27. Simplify: $4x\sqrt{12x^2y}+\sqrt{3x^4y}-x^2\sqrt{27y}$

28. The velocity of the wind determines the amount of power generated by a windmill. A typical equation for this relationship is $v=4.05\sqrt[3]{P}$, where v is the velocity in miles per hour and P is the power in watts. Find the amount of power generated by a 20-mph wind. Round to the nearest whole number.

29. Find the distance required for a car to reach a velocity of 88 ft/s when the acceleration is 16 ft/s². Use the equation $v=\sqrt{2as}$, where v is the velocity in feet per second, a is the acceleration, and s is the distance in feet.

30. A 12-foot ladder is leaning against a building. How far from the building is the bottom of the ladder when the top of the ladder touches the building 10 ft above the ground? Round to the nearest hundredth.

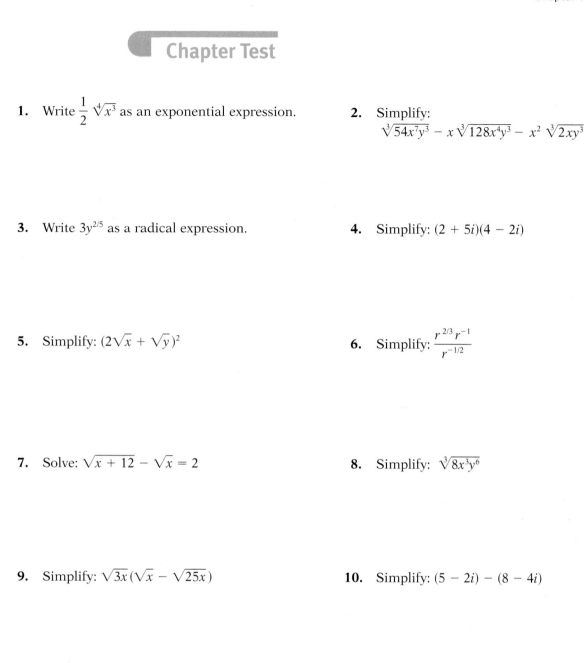

Chapter Test

1. Write $\dfrac{1}{2}\sqrt[4]{x^3}$ as an exponential expression.

2. Simplify:
$\sqrt[3]{54x^7y^3} - x\sqrt[3]{128x^4y^3} - x^2\sqrt[3]{2xy^3}$

3. Write $3y^{2/5}$ as a radical expression.

4. Simplify: $(2 + 5i)(4 - 2i)$

5. Simplify: $(2\sqrt{x} + \sqrt{y})^2$

6. Simplify: $\dfrac{r^{2/3}\, r^{-1}}{r^{-1/2}}$

7. Solve: $\sqrt{x + 12} - \sqrt{x} = 2$

8. Simplify: $\sqrt[3]{8x^3y^6}$

9. Simplify: $\sqrt{3x}\left(\sqrt{x} - \sqrt{25x}\right)$

10. Simplify: $(5 - 2i) - (8 - 4i)$

11. Simplify: $\sqrt{32x^4y^7}$

12. Simplify: $(2\sqrt{3} + 4)(3\sqrt{3} - 1)$

13. Simplify: $(2 + i) + (2 - i)(3 + 2i)$

14. Simplify: $\dfrac{4 - 2\sqrt{5}}{2 - \sqrt{5}}$

15. Simplify: $\sqrt{18a^3} + a\sqrt{50a}$

16. Simplify: $(\sqrt{a} - 3\sqrt{b})(2\sqrt{a} + 5\sqrt{b})$

17. Simplify: $\dfrac{(2x^{1/3}y^{-2/3})^6}{(x^{-4}y^8)^{1/4}}$

18. Simplify: $\dfrac{\sqrt{x}}{\sqrt{x} - \sqrt{y}}$

19. Simplify: $\dfrac{2 + 3i}{1 - 2i}$

20. Solve: $\sqrt[3]{2x - 2} + 4 = 2$

21. Simplify: $\left(\dfrac{4a^4}{b^2}\right)^{-3/2}$

22. Simplify: $\sqrt[3]{27a^4b^3c^7}$

23. Simplify: $\dfrac{\sqrt{32x^5y}}{\sqrt{2xy^3}}$

24. Simplify: $(\sqrt{-8})(\sqrt{-2})$

25. An object is dropped from a high building. Find the distance the object has fallen when its speed reaches 192 ft/s. Use the equation $v = \sqrt{64d}$, where v is the speed of the object in feet per second and d is the distance in feet.

Cumulative Review

1. Identify the property that justifies the statement $(a + 2)b = ab + 2b$.

2. Given $f(x) = 3x^2 - 2x + 1$, evaluate $f(-3)$.

3. Solve: $5 - \dfrac{2}{3}x = 4$

4. Solve: $2[4 - 2(3 - 2x)] = 4(1 - x)$

5. Solve: $2 + |4 - 3x| = 5$

6. Solve: $6x - 3(2x + 2) > 3 - 3(x + 2)$

7. Solve: $|2x + 3| \le 9$

8. Factor: $81x^2 - y^2$

9. Factor: $x^5 + 2x^3 - 3x$

10. Find the equation of the line that passes through the points $(2, 3)$ and $(-1, 2)$.

11. Find the value of the determinant:
$$\begin{vmatrix} 1 & 2 & -3 \\ 0 & -1 & 2 \\ 3 & 1 & -2 \end{vmatrix}$$

12. Solve $P = \dfrac{R - C}{n}$ for C.

13. Simplify: $(2^{-1}x^2y^{-6})(2^{-1}y^{-4})^{-2}$

14. Simplify: $\dfrac{x^2y^3}{x^2 + 2x - 8} \cdot \dfrac{2x^2 - 7x + 6}{xy^4}$

15. Simplify: $\sqrt{40x^3} - x\sqrt{90x}$

16. Solve: $\dfrac{x}{x - 2} - 2x = \dfrac{-3}{x - 2}$

17. Graph $3x - 2y = -6$ and find the slope and y-intercept.

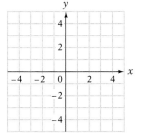

18. Graph the solution set of $3x + 2y \le 4$.

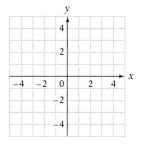

19. Simplify: $\dfrac{2i}{3-i}$

20. Solve: $\sqrt[3]{3x-4}+5=1$

21. Simplify: $\dfrac{x}{2x-3}+\dfrac{4}{x+4}$

22. Solve by using Cramer's Rule:
$$2x - y = 4$$
$$-2x + 3y = 5$$

23. A collection of 30 stamps consists of 13¢ stamps and 18¢ stamps. The total value of the stamps is $4.85. Find the number of 18¢ stamps.

24. A sales executive traveled 25 mi by car and then an additional 625 mi by plane. The rate of the plane was five times faster than the rate of the car. The total time of the trip was 3 h. Find the rate of the plane.

25. How long does it take light to travel to Earth from the moon when the moon is 232,500 mi from Earth? Light travels 1.86×10^5 mi/s.

26. How far would a submarine periscope have to be above the water to locate a ship 7 mi away? The equation for the distance in miles that the lookout can see is $d = \sqrt{1.5h}$, where h is the height in feet above the surface of the water. Round to the nearest tenth of a foot.

27. The graph shows the amount invested and the annual income from an investment. Find the slope of the line between the two points shown on the graph. Then write a sentence that states the meaning of the slope.

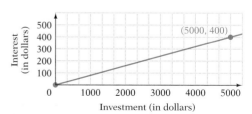

Chapter

8

Quadratic Equations

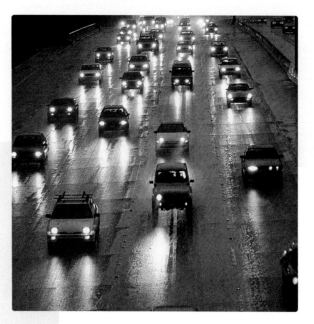

A rainstorm like the one shown here directly affects highway driving conditions. How quickly a car is able to stop once the brakes are applied is determined by many factors, including how wet the road is, the type of pavement, and the speed of the car. Some factors, such as the tread on a car's tires, can be controlled, and good tread can minimize the adverse effects of braking on dangerously wet roads. The braking distance of a car can be modeled by quadratic equations such as the ones in **Exercises 5 and 9 on page 453**.

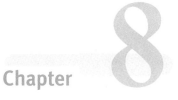

Need help? For on-line student resources, such as section quizzes, visit this textbook's web site at **math.college.hmco.com/students**.

1. Simplify: $\sqrt{18}$

2. Simplify: $\sqrt{-9}$

3. Simplify: $\dfrac{3x - 2}{x - 1} - 1$

4. Evaluate $b^2 - 4ac$ when $a = 2$, $b = -4$, and $c = 1$.

5. Is $4x^2 + 28x + 49$ a perfect square trinomial?

6. Factor: $4x^2 - 4x + 1$

7. Factor: $9x^2 - 4$

8. Graph $\{x \mid x < -1\} \cap \{x \mid x < 4\}$

9. Solve: $x(x - 1) = x + 15$

10. Solve: $\dfrac{4}{x - 3} = \dfrac{16}{x}$

The numerals .AAA... and .BBB... are repeating decimals whose digits are A and B, respectively. Given $\sqrt{.AAA...} = \sqrt{.BBB...}$ and $A = B$, find the value of A.

8.1 Solving Quadratic Equations by Factoring or by Taking Square Roots

Objective A   **To solve a quadratic equation by factoring**   〔 8 〕

Recall that a **quadratic equation** is an equation of the form $ax^2 + bx + c = 0$, where a and b are coefficients, c is a constant, and $a \neq 0$.

$$\text{Quadratic Equations} \begin{cases} 3x^2 - x + 2 = 0, & a = 3, & b = -1, & c = 2 \\ -x^2 + 4 = 0, & a = -1, & b = 0, & c = 4 \\ 6x^2 - 5x = 0, & a = 6, & b = -5, & c = 0 \end{cases}$$

A quadratic equation is in **standard form** when the polynomial is in descending order and equal to zero. Because the degree of the polynomial $ax^2 + bx + c$ is 2, a quadratic equation is also called a **second-degree equation.**

As we discussed earlier, quadratic equations sometimes can be solved by using the Principle of Zero Products. This method is reviewed here.

> **The Principle of Zero Products**
>
> If a and b are real numbers and $ab = 0$, then $a = 0$ or $b = 0$.

The Principle of Zero Products states that if the product of two factors is zero, then at least one of the factors must be zero.

TAKE NOTE

Recall that the steps involved in solving a quadratic equation by factoring are:
1. Write the equation in standard form.
2. Factor.
3. Use the Principle of Zero Products to set each factor equal to 0.
4. Solve each equation.
5. Check the solutions.

⇒ Solve by factoring: $3x^2 = 2 - 5x$

$$\begin{aligned} 3x^2 &= 2 - 5x \\ 3x^2 + 5x - 2 &= 0 \\ (3x - 1)(x + 2) &= 0 \end{aligned}$$

$3x - 1 = 0 \qquad\qquad x + 2 = 0$

$3x = 1 \qquad\qquad x = -2$

$x = \dfrac{1}{3}$

- Write the equation in standard form.
- Factor.
- Use the Principle of Zero Products to write two equations.
- Solve each equation.

$\dfrac{1}{3}$ and -2 check as solutions. The solutions are $\dfrac{1}{3}$ and -2.

TAKE NOTE

When a quadratic equation has two solutions that are the same number, the solution is called a **double root** of the equation. 3 is a double root of $x^2 - 6x = -9$.

⇒ Solve by factoring: $x^2 - 6x = -9$

$$\begin{aligned} x^2 - 6x &= -9 \\ x^2 - 6x + 9 &= 0 \\ (x - 3)(x - 3) &= 0 \end{aligned}$$

$x - 3 = 0 \qquad\qquad x - 3 = 0$

$x = 3 \qquad\qquad x = 3$

- Write the equation in standard form.
- Factor.
- Use the Principle of Zero Products.
- Solve each equation.

3 checks as a solution. The solution is 3.

Example 1
Solve by factoring: $2x(x - 3) = x + 4$

Solution

$$2x(x - 3) = x + 4$$
$$2x^2 - 6x = x + 4$$
$$2x^2 - 7x - 4 = 0$$
$$(2x + 1)(x - 4) = 0$$

$$2x + 1 = 0 \qquad x - 4 = 0$$
$$2x = -1 \qquad x = 4$$
$$x = -\frac{1}{2}$$

The solutions are $-\frac{1}{2}$ and 4.

You Try It 1
Solve by factoring: $2x^2 = 7x - 3$

Your solution

Example 2
Solve for x by factoring: $x^2 - 4ax - 5a^2 = 0$

Solution
This is a literal equation. Solve for x in terms of a.

$$x^2 - 4ax - 5a^2 = 0$$
$$(x + a)(x - 5a) = 0$$
$$x + a = 0 \qquad x - 5a = 0$$
$$x = -a \qquad x = 5a$$

The solutions are $-a$ and $5a$.

You Try It 2
Solve for x by factoring: $x^2 - 3ax - 4a^2 = 0$

Your solution

Solutions on p. S24

Objective B   **To write a quadratic equation given its solutions**   8

As shown below, the solutions of the equation $(x - r_1)(x - r_2) = 0$ are r_1 and r_2.

$$(x - r_1)(x - r_2) = 0$$

$$x - r_1 = 0 \qquad\qquad x - r_2 = 0$$
$$x = r_1 \qquad\qquad x = r_2$$

Check:

$(x - r_1)(x - r_2) = 0$		$(x - r_1)(x - r_2) = 0$	
$(r_1 - r_1)(r_1 - r_2)$	0	$(r_2 - r_1)(r_2 - r_2)$	0
$0 \cdot (r_1 - r_2)$	0	$(r_2 - r_1) \cdot 0$	0
$0 = 0$		$0 = 0$	

Using the equation $(x - r_1)(x - r_2) = 0$ and the fact that r_1 and r_2 are solutions of this equation, it is possible to write a quadratic equation given its solutions.

⇒ Write a quadratic equation that has solutions 4 and -5.

$$(x - r_1)(x - r_2) = 0$$
$$(x - 4)[x - (-5)] = 0$$
$$(x - 4)(x + 5) = 0$$
$$x^2 + x - 20 = 0$$

- Replace r_1 by 4 and r_2 by -5.
- Simplify.
- Multiply.

⇒ Write a quadratic equation with integer coefficients and solutions $\frac{2}{3}$ and $\frac{1}{2}$.

$$(x - r_1)(x - r_2) = 0$$

$$\left(x - \frac{2}{3}\right)\left(x - \frac{1}{2}\right) = 0$$

$$x^2 - \frac{7}{6}x + \frac{1}{3} = 0$$

$$6\left(x^2 - \frac{7}{6}x + \frac{1}{3}\right) = 6 \cdot 0$$

$$6x^2 - 7x + 2 = 0$$

- Replace r_1 by $\frac{2}{3}$ and r_2 by $\frac{1}{2}$.
- Multiply.
- Multiply each side of the equation by the LCM of the denominators.

Example 3

Write a quadratic equation with integer coefficients and solutions $\frac{1}{2}$ and -4.

Solution

$$(x - r_1)(x - r_2) = 0$$

$$\left(x - \frac{1}{2}\right)[x - (-4)] = 0$$

$$\left(x - \frac{1}{2}\right)(x + 4) = 0$$

$$x^2 + \frac{7}{2}x - 2 = 0$$

$$2\left(x^2 + \frac{7}{2}x - 2\right) = 2 \cdot 0$$

$$2x^2 + 7x - 4 = 0$$

You Try It 3

Write a quadratic equation with integer coefficients and solutions 3 and $-\frac{1}{2}$.

Your solution

Solution on p. S24

Objective C   **To solve a quadratic equation by taking square roots**   8

The solution of the quadratic equation $x^2 = 16$ is shown at the right.

$$x^2 = 16$$
$$x^2 - 16 = 0$$
$$(x - 4)(x + 4) = 0$$
$$x - 4 = 0 \qquad x + 4 = 0$$
$$x = 4 \qquad x = -4$$

The solution can also be found by taking the square root of each side of the equation and writing the positive and negative square roots of the radicand. The notation ± 4 means $x = 4$ or $x = -4$.

$$x^2 = 16$$
$$\sqrt{x^2} = \sqrt{16}$$
$$x = \pm 4$$

The solutions are 4 and -4.

⇒ Solve by taking square roots: $3x^2 = 54$

$$3x^2 = 54$$
$$x^2 = 18$$
$$\sqrt{x^2} = \sqrt{18}$$
$$x = \pm\sqrt{18}$$
$$x = \pm 3\sqrt{2}$$

- Solve for x^2.
- Take the square root of each side of the equation.
- Simplify.

The solutions are $3\sqrt{2}$ and $-3\sqrt{2}$.   - $3\sqrt{2}$ and $-3\sqrt{2}$ check as solutions.

Solving a quadratic equation by taking the square root of each side of the equation can lead to solutions that are complex numbers.

➡ Solve by taking square roots: $2x^2 + 18 = 0$

$$2x^2 + 18 = 0$$
$$2x^2 = -18 \qquad \bullet \text{ Solve for } x^2.$$
$$x^2 = -9$$
$$\sqrt{x^2} = \sqrt{-9} \qquad \bullet \text{ Take the square root of each side of the equation.}$$
$$x = \pm\sqrt{-9} \qquad \bullet \text{ Simplify.}$$
$$x = \pm 3i$$

Check:

$$\begin{array}{c|c} 2x^2 + 18 = 0 & 2x^2 + 18 = 0 \\ \hline 2(3i)^2 + 18 \;\big|\; 0 & 2(-3i)^2 + 18 \;\big|\; 0 \\ 2(-9) + 18 \;\big|\; 0 & 2(-9) + 18 \;\big|\; 0 \\ -18 + 18 \;\big|\; 0 & -18 + 18 \;\big|\; 0 \\ 0 = 0 & 0 = 0 \end{array}$$

The solutions are $3i$ and $-3i$.

An equation containing the square of a binomial can be solved by taking square roots.

➡ Solve by taking square roots: $(x + 2)^2 - 24 = 0$

$$(x + 2)^2 - 24 = 0$$
$$(x + 2)^2 = 24 \qquad \bullet \text{ Solve for } (x + 2)^2.$$
$$\sqrt{(x + 2)^2} = \sqrt{24} \qquad \bullet \text{ Take the square root of each}$$
$$x + 2 = \pm\sqrt{24} \qquad \text{side of the equation. Then}$$
$$x + 2 = \pm 2\sqrt{6} \qquad \text{simplify.}$$

$$x + 2 = 2\sqrt{6} \qquad x + 2 = -2\sqrt{6} \qquad \bullet \text{ Solve for } x.$$
$$x = -2 + 2\sqrt{6} \qquad x = -2 - 2\sqrt{6}$$

The solutions are $-2 + 2\sqrt{6}$ and $-2 - 2\sqrt{6}$.

Example 4

Solve by taking square roots.
$3(x - 2)^2 + 12 = 0$

Solution

$$3(x - 2)^2 + 12 = 0$$
$$3(x - 2)^2 = -12$$
$$(x - 2)^2 = -4$$
$$\sqrt{(x - 2)^2} = \sqrt{-4}$$
$$x - 2 = \pm\sqrt{-4}$$
$$x - 2 = \pm 2i$$

$$x - 2 = 2i \qquad x - 2 = -2i$$
$$x = 2 + 2i \qquad x = 2 - 2i$$

The solutions are $2 + 2i$ and $2 - 2i$.

You Try It 4

Solve by taking square roots.
$2(x + 1)^2 - 24 = 0$

Your solution

Solution on p. S24

8.1 Exercises

Objective A

1. Explain why the restriction $a \neq 0$ is necessary in the definition of a quadratic equation.

2. What does the Principle of Zero Products state? How is it used to solve a quadratic equation?

Write the quadratic equation in standard form with the coefficient of x^2 positive. Name the values of a, b, and c.

3. $2x^2 - 4x = 5$ **4.** $x^2 = 3x + 1$ **5.** $5x = 4x^2 + 6$ **6.** $3x^2 = 7$

Solve by factoring.

7. $x^2 - 4x = 0$ **8.** $y^2 + 6y = 0$ **9.** $t^2 - 25 = 0$

10. $p^2 - 81 = 0$ **11.** $s^2 - s - 6 = 0$ **12.** $v^2 + 4v - 5 = 0$

13. $y^2 - 6y + 9 = 0$ **14.** $x^2 + 10x + 25 = 0$ **15.** $9z^2 - 18z = 0$

16. $4y^2 + 20y = 0$ **17.** $r^2 - 3r = 10$ **18.** $p^2 + 5p = 6$

19. $v^2 + 10 = 7v$ **20.** $t^2 - 16 = 15t$ **21.** $2x^2 - 9x - 18 = 0$

22. $3y^2 - 4y - 4 = 0$ **23.** $4z^2 - 9z + 2 = 0$ **24.** $2s^2 - 9s + 9 = 0$

25. $3w^2 + 11w = 4$ **26.** $2r^2 + r = 6$ **27.** $6x^2 = 23x + 18$

28. $6x^2 = 7x - 2$ **29.** $4 - 15u - 4u^2 = 0$ **30.** $3 - 2y - 8y^2 = 0$

31. $x + 18 = x(x - 6)$ **32.** $t + 24 = t(t + 6)$ **33.** $4s(s + 3) = s - 6$

34. $3v(v - 2) = 11v + 6$

35. $u^2 - 2u + 4 = (2u - 3)(u + 2)$

36. $(3v - 2)(2v + 1) = 3v^2 - 11v - 10$

37. $(3x - 4)(x + 4) = x^2 - 3x - 28$

Solve for x by factoring.

38. $x^2 + 14ax + 48a^2 = 0$

39. $x^2 - 9bx + 14b^2 = 0$

40. $x^2 + 9xy - 36y^2 = 0$

41. $x^2 - 6cx - 7c^2 = 0$

42. $x^2 - ax - 20a^2 = 0$

43. $2x^2 + 3bx + b^2 = 0$

44. $3x^2 - 4cx + c^2 = 0$

45. $3x^2 - 14ax + 8a^2 = 0$

46. $3x^2 - 11xy + 6y^2 = 0$

47. $3x^2 - 8ax - 3a^2 = 0$

48. $3x^2 - 4bx - 4b^2 = 0$

49. $4x^2 + 8xy + 3y^2 = 0$

50. $6x^2 - 11cx + 3c^2 = 0$

51. $6x^2 + 11ax + 4a^2 = 0$

52. $12x^2 - 5xy - 2y^2 = 0$

Objective B

Write a quadratic equation that has integer coefficients and has as solutions the given pair of numbers.

53. 2 and 5

54. 3 and 1

55. -2 and -4

56. -1 and -3

57. 6 and -1

58. -2 and 5

59. 3 and -3

60. 5 and -5

61. 4 and 4

62. 2 and 2

63. 0 and 5

64. 0 and -2

65. 0 and 3

66. 0 and -1

67. 3 and $\dfrac{1}{2}$

68. 2 and $\dfrac{2}{3}$

69. $-\dfrac{3}{4}$ and 2

70. $-\dfrac{1}{2}$ and 5

71. $-\dfrac{5}{3}$ and -2

72. $-\dfrac{3}{2}$ and -1

73. $-\dfrac{2}{3}$ and $\dfrac{2}{3}$

74. $-\dfrac{1}{2}$ and $\dfrac{1}{2}$

75. $\dfrac{1}{2}$ and $\dfrac{1}{3}$

76. $\dfrac{3}{4}$ and $\dfrac{2}{3}$

77. $\dfrac{6}{5}$ and $-\dfrac{1}{2}$

78. $\dfrac{3}{4}$ and $-\dfrac{3}{2}$

79. $-\dfrac{1}{4}$ and $-\dfrac{1}{2}$

80. $-\dfrac{5}{6}$ and $-\dfrac{2}{3}$

81. $\dfrac{3}{5}$ and $-\dfrac{1}{10}$

82. $\dfrac{7}{2}$ and $-\dfrac{1}{4}$

Objective C

Solve by taking square roots.

83. $y^2 = 49$

84. $x^2 = 64$

85. $z^2 = -4$

86. $v^2 = -16$

87. $s^2 - 4 = 0$

88. $r^2 - 36 = 0$

89. $4x^2 - 81 = 0$

90. $9x^2 - 16 = 0$

91. $y^2 + 49 = 0$

92. $z^2 + 16 = 0$

93. $v^2 - 48 = 0$

94. $s^2 - 32 = 0$

95. $r^2 - 75 = 0$

96. $u^2 - 54 = 0$

97. $z^2 + 18 = 0$

98. $t^2 + 27 = 0$

99. $(x - 1)^2 = 36$

100. $(x + 2)^2 = 25$

101. $3(y + 3)^2 = 27$

102. $4(s - 2)^2 = 36$

103. $5(z + 2)^2 = 125$

104. $2(y - 3)^2 = 18$

105. $\left(v - \dfrac{1}{2}\right)^2 = \dfrac{1}{4}$

106. $\left(r + \dfrac{2}{3}\right)^2 = \dfrac{1}{9}$

107. $(x + 5)^2 - 6 = 0$

108. $(t - 1)^2 - 15 = 0$

109. $(v - 3)^2 + 45 = 0$

110. $(x + 5)^2 + 32 = 0$

111. $\left(u + \dfrac{2}{3}\right)^2 - 18 = 0$

112. $\left(z - \dfrac{1}{2}\right)^2 - 20 = 0$

APPLYING THE CONCEPTS

Write a quadratic equation that has as solutions the given pair of numbers.

113. $\sqrt{2}$ and $-\sqrt{2}$ **114.** $\sqrt{5}$ and $-\sqrt{5}$ **115.** i and $-i$ **116.** $2i$ and $-2i$

117. $2\sqrt{2}$ and $-2\sqrt{2}$ **118.** $3\sqrt{2}$ and $-3\sqrt{2}$ **119.** $i\sqrt{2}$ and $-i\sqrt{2}$ **120.** $2i\sqrt{3}$ and $-2i\sqrt{3}$

Solve for x.

121. $4a^2x^2 = 36b^2, a > 0, b > 0$ **122.** $3y^2x^2 = 27z^2, y > 0, z > 0$ **123.** $(x + a)^2 - 4 = 0$

124. $(x - b)^2 - 1 = 0$ **125.** $(2x - 1)^2 = (2x + 3)^2$ **126.** $(x - 4)^2 = (x + 2)^2$

127. Show that the solutions of the equation $ax^2 + bx = 0, a > 0, b > 0$, are 0 and $-\dfrac{b}{a}$.

128. Show that the solutions of the equation $ax^2 + c = 0, a > 0, c > 0$, are $\dfrac{\sqrt{ca}}{a}i$ and $-\dfrac{\sqrt{ca}}{a}i$.

8.2 Solving Quadratic Equations by Completing the Square

Objective A   To solve a quadratic equation by completing the square

Recall that a perfect-square trinomial is the square of a binomial.

Perfect-Square Trinomial		Square of a Binomial
$x^2 + 8x + 16$	$=$	$(x + 4)^2$
$x^2 - 10x + 25$	$=$	$(x - 5)^2$
$x^2 + 2ax + a^2$	$=$	$(x + a)^2$

For each perfect-square trinomial, the square of $\frac{1}{2}$ of the coefficient of x equals the constant term.

$$\left(\frac{1}{2} \text{ coefficient of } x\right)^2 = \text{constant term}$$

$x^2 + 8x + 16, \quad \left(\frac{1}{2} \cdot 8\right)^2 = 16$

$x^2 - 10x + 25, \quad \left[\frac{1}{2}(-10)\right]^2 = 25$

$x^2 + 2ax + a^2, \quad \left(\frac{1}{2} \cdot 2a\right)^2 = a^2$

This relationship can be used to write the constant term for a perfect-square trinomial. Adding to a binomial the constant term that makes it a perfect-square trinomial is called **completing the square.**

➡ Complete the square on $x^2 + 12x$. Write the resulting perfect-square trinomial as the square of a binomial.

$\left[\frac{1}{2}(12)\right]^2 = (6)^2 = 36$ • Find the constant term.

$x^2 + 12x + 36$ • Complete the square on $x^2 + 12x$ by adding the constant term.

$x^2 + 12x + 36 = (x + 6)^2$ • Write the resulting perfect-square trinomial as the square of a binomial.

➡ Complete the square on $z^2 - 3z$. Write the resulting perfect-square trinomial as the square of a binomial.

$\left[\frac{1}{2} \cdot (-3)\right]^2 = \left(-\frac{3}{2}\right)^2 = \frac{9}{4}$ • Find the constant term.

$z^2 - 3z + \frac{9}{4}$ • Complete the square on $z^2 - 3z$ by adding the constant term.

$z^2 - 3z + \frac{9}{4} = \left(z - \frac{3}{2}\right)^2$ • Write the resulting perfect-square trinomial as the square of a binomial.

Any quadratic equation can be solved by completing the square. Add to each side of the equation the term that completes the square. Rewrite the equation in the form $(x + a)^2 = b$. Then take the square root of each side of the equation.

Point of Interest

Early attempts to solve quadratic equations were primarily geometric. The Persian mathematician al-Khowarizmi (c. A.D. 800) essentially completed a square of $x^2 + 12x$ as follows.

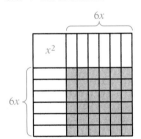

⇒ Solve by completing the square: $x^2 - 6x - 15 = 0$

$$x^2 - 6x - 15 = 0$$

$$x^2 - 6x = 15$$
- Add 15 to each side of the equation.

$$x^2 - 6x + 9 = 15 + 9$$
- Complete the square. Add $\left[\dfrac{1}{2}(-6)\right]^2 = (-3)^2 = 9$ to each side of the equation.

$$(x - 3)^2 = 24$$
- Factor the perfect-square trinomial.

$$\sqrt{(x - 3)^2} = \sqrt{24}$$
- Take the square root of each side of the equation.

$$x - 3 = \pm\sqrt{24}$$
- Solve for x.

$$x - 3 = \pm2\sqrt{6}$$

$$x - 3 = 2\sqrt{6} \qquad\qquad x - 3 = -2\sqrt{6}$$

$$x = 3 + 2\sqrt{6} \qquad\qquad x = 3 - 2\sqrt{6}$$

Check:

$$
\begin{array}{c|c}
x^2 - 6x - 15 = 0 & \\
\hline
(3 + 2\sqrt{6})^2 - 6(3 + 2\sqrt{6}) - 15 & 0 \\
9 + 12\sqrt{6} + 24 - 18 - 12\sqrt{6} - 15 & 0 \\
0 = 0 &
\end{array}
$$

$$
\begin{array}{c|c}
x^2 - 6x - 15 = 0 & \\
\hline
(3 - 2\sqrt{6})^2 - 6(3 - 2\sqrt{6}) - 15 & 0 \\
9 - 12\sqrt{6} + 24 - 18 + 12\sqrt{6} - 15 & 0 \\
0 = 0 &
\end{array}
$$

The solutions are $3 + 2\sqrt{6}$ and $3 - 2\sqrt{6}$.

TAKE NOTE

The exact solutions of the equation $x^2 - 6x - 15 = 0$ are $3 + 2\sqrt{6}$ and $3 - 2\sqrt{6}$. 7.899 and -1.899 are approximate solutions of the equation.

In the example above, the solutions of the equation $x^2 - 6x - 15 = 0$ are $3 + 2\sqrt{6}$ and $3 - 2\sqrt{6}$. These are the exact solutions. However, in some situations it may be preferable to have decimal approximations of the solutions of a quadratic equation. Approximate solutions can be found by using a calculator and then rounding to the desired degree of accuracy.

$$3 + 2\sqrt{6} \approx 7.899 \text{ and } 3 - 2\sqrt{6} \approx -1.899$$

To the nearest thousandth, the approximate solutions of the equation $x^2 - 6x - 15 = 0$ are 7.899 and -1.899.

⇒ Solve $2x^2 - x - 2 = 0$ by completing the square. Find the exact solutions and approximate the solutions to the nearest thousandth.

In order for us to complete the square on an expression, the coefficient of the squared term must be 1. After adding the constant term to each side of the equation, multiply each side of the equation by $\dfrac{1}{2}$.

$$2x^2 - x - 2 = 0$$

$$2x^2 - x = 2$$
- Add 2 to each side of the equation.

$$\dfrac{1}{2}(2x^2 - x) = \dfrac{1}{2} \cdot 2$$
- Multiply each side of the equation by $\dfrac{1}{2}$.

TAKE NOTE

This example illustrates all the steps required to solve a quadratic equation by completing the square.

1. Write the equation in the form $ax^2 + bx = -c$.
2. Multiply both sides of the equation by $\frac{1}{a}$.
3. Complete the square on $x^2 + \frac{b}{a}x$. Add the number that completes the square to both sides of the equation.
4. Factor the perfect-square trinomial.
5. Take the square root of each side of the equation.
6. Solve the resulting equation for x.

$$x^2 - \frac{1}{2}x = 1$$

• The coefficient of x^2 is now 1.

$$x^2 - \frac{1}{2}x + \frac{1}{16} = 1 + \frac{1}{16}$$

• Complete the square. Add $$\left[\frac{1}{2}\left(-\frac{1}{2}\right)\right]^2 = \left(-\frac{1}{4}\right)^2 = \frac{1}{16}$$ to each side of the equation.

$$\left(x - \frac{1}{4}\right)^2 = \frac{17}{16}$$

• Factor the perfect-square trinomial.

$$\sqrt{\left(x - \frac{1}{4}\right)^2} = \sqrt{\frac{17}{16}}$$

• Take the square root of each side of the equation.

$$x - \frac{1}{4} = \pm\frac{\sqrt{17}}{4}$$

• Solve for x.

$$x - \frac{1}{4} = \frac{\sqrt{17}}{4} \qquad x - \frac{1}{4} = -\frac{\sqrt{17}}{4}$$

$$x = \frac{1}{4} + \frac{\sqrt{17}}{4} \qquad x = \frac{1}{4} - \frac{\sqrt{17}}{4}$$

The exact solutions are $\frac{1 + \sqrt{17}}{4}$ and $\frac{1 - \sqrt{17}}{4}$.

$$\frac{1 + \sqrt{17}}{4} \approx 1.281 \qquad \frac{1 - \sqrt{17}}{4} \approx -0.781$$

To the nearest thousandth, the solutions are 1.281 and -0.781.

➡ Solve by completing the square: $x^2 - 3x + 5 = 0$

$$x^2 - 3x + 5 = 0$$

$$x^2 - 3x = -5$$

• Add the opposite of the constant term to each side of the equation.

$$x^2 - 3x + \frac{9}{4} = -5 + \frac{9}{4}$$

• Add to each side of the equation the term that completes the square on $x^2 - 3x$.

$$\left(x - \frac{3}{2}\right)^2 = -\frac{11}{4}$$

• Factor the perfect-square trinomial.

$$\sqrt{\left(x - \frac{3}{2}\right)^2} = \sqrt{-\frac{11}{4}}$$

• Take the square root of each side of the equation.

$$x - \frac{3}{2} = \pm\frac{i\sqrt{11}}{2}$$

• Simplify.

$$x - \frac{3}{2} = \frac{i\sqrt{11}}{2} \qquad x - \frac{3}{2} = -\frac{i\sqrt{11}}{2}$$

• Solve for x.

$$x = \frac{3}{2} + \frac{i\sqrt{11}}{2} \qquad x = \frac{3}{2} - \frac{i\sqrt{11}}{2}$$

$\frac{3}{2} + \frac{\sqrt{11}}{2}i$ and $\frac{3}{2} - \frac{\sqrt{11}}{2}i$ check as solutions.

The solutions are $\frac{3}{2} + \frac{\sqrt{11}}{2}i$ and $\frac{3}{2} - \frac{\sqrt{11}}{2}i$.

Point of Interest

Mathematicians have studied quadratic equations for centuries. Many of the initial equations were a result of trying to solve a geometry problem. One of the most famous, which dates from around 500 B.C., is "squaring the circle." The question was "Is it possible to construct a square whose area is that of a given circle?" For these early mathematicians, to *construct* meant to draw with only a straightedge and a compass. It was approximately 2300 years later that mathematicians were able to prove that such a construction is impossible.

Example 1

Solve by completing the square:
$4x^2 - 8x + 1 = 0$

Solution

$$4x^2 - 8x + 1 = 0$$
$$4x^2 - 8x = -1$$
$$\frac{1}{4}(4x^2 - 8x) = \frac{1}{4}(-1)$$
$$x^2 - 2x = -\frac{1}{4}$$
$$x^2 - 2x + 1 = -\frac{1}{4} + 1 \qquad \text{• Complete the square.}$$
$$(x - 1)^2 = \frac{3}{4}$$
$$\sqrt{(x - 1)^2} = \sqrt{\frac{3}{4}}$$
$$x - 1 = \pm\frac{\sqrt{3}}{2}$$

$$x - 1 = \frac{\sqrt{3}}{2} \qquad\qquad x - 1 = -\frac{\sqrt{3}}{2}$$
$$x = 1 + \frac{\sqrt{3}}{2} \qquad\qquad x = 1 - \frac{\sqrt{3}}{2}$$
$$= \frac{2 + \sqrt{3}}{2} \qquad\qquad = \frac{2 - \sqrt{3}}{2}$$

The solutions are $\dfrac{2 + \sqrt{3}}{2}$ and $\dfrac{2 - \sqrt{3}}{2}$.

You Try It 1

Solve by completing the square:
$4x^2 - 4x - 1 = 0$

Your solution

Example 2

Solve by completing the square:
$x^2 + 4x + 5 = 0$

Solution

$$x^2 + 4x + 5 = 0$$
$$x^2 + 4x = -5$$
$$x^2 + 4x + 4 = -5 + 4 \qquad \text{• Complete the square.}$$
$$(x + 2)^2 = -1$$
$$\sqrt{(x + 2)^2} = \sqrt{-1}$$
$$x + 2 = \pm i$$

$$x + 2 = i \qquad\qquad x + 2 = -i$$
$$x = -2 + i \qquad\qquad x = -2 - i$$

The solutions are $-2 + i$ and $-2 - i$.

You Try It 2

Solve by completing the square:
$2x^2 + x - 5 = 0$

Your solution

Solutions on pp. S24–S25

8.2 Exercises

Objective A

Solve by completing the square.

1. $x^2 - 4x - 5 = 0$

2. $y^2 + 6y + 5 = 0$

3. $v^2 + 8v - 9 = 0$

4. $w^2 - 2w - 24 = 0$

5. $z^2 - 6z + 9 = 0$

6. $u^2 + 10u + 25 = 0$

7. $r^2 + 4r - 7 = 0$

8. $s^2 + 6s - 1 = 0$

9. $x^2 - 6x + 7 = 0$

10. $y^2 + 8y + 13 = 0$

11. $z^2 - 2z + 2 = 0$

12. $t^2 - 4t + 8 = 0$

13. $s^2 - 5s - 24 = 0$

14. $v^2 + 7v - 44 = 0$

15. $x^2 + 5x - 36 = 0$

16. $y^2 - 9y + 20 = 0$

17. $p^2 - 3p + 1 = 0$

18. $r^2 - 5r - 2 = 0$

19. $t^2 - t - 1 = 0$

20. $u^2 - u - 7 = 0$

21. $y^2 - 6y = 4$

22. $w^2 + 4w = 2$

23. $x^2 = 8x - 15$

24. $z^2 = 4z - 3$

25. $v^2 = 4v - 13$

26. $x^2 = 2x - 17$

27. $p^2 + 6p = -13$

28. $x^2 + 4x = -20$

29. $y^2 - 2y = 17$

30. $x^2 + 10x = 7$

31. $z^2 = z + 4$

32. $r^2 = 3r - 1$

33. $x^2 + 13 = 2x$

34. $6v^2 - 7v = 3$

35. $4x^2 - 4x + 5 = 0$

36. $4t^2 - 4t + 17 = 0$

37. $9x^2 - 6x + 2 = 0$

38. $9y^2 - 12y + 13 = 0$

39. $2s^2 = 4s + 5$

40. $3u^2 = 6u + 1$

41. $2r^2 = 3 - r$

42. $2x^2 = 12 - 5x$

43. $y - 2 = (y - 3)(y + 2)$

44. $8s - 11 = (s - 4)(s - 2)$

45. $6t - 2 = (2t - 3)(t - 1)$

46. $2z + 9 = (2z + 3)(z + 2)$

47. $(x - 4)(x + 1) = x - 3$

48. $(y - 3)^2 = 2y + 10$

Solve by completing the square. Approximate the solutions to the nearest thousandth.

49. $z^2 + 2z = 4$

50. $t^2 - 4t = 7$

51. $2x^2 = 4x - 1$

52. $3y^2 = 5y - 1$

53. $4z^2 + 2z = 1$

54. $4w^2 - 8w = 3$

APPLYING THE CONCEPTS

Solve for x by completing the square.

55. $x^2 - ax - 2a^2 = 0$

56. $x^2 + 3ax - 4a^2 = 0$

57. $x^2 + 3ax - 10a^2 = 0$

58. After a baseball is hit, the height h (in feet) of the ball above the ground t seconds after it is hit can be approximated by the equation $h = -16t^2 + 70t + 4$. Using this equation, determine when the ball will hit the ground. (*Hint:* The ball hits the ground when $h = 0$.)

59. After a baseball is hit, there are two equations that can be considered. One gives the height h (in feet) the ball is above the ground t seconds after it is hit. The second is the distance s (in feet) the ball is from home plate t seconds after it is hit. A model of this situation is given by $h = -16t^2 + 70t + 4$ and $s = 44.5t$. Using this model, determine whether the ball will clear a fence 325 ft from home plate.

60. Explain how to complete the square on $x^2 + bx$.

8.3

Solving Quadratic Equations by Using the Quadratic Formula

Objective A

To solve a quadratic equation by using the quadratic formula

A general formula known as the **quadratic formula** can be derived by applying the method of completing the square to the standard form of a quadratic equation. This formula can be used to solve any quadratic equation. The equation $ax^2 + bx + c = 0$ is solved by completing the square as follows.

Add the opposite of the constant term to each side of the equation.

$$ax^2 + bx + c = 0$$
$$ax^2 + bx + c + (-c) = 0 + (-c)$$

$$ax^2 + bx = -c$$

Multiply each side of the equation by the reciprocal of a, the coefficient of x^2.

$$\frac{1}{a}(ax^2 + bx) = \frac{1}{a}(-c)$$

$$x^2 + \frac{b}{a}x = -\frac{c}{a}$$

Complete the square by adding $\left(\frac{1}{2} \cdot \frac{b}{a}\right)^2 = \frac{b^2}{4a^2}$ to each side of the equation.

$$x^2 + \frac{b}{a}x + \frac{b^2}{4a^2} = \frac{b^2}{4a^2} - \frac{c}{a}$$

Simplify the right side of the equation.

$$x^2 + \frac{b}{a}x + \frac{b^2}{4a^2} = \frac{b^2}{4a^2} - \left(\frac{c}{a} \cdot \frac{4a}{4a}\right)$$

$$x^2 + \frac{b}{a}x + \frac{b^2}{4a^2} = \frac{b^2}{4a^2} - \frac{4ac}{4a^2}$$

$$x^2 + \frac{b}{a}x + \frac{b^2}{4a^2} = \frac{b^2 - 4ac}{4a^2}$$

Factor the perfect-square trinomial on the left side of the equation.

$$\left(x + \frac{b}{2a}\right)^2 = \frac{b^2 - 4ac}{4a^2}$$

Take the square root of each side of the equation.

$$\sqrt{\left(x + \frac{b}{2a}\right)^2} = \sqrt{\frac{b^2 - 4ac}{4a^2}}$$

$$x + \frac{b}{2a} = \pm\frac{\sqrt{b^2 - 4ac}}{2a}$$

Solve for x.

$$x + \frac{b}{2a} = \frac{\sqrt{b^2 - 4ac}}{2a} \qquad x + \frac{b}{2a} = -\frac{\sqrt{b^2 - 4ac}}{2a}$$

$$x = -\frac{b}{2a} + \frac{\sqrt{b^2 - 4ac}}{2a} \qquad x = -\frac{b}{2a} - \frac{\sqrt{b^2 - 4ac}}{2a}$$

$$= \frac{-b + \sqrt{b^2 - 4ac}}{2a} \qquad = \frac{-b - \sqrt{b^2 - 4ac}}{2a}$$

Point of Interest

Although mathematicians have studied quadratic equations since around 500 B.C., it was not until the 18th century that the formula was written as it is today. Of further note, the word *quadratic* has the same Latin root as does the word *square*.

The Quadratic Formula

The solutions of $ax^2 + bx + c = 0$, $a \neq 0$, are

$$\frac{-b + \sqrt{b^2 - 4ac}}{2a} \quad \text{and} \quad \frac{-b - \sqrt{b^2 - 4ac}}{2a}$$

The quadratic formula is frequently written as $x = \dfrac{-b \pm \sqrt{b^2 - 4ac}}{2a}$.

⇨ Solve by using the quadratic formula: $2x^2 + 5x + 3 = 0$

$$x = \dfrac{-b \pm \sqrt{b^2 - 4ac}}{2a}$$

$$= \dfrac{-(5) \pm \sqrt{(5)^2 - 4(2)(3)}}{2(2)}$$

$$= \dfrac{-5 \pm \sqrt{25 - 24}}{4}$$

$$= \dfrac{-5 \pm \sqrt{1}}{4} = \dfrac{-5 \pm 1}{4}$$

- The equation $2x^2 + 5x + 3 = 0$ is in standard form. $a = 2$, $b = 5$, $c = 3$
- Replace a, b, and c in the quadratic formula wth these values.

$$x = \dfrac{-5 + 1}{4} = \dfrac{-4}{4} = -1 \qquad x = \dfrac{-5 - 1}{4} = \dfrac{-6}{4} = -\dfrac{3}{2}$$

The solutions are -1 and $-\dfrac{3}{2}$.

⇨ Solve $3x^2 = 4x + 6$ by using the quadratic formula. Find the exact solutions and approximate the solutions to the nearest thousandth.

$$3x^2 = 4x + 6$$
$$3x^2 - 4x - 6 = 0$$

$$x = \dfrac{-b \pm \sqrt{b^2 - 4ac}}{2a}$$

$$= \dfrac{-(-4) \pm \sqrt{(-4)^2 - 4(3)(-6)}}{2(3)}$$

$$= \dfrac{4 \pm \sqrt{16 - (-72)}}{6}$$

$$= \dfrac{4 \pm \sqrt{88}}{6} = \dfrac{4 \pm 2\sqrt{22}}{6}$$

$$= \dfrac{\cancel{2}(2 \pm \sqrt{22})}{\cancel{2} \cdot 3} = \dfrac{2 \pm \sqrt{22}}{3}$$

- Write the equation in standard form. Subtract $4x$ and 6 from each side of the equation.
  $a = 3$, $b = -4$, $c = -6$
- Replace a, b, and c in the quadratic formula with these values.

Check:

$$3x^2 = 4x + 6$$

$3\left(\dfrac{2 + \sqrt{22}}{3}\right)^2$	$4\left(\dfrac{2 + \sqrt{22}}{3}\right) + 6$
$3\left(\dfrac{4 + 4\sqrt{22} + 22}{9}\right)$	$\dfrac{8}{3} + \dfrac{4\sqrt{22}}{3} + \dfrac{18}{3}$
$3\left(\dfrac{26 + 4\sqrt{22}}{9}\right)$	$\dfrac{26}{3} + \dfrac{4\sqrt{22}}{3}$
$\dfrac{26 + 4\sqrt{22}}{3}$	$= \dfrac{26 + 4\sqrt{22}}{3}$

$$3x^2 = 4x + 6$$

$3\left(\dfrac{2 - \sqrt{22}}{3}\right)^2$	$4\left(\dfrac{2 - \sqrt{22}}{3}\right) + 6$
$3\left(\dfrac{4 - 4\sqrt{22} + 22}{9}\right)$	$\dfrac{8}{3} - \dfrac{4\sqrt{22}}{3} + \dfrac{18}{3}$
$3\left(\dfrac{26 - 4\sqrt{22}}{9}\right)$	$\dfrac{26}{3} - \dfrac{4\sqrt{22}}{3}$
$\dfrac{26 - 4\sqrt{22}}{3}$	$= \dfrac{26 - 4\sqrt{22}}{3}$

The exact solutions are $\dfrac{2 + \sqrt{22}}{3}$ and $\dfrac{2 - \sqrt{22}}{3}$.

$$\dfrac{2 + \sqrt{22}}{3} \approx 2.230 \qquad \dfrac{2 - \sqrt{22}}{3} \approx -0.897$$

To the nearest thousandth, the solutions are 2.230 and -0.897.

⇒ Solve by using the quadratic formula: $4x^2 = 8x - 13$

$$4x^2 = 8x - 13$$
$$4x^2 - 8x + 13 = 0$$ • Write the equation in standard form.

$$x = \frac{-b \pm \sqrt{b^2 - 4ac}}{2a}$$ • Use the quadratic formula.

$$= \frac{-(-8) \pm \sqrt{(-8)^2 - 4 \cdot 4 \cdot 13}}{2 \cdot 4}$$ • $a = 4, b = -8, c = 13$

$$= \frac{8 \pm \sqrt{64 - 208}}{8} = \frac{8 \pm \sqrt{-144}}{8}$$

$$= \frac{8 \pm 12i}{8} = \frac{2 \pm 3i}{2}$$

The solutions are $1 + \frac{3}{2}i$ and $1 - \frac{3}{2}i$.

Of the three preceding examples, the first two had real number solutions; the last one had complex number solutions.

In the quadratic formula, the quantity $b^2 - 4ac$ is called the **discriminant.** When a, b, and c are real numbers, the discriminant determines whether a quadratic equation will have a double root, two real number solutions that are not equal, or two complex number solutions.

The Effect of the Discriminant on the Solutions of a Quadratic Equation

1. If $b^2 - 4ac = 0$, the equation has one real number solution, a double root.

2. If $b^2 - 4ac > 0$, the equation has two unequal real number solutions.

3. If $b^2 - 4ac < 0$, the equation has two complex number solutions.

⇒ Use the discriminant to determine whether $x^2 - 4x - 5 = 0$ has one real number solution, two real number solutions, or two complex number solutions.

$$b^2 - 4ac$$ • Evaluate the discriminant.
$$(-4)^2 - 4(1)(-5) = 16 + 20 = 36$$ $a = 1, b = -4, c = -5$
$$36 > 0$$

Because $b^2 - 4ac > 0$, the equation has two real number solutions.

. . . .

Example 1   Solve by using the quadratic formula: $2x^2 - x + 5 = 0$

You Try It 1   Solve by using the quadratic formula: $x^2 - 2x + 10 = 0$

Solution

$$2x^2 - x + 5 = 0$$
$$a = 2, b = -1, c = 5$$
$$x = \frac{-b \pm \sqrt{b^2 - 4ac}}{2a}$$

$$= \frac{-(-1) \pm \sqrt{(-1)^2 - 4(2)(5)}}{2 \cdot 2}$$

$$= \frac{1 \pm \sqrt{1 - 40}}{4} = \frac{1 \pm \sqrt{-39}}{4}$$

$$= \frac{1 \pm i\sqrt{39}}{4}$$

Your solution

The solutions are $\frac{1}{4} + \frac{\sqrt{39}}{4}i$ and $\frac{1}{4} - \frac{\sqrt{39}}{4}i$.

Solution on p. S25

Example 2

Solve by using the quadratic formula:
$2x^2 = (x - 2)(x - 3)$

Solution

$2x^2 = (x - 2)(x - 3)$

$2x^2 = x^2 - 5x + 6$

$x^2 + 5x - 6 = 0$

$a = 1, b = 5, c = -6$

$$x = \frac{-b \pm \sqrt{b^2 - 4ac}}{2a}$$

$$= \frac{-5 \pm \sqrt{5^2 - 4(1)(-6)}}{2 \cdot 1}$$

$$= \frac{-5 \pm \sqrt{25 + 24}}{2} = \frac{-5 \pm \sqrt{49}}{2}$$

$$= \frac{-5 \pm 7}{2}$$

$$x = \frac{-5 + 7}{2} \qquad x = \frac{-5 - 7}{2}$$

$$= \frac{2}{2} = 1 \qquad = \frac{-12}{2} = -6$$

The solutions are 1 and -6.

You Try It 2

Solve by using the quadratic formula:
$4x^2 = 4x - 1$

Your solution

Example 3

Use the discriminant to determine whether $4x^2 - 2x + 5 = 0$ has one real number solution, two real number solutions, or two complex number solutions.

Solution

$a = 4, b = -2, c = 5$

$b^2 - 4ac = (-2)^2 - 4(4)(5)$

$\qquad\quad = 4 - 80$

$\qquad\quad = -76$

$-76 < 0$

Because the discriminant is less than zero, the equation has two complex number solutions.

You Try It 3

Use the discriminant to determine whether $3x^2 - x - 1 = 0$ has one real number solution, two real number solutions, or two complex number solutions.

Your solution

Solutions on p. S25

8.3 Exercises

Objective A

1. Write the quadratic formula. What does each variable in the formula represent?

2. Write the expression that appears under the radical symbol in the quadratic formula. What is this quantity called? What can it be used to determine?

Solve by using the quadratic formula.

3. $x^2 - 3x - 10 = 0$

4. $z^2 - 4z - 8 = 0$

5. $y^2 + 5y - 36 = 0$

6. $z^2 - 3z - 40 = 0$

7. $w^2 = 8w + 72$

8. $t^2 = 2t + 35$

9. $v^2 = 24 - 5v$

10. $x^2 = 18 - 7x$

11. $2y^2 + 5y - 1 = 0$

12. $4p^2 - 7p + 1 = 0$

13. $8s^2 = 10s + 3$

14. $12t^2 = 5t + 2$

15. $x^2 = 14x - 4$

16. $v^2 = 12v - 24$

17. $2z^2 - 2z - 1 = 0$

18. $6w^2 = 9w - 1$

19. $z^2 + 2z + 2 = 0$

20. $p^2 - 4p + 5 = 0$

21. $y^2 - 2y + 5 = 0$

22. $x^2 + 6x + 13 = 0$

23. $s^2 - 4s + 13 = 0$

24. $t^2 - 6t + 10 = 0$

25. $2w^2 - 2w - 5 = 0$

26. $2v^2 + 8v + 3 = 0$

27. $2x^2 + 6x + 5 = 0$

28. $2y^2 + 2y + 13 = 0$

29. $4t^2 - 6t + 9 = 0$

Solve by using the quadratic formula. Approximate the solutions to the nearest thousandth.

30. $x^2 - 6x - 6 = 0$ **31.** $p^2 - 8p + 3 = 0$ **32.** $r^2 - 2r = 4$

33. $w^2 + 4w = 1$ **34.** $3t^2 = 7t + 1$ **35.** $2y^2 = y + 5$

Use the discriminant to determine whether the quadratic equation has one real number solution, two real number solutions, or two complex number solutions.

36. $2z^2 - z + 5 = 0$ **37.** $3y^2 + y + 1 = 0$ **38.** $9x^2 - 12x + 4 = 0$

39. $4x^2 + 20x + 25 = 0$ **40.** $2v^2 - 3v - 1 = 0$ **41.** $3w^2 + 3w - 2 = 0$

APPLYING THE CONCEPTS

42. Name the three methods of solving a quadratic equation that have been discussed. What are the advantages and disadvantages of each?

For what values of p does the quadratic equation have two real number solutions that are not equal? Write the answer in set-builder notation.

43. $x^2 - 6x + p = 0$ **44.** $x^2 + 10x + p = 0$

For what values of p does the quadratic equation have two complex number solutions? Write the answer in set-builder notation.

45. $x^2 - 2x + p = 0$ **46.** $x^2 + 4x + p = 0$

47. Find all values of x that satisfy the equation $x^2 + ix + 2 = 0$.

48. Show that the equation $x^2 + bx - 1 = 0$ always has real number solutions regardless of the value of b.

49. Show that the equation $2x^2 + bx - 2 = 0$ always has real number solutions regardless of the value of b.

50. Can the quadratic formula be used to solve *any* quadratic equation? If so, explain why. If not, give an example of a quadratic equation that cannot be solved by using the quadratic formula.

51. One of the steps in the derivation of the quadratic formula is $x + \dfrac{b}{2a} = \pm\sqrt{\dfrac{b^2 - 4ac}{2a}}$. Carefully explain the occurrence of the $\pm$ sign. (*Hint:* Recall that $\sqrt{x^2} = |x|$.)

8.4 Solving Equations That Are Reducible to Quadratic Equations

Objective A   **To solve an equation that is quadratic in form**

Certain equations that are not quadratic can be expressed in quadratic form by making suitable substitutions. An equation is **quadratic in form** if it can be written as $au^2 + bu + c = 0$.

The equation $x^4 - 4x^2 - 5 = 0$ is quadratic in form.

$$x^4 - 4x^2 - 5 = 0$$
$$(x^2)^2 - 4(x^2) - 5 = 0$$
$$u^2 - 4u - 5 = 0 \qquad \bullet \text{ Let } x^2 = u.$$

The equation $y - y^{1/2} - 6 = 0$ is quadratic in form.

$$y - y^{1/2} - 6 = 0$$
$$(y^{1/2})^2 - (y^{1/2}) - 6 = 0$$
$$u^2 - u - 6 = 0 \qquad \bullet \text{ Let } y^{1/2} = u.$$

The key to recognizing equations that are quadratic in form is as follows. When the equation is written in standard form, the exponent on one variable term is $\dfrac{1}{2}$ the exponent on the other variable term.

➡ Solve: $z + 7z^{1/2} - 18 = 0$

$$z + 7z^{1/2} - 18 = 0 \qquad \bullet \text{ The equation is quadratic in form.}$$
$$(z^{1/2})^2 + 7(z^{1/2}) - 18 = 0$$
$$u^2 + 7u - 18 = 0 \qquad \bullet \text{ Let } z^{1/2} = u.$$

$$(u - 2)(u + 9) = 0 \qquad \bullet \text{ Solve by factoring.}$$

$$\begin{array}{ll} u - 2 = 0 & u + 9 = 0 \\ u = 2 & u = -9 \end{array}$$

$$\begin{array}{ll} z^{1/2} = 2 & z^{1/2} = -9 \\ \sqrt{z} = 2 & \sqrt{z} = -9 \end{array} \qquad \bullet \text{ Replace } u \text{ by } z^{1/2}.$$

$$\begin{array}{ll} (\sqrt{z})^2 = 2^2 & (\sqrt{z})^2 = (-9)^2 \\ z = 4 & z = 81 \end{array} \qquad \bullet \text{ Solve for } z.$$

Check each solution.

TAKE NOTE
When each side of an equation is squared, the resulting equation may have a solution that is not a solution of the original equation.

Check:

$z + 7z^{1/2} - 18 = 0$		$z + 7z^{1/2} - 18 = 0$	
$4 + 7(4)^{1/2} - 18$	0	$81 + 7(81)^{1/2} - 18$	0
$4 + 7 \cdot 2 - 18$	0	$81 + 7 \cdot 9 - 18$	0
$4 + 14 - 18$	0	$81 + 63 - 18$	0
	$0 = 0$		$126 \neq 0$

4 checks as a solution, but 81 does not check as a solution.

The solution is 4.

Example 1 Solve: $x^4 + x^2 - 12 = 0$

Solution

$$x^4 + x^2 - 12 = 0$$
$$(x^2)^2 + (x^2) - 12 = 0$$
$$u^2 + u - 12 = 0$$
$$(u - 3)(u + 4) = 0$$

$$u - 3 = 0 \qquad u + 4 = 0$$
$$u = 3 \qquad u = -4$$

Replace u by x^2.

$$x^2 = 3 \qquad\qquad x^2 = -4$$
$$\sqrt{x^2} = \sqrt{3} \qquad\qquad \sqrt{x^2} = \sqrt{-4}$$
$$x = \pm\sqrt{3} \qquad\qquad x = \pm 2i$$

The solutions are $\sqrt{3}$, $-\sqrt{3}$, $2i$, and $-2i$.

You Try It 1 Solve: $x - 5x^{1/2} + 6 = 0$

Your solution

Solution on p. S25

Objective B   **To solve a radical equation that is reducible to a quadratic equation**

Certain equations containing radicals can be expressed as quadratic equations.

⇒ Solve: $\sqrt{x + 2} + 4 = x$

$$\sqrt{x + 2} + 4 = x$$

$$\sqrt{x + 2} = x - 4$$ • Solve for the radical expression.

$$(\sqrt{x + 2})^2 = (x - 4)^2$$ • Square each side of the equation.

$$x + 2 = x^2 - 8x + 16$$ • Simplify.

$$0 = x^2 - 9x + 14$$ • Write the equation in standard form.

$$0 = (x - 7)(x - 2)$$ • Solve for x.

$$x - 7 = 0 \qquad x - 2 = 0$$
$$x = 7 \qquad\quad x = 2$$

Check each solution.

Check:

$$\begin{array}{c|c} \sqrt{x + 2} + 4 = x & \\ \hline \sqrt{7 + 2} + 4 & 7 \\ \sqrt{9} + 4 & 7 \\ 3 + 4 & 7 \\ 7 = 7 & \end{array} \qquad \begin{array}{c|c} \sqrt{x + 2} + 4 = x & \\ \hline \sqrt{2 + 2} + 4 & 2 \\ \sqrt{4} + 4 & 2 \\ 2 + 4 & 2 \\ 6 \ne 2 & \end{array}$$

7 checks as a solution, but 2 does not check as a solution.

The solution is 7.

Example 2

Solve: $\sqrt{7y - 3} + 3 = 2y$

Solution

$$\sqrt{7y - 3} + 3 = 2y$$
$$\sqrt{7y - 3} = 2y - 3$$
$$(\sqrt{7y - 3})^2 = (2y - 3)^2$$
$$7y - 3 = 4y^2 - 12y + 9$$
$$0 = 4y^2 - 19y + 12$$
$$0 = (4y - 3)(y - 4)$$

$$4y - 3 = 0 \qquad y - 4 = 0$$
$$4y = 3 \qquad\quad y = 4$$
$$y = \frac{3}{4}$$

4 checks as a solution.

$\dfrac{3}{4}$ does not check as a solution.

The solution is 4.

You Try It 2

Solve: $\sqrt{2x + 1} + x = 7$

Your solution

Example 3

Solve: $\sqrt{2y + 1} - \sqrt{y} = 1$

Solution

$$\sqrt{2y + 1} - \sqrt{y} = 1$$

Solve for one of the radical expressions.

$$\sqrt{2y + 1} = \sqrt{y} + 1$$
$$(\sqrt{2y + 1})^2 = (\sqrt{y} + 1)^2$$
$$2y + 1 = y + 2\sqrt{y} + 1$$
$$y = 2\sqrt{y}$$

Square each side of the equation.

$$y^2 = (2\sqrt{y})^2$$
$$y^2 = 4y$$
$$y^2 - 4y = 0$$
$$y(y - 4) = 0$$
$$y = 0 \qquad y - 4 = 0$$
$$y = 4$$

0 and 4 check as solutions.

The solutions are 0 and 4.

You Try It 3

Solve: $\sqrt{2x - 1} + \sqrt{x} = 2$

Your solution

Solutions on pp. S25–S26

Objective C **To solve a fractional equation that is reducible to a quadratic equation**

After each side of a fractional equation has been multiplied by the LCM of the denominators, the resulting equation may be a quadratic equation.

➡ Solve: $\dfrac{1}{r} + \dfrac{1}{r+1} = \dfrac{3}{2}$

$$\dfrac{1}{r} + \dfrac{1}{r+1} = \dfrac{3}{2}$$

$$2r(r+1)\left(\dfrac{1}{r} + \dfrac{1}{r+1}\right) = 2r(r+1) \cdot \dfrac{3}{2}$$ • Multiply each side of the equation by the LCM of the denominators.

$$2(r+1) + 2r = r(r+1) \cdot 3$$
$$2r + 2 + 2r = 3r(r+1)$$
$$4r + 2 = 3r^2 + 3r$$
$$0 = 3r^2 - r - 2$$ • Write the equation in standard form.
$$0 = (3r+2)(r-1)$$ • Solve for r by factoring.

$$3r + 2 = 0 \qquad r - 1 = 0$$
$$3r = -2 \qquad\quad r = 1$$
$$r = -\dfrac{2}{3}$$

$-\dfrac{2}{3}$ and 1 check as solutions.

The solutions are $-\dfrac{2}{3}$ and 1.

Example 4

Solve: $\dfrac{9}{x-3} = 2x + 1$

Solution

$$\dfrac{9}{x-3} = 2x + 1$$
$$(x-3)\dfrac{9}{x-3} = (x-3)(2x+1)$$
$$9 = 2x^2 - 5x - 3$$
$$0 = 2x^2 - 5x - 12$$
$$0 = (2x+3)(x-4)$$

$$2x + 3 = 0 \qquad x - 4 = 0$$
$$2x = -3 \qquad\quad x = 4$$
$$x = -\dfrac{3}{2}$$

The solutions are $-\dfrac{3}{2}$ and 4.

You Try It 4

Solve: $3y + \dfrac{25}{3y-2} = -8$

Your solution

Solution on p. S26

8.4 Exercises

Objective A

Solve.

1. $x^4 - 13x^2 + 36 = 0$

2. $y^4 - 5y^2 + 4 = 0$

3. $z^4 - 6z^2 + 8 = 0$

4. $t^4 - 12t^2 + 27 = 0$

5. $p - 3p^{1/2} + 2 = 0$

6. $v - 7v^{1/2} + 12 = 0$

7. $x - x^{1/2} - 12 = 0$

8. $w - 2w^{1/2} - 15 = 0$

9. $z^4 + 3z^2 - 4 = 0$

10. $y^4 + 5y^2 - 36 = 0$

11. $x^4 + 12x^2 - 64 = 0$

12. $x^4 - 81 = 0$

13. $p + 2p^{1/2} - 24 = 0$

14. $v + 3v^{1/2} - 4 = 0$

15. $y^{2/3} - 9y^{1/3} + 8 = 0$

16. $z^{2/3} - z^{1/3} - 6 = 0$

17. $9w^4 - 13w^2 + 4 = 0$

18. $4y^4 - 7y^2 - 36 = 0$

Objective B

Solve.

19. $\sqrt{x + 1} + x = 5$

20. $\sqrt{x - 4} + x = 6$

21. $x = \sqrt{x} + 6$

22. $\sqrt{2y - 1} = y - 2$

23. $\sqrt{3w + 3} = w + 1$

24. $\sqrt{2s + 1} = s - 1$

25. $\sqrt{4y + 1} - y = 1$

26. $\sqrt{3s + 4} + 2s = 12$

27. $\sqrt{10x + 5} - 2x = 1$

28. $\sqrt{t + 8} = 2t + 1$

29. $\sqrt{p + 11} = 1 - p$

30. $x - 7 = \sqrt{x - 5}$

31. $\sqrt{x - 1} - \sqrt{x} = -1$

32. $\sqrt{y} + 1 = \sqrt{y + 5}$

33. $\sqrt{2x - 1} = 1 - \sqrt{x - 1}$

34. $\sqrt{x + 6} + \sqrt{x + 2} = 2$

35. $\sqrt{t + 3} + \sqrt{2t + 7} = 1$

36. $\sqrt{5 - 2x} = \sqrt{2 - x} + 1$

Objective C

Solve.

37. $x = \dfrac{10}{x - 9}$

38. $z = \dfrac{5}{z - 4}$

39. $\dfrac{t}{t + 1} = \dfrac{-2}{t - 1}$

40. $\dfrac{2v}{v - 1} = \dfrac{5}{v + 2}$

41. $\dfrac{y - 1}{y + 2} + y = 1$

42. $\dfrac{2p - 1}{p - 2} + p = 8$

43. $\dfrac{3r + 2}{r + 2} - 2r = 1$

44. $\dfrac{2v + 3}{v + 4} + 3v = 4$

45. $\dfrac{2}{2x + 1} + \dfrac{1}{x} = 3$

46. $\dfrac{3}{s} - \dfrac{2}{2s - 1} = 1$

47. $\dfrac{16}{z - 2} + \dfrac{16}{z + 2} = 6$

48. $\dfrac{2}{y + 1} + \dfrac{1}{y - 1} = 1$

49. $\dfrac{t}{t - 2} + \dfrac{2}{t - 1} = 4$

50. $\dfrac{4t + 1}{t + 4} + \dfrac{3t - 1}{t + 1} = 2$

51. $\dfrac{5}{2p - 1} + \dfrac{4}{p + 1} = 2$

APPLYING THE CONCEPTS

52. Solve: $(x^2 - 7)^{1/2} = (x - 1)^{1/2}$

53. Solve: $(\sqrt{x} - 2)^2 - 5\sqrt{x} + 14 = 0$ (*Hint:* Let $u = \sqrt{x} - 2$.)

54. Solve: $(\sqrt{x} + 3)^2 - 4\sqrt{x} - 17 = 0$ (*Hint:* Let $u = \sqrt{x} + 3$.)

55. The fourth power of a number is twenty-five less than ten times the square of the number. Find the number.

8.5 Quadratic Inequalities and Rational Inequalities

Objective A   **To solve a nonlinear inequality**   (8)

A **quadratic inequality** is one that can be written in the form $ax^2 + bx + c < 0$ or $ax^2 + bx + c > 0$, where $a \neq 0$. The symbols $\leq$ and $\geq$ can also be used. The solution set of a quadratic inequality can be found by solving a compound inequality.

To solve $x^2 - 3x - 10 > 0$, first factor the trinomial.

$$x^2 - 3x - 10 > 0$$
$$(x + 2)(x - 5) > 0$$

There are two cases for which the product of the factors will be positive: (1) both factors are positive, or (2) both factors are negative.

(1) $x + 2 > 0$ and $x - 5 > 0$
(2) $x + 2 < 0$ and $x - 5 < 0$

Solve each pair of compound inequalities.

(1) $x + 2 > 0$ and $x - 5 > 0$
     $x > -2$ $x > 5$
$\{x \mid x > -2\} \cap \{x \mid x > 5\} = \{x \mid x > 5\}$

(2) $x + 2 < 0$ and $x - 5 < 0$
     $x < -2$ $x < 5$
$\{x \mid x < -2\} \cap \{x \mid x < 5\} = \{x \mid x < -2\}$

Because the two cases for which the product will be positive are connected by *or,* the solution set is the union of the solution sets of the individual inequalities.

$$\{x \mid x > 5\} \cup \{x \mid x < -2\} = \{x \mid x > 5 \text{ or } x < -2\}$$

Although the solution set of any quadratic inequality can be found by using the method outlined above, a graphical method is often easier to use.

⇒ Solve and graph the solution set of $x^2 - x - 6 < 0$.

Factor the trinomial.

$$x^2 - x - 6 < 0$$
$$(x - 3)(x + 2) < 0$$

On a number line, draw lines indicating the numbers that make each factor equal to zero.

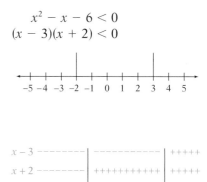

$x - 3 = 0$ $x + 2 = 0$
    $x = 3$ $x = -2$

For each factor, place plus signs above the number line for those regions where the factor is positive and minus signs where the factor is negative.

Because $x^2 - x - 6 < 0$, the solution set will be the regions where one factor is positive and the other factor is negative.

Write the solution set.

$$\{x \mid -2 < x < 3\}$$

The graph of the solution set of $x^2 - x - 6 < 0$ is shown at the right.

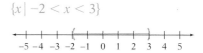

➡ Solve and graph the solution set of $(x - 2)(x + 1)(x - 4) > 0$.

On a number line, identify for each factor the regions where the factor is positive and those where the factor is negative.

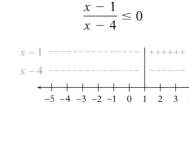

There are two regions where the product of the three factors is positive.

Write the solution set.

$\{x \mid -1 < x < 2 \text{ or } x > 4\}$

The graph of the solution set of $(x - 2)(x + 1)(x - 4) > 0$ is shown at the right.

➡ Solve: $\dfrac{2x - 5}{x - 4} \le 1$

$$\dfrac{2x - 5}{x - 4} \le 1$$

Rewrite the inequality so that 0 appears on the right side of the inequality.

$$\dfrac{2x - 5}{x - 4} - 1 \le 0$$

Simplify.

$$\dfrac{2x - 5}{x - 4} - \dfrac{x - 4}{x - 4} \le 0$$

$$\dfrac{x - 1}{x - 4} \le 0$$

On a number line, identify for each factor of the numerator and each factor of the denominator the regions where the factor is positive and those where the factor is negative.

The region where the quotient of the two factors is negative is between 1 and 4.

Write the solution set.

$\{x \mid 1 \le x < 4\}$

Note that 1 is part of the solution set but 4 is not because the denominator of the rational expression is zero when $x = 4$.

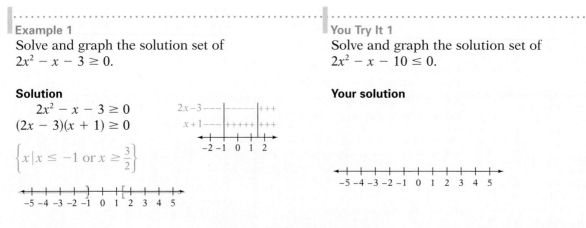

Example 1
Solve and graph the solution set of $2x^2 - x - 3 \ge 0$.

Solution
$$2x^2 - x - 3 \ge 0$$
$$(2x - 3)(x + 1) \ge 0$$

$$\left\{ x \mid x \le -1 \text{ or } x \ge \dfrac{3}{2} \right\}$$

You Try It 1
Solve and graph the solution set of $2x^2 - x - 10 \le 0$.

Your solution

Solution on p. S26

8.5 Exordises

Objective A

1. If $(x - 3)(x - 5) > 0$, what must be true of the values of $x - 3$ and $x - 5$?

2. For the inequality $\dfrac{x - 2}{x - 3} \le 1$, which of the values 1, 2, and 3 is not a possible element of the solution set? Why?

Solve and graph the solution set.

3. $(x - 4)(x + 2) > 0$

$$-5\ -4\ -3\ -2\ -1\ \ 0\ \ 1\ \ 2\ \ 3\ \ 4\ \ 5$$

4. $(x + 1)(x - 3) > 0$

$$-5\ -4\ -3\ -2\ -1\ \ 0\ \ 1\ \ 2\ \ 3\ \ 4\ \ 5$$

5. $x^2 - 3x + 2 \ge 0$

$$-5\ -4\ -3\ -2\ -1\ \ 0\ \ 1\ \ 2\ \ 3\ \ 4\ \ 5$$

6. $x^2 + 5x + 6 > 0$

$$-5\ -4\ -3\ -2\ -1\ \ 0\ \ 1\ \ 2\ \ 3\ \ 4\ \ 5$$

7. $x^2 - x - 12 < 0$

$$-5\ -4\ -3\ -2\ -1\ \ 0\ \ 1\ \ 2\ \ 3\ \ 4\ \ 5$$

8. $x^2 + x - 20 < 0$

$$-5\ -4\ -3\ -2\ -1\ \ 0\ \ 1\ \ 2\ \ 3\ \ 4\ \ 5$$

9. $(x - 1)(x + 2)(x - 3) < 0$

$$-5\ -4\ -3\ -2\ -1\ \ 0\ \ 1\ \ 2\ \ 3\ \ 4\ \ 5$$

10. $(x + 4)(x - 2)(x + 1) > 0$

$$-5\ -4\ -3\ -2\ -1\ \ 0\ \ 1\ \ 2\ \ 3\ \ 4\ \ 5$$

11. $(x + 4)(x - 2)(x - 1) \ge 0$

$$-5\ -4\ -3\ -2\ -1\ \ 0\ \ 1\ \ 2\ \ 3\ \ 4\ \ 5$$

12. $(x - 1)(x + 5)(x - 2) \le 0$

$$-5\ -4\ -3\ -2\ -1\ \ 0\ \ 1\ \ 2\ \ 3\ \ 4\ \ 5$$

13. $\dfrac{x - 4}{x + 2} > 0$

$$-5\ -4\ -3\ -2\ -1\ \ 0\ \ 1\ \ 2\ \ 3\ \ 4\ \ 5$$

14. $\dfrac{x + 2}{x - 3} > 0$

$$-5\ -4\ -3\ -2\ -1\ \ 0\ \ 1\ \ 2\ \ 3\ \ 4\ \ 5$$

15. $\dfrac{x - 3}{x + 1} \le 0$

$$-5\ -4\ -3\ -2\ -1\ \ 0\ \ 1\ \ 2\ \ 3\ \ 4\ \ 5$$

16. $\dfrac{x - 1}{x} > 0$

$$-5\ -4\ -3\ -2\ -1\ \ 0\ \ 1\ \ 2\ \ 3\ \ 4\ \ 5$$

17. $\dfrac{(x-1)(x+2)}{x-3} \le 0$

18. $\dfrac{(x+3)(x-1)}{x-2} \ge 0$

Solve.

19. $x^2 - 16 > 0$

20. $x^2 - 4 \ge 0$

21. $x^2 - 9x \le 36$

22. $x^2 + 4x > 21$

23. $4x^2 - 8x + 3 < 0$

24. $2x^2 + 11x + 12 \ge 0$

25. $\dfrac{3}{x-1} < 2$

26. $\dfrac{x}{(x-1)(x+2)} \ge 0$

27. $\dfrac{x-2}{(x+1)(x-1)} \le 0$

28. $\dfrac{1}{x} < 2$

29. $\dfrac{x}{2x-1} \ge 1$

30. $\dfrac{x}{2x-3} \le 1$

31. $\dfrac{x}{2-x} \le -3$

32. $\dfrac{3}{x-2} > \dfrac{2}{x+2}$

33. $\dfrac{3}{x-5} > \dfrac{1}{x+1}$

APPLYING THE CONCEPTS

Graph the solution set.

34. $(x+2)(x-3)(x+1)(x+4) > 0$

35. $(x-1)(x+3)(x-2)(x-4) \ge 0$

36. $(x^2 + 2x - 8)(x^2 - 2x - 3) < 0$

37. $(x^2 + 2x - 3)(x^2 + 3x + 2) \ge 0$

38. $(x^2 + 1)(x^2 - 3x + 2) > 0$

39. $\dfrac{x^2(3-x)(2x+1)}{(x+4)(x+2)} \ge 0$

8.6 Applications of Quadratic Equations

Objective A   **To solve application problems**   ⟨ 8 ⟩

The application problems in this section are similar to problems solved earlier in the text. Each of the strategies for the problems in this section will result in a quadratic equation.

⟹ A small pipe takes 16 min longer to empty a tank than does a larger pipe. Working together, the pipes can empty the tank in 6 min. How long would it take each pipe working alone to empty the tank?

Strategy for Solving an Application Problem

1. Determine the type of problem. Is it a uniform motion problem, a geometry problem, an integer problem, or a work problem?

The problem is a work problem.

2. Choose a variable to represent the unknown quantity. Write numerical or variable expressions for all the remaining quantities. These results can be recorded in a table.

The unknown time of the larger pipe: t
The unknown time of the smaller pipe: $t + 16$

	Rate of Work	·	Time Worked	=	Part of Task Completed
Larger pipe	$\dfrac{1}{t}$	·	6	=	$\dfrac{6}{t}$
Smaller pipe	$\dfrac{1}{t + 16}$	·	6	=	$\dfrac{6}{t + 16}$

3. Determine how the quantities are related.

$$\frac{6}{t} + \frac{6}{t + 16} = 1$$ • The sum of the parts of the task completed must equal 1.

$$t(t + 16)\left(\frac{6}{t} + \frac{6}{t + 16}\right) = t(t + 16) \cdot 1$$
$$(t + 16)6 + 6t = t^2 + 16t$$
$$6t + 96 + 6t = t^2 + 16t$$
$$0 = t^2 + 4t - 96$$
$$0 = (t + 12)(t - 8)$$

$$t + 12 = 0 \qquad t - 8 = 0$$
$$t = -12 \qquad t = 8$$

Because time cannot be negative, the solution $t = -12$ is not possible.

The time for the smaller pipe is $t + 16$.

$$t + 16 = 8 + 16 = 24$$ • Replace t by 8 and evaluate.

The larger pipe requires 8 min to empty the tank.
The smaller pipe requires 24 min to empty the tank.

....

Example 1

In 8 h, two campers rowed 15 mi down a river and then rowed back to their campsite. The rate of the river's current was 1 mph. Find the rate at which the campers rowed.

You Try It 1

The length of a rectangle is 3 m more than the width. The area is 54 m². Find the length of the rectangle.

Strategy

- This is a uniform motion problem.
- Unknown rowing rate of the campers: r

	Distance	Rate	Time
Down river	15	$r + 1$	$\dfrac{15}{r+1}$
Up river	15	$r - 1$	$\dfrac{15}{r-1}$

- The total time of the trip was 8 h.

Your strategy

Solution

$$\frac{15}{r+1} + \frac{15}{r-1} = 8$$

$$(r+1)(r-1)\left(\frac{15}{r+1} + \frac{15}{r-1}\right) = (r+1)(r-1)8$$

$$(r-1)15 + (r+1)15 = (r^2 - 1)8$$

$$15r - 15 + 15r + 15 = 8r^2 - 8$$

$$30r = 8r^2 - 8$$

$$0 = 8r^2 - 30r - 8$$

$$0 = 2(4r^2 - 15r - 4)$$

$$0 = 2(4r + 1)(r - 4)$$

$$
\begin{array}{ll}
4r + 1 = 0 & r - 4 = 0 \\
4r = -1 & r = 4 \\
r = -\dfrac{1}{4} &
\end{array}
$$

The solution $r = -\dfrac{1}{4}$ is not possible, because the rate cannot be a negative number.

The rowing rate was 4 mph.

Your solution

Solution on p. S26

8.6 Exercises

Objective A   *Application Problems*

1. The length of the base of a triangle is 1 cm less than five times the height of the triangle. The area of the triangle is 21 cm². Find the height and the length of the base of the triangle.

2. The length of a rectangle is 2 ft less than three times the width of the rectangle. The area of the rectangle is 65 ft². Find the length and width of the rectangle.

3. The state of Colorado is almost perfectly rectangular, with its north border 111 mi longer than its west border. If the state encompasses 104,000 mi², estimate the dimensions of Colorado. Round to the nearest mile.

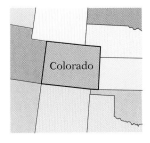

4. A square piece of cardboard is formed into a box by cutting 10-centimeter squares from each of the four corners and then folding up the sides, as shown in the figure. If the volume, V, of the box is to be 49,000 cm³, what size square piece of cardboard is needed? Recall that $V = LWH$.

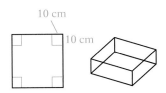

5. A car with good tire tread can stop in less distance than a car with poor tread. The formula for the stopping distance d, in feet, of a car with good tread on dry cement is approximated by $d = 0.04v^2 + 0.5v$, where v is the speed of the car. If the driver must be able to stop within 60 ft, what is the maximum safe speed, to the nearest mile per hour, of the car?

6. A model rocket is launched with an initial velocity of 200 ft/s. The height h, in feet, of the rocket t seconds after the launch is given by $h = -16t^2 + 200t$. How many seconds after the launch will the rocket be 300 ft above the ground? Round to the nearest hundredth of a second.

7. The height of a projectile fired upward is given by the formula $s = v_0t - 16t^2$, where s is the height in feet, v_0 is the initial velocity, and t is the time in seconds. Find the time for a projectile to return to Earth if it has an initial velocity of 200 ft/s.

8. The height of a projectile fired upward is given by the formula $s = v_0t - 16t^2$, where s is the height in feet, v_0 is the initial velocity, and t is the time in seconds. Find the time for a projectile to reach a height of 64 ft if it has an initial velocity of 128 ft/s. Round to the nearest hundredth of a second.

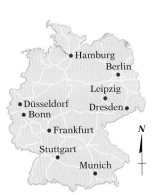

9. In Germany, there is no speed limit on some portions of the autobahn (highway). Other portions have a speed limit of 180 km/h (approximately 112 mph). The distance d, in meters, required to stop a car traveling at v kilometers per hour is $d = 0.019v^2 + 0.69v$. Approximate, to the nearest tenth, the maximum speed a driver can be traveling and still be able to stop within 150 m.

German Autobahn System

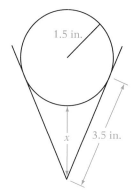

10. A perfectly spherical scoop of mint chocolate chip ice cream is placed in a cone, as shown in the figure. How far is the bottom of the scoop of ice cream from the bottom of the cone? Round to the nearest tenth. (*Hint:* A line segment from the center of the ice cream to the point at which the ice cream touches the cone is perpendicular to the edge of the cone.)

11. A small pipe can fill a tank in 6 min more time than it takes a larger pipe to fill the same tank. Working together, both pipes can fill the tank in 4 min. How long would it take each pipe working alone to fill the tank?

12. A cruise ship made a trip of 100 mi in 8 h. The ship traveled the first 40 mi at a constant rate before increasing its speed by 5 mph. Then it traveled another 60 mi at the increased speed. Find the rate of the cruise ship for the first 40 mi.

13. The Concorde's speed in calm air is 1320 mph. Flying with the wind, the Concorde can fly from New York to London, a distance of approximately 4000 mi, in 0.5 h less than the time required to make the return trip. Find the rate of the wind to the nearest mile per hour.

14. A car travels 120 mi. A second car, traveling 10 mph faster than the first car, makes the same trip in 1 h less time. Find the speed of each car.

15. For a portion of the Green River in Utah, the rate of the river's current is 4 mph. A tour guide can row 5 mi down this river and back in 3 h. Find the rowing rate of the guide in calm water.

16. The height h, in feet, of an arch is given by the equation $h(x) = -\dfrac{3}{64}x^2 + 27$, where $|x|$ is the distance in feet from the center of the arch.

    a. What is the maximum height of the arch?
    b. What is the height of the arch 8 ft to the right of the center?
    c. How far from the center is the arch 8 ft tall?

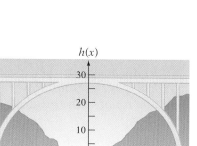

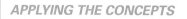
APPLYING THE CONCEPTS

17. The surface area of the ice cream cone shown at the right is given by $A = \pi r^2 + \pi rs$, where r is the radius of the circular top of the cone and s is the slant height of the cone. If the surface area of the cone is 11.25π in² and the slant height is 6 in., find the radius of the cone.

18. Using Torricelli's Principle, it can be shown that the depth, d, of a liquid in a bottle with a hole of area 0.5 cm² in its side can be approximated by $d = 0.0034t^2 - 0.52518t + 20$, where t is the time since a stopper was removed from the hole. When will the depth be 10 cm? Round to the nearest tenth of a second.

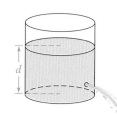

Focus on Problem Solving

Inductive and Deductive Reasoning

Consider the following sums of odd positive integers.

$1 + 3 = 4 = 2^2$ Sum of the first two odd numbers is 2^2.

$1 + 3 + 5 = 9 = 3^2$ Sum of the first three odd numbers is 3^2.

$1 + 3 + 5 + 7 = 16 = 4^2$ Sum of the first four odd numbers is 4^2.

$1 + 3 + 5 + 7 + 9 = 25 = 5^2$ Sum of the first five odd numbers is 5^2.

1. Make a conjecture about the value of the sum

$$1 + 3 + 5 + 7 + 9 + 11 + 13 + 15$$

without adding the numbers.

If the pattern continues, a possible conjecture is that the sum of the first eight odd positive integers is $8^2 = 64$. By adding the numbers, you can verify that this is correct. Inferring that the pattern established by a few cases is valid for all cases is an example of **inductive reasoning.**

2. Use inductive reasoning to find the next figure in the sequence below.

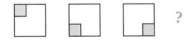   ?

The fact that a pattern appears to be true does not *prove* that it is true for all cases. For example, consider the polynomial $n^2 - n + 41$. If we begin substituting positive integer values for n and evaluating the expression, we produce the following table.

n	$n^2 - n + 41$	
1	$1^2 - 1 + 41 = 41$	41 is a prime number.
2	$2^2 - 2 + 41 = 43$	43 is a prime number.
3	$3^2 - 3 + 41 = 47$	47 is a prime number.
4	$4^2 - 4 + 41 = 53$	53 is a prime number.

Even if we try a number like 30, we have $30^2 - 30 + 41 = 911$ and 911 is a prime number. Thus it appears that the value of this polynomial is a prime number for any value of n. However, the conjecture is not true. For instance, if $n = 41$, we have $41^2 - 41 + 41 = 41^2 = 1681$, which is not a prime number because it is divisible by 41. This illustrates that inductive reasoning may lead to incorrect conclusions and that an inductive proof must be available to prove conjectures. There is such a proof, called *mathematical induction*, that you may study in a future math course.

Now consider the true statement that the sum of the measures of the interior angles of a triangle is 180°. The figure at the left is a triangle; therefore the sum of the measures of the interior angles must be 180°. This is an example of *deductive reasoning.* **Deductive reasoning** uses a rule or statement of fact to reach a conclusion.

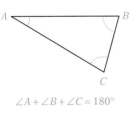

$\angle A + \angle B + \angle C = 180°$

3. Use deductive reasoning to complete the following sentence. All even numbers are divisible by 2. Because 14,386 is an even number, . . .

For Exercises 4 and 5 on the next page, determine whether inductive or deductive reasoning is being used.

4. The tenth number in the list 1, 4, 9, 16, 25, . . . is 100.

5. All quadrilaterals have four sides. A square is a quadrilateral. Therefore a square has four sides.

6. Explain the difference between inductive and deductive reasoning.

Projects and Group Activities

Completing the Square

Essentially all of the investigations into mathematics before the Renaissance were geometric. The solutions of quadratic equations were calculated from a construction of a certain area. Proofs of theorems, even theorems about numbers, were based entirely on geometry. In this project, we will examine the solution of a quadratic equation.

⟹ Solve: $x^2 + 6x = 7$

Begin with a line of unknown length, x, and one of length 6, the coefficient of x. Using these lines, construct a rectangle as shown in Figure 1.

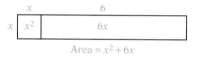

Area $= x^2 + 6x$

Figure 1

Now draw another area that has exactly the same area as Figure 1 by cutting one-half of the rectangle off and placing it on the bottom of the square. See Figure 2.

The unshaded area in Figure 2 has exactly the same area as Figure 1. However, when the shaded area is added to Figure 2 to make a square, the area of the square is 9 square units larger than that of Figure 1. In equation form,

(Area of Figure 1) + 9 = area of Figure 2

or

$$x^2 + 6x + 9 = (x + 3)^2$$

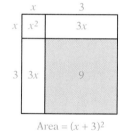

Area $= (x + 3)^2$

Figure 2

From the original equation, $x^2 + 6x = 7$. Thus,

$$x^2 + 6x + 9 = (x + 3)^2$$
$$7 + 9 = (x + 3)^2 \qquad \bullet \ x^2 + 6x = 7$$
$$16 = (x + 3)^2$$
$$4 = x + 3 \qquad \bullet \ \text{See note below.}$$
$$1 = x$$

Note: Although early mathematicians knew that a quadratic equation may have two solutions, both solutions were allowed only if they were positive. After all, a geometric construction could not have a negative length. Therefore, the solution of this equation was 1; the solution −7 would have been dismissed as *fictitious*, the actual word that was frequently used through the 15th century for negative-number solutions of an equation.

Try to solve the quadratic equation $x^2 + 4x = 12$ by geometrically completing the square.

Using a Graphing Calculator to Solve a Quadratic Equation

Recall that an x-intercept of the graph of an equation is a point at which the graph crosses the x-axis. For the graph in Figure 3, the x-intercepts are $(-2, 0)$ and $(3, 0)$.

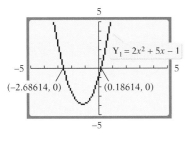

Figure 3

Recall also that to find the x-intercept of a graph, set $y = 0$ and then solve for x. For the equation in Figure 3, if we set $y = 0$, the resulting equation is $0 = x^2 - x - 6$, which is a quadratic equation. Solving this equation by factoring, we have

$$0 = x^2 - x - 6$$
$$0 = (x + 2)(x - 3)$$

$$x + 2 = 0 \qquad\qquad x - 3 = 0$$
$$x = -2 \qquad\qquad x = 3$$

Thus the solutions of the equation are the x-coordinates of the x-intercepts of the graph.

This connection between the solutions of an equation and the x-intercepts of its graph allows us to graphically find approximations of the real number solutions of an equation. For example, to use a TI-83 to graphically approximate the solutions of $2x^2 + 5x - 1 = 0$, use the following keystrokes to graph $y = 2x^2 + 5x - 1$ and find the decimal approximations of the x-coordinates of the x-intercepts.

$\boxed{\text{Y=}}$ $\boxed{\text{CLEAR}}$ 2 $\boxed{\text{X,T,}\theta,n}$ $\boxed{x^2}$ $\boxed{+}$ 5 $\boxed{\text{X,T,}\theta,n}$ $\boxed{-}$ 1
$\boxed{\text{ZOOM}}$ 6 $\boxed{\text{2nd}}$ $\boxed{\text{CALC}}$ 2

Figure 4

Use the arrow keys to move to the left of the leftmost x-intercept. Press ENTER. Use the arrow keys to move to a point just to the right of the leftmost x-intercept. Press ENTER twice. The x-coordinate at the bottom of the screen is the approximation of one solution of the equation. To find the other solution, use the same procedure but move the cursor first to the left and then to the right of the rightmost x-intercept.

Chapter Summary

Key Words

A *quadratic equation* is an equation of the form $ax^2 + bx + c = 0$, where a and b are coefficients, c is a constant, and $a \neq 0$. A quadratic equation is also called a *second-degree equation*. A quadratic equation is in *standard form* when the polynomial is in descending order and equal to zero. [p. 421]

When a quadratic equation has two solutions that are the same number, the solution is called a *double root* of the equation. [p. 421]

Methods of Solving a Quadratic Equation:

1. **Factoring** [p. 421]
   This method is used only when the polynomial $ax^2 + bx + c$ is factorable.
2. **Taking square roots** [pp. 423–424]
   This method is used only when the value of b in $ax^2 + bx + c$ is 0. It is also used when an equation can be written in the form binomial^2 = constant.
3. **Completing the square** [pp. 429–431]
   This method can be used to solve any quadratic equation.
4. **Quadratic formula** [p. 436]
   This method can be used to solve any quadratic equation.

Adding to a binomial the constant term that makes it a perfect-square trinomial is called *completing the square*. [p. 429]

For an equation of the form $ax^2 + bx + c = 0$, the quantity $b^2 - 4ac$ is called the *discriminant*. [p. 437]

An equation is *quadratic in form* if it can be written as $au^2 + bu + c = 0$. [p. 441]

A *quadratic inequality* is one that can be written in the form $ax^2 + bx + c > 0$ or $ax^2 + bx + c < 0$, where $a \neq 0$. The symbols $\leq$ or $\geq$ can also be used. [p. 447]

Essential Rules

The Principle of Zero Products

If $ab = 0$, then $a = 0$ or $b = 0$. [p. 421]

To Solve a Quadratic Equation by Factoring:

1. Write the equation in standard form.
2. Factor.
3. Use the Principle of Zero Products to set each factor equal to 0.
4. Solve each equation.
5. Check the solutions. [p. 421]

To Write a Quadratic Equation Given Its Solutions:

Use the equation $(x - r_1)(x - r_2) = 0$. Replace r_1 with one solution and r_2 with the other solution. Then multiply the two factors. [p. 422]

To Complete the Square:

Add to a binomial of the form $x^2 + bx$ the square of $\frac{1}{2}$ of the coefficient of x. [p. 429]

To Solve a Quadratic Equation by Completing the Square:

1. Write the equation in the form $ax^2 + bx = -c$.

2. Multiply both sides of the equation by $\frac{1}{a}$.

3. Complete the square on $x^2 + \frac{b}{a}x$. Add the number that completes the square to both sides of the equation.

4. Factor the perfect-square trinomial.
5. Take the square root of each side of the equation.
6. Solve the resulting equation for x.
7. Check the solutions. [p. 431]

The Quadratic Formula

The solutions of $ax^2 + bx + c = 0$, $a \neq 0$, are $x = \dfrac{-b \pm \sqrt{b^2 - 4ac}}{2a}$. [p. 435]

The Effect of the Discriminant on the Solutions of a Quadratic Equation

1. If $b^2 - 4ac = 0$, the equation has one real number solution, a double root.
2. If $b^2 - 4ac > 0$, the equation has two unequal real number solutions.
3. If $b^2 - 4ac < 0$, the equation has two complex number solutions. [p. 437]

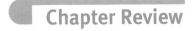 Chapter Review

1. Solve by factoring: $2x^2 - 3x = 0$

2. Solve by factoring: $6x^2 + 9cx = 6c^2$

3. Solve by taking square roots:
   $x^2 = 48$

4. Solve by taking square roots:
   $\left(x + \dfrac{1}{2}\right)^2 + 4 = 0$

5. Solve by completing the square:
   $x^2 + 4x + 3 = 0$

6. Solve by completing the square:
   $7x^2 - 14x + 3 = 0$

7. Solve by using the quadratic formula:
   $12x^2 - 25x + 12 = 0$

8. Solve by using the quadratic formula:
   $x^2 - x + 8 = 0$

9. Write a quadratic equation that has integer coefficients and has solutions 0 and -3.

10. Write a quadratic equation that has integer coefficients and has solutions $\dfrac{3}{4}$ and $-\dfrac{2}{3}$.

11. Solve by completing the square:
    $x^2 - 2x + 8 = 0$

12. Solve by completing the square:
    $(x - 2)(x + 3) = x - 10$

13. Solve by using the quadratic formula:
    $3x(x - 3) = 2x - 4$

14. Use the discriminant to determine whether $3x^2 - 5x + 1 = 0$ has one real number solution, two real number solutions, or two complex number solutions.

15. Solve: $(x + 3)(2x - 5) < 0$

16. Solve: $(x - 2)(x + 4)(2x + 3) \le 0$

17. Solve: $x^{2/3} + x^{1/3} - 12 = 0$

18. Solve: $2(x - 1) + 3\sqrt{x - 1} - 2 = 0$

19. Solve: $3x = \dfrac{9}{x - 2}$

20. Solve: $\dfrac{3x + 7}{x + 2} + x = 3$

21. Solve and graph the solution set:
$\dfrac{x - 2}{2x - 3} \geq 0$

22. Solve and graph the solution set:
$\dfrac{(2x - 1)(x + 3)}{x - 4} \leq 0$

23. Solve: $x = \sqrt{x} + 2$

24. Solve: $2x = \sqrt{5x + 24} + 3$

25. Solve: $\dfrac{x - 2}{2x + 3} - \dfrac{x - 4}{x} = 2$

26. Solve: $1 - \dfrac{x + 4}{2 - x} = \dfrac{x - 3}{x + 2}$

27. The length of a rectangle is two more than twice the width. The area of the rectangle is 60 cm². Find the length and width of the rectangle.

28. The sum of the squares of three consecutive even integers is fifty-six. Find the three integers.

29. An older computer requires 12 min longer to print the payroll than does a newer computer. Together the computers can print the payroll in 8 min. Find the time for the new computer working alone to print the payroll.

30. A car travels 200 mi. A second car, making the same trip, travels 10 mph faster than the first car and makes the trip in 1 h less time. Find the speed of each car.

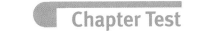

Chapter Test

1. Solve by factoring: $3x^2 + 10x = 8$

2. Solve by factoring: $6x^2 - 5x - 6 = 0$

3. Write a quadratic equation that has integer coefficients and has solutions 3 and -3.

4. Write a quadratic equation that has integer coefficients and has solutions $\frac{1}{2}$ and -4.

5. Solve by taking square roots:
$3(x - 2)^2 - 24 = 0$

6. Solve by completing the square:
$x^2 - 6x - 2 = 0$

7. Solve by completing the square:
$3x^2 - 6x = 2$

8. Solve by using the quadratic formula:
$2x^2 - 2x = 1$

9. Solve by using the quadratic formula:
$x^2 + 4x + 12 = 0$

10. Use the discriminant to determine whether $3x^2 - 4x = 1$ has one real number solution, two real number solutions, or two complex number solutions.

11. Use the discriminant to determine whether $x^2 - 6x = -15$ has one real number solution, two real number solutions, or two complex number solutions.

12. Solve: $2x + 7x^{1/2} - 4 = 0$

13. Solve: $x^4 - 4x^2 + 3 = 0$

14. Solve: $\sqrt{2x + 1} + 5 = 2x$

15. Solve: $\sqrt{x - 2} = \sqrt{x} - 2$

16. Solve: $\dfrac{2x}{x - 3} + \dfrac{5}{x - 1} = 1$

17. Solve and graph the solution set of $(x - 2)(x + 4)(x - 4) < 0$.

18. Solve and graph the solution set of $\dfrac{2x - 3}{x + 4} \leq 0$.

19. The length of the base of a triangle is 3 ft more than three times the height. The area of the triangle is 30 ft². Find the length of the base and the height of the triangle.

20. The rate of a river's current is 2 mph. A canoe was rowed 6 mi down the river and back in 4 h. Find the rowing rate in calm water.

Cumulative Review

1. Evaluate $2a^2 - b^2 \div c^2$ when $a = 3$, $b = -4$, and $c = -2$.

2. Solve: $\dfrac{2x - 3}{4} - \dfrac{x + 4}{6} = \dfrac{3x - 2}{8}$

3. Find the slope of the line containing the points $(3, -4)$ and $(-1, 2)$.

4. Find the equation of the line that contains the point $(1, 2)$ and is parallel to the line $x - y = 1$.

5. Factor: $-3x^3y + 6x^2y^2 - 9xy^3$

6. Factor: $6x^2 - 7x - 20$

7. Factor: $a^n x + a^n y - 2x - 2y$

8. Divide: $(3x^3 - 13x^2 + 10) \div (3x - 4)$

9. Simplify: $\dfrac{x^2 + 2x + 1}{8x^2 + 8x} \cdot \dfrac{4x^3 - 4x^2}{x^2 - 1}$

10. Find the distance between the points $(-2, 3)$ and $(2, 5)$.

11. Solve $S = \dfrac{n}{2}(a + b)$ for b.

12. Simplify: $-2i(7 - 4i)$

13. Simplify: $a^{-1/2}(a^{1/2} - a^{3/2})$

14. Simplify: $\dfrac{\sqrt[3]{8x^4y^5}}{\sqrt[3]{16xy^6}}$

15. Solve: $\dfrac{x}{x + 2} - \dfrac{4x}{x + 3} = 1$

16. Solve: $\dfrac{x}{2x + 3} - \dfrac{3}{4x^2 - 9} = \dfrac{x}{2x - 3}$

17. Solve: $x^4 - 6x^2 + 8 = 0$

18. Solve: $\sqrt{3x + 1} - 1 = x$

19. Solve: $|3x - 2| < 8$

20. Find the x- and y-intercepts of the graph of $6x - 5y = 15$.

21. Graph the solution set:
$$x + y \leq 3$$
$$2x - y < 4$$

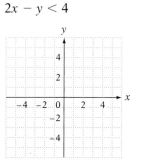

22. Solve the system of equations.
$$x + y + z = 2$$
$$-x + 2y - 3z = -9$$
$$x - 2y - 2z = -1$$

23. Given $f(x) = \dfrac{2x - 3}{x^2 - 1}$, find $f(-2)$.

24. Find the domain of the function
$$f(x) = \frac{x - 2}{x^2 - 2x - 15}.$$

25. Solve and graph the solution set of $x^3 + x^2 - 6x < 0$.

$$\xleftarrow{\quad}\overset{\;-5\;-4\;-3\;-2\;-1\quad0\quad1\quad2\quad3\quad4\quad5}{\rule{6cm}{0.4pt}}\xrightarrow{\quad}$$

26. Solve and graph the solution set of
$$\frac{(x - 1)(x - 5)}{x + 3} \geq 0.$$

$$\xleftarrow{\quad}\overset{\;-5\;-4\;-3\;-2\;-1\quad0\quad1\quad2\quad3\quad4\quad5}{\rule{6cm}{0.4pt}}\xrightarrow{\quad}$$

27. A piston rod for an automobile is $9\dfrac{3}{8}$ in. with a tolerance of $\dfrac{1}{64}$ in. Find the lower and upper limits of the length of the piston rod.

28. The length of the base of a triangle is $(x + 8)$ ft. The height is $(2x - 4)$ ft. Find the area of the triangle in terms of the variable x.

29. Use the discriminant to determine whether $2x^2 + 4x + 3 = 0$ has one real number solution, two real number solutions, or two complex number solutions.

30. The graph shows the relationship between the cost of a building and the depreciation allowed for income tax purposes. Find the slope of the line between the two points shown on the graph. Write a sentence that states the meaning of the slope.

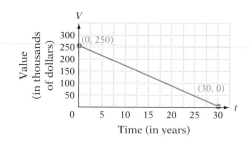

Chapter 9

Functions and Relations

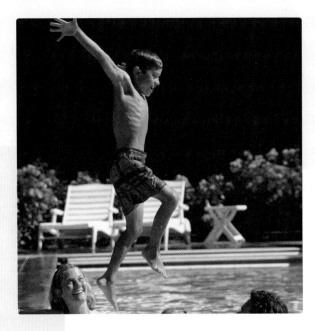

Objectives

Section 9.1

A To graph a quadratic function
B To find the *x*-intercepts of a parabola
C To find the minimum or maximum of a quadratic function
D To solve application problems

Section 9.2

A To graph functions

Section 9.3

A To perform operations on functions
B To find the composition of two functions

Section 9.4

A To determine whether a function is one-to-one
B To find the inverse of a function

Personal swimming pools should be as clean and safe as possible for families' and friends' enjoyment. Basic pool upkeep includes scheduled chemical treatments to reduce any algae in the water. The treatment eventually reaches a level of maximum efficiency, which results in a minimum amount of algae in the pool. **Exercise 94 on page 481** provides a function that is used to determine when a treated pool will have the least amount of algae.

Need help? For on-line student resources, such as section quizzes, visit this textbook's web site at **math.college.hmco.com/students**.

1. Evaluate $-\dfrac{b}{2a}$ for $b = -4$ and $a = 2$.

2. Given $y = -x^2 + 2x + 1$, find the value of y when $x = -2$.

3. Given $f(x) = x^2 - 3x + 2$, find $f(-4)$.

4. Evaluate $p(r) = r^2 - 5$ when $r = 2 + h$.

5. Solve: $0 = 3x^2 - 7x - 6$

6. Solve: $0 = x^2 - 4x + 1$

7. Solve $x = 2y + 4$ for y.

8. Find the domain and range of the relation $\{(-2, 4), (3, 5), (4, 6), (6, 5)\}$. Is the relation a function?

9. What value is excluded from the domain of $f(x) = \dfrac{3}{x - 8}$?

10. Graph: $x = -2$

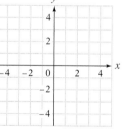

Go Figure

Each time the two hands of a certain standard, analog, 12-hour clock form a 180° angle, a bell chimes once. From noon today until noon tomorrow, how many chimes will be heard?

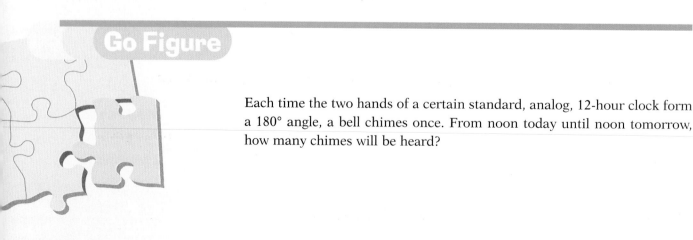

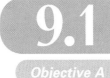

Properties of Quadratic Functions

Objective A   **To graph a quadratic function**   < 9 >

Point of Interest

The suspension cables for some bridges, such as the Golden Gate bridge, have the shape of a parabola. Parabolic shapes are also used for the mirrors in telescopes and in the design of certain antennas.

Recall that a linear function is one that can be expressed by the equation $f(x) = mx + b$. The graph of a linear function has certain characteristics. It is a straight line with slope m and y-intercept $(0, b)$. A **quadratic function** is one that can be expressed by the equation $f(x) = ax^2 + bx + c$, $a \neq 0$. The graph of this function, called a **parabola,** also has certain characteristics. The graph of a quadratic function can be drawn by finding ordered pairs that belong to the function.

⟹ Graph $f(x) = x^2 - 2x - 3$.

By evaluating the function for various values of x, find enough ordered pairs to determine the shape of the graph.

TAKE NOTE

Sometimes the value of the independent variable is called the **input** because it is *put in* place of the independent variable. The result of evaluating the function is called the **output.**

An **input/output table** shows the results of evaluating a function for various values of the independent variable. An input/output table for $f(x) = x^2 - 2x - 3$ is shown at the right.

x	$f(x) = x^2 - 2x - 3$	$f(x)$	(x, y)
-2	$f(-2) = (-2)^2 - 2(-2) - 3$	5	$(-2, 5)$
-1	$f(-1) = (-1)^2 - 2(-1) - 3$	0	$(-1, 0)$
0	$f(0) = (0)^2 - 2(0) - 3$	-3	$(0, -3)$
1	$f(1) = (1)^2 - 2(1) - 3$	-4	$(1, -4)$
2	$f(2) = (2)^2 - 2(2) - 3$	-3	$(2, -3)$
3	$f(3) = (3)^2 - 2(3) - 3$	0	$(3, 0)$
4	$f(4) = (4)^2 - 2(4) - 3$	5	$(4, 5)$

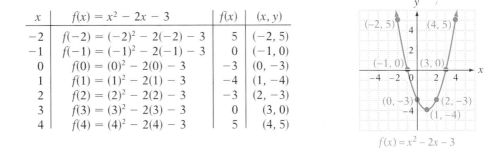

$f(x) = x^2 - 2x - 3$

Because the value of $f(x) = x^2 - 2x - 3$ is a real number for all values of x, the domain of f is all real numbers. From the graph, it appears that no value of y is less than -4. Thus the range is $\{y \,|\, y \geq -4\}$. The range can also be determined algebraically, as shown below, by completing the square.

TAKE NOTE

In completing the square, 1 is both added and subtracted. Because $1 - 1 = 0$, the expression $x^2 - 2x - 3$ is not changed. Note that

$(x - 1)^2 - 4$
$= (x^2 - 2x + 1) - 4$
$= x^2 - 2x - 3$

which is the original expression.

$f(x) = x^2 - 2x - 3$

$\quad = (x^2 - 2x) - 3$ • Group the variable terms.

$\quad = (x^2 - 2x + 1) - 1 - 3$ • Complete the square on $x^2 - 2x$. Add and subtract $\left[\dfrac{1}{2}(-2)\right]^2 = 1$ to and from $x^2 - 2x$.

$\quad = (x - 1)^2 - 4$ • Factor and combine like terms.

Because the square of a positive number is always positive, we have

$(x - 1)^2 \geq 0$

$(x - 1)^2 - 4 \geq -4$ • Subtract 4 from each side of the inequality.

$\qquad f(x) \geq -4$ • $f(x) = (x - 1)^2 - 4$

$\qquad y \geq -4$

From the last inequality, the range is $\{y \,|\, y \geq -4\}$.

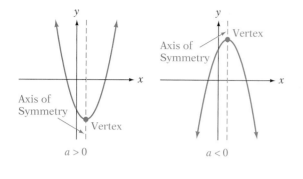

$a > 0$ $a < 0$

In general, the graph of $f(x) = ax^2 + bx + c$, $a \neq 0$, resembles a "cup" shape as shown at the left. When $a > 0$, the parabola opens up and the **vertex** of the parabola is the point with the smallest y-coordinate. When $a < 0$, the parabola opens down and the vertex is the point with the largest y-coordinate. The **axis of symmetry** is the vertical line that passes through the vertex of the parabola and is parallel to the y-axis. To understand the axis of symmetry, think of folding the graph along that vertical line. The two portions of the graph will match up.

The vertex and axis of symmetry of a parabola can be found by completing the square.

⇨ Find the vertex and axis of symmetry of the graph of $F(x) = x^2 + 4x + 3$.

To find the coordinates of the vertex, complete the square.

$$F(x) = x^2 + 4x + 3$$
$$= (x^2 + 4x) + 3 \qquad \bullet \text{ Group the variable terms.}$$
$$= (x^2 + 4x + 4) - 4 + 3 \qquad \bullet \text{ Complete the square on } x^2 + 4x. \text{ Add and}$$
$$\text{subtract } \left[\frac{1}{2}(4)\right]^2 = 4 \text{ to and from } x^2 + 4x.$$
$$= (x + 2)^2 - 1 \qquad \bullet \text{ Factor and combine like terms.}$$

Because a, the coefficient of x^2, is positive ($a = 1$), the parabola opens up and the vertex is the point with the smallest y-coordinate. Because $(x + 2)^2 \geq 0$ for all values of x, the smallest y-coordinate occurs when $(x + 2)^2 = 0$. The quantity $(x + 2)^2$ is equal to zero when $x = -2$. Therefore, the x-coordinate of the vertex is -2.

To find the y-coordinate of the vertex, evaluate the function at $x = -2$.

$$F(x) = (x + 2)^2 - 1$$
$$F(-2) = (-2 + 2)^2 - 1 = -1$$

The y-coordinate of the vertex is -1.

From the results above, the coordinates of the vertex are $(-2, -1)$.

The axis of symmetry is the vertical line that passes through the vertex. The equation of the vertical line that passes through the point $(-2, -1)$ is $x = -2$. The axis of symmetry is the line $x = -2$.

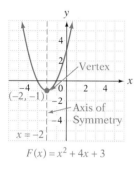

$F(x) = x^2 + 4x + 3$

By following the process illustrated in the preceding example and completing the square on $f(x) = ax^2 + bx + c$, we can find a formula for the coordinates of the vertex and the equation of the axis of symmetry of a parabola.

Vertex and Axis of Symmetry of a Parabola

Let $f(x) = ax^2 + bx + c$ be the equation of a parabola. The coordinates of the vertex are $\left(-\dfrac{b}{2a}, f\left(-\dfrac{b}{2a}\right)\right)$. The equation of the axis of symmetry is $x = -\dfrac{b}{2a}$.

➡ Find the vertex of the parabola whose equation is $g(x) = -2x^2 + 3x + 1$. Then graph the equation.

x-coordinate of the vertex: $-\dfrac{b}{2a} = -\dfrac{3}{2(-2)} = \dfrac{3}{4}$

• From the equation $g(x) = -2x^2 + 3x + 1$, $a = -2$, $b = 3$.

y-coordinate of the vertex: $g(x) = -2x^2 + 3x + 1$

$g\left(\dfrac{3}{4}\right) = -2\left(\dfrac{3}{4}\right)^2 + 3\left(\dfrac{3}{4}\right) + 1$

$g\left(\dfrac{3}{4}\right) = \dfrac{17}{8}$

• Evaluate the function at the value of the x-coordinate of the vertex.

The vertex is $\left(\dfrac{3}{4}, \dfrac{17}{8}\right)$.

Because a is negative, the graph opens down. Find a few ordered pairs that belong to the function and then sketch the graph, as shown at the left.

TAKE NOTE

Once the coordinates of the vertex are found, the range of the quadratic function can be determined.

Once the y-coordinate of the vertex is known, the range of the function can be determined. Here, the graph of g opens down, so the y-coordinate of the vertex is the largest value of y. Therefore the range of g is $\left\{y \mid y \le \dfrac{17}{8}\right\}$. The domain is $\{x \mid x \in \text{ real numbers}\}$.

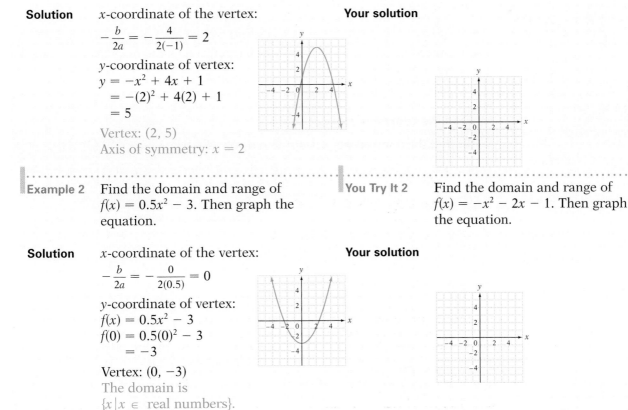

Example 1   Find the vertex and axis of symmetry of the parabola whose equation is $y = -x^2 + 4x + 1$. Then graph the equation.

Solution x-coordinate of the vertex:

$-\dfrac{b}{2a} = -\dfrac{4}{2(-1)} = 2$

y-coordinate of vertex:
$y = -x^2 + 4x + 1$
$= -(2)^2 + 4(2) + 1$
$= 5$

Vertex: $(2, 5)$
Axis of symmetry: $x = 2$

You Try It 1   Find the vertex and axis of symmetry of the parabola whose equation is $y = 4x^2 + 4x + 1$. Then graph the equation.

Your solution

Example 2   Find the domain and range of $f(x) = 0.5x^2 - 3$. Then graph the equation.

Solution x-coordinate of the vertex:

$-\dfrac{b}{2a} = -\dfrac{0}{2(0.5)} = 0$

y-coordinate of vertex:
$f(x) = 0.5x^2 - 3$
$f(0) = 0.5(0)^2 - 3$
$= -3$

Vertex: $(0, -3)$
The domain is
$\{x \mid x \in \text{ real numbers}\}$.
The range is $\{y \mid y \ge -3\}$.

You Try It 2   Find the domain and range of $f(x) = -x^2 - 2x - 1$. Then graph the equation.

Your solution

Solutions on p. S26

Objective B **To find the *x*-intercepts of a parabola**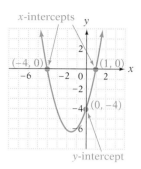

Recall that a point at which a graph crosses the *x*- or *y*-axis is called an *intercept* of the graph. The *x*-intercepts of the graph of an equation occur when $y = 0$; the *y*-intercepts occur when $x = 0$.

The graph of $y = x^2 + 3x - 4$ is shown at the right. The points whose coordinates are $(-4, 0)$ and $(1, 0)$ are the *x*-intercepts of the graph. The point whose coordinates are $(0, -4)$ is the *y*-intercept of the graph.

➡ Find the *x*-intercepts of the parabola whose equation is $y = 4x^2 - 4x + 1$.

To find the *x*-intercepts, let $y = 0$ and then solve for x.

$$y = 4x^2 - 4x + 1$$
$$0 = 4x^2 - 4x + 1$$
   • Let **y = 0**.

$$0 = (2x - 1)(2x - 1)$$
   • Solve for **x** by factoring and using the Principle of Zero Products.

$$
\begin{array}{ll}
2x - 1 = 0 & \quad 2x - 1 = 0 \\
2x = 1 & \quad 2x = 1 \\
x = \dfrac{1}{2} & \quad x = \dfrac{1}{2}
\end{array}
$$

The *x*-intercept is $\left(\dfrac{1}{2}, 0\right)$.

(graph at left labeled) $y = 4x^2 - 4x + 1$

In the preceding example, the parabola has only one *x*-intercept. In this case, the parabola is said to be **tangent** to the *x*-axis at $x = \dfrac{1}{2}$.

➡ Find the *x*-intercepts of $y = 2x^2 - x - 6$.

To find the *x*-intercepts, let $y = 0$ and solve for x.

$$y = 2x^2 - x - 6$$
$$0 = 2x^2 - x - 6$$
   • Let **y = 0**.
$$0 = (2x + 3)(x - 2)$$
   • Solve for **x** by factoring and using the Principle of Zero Products.

$$
\begin{array}{ll}
2x + 3 = 0 & \qquad x - 2 = 0 \\
x = -\dfrac{3}{2} & \qquad x = 2
\end{array}
$$

The *x*-intercepts are $\left(-\dfrac{3}{2}, 0\right)$ and $(2, 0)$.

TAKE NOTE

A zero of a function is the *x*-coordinate of the *x*-intercept of the graph of the function. Because the *x*-intercepts of the graph of $f(x) = 2x^2 - x - 6$ are $\left(-\dfrac{3}{2}, 0\right)$ and $(2, 0)$, the zeros are $-\dfrac{3}{2}$ and 2.

If the equation above, $y = 2x^2 - x - 6$, is written in functional notation as $f(x) = 2x^2 - x - 6$, then to find the *x*-intercepts you would let $f(x) = 0$ and solve for x. A value of x for which $f(x) = 0$ is a **zero** of the function. Thus, $-\dfrac{3}{2}$ and and 2 are *zeros* of $f(x) = 2x^2 - x - 6$.

⟹ Find the zeros of $f(x) = x^2 - 2x - 1$.

To find the zeros, let $f(x) = 0$ and solve for x.

$$f(x) = x^2 - 2x - 1$$
$$0 = x^2 - 2x - 1$$

$$x = \frac{-b \pm \sqrt{b^2 - 4ac}}{2a}$$

- Because $x^2 - 2x - 1$ does not easily factor, use the quadratic formula to solve for x.

$$= \frac{-(-2) \pm \sqrt{(-2)^2 - 4(1)(-1)}}{2(1)}$$

- $a = 1, b = -2, c = -1$

$$= \frac{2 \pm \sqrt{4 + 4}}{2} = \frac{2 \pm \sqrt{8}}{2}$$

$$= \frac{2 \pm 2\sqrt{2}}{2}$$

$$= 1 \pm \sqrt{2}$$

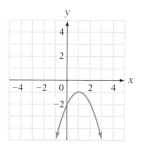

The zeros of the function are $1 - \sqrt{2}$ and $1 + \sqrt{2}$.

The preceding examples suggest that there is a relationship among the x-intercepts of the graph of a function, the zeros of a function, and the solutions of an equation. In fact, those three concepts are different ways of discussing the same number. The choice depends on the focus of the discussion. If we are discussing graphing, then the intercept is our focus; if we are discussing functions, then the zero of the function is our focus; and if we are discussing equations, the solution of the equation is our focus.

The graph of a parabola may not have x-intercepts. The graph of $y = -x^2 + 2x - 2$ is shown at the right. Note that the graph does not pass through the x-axis and thus there are no x-intercepts. This means there are no real number zeros of $f(x) = -x^2 + 2x - 2$ and there are no real number solutions of $-x^2 + 2x - 2 = 0$.

Using the quadratic formula, the solutions of the equation $-x^2 + 2x - 2 = 0$ are $1 - i$ and $1 + i$. Thus the zeros of $f(x) = -x^2 + 2x - 2$ are the complex numbers $1 - i$ and $1 + i$.

Recall that the *discriminant* of $ax^2 + bx + c$ is the expression $b^2 - 4ac$ and that this expression can be used to determine whether $ax^2 + bx + c = 0$ has zero, one, or two real number solutions. Because there is a connection between the solutions of $ax^2 + bx + c = 0$ and the x-intercepts of the graph of $y = ax^2 + bx + c$, the discriminant can be used to determine the number of x-intercepts of a parabola.

The Effect of the Discriminant on the Number of x-Intercepts of a Parabola

1. If $b^2 - 4ac = 0$, the parabola has one x-intercept.
2. If $b^2 - 4ac > 0$, the parabola has two x-intercepts.
3. If $b^2 - 4ac < 0$, the parabola has no x-intercepts.

⇒ Use the discriminant to determine the number of x-intercepts of the parabola whose equation is $y = 2x^2 - x + 2$.

$b^2 - 4ac$ • Evaluate the discriminant.

$(-1)^2 - 4(2)(2) = 1 - 16 = -15$ • $a = 2, b = -1, c = 2$

$-15 < 0$

The discriminant is less than zero. Therefore, the parabola has no x-intercepts.

Example 3

Find the x-intercepts of $y = 2x^2 - 5x + 2$.

Solution

$y = 2x^2 - 5x + 2$

$0 = 2x^2 - 5x + 2$ • Let $y = 0$.

$0 = (2x - 1)(x - 2)$ • Solve for x by factoring.

$2x - 1 = 0 \qquad x - 2 = 0$

$\qquad 2x = 1 \qquad\qquad x = 2$

$\qquad\quad x = \dfrac{1}{2}$

The x-intercepts are $\left(\dfrac{1}{2}, 0\right)$ and $(2, 0)$.

You Try It 3

Find the x-intercepts of $y = x^2 + 3x + 4$.

Your solution

Example 4

Find the zeros of $f(x) = x^2 + 4x + 5$.

Solution

$f(x) = x^2 + 4x + 5$

$\quad 0 = x^2 + 4x + 5$ • Let $f(x) = 0$.

$x = \dfrac{-b \pm \sqrt{b^2 - 4ac}}{2a}$ • Use the quadratic formula.

$\quad = \dfrac{-4 \pm \sqrt{4^2 - 4(1)(5)}}{2(1)}$ • $a = 1, b = 4, c = 5$

$\quad = \dfrac{-4 \pm \sqrt{16 - 20}}{2} = \dfrac{-4 \pm \sqrt{-4}}{2}$

$\quad = \dfrac{-4 \pm 2i}{2} = -2 \pm i$

The zeros of the function are $-2 + i$ and $-2 - i$.

You Try It 4

Find the zeros of $g(x) = x^2 - x + 6$.

Your solution

Example 5

Use the discriminant to determine the number of x-intercepts of $y = x^2 - 6x + 9$.

Solution

$b^2 - 4ac$ • Evaluate the discriminant.

$(-6)^2 - 4(1)(9)$ • $a = 1, b = -6, c = 9$

$\quad = 36 - 36 = 0$

The discriminant is equal to zero.

The parabola has one x-intercept.

You Try It 5

Use the discriminant to determine the number of x-intercepts of $y = x^2 - x - 6$.

Your solution

Solutions on pp. S26–S27

Objective C   **To find the minimum or maximum of a quadratic function**   < 9 >

The graph of $f(x) = x^2 - 2x + 3$ is shown at the right. Because a is positive, the parabola opens up. The vertex of the parabola is the lowest point on the parabola. It is the point that has the minimum y-coordinate. Therefore, the value of the function at this point is a **minimum.**

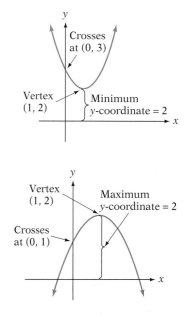

Crosses at (0, 3)

Vertex (1, 2)

Minimum y-coordinate = 2

The graph of $f(x) = -x^2 + 2x + 1$ is shown at the right. Because a is negative, the parabola opens down. The vertex of the parabola is the highest point on the parabola. It is the point that has the maximum y-coordinate. Therefore, the value of the function at this point is a **maximum.**

Vertex (1, 2)

Maximum y-coordinate = 2

Crosses at (0, 1)

To find the minimum or maximum value of a quadratic function, first find the x-coordinate of the vertex. Then evaluate the function at that value.

➡ Find the maximum or minimum value of the function $f(x) = -2x^2 + 4x + 3$.

$$x = -\frac{b}{2a} = -\frac{4}{2(-2)} = 1$$

• Find the x-coordinate of the vertex.
  $a = -2, b = 4$

$$f(x) = -2x^2 + 4x + 3$$
$$f(1) = -2(1)^2 + 4(1) + 3$$
$$= -2 + 4 + 3$$
$$= 5$$

• Evaluate the function at $x = 1$.

The maximum value of the function is 5.

• Because $a < 0$, the graph of f opens down. Therefore, the function has a maximum value.

· ·

Example 6

Find the maximum or minimum value of $f(x) = 2x^2 - 3x + 1$.

Solution

$$x = -\frac{b}{2a} = -\frac{-3}{2(2)} = \frac{3}{4}$$

$$f(x) = 2x^2 - 3x + 1$$

$$f\left(\frac{3}{4}\right) = 2\left(\frac{3}{4}\right)^2 - 3\left(\frac{3}{4}\right) + 1 = \frac{9}{8} - \frac{9}{4} + 1$$

$$= -\frac{1}{8}$$

Because a is positive, the graph opens up. The function has a minimum value.

The minimum value of the function is $-\frac{1}{8}$.

You Try It 6

Find the maximum or minimum value of $f(x) = -3x^2 + 4x - 1$.

Your solution

Solution on p. S27

Objective D   **To solve application problems**   〈 9 〉

➡ A mason is forming a rectangular floor for a storage shed. The perimeter of the rectangle is 44 ft. What dimensions of the rectangle would give the floor a maximum area? What is the maximum area?

We are given the perimeter of the rectangle, and we want to find the dimensions of the rectangle that will yield the maximum area for the floor. Use the equation for the perimeter of a rectangle.

$$P = 2L + 2W$$
$$44 = 2L + 2W$$
$$22 = L + W$$
$$22 - L = W$$

- $P = 44$
- Divide both sides of the equation by 2.
- Solve the equation for W.

Now use the equation for the area of a rectangle. Use substitution to express the area in terms of L.

$$A = LW$$
$$A = L(22 - L)$$

$$A = 22L - L^2$$

- From the equation above, $W = 22 - L$. Substitute $22 - L$ for W.
- The area of the rectangle is $22L - L^2$.

To find the length of the rectangle, find the L-coordinate of the vertex of the function $f(L) = -L^2 + 22L$.

$$L = -\frac{b}{2a} = -\frac{22}{2(-1)} = 11$$

- For the equation $f(L) = -L^2 + 22L$, $a = -1$ and $b = 22$.

The length of the rectangle is 11 ft.

To find the width, replace L in $22 - L$ by the L-coordinate of the vertex and evaluate.

$$W = 22 - L$$
$$W = 22 - 11 = 11$$

- Replace L by 11 and evaluate.

The width of the rectangle is 11 ft.

The dimensions of the rectangle that would give the floor a maximum area are 11 ft by 11 ft.

To find the maximum area of the floor, evaluate $f(L) = -L^2 + 22L$ at the L-coordinate of the vertex.

$$f(L) = -L^2 + 22L$$
$$f(11) = -(11)^2 + 22(11)$$
$$= -121 + 242 = 121$$

- Evaluate the function at 11.

The maximum area of the floor is 121 ft².

The graph of the function $f(L) = -L^2 + 22L$ is shown at the right. Note that the vertex of the parabola is $(11, 121)$. For any value of L less than 11, the area of the floor will be less than 121 ft². For any value of L greater than 11, the area of the floor will be less than 121 ft². The maximum value of the function is 121, and the maximum value occurs when $L = 11$.

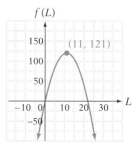

Example 7

A mining company has determined that the cost c, in dollars per ton, of mining a mineral is given by $c(x) = 0.2x^2 - 2x + 12$, where x is the number of tons of the mineral mined. Find the number of tons of the mineral that should be mined to minimize the cost. What is the minimum cost per ton?

Strategy

- To find the number of tons of the mineral that should be mined to minimize the cost, find the x-coordinate of the vertex.
- To find the minimum cost, evaluate $c(x)$ at the x-coordinate of the vertex.

Solution

$$c(x) = 0.2x^2 - 2x + 12 \qquad \bullet \; a = 0.2, \, b = -2,$$
$$c = 12$$

$$x = -\frac{b}{2a} = -\frac{(-2)}{2(0.2)} = 5$$

To minimize the cost, 5 tons of the mineral should be mined.

$$c(x) = 0.2x^2 - 2x + 12$$
$$c(5) = 0.2(5)^2 - 2(5) + 12$$
$$= 5 - 10 + 12$$
$$= 7$$

The minimum cost is $7 per ton.

Note: The graph of the function $c(x) = 0.2x^2 - 2x + 12$ is shown below. The vertex of the parabola is $(5, 7)$. For any value of x less than 5, the cost per ton is greater than $7. For any value of x greater than 5, the cost per ton is greater than $7. 7 is the minimum value of the function, and the minimum value occurs when $x = 5$.

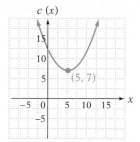

You Try It 7

The height s, in feet, of a ball thrown straight up is given by $s(t) = -16t^2 + 64t$, where t is the time in seconds. Find the time it takes the ball to reach its maximum height. What is the maximum height?

Your strategy

Your solution

Solution on p. S27

Example 8

Find two numbers whose difference is 10 and whose product is a minimum. What is the minimum product of the two numbers?

You Try It 8

A rectangular fence is being constructed along a stream to enclose a picnic area. If there are 100 ft of fence available, what dimensions of the rectangle will produce the maximum area for picnicking?

Strategy

- Let x represent one number. Because the difference between the two numbers is 10,

$$x + 10$$

represents the other number.
[*Note:* $(x + 10) - (x) = 10$]
Then their product is represented by

$$x(x + 10) = x^2 + 10x$$

- To find one of the two numbers, find the x-coordinate of the vertex of $f(x) = x^2 + 10x$.
- To find the other number, replace x in $x + 10$ by the x-coordinate of the vertex and evaluate.
- To find the minimum product, evaluate the function at the x-coordinate of the vertex.

Your strategy

Solution

$f(x) = x^2 + 10x$

$x = -\dfrac{b}{2a} = -\dfrac{10}{2(1)} = -5$

- $a = 1, b = 10, c = 0$
- One number is -5.

$\begin{aligned} &x + 10 \\ &-5 + 10 = 5 \end{aligned}$

- The other number is 5.

The numbers are -5 and 5.

$\begin{aligned} f(x) &= x^2 + 10x \\ f(-5) &= (-5)^2 + 10(-5) \\ &= 25 - 50 \\ &= -25 \end{aligned}$

The minimum product of the two numbers is -25.

Your solution

Solution on p. S27

9.1 Exercises

Objective A

1. What is a quadratic function?

2. Describe the graph of a parabola.

3. What is the vertex of a parabola?

4. What is the axis of symmetry of the graph of a parabola?

5. The axis of symmetry of a parabola is the line $x = -5$. What is the x-coordinate of the vertex of the parabola?

6. The axis of symmetry of a parabola is the line $x = 8$. What is the x-coordinate of the vertex of the parabola?

7. The vertex of a parabola is $(7, -9)$. What is the equation of the axis of symmetry of the parabola?

8. The vertex of a parabola is $(-4, 10)$. What is the equation of the axis of symmetry of the parabola?

Find the vertex and axis of symmetry of the parabola given by each equation. Then graph the equation.

9. $y = x^2 - 2x - 4$ 10. $y = x^2 + 4x - 4$ 11. $y = -x^2 + 2x - 3$

12. $y = -x^2 + 4x - 5$ 13. $f(x) = x^2 - x - 6$ 14. $G(x) = x^2 - x - 2$

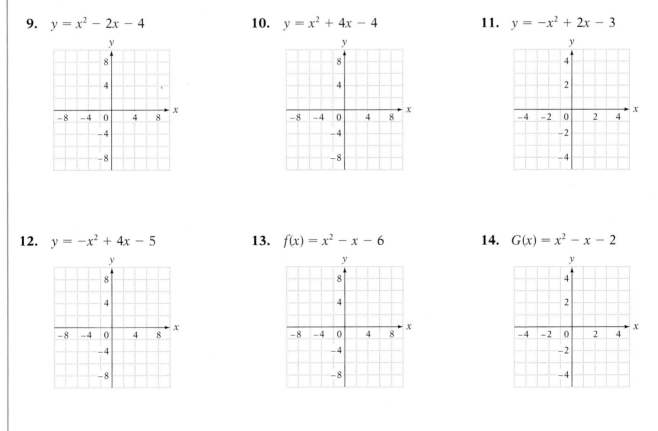

15. $F(x) = x^2 - 3x + 2$

16. $y = 2x^2 - 4x + 1$

17. $y = -2x^2 + 6x$

18. $y = \dfrac{1}{2}x^2 + 4$

19. $y = -\dfrac{1}{4}x^2 - 1$

20. $h(x) = \dfrac{1}{2}x^2 - x + 1$

21. $P(x) = -\dfrac{1}{2}x^2 + 2x - 3$

22. $y = \dfrac{1}{2}x^2 + 2x - 6$

23. $y = -\dfrac{1}{2}x^2 + x - 3$

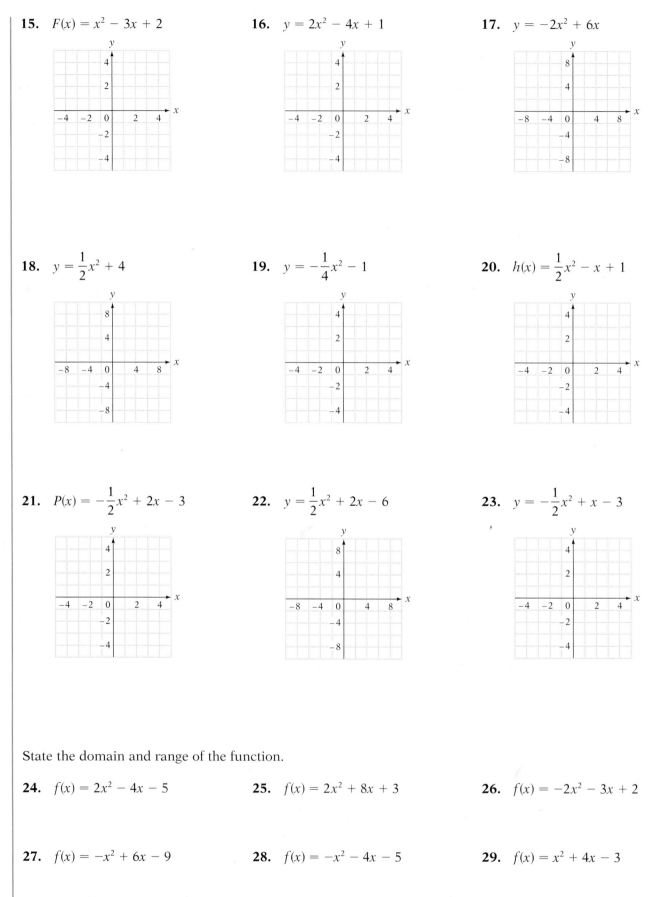

State the domain and range of the function.

24. $f(x) = 2x^2 - 4x - 5$

25. $f(x) = 2x^2 + 8x + 3$

26. $f(x) = -2x^2 - 3x + 2$

27. $f(x) = -x^2 + 6x - 9$

28. $f(x) = -x^2 - 4x - 5$

29. $f(x) = x^2 + 4x - 3$

Objective B

30. **a.** What is an x-intercept of the graph of a parabola?
   b. How many x-intercepts can the graph of a parabola have?

31. **a.** What is the y-intercept of the graph of a parabola?
   b. How many y-intercepts can the graph of a parabola have?

Find the x-intercepts of the parabola given by each equation.

32. $y = x^2 - 4$

33. $y = x^2 - 9$

34. $y = 2x^2 - 4x$

35. $y = 3x^2 + 6x$

36. $y = x^2 - x - 2$

37. $y = x^2 - 2x - 8$

38. $y = 2x^2 - x - 1$

39. $y = 2x^2 - 5x - 3$

40. $y = x^2 + 2x - 1$

41. $y = x^2 + 4x - 3$

42. $y = x^2 + 6x + 10$

43. $y = -x^2 - 4x - 5$

Find the zeros of the function.

44. $f(x) = x^2 + 3x + 2$

45. $f(x) = x^2 - 6x + 9$

46. $f(x) = -x^2 + 4x - 5$

47. $f(x) = -x^2 + 3x + 8$

48. $f(x) = 2x^2 - 3x$

49. $f(x) = -3x^2 + 4x$

50. $f(x) = 2x^2 - 4$

51. $f(x) = 3x^2 + 6$

52. $f(x) = 2x^2 + 3x + 2$

53. $f(x) = 3x^2 - x + 4$

54. $f(x) = -3x^2 - 4x + 1$

55. $f(x) = -2x^2 + x + 5$

Use the discriminant to determine the number of x-intercepts of the graph.

56. $y = 2x^2 + 2x - 1$

57. $y = -x^2 - x + 3$

58. $y = x^2 - 8x + 16$

59. $y = x^2 - 10x + 25$ **60.** $y = -3x^2 - x - 2$ **61.** $y = -2x^2 + x - 1$

62. $y = -2x^2 + x + 1$ **63.** $y = 4x^2 - x - 2$ **64.** $y = 2x^2 + x + 1$

65. $y = 2x^2 + x + 4$ **66.** $y = -3x^2 + 2x - 8$ **67.** $y = 4x^2 + 2x - 5$

68. The zeros of the function $f(x) = x^2 - 2x - 3$ are -1 and 3. What are the x-intercepts of the graph of the equation $y = x^2 - 2x - 3$?

69. The zeros of the function $f(x) = x^2 - x - 20$ are -4 and 5. What are the x-intercepts of the graph of the equation $y = x^2 - x - 20$?

Objective C

70. What is the minimum or maximum value of a quadratic function?

71. Describe how to find the minimum or maximum value of a quadratic function.

72. Does the function have a minimum or a maximum value?
  a. $f(x) = -x^2 + 6x - 1$ **b.** $f(x) = 2x^2 - 4$ **c.** $f(x) = -5x^2 + x$

73. Does the function have a minimum or a maximum value?
  a. $f(x) = 3x^2 - 2x + 4$ **b.** $f(x) = -x^2 + 9$ **c.** $f(x) = 6x^2 - 3x$

Find the minimum or maximum value of the quadratic function.

74. $f(x) = x^2 - 2x + 3$ **75.** $f(x) = 2x^2 + 4x$ **76.** $f(x) = -2x^2 + 4x - 3$

77. $f(x) = -2x^2 + 4x - 5$ **78.** $f(x) = -2x^2 - 3x + 4$ **79.** $f(x) = -2x^2 - 3x$

80. $f(x) = 2x^2 + 3x - 8$ **81.** $f(x) = 3x^2 + 3x - 2$ **82.** $f(x) = -3x^2 + x - 6$

83. $f(x) = -x^2 - x + 2$ **84.** $f(x) = x^2 - 5x + 3$ **85.** $f(x) = 3x^2 + 5x + 2$

86. Which of the following parabolas has the greatest minimum value?
a. $y = x^2 - 2x - 3$ **b.** $y = x^2 - 10x + 20$ **c.** $y = 3x^2 - 6$

87. Which of the following parabolas has the greatest maximum value?
a. $y = -2x^2 + 2x - 1$ **b.** $y = -x^2 + 8x - 2$ **c.** $y = -4x^2 + 3$

88. The vertex of a parabola that opens up is $(-4, 7)$. Does the function have a maximum or a minimum value? What is the maximum or minimum value of the function?

89. The vertex of a parabola that opens down is $(3, -5)$. Does the function have a maximum or a minimum value? What is the maximum or minimum value of the function?

Objective D

90. The height s, in feet, of a rock thrown upward at an initial speed of 64 ft/s from a cliff 50 ft above an ocean beach is given by the function $s(t) = -16t^2 + 64t + 50$, where t is the time in seconds. Find the maximum height above the beach that the rock will attain.

91. An event in the Summer Olympics is 10-meter springboard diving. In this event, the height s, in meters, of a diver above the water t seconds after jumping is given by $s(t) = -4.9t^2 + 7.8t + 10$. What is the maximum height that the diver will be above the water? Round to the nearest tenth.

92. A tour operator believes that the profit P, in dollars, from selling x tickets is given by $P(x) = 40x - 0.25x^2$. Using this model, what is the maximum profit the tour operator can expect?

93. A manufacturer of microwave ovens believes that the revenue R, in dollars, the company receives is related to the price P, in dollars, of an oven by the function $R(P) = 125P - 0.25P^2$. What price will give the maximum revenue?

94. A pool is treated with a chemical to reduce the amount of algae in the pool. The amount of algae in the pool t days after the treatment can be approximated by the function $A(t) = 40t^2 - 400t + 500$. How many days after the treatment will the pool have the least amount of algae?

95. The suspension cable that supports a small footbridge hangs in the shape of a parabola. The height h, in feet, of the cable above the bridge is given by the function $h(x) = 0.25x^2 - 0.8x + 25$, where x is the distance in feet from one end of the bridge. What is the minimum height of the cable above the bridge?

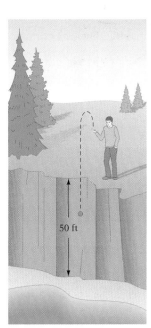

50 ft

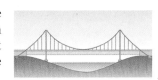

96. An equation that models the thickness h, in inches, of the mirror at the Palomar Mountain Observatory is given by $h(x) = 0.000379x^2 - 0.0758x + 24$, where x is measured in inches from the edge of the mirror. Find the minimum thickness of the mirror.

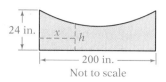

97. The Buningham Fountain in Chicago shoots water from a nozzle at the base of the fountain. The height h, in feet, of the water above the ground t seconds after it leaves the nozzle is given by the function $h(t) = -16t^2 + 90t + 15$. What is the maximum height of the water? Round to the nearest tenth.

98. The height s, in feet, of water squirting from a fire hose nozzle is given by the equation $s(x) = -\dfrac{1}{30}x^2 + 2x + 5$, where x is the horizontal distance, in feet, from the nozzle. How high on a building 40 ft from the fire hose will the water land?

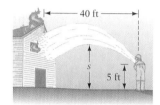

99. On wet concrete the stopping distance s, in feet, of a car traveling v miles per hour is given by the function $s(v) = 0.055v^2 + 1.1v$. What is the maximum speed at which a car could be traveling and still stop at a stop sign 44 ft away?

100. Find two numbers whose sum is 20 and whose product is a maximum.

101. Find two numbers whose difference is 14 and whose product is a minimum.

102. A rancher has 200 ft of fencing to build a rectangular corral alongside an existing fence. Determine the dimensions of the corral that will maximize the enclosed area.

APPLYING THE CONCEPTS

103. One root of the quadratic equation $2x^2 - 5x + k = 0$ is 4. What is the other root?

104. What is the value of k if the vertex of the parabola $y = x^2 - 8x + k$ is a point on the x-axis?

105. The roots of the function $f(x) = mx^2 + nx + 1$ are -2 and 3. What are the roots of the function $g(x) = nx^2 + mx - 1$?

106. Show that the minimum value of
$$S(x) = (2 - x)^2 + (5 - x)^2 + (4 - x)^2 + (7 - x)^2$$
occurs when x is the average of the numbers 2, 5, 4, and 7.

9.2 Graphs of Functions

Objective A   **To graph functions**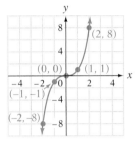

The graphs of the polynomial functions $L(x) = mx + b$ (a straight line) and $Q(x) = ax^2 + bx + c$, $a \neq 0$ (a parabola) were discussed in previous objectives. The graphs of other functions can also be drawn by finding ordered pairs that belong to the function, plotting the points that correspond to the ordered pairs, and then drawing a curve through the points.

➡ Graph: $F(x) = x^3$

Select several values of x and evaluate the function.

x	$F(x) = x^3$	$F(x)$	(x, y)
-2	$(-2)^3$	-8	$(-2, -8)$
-1	$(-1)^3$	-1	$(-1, -1)$
0	0^3	0	$(0, 0)$
1	1^3	1	$(1, 1)$
2	2^3	8	$(2, 8)$

> **TAKE NOTE**
> The units along the x-axis are different from those along the y-axis. If the units along the x-axis were the same as those along the y-axis, the graph would appear narrower than the one shown.

Plot the ordered pairs and draw a graph through the points.

➡ Graph: $g(x) = x^3 - 4x + 5$

Select several values of x and evaluate the function.

x	$g(x) = x^3 - 4x + 5$	$g(x)$	(x, y)
-3	$(-3)^3 - 4(-3) + 5$	-10	$(-3, -10)$
-2	$(-2)^3 - 4(-2) + 5$	5	$(-2, 5)$
-1	$(-1)^3 - 4(-1) + 5$	8	$(-1, 8)$
0	$(0)^3 - 4(0) + 5$	5	$(0, 5)$
1	$(1)^3 - 4(1) + 5$	2	$(1, 2)$
2	$(2)^3 - 4(2) + 5$	5	$(2, 5)$

Plot the ordered pairs and draw a graph through the points.

Note from the graphs of the two cubic functions above that the shapes of the graphs can be quite different. The following graphs of typical cubic polynomial functions show the general shape of a cubic polynomial.

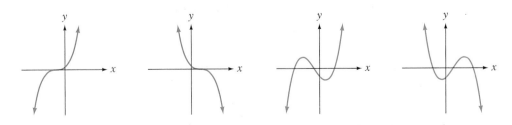

As the degree of a polynomial increases, the graph of the polynomial function can change significantly. In such cases, it may be necessary to plot many points before an accurate graph can be drawn. Only polynomials of degree 3 are considered here.

⇒ Graph: $f(x) = |x + 2|$

This is an **absolute-value function.**

x	$f(x) = \lvert x + 2 \rvert$	$f(x)$	(x, y)
-3	$\lvert -3 + 2 \rvert$	1	$(-3, 1)$
-2	$\lvert -2 + 2 \rvert$	0	$(-2, 0)$
-1	$\lvert -1 + 2 \rvert$	1	$(-1, 1)$
0	$\lvert 0 + 2 \rvert$	2	$(0, 2)$
1	$\lvert 1 + 2 \rvert$	3	$(1, 3)$
2	$\lvert 2 + 2 \rvert$	4	$(2, 4)$

In general, the graph of the absolute value of a linear polynomial is V-shaped.

⇒ Graph: $R(x) = \sqrt{2x - 4}$

This is a **radical function.** Because the square root of a negative number is not a real number, the domain of this function requires that $2x - 4 \geq 0$. Solve this inequality for x.

$$2x - 4 \geq 0$$
$$2x \geq 4$$
$$x \geq 2$$

The domain is $\{x \mid x \geq 2\}$. This means that only values of x that are greater than or equal to 2 can be chosen as values at which to evaluate the function. In this case, some of the y-coordinates must be approximated.

x	$R(x) = \sqrt{2x - 4}$	$R(x)$	(x, y)
2	$\sqrt{2(2) - 4}$	0	$(2, 0)$
3	$\sqrt{2(3) - 4}$	1.41	$(3, 1.41)$
4	$\sqrt{2(4) - 4}$	2	$(4, 2)$
5	$\sqrt{2(5) - 4}$	2.45	$(5, 2.45)$
6	$\sqrt{2(6) - 4}$	2.83	$(6, 2.83)$

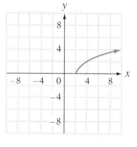

Recall that a function is a special type of relation, one for which no two ordered pairs have the same first coordinate. Graphically, this means that the graph of a function cannot pass through two points that have the same x-coordinate and different y-coordinates. For instance, the graph at the right is not the graph of a function because there are ordered pairs with the same x-coordinate and different y-coordinates.

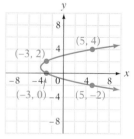

The last graph illustrates a general statement that can be made about whether a graph defines a function. It is called the **vertical-line test.**

> **Vertical-Line Test**
>
> A graph defines a function if any vertical line intersects the graph at no more than one point.

For example, the graph of a nonvertical straight line is the graph of a function. Any vertical line intersects the graph no more than once. The graph of a circle, however, is not the graph of a function. There are vertical lines that intersect the graph at more than one point.

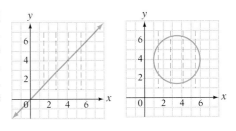

Some practical applications can be modeled by a graph that is not the graph of a function. Below is an example of such an application.

One of the causes of smog is an inversion layer of air where temperatures at higher altitudes are warmer than those at lower altitudes. The graph at the right shows the altitudes at which various temperatures were recorded. As shown by the dashed lines in the graph, there are two altitudes at which the temperature was 25°C. This means that there are two ordered pairs (shown in the graph) with the same first coordinate but different second coordinates. The graph does not define a function.

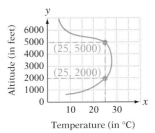

When a graph does define a function, the domain and range can be estimated from the graph.

TAKE NOTE

To determine the domain, think of collapsing the graph onto the *x*-axis and determining the interval on the *x*-axis that the graph covers. To determine the range, think of collapsing the graph onto the *y*-axis and determining the interval on the *y*-axis that the graph covers.

➡ Determine the domain and range of the function given by the graph at the right. Write the answer in set-builder notation.

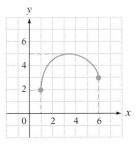

The solid dots on the graph indicate its beginning and ending points.

The domain is the set of *x*-coordinates.
Domain: $\{x \mid 1 \le x \le 6\}$

The range is the set of *y*-coordinates.
Range: $\{y \mid 2 \le y \le 5\}$

TAKE NOTE

In the first example, the domain and range are given in set-builder notation. In the second example, the domain and range are given in interval notation. Either method may be used.

➡ Determine the domain and range of the function given by the graph at the right. Write the answer in interval notation.

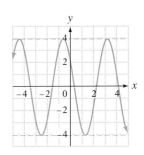

The arrows on the graph indicate that the graph continues in the same manner.

The domain is the set of *x*-coordinates.
Domain: $(-\infty, \infty)$

The range is the set of *y*-coordinates.
Range: $[-4, 4]$

Example 1

Use the vertical-line test to determine whether the graph shown is the graph of a function.

Solution

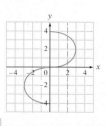

A vertical line intersects the graph more than once. The graph is not the graph of a function.

Example 2

Graph and write in set-builder notation the domain and range of the function defined by $f(x) = x^3 - 3x$.

Solution

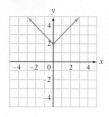

Domain: $\{x | x \in \text{real numbers}\}$
Range: $\{y | y \in \text{real numbers}\}$

Example 3

Graph and write in interval notation the domain and range of the function defined by $f(x) = |x| + 2$.

Solution

Domain: $(-\infty, \infty)$
Range: $[2, \infty)$

Example 4

Graph and write in set-builder notation the domain and range of the function defined by $f(x) = \sqrt{2 - x}$.

Solution

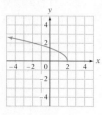

Domain: $\{x | x \leq 2\}$
Range: $\{y | y \geq 0\}$

You Try It 1

Use the vertical-line test to determine whether the graph shown is the graph of a function.

Your solution

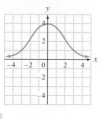

You Try It 2

Graph and write in interval notation the domain and range of the function defined by $f(x) = -\frac{1}{2}x^3 + 2x$.

Your solution

You Try It 3

Graph and write in set-builder notation the domain and range of the function defined by $f(x) = |x - 3|$.

Your solution

You Try It 4

Graph and write in interval notation the domain and range of the function defined by $f(x) = -\sqrt{x - 1}$.

Your solution

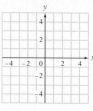

Solutions on pp. S27–S28

9.2 Exercises

Objective A

Use the vertical-line test to determine whether the graph is the graph of a function.

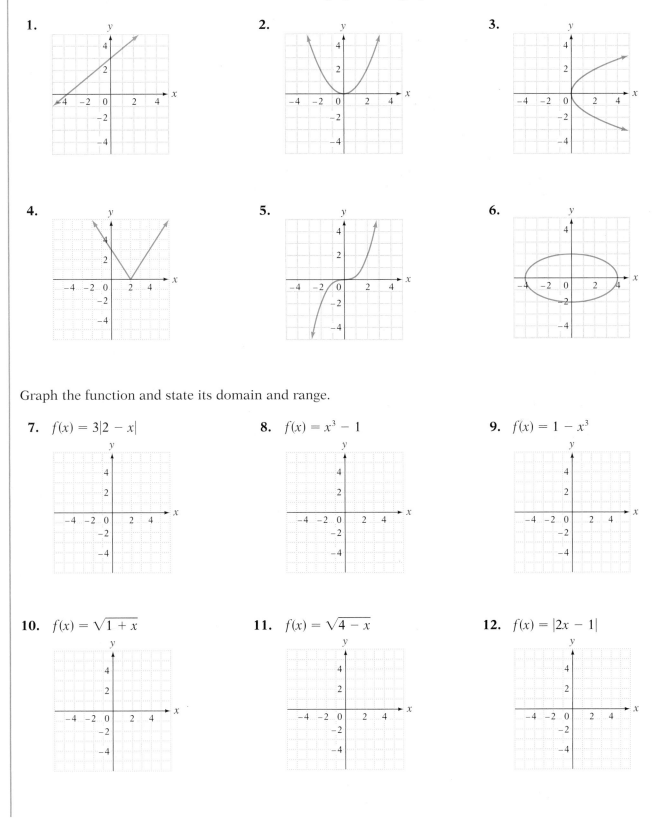

1.

2.

3.

4.

5.

6.

Graph the function and state its domain and range.

7. $f(x) = 3|2 - x|$

8. $f(x) = x^3 - 1$

9. $f(x) = 1 - x^3$

10. $f(x) = \sqrt{1 + x}$

11. $f(x) = \sqrt{4 - x}$

12. $f(x) = |2x - 1|$

13. $f(x) = x^3 + 4x^2 + 4x$ **14.** $f(x) = x^3 - x^2 - x + 1$ **15.** $f(x) = -\sqrt{x + 2}$

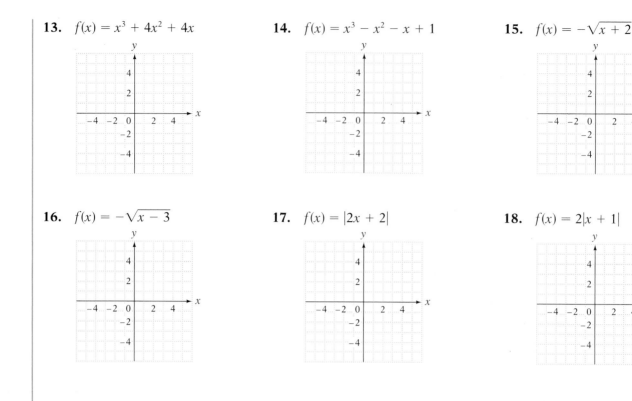

16. $f(x) = -\sqrt{x - 3}$ **17.** $f(x) = |2x + 2|$ **18.** $f(x) = 2|x + 1|$

APPLYING THE CONCEPTS

19. If $f(x) = \sqrt{x - 2}$ and $f(a) = 4$, find a.

20. If $f(x) = \sqrt{x + 5}$ and $f(a) = 3$, find a.

21. $f(a, b) =$ the sum of a and b
$g(a, b) =$ the product of a and b
Find $f(2, 5) + g(2, 5)$.

22. The graph of the function f is shown at the right. For this function, which of the following are true?
  a. $f(4) = 1$ **b.** $f(0) = 3$ **c.** $f(-3) = 2$

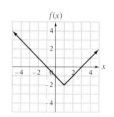

23. Let $f(x)$ be the digit in the xth decimal place of the repeating decimal $0.\overline{387}$. For example, $f(3) = 7$ because 7 is the digit in the third decimal place. Find $f(14)$.

24. Given $f(x) = (x + 1)(x - 1)$, for what values of x is $f(x)$ negative? Write your answer in set-builder notation.

25. Given $f(x) = (x + 2)(x - 2)$, for what values of x is $f(x)$ negative? Write your answer in set-builder notation.

26. Given $f(x) = -|x + 3|$, for what value of x is $f(x)$ greatest?

27. Given $f(x) = |2x - 2|$, for what value of x is $f(x)$ smallest?

9.3   Algebra of Functions

Objective A   **To perform operations on functions**

The operations of addition, subtraction, multiplication, and division of functions are defined as follows.

> **Operations on Functions**
>
> If f and g are functions and x is an element of the domain of each function, then
>
> $(f + g)(x) = f(x) + g(x)$ $\qquad$ $(f \cdot g)(x) = f(x) \cdot g(x)$
>
> $(f - g)(x) = f(x) - g(x)$ $\qquad$ $\left(\dfrac{f}{g}\right)(x) = \dfrac{f(x)}{g(x)}, g(x) \neq 0$

⟹ Given $f(x) = x^2 + 1$ and $g(x) = 3x - 2$, find $(f + g)(3)$ and $(f \cdot g)(-1)$.

$$
\begin{aligned}
(f + g)(3) &= f(3) + g(3)\\
&= [(3)^2 + 1] + [3(3) - 2]\\
&= 10 + 7 = 17
\end{aligned}
$$

$$
\begin{aligned}
(f \cdot g)(-1) &= f(-1) \cdot g(-1)\\
&= [(-1)^2 + 1] \cdot [3(-1) - 2]\\
&= 2 \cdot (-5) = -10
\end{aligned}
$$

Consider the functions f and g from the last example. Let $S(x)$ be the sum of the two functions. Then

$$
\begin{aligned}
S(x) = (f + g)(x) &= f(x) + g(x)\\
&= [x^2 + 1] + [3x - 2]\\
S(x) &= x^2 + 3x - 1
\end{aligned}
$$

• The definition of addition of functions

• $f(x) = x^2 + 1, g(x) = 3x - 2$

Now evaluate $S(3)$.

$$
\begin{aligned}
S(x) &= x^2 + 3x - 1\\
S(3) &= (3)^2 + 3(3) - 1\\
&= 9 + 9 - 1\\
&= 17 = (f + g)(3)
\end{aligned}
$$

Note that $S(3) = 17$ and $(f + g)(3) = 17$. This shows that adding $f(x) + g(x)$ and then evaluating is the same as evaluating $f(x)$ and $g(x)$ and then adding. The same is true for the other operations on functions. For instance, let $P(x)$ be the product of the functions f and g. Then

$$
\begin{aligned}
P(x) = (f \cdot g)(x) &= f(x) \cdot g(x)\\
&= (x^2 + 1)(3x - 2)\\
&= 3x^3 - 2x^2 + 3x - 2\\
P(-1) &= 3(-1)^3 - 2(-1)^2 + 3(-1) - 2\\
&= -3 - 2 - 3 - 2\\
&= -10
\end{aligned}
$$

• Note the $P(-1) = -10$ and $(f \cdot g)(-1) = -10$.

➡ Given $f(x) = 2x^2 - 5x + 3$ and $g(x) = x^2 - 1$, find $\left(\dfrac{f}{g}\right)(1)$.

$$\left(\dfrac{f}{g}\right)(1) = \dfrac{f(1)}{g(1)}$$

$$= \dfrac{2(1)^2 - 5(1) + 3}{(1)^2 - 1} = \dfrac{0}{0} \qquad \bullet \text{ Not a real number}$$

Because $\dfrac{0}{0}$ is not defined, the expression $\left(\dfrac{f}{g}\right)(1)$ cannot be evaluated.

Example 1

Given $f(x) = x^2 - x + 1$ and $g(x) = x^3 - 4$, find $(f - g)(3)$.

Solution

$$\begin{aligned}(f - g)(3) &= f(3) - g(3) \\ &= (3^2 - 3 + 1) - (3^3 - 4) \\ &= 7 - 23 \\ &= -16\end{aligned}$$

$(f - g)(3) = -16$

You Try It 1

Given $f(x) = x^2 + 2x$ and $g(x) = 5x - 2$, find $(f + g)(-2)$.

Your solution

Example 2

Given $f(x) = x^2 + 2$ and $g(x) = 2x + 3$, find $(f \cdot g)(-2)$.

Solution

$$\begin{aligned}(f \cdot g)(-2) &= f(-2) \cdot g(-2) \\ &= [(-2)^2 + 2] \cdot [2(-2) + 3] \\ &= 6(-1) \\ &= -6\end{aligned}$$

$(f \cdot g)(-2) = -6$

You Try It 2

Given $f(x) = 4 - x^2$ and $g(x) = 3x - 4$, find $(f \cdot g)(3)$.

Your solution

Example 3

Given $f(x) = x^2 + 4x + 4$ and $g(x) = x^3 - 2$, find $\left(\dfrac{f}{g}\right)(3)$.

Solution

$$\begin{aligned}\left(\dfrac{f}{g}\right)(3) &= \dfrac{f(3)}{g(3)} \\ &= \dfrac{3^2 + 4(3) + 4}{3^3 - 2} \\ &= \dfrac{25}{25} \\ &= 1\end{aligned}$$

$\left(\dfrac{f}{g}\right)(3) = 1$

You Try It 3

Given $f(x) = x^2 - 4$ and $g(x) = x^2 + 2x + 1$, find $\left(\dfrac{f}{g}\right)(4)$.

Your solution

Solutions on p. S28

Objective B   **To find the composition of two functions**

A function can be evaluated at the value of another function. Consider

$$f(x) = 2x + 7 \qquad \text{and} \qquad g(x) = x^2 + 1$$

The expression $f[g(-2)]$ means to evaluate the function f at $g(-2)$.

$$g(-2) = (-2)^2 + 1 = 4 + 1 = 5$$
$$f[g(-2)] = f(5) = 2(5) + 7 = 10 + 7 = 17$$

Definition of the Composition of Two Functions

Let f and g be two functions such that $g(x)$ is in the domain of f for all x in the domain of g. Then the **composition** of the two functions, denoted by $f \circ g$, is the function whose value at x is given by $(f \circ g)(x) = f[g(x)]$.

The function defined by $f[g(x)]$ is called the **composite** of f and g.

The function machine at the right illustrates the composition of the two functions $g(x) = x^2$ and $f(x) = 2x$. Note that a composite function combines two functions. First one function pairs an input with an output. Then that output is used as the input for a second function, which in turn produces a final output.

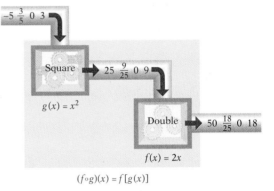

$(f \circ g)(x) = f[g(x)]$

The requirement in the definition of the composition of two functions that $g(x)$ be in the domain of f for all x in the domain of g is important. For instance, let

$$f(x) = \frac{1}{x - 1} \qquad \text{and} \qquad g(x) = 3x - 5$$

When $x = 2$,

$$g(2) = 3(2) - 5 = 1$$

$$f[g(2)] = f(1) = \frac{1}{1 - 1} = \frac{1}{0} \qquad \bullet \text{ This is not a real number.}$$

In this case, $g(2)$ is not in the domain of f. Thus the composition is not defined at 2.

⟹ Given $f(x) = x^3 - x + 1$ and $g(x) = 2x^2 - 10$, evaluate $(g \circ f)(2)$.

$$f(2) = (2)^3 - (2) + 1$$
$$= 7$$
$$(g \circ f)(2) = g[f(2)]$$
$$= g(7)$$
$$= 2(7)^2 - 10 = 88$$
$$(g \circ f)(2) = 88$$

⇒ Given $f(x) = 3x - 2$ and $g(x) = x^2 - 2x$, find $(f \circ g)(x)$.

$$(f \circ g)(x) = f[g(x)]$$
$$= 3(x^2 - 2x) - 2$$
$$= 3x^2 - 6x - 2$$

When evaluating compositions of functions, the order in which the functions are applied is important. In the two diagrams below, the order of the *square* function, $g(x) = x^2$, and the *double* function, $f(x) = 2x$, is interchanged. Notice that the final outputs are different. Therefore, $(f \circ g)(x) \neq (g \circ f)(x)$.

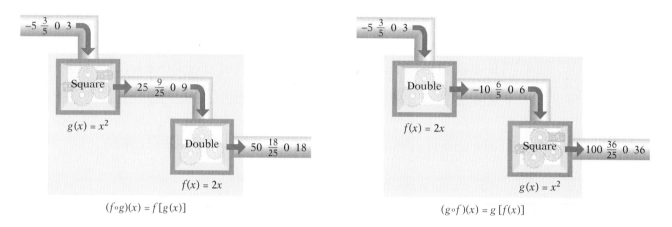

$$(f \circ g)(x) = f[g(x)] \qquad\qquad (g \circ f)(x) = g[f(x)]$$

Example 4

Given $f(x) = x^2 - x$ and $g(x) = 3x - 2$, find $f[g(3)]$.

Solution

$$g(x) = 3x - 2$$
$$g(3) = 3(3) - 2 = 9 - 2 = 7$$
$$f(x) = x^2 - x$$
$$f[g(3)] = f(7) = 7^2 - 7 = 42$$

You Try It 4

Given $f(x) = 1 - 2x$ and $g(x) = x^2$, find $f[g(-1)]$.

Your solution

Example 5

Given $s(t) = t^2 + 3t - 1$ and $v(t) = 2t + 1$, determine $s[v(t)]$.

Solution

$$s(t) = t^2 + 3t - 1$$
$$s[v(t)] = (2t + 1)^2 + 3(2t + 1) - 1$$
$$= (4t^2 + 4t + 1) + 6t + 3 - 1$$
$$= 4t^2 + 10t + 3$$

You Try It 5

Given $L(s) = s + 1$ and $M(s) = s^3 + 1$, determine $M[L(s)]$.

Your solution

Solutions on p. S28

9.3 Exercises

Objective A

For $f(x) = 2x^2 - 3$ and $g(x) = -2x + 4$, find:

1. $f(2) - g(2)$

2. $f(3) - g(3)$

3. $f(0) + g(0)$

4. $f(1) + g(1)$

5. $(f \cdot g)(2)$

6. $(f \cdot g)(-1)$

7. $\left(\dfrac{f}{g}\right)(4)$

8. $\left(\dfrac{f}{g}\right)(-1)$

9. $\left(\dfrac{g}{f}\right)(-3)$

For $f(x) = 2x^2 + 3x - 1$ and $g(x) = 2x - 4$, find:

10. $f(-3) + g(-3)$

11. $f(1) + g(1)$

12. $f(-2) - g(-2)$

13. $f(4) - g(4)$

14. $(f \cdot g)(-2)$

15. $(f \cdot g)(1)$

16. $\left(\dfrac{f}{g}\right)(2)$

17. $\left(\dfrac{f}{g}\right)(-3)$

18. $(f \cdot g)\left(\dfrac{1}{2}\right)$

For $f(x) = x^2 + 3x - 5$ and $g(x) = x^3 - 2x + 3$, find:

19. $f(2) - g(2)$

20. $(f \cdot g)(-3)$

21. $\left(\dfrac{f}{g}\right)(-2)$

Objective B

Given $f(x) = 2x - 3$ and $g(x) = 4x - 1$, evaluate the composite function.

22. $f[g(0)]$

23. $g[f(0)]$

24. $f[g(2)]$

25. $g[f(-2)]$

26. $f[g(x)]$

27. $g[f(x)]$

Given $h(x) = 2x + 4$ and $f(x) = \frac{1}{2}x + 2$, evaluate the composite function.

28. $h[f(0)]$ **29.** $f[h(0)]$ **30.** $h[f(2)]$

31. $f[h(-1)]$ **32.** $h[f(x)]$ **33.** $f[h(x)]$

Given $g(x) = x^2 + 3$ and $h(x) = x - 2$, evaluate the composite function.

34. $g[h(0)]$ **35.** $h[g(0)]$ **36.** $g[h(4)]$

37. $h[g(-2)]$ **38.** $g[h(x)]$ **39.** $h[g(x)]$

Given $f(x) = x^2 + x + 1$ and $h(x) = 3x + 2$, evaluate the composite function.

40. $f[h(0)]$ **41.** $h[f(0)]$ **42.** $f[h(-1)]$

43. $h[f(-2)]$ **44.** $f[h(x)]$ **45.** $h[f(x)]$

Given $f(x) = x - 2$ and $g(x) = x^3$, evaluate the composite function.

46. $f[g(2)]$ **47.** $f[g(-1)]$ **48.** $g[f(2)]$

49. $g[f(-1)]$ **50.** $f[g(x)]$ **51.** $g[f(x)]$

APPLYING THE CONCEPTS

For the function $g(x) = x^2 - 1$, find:

52. $g(2 + h)$ **53.** $g(3 + h) - g(3)$ **54.** $g(-1 + h) - g(-1)$

55. $\dfrac{g(1 + h) - g(1)}{h}$ **56.** $\dfrac{g(-2 + h) - g(-2)}{h}$ **57.** $\dfrac{g(a + h) - g(a)}{h}$

Given $f(x) = 2x$, $g(x) = 3x - 1$, and $h(x) = x - 2$, find:

58. $f(g[h(2)])$ **59.** $g(h[f(1)])$ **60.** $h(g[f(-1)])$

61. $f(h[g(0)])$ **62.** $f(g[h(x)])$ **63.** $g(f[h(x)])$

9.4 One-to-One and Inverse Functions

Objective A   **To determine whether a function is one-to-one**

Recall that a function is a set of ordered pairs in which no two ordered pairs that have the same first coordinate have different second coordinates. This means that given any x, there is only one y that can be paired with that x. A **one-to-one function** satisfies the additional condition that given any y, there is only one x that can be paired with the given y. One-to-one functions are commonly written as 1–1.

> **One-to-One Function**
>
> A function f is a 1–1 function if for any a and b in the domain of f, $f(a) = f(b)$ implies that $a = b$.

This definition states that if the y-coordinates of an ordered pair are equal, $f(a) = f(b)$, then the x-coordinates must be equal, $a = b$.

The function defined by $f(x) = 2x + 1$ is a 1–1 function. To show this, determine $f(a)$ and $f(b)$. Then form the equation $f(a) = f(b)$.

$$f(a) = 2a + 1 \qquad f(b) = 2b + 1$$

$$
\begin{aligned}
f(a) &= f(b) \\
2a + 1 &= 2b + 1 \\
2a &= 2b \qquad &&\bullet \text{ Subtract 1 from each side of the equation.} \\
a &= b \qquad &&\bullet \text{ Divide each side of the equation by 2.}
\end{aligned}
$$

Because $f(a) = f(b)$ implies that $a = b$, the function is a 1–1 function.

Consider the function defined by $g(x) = x^2 - x$. Evaluate the function at -2 and 3.

$$g(-2) = (-2)^2 - (-2) = 6 \qquad g(3) = 3^2 - 3 = 6$$

From this evaluation, $g(-2) = 6$ and $g(3) = 6$, but $-2 \neq 3$. Thus g is not a 1–1 function.

The graphs of $f(x) = 2x + 1$ and $g(x) = x^2 - x$ are shown below. Note that a horizontal line intersects the graph of f at no more than one point. However, a horizontal line intersects the graph of g at more than one point.

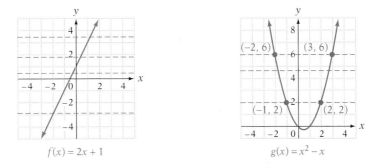

Looking at the graph of f on the previous page, note that for each y-coordinate there is only one x-coordinate. Thus f is a 1–1 function. From the graph of g, however, there are *two* x-coordinates for a given y-coordinate. For instance, $(-2, 6)$ and $(3, 6)$ are the coordinates of two points on the graph for which the y-coordinates are the same and the x-coordinates are different. Therefore, g is not a 1–1 function.

Horizontal-Line Test

The graph of a function represents the graph of a 1–1 function if any horizontal line intersects the graph at no more than one point.

Example 1

Determine whether the graph is the graph of a 1–1 function.

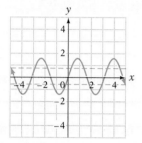

You Try It 1

Determine whether the graph is the graph of a 1–1 function.

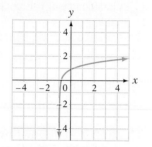

Solution

Because a horizontal line intersects the graph more than once, the graph is not the graph of a 1–1 function.

Your solution

Solution on p. S28

Objective B **To find the inverse of a function**

The **inverse of a function** is the set of ordered pairs formed by reversing the coordinates of each ordered pair of the function.

For example, the set of ordered pairs of the function defined by $f(x) = 2x$ with domain $\{-2, -1, 0, 1, 2\}$ is $\{(-2, -4), (-1, -2), (0, 0), (1, 2), (2, 4)\}$. The set of ordered pairs of the inverse function is $\{(-4, -2), (-2, -1), (0, 0), (2, 1), (4, 2)\}$.

From the ordered pairs of f, we have

Domain $= \{-2, -1, 0, 1, 2\}$ and   range $= \{-4, -2, 0, 2, 4\}$

From the ordered pairs of the inverse function, we have

Domain $= \{-4, -2, 0, 2, 4\}$ and   range $= \{-2, -1, 0, 1, 2\}$

Note that the domain of the inverse function is the range of the original function, and the range of the inverse function is the domain of the original function.

Now consider the function defined by $g(x) = x^2$ with domain $\{-2, -1, 0, 1, 2\}$. The set of ordered pairs of this function is $\{(-2, 4), (-1, 1), (0, 0), (1, 1), (2, 4)\}$. Reversing the ordered pairs gives $\{(4, -2), (1, -1), (0, 0), (1, 1), (4, 2)\}$. These ordered pairs do not satisfy the condition of a function, because there are ordered pairs with the same first coordinate and different second coordinates. This example illustrates that not all functions have an inverse function.

The graphs of $f(x) = 2x$ and $g(x) = x^2$, with the set of real numbers as the domain, are shown below.

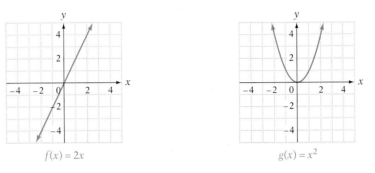

$f(x) = 2x$ $g(x) = x^2$

By the horizontal-line test, f is a 1–1 function but g is not.

Condition for an Inverse Function

A function f has an inverse function if and only if f is a 1–1 function.

The symbol f^{-1} is used to denote the inverse of the function f. The symbol $f^{-1}(x)$ is read "f inverse of x."

$f^{-1}(x)$ is *not* the reciprocal of $f(x)$, but rather is the notation for the inverse of a 1–1 function.

To find the inverse of a function, interchange x and y. Then solve for y.

⟹ Find the inverse of the function defined by $f(x) = 3x + 6$.

$$f(x) = 3x + 6$$
$$y = 3x + 6 \qquad \bullet \text{ Replace } f(x) \text{ by } y.$$
$$x = 3y + 6 \qquad \bullet \text{ Interchange } x \text{ and } y.$$
$$x - 6 = 3y \qquad \bullet \text{ Solve for } y.$$
$$\frac{1}{3}x - 2 = y$$
$$f^{-1}(x) = \frac{1}{3}x - 2 \qquad \bullet \text{ Replace } y \text{ by } f^{-1}(x).$$

The inverse of the function $f(x) = 3x + 6$ is $f^{-1}(x) = \frac{1}{3}x - 2$.

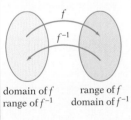

domain of f range of f
range of f^{-1} domain of f^{-1}

The fact that the ordered pairs of the inverse of a function are the reverse of those of the original function has a graphical interpretation. In the graph in the middle, the points from the figure on the left are plotted with the coordinates reversed. The inverse function is graphed by drawing a smooth curve through those points, as shown in the figure on the right.

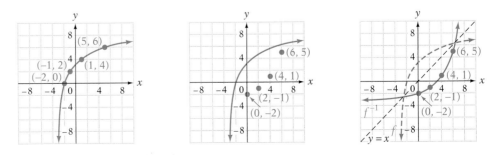

Note that the dashed graph of $y = x$ is shown in the figure on the bottom right of the preceding page. If two functions are inverses of each other, their graphs are mirror images with respect to the graph of the line $y = x$.

The composition of a function and its inverse has a special property.

Composition of Inverse Functions Property

$$f^{-1}[f(x)] = x \quad \text{and} \quad f[f^{-1}(x)] = x$$

This property can be used to determine whether two functions are inverses of each other.

➡ Are $f(x) = 2x - 4$ and $g(x) = \frac{1}{2}x + 2$ inverses of each other?

To determine whether the functions are inverses, use the Composition of Inverse Functions Property.

$$f[g(x)] = 2\left(\frac{1}{2}x + 2\right) - 4 \qquad\qquad g[f(x)] = \frac{1}{2}(2x - 4) + 2$$
$$= x + 4 - 4 \qquad\qquad\qquad = x - 2 + 2$$
$$= x \qquad\qquad\qquad\qquad\quad = x$$

Because $f[g(x)] = x$ and $g[f(x)] = x$, the functions are inverses of each other.

Example 2

Find the inverse of the function defined by $f(x) = 2x - 3$.

Solution

$$f(x) = 2x - 3$$
$$y = 2x - 3 \qquad \bullet \text{ Replace } f(x) \text{ by } y.$$
$$x = 2y - 3 \qquad \bullet \text{ Interchange } x \text{ and } y.$$
$$x + 3 = 2y \qquad \bullet \text{ Solve for } y.$$
$$\frac{x}{2} + \frac{3}{2} = y$$
$$f^{-1}(x) = \frac{x}{2} + \frac{3}{2} \qquad \bullet \text{ Replace } y \text{ by } f^{-1}(x).$$

The inverse of the function is given by
$$f^{-1}(x) = \frac{x}{2} + \frac{3}{2}.$$

You Try It 2

Find the inverse of the function defined by
$$f(x) = \frac{1}{2}x + 4.$$

Your solution

Example 3

Are $f(x) = 3x - 6$ and $g(x) = \frac{1}{3}x + 2$ inverses of each other?

Solution

$$f[g(x)] = 3\left(\frac{1}{3}x + 2\right) - 6 = x + 6 - 6 = x$$

$$g[f(x)] = \frac{1}{3}(3x - 6) + 2 = x - 2 + 2 = x$$

Yes, the functions are inverses of each other.

You Try It 3

Are $f(x) = 2x - 6$ and $g(x) = \frac{1}{2}x - 3$ inverses of each other?

Your solution

Solutions on p. S28

9.4 Exercises

Objective A

1. What is a 1–1 function?

2. What is the horizontal-line test?

Determine whether the graph represents the graph of a 1–1 function.

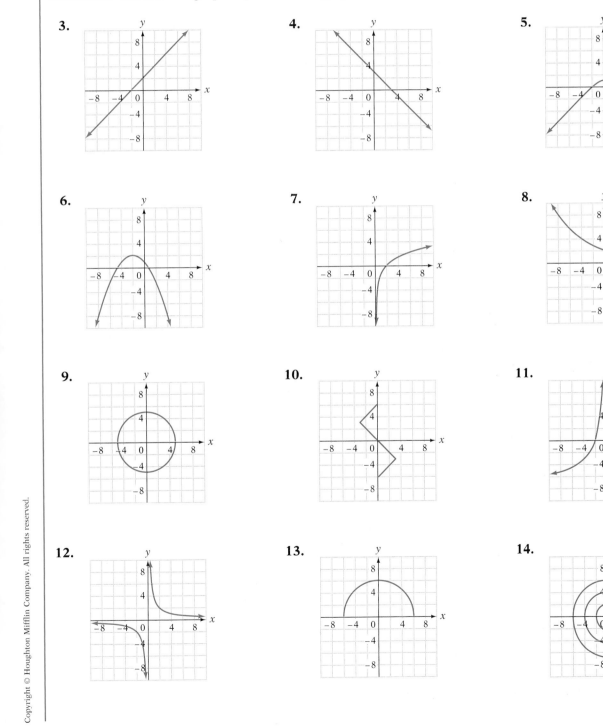

15. How are the ordered pairs of the inverse of a function related to the function?

16. Why is it that not all functions have an inverse function?

Find the inverse of the function. If the function does not have an inverse function, write "no inverse."

17. $\{(1, 0), (2, 3), (3, 8), (4, 15)\}$

18. $\{(1, 0), (2, 1), (-1, 0), (-2, 0)\}$

19. $\{(3, 5), (-3, -5), (2, 5), (-2, -5)\}$

20. $\{(-5, -5), (-3, -1), (-1, 3), (1, 7)\}$

21. $\{(0, -2), (-1, 5), (3, 3), (-4, 6)\}$

22. $\{(-2, -2), (0, 0), (2, 2), (4, 4)\}$

23. $\{(-2, -3), (-1, 3), (0, 3), (1, 3)\}$

24. $\{(2, 0), (1, 0), (3, 0), (4, 0)\}$

Find $f^{-1}(x)$.

25. $f(x) = 4x - 8$

26. $f(x) = 3x + 6$

27. $f(x) = 2x + 4$

28. $f(x) = x - 5$

29. $f(x) = \dfrac{1}{2}x - 1$

30. $f(x) = \dfrac{1}{3}x + 2$

31. $f(x) = -2x + 2$

32. $f(x) = -3x - 9$

33. $f(x) = \dfrac{2}{3}x + 4$

34. $f(x) = \dfrac{3}{4}x - 4$

35. $f(x) = -\dfrac{1}{3}x + 1$

36. $f(x) = -\dfrac{1}{2}x + 2$

37. $f(x) = 2x - 5$

38. $f(x) = 3x + 4$

39. $f(x) = 5x - 2$

40. $f(x) = 4x - 2$

41. $f(x) = 6x - 3$

42. $f(x) = -8x + 4$

Given $f(x) = 3x - 5$, find:

43. $f^{-1}(0)$

44. $f^{-1}(2)$

45. $f^{-1}(4)$

State whether the graph is the graph of a function. If it is the graph of a function, does the function have an inverse?

46.   **47.**   **48.**

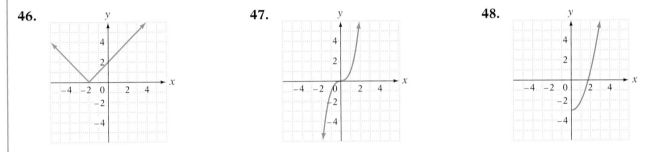

Use the Composition of Inverse Functions Property to determine whether the functions are inverses.

49. $f(x) = 4x; g(x) = \dfrac{x}{4}$

50. $g(x) = x + 5; h(x) = x - 5$

51. $f(x) = 3x; h(x) = \dfrac{1}{3x}$

52. $h(x) = x + 2; g(x) = 2 - x$

53. $g(x) = 3x + 2; f(x) = \dfrac{1}{3}x - \dfrac{2}{3}$

54. $h(x) = 4x - 1; f(x) = \dfrac{1}{4}x + \dfrac{1}{4}$

55. $f(x) = \dfrac{1}{2}x - \dfrac{3}{2}; g(x) = 2x + 3$

56. $g(x) = -\dfrac{1}{2}x - \dfrac{1}{2}; h(x) = -2x + 1$

APPLYING THE CONCEPTS

Given the graph of the 1–1 function, draw the graph of the inverse of the function by using the technique shown in Objective B of this section.

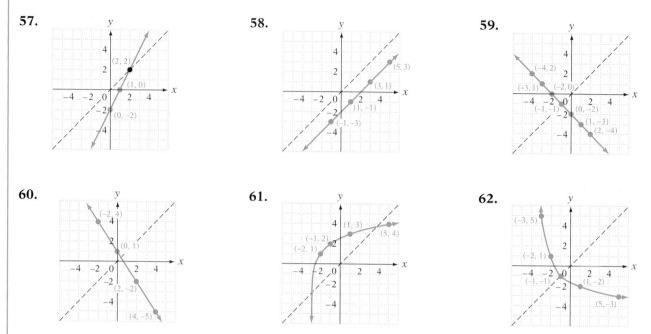

57. **58.** **59.**

60. **61.** **62.**

Each of the tables below defines a function. Is the inverse of the function a function? Explain your answer.

63. Grading Scale

Score	Grade
90–100	A
80–89	B
70–79	C
60–69	D
0–59	F

64. First-Class Postage Rates

Weight	Cost
$0 < w < 1$	\$.34
$1 < w \le 2$	\$.55
$2 < w \le 3$	\$.76
$3 < w \le 4$	\$.97

If f is a 1–1 function and $f(0) = 1, f(3) = -1$, and $f(5) = -3$, find:

65. $f^{-1}(-3)$ **66.** $f^{-1}(-1)$ **67.** $f^{-1}(1)$

If f is a 1–1 function and $f(-3) = 3, f(-4) = 7$, and $f(0) = 8$, find:

68. $f^{-1}(3)$ **69.** $f^{-1}(7)$ **70.** $f^{-1}(8)$

71. Is the inverse of a constant function a function? Explain your answer.

72. The graphs of all functions given by $f(x) = mx + b, m \ne 0$, are straight lines. Are all of these functions 1–1 functions? If so, explain why. If not, give an example of a linear function that is not 1–1.

Focus On Problem Solving

Proof in Mathematics

Mathematics is based on the concept of proof, which in turn is based on principles of logic. It was Aristotle (384 B.C. to 322 B.C.) who took the logical principles that had been developed over time and set them in a structured form. From this structured form, several methods of proof were developed.

implies   implies
A true → B true → C true
implies
A true → C true

A **deductive proof** is one that proceeds in a carefully ordered sequence of steps. It is the type of proof you may have seen in a geometry class. Deductive proofs can be modeled as shown at the left.

Here is an example of a deductive proof. It proves that the product of a negative number and a positive number is a negative number. The proof is based on the fact that every real number has exactly one additive inverse. For instance, the additive inverse of 4 is -4 and the additive inverse of $-\frac{2}{3}$ is $\frac{2}{3}$.

Theorem: If a and b are positive real numbers, then $(-a)b = -(ab)$.

Proof:
$$
\begin{aligned}
(-a)b + ab &= (-a + a)b \\
&= 0 \cdot b \\
&= 0
\end{aligned}
$$

- Distributive Property
- Inverse Property of Addition
- Multiplication Property of Zero

Since $(-a)b + ab = 0$, $(-a)b$ is the additive inverse of ab. But $-(ab)$ is also the additive inverse of ab and because every number has exactly one additive inverse, $(-a)b = -(ab)$.

Here is another example of a deductive proof. It is a proof of part of the Pythagorean Theorem.

Theorem: If a and b are the lengths of the legs of a right triangle and c is the length of the hypotenuse, then $a^2 + b^2 = c^2$.

Proof: Let ΔDBA be a right triangle and let BC be the altitude to the hypotenuse of ΔDBA, as shown at the left. Then ΔBCD and ΔBCA are right triangles.

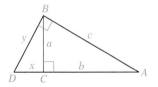

1. ΔDBA is a right triangle with height BC.

   Given.

2. ΔDBC is similar to ΔBAC and ΔDBA is similar to ΔBCA.

   Theorem from geometry: the altitude to the hypotenuse of a right triangle forms triangles that are similar to each other.

3. $\dfrac{a}{b} = \dfrac{x}{a}$

   The ratios of corresponding sides of similar triangles are equal.

4. $a^2 = bx$

   Multiply each side by ab.

5. $\dfrac{x + b}{c} = \dfrac{c}{b}$

   The ratios of corresponding sides of similar triangles are equal.

6. $bx + b^2 = c^2$

   Multiply each side by cb.

7. $a^2 + b^2 = c^2$

   Substitution Property from (**4**).

Because the first statement of the proof is true and each subsequent statement of the proof is true, the first statement implies that the last statement is true. Therefore we have proved: If a and b are the lengths of the legs of a right triangle and c is the length of the hypotenuse, then $a^2 + b^2 = c^2$.

Projects and Group Activities

Properties of Polynomials

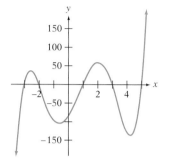

Recall that the x-coordinate of an x-intercept of the graph of $y = f(x)$ is a real number solution of the equation $f(x) = 0$. For instance, the graph of $P(x) = x^5 - 4x^4 - 16x^3 + 46x^2 + 63x - 90$ is shown at the right. The x-intercepts of the graph are $(-3, 0)$, $(-2, 0)$, $(1, 0)$, $(3, 0)$, and $(5, 0)$, and the roots of the equation $x^5 - 4x^4 - 16x^3 + 46x^2 + 63x - 90 = 0$ are -3, -2, 1, 3, and 5.

1. Graph $P(x) = x^5 + x^4 - 7x^3 - x^2 + 6x$ and determine the x-intercepts. Use the viewing window Xmin $= -4$, Xmax $= 4$, Xscl $= 1$, Ymin $= -10$, Ymax $= 40$, Yscl $= 10$.

2. What are the real number solutions of $x^5 + x^4 - 7x^3 - x^2 + 6x = 0$?

3. Graph $P(x) = x^5 - 2x^4 + 5x^3 - 10x^2 + 4x - 8$ and determine the x-intercepts. Use the viewing window Xmin $= -4$, Xmax $= 4$, Xscl $= 1$, Ymin $= -100$, Ymax $= 100$, Yscl $= 50$.

4. What are the real number solutions of $x^5 - 2x^4 + 5x^3 - 10x^2 + 4x - 8 = 0$?

5. On the basis of the graph shown above and your answers to Exercises 2 and 4, make a conjecture as to the number of real number solutions a fifth-degree equation can have.

6. Make up a few more fifth-degree equations and produce their graphs. Do the numbers of real number solutions of these equations satisfy your conjecture?

A **turning point** in the graph of a function is a point at which the graph changes direction. The graph of a fifth-degree polynomial with four turning points is shown at the right. In general, the number of turning points of the graph of a polynomial function depends on the degree of the polynomial.

Use a graphing calculator to graph each of the following equations. Try a viewing window of Xmin $= -10$, Xmax $= 10$, Xscl $= 1$, Ymin $= -80$, Ymax $= -80$, and Yscl $= 10$. Record the degree of the equation and the number of turning points.

$y = x^3 + 4x^2 - 2x - 14$
$y = x^3 + 2x^2 - 5x - 6$
$y = x^4 - x^3 - 11x^2 - x - 12$
$y = x^4 - 2x^3 - 13x^2 + 14x + 24$
$y = x^5 + x^4 - 9x^3 + 8x$
$y = x^5 - 3x^4 - 11x^3 + 27x^2 + 10x - 24$

7. Make a conjecture as to the relationship between the degree of a polynomial and the number of turning points of its graph. Graph a few more polynomials of your choosing and see if your conjecture is valid for those graphs. If not, refine your conjecture and test it again.

Finding the Maximum or Minimum of a Function Using a Graphing Calculator

Recall that, for a quadratic function, the maximum or minimum value of the function is the y-coordinate of the vertex of the graph of the function. We determined algebraically that the coordinates of the vertex are $\left(-\frac{b}{2a}, f\left(-\frac{b}{2a}\right)\right)$. It is also possible to use a graphing calculator to approximate a maximum or minimum value. The method used to find the value depends on the calculator model. However, it is generally true for most calculators that the point at which the function has a maximum or minimum must be shown on the screen. If it is not, you must select a different viewing window.

For a TI-83 graphing calculator, press 2nd CALC 3 to determine a minimum value or 2nd CALC 4 to determine a maximum value. Move the cursor to a point on the curve that is to the left of the minimum (or maximum). Press ENTER. Move the cursor to a point on the curve that is to the right of the minimum (or maximum). Press ENTER twice. The minimum (or maximum) value is shown as the y-coordinate at the bottom of the screen.

1. Find the maximum or minimum value of $f(x) = -x^2 - 3x + 2$ by using a graphing calculator.

2. Find the maximum or minimum value of $f(x) = x^2 - 4x - 2$ by using a graphing calculator.

3. Find the maximum or minimum value of $f(x) = -\sqrt{2}x^2 - \pi x + 2$ by using a graphing calculator.

4. Find the maximum or minimum value of $f(x) = \frac{x^2}{\sqrt{5}} - \sqrt{7}x - \sqrt{11}$ by using a graphing calculator.

The project on the preceding page described the turning points of a graph. These points are also called **local maximums** or **local minimums.** For the graph at the right, there is a local maximum at A and a local minimum at B. These points are called *local* because they are not really the maximum or minimum values of the function. For instance, the value of the function at C is greater than the value of the function at A, and the value of the function at D is less than the value of the function at B.

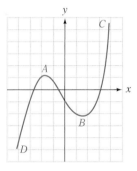

In future math courses you may study methods that will allow you to exactly determine the local maximums or local minimums of a function. However, by using a graphing calculator, it is possible to approximate these values.

For the following exercises, use the viewing window Xmin $= -10$, Xmax $= 10$, Xscl $= 1$, Ymin $= -100$, Ymax $= 100$, Yscl $= 10$.

5. Find the local maximum and minimum of $f(x) = 2x^3 - 3x^2 - 36x + 10$.

6. Find the local maximum and minimum of $f(x) = -x^3 + 4x^2 + 4x - 16$.

7. Find the local maximum and minimum of $f(x) = x^4 - 3x^3 - 7x^2 + x + 6$.

Chapter Summary

Key Words

A *quadratic function* is one of the form $f(x) = ax^2 + bx + c$, where $a \neq 0$. The graph of a quadratic function is a *parabola*. The *vertex* of a parabola is the point with the smallest y-coordinate or the largest y-coordinate. When $a > 0$, the parabola opens up and the vertex of the parabola is the point with the smallest y-coordinate. The value of the function at this point is a *minimum*. When $a < 0$, the parabola opens down and the vertex is the point with the largest y-coordinate. The value of the function at this point is a *maximum*. [pp. 467, 468, 473]

The *axis of symmetry* of a parabola is the vertical line that passes through the vertex of the parabola and is parallel to the y-axis. [p. 468]

A point at which a graph crosses the x-axis is called an *x-intercept* of the graph. [p. 470]

The *inverse* of a function is the set of ordered pairs formed by reversing the coordinates of each ordered pair of the function. A function f has an inverse function if and only if f is a 1–1 function. [pp. 496–497]

Essential Rules

Vertex and Axis of Symmetry of a Parabola

Let $f(x) = ax^2 + bx + c$ be the equation of a parabola. The coordinates of the vertex are $\left(-\dfrac{b}{2a}, f\left(-\dfrac{b}{2a}\right)\right)$. The equation of the axis of symmetry is $x = -\dfrac{b}{2a}$. [p. 468]

The Effect of the Discriminant on the Number of x-Intercepts of a Parabola
1. If $b^2 - 4ac = 0$, the parabola has one x-intercept.
2. If $b^2 - 4ac > 0$, the parabola has two x-intercepts.
3. If $b^2 - 4ac < 0$, the parabola has no x-intercepts. [p. 471]

To Find the Minimum or Maximum Value of a Quadratic Function:
Find the x-coordinate of the vertex. Then evaluate the function at that value. [p. 473]

Vertical-Line Test
A graph defines the graph of a function if any vertical line intersects the graph at no more than one point. [p. 484]

Operations on Functions
If f and g are functions and x is an element of the domain of each function, then

$$(f + g)(x) = f(x) + g(x) \qquad\qquad (f \cdot g)(x) = f(x) \cdot g(x)$$

$$(f - g)(x) = f(x) - g(x) \qquad\qquad \left(\frac{f}{g}\right)(x) = \frac{f(x)}{g(x)}, \, g(x) \neq 0 \text{ [p. 489]}$$

Composition of Functions
The composition of two functions f and g, symbolized by $f \circ g$, is the function whose value at x is given by $(f \circ g)(x) = f[g(x)]$. [p. 491]

One-to-One Functions
A function is a 1–1 function if, for any a and b in the domain of f, $f(a) = f(b)$ implies $a = b$. [p. 495]

Horizontal-Line Test
The graph of a function represents the graph of a 1–1 function if any horizontal line intersects the graph at no more than one point. [p. 496]

Composition of Inverse Functions Property
$f^{-1}[f(x)] = x$ and $f[f^{-1}(x)] = x$. [p. 498]

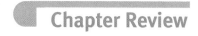

Chapter Review

1. Is the graph below the graph of a function?

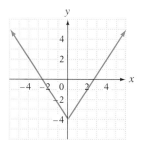

2. Is the graph below the graph of a 1–1 function?

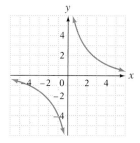

3. Graph $f(x) = 3x^3 - 2$. State the domain and range.

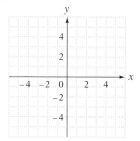

4. Graph $f(x) = \sqrt{x + 4}$. State the domain and range.

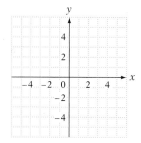

5. Use the discriminant to determine the number of x-intercepts of $y = -3x^2 + 4x + 6$.

6. Find the x-intercepts of $y = 3x^2 + 9x$.

7. Find the zeros of $f(x) = 3x^2 + 2x + 2$.

8. Find the maximum value of $f(x) = -2x^2 + 4x + 1$.

9. Find the minimum value of $f(x) = x^2 - 7x + 8$.

10. Given $f(x) = x^2 + 4$ and $g(x) = 4x - 1$, find $f[g(0)]$.

11. Given $f(x) = 6x + 8$ and $g(x) = 4x + 2$, find $g[f(-1)]$.

12. Given $f(x) = 3x^2 - 4$ and $g(x) = 2x + 1$, find $f[g(x)]$.

13. Are the functions given by $f(x) = -\frac{1}{4}x + \frac{5}{4}$ and $g(x) = -4x + 5$ inverses of each other?

14. Graph $f(x) = x^2 + 2x - 4$. State the domain and range.

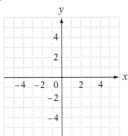

15. Find the vertex and axis of symmetry of the parabola whose equation is $y = x^2 - 2x + 3$.

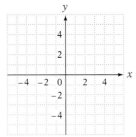

16. Graph $f(x) = |x| - 3$. State the domain and range.

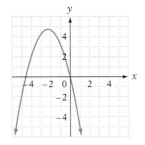

17. Is the graph below the graph of a 1–1 function?

For problems 18–21, use $f(x) = x^2 + 2x - 3$ and $g(x) = x^2 - 2$.

18. Evaluate: $(f + g)(2)$

19. Evaluate: $(f - g)(-4)$

20. Evaluate: $(f \cdot g)(-4)$

21. Evaluate: $\left(\dfrac{f}{g}\right)(3)$

22. Given $f(x) = 2x^2 + x - 5$ and $g(x) = 3x - 1$, find $g[f(x)]$.

23. Find the inverse of $f(x) = -6x + 4$.

24. Find the inverse of $f(x) = \dfrac{2}{3}x - 12$.

25. Find the inverse of $f(x) = \dfrac{1}{2}x + 8$.

26. The perimeter of a rectangle is 28 ft. What dimensions would give the rectangle a maximum area?

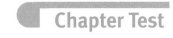

Chapter Test

1. Find the zeros of the function
   $f(x) = 2x^2 - 3x + 4$.

2. Find the x-intercepts of $y = x^2 + 3x - 8$.

3. Use the discriminant to determine the number of zeros of the parabola
   $y = 3x^2 + 2x - 4$.

4. Given $f(x) = x^2 + 2x - 3$ and $g(x) = x^3 - 1$,
   find $(f - g)(2)$.

5. Given $f(x) = x^3 + 1$ and $g(x) = 2x - 3$, find
   $(f \cdot g)(-3)$.

6. Given $f(x) = 4x - 5$ and $g(x) = x^2 + 3x + 4$,
   find $\left(\dfrac{f}{g}\right)(-2)$.

7. Given $f(x) = x^2 + 4$ and $g(x) = 2x^2 + 2x + 1$,
   find $(f - g)(-4)$.

8. Given $f(x) = 4x + 2$ and $g(x) = \dfrac{x}{x + 1}$, find
   $f[g(3)]$.

9. Given $f(x) = 2x^2 - 7$ and $g(x) = x - 1$, find
   $f[g(x)]$.

10. Find the maximum value of the function
    $f(x) = -x^2 + 8x - 7$.

11. Find the inverse of the function
    $f(x) = 4x - 2$.

12. Find the inverse of the function
$f(x) = \frac{1}{4}x - 4$.

13. Find the inverse of the function
$\{(2, 6), (3, 5), (4, 4), (5, 3)\}$.

14. Are the functions $f(x) = \frac{1}{2}x + 2$ and
$g(x) = 2x - 4$ inverses of each other?

15. Find the maximum product of two numbers whose sum is 20.

16. Graph $f(x) = -\sqrt{3 - x}$. State the domain and range.

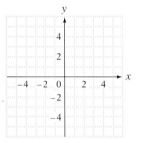

17. Graph $f(x) = \left| \frac{1}{2}x \right| - 2$. State the domain and range.

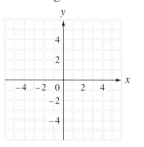

18. Graph $f(x) = x^3 - 3x + 2$. State the domain and range.

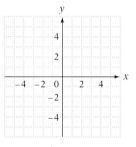

19. Determine whether the graph is the graph of a 1–1 function.

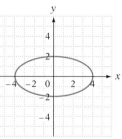

20. The perimeter of a rectangle is 200 cm. What dimensions would give the rectangle a maximum area? What is the maximum area?

Cumulative Review

1. Evaluate $-3a + \left| \dfrac{3b - ab}{3b - c} \right|$ when $a = 2$, $b = 2$, and $c = -2$.

2. Graph $\{x \mid x < -3\} \cap \{x \mid x > -4\}$.

$$-5 \ -4 \ -3 \ -2 \ -1 \ \ 0 \ \ 1 \ \ 2 \ \ 3 \ \ 4 \ \ 5$$

3. Solve: $\dfrac{3x - 1}{6} - \dfrac{5 - x}{4} = \dfrac{5}{6}$

4. Solve: $4x - 2 < -10$ or $3x - 1 > 8$

5. Solve: $|8 - 2x| \geq 0$

6. Simplify: $\left(\dfrac{3a^3 b}{2a} \right)^2 \left(\dfrac{a^2}{-3b^2} \right)^3$

7. Simplify: $(x - 4)(2x^2 + 4x - 1)$

8. Factor: $a^4 - 2a^2 - 8$

9. Factor: $x^3 y + x^2 y^2 - 6xy^3$

10. Solve: $(b + 2)(b - 5) = 2b + 14$

11. Solve: $x^2 - 2x > 15$

12. Simplify: $\dfrac{x^2 + 4x - 5}{2x^2 - 3x + 1} - \dfrac{x}{2x - 1}$

13. Solve: $\dfrac{5}{x^2 + 7x + 12} = \dfrac{9}{x + 4} - \dfrac{2}{x + 3}$

14. Simplify: $\dfrac{4 - 6i}{2i}$

15. Graph $f(x) = \dfrac{1}{4} x^2$. Find the vertex and axis of symmetry.

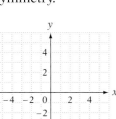

16. Graph the solution set of $3x - 4y \geq 8$.

17. Find the equation of the line containing the points $(-3, 4)$ and $(2, -6)$.

18. Find the equation of the line that contains the point $(-3, 1)$ and is perpendicular to the line $2x - 3y = 6$.

19. Solve: $3x^2 = 3x - 1$

20. Solve: $\sqrt{8x + 1} = 2x - 1$

21. Find the minimum value of the function $f(x) = 2x^2 - 3$.

22. Find the range of $f(x) = |3x - 4|$ if the domain is $\{0, 1, 2, 3\}$.

23. Is this set of ordered pairs a function? $\{(-3, 0), (-2, 0), (-1, 1), (0, 1)\}$

24. Solve: $\sqrt[3]{5x - 2} = 2$

25. Given $g(x) = 3x - 5$ and $h(x) = \frac{1}{2}x + 4$, find $g[h(2)]$.

26. Find the inverse of the function given by $f(x) = -3x + 9$.

27. Find the cost per pound of a tea mixture made from 30 lb of tea costing $4.50 per pound and 45 lb of tea costing $3.60 per pound.

28. How many pounds of an 80% copper alloy must be mixed with 50 lb of a 20% copper alloy to make an alloy that is 40% copper?

29. Six ounces of an insecticide are mixed with 16 gal of water to make a spray for spraying an orange grove. How much additional insecticide is required if it is to be mixed with 28 gal of water?

30. A large pipe can fill a tank in 8 min less time than it takes a smaller pipe to fill the same tank. Working together, both pipes can fill the tank in 3 min. How long would it take the larger pipe working alone to fill the tank?

31. The distance, d, a spring stretches varies directly as the force, f, used to stretch the spring. If a force of 50 lb can stretch a spring 30 in., how far can a force of 40 lb stretch the spring?

32. The frequency of vibration, f, in an open pipe organ varies inversely as the length, L, of the pipe. If the air in a pipe 2 m long vibrates 60 times per minute, find the frequency in a pipe that is 1.5 m long.

Chapter 10

Exponential and Logarithmic Functions

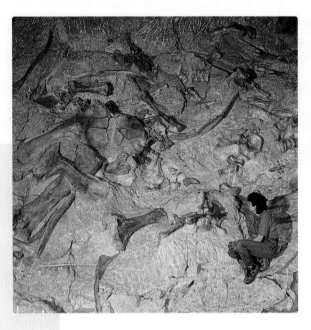

This researcher is examining sauropod bones in Jurassic sediments, hoping to determine the age of these long-necked dinosaurs' bones. Archaeologists use the carbon-dating method, which involves an exponential function, to calculate just how old the bones are. Instruments are used to detect the amount of carbon-14 left in an object. Since carbon-14 occurs naturally in living things and gradually decays after death, the measure of the amount of carbon-14 remaining reveals the object's age. An example of the process of using carbon dating is given in the **problem on page 544**.

Need help? For on-line student resources, such as section quizzes, visit this textbook's web site at **math.college.hmco.com/students**.

1. Simplify: 3^{-2}

2. Simplify: $\left(\dfrac{1}{2}\right)^{-4}$

3. Complete: $\dfrac{1}{8} = 2^{?}$

4. Evaluate $f(x) = x^4 + x^3$ for $x = -1$ and $x = 3$.

5. Solve: $3x + 7 = x - 5$

6. Solve: $16 = x^2 - 6x$

7. Evaluate $A(1 + i)^n$ for $A = 5000$, $i = 0.04$, and $n = 6$. Round to the nearest hundredth.

8. Graph: $f(x) = x^2 - 1$

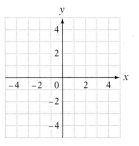

Go Figure

What is the ones digit of $1 + 9 + 9^2 + 9^3 + 9^4 + \ldots + 9^{2000} + 9^{2001}$?

10.1 Exponential Functions

Objective A   **To evaluate an exponential function**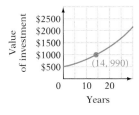

The growth of a $500 savings account that earns 5% annual interest compounded daily is shown in the graph at the right. In approximately 14 years, the savings account contains approximately $1000, twice the initial amount. The growth of this savings account is an example of an exponential function.

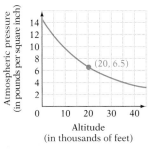

The pressure of the atmosphere at a certain height is shown in the graph at the right. This is another example of an exponential function. From the graph, we read that the air pressure is approximately 6.5 lb/in² at an altitude of 20,000 ft.

Definition of an Exponential Function

The **exponential function** with base b is defined by

$$f(x) = b^x$$

where $b > 0$, $b \neq 1$, and x is any real number.

In the definition of an exponential function, b, the base, is required to be positive. If the base were a negative number, the value of the function would be a complex number for some values of x. For instance, the value of $f(x) = (-4)^x$ when $x = \frac{1}{2}$ is $f\left(\frac{1}{2}\right) = (-4)^{1/2} = \sqrt{-4} = 2i$. To avoid complex number values of a function, the base of the exponential function is always a positive number.

⇒ Evaluate $f(x) = 2^x$ at $x = 3$ and $x = -2$.

$f(3) = 2^3 = 8$ • Substitute 3 for x and simplify.

$f(-2) = 2^{-2} = \dfrac{1}{2^2} = \dfrac{1}{4}$ • Substitute −2 for x and simplify.

To evaluate an exponential expression for an irrational number such as $\sqrt{2}$, we obtain an approximation to the value of the function by approximating the irrational number. For instance, the value of $f(x) = 4^x$ when $x = \sqrt{2}$ can be approximated by using an approximation of $\sqrt{2}$.

$$f(\sqrt{2}) = 4^{\sqrt{2}} \approx 4^{1.4142} \approx 7.1029$$

Because $f(x) = b^x$ ($b > 0$, $b \neq 1$) can be evaluated at both rational and irrational numbers, the domain of f is all real numbers. And because $b^x > 0$ for all values of x, the range of f is the positive real numbers.

A frequently used base in applications of exponential functions is an irrational number designated by e. The number e is approximately 2.71828183. It is an irrational number, so it has a nonterminating, nonrepeating decimal representation.

Natural Exponential Function

The function defined by $f(x) = e^x$ is called the **natural exponential function.**

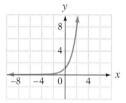

The e^x key on a calculator can be used to evaluate the natural exponential function. The graph of $y = e^x$ is shown at the left.

Example 1

Evaluate $f(x) = \left(\frac{1}{2}\right)^x$ at $x = 2$ and $x = -3$.

Solution

$$f(x) = \left(\frac{1}{2}\right)^x$$

$$f(2) = \left(\frac{1}{2}\right)^2 = \frac{1}{4}$$

$$f(-3) = \left(\frac{1}{2}\right)^{-3} = 2^3 = 8$$

You Try It 1

Evaluate $f(x) = \left(\frac{2}{3}\right)^x$ at $x = 3$ and $x = -2$.

Your solution

Example 2

Evaluate $f(x) = 2^{3x-1}$ at $x = 1$ and $x = -1$.

Solution

$$f(x) = 2^{3x-1}$$
$$f(1) = 2^{3(1)-1} = 2^2 = 4$$
$$f(-1) = 2^{3(-1)-1} = 2^{-4} = \frac{1}{2^4} = \frac{1}{16}$$

You Try It 2

Evaluate $f(x) = 2^{2x+1}$ at $x = 0$ and $x = -2$.

Your solution

Example 3

Evaluate $f(x) = e^{2x}$ at $x = 1$ and $x = -1$. Round to the nearest ten-thousandth.

Solution

$$f(x) = e^{2x}$$
$$f(1) = e^{2 \cdot 1} = e^2 \approx 7.3891$$
$$f(-1) = e^{2 \cdot (-1)} = e^{-2} \approx 0.1353$$

You Try It 3

Evaluate $f(x) = e^{2x-1}$ at $x = 2$ and $x = -2$. Round to the nearest ten-thousandth.

Your solution

Solutions on p. S29

Objective B **To graph an exponential function**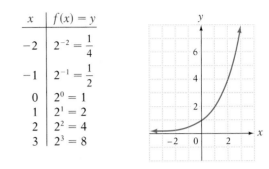

Some properties of an exponential function can be seen from its graph.

➡ Graph $f(x) = 2^x$.

Think of this as the equation $y = 2^x$.

Choose values of x and find the corresponding values of y. The results can be recorded in a table.

Graph the ordered pairs on a rectangular coordinate system.

x	$f(x) = y$
-2	$2^{-2} = \dfrac{1}{4}$
-1	$2^{-1} = \dfrac{1}{2}$
0	$2^0 = 1$
1	$2^1 = 2$
2	$2^2 = 4$
3	$2^3 = 8$

Connect the points with a smooth curve.

Note that any vertical line would intersect the graph at only one point. Therefore, by the vertical-line test, the graph of $f(x) = 2^x$ is the graph of a function. Also note that any horizontal line would intersect the graph at only one point. Therefore, the graph of $f(x) = 2^x$ is the graph of a one-to-one function.

➡ Graph $f(x) = \left(\dfrac{1}{2}\right)^x$.

Think of this as the equation $y = \left(\dfrac{1}{2}\right)^x$.

Choose values of x and find the corresponding values of y.

Graph the ordered pairs on a rectangular coordinate system.

Connect the points with a smooth curve.

x	$f(x) = y$
-3	$\left(\dfrac{1}{2}\right)^{-3} = 8$
-2	$\left(\dfrac{1}{2}\right)^{-2} = 4$
-1	$\left(\dfrac{1}{2}\right)^{-1} = 2$
0	$\left(\dfrac{1}{2}\right)^{0} = 1$
1	$\left(\dfrac{1}{2}\right)^{1} = \dfrac{1}{2}$
2	$\left(\dfrac{1}{2}\right)^{2} = \dfrac{1}{4}$

Applying the vertical-line and horizontal-line tests reveals that the graph of $f(x) = \left(\dfrac{1}{2}\right)^x$ is also the graph of a one-to-one function.

➡ Graph $f(x) = 2^{-x}$.

Think of this as the equation $y = 2^{-x}$.

Choose values of x and find the corresponding values of y.

Graph the ordered pairs on a rectangular coordinate system.

Connect the points with a smooth curve.

x	y
-3	8
-2	4
-1	2
0	1
1	$\dfrac{1}{2}$
2	$\dfrac{1}{4}$

TAKE NOTE

Note that because $2^{-x} = (2^{-1})^x = \left(\dfrac{1}{2}\right)^x$, the graphs of $f(x) = 2^{-x}$ and $f(x) = \left(\dfrac{1}{2}\right)^x$ are the same.

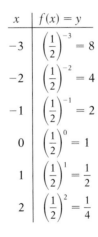
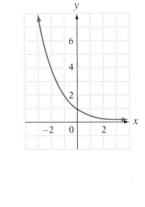
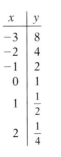
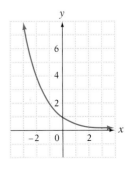

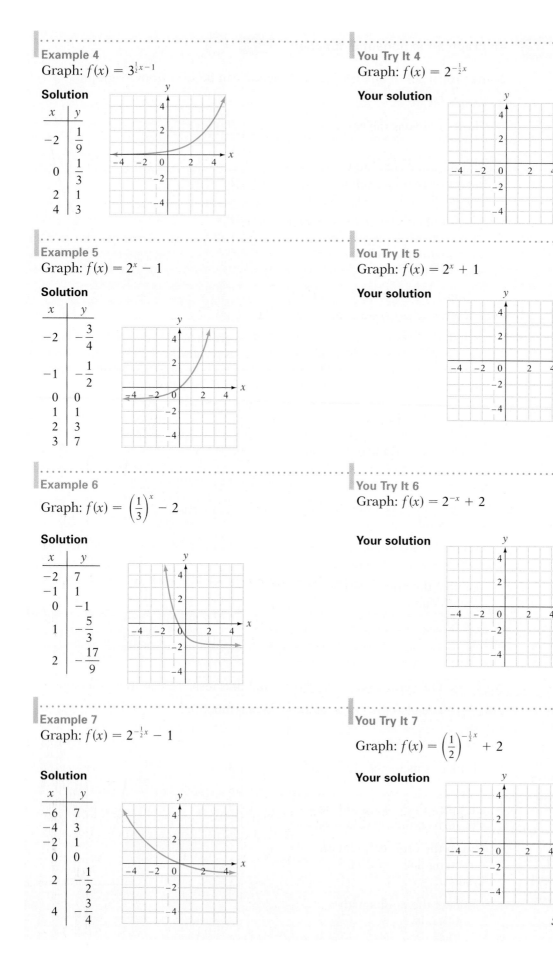

Example 4

Graph: $f(x) = 3^{\frac{1}{2}x-1}$

Solution

x	y
-2	$\frac{1}{9}$
0	$\frac{1}{3}$
2	1
4	3

You Try It 4

Graph: $f(x) = 2^{-\frac{1}{2}x}$

Your solution

Example 5

Graph: $f(x) = 2^x - 1$

Solution

x	y
-2	$-\frac{3}{4}$
-1	$-\frac{1}{2}$
0	0
1	1
2	3
3	7

You Try It 5

Graph: $f(x) = 2^x + 1$

Your solution

Example 6

Graph: $f(x) = \left(\frac{1}{3}\right)^x - 2$

Solution

x	y
-2	7
-1	1
0	-1
1	$-\frac{5}{3}$
2	$-\frac{17}{9}$

You Try It 6

Graph: $f(x) = 2^{-x} + 2$

Your solution

Example 7

Graph: $f(x) = 2^{-\frac{1}{2}x} - 1$

Solution

x	y
-6	7
-4	3
-2	1
0	0
2	$-\frac{1}{2}$
4	$-\frac{3}{4}$

You Try It 7

Graph: $f(x) = \left(\frac{1}{2}\right)^{-\frac{1}{2}x} + 2$

Your solution

Solutions on p. S29

31. Which of the following functions have the same graph?

   a. $f(x) = 3^x$ **b.** $f(x) = \left(\dfrac{1}{3}\right)^x$ **c.** $f(x) = x^3$ **d.** $f(x) = 3^{-x}$

32. Which of the following functions have the same graph?

   a. $f(x) = x^4$ **b.** $f(x) = 4^{-x}$ **c.** $f(x) = 4^x$ **d.** $f(x) = \left(\dfrac{1}{4}\right)^x$

33. Graph $f(x) = 3^x$ and $f(x) = 3^{-x}$ and find the point of intersection of the two graphs.

34. Graph $f(x) = 2^{x+1}$ and $f(x) = 2^{-x+1}$ and find the point of intersection of the two graphs.

35. Graph $f(x) = \left(\dfrac{1}{3}\right)^x$. What are the x- and y-intercepts of the graph of the function?

36. Graph $f(x) = \left(\dfrac{1}{3}\right)^{-x}$. What are the x- and y-intercepts of the graph of the function?

APPLYING THE CONCEPTS

Use a graphing calculator to graph the function.

37. $P(x) = \left(\sqrt{3}\right)^x$ **38.** $F(x) = \left(\sqrt{5}\right)^x$ **39.** $Q(x) = \left(\sqrt{3}\right)^{-x}$

40. $A(x) = \left(\sqrt{2}\right)^{-x}$ **41.** $f(x) = \pi^x$ **42.** $g(x) = \pi^{-x}$

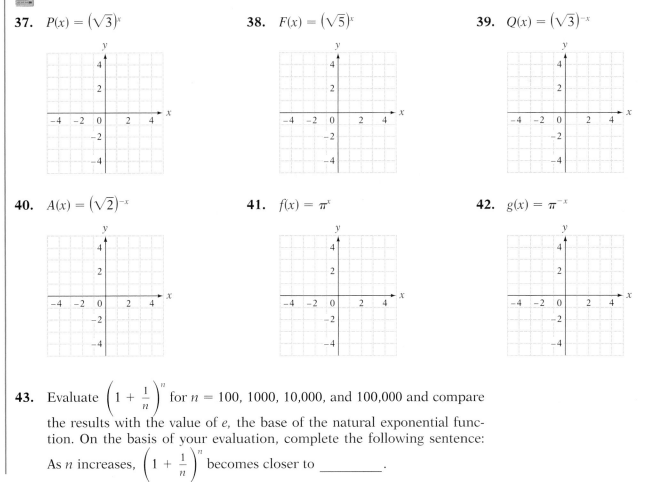

43. Evaluate $\left(1 + \dfrac{1}{n}\right)^n$ for $n = 100, 1000, 10{,}000,$ and $100{,}000$ and compare the results with the value of e, the base of the natural exponential function. On the basis of your evaluation, complete the following sentence:

As n increases, $\left(1 + \dfrac{1}{n}\right)^n$ becomes closer to _____.

44. Evaluate $(1 + x)^{1/x}$ for $x = 0.1$, 0.01, 0.001, 0.00001, and 0.0000001 and compare the results with the value of e, the base of the natural exponential function. On the basis of your evaluation, complete the following sentence: As x gets closer to 0, $(1 + x)^{1/x}$ becomes closer to _____ .

45. According to population studies, the population of China can be approximated by the equation $P(t) = 1.17(1.012)^t$, where $t = 0$ corresponds to 1993 and $P(t)$ is the population, in billions, of China in t years.

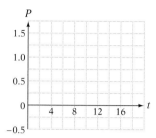

  a. Graph this equation. *Suggestion:* Use Xmin = 0, Xmax = 19, Ymin = −0.5, Ymax = 1.75, and Yscl = 0.25.

  b. The point whose approximate coordinates are (9, 1.303) is on this graph. Write a sentence that explains the meaning of these coordinates.

46. The growth of the population of India can be approximated by the equation $P(t) = 0.883(1.017)^t$, where $t = 0$ corresponds to 1993 and $P(t)$ is the population, in billions, of India in t years.

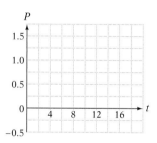

  a. Graph this equation. *Suggestion:* Use Xmin = 0, Xmax = 20, Ymin = −0.5, Ymax = 1.75, and Yscl = 0.25.

  b. The point whose approximate coordinates are (8, 1.01) is on this graph. Write a sentence that explains the meaning of these coordinates.

47. If air resistance is ignored, the speed v, in feet per second, of an object t seconds after it has been dropped is given by $v = 32t$. However, if air resistance is considered, then the speed depends on the mass (and on other things). For a certain mass, the speed t seconds after it has been dropped is given by $v = 32(1 - e^{-t})$.

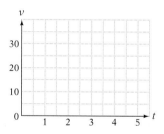

  a. Graph this equation. *Suggestion:* Use Xmin = 0, Xmax = 5.5, Ymin = 0, Ymax = 40, and Yscl = 5.

  b. The point whose approximate coordinates are (2, 27.7) is on this graph. Write a sentence that explains the meaning of these coordinates.

48. If air resistance is ignored, the speed v, in feet per second, of an object t seconds after it has been dropped is given by $v = 32t$. However, if air resistance is considered, then the speed depends on the mass (and on other things). For a certain mass, the speed t seconds after it has been dropped is given by $v = 64(1 - e^{-t/2})$.

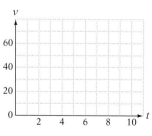

  a. Graph this equation. *Suggestion:* Use Xmin = 0, Xmax = 11, Ymin = 0, Ymax = 80, and Yscl = 10.

  b. The point whose approximate coordinates are (4, 55.3) is on this graph. Write a sentence that explains the meaning of these coordinates.

10.2 Introduction to Logarithms

Objective A   **To write equivalent exponential and logarithmic equations**   10

Because the exponential function is a 1–1 function, it has an inverse function, which is called a **logarithm**. A logarithm is used to answer a question similar to the following: "If $16 = 2^y$, what is the value of y?" Because $16 = 2^4$, the logarithm, base 2, of 16 is 4. This is written $\log_2 16 = 4$. Note that a logarithm is an exponent that solves a certain equation.

Definition of Logarithm

For $b > 0$, $b \neq 1$, $y = \log_b x$ is equivalent to $x = b^y$.

Read $\log_b x$ as "the logarithm of x, base b" or "log base b of x."

The table at the right shows equivalent statements written in both exponential and logarithmic form.

Exponential Form	Logarithmic Form
$2^4 = 16$	$\log_2 16 = 4$
$\left(\dfrac{2}{3}\right)^2 = \dfrac{4}{9}$	$\log_{2/3}\left(\dfrac{4}{9}\right) = 2$
$10^{-1} = 0.1$	$\log_{10}(0.1) = -1$

⇒ Write $\log_3 81 = 4$ in exponential form.

$\log_3 81 = 4$ is equivalent to $3^4 = 81$.

⇒ Write $10^{-2} = 0.01$ in logarithmic form.

$10^{-2} = 0.01$ is equivalent to $\log_{10}(0.01) = -2$.

The 1–1 property of exponential functions can be used to evaluate some logarithms.

1–1 Property of Exponential Functions

For $b > 0$, $b \neq 1$, if $b^u = b^v$, then $u = v$.

⇒ Evaluate $\log_2 8$.

$\log_2 8 = x$	• Write an equation.
$8 = 2^x$	• Write the equation in its equivalent exponential form.
$2^3 = 2^x$	• Write 8 as 2^3.
$3 = x$	• Use the 1–1 Property of Exponential Functions.
$\log_2 8 = 3$	

⇒ Solve $\log_4 x = -2$ for x.

$\log_4 x = -2$

$4^{-2} = x$ • Write the equation in its equivalent exponential form.

$\dfrac{1}{16} = x$ • Simplify the negative exponent.

The solution is $\dfrac{1}{16}$.

In this example, $\dfrac{1}{16}$ is called the *antilogarithm* base 4, of negative 2. In general, if $\log_b M = N$, then M is the antilogarithm base b of N. The antilogarithm of a number can be determined by rewriting the logarithmic expression in exponential form. For instance, the exponential form of $\log_5 x = 3$ is $5^3 = x$; x is the antilogarithm base 5 of 3.

Definition of Antilogarithm

If $\log_b M = N$, the **antilogarithm,** base b, of N is M. In exponential form,

$$M = b^N$$

Logarithms base 10 are called **common logarithms.** Usually the base, 10, is omitted when writing the common logarithm of a number. Therefore, $\log_{10} x$ is written $\log x$. To find the common logarithm of most numbers, a calculator is necessary. A calculator was used to find the value of $\log 384$, shown below.

CALCULATOR NOTE

The logarithms of most numbers are irrational numbers. Therefore, the value displayed by a calculator is an approximation.

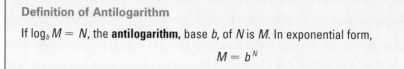

$$\log 384 \approx 2.5843312$$

The decimal part of a common logarithm is called the **mantissa;** the integer part is called the **characteristic.**

When e (the base of the natural exponential function) is used as the base of a logarithm, the logarithm is referred to as the **natural logarithm** and is abbreviated $\ln x$. This is read "el en x." Using a calculator, we learn that

$$\ln 23 \approx 3.135494216$$

The integer and decimal parts of a natural logarithm do not have special names.

Example 1

Evaluate: $\log_3\left(\dfrac{1}{9}\right)$

Solution

$\log_3\left(\dfrac{1}{9}\right) = x$

$\dfrac{1}{9} = 3^x$

$3^{-2} = 3^x$

$-2 = x$

$\log_3\left(\dfrac{1}{9}\right) = -2$

You Try It 1

Evaluate: $\log_4 64$

Your solution

Solution on p. S29

Example 2   Solve for x: $\log_5 x = 2$

Solution $\log_5 x = 2$
$5^2 = x$
$25 = x$

The solution is 25.

You Try It 2   Solve for x: $\log_2 x = -4$

Your solution

Example 3   Solve $\log x = -1.5$ for x. Round to the nearest ten-thousandth.

Solution $\log x = -1.5$
$10^{-1.5} = x$

$0.0316 \approx x$ • **Use a calculator.**

You Try It 3   Solve $\ln x = 3$ for x. Round to the nearest ten-thousandth.

Your solution

Solutions on p. S29

Objective B   **To use the Properties of Logarithms to simplify expressions containing logarithms**

Because a logarithm is a special kind of exponent, the Properties of Logarithms are similar to the Properties of Exponents.

The property of logarithms that states that the logarithm of the product of two numbers equals the sum of the logarithms of the two numbers is similar to the property of exponents that states that to multiply two exponential expressions with the same base, we add the exponents.

The Logarithm Property of the Product of Two Numbers

For any positive real numbers x, y, and b, $b \neq 1$, $\log_b(xy) = \log_b x + \log_b y$.

TAKE NOTE

Pay close attention to this theorem. Note, for instance, that this theorem states that $\log_3(4 \cdot p) = \log_3 4 + \log_3 p$. It also states that $\log_5 9 + \log_5 z = \log_5(9z)$. It does *not* state any relationship regarding the expression $\log_b(x + y)$. **This expression cannot be simplified.**

A proof of this property can be found in the Appendix.

▷ Write $\log_b(6z)$ in expanded form.

$\log_b(6z) = \log_b 6 + \log_b z$ • **Use the Logarithm Property of Products.**

▷ Write $\log_b 12 + \log_b r$ as a single logarithm.

$\log_b 12 + \log_b r = \log_b(12r)$ • **Use the Logarithm Property of Products.**

The Logarithm Property of Products can be extended to include the logarithm of the product of more than two factors. For instance,

$$\log_b(xyz) = \log_b x + \log_b y + \log_b z$$

$$\log_b(7rt) = \log_b 7 + \log_b r + \log_b t$$

A second property of logarithms involves the logarithm of the quotient of two numbers. This property of logarithms is also based on the fact that a logarithm is an exponent and that to divide two exponential expressions with the same base, we subtract the exponents.

> **The Logarithm Property of the Quotient of Two Numbers**
>
> For any positive real numbers x, y, and b, $b \neq 1$,
>
> $$\log_b \frac{x}{y} = \log_b x - \log_b y.$$

A proof of this property can be found in the Appendix.

➡ Write $\log_b \dfrac{p}{8}$ in expanded form.

$$\log_b \frac{p}{8} = \log_b p - \log_b 8 \qquad \bullet \text{ Use the Logarithm Property of Quotients.}$$

➡ Write $\log_b y - \log_b v$ as a single logarithm.

$$\log_b y - \log_b v = \log_b \frac{y}{v} \qquad \bullet \text{ Use the Logarithm Property of Quotients.}$$

A third property of logarithms is used to simplify powers of a number.

Point of Interest

Logarithms were developed independently by Jobst Burgi (1552–1632) and John Napier (1550–1617) as a means of simplifying the calculations of astronomers. The idea was to devise a method by which two numbers could be multiplied by performing additions. Napier is usually given credit for logarithms because he published his results first.

In Napier's original work, the logarithm of 10,000,000 was 0. After this work was published, Napier, in discussions with Henry Briggs (1561–1631), decided that tables of logarithms would be easier to use if the logarithm of 1 were 0. Napier died before new tables could be determined, and Briggs took on the task. His table consisted of logarithms accurate to 30 decimal places, all accomplished without a calculator!

The logarithms Briggs calculated are the common logarithms mentioned earlier.

> **The Logarithm Property of the Power of a Number**
>
> For any positive real numbers x and b, $b \neq 1$, and for any real number r,
> $$\log_b x^r = r \log_b x.$$

A proof of this property can be found in the Appendix.

➡ Rewrite $\log_b x^3$ in terms of $\log_b x$.

$$\log_b x^3 = 3 \log_b x \qquad \bullet \text{ Use the Logarithm Property of Powers.}$$

➡ Rewrite $\dfrac{2}{3} \log_b x$ with a coefficient of 1.

$$\frac{2}{3} \log_b x = \log_b x^{2/3} \qquad \bullet \text{ Use the Logarithm Property of Powers.}$$

The following table summarizes the properties of logarithms that we have discussed, along with three other properties.

> **Summary of the Properties of Logarithms**
>
> Let x, y, and b be positive real numbers with $b \neq 1$. Then
>
> | | |
> |---|---|
> | Product Property | $\log_b(x \cdot y) = \log_b x + \log_b y$ |
> | Quotient Property | $\log_b\left(\dfrac{x}{y}\right) = \log_b x - \log_b y$ |
> | Power Property | $\log_b x^r = r \log_b x$, r a real number |
> | Logarithm of One | $\log_b 1 = 0$ |
> | Inverse Property | $b^{\log_b x} = x$ and $\log_b b^x = x$ |
> | 1–1 Property | If $\log_b x = \log_b y$, then $x = y$. |

➡ Write $\log_b \frac{xy}{z}$ in expanded form.

$$\log_b \frac{xy}{z} = \log_b (xy) - \log_b z$$ • **Use the Logarithm Property of Quotients.**

$$= \log_b x + \log_b y - \log_b z$$ • **Use the Logarithm Property of Products.**

➡ Write $\log_b \frac{x^2}{y^3}$ in expanded form.

$$\log_b \frac{x^2}{y^3} = \log_b x^2 - \log_b y^3$$ • **Use the Logarithm Property of Quotients.**

$$= 2 \log_b x - 3 \log_b y$$ • **Use the Logarithm Property of Powers.**

➡ Write $2 \log_b x + 4 \log_b y$ as a single logarithm with a coefficient of 1.

$$2 \log_b x + 4 \log_b y = \log_b x^2 + \log_b y^4$$ • **Use the Logarithm Property of Powers.**

$$= \log_b x^2 y^4$$ • **Use the Logarithm Property of Products.**

- -

Example 4

Write $\log \sqrt{x^3 y}$ in expanded form.

Solution

$$\log \sqrt{x^3 y} = \log (x^3 y)^{1/2} = \frac{1}{2} \log (x^3 y)$$

$$= \frac{1}{2} (\log x^3 + \log y)$$

$$= \frac{1}{2} (3 \log x + \log y)$$

$$= \frac{3}{2} \log x + \frac{1}{2} \log y$$

You Try It 4

Write $\log_8 \sqrt[3]{x y^2}$ in expanded form.

Your solution

- -

Example 5

Write $\frac{1}{2} (\log_3 x - 3 \log_3 y + \log_3 z)$ as a single logarithm with a coefficient of 1.

Solution

$$\frac{1}{2} (\log_3 x - 3 \log_3 y + \log_3 z)$$

$$= \frac{1}{2} (\log_3 x - \log_3 y^3 + \log_3 z)$$

$$= \frac{1}{2} \left(\log_3 \frac{x}{y^3} + \log_3 z \right)$$

$$= \frac{1}{2} \left(\log_3 \frac{xz}{y^3} \right) = \log_3 \left(\frac{xz}{y^3} \right)^{1/2} = \log_3 \sqrt{\frac{xz}{y^3}}$$

You Try It 5

Write $\frac{1}{3} (\log_4 x - 2 \log_4 y + \log_4 z)$ as a single logarithm with a coefficient of 1.

Your solution

- -

Example 6

Find $\log_4 4^7$.

Solution

$$\log_4 4^7 = 7$$ • **Inverse Property**

You Try It 6

Find $\log_9 1$.

Your solution

Solutions on pp. S29–S30

Objective C **To use the Change-of-Base Formula** ◂ 10 ▸

Although only common logarithms and natural logarithms are programmed into calculators, the logarithms for other positive bases can be found.

➡ Evaluate $\log_5 22$.

$$\log_5 22 = x$$
$$5^x = 22$$
$$\log 5^x = \log 22$$
$$x \log 5 = \log 22$$
$$x = \frac{\log 22}{\log 5}$$
$$x \approx 1.92057$$
$$\log_5 22 \approx 1.92057$$

- Write an equation.
- Write the equation in its equivalent exponential form.
- Apply the common logarithm to each side of the equation.
- Use the Power Property of Logarithms.
- Exact answer
- Approximate answer

In the third step above, the natural logarithm, instead of the common logarithm, could have been applied to each side of the equation. The same result would have been obtained.

Using a procedure similar to the one used to evaluate $\log_5 22$, a formula for changing bases can be derived.

Change-of-Base Formula

$$\log_a N = \frac{\log_b N}{\log_b a}$$

➡ Evaluate $\log_2 14$.

$$\log_2 14 = \frac{\log 14}{\log 2}$$

- Use the Change-of-Base Formula with $N = 14$, $a = 2$, $b = 10$.

$$\approx 3.80735$$

In the last example, common logarithms were used. Here is the same example using natural logarithms. Note that the answers are the same.

$$\log_2 14 = \frac{\ln 14}{\ln 2} \approx 3.80735$$

Example 7
Evaluate $\log_8 0.137$ by using natural logarithms.

Solution
$$\log_8 0.137 = \frac{\ln 0.137}{\ln 8} \approx -0.95592$$

Example 8
Evaluate $\log_2 90.813$ by using common logarithms.

Solution
$$\log_2 90.813 = \frac{\log 90.813}{\log 2} \approx 6.50483$$

You Try It 7
Evaluate $\log_7 6.45$ by using common logarithms.

Your solution

You Try It 8
Evaluate $\log_3 0.834$ by using natural logarithms.

Your solution

Solutions on p. S30

10.2 Exercises

Objective A

1a. What is a common logarithm?
 b. How is the common logarithm of $4z$ written?

2a. What is a natural logarithm?
 b. How is the natural logarithm of $3x$ written?

Write the exponential expression in logarithmic form.

3. $5^2 = 25$ **4.** $10^3 = 1000$ **5.** $4^{-2} = \dfrac{1}{16}$ **6.** $3^{-3} = \dfrac{1}{27}$

7. $10^y = x$ **8.** $e^y = x$ **9.** $a^x = w$ **10.** $b^y = c$

Write the logarithmic expression in exponential form.

11. $\log_3 9 = 2$ **12.** $\log_2 32 = 5$ **13.** $\log 0.01 = -2$ **14.** $\log_5 \dfrac{1}{5} = -1$

15. $\ln x = y$ **16.** $\log x = y$ **17.** $\log_b u = v$ **18.** $\log_c x = y$

Evaluate.

19. $\log_3 81$ **20.** $\log_7 49$ **21.** $\log_2 128$ **22.** $\log_5 125$

23. $\log 100$ **24.** $\log 0.001$ **25.** $\ln e^3$ **26.** $\ln e^2$

27. $\log_8 1$ **28.** $\log_3 243$ **29.** $\log_5 625$ **30.** $\log_2 64$

Solve for x.

31. $\log_3 x = 2$ **32.** $\log_5 x = 1$ **33.** $\log_4 x = 3$ **34.** $\log_2 x = 6$

35. $\log_7 x = -1$ **36.** $\log_8 x = -2$ **37.** $\log_6 x = 0$ **38.** $\log_4 x = 0$

Solve for x. Round to the nearest hundredth.

39. $\log x = 2.5$ **40.** $\log x = 3.2$ **41.** $\log x = -1.75$ **42.** $\log x = -2.1$

43. $\ln x = 2$ **44.** $\ln x = 1.4$ **45.** $\ln x = -\dfrac{1}{2}$ **46.** $\ln x = -1.7$

Objective B

47. What is the Product Property of Logarithms?

48. What is the Quotient Property of Logarithms?

Express as a single logarithm with a coefficient of 1.

49. $\log_3 x^3 + \log_3 y^2$ **50.** $\log_7 x + \log_7 z^2$ **51.** $\ln x^4 - \ln y^2$

52. $\ln x^2 - \ln y$ **53.** $3\log_7 x$ **54.** $4\log_8 y$

55. $3\ln x + 4\ln y$ **56.** $2\ln x - 5\ln y$ **57.** $2(\log_4 x + \log_4 y)$

58. $3(\log_5 r + \log_5 t)$ **59.** $2\log_3 x - \log_3 y + 2\log_3 z$ **60.** $4\log_5 r - 3\log_5 s + \log_5 t$

61. $\ln x - (2\ln y + \ln z)$ **62.** $2\log_b x - 3(\log_b y + \log_b z)$

63. $\dfrac{1}{2}(\log_6 x - \log_6 y)$ **64.** $\dfrac{1}{3}(\log_8 x - \log_8 y)$

65. $2(\log_4 s - 2\log_4 t + \log_4 r)$ **66.** $3(\log_9 x + 2\log_9 y - 2\log_9 z)$

67. $\ln x - 2(\ln y + \ln z)$ **68.** $\ln t - 3(\ln u + \ln v)$

69. $3\log_2 t - 2(\log_2 r - \log_2 v)$ **70.** $2\log_{10} x - 3(\log_{10} y - \log_{10} z)$

71. $\dfrac{1}{2}(3\log_4 x - 2\log_4 y + \log_4 z)$

72. $\dfrac{1}{3}(4\log_5 t - 5\log_5 u - 7\log_5 v)$

73. $\dfrac{1}{2}(\ln x - 3\ln y)$

74. $\dfrac{1}{3}\ln a + \dfrac{2}{3}\ln b$

75. $\dfrac{1}{2}\log_2 x - \dfrac{2}{3}\log_2 y + \dfrac{1}{2}\log_2 z$

76. $\dfrac{2}{3}\log_3 x + \dfrac{1}{3}\log_3 y - \dfrac{1}{2}\log_3 z$

Write the logarithm in expanded form.

77. $\log_8(xz)$

78. $\log_7(rt)$

79. $\log_3 x^5$

80. $\log_2 y^7$

81. $\log_b\left(\dfrac{r}{s}\right)$

82. $\log_c\left(\dfrac{z}{4}\right)$

83. $\log_3(x^2 y^6)$

84. $\log_4(t^4 u^2)$

85. $\log_7\left(\dfrac{u^3}{v^4}\right)$

86. $\log_{10}\left(\dfrac{s^5}{t^2}\right)$

87. $\log_2(rs)^2$

88. $\log_3(x^2 y)^3$

89. $\ln(x^2 yz)$

90. $\ln(xy^2 z^3)$

91. $\log_5\left(\dfrac{xy^2}{z^4}\right)$

92. $\log_b\left(\dfrac{r^2 s}{t^3}\right)$

93. $\log_8\left(\dfrac{x^2}{yz^2}\right)$

94. $\log_9\left(\dfrac{x}{y^2 z^3}\right)$

95. $\log_4\sqrt{x^3 y}$

96. $\log_3\sqrt{x^5 y^3}$

97. $\log_7\sqrt{\dfrac{x^3}{y}}$

98. $\log_b \sqrt[3]{\dfrac{r^2}{t}}$ **99.** $\log_3 \dfrac{t}{\sqrt{x}}$ **100.** $\log_4 \dfrac{x}{\sqrt{y^2 z}}$

Objective C

Evaluate. Round to the nearest ten-thousandth.

101. $\log_{10} 7$ **102.** $\log_{10} 9$ **103.** $\log_{10}\left(\dfrac{3}{5}\right)$ **104.** $\log_{10}\left(\dfrac{13}{3}\right)$

105. $\ln 4$ **106.** $\ln 6$ **107.** $\ln\left(\dfrac{17}{6}\right)$ **108.** $\ln\left(\dfrac{13}{17}\right)$

109. $\log_8 6$ **110.** $\log_4 8$ **111.** $\log_5 30$ **112.** $\log_6 28$

113. $\log_3 0.5$ **114.** $\log_5 0.6$ **115.** $\log_7 1.7$ **116.** $\log_6 3.2$

117. $\log_5 15$ **118.** $\log_3 25$ **119.** $\log_{12} 120$ **120.** $\log_9 90$

121. $\log_4 2.55$ **122.** $\log_8 6.42$ **123.** $\log_5 67$ **124.** $\log_8 35$

APPLYING THE CONCEPTS

125. For each of the following, answer True or False. Assume all variables represent positive numbers.

a. $\log_3(-9) = -2$ **b.** $x^y = z$ and $\log_x z = y$ are equivalent equations.

c. $\log(x^{-1}) = \dfrac{1}{\log x}$ **d.** $\log\left(\dfrac{x}{y}\right) = \log x - \log y$

e. $\log(x \cdot y) = \log x \cdot \log y$ **f.** If $\log x = \log y$, then $x = y$.

126. Complete each statement using the equation $\log_a b = c$.

a. $a^c = $ _____   **b.** $\text{antilog}_a(\log_a b) = $ _____

10.3 Graphs of Logarithmic Functions

Objective A **To graph a logarithmic function**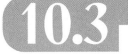

The graph of a logarithmic function can be drawn by using the relationship between the exponential and logarithmic functions.

⇒ Graph: $f(x) = \log_2 x$

Think of this as the equation $y = \log_2 x$.

$$f(x) = \log_2 x$$
$$y = \log_2 x$$

Write the equivalent exponential equation.

$$x = 2^y$$

Point of Interest

Although logarithms were originally developed to assist with computations, logarithmic functions have a much broader use today. These functions occur in geology, acoustics, chemistry, and economics, for example.

Because the equation is solved for x in terms of y, it is easier to choose values of y and find the corresponding values of x. The results can be recorded in a table.

Graph the ordered pairs on a rectangular coordinate system.

Connect the points with a smooth curve.

x	y
$\frac{1}{4}$	-2
$\frac{1}{2}$	-1
1	0
2	1
4	2

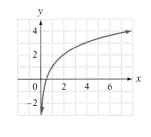

Applying the vertical-line and horizontal-line tests reveals that $f(x) = \log_2 x$ is a one-to-one function.

CALCULATOR NOTE

See the Projects and Group Activities at the end of this chapter for suggestions on graphing logarithmic functions using a graphing calculator.

⇒ Graph: $f(x) = \log_2(x) + 1$

Think of $f(x) = \log_2(x) + 1$ as the equation $y = \log_2(x) + 1$.

$$f(x) = \log_2(x) + 1$$
$$y = \log_2(x) + 1$$

Solve for $\log_2(x)$.

$$y - 1 = \log_2(x)$$

Write the equivalent exponential equation.

$$2^{y-1} = x$$

Choose values of y and find the corresponding values of x.

Graph the ordered pairs on a rectangular coordinate system.

Connect the points with a smooth curve.

x	y
$\frac{1}{4}$	-1
$\frac{1}{2}$	0
1	1
2	2
4	3

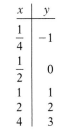

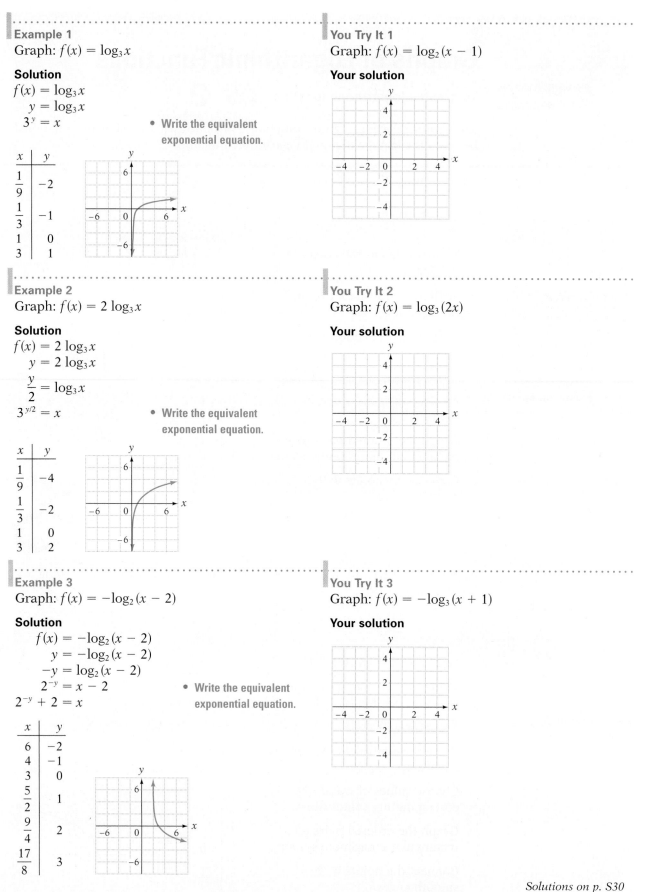

Example 1
Graph: $f(x) = \log_3 x$

Solution
$f(x) = \log_3 x$
$y = \log_3 x$
$3^y = x$ • Write the equivalent exponential equation.

x	y
$\frac{1}{9}$	-2
$\frac{1}{3}$	-1
1	0
3	1

You Try It 1
Graph: $f(x) = \log_2(x - 1)$

Your solution

Example 2
Graph: $f(x) = 2\log_3 x$

Solution
$f(x) = 2\log_3 x$
$y = 2\log_3 x$
$\frac{y}{2} = \log_3 x$
$3^{y/2} = x$ • Write the equivalent exponential equation.

x	y
$\frac{1}{9}$	-4
$\frac{1}{3}$	-2
1	0
3	2

You Try It 2
Graph: $f(x) = \log_3(2x)$

Your solution

Example 3
Graph: $f(x) = -\log_2(x - 2)$

Solution
$f(x) = -\log_2(x - 2)$
$y = -\log_2(x - 2)$
$-y = \log_2(x - 2)$
$2^{-y} = x - 2$
$2^{-y} + 2 = x$ • Write the equivalent exponential equation.

x	y
6	-2
4	-1
3	0
$\frac{5}{2}$	1
$\frac{9}{4}$	2
$\frac{17}{8}$	3

You Try It 3
Graph: $f(x) = -\log_3(x + 1)$

Your solution

Solutions on p. S30

10.3 Exercises

Objective A

1. Is the function $f(x) = \log x$ a 1–1 function? Why and why not?

2. Name two characteristics of the graph of $y = \log_b x$, $b > 1$.

3. What is the relationship between the graphs of $x = 3^y$ and $y = \log_3 x$?

4. What is the relationship between the graphs of $y = 3^x$ and $y = \log_3 x$?

Graph.

5. $f(x) = \log_4 x$

6. $f(x) = \log_2(x + 1)$

7. $f(x) = \log_3(2x - 1)$

8. $f(x) = \log_2\left(\dfrac{1}{2}x\right)$

9. $f(x) = 3 \log_2 x$

10. $f(x) = \dfrac{1}{2}\log_2 x$

11. $f(x) = -\log_2 x$

12. $f(x) = -\log_3 x$

13. $f(x) = \log_2(x - 1)$

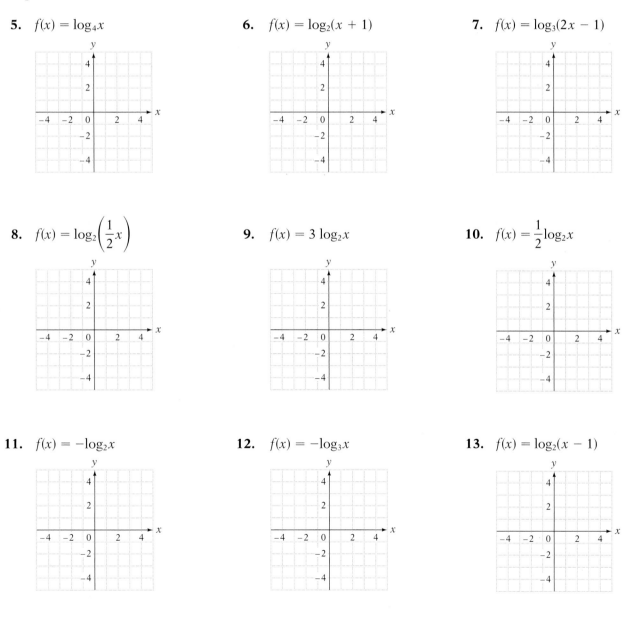

14. $f(x) = \log_3(2 - x)$

15. $f(x) = -\log_2(x - 1)$

16. $f(x) = -\log_2(1 - x)$

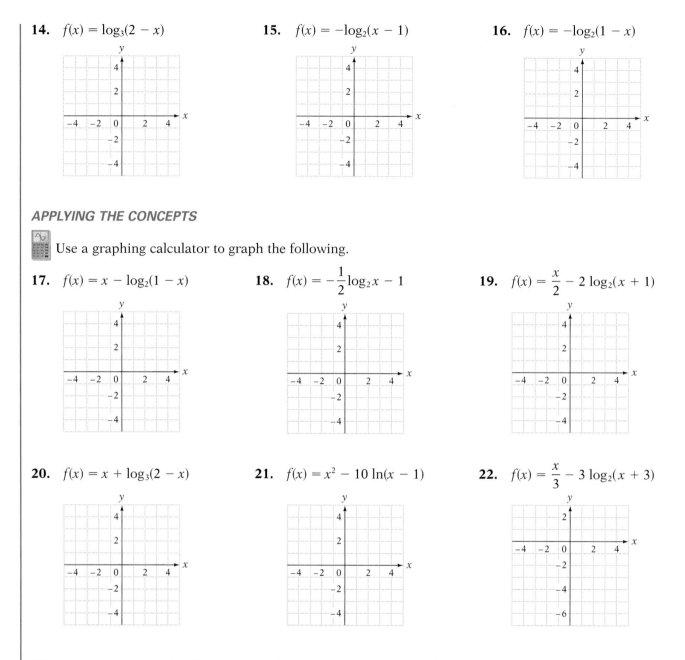

APPLYING THE CONCEPTS

Use a graphing calculator to graph the following.

17. $f(x) = x - \log_2(1 - x)$

18. $f(x) = -\dfrac{1}{2}\log_2 x - 1$

19. $f(x) = \dfrac{x}{2} - 2\log_2(x + 1)$

20. $f(x) = x + \log_3(2 - x)$

21. $f(x) = x^2 - 10\ln(x - 1)$

22. $f(x) = \dfrac{x}{3} - 3\log_2(x + 3)$

23. Astronomers use the *distance modulus* of a star as a method of determining the star's distance from Earth. The formula is $M = 5\log s - 5$, where M is the distance modulus and s is the star's distance from the Earth in parsecs. (One parsec $\approx 1.9 \times 10^{13}$ mi)
   a. Graph the equation.
   b. The point with coordinates (25.1, 2) is on the graph. Write a sentence that describes the meaning of this ordered pair.

24. Without practice, the proficiency of a typist decreases. The equation $S = 60 - 7\ln(t + 1)$, where S is the typing speed in words per minute and t is the number of months without typing, approximates this decrease.
   a. Graph the equation.
   b. The point with coordinates (4, 49) is on the graph. Write a sentence that describes the meaning of this ordered pair.

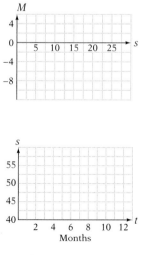

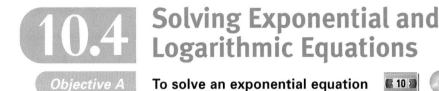

Solving Exponential and Logarithmic Equations

Objective A   **To solve an exponential equation** ◖ 10 ◗ ◉

An **exponential equation** is one in which the variable occurs in the exponent. The examples at the right are exponential equations.

$$6^{2x+1} = 6^{3x-2}$$
$$4^x = 3$$
$$2^{x+1} = 7$$

An exponential equation in which each side of the equation can be expressed in terms of the same base can be solved by using the 1-1 Property of Exponential Functions. Recall that the 1-1 Property of Exponential Functions states that for $b > 0, b \neq 1$,

$$\text{if } b^x = b^y, \text{ then } x = y.$$

In the two examples below, this property is used in solving exponential equations.

➡ Solve: $10^{3x+5} = 10^{x-3}$

$10^{3x+5} = 10^{x-3}$

$3x + 5 = x - 3$ 　　　• Use the 1–1 Property of Exponential Functions to equate the exponents.
　　　　　　　　　　　• Solve the resulting equation.

$2x + 5 = -3$

$2x = -8$

$x = -4$

Check: 　$\dfrac{10^{3x+5} = 10^{x-3}}{\begin{array}{c|c} 10^{3(-4)+5} & 10^{-4-3} \\ 10^{-12+5} & 10^{-7} \\ 10^{-7} = & 10^{-7} \end{array}}$

The solution is -4.

➡ Solve: $9^{x+1} = 27^{x-1}$

$9^{x+1} = 27^{x-1}$

$(3^2)^{x+1} = (3^3)^{x-1}$ 　　　• $3^2 = 9; 3^3 = 27$

$3^{2x+2} = 3^{3x-3}$

$2x + 2 = 3x - 3$ 　　　• Use the 1–1 Property of Exponential Functions to equate the exponents.
　　　　　　　　　　　• Solve for x.

$2 = x - 3$

$5 = x$

Check: 　$\dfrac{9^{x+1} = 27^{x-1}}{\begin{array}{c|c} 9^{5+1} & 27^{5-1} \\ 9^6 & 27^4 \\ 531,441 = & 531,441 \end{array}}$

The solution is 5.

To evaluate $\dfrac{\log 7}{\log 4}$ on a scientific calculator, use the keystrokes

7 [log] [÷] 4 [log] [=]

The display should read 1.4036775.

When each side of an exponential equation cannot easily be expressed in terms of the same base, logarithms are used to solve the exponential equation.

➡ Solve: $4^x = 7$

$$4^x = 7$$
$$\log 4^x = \log 7$$

$$x \log 4 = \log 7$$

$$x = \frac{\log 7}{\log 4} \approx 1.4037$$

The solution is 1.4037.

- Take the common logarithm of each side of the equation.
- Rewrite the equation using the Properties of Logarithms.
- Solve for x.
- Note that $\dfrac{\log 7}{\log 4} \neq \log 7 - \log 4$.

➡ Solve: $3^{x+1} = 5$

$$3^{x+1} = 5$$
$$\log 3^{x+1} = \log 5$$

$$(x + 1)\log 3 = \log 5$$

$$x + 1 = \frac{\log 5}{\log 3}$$
$$x + 1 \approx 1.4650$$
$$x \approx 0.4650$$

The solution is 0.4650.

- Take the common logarithm of each side of the equation.
- Rewrite the equation using the Properties of Logarithms.
- Solve for x.

Example 1
Solve for n: $(1.1)^n = 2$

Solution
$$(1.1)^n = 2$$
$$\log (1.1)^n = \log 2$$
$$n \log 1.1 = \log 2$$
$$n = \frac{\log 2}{\log 1.1}$$
$$n \approx 7.2725$$

The solution is 7.2725.

You Try It 1
Solve for n: $(1.06)^n = 1.5$

Your solution

Example 2
Solve for x: $3^{2x} = 4$

Solution
$$3^{2x} = 4$$
$$\log 3^{2x} = \log 4$$
$$2x \log 3 = \log 4$$
$$2x = \frac{\log 4}{\log 3}$$
$$x = \frac{\log 4}{2 \log 3}$$
$$x \approx 0.6309$$

The solution is 0.6309.

You Try It 2
Solve for x: $4^{3x} = 25$

Your solution

Solutions on p. S30

Objective B **To solve a logarithmic equation** 〈 10 〉

A logarithmic equation can be solved by using the Properties of Logarithms. Here are two examples.

⇒ Solve: $\log_9 x + \log_9(x - 8) = 1$

$$\log_9 x + \log_9(x - 8) = 1$$
$$\log_9[x(x - 8)] = 1$$

- Use the Logarithm Property of Products to rewrite the left side of the equation.

$$9^1 = x(x - 8)$$
$$9 = x^2 - 8x$$
$$0 = x^2 - 8x - 9$$
$$0 = (x - 9)(x + 1)$$

- Write the equation in exponential form.
- Simplify.
- Solve for x.
- Factor and use the Principle of Zero Products.

$$x - 9 = 0 \qquad x + 1 = 0$$
$$x = 9 \qquad\quad x = -1$$

Replacing x by 9 in the original equation reveals that 9 checks as a solution. Replacing x by -1 in the original equation results in the expression $\log_9(-1)$. Because the logarithm of a negative number is not a real number, -1 does not check as a solution.

The solution of the equation is 9.

⇒ Solve: $\log_3 6 - \log_3(2x + 3) = \log_3(x + 1)$

$$\log_3 6 - \log_3(2x + 3) = \log_3(x + 1)$$
$$\log_3 \frac{6}{2x + 3} = \log_3(x + 1)$$

- Use the Quotient Property of Logarithms to rewrite the left side of the equation.

$$\frac{6}{2x + 3} = x + 1$$

- Use the 1–1 Property of Logarithms.

$$6 = (2x + 3)(x + 1)$$
$$6 = 2x^2 + 5x + 3$$
$$0 = 2x^2 + 5x - 3$$
$$0 = (2x - 1)(x + 3)$$

- Write in standard form.
- Solve for x.

$$2x - 1 = 0 \qquad x + 3 = 0$$
$$x = \frac{1}{2} \qquad\quad x = -3$$

Replacing x by $\frac{1}{2}$ in the original equation reveals that $\frac{1}{2}$ checks as a solution. Replacing x by -3 in the original equation results in the expression $\log_3(-2)$. Because the logarithm of a negative number is not a real number, -3 does not check as a solution.

The solution of the equation is $\frac{1}{2}$.

Example 3

Solve for x: $\log_3(2x - 1) = 2$

Solution

$\log_3(2x - 1) = 2$

$\quad\quad 3^2 = 2x - 1$ • Write in exponential form.

$\quad\quad\quad 9 = 2x - 1$

$\quad\quad 10 = 2x$

$\quad\quad\quad 5 = x$

The solution is 5.

You Try It 3

Solve for x: $\log_4(x^2 - 3x) = 1$

Your solution

Example 4

Solve for x: $\log_2 x - \log_2(x - 1) = \log_2 2$

Solution

$\log_2 x - \log_2(x - 1) = \log_2 2$

$\log_2\left(\dfrac{x}{x-1}\right) = \log_2 2$ • Use the Quotient Property of Logarithms.

$\dfrac{x}{x-1} = 2$ • Use the 1–1 Property of Logarithms.

$(x - 1)\left(\dfrac{x}{x-1}\right) = (x - 1)2$

$\quad\quad\quad x = 2x - 2$

$\quad\quad -x = -2$

$\quad\quad\quad x = 2$

The solution is 2.

You Try It 4

Solve for x: $\log_3 x + \log_3(x + 3) = \log_3 4$

Your solution

Example 5

Solve for x:

$\log_2(3x + 8) = \log_2(2x + 2) + \log_2(x - 2)$

Solution

$\log_2(3x + 8) = \log_2(2x + 2) + \log_2(x - 2)$

$\log_2(3x + 8) = \log_2[(2x + 2)(x - 2)]$

$\log_2(3x + 8) = \log_2(2x^2 - 2x - 4)$

$3x + 8 = 2x^2 - 2x - 4$ • Use the 1–1 Property of Logarithms.

$\quad\quad 0 = 2x^2 - 5x - 12$

$\quad\quad 0 = (2x + 3)(x - 4)$

$2x + 3 = 0 \quad\quad\quad x - 4 = 0$

$\quad x = -\dfrac{3}{2} \quad\quad\quad\quad x = 4$

$-\dfrac{3}{2}$ does not check as a solution;

4 checks as a solution. The solution is 4.

You Try It 5

Solve for x: $\log_3 x + \log_3(x + 6) = 3$

Your solution

Solutions on p. S30

10.4 Exercises

Objective A

1. What is an exponential equation?

2a. What does the 1–1 Property of Exponential Functions state?
 b. Provide an example of when you would use this property.

Solve for x. For Exercises 15–26, round to the nearest ten-thousandth.

3. $5^{4x-1} = 5^{x-2}$

4. $7^{4x-3} = 7^{2x+1}$

5. $8^{x-4} = 8^{5x+8}$

6. $10^{4x-5} = 10^{x+4}$

7. $9^x = 3^{x+1}$

8. $2^{x-1} = 4^x$

9. $8^{x+2} = 16^x$

10. $9^{3x} = 81^{x-4}$

11. $16^{2-x} = 32^{2x}$

12. $27^{2x-3} = 81^{4-x}$

13. $25^{3-x} = 125^{2x-1}$

14. $8^{4x-7} = 64^{x-3}$

15. $5^x = 6$

16. $7^x = 10$

17. $e^x = 3$

18. $e^x = 2$

19. $10^x = 21$

20. $10^x = 37$

21. $2^{-x} = 7$

22. $3^{-x} = 14$

23. $2^{x-1} = 6$

24. $4^{x+1} = 9$

25. $3^{2x-1} = 4$

26. $4^{-x+2} = 12$

Objective B

27. What is a logarithmic equation?

28. What does the 1–1 Property of Logarithms state?

Solve for x.

29. $\log_2(2x - 3) = 3$

30. $\log_4(3x + 1) = 2$

31. $\log_2(x^2 + 2x) = 3$

32. $\log_3(x^2 + 6x) = 3$

33. $\log_5\left(\dfrac{2x}{x-1}\right) = 1$

34. $\log_6\left(\dfrac{3x}{x+1}\right) = 1$

35. $\log x = \log (1 - x)$

36. $\ln(3x - 2) = \ln(x + 1)$

37. $\ln 5 = \ln(4x - 13)$

38. $\log_3(x - 2) = \log_3(2x)$

39. $\ln(3x + 2) = 4$

40. $\ln(2x + 3) = -1$

41. $\log_2(8x) - \log_2(x^2 - 1) = \log_2 3$

42. $\log_5(3x) - \log_5(x^2 - 1) = \log_5 2$

43. $\log_9 x + \log_9(2x - 3) = \log_9 2$

44. $\log_6 x + \log_6(3x - 5) = \log_6 2$

45. $\log_8(6x) = \log_8 2 + \log_8(x - 4)$

46. $\log_7(5x) = \log_7 3 + \log_7(2x + 1)$

APPLYING THE CONCEPTS

Solve each equation with a graphing calculator. Round answers to the nearest hundredth.

47. $3^x = -x$

48. $2^x = -x + 2$

49. $2^{-x} = x - 1$

50. $3^{-x} = 2x$

51. $\ln x = x^2$

52. $2 \ln x = -x + 1$

53. $\log_3 x = -2x - 2$

54. $\log_5(2x) = -2x + 1$

55. A model for the distance s (in feet) that an object experiencing air resistance will fall in t seconds is given by $s = 312.5 \ln\left(\dfrac{e^{0.32t} + e^{-0.32t}}{2}\right)$.

    a. Graph this equation. *Suggestion:* Use Xmin = 0, Xmax = 4.5, Ymin = 0, Ymax = 140, and Yscl = 20.

    b. Determine, to the nearest hundredth of a second, the time it takes the object to travel 100 ft.

56. A model for the distance s (in feet) that an object experiencing air resistance will fall in t seconds is given by $s = 78 \ln\left(\dfrac{e^{0.8t} + e^{-0.8t}}{2}\right)$.

    a. Graph this equation. *Suggestion:* Use Xmin = 0, Xmax = 4.5, Ymin = 0, Ymax = 140, and Yscl = 20.

    b. Determine, to the nearest hundredth of a second, the time it takes the object to travel 125 ft.

57. The following "proof" shows that $0.5 < 0.25$. Explain the error.

$$1 < 2$$
$$1 \cdot \log 0.5 < 2 \cdot \log 0.5$$
$$\log 0.5 < \log (0.5)^2$$
$$0.5 < (0.5)^2$$
$$0.5 < 0.25$$

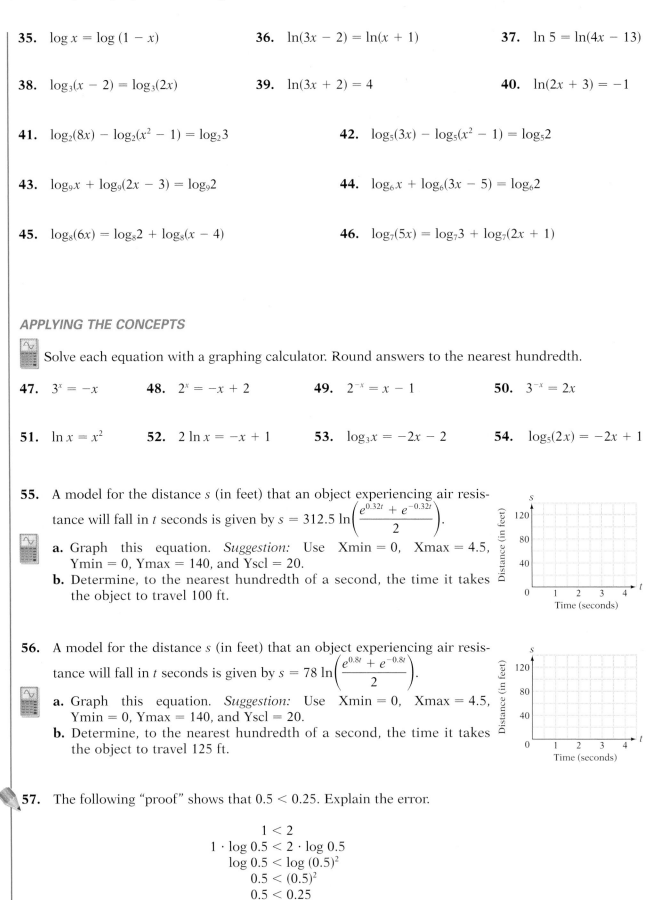

10.5 Applications of Exponential and Logarithmic Functions

Objective A **To solve application problems** ⟨ 10 ⟩

A biologist places one single-celled bacterium in a culture, and each hour that particular species of bacteria divides into two bacteria. After one hour there will be two bacteria. After two hours, each of the two bacteria will divide and there will be four bacteria. After three hours, each of the four bacteria will divide and there will be eight bacteria.

The table at the right shows the number of bacteria in the culture after various intervals of time, t, in hours. Values in this table could also be found by using the exponential equation $N = 2^t$.

Time, t	Number of Bacteria, N
0	1
1	2
2	4
3	8
4	16

Point of Interest

Parkinson's Law (C. Northcote Parkinson) is sometimes stated as "A job will expand to fill the time alloted for the job." However, Parkinson actually said that in any new government administration, administrative employees will be added at the rate of about 5% to 6% per year. This is an example of exponential growth and means that a staff of 500 will grow to approximately 630 by the end of a 4-year term.

The equation $N = 2^t$ is an example of an **exponential growth equation.** In general, any equation that can be written in the form $A = A_0 b^{kt}$, where A is the size at time t, A_0 is the initial size, $b > 1$, and k is a positive real number, is an exponential growth equation. These equations are important not only in population growth studies but also in physics, chemistry, psychology, and economics.

Recall that interest is the amount of money paid (or received) when borrowing (or investing) money. **Compound interest** is interest that is computed not only on the original principal, but also on the interest already earned. The compound interest formula is an exponential equation.

The **compound interest formula** is $P = A(1 + i)^n$, where A is the original value of an investment, i is the interest rate per compounding period, n is the total number of compounding periods, and P is the value of the investment after n periods.

⇒ An investment broker deposits $1000 into an account that earns 12% annual interest compounded quarterly. What is the value of the investment after 2 years?

$i = \dfrac{12\%}{4} = \dfrac{0.12}{4} = 0.03$
- Find i, the interest rate per quarter. The quarterly rate is the annual rate divided by 4, the number of quarters in one year.

$n = 4 \cdot 2 = 8$
- Find n, the number of compounding periods. The investment is compounded quarterly, 4 times a year, for 2 years.

$P = A(1 + i)^n$
- Use the compound interest formula.

$P = 1000(1 + 0.03)^8$
- Replace A, i, and n by their values.

$P \approx 1267$
- Solve for P.

The value of the investment after 2 years is approximately $1267.

Exponential decay offers another example of an exponential equation. One of the most common illustrations of exponential decay is the decay of a radioactive substance. For instance, tritium, a radioactive nucleus of hydrogen that has been used in luminous watch dials, has a half-life of approximately 12 years. This means that one-half of any given amount of tritium will disintegrate in 12 years.

The table at the right indicates the amount of an initial 10-microgram sample of tritium that remains after various intervals of time, t, in years. Values in this table could also be found by using the exponential equation $A = 10(0.5)^{t/12}$.

Time, t	Amount, A
0	10
12	5
24	2.5
36	1.25
48	0.625

The equation $A = 10(0.5)^{t/12}$ is an example of an **exponential decay equation.** Comparing this equation to the exponential growth equation, note that for exponential growth, the base of the exponential equation is greater than 1, whereas for exponential decay, the base is between 0 and 1.

A method by which an archaeologist can measure the age of a bone is called carbon dating. Carbon dating is based on a radioactive isotope of carbon called carbon-14, which has a half-life of approximately 5570 years. The exponential decay equation is given by $A = A_0(0.5)^{t/5570}$, where A_0 is the original amount of carbon-14 present in the bone, t is the age of the bone in years, and A is the amount of carbon-14 present after t years.

⇒ A bone that originally contained 100 mg of carbon-14 now has 70 mg of carbon-14. What is the approximate age of the bone?

$$A = A_0(0.5)^{t/5570}$$ • Use the exponential decay equation.

$$70 = 100(0.5)^{t/5570}$$ • Replace A by 70 and A_0 by 100 and solve for t.

$$0.7 = (0.5)^{t/5570}$$ • Divide each side by 100.

$$\log 0.7 = \log(0.5)^{t/5570}$$ • Take the common logarithm of each side of the equation. Then simplify.

$$\log 0.7 = \frac{t}{5570} \log 0.5$$

$$\frac{5570 \log 0.7}{\log 0.5} = t$$

$$2866 \approx t$$

The bone is approximately 2866 years old.

A chemist measures the acidity or alkalinity of a solution by measuring the concentration of hydrogen ions, H^+, in the solution using the formula $pH = -\log(H^+)$. A neutral solution such as distilled water has a pH of 7, acids have a pH less than 7, and alkaline solutions (also called basic solutions) have a pH greater than 7.

Point of Interest

The pH scale was created by the Danish biochemist Søren Sørensen in 1909 to measure the acidity of water used in the brewing of beer. pH is an abbreviation for *pondus hydrogenii*, which translates as "potential hydrogen."

⇒ Find the pH of orange juice that has a hydrogen ion concentration, H^+, of 2.9×10^{-4}. Round to the nearest tenth.

$$pH = -\log(H^+)$$

$$= -\log(2.9 \times 10^{-4})$$ • $H^+ = 2.9 \times 10^{-4}$

$$\approx 3.5376$$

The pH of the orange juice is approximately 3.5.

Logarithmic functions are used to scale very large or very small numbers into numbers that are easier to comprehend. For instance, the *Richter scale magnitude* of an earthquake uses a logarithmic function to convert the intensity of shock waves I into a number M, which for most earthquakes is in the range of 0 to 10. The intensity I of an earthquake is often given in terms of the constant I_0, where I_0 is the intensity of the smallest earthquake, called a *zero-level earthquake,* that can be measured on a seismograph near the earthquake's epicenter. An earthquake with an intensity I has a Richter scale magnitude of $M = \log\left(\dfrac{I}{I_0}\right)$, where I_0 is the measure of the intensity of a zero-level earthquake.

⇒ Find the Richter scale magnitude of the 1999 Joshua Tree, California earthquake that had an intensity I of $12,589,254 I_0$. Round to the nearest tenth.

$$M = \log\left(\frac{I}{I_0}\right)$$

$$M = \log\left(\frac{12,589,254 I_0}{I_0}\right) \qquad \bullet \ I = 12,589,254 I_0$$

$$M = \log 12,589,254 \qquad \bullet \ \text{Divide the numerator and denominator by } I_0.$$

$$M \approx 7.1 \qquad \bullet \ \text{Evaluate log 12,589,254.}$$

The 1999 Joshua Tree earthquake had a Richter scale magnitude of 7.1.

If you know the Richter scale magnitude of an earthquake, you can determine the intensity of the earthquake.

⇒ Find the intensity of the 1999 Taiwan earthquake that measured 7.6 on the Richter scale. Write the answer in terms of I_0.

$$M = \log\left(\frac{I}{I_0}\right)$$

$$7.6 = \log\left(\frac{I}{I_0}\right) \qquad \bullet \ \text{Replace } M \text{ by 7.6.}$$

$$10^{7.6} = \frac{I}{I_0} \qquad \bullet \ \text{Write in exponential form.}$$

$$10^{7.6} I_0 = I \qquad \bullet \ \text{Multiply both sides by } I_0.$$

$$39,810,717 I_0 \approx I \qquad \bullet \ \text{Evaluate } 10^{7.6}.$$

The 1999 Taiwan earthquake had an intensity that was approximately 39,811,000 times the intensity of a zero-level earthquake.

The percent of light that will pass through a substance is given by $\log P = -kd$, where P is the percent of light passing through the substance, k is a constant depending on the substance, and d is the thickness of the substance in meters.

⇒ For certain parts of the ocean, $k = 0.03$. Using this value, at what depth will the percent of light be 50% of the light at the surface of the ocean? Round to the nearest meter.

$$\log P = -kd$$

$$\log(0.5) = -0.03 d \qquad \bullet \ \text{Replace } P \text{ by 0.5 (50\%) and } k \text{ by 0.03.}$$

$$\frac{\log(0.5)}{-0.03} = d \qquad \bullet \ \text{Solve for } d.$$

$$10.0343 \approx d$$

At a depth of about 10 m, the light will be 50% of the light at the surface.

Example 1

An investment of $3000 is placed into an account that earns 12% annual interest compounded monthly. In approximately how many years will the investment be worth twice the original amount?

Strategy

To find the time, solve the compound interest formula for n. Use $P = 6000$, $A = 3000$, and $i = \frac{12\%}{12} = \frac{0.12}{12} = 0.01$.

Solution

$$P = A(1 + i)^n$$
$$6000 = 3000(1 + 0.01)^n$$
$$6000 = 3000(1.01)^n$$
$$2 = (1.01)^n$$
$$\log 2 = \log(1.01)^n$$
$$\log 2 = n \log 1.01$$
$$\frac{\log 2}{\log 1.01} = n$$
$$70 \approx n$$

70 months ÷ 12 ≈ 5.8 years

In approximately 6 years, the investment will be worth $6000.

You Try It 1

Find the hydrogen ion concentration, H^+, of vinegar that has a pH of 2.9.

Your strategy

Your solution

Example 2

The number of words per minute that a student can type will increase with practice and can be approximated by the equation $N = 100[1 - (0.9)^t]$, where N is the number of words typed per minute after t days of instruction. Find the number of words a student will type per minute after 8 days of instruction.

Strategy

To find the number of words per minute, replace t in the equation by its given value and solve for N.

Solution

$$N = 100[1 - (0.9)^t]$$
$$= 100[1 - (0.9)^8]$$
$$\approx 56.95$$

After 8 days of instruction, a student will type approximately 57 words per minute.

You Try It 2

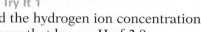 On September 3, 2000, an earthquake measuring 5.2 on the Richter scale struck the Napa Valley, located 50 mi north of San Francisco. Find the intensity of the quake in terms of I_0.

Your strategy

Your solution

Solutions on p. S31

10.5 Exercises

Objective A

For Exercises 1–4, use the compound interest formula $P = A(1 + i)^n$, where A is the original value of an investment, i is the interest rate per compounding period, n is the total number of compounding periods, and P is the value of the investment after n periods.

1. An investment broker deposits $1000 into an account that earns 8% annual interest compounded quarterly. What is the value of the investment after 2 years? Round to the nearest dollar.

2. A financial advisor recommends that a client deposit $2500 into a fund that earns 7.5% annual interest compounded monthly. What will be the value of the investment after 3 years? Round to the nearest cent.

3. To save for college tuition, the parents of a preschooler invest $5000 in a bond fund that earns 6% annual interest compounded monthly. In approximately how many years will the investment be worth $15,000?

4. A hospital administrator deposits $10,000 into an account that earns 9% annual interest compounded monthly. In approximately how many years will the investment be worth $15,000?

For Exercises 5–8, use the exponential decay equation $A = A_0\left(\dfrac{1}{2}\right)^{t/k}$, where A is the amount of a radioactive material present after time t, k is the half-life of the radioactive substance, and A_0 is the original amount of the radioactive substance. Round to the nearest tenth.

5. An isotope of technetium is used to prepare images of internal body organs. This isotope has a half-life of approximately 6 h. A patient is injected with 30 mg of this isotope.
   a. What is the technetium level in the patient after 3 h?
   b. How long (in hours) will it take for the technetium level to reach 20 mg?

6. Iodine-131 is an isotope that is used to study the functioning of the thyroid gland. This isotope has a half-life of approximately 8 days. A patient is given an injection that contains 8 micrograms of iodine-131.
   a. What is the iodine level in the patient after 5 days?
   b. How long (in days) will it take for the iodine level to reach 5 micrograms?

7. A sample of promethium-147 (used in some luminous paints) weighs 25 mg. One year later, the sample weighs 18.95 mg. What is the half-life of promethium-147, in years?

8. Francium-223 is a very rare radioactive isotope discovered in 1939 by Marguerite Percy. A 3-microgram sample of francium-223 decays to 2.54 micrograms in 5 min. What is the half-life of francium-223, in minutes?

9. Earth's atmospheric pressure changes as you rise above its surface. At an altitude of h kilometers, where $0 < h < 80$, the pressure P in newtons per square centimeter is approximately modeled by the equation $P(h) = 10.13e^{-0.116h}$.

    a. What is the approximate pressure at 40 km above Earth's surface?

    b. What is the approximate pressure on Earth's surface?

    c. Does atmospheric pressure increase or decrease as you rise above Earth's surface?

10. The U.S. Census Bureau provides information about various segments of the population in the United States. The following table gives the number of people, in millions, age 80 and older at the beginning of each decade from 1910 to 2000. An equation that approximately models the data is $y = 0.18808(1.0365)^x$, where x is the number of years since 1900 and y is the population, in millions, of people age 80 and over.

Year	1910	1920	1930	1940	1950	1960	1970	1980	1990	2000
Number of people age 80 and over (in millions)	0.3	0.4	0.5	0.8	1.1	1.6	2.3	2.9	3.9	9.3

    a. According to the model, what is the predicted population of this age group in the year 2020? Round to the nearest tenth of a million. (*Hint:* You will need to determine the x-value for the year 2020.)

    b. In what year does this model predict that the population of this age group will be 15 million? Round to the nearest year.

For Exercises 11 and 12, use the equation $pH = -\log(H^+)$, where H^+ is the hydrogen ion concentration of a solution. Round to the nearest tenth.

11. Find the pH of milk, which has a hydrogen ion concentration of 3.97×10^{-7}.

12. Find the pH of a baking soda solution for which the hydrogen ion concentration is 3.98×10^{-9}.

For Exercises 13 and 14, use the equation $\log P = -kd$, which gives the relationship between the percent P, as a decimal, of light passing through a substance of thickness d, in meters.

13. The value of k for a swimming pool is approximately 0.05. At what depth, in meters, will the percent of light be 75% of the light at the surface of the pool?

14. The constant k for a piece of blue stained glass is 20. What percent of light will pass through a piece of this glass that is 0.005 m thick?

For Exercises 15 and 16, use the equation $D = 10(\log I + 16)$, where D is the number of decibels of a sound and I is the power of the sound measured in watts. Round to the nearest whole number.

15. Find the number of decibels of normal conversation. The power of the sound of normal conversation is approximately 3.2×10^{-10} watts.

16. The loudest sound made by any animal is made by the blue whale and can be heard from more than 500 mi away. The power of the sound is 630 watts. Find the number of decibels of sound emitted by the blue whale.

17. During the 1980s and 1990s, the average time T to play a Major League baseball game increased each year. If the year 1981 is represented by $x = 1$, then the function $T(x) = 149.57 + 7.63 \ln x$ approximates the time T, in minutes, to play a Major League baseball game for the years $x = 1$ to $x = 19$. By how many minutes did the average time of a Major League baseball game increase from 1981 to 1999? Round to the nearest tenth.

18. In 1962, the cost of a first-class postage stamp was $.04. In 2001, the cost was $.34. The increase in cost can be modeled by the equation $C = 0.04e^{0.057t}$, where C is the cost and t is the number of years after 1962. According to this model, in what year did a first-class postage stamp cost $.22?

19. The intensity I of an x-ray after it has passed through a material that is x centimeters thick is given by $I = I_0 e^{-kx}$, where I_0 is the initial intensity of the x-ray and k is a number that depends on the material. The constant k for copper is 3.2. Find the thickness of copper that is needed so that the intensity of an x-ray after passing through the copper is 25% of the original intensity. Round to the nearest tenth.

20. One model for the time it will take for the world's oil supply to be depleted is given by the equation $T = 14.29 \ln (0.00411r + 1)$, where r is the estimated world oil reserves in billions of barrels and T is the time, in years, before that amount of oil is depleted. Use this equation to determine how many barrels of oil are necessary to meet the world demand for 20 years. Round to the nearest tenth.

For Exercises 21–24, use the Richter scale equation $M = \log \dfrac{I}{I_0}$, where M is the magnitude of an earthquake, I is the intensity of the shock waves, and I_0 is the measure of the intensity of a zero-level earthquake.

21. On July 14, 2000, an earthquake struck the Kodiak Island region of Alaska. The earthquake had an intensity of $I = 6,309,573I_0$. Find the Richter scale magnitude of the earthquake. Round to the nearest tenth.

22. The earthquake on January 26, 2000 near Gujarat, India had an intensity of $I = 50{,}118{,}723 I_0$. Find the Richter scale magnitude of the earthquake. Round to the nearest tenth.

23. An earthquake in Japan on March 2, 1933 measured 8.9 on the Richter scale. Find the intensity of the earthquake in terms of I_0. Round to the nearest whole number.

24. An earthquake that occurred in China in 1978 measured 8.2 on the Richter scale. Find the intensity of the earthquake in terms of I_0. Round to the nearest whole number.

Shown at the right is a *seismogram,* which is used to measure the magnitude of an earthquake. The magnitude is determined by the amplitude A of a shock wave and the difference in time t, in seconds, between the occurrences of two types of waves called *primary waves* and *secondary waves.* As you can see on the graph, a primary wave is abbreviated *p-wave* and a secondary wave is abbreviated *s-wave.* The amplitude A of a wave is one-half the difference between its highest and lowest points. For this graph, A is 23 mm. The equation is $M = \log A + 3 \log 8t - 2.92$.

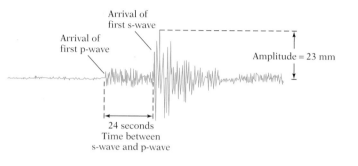

25. Determine the magnitude of the earthquake for the seismogram given in the figure. Round to the nearest tenth.

26. Find the magnitude of an earthquake that has a seismogram with an amplitude of 30 mm and for which t is 21 s.

27. Find the magnitude of an earthquake that has a seismogram with an amplitude of 28 mm and for which t is 28 s.

APPLYING THE CONCEPTS

28. The value of an investment in an account that earns an annual interest rate of 10% compounded daily grows according to the equation $A = A_0 \left(1 + \dfrac{0.10}{365}\right)^{365t}$, where A_0 is the original value of an investment and t is the time in years. Find the time for the investment to double in value. Round to the nearest year.

29. Some banks now use continuous compounding of an amount invested. In this case, the equation that relates the value of an initial investment of A dollars after t years earning an annual interest rate of $r\%$ is given by $P = Ae^{rt}$. Using this equation, find the value after 5 years of an investment of $2500 in an account that earns 5% annual interest.

Focus on Problem Solving

Proof by Contradiction

The four-step plan for solving problems that we have used before is restated here.

1. Understand the problem.
2. Devise a plan.
3. Carry out the plan.
4. Review the solution.

One of the techniques that can be used in the second step is a method called *proof by contradiction*. In this method you assume that the conditions of the problem you are trying to solve can be met and then show that your assumption leads to a condition you already know is not true.

To illustrate this method, suppose we try to prove that $\sqrt{2}$ is a rational number. We begin by recalling that a rational number is one that can be written as the quotient of integers. Therefore, let a and b be two integers with no common factors. If $\sqrt{2}$ is a rational number, then

$$\sqrt{2} = \frac{a}{b}$$

$$(\sqrt{2})^2 = \left(\frac{a}{b}\right)^2 \qquad \bullet \text{ Square each side.}$$

$$2 = \frac{a^2}{b^2}$$

$$2b^2 = a^2 \qquad \bullet \text{ Multiply each side by } b^2.$$

From the last equation, a^2 is an even number. Because a^2 is an even number, a is an even number. Now divide each side of the last equation by 2.

$$2b^2 = a^2 = a \cdot a$$

$$2b^2 = a \cdot a$$

$$b^2 = a \cdot x \qquad \qquad \bullet \ x = \frac{a}{2}$$

Because a is an even number, $a \cdot x$ is an even number. Because $a \cdot x$ is an even number, b^2 is an even number, and this in turn means that b is an even number. This result, however, contradicts the assumption that a and b are two integers with no common factors. Because this assumption is now known not to be true, we conclude that our original assumption, that $\sqrt{2}$ is a rational number, is false. This proves that $\sqrt{2}$ is an irrational number.

Try a proof by contradiction for the following problem: "Is it possible to write numbers using each of the digits 0, 1, 2, 3, 4, 5, 6, 7, 8, and 9 exactly once such that the sum of the numbers is exactly 100?"* Here are some suggestions. First note that the sum of the 10 digits is 45. This means that some of the digits used must be tens digits. Let x be the sum of those digits.

a. What is the sum of the remaining units digits?

b. Express "the sum of the units digits and the tens digits equals 100" as an equation.

c. Solve the equation for x.

d. Explain why this result means that it is impossible to satisfy both conditions of the problem.

*G. Polya, *How to Solve It: A New Aspect of Mathematical Method*. Copyright © 1945, renewed 1973 by Princeton University Press. Reprinted by permission of Princeton University Press.

Projects and Group Activities

Fractals*   Fractals have a wide variety of applications. They have been used to create special effects for movies such as *Star Trek II: The Wrath of Khan* and to explain the behavior of some biological and economic systems. One aspect of fractals that has fascinated mathematicians is that they apparently have *fractional dimension*.

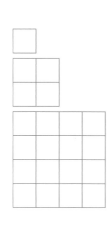

To understand the idea of fractional dimension, one must first understand the terms "scale factor" and "size." Consider a unit square (a square of length 1). By joining four of these squares, we can create another square, the length of which is 2, and the size of which is 4. (Here, size = number of square units.) Four of these larger squares can in turn be put together to make a third square of length 4 and size 16. This process of grouping together four squares can in theory be done an infinite number of times; yet at each step, the following quantities will be the same:

$$\text{scale factor} = \frac{\text{new length}}{\text{old length}} \qquad \text{size ratio} = \frac{\text{new size}}{\text{old size}}$$

Consider the unit square as Step 1, the four unit squares as Step 2, etc.

1. Calculate the scale factor going from **a.** Step 1 to Step 2, **b.** Step 2 to Step 3, and **c.** Step 3 to Step 4.

2. Calculate the size ratio going from **a.** Step 1 to Step 2, **b.** Step 2 to Step 3, and **c.** Step 3 to Step 4.

3. What is **a.** the scale factor and **b.** the size ratio going from Step n to Step $n + 1$?

Mathematicians have defined dimension using the formula $d = \dfrac{\log(\text{size ratio})}{\log(\text{scale factor})}$.

For the squares discussed above, $d = \dfrac{\log(\text{size ratio})}{\log(\text{scale factor})} = \dfrac{\log 4}{\log 2} = 2$.

So by this definition of dimension, squares are two-dimensional figures.

Now consider a unit cube (Step 1). Group eight unit cubes to form a cube that is 2 units on each side (Step 2). Group eight of the cubes from Step 2 to form a cube that is 4 units on each side (Step 3).

4. Calculate **a.** the scale factor and **b.** the size ratio for this process.

5. Show that the cubes are three-dimensional figures.

In each of the above examples, if the process is continued indefinitely, we still have a square or a cube. Consider a process that is more difficult to envision. Let Step 1 be an equilateral triangle whose base has length 1 unit, and let Step 2 be a grouping of three of these equilateral triangles, such that the space between them is another equilateral triangle with a base of length 1 unit. Three shapes from Step 2 are arranged with an equilateral triangle in their center, and so on. It is hard to imagine the result if this is done an infinite number of times, but mathematicians have shown that the result is a single figure of fractional dimension. (Similar processes have been used to create fascinating artistic patterns and to explain scientific phenomena.)

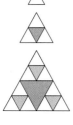

6. Show that for this process **a.** the scale factor is 2 and **b.** the size ratio is 3.

7. Calculate the dimension of the fractal. (Note that it is a *fractional* dimension!)

*Adapted with permission from Tami Martin, *Student Math Notes*, by the National Council of Teachers of Mathematics.

Solving Exponential and Logarithmic Equations Using a Graphing Calculator

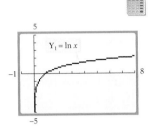

A graphing calculator can be used to draw the graphs of logarithmic functions. Note that there are two logarithmic keys, LOG and LN, on a graphing calculator. The first key gives the values of common logarithms (base 10), and the second gives the values of natural logarithms (base e).

To graph $y = \ln x$, press the Y= key. Clear any equations already entered. Press LN X,T,θ,n) GRAPH. The graph is shown at the left with a viewing window of Xmin $= -1$, Xmax $= 8$, Ymin $= -5$, and Ymax $= 5$.

Some exponential and logarithmic equations cannot be solved algebraically. In these cases, a graphical approach may be appropriate. Here is an example.

⇒ Solve $\ln (2x + 4) = x^2$ for x. Round to the nearest hundredth.

Rewrite the equation by subtracting x^2 from each side.

$$\ln (2x + 4) = x^2$$
$$\ln (2x + 4) - x^2 = 0$$

The zeros of $f(x) = \ln (2x + 4) - x^2$ are the solutions of $\ln (2x + 4) - x^2 = 0$.

Graph f and use the zero feature of the graphing calculator to estimate the solutions to the nearest hundredth.

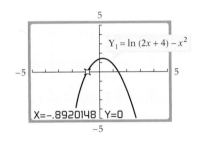

The zeros are approximately -0.89 and 1.38. The solutions are -0.89 and 1.38.

Solve for x by graphing. Round to the nearest hundredth.

1. $2^x = 2x + 4$ **2.** $3^x = -x - 1$
3. $e^x = -2x - 2$ **4.** $e^x = 3x + 4$
5. $\log (2x - 1) = -x + 3$ **6.** $\log (x + 4) = -2x + 1$
7. $\ln (x + 2) = x^2 - 3$ **8.** $\ln x = -x^2 + 1$

To graph a logarithmic function with a base other than base 10 or base e, the Change-of-Base Formula is used. Here is the Change-of-Base Formula, discussed earlier.

$$\log_a N = \frac{\log_b N}{\log_b a}$$

To graph $y = \log_2 x$, first change from base 2 logarithms to the equivalent base 10 logarithms or natural logarithms. Base 10 is shown here.

$$y = \log_2 x = \frac{\log_{10} x}{\log_{10} 2} \qquad \bullet \ a = 2, \ b = 10, \ N = x$$

Then graph the equivalent equation $y = \dfrac{\log x}{\log 2}$. The graph is shown at the left with a viewing window of Xmin $= -1$, Xmax $= 8$, Ymin $= -5$, and Ymax $= 5$.

The graphs of more complicated functions involving logarithms are produced in much the same way as outlined above. To graph $y = -2 \log_3(2x + 5) + 1$, enter into the calculator the equivalent equation $y = \dfrac{-2 \log(2x + 5)}{\log 3} + 1$.

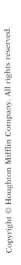

Chapter Summary

Key Words

A function of the form $f(x) = b^x$, where b is a positive real number not equal to 1, is an *exponential function*. The number b is the *base* of the exponential function. [p. 515]

The function defined by $f(x) = e^x$ is called the *natural exponential function*. [p. 516]

Because the exponential function is a 1–1 function, it has an inverse function that is called a *logarithm*. The definition of logarithm is: For $b > 0$, $b \neq 1$, $y = \log_b x$ is equivalent to $x = b^y$. [p. 523]

If $\log_b M = N$, then M is the *antilogarithm* base b of N. In exponential form, $M = b^N$. [p. 524]

Logarithms with base 10 are called *common logarithms*. The base, 10, is usually omitted when the common logarithm of a number is written. The decimal part of a common logarithm is called the *mantissa*; the integer part is called the *characteristic*. [p. 524]

When e (the base of the natural exponential function) is used as the base of a logarithm, the logarithm is referred to as the *natural logarithm* and is abbreviated $\ln x$. [p. 524]

An *exponential equation* is one in which a variable occurs in the exponent. [p. 537]

An *exponential growth equation* is an equation that can be written in the form $A = A_0 b^{kt}$, where A is the size at time t, A_0 is the initial size, $b > 1$, and k is a positive real number. In an *exponential decay equation*, the value of b is between 0 and 1. [pp. 543–544]

Essential Rules

The One-to-One Property of Exponential Functions
For $b > 0$, $b \neq 1$, if $b^u = b^v$, then $u = v$. [p. 523]

The Logarithm Property of the Product of Two Numbers
For any positive real numbers x, y, and b, $b \neq 1$, $\log_b(xy) = \log_b x + \log_b y$. [p. 525]

The Logarithm Property of the Quotient of Two Numbers
For any positive real numbers x, y, and b, $b \neq 1$, $\log_b \dfrac{x}{y} = \log_b x - \log_b y$. [p. 526]

The Logarithm Property of the Power of a Number
For any positive real numbers x and b, $b \neq 1$, and for any real number r, $\log_b x^r = r \log_b x$. [p. 526]

Additional Properties of Logarithms
For any positive real numbers x, y, and b, $b \neq 1$:
$$\log_b 1 = 0$$
$$b^{\log_b x} = x$$
$$\log_b b^x = x \text{ [p. 526]}$$

Change-of-Base Formula
$$\log_a N = \frac{\log_b N}{\log_b a} \text{ [p. 528]}$$

To Solve Numerical Problems Using Logarithms
For any positive real numbers x, y, and b, $b \neq 1$, if $\log_b x = \log_b y$, then $x = y$. [p. 526]

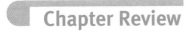

Chapter Review

1. Evaluate $f(x) = e^{x-2}$ at $x = 2$.

2. Write $\log_5 25 = 2$ in exponential form.

3. Graph: $f(x) = 3^{-x} + 2$

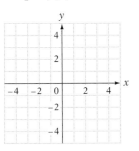

4. Graph: $f(x) = \log_3(x - 1)$

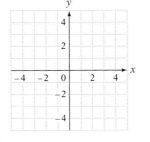

5. Write $\log_3 \sqrt[5]{x^2 y^4}$ in expanded form.

6. Write $2 \log_3 x - 5 \log_3 y$ as a single logarithm with a coefficient of 1.

7. Solve: $27^{2x+4} = 81^{x-3}$

8. Solve: $\log_5 \dfrac{7x + 2}{3x} = 1$

9. Find $\log_6 22$. Round to the nearest ten-thousandth.

10. Solve: $\log_2 x = 5$

11. Solve: $\log_3(x + 2) = 4$

12. Solve: $\log_{10} x = 3$

13. Write $\dfrac{1}{3}(\log_7 x + 4 \log_7 y)$ as a single logarithm with a coefficient of 1.

14. Write $\log_8 \sqrt{\dfrac{x^5}{y^3}}$ in expanded form.

15. Write $2^5 = 32$ in logarithmic form.

16. Find $\log_3 1.6$. Round to the nearest ten-thousandth.

17. Solve $3^{x+2} = 5$. Round to the nearest thousandth.

18. Evaluate $f(x) = \left(\dfrac{2}{3}\right)^{x+2}$ at $x = -3$.

19. Solve: $\log_8(x + 2) - \log_8 x = \log_8 4$

20. Solve: $\log_6(2x) = \log_6 2 + \log_6(3x - 4)$

21. Graph: $f(x) = \left(\dfrac{2}{3}\right)^{x+1}$

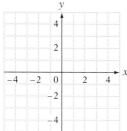

22. Graph: $f(x) = \log_2(2x - 1)$

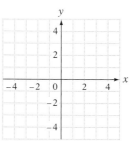

23. Use the compound interest formula $P = A(1 + i)^n$, where A is the original value of an investment, i is the interest rate per compounding period, n is the number of compounding periods, and P is the value of the investment after n periods, to find the value of an investment after 2 years. The amount of the investment is \$4000, and it is invested at 8% compounded monthly. Round to the nearest dollar.

24. An earthquake off the coast of Central America in January, 2001 had an intensity of $I = 39{,}810{,}717I_0$. Find the Richter scale magnitude of the earthquake. Use the Richter scale equation $M = \log\dfrac{I}{I_0}$, where M is the magnitude of an earthquake, I is the intensity of the shock waves, and I_0 is the measure of the intensity of a zero-level earthquake. Round to the nearest tenth.

25. Use the exponential decay equation $A = A_0\left(\dfrac{1}{2}\right)^{t/k}$, where A is the amount of a radioactive material present after time t, k is the half-life of the radioactive material, and A_0 is the original amount of radioactive material, to find the half-life of a material that decays from 25 mg to 15 mg in 20 days. Round to the nearest whole number.

26. The number of decibels, D, of a sound can be given by the equation $D = 10(\log I + 16)$, where I is the power of the sound measured in watts. Find the number of decibels of sound emitted from a busy street corner for which the power of the sound is 5×10^{-6} watts.

Chapter Test

1. Evaluate $f(x) = \left(\dfrac{2}{3}\right)^x$ at $x = 0$.

2. Evaluate $f(x) = 3^{x+1}$ at $x = -2$.

3. Graph: $f(x) = 2^x - 3$

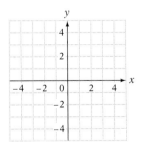

4. Graph: $f(x) = 2^x + 2$

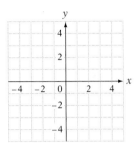

5. Evaluate: $\log_4 16$

6. Solve for x: $\log_3 x = -2$

7. Graph: $f(x) = \log_2(2x)$

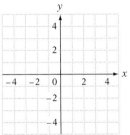

8. Graph: $f(x) = \log_3(x + 1)$

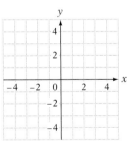

9. Write $\log_6 \sqrt{xy^3}$ in expanded form.

10. Write $\dfrac{1}{2}(\log_3 x - \log_3 y)$ as a single logarithm with a coefficient of 1.

11. Write $\ln\left(\dfrac{x}{\sqrt{z}}\right)$ in expanded form.

12. Write $3 \ln x - \ln y - \dfrac{1}{2} \ln z$ as a single logarithm with a coefficient of 1.

13. Solve for x: $3^{7x+1} = 3^{4x-5}$

14. Solve for x: $8^x = 2^{x-6}$

15. Solve for x: $3^x = 17$

16. Solve for x: $\log x + \log(x - 4) = \log 12$

17. Solve for x: $\log_6 x + \log_6(x - 1) = 1$

18. Find $\log_5 9$.

19. Find $\log_3 19$.

20. Use the exponential decay equation $A = A_0\left(\dfrac{1}{2}\right)^{t/k}$, where A is the amount of a radioactive material present after time t, k is the half-life of the material, and A_0 is the original amount of radioactive material, to find the half-life of a material that decays from 10 mg to 9 mg in 5 h. Round to the nearest whole number.

Cumulative Review

1. Solve: $4 - 2[x - 3(2 - 3x) - 4x] = 2x$

2. Find the equation of the line that contains the point $(2, -2)$ and is parallel to the line $2x - y = 5$.

3. Factor: $4x^{2n} + 7x^n + 3$

4. Simplify: $\dfrac{1 - \dfrac{5}{x} + \dfrac{6}{x^2}}{1 + \dfrac{1}{x} - \dfrac{6}{x^2}}$

5. Simplify: $\dfrac{\sqrt{xy}}{\sqrt{x} - \sqrt{y}}$

6. Solve by completing the square: $x^2 - 4x - 6 = 0$

7. Write a quadratic equation that has integer coefficients and has solutions $\dfrac{1}{3}$ and -3.

8. Graph the solution set: $2x - y < 3$
$$x + y < 1$$

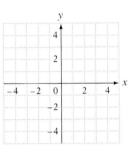

9. Solve by the addition method:
$$3x - y + z = 3$$
$$x + y + 4z = 7$$
$$3x - 2y + 3z = 8$$

10. Simplify: $\dfrac{x - 4}{2 - x} - \dfrac{1 - 6x}{2x^2 - 7x + 6}$

11. Solve: $x^2 + 4x - 5 \le 0$

12. Solve: $|2x - 5| \le 3$

13. Graph: $f(x) = \left(\dfrac{1}{2}\right)^x + 1$

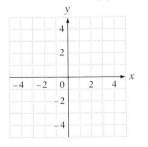

14. Graph: $f(x) = \log_2 x - 1$

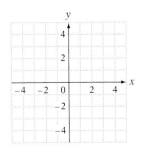

15. Evaluate the function $f(x) = 2^{-x-1}$ at $x = -3$.

16. Solve for x: $\log_5 x = 3$

17. Write $3 \log_b x - 5 \log_b y$ as a single logarithm with a coefficient of 1.

18. Find $\log_3 7$.

19. Solve for x: $4^{5x-2} = 4^{3x+2}$

20. Solve for x: $\log x + \log(2x + 3) = \log 2$

21. A bank offers two types of checking accounts. One account has a charge of $5 per month plus 2 cents per check. The second account has a charge of $2 per month plus 8 cents per check. How many checks can a customer who has the second type of account write if it is to cost the customer less than the first type of checking account?

22. Find the cost per pound of a mixture made from 16 lb of chocolate that costs $4.00 per pound and 24 lb of chocolate that costs $2.50 per pound.

23. A plane can fly at a rate of 225 mph in calm air. Traveling with the wind, the plane flew 1000 mi in the same amount of time that it took to fly 800 mi against the wind. Find the rate of the wind.

24. The distance, d, that a spring stretches varies directly as the force, f, used to stretch the spring. If a force of 20 lb stretches a spring 6 in., how far will a force of 34 lb stretch the spring?

25. A carpenter purchased 80 ft of redwood and 140 ft of fir for a total cost of $67. A second purchase, at the same prices, included 140 ft of redwood and 100 ft of fir for a total cost of $81. Find the cost of redwood and of fir.

26. The compound interest formula is $P = A(1 + i)^n$, where A is the original value of an investment, i is the interest rate per compounding period, n is the total number of compounding periods, and P is the value of the investment after n periods. Use the compound interest formula to find the number of years in which an investment of $5000 will double in value. The investment earns 9% annual interest and is compounded semiannually.

Chapter 11 Conic Sections

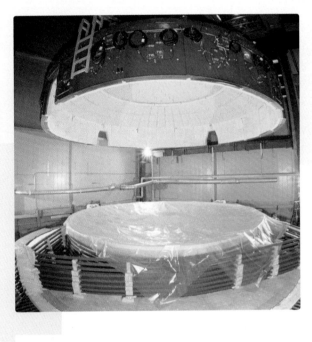

When creating a parabolic-shaped mirror for use in a telescope, such as the one in this photo, there is little room for error. The multi-million-dollar Hubble Telescope made international headlines when it was discovered that the outer edge of its primary mirror was ground too flat by one fiftieth the thickness of a human hair! This flaw (which was later corrected) caused Hubble to transmit fuzzy images because the light from viewed objects was scattering instead of reflecting to a single point. This single point is called the focus, as explained in **Exercises 25 and 26 on page 568**.

Need help? For on-line student resources, such as section quizzes, visit this textbook's web site at **math.college.hmco.com/students**.

1. Find the distance between the points whose coordinates are $(-2, 3)$ and $(4, -1)$. Round to the nearest hundredth.

2. Complete the square on $x^2 - 8x$. Write the resulting perfect-square trinomial as the square of a binomial.

3. Solve $\dfrac{x^2}{16} + \dfrac{y^2}{9} = 1$ for x when $y = 3$ and when $y = 0$.

4. Solve by the substitution method:
$$7x + 4y = 3$$
$$y = x - 2$$

5. Solve by the addition method: $\quad 4x - y = 9$
$$2x + 3y = -13$$

6. Find the axis of symmetry and the vertex of the graph of $y = x^2 - 4x + 2$.

7. Graph: $f(x) = -2x^2 + 4x$

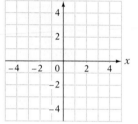

8. Graph the solution set of $5x - 2y > 10$.

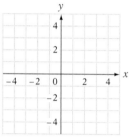

9. Graph the solution set:
$$x + 2y \le 4$$
$$x - y \le 2$$

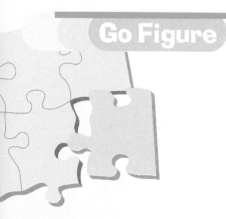

Go Figure

What is the minimum number of times a paper must be folded in half so that its thickness is at least the distance from Earth to the moon? Assume the paper can be folded in half indefinitely and that the paper is 0.01 in. thick. Use a distance from Earth to the moon of 240,000 mi. (*Hint:* 5280 ft = 1 mi)

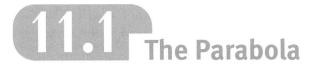

The Parabola

Objective A   **To graph a parabola**   11

The **conic sections** are curves that can be constructed from the intersection of a plane and a right circular cone. The parabola, which was introduced earlier, is one of these curves. Here we will review some of that previous discussion and look at equations of parabolas that were not discussed before.

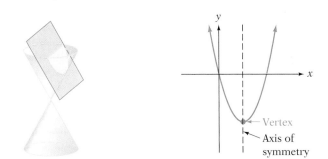

Every parabola has an axis of symmetry and a vertex that is on the axis of symmetry. To understand the axis of symmetry, think of folding the paper along that axis. The two halves of the curve will match up.

The graph of the equation $y = ax^2 + bx + c$, $a \neq 0$, is a parabola with the axis of symmetry parallel to the y-axis. The parabola opens up when $a > 0$ and opens down when $a < 0$. When the parabola opens up, the vertex is the lowest point on the parabola. When the parabola opens down, the vertex is the highest point on the parabola.

The coordinates of the vertex can be found by completing the square.

→ Find the vertex of the parabola whose equation is $y = x^2 - 4x + 5$.

$y = x^2 - 4x + 5$
$y = (x^2 - 4x) + 5$
$y = (x^2 - 4x + 4) - 4 + 5$

- Group the terms involving x.
- Complete the square on $x^2 - 4x$. Note that 4 is added and subtracted. Because $4 - 4 = 0$, the equation is not changed.
- Factor the trinomial and combine like terms.

$y = (x - 2)^2 + 1$

The coefficient of x^2 is positive, so the parabola opens up. The vertex is the lowest point on the parabola, or the point that has the least y-coordinate.

Because $(x - 2)^2 \geq 0$ for all x, the least y-coordinate occurs when $(x - 2)^2 = 0$, which occurs when $x = 2$. This means the x-coordinate of the vertex is 2.

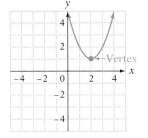

To find the y-coordinate of the vertex, replace x in $y = (x - 2)^2 + 1$ by 2 and solve for y.

$y = (x - 2)^2 + 1$
$ = (2 - 2)^2 + 1 = 1$

The vertex is (2, 1).

By following the procedure of the last example and completing the square on the equation $y = ax^2 + bx + c$, we find that the **x-coordinate of the vertex** is $-\dfrac{b}{2a}$. The y-coordinate of the vertex can then be determined by substituting this value of x into $y = ax^2 + bx + c$ and solving for y.

Because the axis of symmetry is parallel to the y-axis and passes through the vertex, the equation of the **axis of symmetry** is $x = -\dfrac{b}{2a}$.

⇒ Find the vertex and axis of symmetry of the parabola whose equation is $y = -3x^2 + 6x + 1$. Then sketch its graph.

x-coordinate: $-\dfrac{b}{2a} = -\dfrac{6}{2(-3)} = 1$

• Find the x-coordinate of the vertex and the axis of symmetry. $a = -3$, $b = 6$

The x-coordinate of the vertex is 1.
The axis of symmetry is the line $x = 1$.

To find the y-coordinate of the vertex, replace x by 1 and solve for y.

$y = -3x^2 + 6x + 1$
$\quad = -3(1)^2 + 6(1) + 1 = 4$

The vertex is (1, 4).

Because a is negative, the parabola opens down.

Find a few ordered pairs and use symmetry to sketch the graph.

⇒ Find the vertex and axis of symmetry of the parabola whose equation is $y = x^2 - 2$. Then sketch its graph.

x-coordinate: $-\dfrac{b}{2a} = -\dfrac{0}{2(1)} = 0$

• Find the x-coordinate of the vertex and the axis of symmetry. $a = 1$, $b = 0$

The x-coordinate of the vertex is 0.
The axis of symmetry is the line $x = 0$.

To find the y-coordinate of the vertex, replace x by 0 and solve for y.

$y = x^2 - 2$
$\quad = 0^2 - 2 = -2$

The vertex is (0, -2).

Because a is positive, the parabola opens up.

Find a few ordered pairs and use symmetry to sketch the graph.

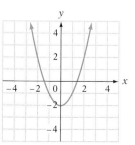

The graph of an equation of the form $x = ay^2 + by + c$, $a \neq 0$, is also a parabola. In this case, the parabola opens to the right when a is positive and opens to the left when a is negative.

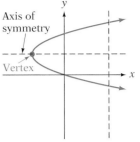

For a parabola of this form, the **y-coordinate of the vertex** is $-\dfrac{b}{2a}$. The **axis of symmetry** is the line $y = -\dfrac{b}{2a}$.

Using the vertical line test, the graph of a parabola of this form is not the graph of a function. The graph of $x = ay^2 + by + c$ is a relation.

➡ Find the vertex and axis of symmetry of the parabola whose equation is $x = 2y^2 - 8y + 5$. Then sketch its graph.

y-coordinate: $-\dfrac{b}{2a} = -\dfrac{-8}{2(2)} = 2$ • Find the y-coordinate of the vertex and the axis of symmetry. $a = 2$, $b = -8$

The y-coordinate of the vertex is 2.
The axis of symmetry is the line $y = 2$.

To find the x-coordinate of the vertex, replace y by 2 and solve for x.

$x = 2y^2 - 8y + 5$
$ = 2(2)^2 - 8(2) + 5 = -3$

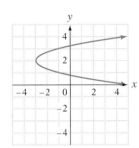

The vertex is $(-3, 2)$.

Since a is positive, the parabola opens to the right.

Find a few ordered pairs and use symmetry to sketch the graph.

➡ Find the vertex and axis of symmetry of the parabola whose equation is $x = -2y^2 - 4y - 3$. Then sketch its graph.

y-coordinate: $-\dfrac{b}{2a} = -\dfrac{-4}{2(-2)} = -1$ • Find the y-coordinate of the vertex and the axis of symmetry. $a = -2$, $b = -4$

The y-coordinate of the vertex is -1.
The axis of symmetry is the line $y = -1$.

To find the x-coordinate of the vertex, replace y by -1 and solve for x.

$x = -2y^2 - 4y - 3$
$ = -2(-1)^2 - 4(-1) - 3 = -1$

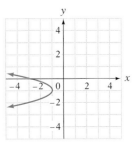

The vertex is $(-1, -1)$.

Because a is negative, the parabola opens to the left.

Find a few ordered pairs and use symmetry to sketch the graph.

Example 1

Find the vertex and axis of symmetry of the parabola whose equation is $y = x^2 - 4x + 3$. Then sketch its graph.

Solution

$$-\frac{b}{2a} = -\frac{-4}{2(1)} = 2$$

Axis of symmetry:
$$x = 2$$

$$y = 2^2 - 4(2) + 3$$
$$= -1$$

Vertex: $(2, -1)$

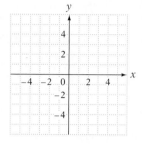

You Try It 1

Find the vertex and axis of symmetry of the parabola whose equation is $y = x^2 + 2x + 1$. Then sketch its graph.

Your solution

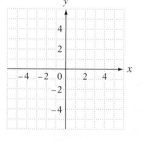

Example 2

Find the vertex and axis of symmetry of the parabola whose equation is $x = 2y^2 - 4y + 1$. Then sketch its graph.

Solution

$$-\frac{b}{2a} = -\frac{-4}{2(2)} = 1$$

Axis of symmetry:
$$y = 1$$

$$x = 2(1)^2 - 4(1) + 1$$
$$= -1$$

Vertex: $(-1, 1)$

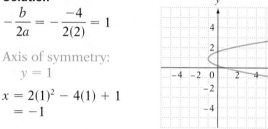

You Try It 2

Find the vertex and axis of symmetry of the parabola whose equation is $x = -y^2 - 2y + 2$. Then sketch its graph.

Your solution

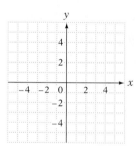

Example 3

Find the vertex and axis of symmetry of the parabola whose equation is $y = x^2 + 1$. Then sketch its graph.

Solution

$$-\frac{b}{2a} = -\frac{0}{2(1)} = 0$$

Axis of symmetry:
$$x = 0$$

$$y = 0^2 + 1$$
$$= 1$$

Vertex: $(0, 1)$

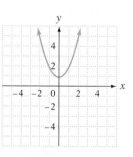

You Try It 3

Find the vertex and axis of symmetry of the parabola whose equation is $y = x^2 - 2x - 1$. Then sketch its graph.

Your solution

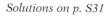

Solutions on p. S31

11.1 Exercises

Objective A

State **a.** whether the axis of symmetry is a vertical or horizontal line and **b.** the direction in which the parabola opens.

1. $y = 3x^2 - 4x + 7$

2. $y = -x^2 + 5x - 2$

3. $x = y^2 + 2y - 8$

4. $x = -3y^2 - y + 9$

5. $x = -\dfrac{1}{2}y^2 - 4y - 7$

6. $y = \dfrac{1}{4}x^2 + 6x - 1$

Find the vertex and axis of symmetry of the parabola given by the equation. Then sketch its graph.

7. $x = y^2 - 3y - 4$

8. $y = x^2 - 2$

9. $y = x^2 + 2$

10. $x = -\dfrac{1}{2}y^2 + 4$

11. $x = -\dfrac{1}{4}y^2 - 1$

12. $x = \dfrac{1}{2}y^2 - y + 1$

13. $x = -\dfrac{1}{2}y^2 + 2y - 3$

14. $y = -\dfrac{1}{2}x^2 + 2x + 6$

15. $y = \dfrac{1}{2}x^2 + x - 3$

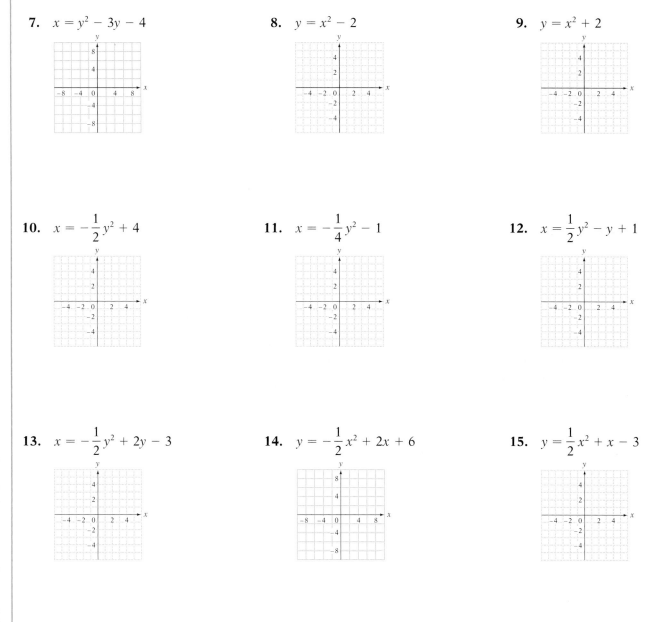

16. $x = y^2 + 6y + 5$

17. $x = y^2 - y - 6$

18. $x = -y^2 + 2y - 3$

19. $y = 2x^2 + 4x - 5$

20. $y = 2x^2 - x - 5$

21. $y = 2x^2 - x - 3$

22. $y = x^2 - 5x + 4$

23. $y = x^2 + 5x + 6$

24. $x = y^2 - 2y - 5$

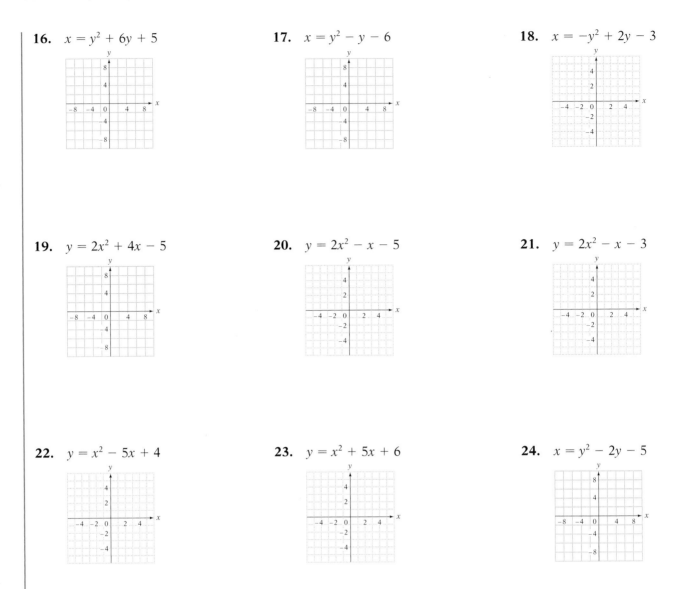

APPLYING THE CONCEPTS

Parabolas have a unique property that is important in the design of telescopes and antennas. If a parabola has a mirrored surface, then all light rays parallel to the axis of symmetry of the parabola are reflected to a single point called the **focus** of the parabola. The location of this point is p units from the vertex on the axis of symmetry. The value of p is given by $p = \dfrac{1}{4a}$, where $y = ax^2$ is the equation of a parabola with vertex at the origin. For the graph of $y = \dfrac{1}{4}x^2$ shown at the right, the coordinates of the focus are $(0, 1)$. For Exercises 25 and 26, find the coordinates of the focus of the parabola given by the equation.

25. $y = 2x^2$

26. $y = \dfrac{1}{10}x^2$

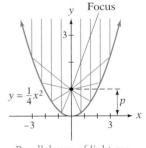

Parallel rays of light are reflected to the focus

$a = \dfrac{1}{4}$

$p = \dfrac{1}{4a} = \dfrac{1}{4(1/4)} = 1$

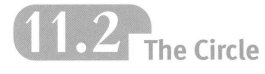

The Circle

Objective A

To find the equation of a circle and then graph the circle

A **circle** is a conic section formed by the intersection of a cone and a plane parallel to the base of the cone.

TAKE NOTE
As the angle of the plane that intersects the cone changes, different conic sections are formed. For a parabola, the plane was *parallel to the side* of the cone. For a circle, the plane is *parallel to the base* of the cone.

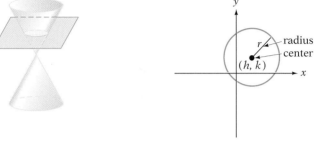

A circle can be defined as all points (x, y) in the plane that are a fixed distance from a given point (h, k), called the **center.** The fixed distance is the **radius** of the circle.

Using the distance formula, the equation of a circle can be determined.

Let (h, k) be the coordinates of the center of the circle, let r be the radius, and let (x, y) be any point on the circle. Then, by the distance formula,

$$r = \sqrt{(x - h)^2 + (y - k)^2}$$

Squaring each side of the equation gives the equation of a circle.

$$r^2 = [\sqrt{(x - h)^2 + (y - k)^2}]^2$$
$$r^2 = (x - h)^2 + (y - k)^2$$

The Standard Form of the Equation of a Circle

Let r be the radius of a circle and let (h, k) be the coordinates of the center of the circle. Then the equation of the circle is given by

$$(x - h)^2 + (y - k)^2 = r^2$$

➡ Sketch a graph of $(x - 1)^2 + (y + 2)^2 = 9$.

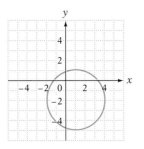

- The standard form of the equation is
  $(x - 1)^2 + [\, y - (-2)]^2 = 3^2$.
  Center: $(1, -2)$ Radius: 3

⇨ Find the equation of the circle with radius 4 and center $(-1, 2)$. Then sketch its graph.

$$(x - h)^2 + (y - k)^2 = r^2$$

$$[x - (-1)]^2 + (y - 2)^2 = 4^2$$
$$(x + 1)^2 + (y - 2)^2 = 16$$

- Use the standard form of the equation of a circle.
- Replace r by 4, h by -1, and k by 2.

- Sketch the graph by drawing a circle with center $(-1, 2)$ and radius 4.

TAKE NOTE

Applying the vertical-line test reveals that the graph of a circle is not the graph of a function. The graph of a circle is the graph of a relation.

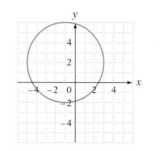

⇨ Find the equation of the circle that passes through the point whose coordinates are $(3, -4)$ and whose center is $(2, 1)$.

A point on the circle is known and the center is known. Use the distance formula to find the radius of the circle.

$$r = \sqrt{(x_1 - x_2)^2 + (y_1 - y_2)^2}$$
$$= \sqrt{(3 - 2)^2 + (-4 - 1)^2}$$
$$= \sqrt{1^2 + (-5)^2}$$
$$= \sqrt{1 + 25}$$
$$= \sqrt{26}$$

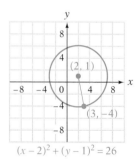

$$(x - 2)^2 + (y - 1)^2 = 26$$

Using $r = \sqrt{26}$ and $(h, k) = (2, 1)$, write the equation of the circle.
Note that $r^2 = (\sqrt{26})^2 = 26$.

$$(x - h)^2 + (y - k)^2 = r^2$$
$$(x - 2)^2 + (y - 1)^2 = 26$$

. . . .

Example 1

Sketch a graph of $(x + 2)^2 + (y - 1)^2 = 4$.

Solution

Center: $(-2, 1)$
Radius: $\sqrt{4} = 2$

You Try It 1

Sketch a graph of $(x - 2)^2 + (y + 3)^2 = 9$.

Your solution

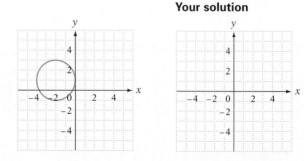

Solution on p. S31

Example 2

Find the equation of the circle with radius 5 and center $(-1, 3)$. Then sketch its graph.

Solution

$$(x - h)^2 + (y - k)^2 = r^2$$
$$[x - (-1)]^2 + (y - 3)^2 = 5^2$$
$$(x + 1)^2 + (y - 3)^2 = 25$$

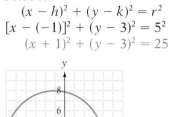

You Try It 2

Find the equation of the circle with radius 4 and center $(2, -3)$. Then sketch its graph.

Your solution

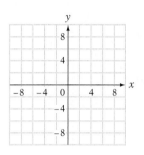

Solution on p. S31

Objective B   To write the equation of a circle in standard form

The equation of a circle can also be expressed as $x^2 + y^2 + ax + by + c = 0$. This is called the **general form** of the equation of a circle. By completing the square, an equation in general form can be written in standard form.

→ Find the center and the radius of a circle whose equation is given by $x^2 + y^2 + 4x + 2y + 1 = 0$. Then sketch its graph.

Write the equation in standard form by completing the square on x and y.

$$x^2 + y^2 + 4x + 2y + 1 = 0$$
$$x^2 + y^2 + 4x + 2y = -1$$

- Subtract the constant term from each side of the equation.

$$(x^2 + 4x) + (y^2 + 2y) = -1$$

- Rewrite the equation by grouping x terms and y terms.

$$(x^2 + 4x + 4) + (y^2 + 2y + 1) = -1 + 4 + 1$$

- Complete the square on $x^2 + 4x$ and $y^2 + 2y$.

$$(x + 2)^2 + (y + 1)^2 = 4$$

- Factor the trinomials.

The center is $(-2, -1)$; the radius is 2.

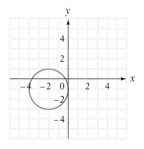

Example 3

Write the equation of the circle
$x^2 + y^2 - 4x + 2y + 4 = 0$ in standard form.
Then sketch its graph.

Solution

$$x^2 + y^2 - 4x + 2y + 4 = 0$$
$$(x^2 - 4x) + (y^2 + 2y) = -4$$
$$(x^2 - 4x + 4) + (y^2 + 2y + 1) = -4 + 4 + 1$$
$$(x - 2)^2 + (y + 1)^2 = 1$$

Center: $(2, -1)$;   Radius: 1

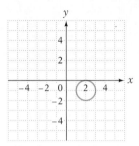

You Try It 3

Write the equation of the circle
$x^2 + y^2 - 2x - 15 = 0$ in standard form.
Then sketch its graph.

Your solution

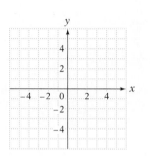

Example 4

Write the equation of the circle
$x^2 + y^2 + 3x - 2y - 1 = 0$ in standard form.
Then sketch its graph.

Solution

$$x^2 + y^2 + 3x - 2y - 1 = 0$$
$$x^2 + y^2 + 3x - 2y = 1$$
$$(x^2 + 3x) + (y^2 - 2y) = 1$$
$$\left(x^2 + 3x + \frac{9}{4}\right) + (y^2 - 2y + 1) = 1 + \frac{9}{4} + 1$$
$$\left(x + \frac{3}{2}\right)^2 + (y - 1)^2 = \frac{17}{4}$$

Center: $\left(-\frac{3}{2}, 1\right)$;   Radius: $\sqrt{\dfrac{17}{4}} = \dfrac{\sqrt{17}}{2}$

You Try It 4

Write the equation of the circle
$x^2 + y^2 - 4x + 8y + 15 = 0$ in standard form.
Then sketch its graph.

Your solution

Solutions on pp. S31–S32

11.2 Exercises

Objective A

Sketch a graph of the circle given by the equation.

1. $(x - 2)^2 + (y + 2)^2 = 9$ **2.** $(x + 2)^2 + (y - 3)^2 = 16$ **3.** $(x + 3)^2 + (y - 1)^2 = 25$

4. $(x - 2)^2 + (y + 3)^2 = 4$ **5.** $(x + 2)^2 + (y + 2)^2 = 4$ **6.** $(x - 1)^2 + (y - 2)^2 = 25$

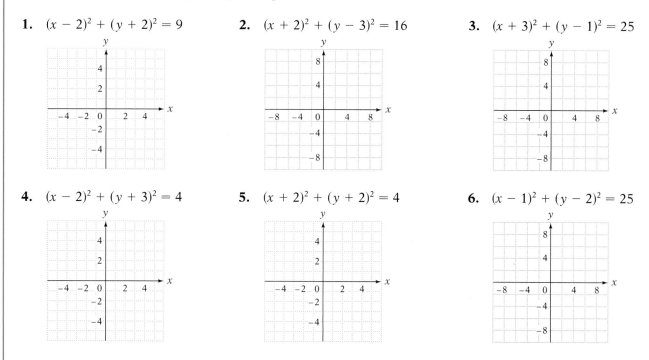

7. Find the equation of the circle with radius 2 and center $(2, -1)$. Then sketch its graph.

8. Find the equation of the circle with radius 3 and center $(-1, -2)$. Then sketch its graph.

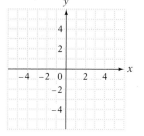

9. Find the equation of the circle that passes through the point $(1, 2)$ and whose center is $(-1, 1)$. Then sketch its graph.

10. Find the equation of the circle that passes through the point $(-1, 3)$ and whose center is $(-2, 1)$. Then sketch its graph.

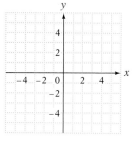

Objective B

Write the equation of the circle in standard form. Then sketch its graph.

11. $x^2 + y^2 - 2x + 4y - 20 = 0$

12. $x^2 + y^2 - 4x + 8y + 4 = 0$

13. $x^2 + y^2 + 6x + 8y + 9 = 0$

14. $x^2 + y^2 - 6x + 10y + 25 = 0$

15. $x^2 + y^2 - 2x + 2y - 23 = 0$

16. $x^2 + y^2 - 6x + 4y + 9 = 0$

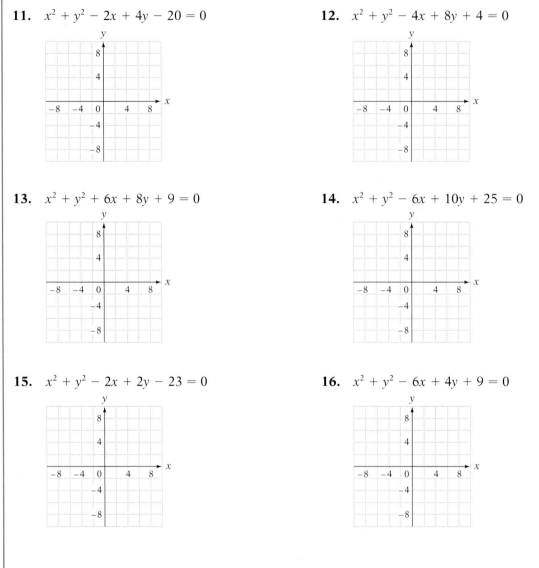

APPLYING THE CONCEPTS

17. Find the equation of a circle that has center $(4, 0)$ and passes through the origin.

18. Find the equation of a circle that has a diameter with endpoints $(-2, 4)$ and $(2, -2)$.

19. The radius of a sphere is 12 in. What is the radius of the circle that is formed by the intersection of the sphere and a plane that is 6 in. from the center of the sphere?

20. Find the equation of the circle with center at $(3, 3)$ if the circle is tangent to the x-axis.

21. Is $x^2 + y^2 + 4x + 8y + 24 = 0$ the equation of a circle? If not, explain why not. If so, find the radius and the coordinates of the center.

11.3 The Ellipse and the Hyperbola

Objective A   **To graph an ellipse with center at the origin**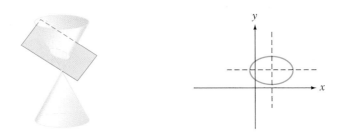

The orbits of the planets around the sun are "oval" shaped. This oval shape can be described as an **ellipse,** which is another of the conic sections.

There are two **axes of symmetry** for an ellipse. The intersection of these two axes is the **center** of the ellipse.

An ellipse with center at the origin is shown at the right. Note that there are two x-intercepts and two y-intercepts.

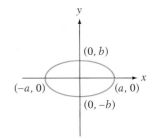

Point of Interest

The word *ellipse* comes from the Greek word *ellipsis*, which means "deficient." The method by which the early Greeks analyzed the conics caused a certain area in the construction of the ellipse to be less than another area (deficient). The word *ellipsis* in English, meaning "omission," has the same Greek root as the word *ellipse*.

The Standard Form of the Equation of an Ellipse with Center at the Origin

The equation of an ellipse with center at the origin is $\dfrac{x^2}{a^2} + \dfrac{y^2}{b^2} = 1$.

The x-intercepts are $(a, 0)$ and $(-a, 0)$. The y-intercepts are $(0, b)$ and $(0, -b)$.

By finding the x- and y-intercepts of an ellipse and using the fact that the ellipse is oval-shaped, we can sketch a graph of the ellipse.

⇒ Sketch the graph of the ellipse whose equation is $\dfrac{x^2}{9} + \dfrac{y^2}{4} = 1$.

The x-intercepts are $(3, 0)$ and $(-3, 0)$.
The y-intercepts are $(0, 2)$ and $(0, -2)$.

- $a^2 = 9$, $b^2 = 4$

- Use the intercepts and symmetry to sketch the graph of the ellipse.

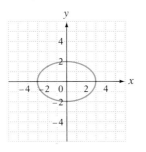

Using the vertical-line test, we find that the graph of an ellipse is not the graph of a function. The graph of an ellipse is the graph of a relation.

⟹ Sketch a graph of the ellipse whose equation is $\dfrac{x^2}{16} + \dfrac{y^2}{16} = 1$.

The x-intercepts are $(4, 0)$ and $(-4, 0)$.
The y-intercepts are $(0, 4)$ and $(0, -4)$.

• $a^2 = 16$, $b^2 = 16$

Point of Interest

For a circle, $a = b$ and thus $\dfrac{a}{b} = 1$. Early Greek astronomers thought that each planet had a circular orbit. Today we know that the planets have elliptical orbits. However, in most cases, the ellipse is very nearly a circle.

For Earth, $\dfrac{a}{b} \approx 1.00014$. The most elliptical orbit is Pluto's, for which $\dfrac{a}{b} \approx 1.0328$.

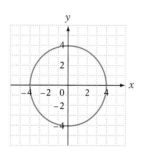

• Use the intercepts and symmetry to sketch the graph of the ellipse.

The graph in this last example is the graph of a circle. A circle is a special case of an ellipse. It occurs when $a^2 = b^2$ in the equation $\dfrac{x^2}{a^2} + \dfrac{y^2}{b^2} = 1$.

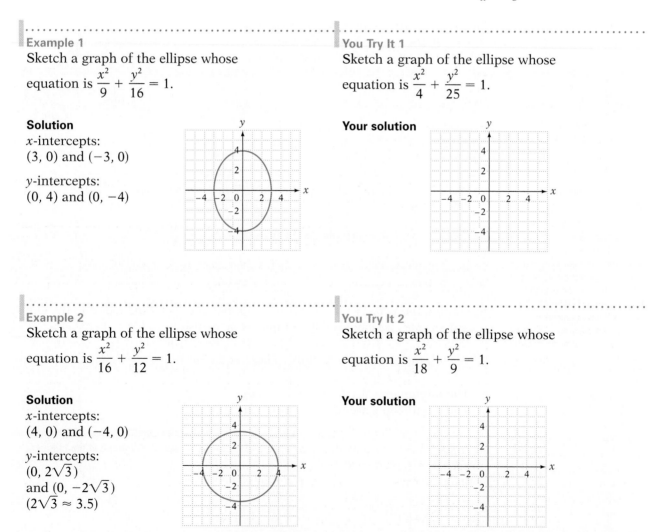

Example 1
Sketch a graph of the ellipse whose equation is $\dfrac{x^2}{9} + \dfrac{y^2}{16} = 1$.

Solution
x-intercepts:
$(3, 0)$ and $(-3, 0)$

y-intercepts:
$(0, 4)$ and $(0, -4)$

You Try It 1
Sketch a graph of the ellipse whose equation is $\dfrac{x^2}{4} + \dfrac{y^2}{25} = 1$.

Your solution

Example 2
Sketch a graph of the ellipse whose equation is $\dfrac{x^2}{16} + \dfrac{y^2}{12} = 1$.

Solution
x-intercepts:
$(4, 0)$ and $(-4, 0)$

y-intercepts:
$(0, 2\sqrt{3})$
and $(0, -2\sqrt{3})$
$(2\sqrt{3} \approx 3.5)$

You Try It 2
Sketch a graph of the ellipse whose equation is $\dfrac{x^2}{18} + \dfrac{y^2}{9} = 1$.

Your solution

Solutions on p. S32

Objective B **To graph a hyperbola with center at the origin**

A **hyperbola** is a conic section that can be formed by the intersection of a cone and a plane perpendicular to the base of the cone.

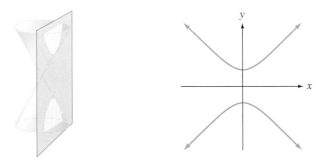

The hyperbola has two **vertices** and an **axis of symmetry** that passes through the vertices. The **center** of a hyperbola is the point halfway between the two vertices.

The graphs at the right show two possible graphs of a hyperbola with center at the origin.

In the first graph, an axis of symmetry is the *x*-axis and the vertices are *x*-intercepts.

In the second graph, an axis of symmetry is the *y*-axis and the vertices are *y*-intercepts.

Note that in either case, the graph of a hyperbola is not the graph of a function. The graph of a hyperbola is the graph of a relation.

Point of Interest

The word *hyperbola* comes from the Greek word *yperboli*, which means "exceeding." The method by which the early Greeks analyzed the conics caused a certain area in the construction of the hyperbola to be greater than (to exceed) another area. The word *hyperbole* in English, meaning "exaggeration," has the same Greek root as the word *hyperbola*.

The word *asymptote* comes from the Greek word *asymptotos*, which means "not capable of meeting."

The Standard Form of the Equation of a Hyperbola with Center at the Origin

The equation of a hyperbola for which an axis of symmetry is the *x*-axis is $\dfrac{x^2}{a^2} - \dfrac{y^2}{b^2} = 1$. The vertices are $(a, 0)$ and $(-a, 0)$.

The equation of a hyperbola for which an axis of symmetry is the *y*-axis is $\dfrac{y^2}{b^2} - \dfrac{x^2}{a^2} = 1$. The vertices are $(0, b)$ and $(0, -b)$.

To sketch a hyperbola, it is helpful to draw two lines that are "approached" by the hyperbola. These two lines are called **asymptotes.** As a point on the hyperbola gets farther from the origin, the hyperbola "gets closer to" the asymptotes.

Because the asymptotes are straight lines, their equations are linear equations. The equations of the asymptotes for a hyperbola with center at the origin are $y = \dfrac{b}{a}x$ and $y = -\dfrac{b}{a}x$.

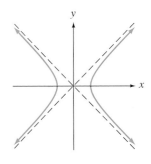

Point of Interest

Hyperbolas are used in LORAN (LOng RAnge Navigation) as a method by which a ship's navigator can determine the position of the ship, as shown in the figure below. They are also used as mirrors in some telescopes to focus incoming light.

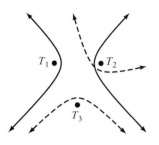

➡ Sketch a graph of the hyperbola whose equation is $\dfrac{y^2}{9} - \dfrac{x^2}{4} = 1$.

An axis of symmetry is the y-axis.
The vertices are $(0, 3)$ and $(0, -3)$.

The asymptotes are $y = \dfrac{3}{2}x$ and $y = -\dfrac{3}{2}x$.

- $b^2 = 9$, $a^2 = 4$

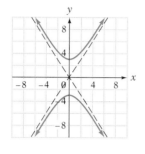

- Sketch the asymptotes. Use symmetry and the fact that the hyperbola will approach the asymptotes to sketch the graph.

Example 3

Sketch a graph of the hyperbola whose equation is $\dfrac{x^2}{16} - \dfrac{y^2}{4} = 1$.

Solution
Axis of symmetry:
x-axis

Vertices:
$(4, 0)$ and $(-4, 0)$

Asymptotes:
$y = \dfrac{1}{2}x$ and $y = -\dfrac{1}{2}x$

You Try It 3

Sketch a graph of the hyperbola whose equation is $\dfrac{x^2}{9} - \dfrac{y^2}{25} = 1$.

Your solution

Example 4

Sketch a graph of the hyperbola whose equation is $\dfrac{y^2}{16} - \dfrac{x^2}{25} = 1$.

Solution
Axis of symmetry:
y-axis

Vertices:
$(0, 4)$ and $(0, -4)$

Asymptotes:
$y = \dfrac{4}{5}x$ and $y = -\dfrac{4}{5}x$

You Try It 4

Sketch a graph of the hyperbola whose equation is $\dfrac{y^2}{9} - \dfrac{x^2}{9} = 1$.

Your solution

Solutions on p. S32

11.3 Exercises

Objective A

Sketch a graph of the ellipse given by the equation.

1. $\dfrac{x^2}{4} + \dfrac{y^2}{9} = 1$

2. $\dfrac{x^2}{25} + \dfrac{y^2}{16} = 1$

3. $\dfrac{x^2}{25} + \dfrac{y^2}{9} = 1$

4. $\dfrac{x^2}{16} + \dfrac{y^2}{9} = 1$

5. $\dfrac{x^2}{36} + \dfrac{y^2}{16} = 1$

6. $\dfrac{x^2}{49} + \dfrac{y^2}{64} = 1$

7. $\dfrac{x^2}{16} + \dfrac{y^2}{49} = 1$

8. $\dfrac{x^2}{25} + \dfrac{y^2}{36} = 1$

9. $\dfrac{x^2}{4} + \dfrac{y^2}{25} = 1$

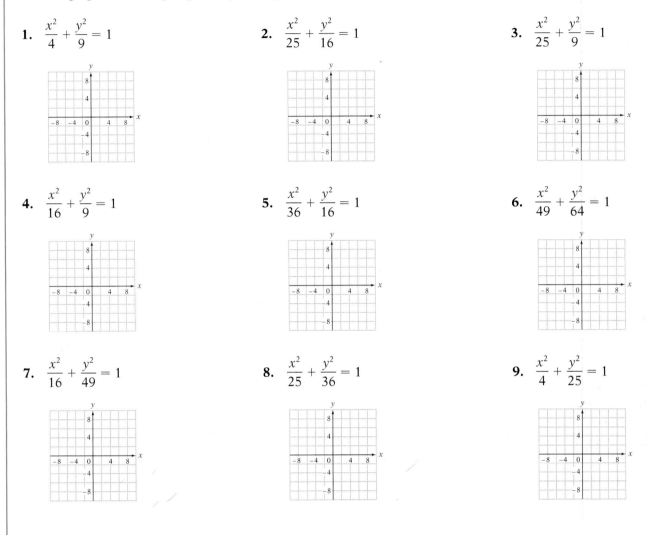

Objective B

Sketch a graph of the hyperbola given by the equation.

10. $\dfrac{x^2}{9} - \dfrac{y^2}{16} = 1$

11. $\dfrac{x^2}{25} - \dfrac{y^2}{4} = 1$

12. $\dfrac{y^2}{16} - \dfrac{x^2}{9} = 1$

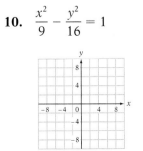

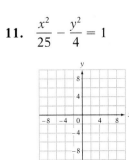

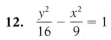

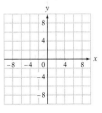

13. $\dfrac{y^2}{16} - \dfrac{x^2}{25} = 1$

14. $\dfrac{x^2}{16} - \dfrac{y^2}{4} = 1$

15. $\dfrac{x^2}{9} - \dfrac{y^2}{49} = 1$

16. $\dfrac{y^2}{25} - \dfrac{x^2}{9} = 1$

17. $\dfrac{y^2}{4} - \dfrac{x^2}{16} = 1$

18. $\dfrac{x^2}{4} - \dfrac{y^2}{25} = 1$

19. $\dfrac{x^2}{36} - \dfrac{y^2}{9} = 1$

20. $\dfrac{y^2}{9} - \dfrac{x^2}{36} = 1$

21. $\dfrac{y^2}{25} - \dfrac{x^2}{4} = 1$

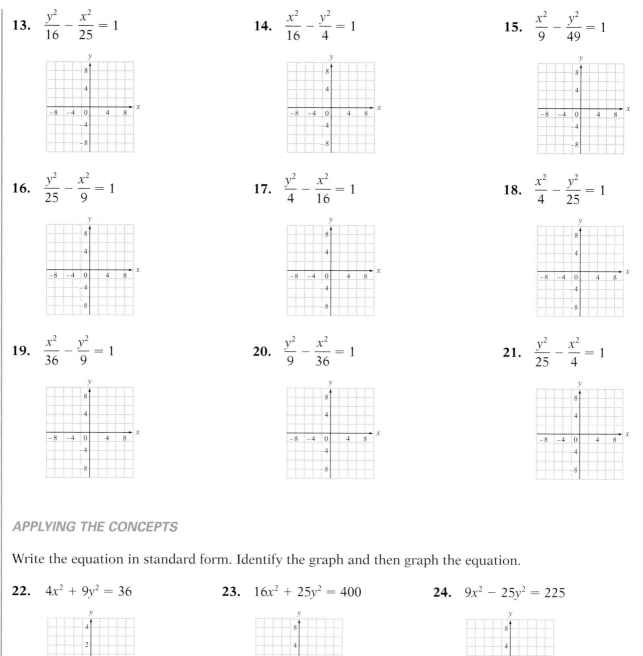

APPLYING THE CONCEPTS

Write the equation in standard form. Identify the graph and then graph the equation.

22. $4x^2 + 9y^2 = 36$

23. $16x^2 + 25y^2 = 400$

24. $9x^2 - 25y^2 = 225$

25. $25y^2 - 4x^2 = -100$

26. $9y^2 - 16x^2 = 144$

27. $4y^2 - x^2 = 36$

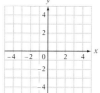

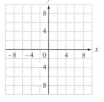

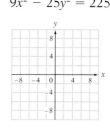

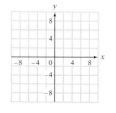

11.4 Solving Nonlinear Systems of Equations

Objective A   **To solve a nonlinear system of equations**

A **nonlinear system of equations** is one in which one or more of the equations are not linear equations. A nonlinear system of equations can be solved by using either a substitution method or an addition method.

➡ Solve: $2x - y = 4$ (1)
$\quad\quad\quad\;\; y^2 = 4x$ (2)

When a nonlinear system of equations contains a linear equation, the substitution method is used.

Solve Equation (1) for y.

$$2x - y = 4$$
$$-y = -2x + 4$$
$$y = 2x - 4$$

Substitute $2x - 4$ for y into equation (2).

$$y^2 = 4x$$
$$(2x - 4)^2 = 4x \qquad \bullet\; y = 2x - 4$$

$$4x^2 - 16x + 16 = 4x \qquad \bullet\; \text{Solve for } x.$$
$$4x^2 - 20x + 16 = 0$$

$$4(x^2 - 5x + 4) = 0$$
$$4(x - 4)(x - 1) = 0$$

$$x - 4 = 0 \qquad x - 1 = 0$$
$$x = 4 \qquad\quad x = 1$$

Substitute the values of x into the equation $y = 2x - 4$ and solve for y.

$$y = 2x - 4 \qquad\qquad y = 2x - 4$$
$$y = 2(4) - 4 \quad \bullet\; x = 4 \qquad y = 2(1) - 4 \quad \bullet\; x = 1$$
$$y = 4 \qquad\qquad\qquad\qquad y = -2$$

The solutions are $(4, 4)$ and $(1, -2)$.

The graph of this system of equations is shown at the right. The graph of the line intersects the parabola at two points. The coordinates of these points are the solutions of the system of equations.

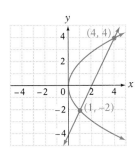

⟹ Solve: $x^2 + y^2 = 4$ (1)

$\qquad\qquad y = x + 3$ (2)

Use the substitution method.

Substitute the expression for y into Equation (1).

$$x^2 + y^2 = 4$$
$$x^2 + (x + 3)^2 = 4 \qquad \bullet \ y = x + 3$$
$$x^2 + x^2 + 6x + 9 = 4 \qquad \bullet \ \text{Solve for } x.$$
$$2x^2 + 6x + 5 = 0$$

Consider the discriminant of $2x^2 + 6x + 5 = 0$.

$$b^2 - 4ac$$
$$6^2 - 4 \cdot 2 \cdot 5 = 36 - 40 = -4 \qquad \bullet \ a = 2, b = 6, c = 5$$

Because the discriminant is less than zero, the equation $2x^2 + 6x + 5 = 0$ has two complex number solutions. Therefore, the system of equations has no real number solutions.

The graphs of the equations of this system are shown at the right. Note that the two graphs do not intersect.

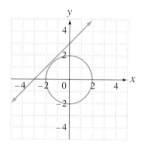

⟹ Solve: $4x^2 + y^2 = 16$ (1)

$\qquad\qquad x^2 + y^2 = 4$ (2)

Use the addition method.

$$\begin{array}{l} 4x^2 + y^2 = 16 \\ \underline{-x^2 - y^2 = -4} \\ \qquad 3x^2 = 12 \end{array}$$

   • **Multiply Equation (2) by −1.**

   • **Add the equations.**

Solve for x.

$$x^2 = 4$$
$$x = \pm 2$$

Substitute the values of x into Equation (2) and solve for y.

$$\begin{aligned} x^2 + y^2 &= 4 & \qquad x^2 + y^2 &= 4 \\ 2^2 + y^2 &= 4 \quad \bullet \ x = 2 & (-2)^2 + y^2 &= 4 \quad \bullet \ x = -2 \\ y^2 &= 0 & y^2 &= 0 \\ y &= 0 & y &= 0 \end{aligned}$$

The solutions are $(2, 0)$ and $(-2, 0)$.

The graphs of the equations in this system are shown at the right. Note that the graphs intersect at the points whose coordinates are $(2, 0)$ and $(-2, 0)$.

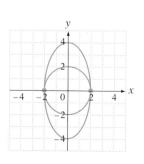

➡ Solve: $2x^2 - y^2 = 1$ (1)
$\;x^2 + 2y^2 = 18$ (2)

Use the addition method.

$$4x^2 - 2y^2 = 2$$
$$\underline{x^2 + 2y^2 = 18}$$
$$5x^2 = 20$$
$$x^2 = 4$$
$$x = \pm 2$$

• Multiply Equation (1) by 2.

• Add the equations.
• Solve for *x*.

Substitute the values of *x* into one of the equations and solve for *y*. Equation (2) is used here.

$$x^2 + 2y^2 = 18$$
$$2^2 + 2y^2 = 18$$
$$2y^2 = 14$$
$$y^2 = 7$$
$$y = \pm\sqrt{7}$$

• *x* = 2

$$x^2 + 2y^2 = 18$$
$$(-2)^2 + 2y^2 = 18$$
$$2y^2 = 14$$
$$y^2 = 7$$
$$y = \pm\sqrt{7}$$

• *x* = −2

The solutions are $(2, \sqrt{7})$, $(2, -\sqrt{7})$, $(-2, \sqrt{7})$, and $(-2, -\sqrt{7})$.

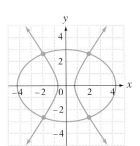

The graphs of the equations in this system are shown at the left. Note that there are four points of intersection.

Note from the preceding three examples that the number of points at which the graphs of the equations of the system intersect is the same as the number of real number solutions of the system of equations. Here are three more examples.

$$x^2 + y^2 = 4$$
$$y = x^2 + 2$$

The graphs intersect at one point.
The system of equations has one solution.

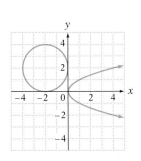

$$y = x^2$$
$$y = -x + 2$$

The graphs intersect at two points.
The system of equations has two solutions.

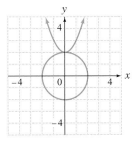

$$(x + 2)^2 + (y - 2)^2 = 4$$
$$x = y^2$$

The graphs do not intersect.
The system of equations has no solution.

Example 1 Solve: (1) $y = 2x^2 - 3x - 1$
(2) $y = x^2 - 2x + 5$

Solution Use the substitution method.

$$y = x^2 - 2x + 5$$
$$2x^2 - 3x - 1 = x^2 - 2x + 5$$
$$x^2 - x - 6 = 0$$
$$(x - 3)(x + 2) = 0$$

$$x - 3 = 0 \qquad x + 2 = 0$$
$$x = 3 \qquad\qquad x = -2$$

Substitute each value of x into Equation (1).

$$y = 2x^2 - 3x - 1$$
$$y = 2(3)^2 - 3(3) - 1$$
$$y = 18 - 9 - 1$$
$$y = 8$$

$$y = 2x^2 - 3x - 1$$
$$y = 2(-2)^2 - 3(-2) - 1$$
$$y = 8 + 6 - 1$$
$$y = 13$$

The solutions are (3, 8) and (−2, 13).

You Try It 1 Solve: $y = 2x^2 + x - 3$
$y = 2x^2 - 2x + 9$

Your solution

Example 2 Solve: (1) $3x^2 - 2y^2 = 26$
(2) $x^2 - y^2 = 5$

Solution Use the addition method. Multiply Equation (2) by −2.

$$-2x^2 + 2y^2 = -10$$
$$3x^2 - 2y^2 = 26$$
$$x^2 = 16$$
$$x = \pm\sqrt{16} = \pm 4$$

Substitute each value of x into Equation (2).

$$x^2 - y^2 = 5 \qquad\qquad x^2 - y^2 = 5$$
$$4^2 - y^2 = 5 \qquad\qquad (-4)^2 - y^2 = 5$$
$$16 - y^2 = 5 \qquad\qquad 16 - y^2 = 5$$
$$-y^2 = -11 \qquad\qquad -y^2 = -11$$
$$y^2 = 11 \qquad\qquad y^2 = 11$$
$$y = \pm\sqrt{11} \qquad\qquad y = \pm\sqrt{11}$$

The solutions are (4, $\sqrt{11}$), (4, $-\sqrt{11}$), (−4, $\sqrt{11}$), and (−4, $-\sqrt{11}$).

You Try It 2 Solve: $x^2 - y^2 = 10$
$x^2 + y^2 = 8$

Your solution

Solutions on p. S32

11.4 Exercises

Solve.

1. $y = x^2 - x - 1$
$y = 2x + 9$

2. $y = x^2 - 3x + 1$
$y = x + 6$

3. $y^2 = -x + 3$
$x - y = 1$

4. $y^2 = 4x$
$x - y = -1$

5. $y^2 = 2x$
$x + 2y = -2$

6. $y^2 = 2x$
$x - y = 4$

7. $x^2 + 2y^2 = 12$
$2x - y = 2$

8. $x^2 + 4y^2 = 37$
$x - y = -4$

9. $x^2 + y^2 = 13$
$x + y = 5$

10. $x^2 + y^2 = 16$
$x - 2y = -4$

11. $4x^2 + y^2 = 12$
$y = 4x^2$

12. $2x^2 + y^2 = 6$
$y = 2x^2$

13. $y = x^2 - 2x - 3$
$y = x - 6$

14. $y = x^2 + 4x + 5$
$y = -x - 3$

15. $3x^2 - y^2 = -1$
$x^2 + 4y^2 = 17$

16. $x^2 + y^2 = 10$
$x^2 + 9y^2 = 18$

17. $2x^2 + 3y^2 = 30$
$x^2 + y^2 = 13$

18. $x^2 + y^2 = 61$
$x^2 - y^2 = 11$

19. $y = 2x^2 - x + 1$
$y = x^2 - x + 5$

20. $y = -x^2 + x - 1$
$y = x^2 + 2x - 2$

21. $2x^2 + 3y^2 = 24$
$x^2 - y^2 = 7$

22. $2x^2 + 3y^2 = 21$
$x^2 + 2y^2 = 12$

23. $x^2 + y^2 = 36$
$4x^2 + 9y^2 = 36$

24. $2x^2 + 3y^2 = 12$
$x^2 - y^2 = 25$

25. $11x^2 - 2y^2 = 4$
$3x^2 + y^2 = 15$

26. $x^2 + 4y^2 = 25$
$x^2 - y^2 = 5$

27. $2x^2 - y^2 = 7$
$2x - y = 5$

28. $3x^2 + 4y^2 = 7$
$x - 2y = -3$

29. $y = 3x^2 + x - 4$
$y = 3x^2 - 8x + 5$

30. $y = 2x^2 + 3x + 1$
$y = 2x^2 + 9x + 7$

31. $x = y + 3$
$x^2 + y^2 = 5$

32. $x - y = -6$
$x^2 + y^2 = 4$

33. $y = x^2 + 4x + 4$
$x + 2y = 4$

APPLYING THE CONCEPTS

34. Give the maximum number of points in which the following pairs of graphs can intersect.
a. A line and an ellipse.
b. A line and a parabola.
c. Two ellipses with centers at the origin.
d. Two different circles with centers at the origin.
e. Two hyperbolas with centers at the origin.

Solve by graphing. Approximate the solutions of the systems to the nearest thousandth.

35. $y = 2^x$
$x + y = 3$

36. $y = 3^{-x}$
$x^2 + y^2 = 9$

37. $y = \log_2 x$
$\dfrac{x^2}{9} + \dfrac{y^2}{1} = 1$

38. $y = \log_3 x$
$x^2 + y^2 = 4$

39. $y = -\log_3 x$
$x + y = 4$

40. $y = \left(\dfrac{1}{2}\right)^x$
$\dfrac{x^2}{9} + \dfrac{y^2}{4} = 1$

11.5 Quadratic Inequalities and Systems of Inequalities

Objective A

To graph the solution set of a quadratic inequality in two variables 〔11〕

The **graph of a quadratic inequality in two variables** is a region of the plane that is bounded by one of the conic sections (parabola, circle, ellipse, or hyperbola). When graphing an inequality of this type, first replace the inequality symbol with an equals sign. Graph the resulting conic using a dashed curve when the original inequality is less than ($<$) or greater than ($>$). Use a solid curve when the original inequality is $\leq$ or $\geq$. Use the point $(0, 0)$ to determine which region of the plane to shade. If $(0, 0)$ is a solution of the inequality, then shade the region of the plane containing $(0, 0)$. If not, shade the other portion of the plane.

⇒ Graph the solution set: $x^2 + y^2 > 9$

Change the inequality to an equality.

$x^2 + y^2 = 9$

This is the equation of a circle with center $(0, 0)$ and radius 3.

The original inequality is $>$, so the graph is drawn as a dashed circle.

Substitute the point $(0, 0)$ in the inequality. Because $0^2 + 0^2 > 9$ is not true, the point $(0, 0)$ should not be in the shaded region.

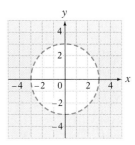

Not all inequalities have a solution set that can be graphed by shading a region of the plane. For example, the inequality

$$\frac{x^2}{9} + \frac{y^2}{4} < -1$$

has no ordered-pair solutions. The solution set is the empty set, and no region of the plane is shaded.

⇒ Graph the solution set: $y \leq x^2 + 2x + 2$

Change the inequality to an equality.

$y = x^2 + 2x + 2$

This is the equation of a parabola that opens up. The vertex is $(-1, 1)$ and the axis of symmetry is the line $x = -1$.

The original inequality is $\leq$, so the graph is drawn as a solid curve.

Substitute the point $(0, 0)$ into the inequality. Because $0 \leq 0^2 + 2(0) + 2$ is true, the point $(0, 0)$ should be in the shaded region.

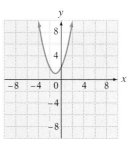

⇒ Graph the solution set: $\dfrac{y^2}{9} - \dfrac{x^2}{4} \geq 1$

Write the inequality as an equality.

$$\dfrac{y^2}{9} - \dfrac{x^2}{4} = 1$$

This is the equation of a hyperbola. The vertices are $(0, -3)$ and $(0, 3)$. The equations of the asymptotes are $y = \dfrac{3}{2}x$ and $y = -\dfrac{3}{2}x$.

The original inequality is ≥, so the graph is drawn as a solid curve.

Substitute the point $(0, 0)$ into the inequality. Because $\dfrac{0^2}{9} - \dfrac{0^2}{4} \geq 1$ is not true, the point $(0, 0)$ should not be in the shaded region.

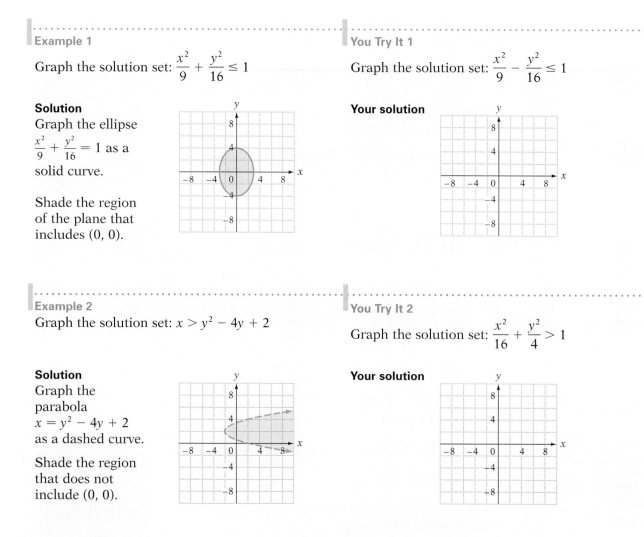

Example 1

Graph the solution set: $\dfrac{x^2}{9} + \dfrac{y^2}{16} \leq 1$

Solution
Graph the ellipse $\dfrac{x^2}{9} + \dfrac{y^2}{16} = 1$ as a solid curve.

Shade the region of the plane that includes $(0, 0)$.

You Try It 1

Graph the solution set: $\dfrac{x^2}{9} - \dfrac{y^2}{16} \leq 1$

Your solution

Example 2

Graph the solution set: $x > y^2 - 4y + 2$

Solution
Graph the parabola $x = y^2 - 4y + 2$ as a dashed curve.

Shade the region that does not include $(0, 0)$.

You Try It 2

Graph the solution set: $\dfrac{x^2}{16} + \dfrac{y^2}{4} > 1$

Your solution

Solutions on p. S33

Objective B

To graph the solution set of a nonlinear system of inequalities

Recall that the solution set of a system of inequalities is the intersection of the solution sets of the individual inequalities. To graph the solution set of a system of inequalities, first graph the solution set of each inequality. The solution set of the system of inequalities is the region of the plane represented by the intersection of the two shaded regions.

⇒ Graph the solution set: $x^2 + y^2 \leq 16$
$ y \geq x^2$

Graph the solution set of each inequality.

$x^2 + y^2 = 16$ is the equation of a circle with center at the origin and a radius of 4. Because the original inequality is ≤, use a solid curve.

$y = x^2$ is the equation of a parabola that opens up. The vertex is (0, 0). Because the original inequality is ≥, use a solid curve.

Substitute the point (0, 0) into the inequality $x^2 + y^2 \leq 16$. $0^2 + 0^2 \leq 16$ is a true statement. Shade inside the circle.

Substitute (2, 0) into the inequality $y \geq x^2$. [We cannot use (0, 0), a point on the parabola.] $0 > 4$ is a false statement. Shade inside the parabola.

The solution is the intersection of the two shaded regions.

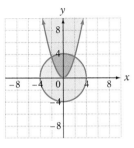

Example 3

Graph the solution set: $\dfrac{x^2}{9} + \dfrac{y^2}{4} \geq 1$
$ \dfrac{x^2}{4} - \dfrac{y^2}{9} \geq 1$

Solution

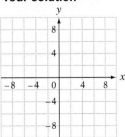

• Draw the ellipse $\dfrac{x^2}{9} + \dfrac{y^2}{4} = 1$ as a solid curve. Shade outside the ellipse.

• Draw the hyperbola $\dfrac{x^2}{4} - \dfrac{y^2}{9} = 1$ as a solid curve. (0, 0) should not be in the shaded region.

You Try It 3

Graph the solution set: $x^2 + y^2 < 16$
$ y^2 > x$

Your solution

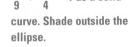

Solution on p. S33

Example 4

Graph the solution set: $y \geq x^2$
$$y \leq x + 2$$

Solution

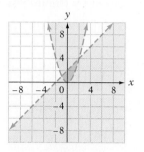

- Draw the parabola $y = x^2$ as a solid curve. Shade inside the parabola.
- Draw the line $y = x + 2$ as a solid line. Shade below the line.

You Try It 4

Graph the solution set: $\dfrac{x^2}{4} + \dfrac{y^2}{9} \leq 1$
$$x > y^2 - 2$$

Your solution

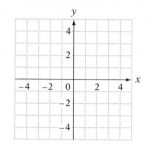

Example 5

Graph the solution set: $\dfrac{x^2}{9} - \dfrac{y^2}{16} > 1$
$$x^2 + y^2 < 4$$

Solution

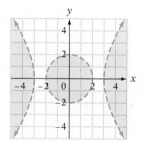

- Draw the hyperbola $\dfrac{x^2}{9} - \dfrac{y^2}{16} = 1$ as a dashed curve. $(0, 0)$ should not be in the shaded region.
- Draw the circle $x^2 + y^2 = 4$ as a dashed curve. Shade inside the circle.

The solution sets of the two inequalities do not intersect. This system of inequalities has no real number solution.

You Try It 5

Graph the solution set: $\dfrac{x^2}{16} + \dfrac{y^2}{25} \geq 1$
$$x^2 + y^2 < 9$$

Your solution

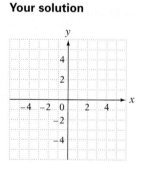

Solutions on p. S33

11.5 Exercises

Objective A

1. How is the graph of the solution set of a nonlinear inequality that uses $\leq$ or $\geq$ different from the graph of the solution set of a nonlinear inequality that uses $<$ or $>$?

2. How can you determine which region to shade when graphing the solution set of a nonlinear inequality?

Graph the solution set.

3. $y \leq x^2 - 4x + 3$

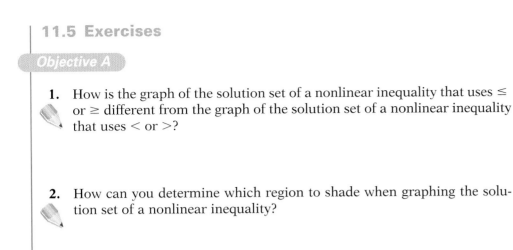

4. $y < x^2 - 2x - 3$

5. $(x - 1)^2 + (y + 2)^2 \leq 9$

6. $(x + 2)^2 + (y - 3)^2 > 4$

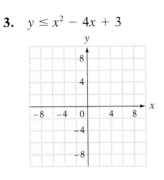

7. $(x + 3)^2 + (y - 2)^2 \geq 9$

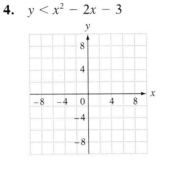

8. $(x - 2)^2 + (y + 1)^2 \leq 16$

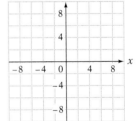

9. $\dfrac{x^2}{16} + \dfrac{y^2}{25} < 1$

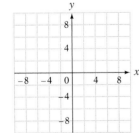

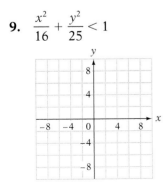

10. $\dfrac{x^2}{9} + \dfrac{y^2}{4} \geq 1$

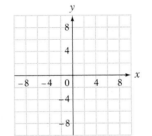

11. $\dfrac{x^2}{25} - \dfrac{y^2}{9} \leq 1$

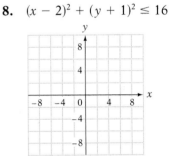

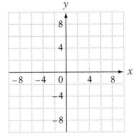

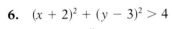

12. $\dfrac{y^2}{25} - \dfrac{x^2}{36} > 1$

13. $\dfrac{x^2}{4} + \dfrac{y^2}{16} \geq 1$

14. $\dfrac{x^2}{4} - \dfrac{y^2}{16} \leq 1$

15. $y \leq x^2 - 2x + 3$

16. $x \leq y^2 + 2y + 1$

17. $\dfrac{y^2}{9} - \dfrac{x^2}{16} \leq 1$

18. $\dfrac{x^2}{16} - \dfrac{y^2}{4} < 1$

19. $\dfrac{x^2}{9} + \dfrac{y^2}{1} \leq 1$

20. $\dfrac{x^2}{16} + \dfrac{y^2}{4} > 1$

21. $(x - 1)^2 + (x + 3)^2 \leq 25$

22. $(x + 1)^2 + (y - 2)^2 \geq 16$

23. $\dfrac{y^2}{25} - \dfrac{x^2}{4} \leq 1$

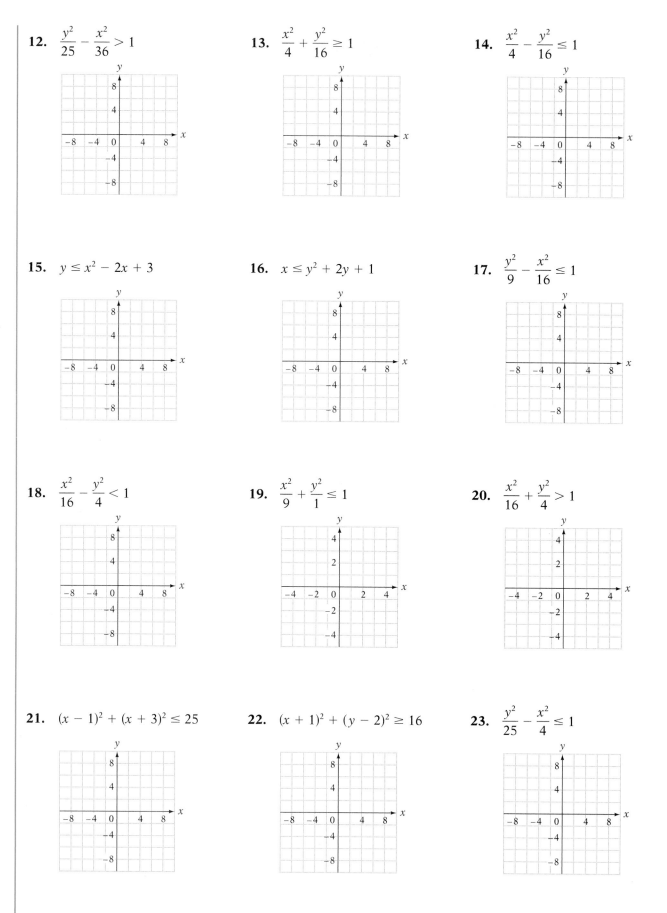

24. $\dfrac{x^2}{9} - \dfrac{y^2}{25} \geq 1$

25. $\dfrac{x^2}{25} + \dfrac{y^2}{9} \leq 1$

26. $\dfrac{x^2}{36} + \dfrac{y^2}{4} \leq 1$

Objective B

Graph the solution set.

27. $y \leq (x - 2)^2$
$y + x > 4$

28. $x^2 + y^2 < 1$
$x + y \geq 4$

29. $x^2 + y^2 < 16$
$y > x + 1$

30. $y > x^2 - 4$
$y < x - 2$

31. $\dfrac{x^2}{4} + \dfrac{y^2}{16} \leq 1$

$y \leq -\dfrac{1}{2}x + 2$

32. $\dfrac{y^2}{4} - \dfrac{x^2}{25} \geq 1$

$y \leq \dfrac{2}{3}x + 4$

33. $x \geq y^2 - 3y + 2$
$y \geq 2x - 2$

34. $x^2 + y^2 \leq 25$

$y \leq -\dfrac{1}{3}x + 2$

35. $x^2 + y^2 < 25$
$\dfrac{x^2}{9} + \dfrac{y^2}{36} < 1$

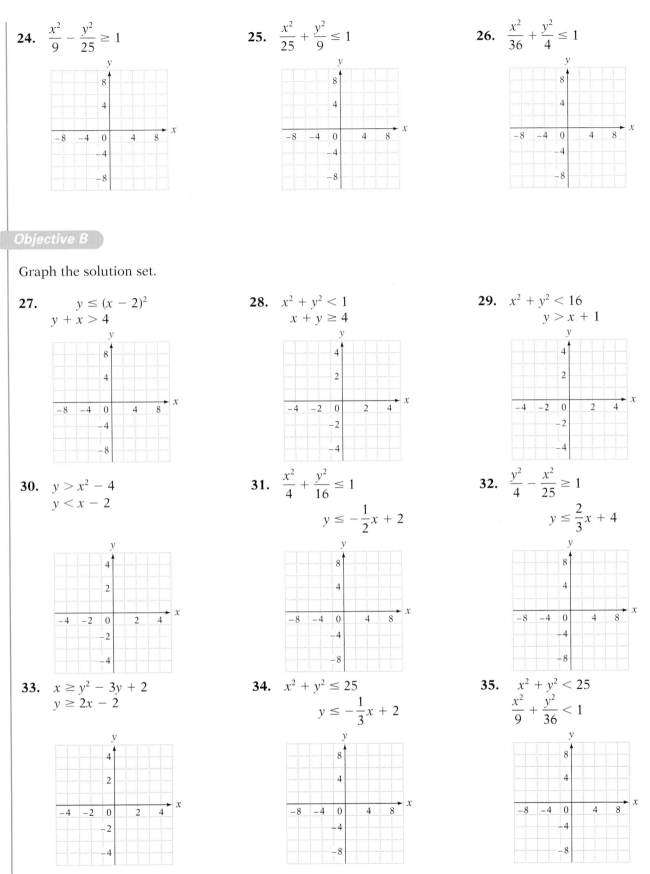

36. $\dfrac{x^2}{9} - \dfrac{y^2}{4} < 1$

$\dfrac{x^2}{25} + \dfrac{y^2}{9} < 1$

37. $x^2 + y^2 > 4$

$x^2 + y^2 < 25$

38. $\dfrac{x^2}{25} + \dfrac{y^2}{16} \le 1$

$\dfrac{x^2}{4} + \dfrac{y^2}{4} \ge 1$

APPLYING THE CONCEPTS

Graph the solution set.

39. $y > x^2 - 3$
$y < x + 3$
$x \le 0$

40. $x^2 + y^2 \le 25$
$y > x + 1$
$x \ge 0$

41. $x^2 + y^2 < 3$
$x > y^2 - 1$
$y \ge 0$

42. $\dfrac{x^2}{4} - \dfrac{y^2}{25} \le 1$

$\dfrac{x^2}{4} + \dfrac{y^2}{4} \le 1$

$y \ge 0$

43. $\dfrac{x^2}{16} + \dfrac{y^2}{4} \le 1$

$x^2 + y^2 \le 4$

$x \ge 0$

$y \le 0$

44. $\dfrac{x^2}{4} + \dfrac{y^2}{25} \le 1$

$x > y^2 - 4$

$x \le 0$

$y \ge 0$

45. $y > 2^x$
$x + y < 4$

46. $y < \left(\dfrac{1}{2}\right)^x$
$2x - y \ge 2$

47. $y \ge \log_2 x$
$x^2 + y^2 < 9$

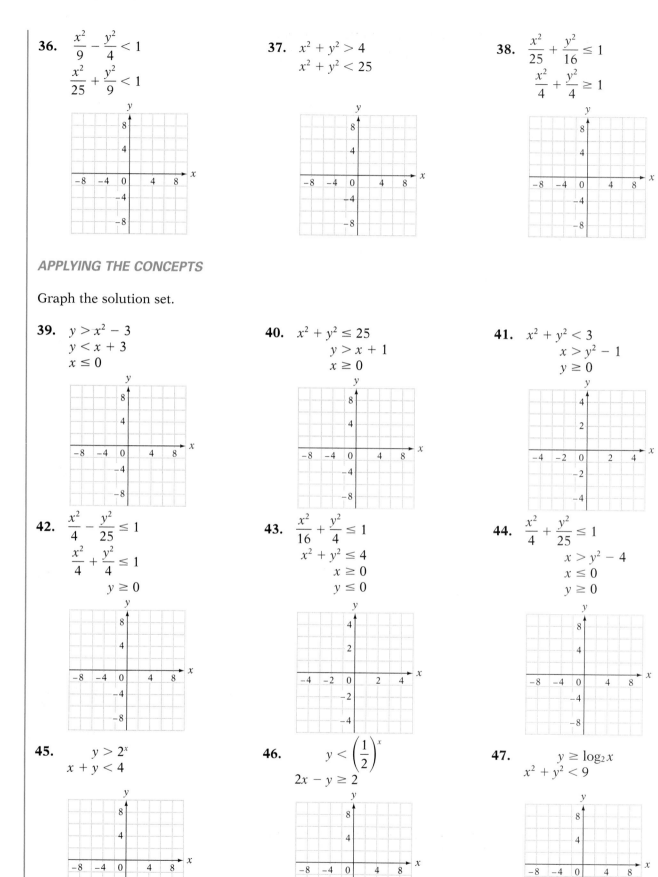

Focus on Problem Solving

Using a Variety of Problem-Solving Techniques

We have examined several problem-solving strategies throughout the text. See if you can apply those techniques to the following problems.

1. Eight coins look exactly alike, but one is lighter than the others. Explain how the different coin can be found in two weighings on a balance scale.

2. For the sequence of numbers 1, 1, 2, 3, 5, 8, 13, . . . , identify a possible pattern and then use that pattern to determine the next number in the sequence.

3. Arrange the numbers 1, 2, 3, 4, 5, 6, 7, 8, and 9 in the squares at the right such that the sum of any row, column, or diagonal is 15. (*Suggestion:* Note that 1, 5, 9; 2, 5, 8; 3, 5, 7; and 4, 5, 6 all add to 15. Because 5 is part of each sum, this suggests that 5 should be placed in the center of the squares.)

4. A restaurant charges $10.00 for a pizza that has a diameter of 9 in. Determine the selling price of a pizza with a diameter of 18 in. such that the selling price per square inch is the same as for the 9-inch pizza.

5. You have a balance scale and weights of 1 g, 4 g, 8 g, and 16 g. Using only these weights, can you weigh something that weighs 7 g? 9 g? 12 g? 19 g?

6. Can the checkerboard at the right be covered with dominos (which look like 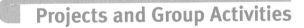) such that every square on the board is covered by a domino? Why or why not? (*Note:* The dominos cannot overlap.)

Projects and Group Activities

The Eccentricity and Foci of an Ellipse

The graph of an ellipse can be long and thin, or it can have a shape that is very close to a circle. The **eccentricity,** *e*, of an ellipse is a measure of its "roundness."

The shapes of ellipses with various eccentricities are shown below.

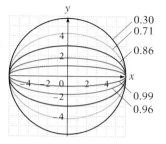

Planet	Eccentricity
Mercury	0.206
Venus	0.007
Earth	0.017
Mars	0.093
Jupiter	0.049
Saturn	0.051
Uranus	0.046
Neptune	0.005
Pluto	0.250

1. Based on the eccentricities of the ellipses shown above, complete the sentence. "As the eccentricity of an ellipse gets closer to 1, the ellipses get *flatter/rounder*."

2. The planets travel around the sun in elliptical orbits. The eccentricities of the orbits of the planets are shown in the table at the left. Which planet has the most nearly circular orbit?

For an ellipse given by the equation $\frac{x^2}{a^2} + \frac{y^2}{b^2} = 1$, $a > b$, a formula for eccentricity is $e = \frac{\sqrt{a^2 - b^2}}{a}$. Use this formula to find the eccentricity of the ellipses in Exercises 3 and 4. If necessary, round to the nearest hundredth.

3. $\frac{x^2}{25} + \frac{y^2}{16} = 1$

4. $\frac{x^2}{9} + \frac{y^2}{4} = 1$

Ellipses have a reflective property that has been used in the design of some buildings. The **foci** of an ellipse are two points on the longer axis, called the **major axis,** of the ellipse.

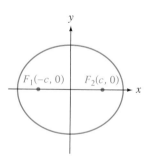

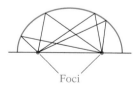

Foci

If light or sound emanates from one focus, it is reflected to the other focus. This phenomenon results in what are called "whispering galleries." The rotunda of the Capitol Building in Washington, D.C. is a whispering gallery. A whisper spoken by a person at one focus can be heard clearly by a person at the other focus.

The foci are c units from the center of an ellipse, where $c = \sqrt{a^2 - b^2}$ for an ellipse whose equation is $\frac{x^2}{a^2} + \frac{y^2}{b^2} = 1$, $a > b$.

5. Find the foci for an ellipse whose equation is $\frac{x^2}{169} + \frac{y^2}{144} = 1$.

Graphing Conic Sections Using a Graphing Calculator

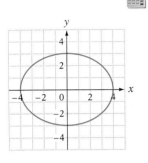

Consider the graph of the ellipse $\dfrac{x^2}{16} + \dfrac{y^2}{9} = 1$ shown at the left. Because a vertical line can intersect the graph at more than one point, the graph is not the graph of a function. And because the graph is not the graph of a function, the equation does not represent a function. Consequently, the equation cannot be entered into a graphing calculator to be graphed. However, by solving the equation for y, we have

$$\frac{y^2}{9} = 1 - \frac{x^2}{16}$$

$$y^2 = 9\left(1 - \frac{x^2}{16}\right)$$

$$y = \pm 3\sqrt{1 - \frac{x^2}{16}}$$

There are two solutions for y, which can be written

$$y_1 = 3\sqrt{1 - \frac{x^2}{16}} \quad \text{and} \quad y_2 = -3\sqrt{1 - \frac{x^2}{16}}$$

Each of these equations is the equation of a function and therefore can be entered into a graphing calculator. The graph of $y_1 = 3\sqrt{1 - \dfrac{x^2}{16}}$ is shown in blue below, and the graph of $y_2 = -3\sqrt{1 - \dfrac{x^2}{16}}$ is shown in red. Note that, together, the graphs are the graph of an ellipse.

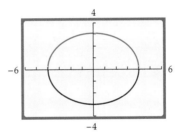

A similar technique can be used to graph a hyperbola. To graph $\dfrac{x^2}{16} - \dfrac{y^2}{4} = 1$, solve for y. Then graph each equation, $y_1 = 2\sqrt{\dfrac{x^2}{16} - 1}$ and $y_2 = -2\sqrt{\dfrac{x^2}{16} - 1}$.

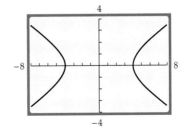

Solve each equation for y. Then graph using a graphing calculator.

1. $\dfrac{x^2}{25} + \dfrac{y^2}{49} = 1$ **2.** $\dfrac{x^2}{4} + \dfrac{y^2}{64} = 1$ **3.** $\dfrac{x^2}{16} - \dfrac{y^2}{4} = 1$ **4.** $\dfrac{x^2}{9} - \dfrac{y^2}{36} = 1$

Chapter Summary

Key Words

The graph of a *conic section* can be represented by the intersection of a plane and a cone. The four conic sections are the *parabola, ellipse, hyperbola,* and *circle.* [pp. 562, 569, 575, 577]

A *circle* is the set of all points (x, y) in the plane that are a fixed distance from a given point (h, k), called the *center.* The fixed distance is the *radius* of the circle. [p. 569]

The *asymptotes* of a hyperbola are the two straight lines that are "approached" by the hyperbola. As the graph of the hyperbola gets farther from the origin, the hyperbola gets "closer to" the asymptotes. [p. 577]

A *nonlinear system of equations* is one in which one or more of the equations are not linear equations. [p. 581]

The *graph of a quadratic inequality in two variables* is a region of the plane that is bounded by one of the conic sections. [p. 587]

Essential Rules

Equations of a Parabola

$y = ax^2 + bx + c$

When $a > 0$, the parabola opens up.
When $a < 0$, the parabola opens down.

The x-coordinate of the vertex is $-\dfrac{b}{2a}$.

The axis of symmetry is the line $x = -\dfrac{b}{2a}$.

$x = ay^2 + by + c$

When $a > 0$, the parabola opens to the right.
When $a < 0$, the parabola opens to the left.

The y-coordinate of the vertex is $-\dfrac{b}{2a}$.

The axis of symmetry is the line $y = -\dfrac{b}{2a}$.

[pp. 563–565]

Equation of a Circle

$(x - h)^2 + (y - k)^2 = r^2$

The center is (h, k) and the radius is r. [p. 569]

Equation of an Ellipse

$\dfrac{x^2}{a^2} + \dfrac{y^2}{b^2} = 1$

The x-intercepts are $(a, 0)$ and $(-a, 0)$.
The y-intercepts are $(0, b)$ and $(0, -b)$. [p. 575]

Equations of a Hyperbola

$\dfrac{x^2}{a^2} - \dfrac{y^2}{b^2} = 1$

An axis of symmetry is the x-axis.
The vertices are $(a, 0)$ and $(-a, 0)$.

The equations of the asymptotes are $y = \pm\dfrac{b}{a}x$.

$\dfrac{y^2}{b^2} - \dfrac{x^2}{a^2} = 1$

An axis of symmetry is the y-axis.
The vertices are $(0, b)$ and $(0, -b)$.

The equations of the asymptotes are $y = \pm\dfrac{b}{a}x$.

[p. 577]

Chapter Review

1. Sketch a graph of $y = -2x^2 + x - 2$.

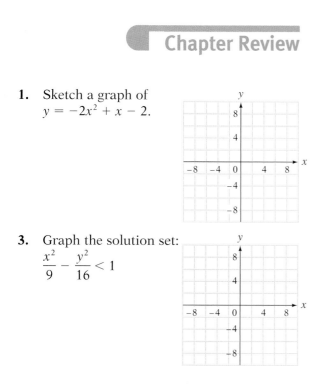

2. Sketch a graph of $\dfrac{x^2}{1} + \dfrac{y^2}{9} = 1$.

3. Graph the solution set: $\dfrac{x^2}{9} - \dfrac{y^2}{16} < 1$

4. Sketch a graph of $(x + 3)^2 + (y + 1)^2 = 1$.

5. Find the vertex and axis of symmetry of the parabola $y = x^2 - 4x + 8$.

6. Solve: $y^2 = 2x^2 - 3x + 6$
$y^2 = 2x^2 + 5x - 2$

7. Graph the solution set:
$\dfrac{x^2}{25} + \dfrac{y^2}{16} \leq 1$
$\dfrac{y^2}{4} - \dfrac{x^2}{4} \geq 1$

8. Sketch a graph of $\dfrac{x^2}{25} - \dfrac{y^2}{1} = 1$.

9. Find the equation of the circle that passes through the point whose coordinates are $(2, -1)$ and whose center is $(-1, 2)$.

10. Find the vertex and axis of symmetry of the parabola $y = -x^2 + 7x - 8$.

11. Solve: $x = 2y^2 - 3y + 1$
$3x - 2y = 0$

12. Find the equation of the circle that passes through the point $(4, 6)$ and whose center is $(0, -3)$.

13. Find the equation of the circle with radius 6 and center $(-1, 5)$.

14. Solve: $2x^2 + y^2 = 19$
$3x^2 - y^2 = 6$

15. Write the equation $x^2 + y^2 + 4x - 2y = 4$ in standard form.

16. Solve: $y = x^2 + 5x - 6$
$\qquad\;\; y = x - 10$

17. Graph the solution set:
$$\frac{x^2}{16} + \frac{y^2}{4} > 1$$

18. Sketch a graph of
$$\frac{y^2}{16} - \frac{x^2}{9} = 1.$$

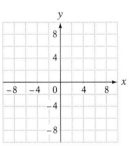

19. Graph the solution set:
$(x - 2)^2 + (y + 1)^2 \le 16$

20. Sketch a graph of
$x^2 + (y - 2)^2 = 9.$

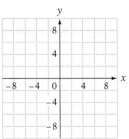

21. Sketch a graph of
$$\frac{x^2}{25} + \frac{y^2}{9} = 1.$$

22. Graph the solution set:
$y \ge -x^2 - 2x + 3$

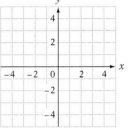

23. Graph the solution set:
$$\frac{x^2}{16} + \frac{y^2}{4} < 1$$
$$x^2 + y^2 > 9$$

24. Graph the solution set:
$y \ge x^2 - 4x + 2$
$$y \le \frac{1}{3}x - 1$$

25. Sketch a graph of
$x = 2y^2 - 6y + 5.$

26. Graph the solution set:
$$\frac{x^2}{9} + \frac{y^2}{1} \ge 1$$
$$\frac{x^2}{4} - \frac{y^2}{1} \le 1$$

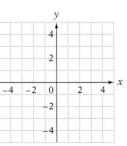

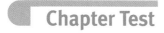

Chapter Test

1. Find the equation of the circle with radius 4 and center $(-3, -3)$.

2. Find the equation of the circle that passes through the point $(2, 5)$ and whose center is $(-2, 1)$.

3. Sketch a graph of $\dfrac{y^2}{25} - \dfrac{x^2}{16} = 1$.

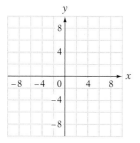

4. Graph the solution set: $x^2 + y^2 < 36$
$$x + y > 4$$

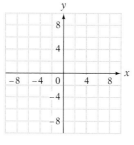

5. Find the axis of symmetry of the parabola $y = -x^2 + 6x - 5$.

6. Solve: $x^2 - y^2 = 24$
$$2x^2 + 5y^2 = 55$$

7. Find the vertex of the parabola $y = -x^2 + 3x - 2$.

8. Sketch a graph of $x = y^2 - y - 2$.

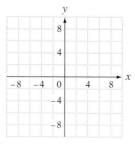

9. Sketch a graph of $\dfrac{x^2}{16} + \dfrac{y^2}{4} = 1$.

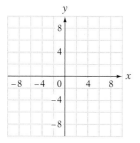

10. Graph the solution set: $\dfrac{x^2}{25} + \dfrac{y^2}{4} \le 1$

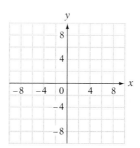

11. Find the equation of the circle with radius 3 and center $(-2, 4)$.

12. Solve: $x = 3y^2 + 2y - 4$
$\quad\quad\quad x = y^2 - 5y$

13. Solve: $x^2 + 2y^2 = 4$
$\quad\quad\quad\quad x + y = 2$

14. Find the equation of the circle that passes through the point $(2, 4)$ and whose center is $(-1, -3)$.

15. Graph the solution set: $\dfrac{x^2}{25} - \dfrac{y^2}{16} \geq 1$
$\quad\quad\quad\quad\quad\quad\quad\quad x^2 + y^2 \leq 9$

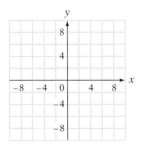

16. Write the equation $x^2 + y^2 - 4x + 2y + 1 = 0$ in standard form and then sketch its graph.

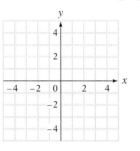

17. Sketch a graph of $y = \dfrac{1}{2}x^2 + x - 4$.

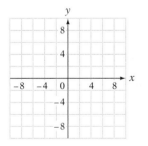

18. Sketch a graph of $(x - 2)^2 + (y + 1)^2 = 9$.

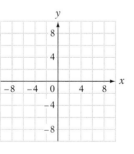

19. Sketch a graph of $\dfrac{x^2}{9} - \dfrac{y^2}{4} = 1$.

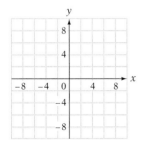

20. Graph the solution set: $\dfrac{x^2}{16} - \dfrac{y^2}{25} < 1$

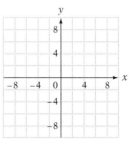

Cumulative Review

1. Graph the solution set.
$\{x \mid x < 4\} \cap \{x \mid x > 2\}$

$$\begin{array}{ccccccccccc} \xleftarrow{\quad} & + & + & + & + & + & + & + & + & + & \xrightarrow{\quad} \\ -5 & -4 & -3 & -2 & -1 & 0 & 1 & 2 & 3 & 4 & 5 \end{array}$$

2. Solve: $\dfrac{5x - 2}{3} - \dfrac{1 - x}{5} = \dfrac{x + 4}{10}$

3. Solve: $4 + |3x + 2| < 6$

4. Find the equation of the line that contains the point $(2, -3)$ and has slope $-\dfrac{3}{2}$.

5. Find the equation of the line that contains the point $(4, -2)$ and is perpendicular to the line $y = -x + 5$.

6. Simplify: $x^{2n}(x^{2n} + 2x^n - 3x)$

7. Factor: $(x - 1)^3 - y^3$

8. Solve: $\dfrac{3x - 2}{x + 4} \leq 1$

9. Simplify: $\dfrac{ax - bx}{ax + ay - bx - by}$

10. Simplify: $\dfrac{x - 4}{3x - 2} - \dfrac{1 + x}{3x^2 + x - 2}$

11. Solve: $\dfrac{6x}{2x - 3} - \dfrac{1}{2x - 3} = 7$

12. Graph the solution set: $5x + 2y > 10$

13. Simplify: $\left(\dfrac{12a^2b^2}{a^{-3}b^{-4}}\right)^{-1}\left(\dfrac{ab}{4^{-1}a^{-2}b^4}\right)^2$

14. Write $2\sqrt[4]{x^3}$ as an exponential expression.

15. Simplify: $\sqrt{18} - \sqrt{-25}$

16. Solve: $2x^2 + 2x - 3 = 0$

17. Solve: $x^2 + y^2 = 20$
$\quad\quad\ \ x^2 - y^2 = 12$

18. Solve: $x - \sqrt{2x - 3} = 3$

19. Evaluate the function $f(x) = -x^2 + 3x - 2$ at $x = -3$.

20. Find the inverse of the function $f(x) = 4x + 8$.

21. Find the maximum value of the function $f(x) = -2x^2 + 4x - 2$.

22. Find the maximum product of two numbers whose sum is 40.

23. Find the distance between the points $(2, 4)$ and $(-1, 0)$.

24. Find the equation of the circle that passes through the point $(3, 1)$ and whose center is $(-1, 2)$.

25. Graph: $x = y^2 - 2y + 3$

26. Graph: $\dfrac{x^2}{25} + \dfrac{y^2}{4} = 1$

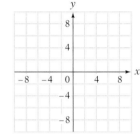

27. Graph: $\dfrac{y^2}{4} - \dfrac{x^2}{25} = 1$

28. Graph the solution set:
$$(x - 1)^2 + y^2 \le 25$$
$$y^2 < x$$

29. Tickets for a school play sold for $4.00 for each adult and $1.50 for each child. The total receipts for the 192 tickets sold were $493. Find the number of adult tickets sold.

30. A motorcycle travels 180 mi in the same amount of time that a car travels 144 mi. The rate of the motorcycle is 12 mph faster than the rate of the car. Find the rate of the motorcycle.

31. The rate of a river's current is 1.5 mph. A rowing crew can row 12 mi downstream and 12 mi back in 6 h. Find the rowing rate of the crew in calm water.

32. The speed, v, of a gear varies inversely as the number of teeth, t. If a gear that has 36 teeth makes 30 revolutions per minute, how many revolutions per minute will a gear that has 60 teeth make?

Chapter 12 Sequences and Series

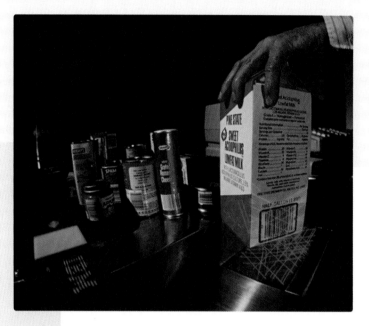

Objectives

Section 12.1

A To write the terms of a sequence
B To find the sum of a series

Section 12.2

A To find the *n*th term of an arithmetic sequence
B To find the sum of an arithmetic series
C To solve application problems

Section 12.3

A To find the *n*th term of a geometric sequence
B To find the sum of a finite geometric series
C To find the sum of an infinite geometric series
D To solve application problems

Section 12.4

A To expand $(a + b)^n$

Grocery stores use lasers, like the one shown here, to scan each product sold. The laser reads a code—on groceries, a Universal Product Code (UPC), and on books, an International Standard Book Number (ISBN)—and interprets the numbers identifying the product. Each product has a different numerical pattern, a series of numbers that is divisible by a particular number. Both ISBN and UPC numbers are explained in the **Project on pages 636 and 637**.

Need help? For on-line student resources, such as section quizzes, visit this textbook's web site at **math.college.hmco.com/students**.

1. Simplify: $[3(1) - 2] + [3(2) - 2] + [3(3) - 2]$

2. Evaluate $f(n) = \dfrac{n}{n + 2}$ for $n = 6$.

3. Evaluate $a_1 + (n - 1)d$ for $a_1 = 2$, $n = 5$, and $d = 4$.

4. Evaluate $a_1 r^{n-1}$ for $a_1 = -3$, $r = -2$, and $n = 6$.

5. Evaluate $\dfrac{a_1(1 - r^n)}{1 - r}$ for $a_1 = -2$, $r = -4$, and $n = 5$.

6. Simplify: $\dfrac{\dfrac{4}{10}}{1 - \dfrac{1}{10}}$

7. Simplify: $(x + y)^2$

8. Simplify: $(x + y)^3$

Go Figure

What is the next number in the following sequence?

1   32   81   64   25   6

12.1 Introduction to Sequences and Series

To write the terms of a sequence ⟨12⟩ ◉

An investor deposits $100 in an account that earns 10% interest compounded annually. The amount of interest earned each year can be determined by using the compound interest formula.

The amount of interest earned in each of the first 4 years of the investment is shown at the right.

Year	1	2	3	4
Interest Earned	$10	$11	$12.10	$13.31

The list of numbers 10, 11, 12.10, 13.31 is called a *sequence*. A **sequence** is an ordered list of numbers. The list 10, 11, 12.10, 13.31 is ordered because the position of a number in this list indicates the year in which that amount of interest was earned.

Each of the numbers of a sequence is called a **term** of the sequence.

For the sequence at the right, the first term is 2, the second term is 4, the third term is 6, and the fourth term is 8.

$2, 4, 6, 8, \ldots$

Examples of other sequences are shown at the right. These sequences are separated into two groups. A **finite sequence** contains a finite number of terms. An **infinite sequence** contains an infinite number of terms.

$1, 1, 2, 3, 5, 8$
$1, 2, 3, 4, 5, 6, 7, 8$ — Finite Sequences
$1, -1, 1, -1$

$1, 3, 5, 7, \ldots$
$1, \frac{1}{2}, \frac{1}{4}, \frac{1}{8}, \ldots$ — Infinite Sequences
$1, 1, 2, 3, 5, 8, \ldots$

A general sequence is shown at the right. The first term is a_1, the second term is a_2, the third term is a_3, and the nth term, also called the **general term** of the sequence, is a_n. Note that each term of the sequence is paired with a natural number.

$a_1, a_2, a_3, \ldots, a_n, \ldots$

Frequently a sequence has a definite pattern that can be expressed by a formula.

Each term of the sequence shown at the right is paired with a natural number by the formula $a_n = 3n$. The first term, a_1, is 3. The second term, a_2, is 6. The third term, a_3, is 9. The nth term, a_n, is $3n$.

$a_n = 3n$

$a_1, \quad a_2, \quad a_3, \quad \ldots, a_n, \ldots$

$3(1), 3(2), 3(3), \ldots, 3(n), \ldots$

$3, \quad 6, \quad 9, \quad \ldots, 3n, \ldots$

⇒ Write the first three terms of the sequence whose nth term is given by the formula $a_n = 4n$.

$a_n = 4n$
$a_1 = 4(1) = 4$ • **Replace *n* by 1.**
$a_2 = 4(2) = 8$ • **Replace *n* by 2.**
$a_3 = 4(3) = 12$ • **Replace *n* by 3.**

The first term is 4, the second term is 8, and the third term is 12.

➡ Find the fifth and seventh terms of the sequence whose nth term is given by the formula $a_n = \dfrac{1}{n^2}$.

$$a_n = \dfrac{1}{n^2}$$

$$a_5 = \dfrac{1}{5^2} = \dfrac{1}{25}$$ • Replace n by 5.

$$a_7 = \dfrac{1}{7^2} = \dfrac{1}{49}$$ • Replace n by 7.

The fifth term is $\dfrac{1}{25}$ and the seventh term is $\dfrac{1}{49}$.

Example 1

Write the first three terms of the sequence whose nth term is given by the formula $a_n = 2n - 1$.

Solution

$a_n = 2n - 1$
$a_1 = 2(1) - 1 = 1$ The first term is 1.
$a_2 = 2(2) - 1 = 3$ The second term is 3.
$a_3 = 2(3) - 1 = 5$ The third term is 5.

You Try It 1

Write the first four terms of the sequence whose nth term is given by the formula $a_n = n(n + 1)$.

Your solution

Example 2

Find the eighth and tenth terms of the sequence whose nth term is given by the formula $a_n = \dfrac{n}{n + 1}$.

Solution

$$a_n = \dfrac{n}{n + 1}$$

$$a_8 = \dfrac{8}{8 + 1} = \dfrac{8}{9}$$ The eighth term is $\dfrac{8}{9}$.

$$a_{10} = \dfrac{10}{10 + 1} = \dfrac{10}{11}$$ The tenth term is $\dfrac{10}{11}$.

You Try It 2

Find the sixth and ninth terms of the sequence whose nth term is given by the formula $a_n = \dfrac{1}{n(n + 2)}$.

Your solution

Solutions on p. S33

Objective B   **To find the sum of a series**

In the last objective, the sequence 10, 11, 12.10, 13.31 was shown to represent the amount of interest earned in each of 4 years of an investment.

10, 11, 12.10, 13.31

The sum of the terms of this sequence represents the total interest earned by the investment over the four-year period.

$10 + 11 + 12.10 + 13.31 = 46.41$

The total interest earned over the four-year period is $46.41.

The indicated sum of the terms of a sequence is called a **series.** Given the sequence 10, 11, 12.10, 13.31, the series $10 + 11 + 12.10 + 13.31$ can be written.

S_n is used to designate the sum of the first n terms of a sequence.

For the preceding example, the sums of the series S_1, S_2, S_3, and S_4 represent the total interest earned for 1, 2, 3, and 4 years, respectively.

$$S_1 = 10 \qquad\qquad\qquad = 10$$
$$S_2 = 10 + 11 \qquad\qquad = 21$$
$$S_3 = 10 + 11 + 12.10 \qquad = 33.10$$
$$S_4 = 10 + 11 + 12.10 + 13.31 = 46.41$$

For the general sequence a_1, a_2, a_3, ... , a_n, the series S_1, S_2, S_3, and S_n are shown at the right.

$$S_1 = a_1$$
$$S_2 = a_1 + a_2$$
$$S_3 = a_1 + a_2 + a_3$$
$$S_n = a_1 + a_2 + a_3 + \cdots + a_n$$

It is convenient to represent a series in a compact form called **summation notation,** or **sigma notation.** The Greek letter sigma, Σ, is used to indicate a sum.

The first four terms of the sequence whose nth term is given by the formula $a_n = 2n$ are 2, 4, 6, and 8. The corresponding series is shown at the right written in summation notation; it is read "the sum from 1 to 4 of $2n$." The letter n is called the **index** of the summation.

$$\sum_{n=1}^{4} 2n$$

To write the terms of the series, replace n by the consecutive integers from 1 to 4.

$$\sum_{n=1}^{4} 2n = 2(1) + 2(2) + 2(3) + 2(4)$$

The series is $2 + 4 + 6 + 8$.

$$= 2 + 4 + 6 + 8$$

The sum of the series is 20.

$$= 20$$

➡ Find the sum of the series $\displaystyle\sum_{n=1}^{6} n^2$.

$$\sum_{n=1}^{6} n^2 = 1^2 + 2^2 + 3^2 + 4^2 + 5^2 + 6^2 \qquad \bullet \text{ Replace } n \text{ by 1, 2, 3, 4, 5, and 6.}$$

$$= 1 + 4 + 9 + 16 + 25 + 36 \qquad \bullet \text{ Simplify.}$$

$$= 91$$

The sum of the series is 91.

TAKE NOTE

The placement of the parentheses in this example is important. For instance,

$$\sum_{i=1}^{3} 2i - 1$$
$$= 2(1) + 2(2) + 2(3) - 1$$
$$= 2 + 4 + 6 - 1 = 11$$

This is *not* the same result as the sum of

$$\sum_{i=1}^{3} (2i - 1).$$

➡ Find the sum of the series $\displaystyle\sum_{i=1}^{3} (2i - 1)$.

$$\sum_{i=1}^{3} (2i - 1) = [2(1) - 1] + [2(2) - 1] + [2(3) - 1] \qquad \bullet \text{ Replace } i \text{ by 1, 2, and 3.}$$

$$= 1 + 3 + 5 \qquad \bullet \text{ Simplify.}$$

$$= 9$$

The sum of the series is 9.

⇒ Find the sum of the series $\sum\limits_{n=3}^{6} \dfrac{1}{2}n$.

$$\sum_{n=3}^{6} \dfrac{1}{2}n = \dfrac{1}{2}(3) + \dfrac{1}{2}(4) + \dfrac{1}{2}(5) + \dfrac{1}{2}(6)$$

• Replace n by 3, 4, 5, and 6.

$$= \dfrac{3}{2} + 2 + \dfrac{5}{2} + 3$$

• Simplify.

$$= 9$$

The sum of the series is 9.

⇒ Write $\sum\limits_{i=1}^{5} x^i$ in expanded form.

This is a variable series.

$$\sum_{i=1}^{5} x^i = x + x^2 + x^3 + x^4 + x^5$$

• Replace i by 1, 2, 3, 4, and 5.

Example 3

Find the sum of the series $\sum\limits_{n=1}^{4} (7 - n)$.

Solution

$$\sum_{n=1}^{4} (7 - n)$$
$$= (7 - 1) + (7 - 2) + (7 - 3) + (7 - 4)$$
$$= 6 + 5 + 4 + 3 = 18$$

You Try It 3

Find the sum of the series $\sum\limits_{i=3}^{6} (i^2 - 2)$.

Your solution

Example 4

Write $\sum\limits_{i=1}^{5} ix$ in expanded form.

Solution

$$\sum_{i=1}^{5} ix = 1 \cdot x + 2 \cdot x + 3 \cdot x + 4 \cdot x + 5 \cdot x$$
$$= x + 2x + 3x + 4x + 5x$$

You Try It 4

Write $\sum\limits_{n=1}^{4} \dfrac{x}{n}$ in expanded form.

Your solution

Solutions on p. S33

12.1 Exercises

Objective A

1. What is a sequence?

2. What is the difference between a finite sequence and an infinite sequence?

3. Name the third term in the sequence 2, 5, 8, 11, 14, ...

4. Name the fourth term in the sequence 1, 2, 4, 8, 16, 32, ...

Write the first four terms of the sequence whose nth term is given by the formula.

5. $a_n = n + 1$

6. $a_n = n - 1$

7. $a_n = 2n + 1$

8. $a_n = 3n - 1$

9. $a_n = 2 - 2n$

10. $a_n = 1 - 2n$

11. $a_n = 2^n$

12. $a_n = 3^n$

13. $a_n = n^2 + 1$

14. $a_n = n - \dfrac{1}{n}$

15. $a_n = n^2 - \dfrac{1}{n}$

16. $a_n = (-1)^{n+1}n$

Find the indicated term of the sequence whose nth term is given by the formula.

17. $a_n = 3n + 4; a_{12}$

18. $a_n = 2n - 5; a_{10}$

19. $a_n = n(n - 1); a_{11}$

20. $a_n = \dfrac{n}{n + 1}; a_{12}$

21. $a_n = (-1)^{n-1}n^2; a_{15}$

22. $a_n = (-1)^{n-1}(n - 1); a_{25}$

23. $a_n = \left(\dfrac{1}{2}\right)^n; a_8$

24. $a_n = \left(\dfrac{2}{3}\right)^n; a_5$

25. $a_n = (n + 2)(n + 3); a_{17}$

26. $a_n = (n + 4)(n + 1); a_7$

27. $a_n = \dfrac{(-1)^{2n-1}}{n^2}; a_6$

28. $a_n = \dfrac{(-1)^{2n}}{n + 4}; a_{16}$

Objective B

Find the sum of the series.

29. $\displaystyle\sum_{n=1}^{5} (2n + 3)$

30. $\displaystyle\sum_{i=1}^{7} (i + 2)$

31. $\displaystyle\sum_{i=1}^{4} 2i$

32. $\displaystyle\sum_{n=1}^{7} n$

33. $\displaystyle\sum_{i=1}^{6} i^2$

34. $\displaystyle\sum_{i=1}^{5} (i^2 + 1)$

35. $\displaystyle\sum_{n=1}^{6} (-1)^n$

36. $\displaystyle\sum_{n=1}^{4} \frac{1}{2n}$

37. $\displaystyle\sum_{i=3}^{6} i^3$

38. $\displaystyle\sum_{i=1}^{4} (-1)^{i-1}(i + 1)$

39. $\displaystyle\sum_{n=3}^{5} \frac{(-1)^{n-1}}{n - 2}$

40. $\displaystyle\sum_{n=4}^{7} \frac{(-1)^{n-1}}{n - 3}$

Write the series in expanded form.

41. $\displaystyle\sum_{n=1}^{5} 2x^n$

42. $\displaystyle\sum_{n=1}^{4} \frac{2n}{x}$

43. $\displaystyle\sum_{i=1}^{5} \frac{x^i}{i}$

44. $\displaystyle\sum_{i=2}^{4} \frac{x^i}{2i - 1}$

45. $\displaystyle\sum_{n=1}^{5} x^{2n}$

46. $\displaystyle\sum_{n=1}^{4} x^{2n-1}$

47. $\displaystyle\sum_{i=1}^{4} \frac{x^i}{i^2}$

48. $\displaystyle\sum_{n=1}^{4} nx^{n-1}$

49. $\displaystyle\sum_{i=1}^{5} x^{-i}$

APPLYING THE CONCEPTS

50. In the first box below, $\frac{1}{2}$ of the box is shaded. In successive boxes the sums $\frac{1}{2} + \frac{1}{4}$ and $\frac{1}{2} + \frac{1}{4} + \frac{1}{8}$ are shown. Can you identify the infinite sum $\frac{1}{2} + \frac{1}{4} + \frac{1}{8} + \frac{1}{16} + \cdots$?

51. Rewrite $1 + \frac{1}{2} + \frac{1}{3} + \cdots + \frac{1}{n}$ using sigma notation.

52. Indicate whether the following statements are true or false.
a. No two terms of a sequence can be equal.
b. The sum of an infinite number of terms is an infinite number.

53. Write a paragraph about the history of the sigma notation used for summation.

54. Write a paragraph explaining the Fibonacci Sequence.

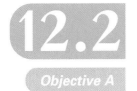

12.2 Arithmetic Sequences and Series

To find the *n*th term of an arithmetic sequence

A company's expenses for training a new employee are quite high. To encourage employees to continue their employment with the company, a company that has a six-month training program offers a starting salary of $1600 per month and then a $200-per-month pay increase each month during the training period.

The sequence at the right shows the employee's monthly salaries during the training period. Each term of the sequence is found by adding $200 to the preceding term.

Month	1	2	3	4	5	6
Salary	1600	1800	2000	2200	2400	2600

> **TAKE NOTE**
> Arithmetic sequences are a special type of sequence, one for which the difference between *any* two successive terms is the same constant. For instance, 5, 10, 15, 20, 25, ... is an arithmetic sequence. The difference between any two successive terms is 5. The sequence 1, 4, 9, 16, ... is not an arithmetic sequence because $4 - 1 \neq 9 - 4$.

The sequence 1600, 1800, 2000, 2200, 2400, 2600 is called an *arithmetic sequence*. An **arithmetic sequence,** or **arithmetic progression,** is one in which the difference between any two consecutive terms is constant. The difference between consecutive terms is called the **common difference** of the sequence.

Each of the sequences shown at the right is an arithmetic sequence. To find the common difference of an arithmetic sequence, subtract the first term from the second term.

$2, 7, 12, 17, 22, \ldots$ Common difference: 5

$3, 1, -1, -3, -5, \ldots$ Common difference: -2

$1, \dfrac{3}{2}, 2, \dfrac{5}{2}, 3, \dfrac{7}{2}$ Common difference: $\dfrac{1}{2}$

Consider an arithmetic sequence in which the first term is a_1 and the common difference is d. Adding the common difference to each successive term of the arithmetic sequence yields a formula for the *n*th term.

The first term is a_1.

$$a_1 = a_1$$

To find the second term, add the common difference d to the first term.

$$a_2 = a_1 + d$$

To find the third term, add the common difference d to the second term.

$$a_3 = a_2 + d = (a_1 + d) + d$$
$$a_3 = a_1 + 2d$$

To find the fourth term, add the common difference d to the third term.

$$a_4 = a_3 + d = (a_1 + 2d) + d$$
$$a_4 = a_1 + 3d$$

Note the relationship between the term number and the number that multiplies d. The multiplier of d is 1 less than the term number.

$$a_n = a_1 + (n - 1)d$$

> **The Formula for the *n*th Term of an Arithmetic Sequence**
>
> The *n*th term of an arithmetic sequence with a common difference of d is given by
> $a_n = a_1 + (n - 1)d$.

➡ Find the 27th term of the arithmetic sequence $-4, -1, 2, 5, 8, \ldots$.

$d = a_2 - a_1 = -1 - (-4) = 3$ • Find the common difference.

$a_n = a_1 + (n - 1)d$
$a_{27} = -4 + (27 - 1)3$
$a_{27} = -4 + (26)3$
$a_{27} = -4 + 78$
$a_{27} = 74$

• Use the Formula for the nth Term of an Arithmetic Sequence to find the 27th term.
$n = 27, a_1 = -4, d = 3$

➡ Find the formula for the nth term of the arithmetic sequence $-5, -2, 1, 4, \ldots$.

$d = a_2 - a_1 = -2 - (-5) = 3$ • Find the common difference.

$a_n = a_1 + (n - 1)d$
$a_n = -5 + (n - 1)3$
$a_n = -5 + 3n - 3$
$a_n = 3n - 8$

• Use the Formula for the nth Term of an Arithmetic Sequence. $a_1 = -5, d = 3$

➡ Find the formula for the nth term of the arithmetic sequence $2, -3, -8, -13, \ldots$.

$d = a_2 - a_1 = -3 - 2 = -5$ • Find the common difference.

$a_n = a_1 + (n - 1)d$
$a_n = 2 + (n - 1)(-5)$
$a_n = 2 - 5n + 5$
$a_n = -5n + 7$

• Use the Formula for the nth Term of an Arithmetic Sequence. $a_1 = 2, d = -5$

Example 1
Find the 10th term of the arithmetic sequence
$1, 6, 11, 16, \ldots$.

Solution
$d = a_2 - a_1 = 6 - 1 = 5$

$a_n = a_1 + (n - 1)d$
$a_{10} = 1 + (10 - 1)5 = 1 + (9)5 = 1 + 45$
$a_{10} = 46$

You Try It 1
Find the 15th term of the arithmetic sequence
$9, 3, -3, -9, \ldots$.

Your solution

Example 2
Find the number of terms in the finite
arithmetic sequence $-3, 1, 5, \ldots, 41$.

Solution
$d = a_2 - a_1 = 1 - (-3) = 4$

$a_n = a_1 + (n - 1)d$
$41 = -3 + (n - 1)4$
$41 = -3 + 4n - 4$
$41 = -7 + 4n$
$48 = 4n$
$12 = n$

There are 12 terms in the sequence.

You Try It 2
Find the number of terms in the finite
arithmetic sequence $7, 9, 11, \ldots, 59$.

Your solution

Solutions on p. S33

Objective B   **To find the sum of an arithmetic series**

The indicated sum of the terms of an arithmetic sequence is called an **arithmetic series.** The sum of an arithmetic series can be found by using a formula. A proof of this formula appears in the Appendix.

Point of Interest

This formula was proven in *Aryabhatiya*, which was written by Aryabhata around 499. The book is the earliest known Indian mathematical work by an identifiable author. Although the proof of the formula appears in that text, the formula was known before Aryabhata's time.

> **The Formula for the Sum of n Terms of an Arithmetic Series**
>
> Let a_1 be the first term of a finite arithmetic sequence, let n be the number of terms, and let a_n be the last term of the sequence.
>
> Then the sum of the series S_n is given by $S_n = \dfrac{n}{2}(a_1 + a_n)$.

⇨ Find the sum of the first 10 terms of the arithmetic sequence 2, 4, 6, 8, ...

$d = a_2 - a_1 = 4 - 2 = 2$ • Find the common difference.

$a_n = a_1 + (n - 1)d$
$a_{10} = 2 + (10 - 1)2$
$a_{10} = 2 + (9)2 = 20$
  • Use the Formula for the nth Term of an Arithmetic Sequence to find the 10th term.
  $n = 10, a_1 = 2, d = 2$

$S_n = \dfrac{n}{2}(a_1 + a_n)$
  • Use the Formula for the Sum of n Terms of an Arithmetic Series. $n = 10, a_1 = 2, a_n = 20$

$S_{10} = \dfrac{10}{2}(2 + 20) = 5(22) = 110$

⇨ Find the sum of the arithmetic series $\displaystyle\sum_{n=1}^{25}(3n + 1)$.

$a_n = 3n + 1$
$a_1 = 3(1) + 1 = 4$ • Find the 1st term.
$a_{25} = 3(25) + 1 = 76$ • Find the 25th term.

$S_n = \dfrac{n}{2}(a_1 + a_n)$
  • Use the Formula for the Sum of n Terms of an Arithmetic Series. $n = 25, a_1 = 4, a_n = 76$

$S_{25} = \dfrac{25}{2}(4 + 76) = \dfrac{25}{2}(80)$

$S_{25} = 1000$

Example 3

Find the sum of the first 20 terms of the arithmetic sequence 3, 8, 13, 18, ...

Solution

$d = a_2 - a_1 = 8 - 3 = 5$

$a_n = a_1 + (n - 1)d$
$a_{20} = 3 + (20 - 1)5 = 3 + (19)5$
$\phantom{a_{20}} = 3 + 95 = 98$

$S_n = \dfrac{n}{2}(a_1 + a_n)$

$S_{20} = \dfrac{20}{2}(3 + 98) = 10(101) = 1010$

You Try It 3

Find the sum of the first 25 terms of the arithmetic sequence −4, −2, 0, 2, 4, ...

Your solution

Solution on p. S34

Example 4

Find the sum of the arithmetic series

$$\sum_{n=1}^{15}(2n + 3).$$

Solution

$a_n = 2n + 3$
$a_1 = 2(1) + 3 = 5$
$a_{15} = 2(15) + 3 = 33$

$S_n = \dfrac{n}{2}(a_1 + a_n)$

$S_{15} = \dfrac{15}{2}(5 + 33) = \dfrac{15}{2}(38) = 285$

You Try It 4

Find the sum of the arithmetic series

$$\sum_{n=1}^{18}(3n - 2).$$

Your solution

Solution on p. S34

Objective C **To solve application problems** ‹ 12 ›

Example 5

The distance a ball rolls down a ramp each second is given by an arithmetic sequence. The distance in feet traveled by the ball during the *n*th second is given by $2n - 1$. Find the distance the ball will travel during the 10th second.

Strategy

To find the distance:

- Find the common difference of the arithmetic sequence.
- Use the Formula for the *n*th Term of an Arithmetic Sequence.

Solution

$a_n = 2n - 1$
$a_1 = 2(1) - 1 = 1$
$a_2 = 2(2) - 1 = 3$
$d = a_2 - a_1 = 3 - 1 = 2$

$a_n = a_1 + (n - 1)d$
$a_{10} = 1 + (10 - 1)2 = 1 + (9)2$
$\quad = 1 + 18 = 19$

The ball will travel 19 ft during the 10th second.

You Try It 5

A contest offers 20 prizes. The first prize is $10,000, and each successive prize is $300 less than the preceding prize. What is the value of the 20th-place prize? What is the total amount of prize money being awarded?

Your strategy

Your solution

Solution on p. S34

12.2 Exercises

Objective A

Find the indicated term of the arithmetic sequence.

1. $1, 11, 21, \ldots; a_{15}$

2. $3, 8, 13, \ldots; a_{20}$

3. $-6, -2, 2, \ldots; a_{15}$

4. $-7, -2, 3, \ldots; a_{14}$

5. $2, \dfrac{5}{2}, 3, \ldots; a_{31}$

6. $1, \dfrac{5}{4}, \dfrac{3}{2}, \ldots; a_{17}$

7. $-4, -\dfrac{5}{2}, -1, \ldots; a_{12}$

8. $-\dfrac{5}{3}, -1, -\dfrac{1}{3}, \ldots; a_{22}$

9. $8, 5, 2, \ldots; a_{40}$

Find the formula for the nth term of the arithmetic sequence.

10. $1, 2, 3, \ldots$

11. $1, 4, 7, \ldots$

12. $6, 2, -2, \ldots$

13. $3, 0, -3, \ldots$

14. $2, \dfrac{7}{2}, 5, \ldots$

15. $7, 4.5, 2, \ldots$

Find the number of terms in the finite arithmetic sequence.

16. $1, 5, 9, \ldots, 81$

17. $3, 8, 13, \ldots, 98$

18. $2, 0, -2, \ldots, -56$

19. $1, -3, -7, \ldots, -75$

20. $\dfrac{5}{2}, 3, \dfrac{7}{2}, \ldots, 13$

21. $\dfrac{7}{3}, \dfrac{13}{3}, \dfrac{19}{3}, \ldots, \dfrac{79}{3}$

22. $1, 0.75, 0.50, \ldots, -4$

23. $3.5, 2, 0.5, \ldots, -25$

24. $-3.4, -2.8, -2.2, \ldots, 11$

Objective B

Find the sum of the indicated number of terms of the arithmetic sequence.

25. $1, 3, 5, \ldots; n = 50$

26. $2, 4, 6, \ldots; n = 25$

27. $20, 18, 16, \ldots; n = 40$

28. $25, 20, 15, \ldots; n = 22$

29. $\dfrac{1}{2}, 1, \dfrac{3}{2}, \ldots; n = 27$

30. $2, \dfrac{11}{4}, \dfrac{7}{2}, \ldots; n = 10$

Find the sum of the arithmetic series.

31. $\displaystyle\sum_{i=1}^{15} (3i - 1)$

32. $\displaystyle\sum_{i=1}^{15} (3i + 4)$

33. $\displaystyle\sum_{n=1}^{17} \left(\frac{1}{2}n + 1\right)$

34. $\displaystyle\sum_{n=1}^{10} (1 - 4n)$

35. $\displaystyle\sum_{i=1}^{15} (4 - 2i)$

36. $\displaystyle\sum_{n=1}^{10} (5 - n)$

Objective C   *Application Problems*

37. An exercise program calls for walking 10 min each day for a week. Each week thereafter, the amount of time spent walking increases by 5 min per day. In how many weeks will a person be walking 60 min each day?

38. A display of cans in a grocery store consists of 24 cans in the bottom row, 21 cans in the next row, and so on in an arithmetic sequence. The top row has 3 cans. Find the number of cans in the display.

39. An object that is dropped from a low-flying airplane will fall 16 ft the first second, 48 ft the next second, 80 ft the third second, and so on in an arithmetic sequence. Find the distance that the object will fall during the fifth second. (Assume that air resistance is negligible.)

40. The loge seating section in a concert hall consists of 27 rows of chairs. There are 73 seats in the first row, 79 seats in the second row, 85 seats in the third row, and so on in an arithmetic sequence. How many seats are in the loge seating section?

41. The salary schedule for a software engineer apprentice program is $1500 for the first month and a $300-per-month salary increase for the next 7 months. Find the monthly salary during the eighth month. Find the total salary for the eight-month period.

APPLYING THE CONCEPTS

42. State whether the following statements are true or false.
a. The sum of the successive terms of an arithmetic sequence always increases in value.
b. Each successive term in an arithmetic series is found by adding a fixed number.

43. Write $\displaystyle\sum_{i=1}^{2} \log 2i$ as a single logarithm with a coefficient of 1.

44. Write a formula for the *n*th term of the sequence of the odd natural numbers

12.3 Geometric Sequences and Series

Objective A   To find the *n*th term of a geometric sequence   12

An ore sample contains 20 mg of a radioactive material with a half-life of one week. The amount of the radioactive material that the sample contains at the beginning of each week can be determined by using an exponential decay equation.

The sequence at the right represents the amount in the sample at the beginning of each week. Each term of the sequence is found by multiplying the preceding term by $\frac{1}{2}$.

Week	1	2	3	4	5
Amount	20	10	5	2.5	1.25

The sequence 20, 10, 5, 2.5, 1.25 is called a *geometric sequence*. A **geometric sequence,** or **geometric progression,** is one in which each successive term of the sequence is the same nonzero constant multiple of the preceding term. The common multiple is called the **common ratio** of the sequence.

> **TAKE NOTE**
> Geometric sequences are different from arithmetic sequences. For a geometric sequence, every two successive terms have the same *ratio*. For an arithmetic sequence, every two successive terms have the same *difference*.

Each of the sequences shown at the right is a geometric sequence. To find the common ratio of a geometric sequence, divide the second term of the sequence by the first term.

3, 6, 12, 24, 48, …   Common ratio: 2

4, −12, 36, −108, 324, …   Common ratio: −3

$6, 4, \frac{8}{3}, \frac{16}{9}, \frac{32}{27}, \ldots$ Common ratio: $\frac{2}{3}$

Consider a geometric sequence in which the first term is a_1 and the common ratio is r. Multiplying each successive term of the geometric sequence by the common ratio yields a formula for the *n*th term.

The first term is a_1.

$$a_1 = a_1$$

To find the second term, multiply the first term by the common ratio r.

$$a_2 = a_1 r$$

To find the third term, multiply the second term by the common ratio r.

$$a_3 = (a_1 r)r$$
$$a_3 = a_1 r^2$$

To find the fourth term, multiply the third term by the common ratio r.

$$a_4 = (a_1 r^2)r$$
$$a_4 = a_1 r^3$$

Note the relationship between the term number and the number that is the exponent on r. The exponent on r is 1 less than the term number.

$$a_n = a_1 r^{n-1}$$

> **The Formula for the *n*th Term of a Geometric Sequence**
>
> The *n*th term of a geometric sequence with first term a_1 and common ratio r is given by $a_n = a_1 r^{n-1}$.

⇒ Find the 6th term of the geometric sequence 3, 6, 12, ...

$$r = \frac{a_2}{a_1} = \frac{6}{3} = 2$$ • Find the common ratio.

$$a_n = a_1 r^{n-1}$$ • Use the Formula for the *n*th Term of a Geometric Sequence
$$a_6 = 3(2)^{6-1}$$ to find the 6th term. $n = 6$, $a_1 = 3$, $r = 2$
$$a_6 = 3(2)^5 = 3(32)$$
$$a_6 = 96$$

⇒ Find a_3 for the geometric sequence 8, a_2, a_3, −27, ...

$$a_n = a_1 r^{n-1}$$ • Use the Formula for the *n*th Term of a Geometric
$$a_4 = a_1 r^{4-1}$$ Sequence to find the common ratio. The 1st and
                          4th terms are known. Substitute 4 for *n*.
$$-27 = 8r^3$$ • $a_4 = -27$, $a_1 = 8$

$$-\frac{27}{8} = r^3$$ • Solve for r^3.

$$-\frac{3}{2} = r$$ • Take the cube root of each side of the equation.

$$a_n = a_1 r^{n-1}$$ • Use the Formula for the *n*th Term of a
$$a_3 = 8\left(-\frac{3}{2}\right)^{3-1} = 8\left(-\frac{3}{2}\right)^2 = 8\left(\frac{9}{4}\right)$$ Geometric Sequence to find a_3.

$n = 3$, $a_1 = 8$, $r = -\frac{3}{2}$

$$a_3 = 18$$

Example 1

Find the 7th term of the geometric sequence 3, −6, 12, ...

Solution

$$r = \frac{a_2}{a_1} = \frac{-6}{3} = -2$$
$$a_n = a_1 r^{n-1}$$
$$a_7 = 3(-2)^{7-1} = 3(-2)^6 = 3(64)$$
$$a_7 = 192$$

You Try It 1

Find the 5th term of the geometric sequence 5, 2, $\frac{4}{5}$, ...

Your solution

Solution on p. S34

Example 2

Find a_2 for the geometric sequence 2, a_2, a_3, 54, ...

Solution

$a_n = a_1 r^{n-1}$

$a_4 = 2r^{4-1}$　　• $n = 4$, $a_1 = 2$

$54 = 2r^{4-1}$　　• $a_4 = 54$

$54 = 2r^3$

$27 = r^3$

$3 = r$　　• The common ratio is 3.

$a_n = a_1 r^{n-1}$

$a_2 = 2(3)^{2-1} = 2(3)$　　• $n = 2$, $r = 3$

$a_2 = 6$

You Try It 2

Find a_3 for the geometric sequence 3, a_2, a_3, -192, ...

Your solution

Solution on p. S34

Objective B

To find the sum of a finite geometric series　

The indicated sum of the terms of a geometric sequence is called a **geometric series.** The sum of a geometric series can be found by a formula. A proof of this formula appears in the Appendix.

Point of Interest

Geometric series are used extensively in the mathematics of finance. Finite geometric series are used to calculate loan balances and monthly payments for amortized loans.

> **The Formula for the Sum of n Terms of a Finite Geometric Series**
>
> Let a_1 be the first term of a finite geometric sequence, let n be the number of terms, and let r be the common ratio. Then the sum of the series S_n is given by $S_n = \dfrac{a_1(1 - r^n)}{1 - r}$.

⇒ Find the sum of the terms of the geometric sequence 2, 8, 32, 128, 512.

$r = \dfrac{a_2}{a_1} = \dfrac{8}{2} = 4$　　• Find the common ratio.

$S_n = \dfrac{a_1(1 - r^n)}{1 - r}$　　• Use the Formula for the Sum of n Terms of a Finite Geometric Series. $n = 5$, $a_1 = 2$, $r = 4$

$S_5 = \dfrac{2(1 - 4^5)}{1 - 4} = \dfrac{2(1 - 1024)}{-3}$

$S_5 = \dfrac{2(-1023)}{-3} = \dfrac{-2046}{-3}$

$S_5 = 682$

⇒ Find the sum of the geometric series $\sum\limits_{n=1}^{10}(-20)(-2)^{n-1}$.

$a_n = (-20)(-2)^{n-1}$

$a_1 = (-20)(-2)^{1-1} = (-20)(-2)^0$ • Find the 1st term.

$\quad = (-20)(1) = -20$

$a_2 = (-20)(-2)^{2-1} = (-20)(-2)^1$ • Find the 2nd term.

$\quad = (-20)(-2) = 40$

$r = \dfrac{a_2}{a_1} = \dfrac{40}{-20} = -2$ • Find the common ratio.

$S_n = \dfrac{a_1(1-r^n)}{1-r}$ • Use the Formula for the Sum of n Terms of a Finite Geometric Series. $n = 10$, $a_1 = -20$, $r = -2$

$S_{10} = \dfrac{-20[1-(-2)^{10}]}{1-(-2)}$

$S_{10} = \dfrac{-20(1-1024)}{3}$

$S_{10} = \dfrac{-20(-1023)}{3} = \dfrac{20{,}460}{3}$

$S_{10} = 6820$

....

Example 3

Find the sum of the terms of the geometric sequence 3, 6, 12, 24, 48, 96.

Solution

$r = \dfrac{a_2}{a_1} = \dfrac{6}{3} = 2$

$S_n = \dfrac{a_1(1-r^n)}{1-r}$

$S_6 = \dfrac{3(1-2^6)}{1-2} = \dfrac{3(1-64)}{-1} = -3(-63) = 189$

You Try It 3

Find the sum of the terms of the geometric sequence $1, -\dfrac{1}{3}, \dfrac{1}{9}, -\dfrac{1}{27}$.

Your solution

....

Example 4

Find the sum of the geometric series $\sum\limits_{n=1}^{4} 4^n$.

Solution

$a_n = 4^n$

$a_1 = 4^1 = 4$

$a_2 = 4^2 = 16$

$r = \dfrac{a_2}{a_1} = \dfrac{16}{4} = 4$

$S_n = \dfrac{a_1(1-r^n)}{1-r}$

$S_4 = \dfrac{4(1-4^4)}{1-4} = \dfrac{4(1-256)}{-3} = \dfrac{4(-255)}{-3} = \dfrac{-1020}{-3}$

$\quad = 340$

You Try It 4

Find the sum of the geometric series $\sum\limits_{n=1}^{5} \left(\dfrac{1}{2}\right)^n$.

Your solution

Solutions on pp. S34–S35

Objective C   **To find the sum of an infinite geometric series**

Point of Interest

The midpoints of the sides of each square in the figure below are connected to form the next smaller square. By the Pythagorean Theorem, if a side of one square has length s, then a side of the next smaller square has length $\dfrac{\sqrt{2}}{2}\,s$.

Therefore, the sides of the consecutively smaller and smaller squares form a sequence in which

$$s_n = \frac{\sqrt{2}}{2}\, s_{n-1}.$$

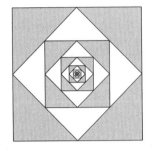

It is from applications such as this that *geometric* sequences got their name.

When the absolute value of the common ratio of a geometric sequence is less than 1, $|r| < 1$, then as n becomes larger, r^n becomes closer to zero.

Examples of geometric sequences for which $|r| < 1$ are shown at the right. Note that as the number of terms increases, the value of the last term listed gets closer to zero.

$$1, \frac{1}{3}, \frac{1}{9}, \frac{1}{27}, \frac{1}{81}, \frac{1}{243}, \cdots$$

$$1, -\frac{1}{2}, \frac{1}{4}, -\frac{1}{8}, \frac{1}{16}, -\frac{1}{32}, \cdots$$

The indicated sum of the terms of an infinite geometric sequence is called an **infinite geometric series.**

An example of an infinite geometric series is shown at the right. The first term is 1. The common ratio is $\dfrac{1}{3}$.

$$1 + \frac{1}{3} + \frac{1}{9} + \frac{1}{27} + \frac{1}{81} + \frac{1}{243} + \cdots$$

The sums of the first 5, 7, 12, and 15 terms, along with the values of r^n, are shown at the right. Note that as n increases, the sum of the terms gets closer to 1.5 and the value of r^n gets closer to zero.

n	S_n	r^n
5	1.4938272	0.0041152
7	1.4993141	0.0004572
12	1.4999972	0.0000019
15	1.4999999	0.0000001

Using the Formula for the Sum of n Terms of a Finite Geometric Series and the fact that r^n approaches zero when $|r| < 1$ and n increases, we can find a formula for the sum of an infinite geometric series.

The sum of the first n terms of a geometric series is shown at the right. If $|r| < 1$, then r^n can be made very close to zero by using larger and larger values of n. Therefore, the sum of the first n terms is approximately $\dfrac{a_1}{1-r}$.

approximately zero

$$S_n = \frac{a_1(1 - r^n)}{1 - r}$$

$$S_n \approx \frac{a_1(1 - 0)}{1 - r} = \frac{a_1}{1 - r}$$

> **The Formula for the Sum of an Infinite Geometric Series**
>
> The sum of an infinite geometric series in which $|r| < 1$ and a_1 is the first term is given by $S = \dfrac{a_1}{1 - r}$.

When $|r| > 1$, the infinite geometric series does not have a sum. For example, the sum of the infinite geometric series $1 + 2 + 4 + 8 + \cdots$ increases without limit.

➡ Find the sum of the terms of the infinite geometric sequence $1, -\frac{1}{2}, \frac{1}{4}, -\frac{1}{8}, \dots$.

The common ratio is $-\frac{1}{2}$. $\left|-\frac{1}{2}\right| < 1$.

$$S = \frac{a_1}{1-r} = \frac{1}{1-\left(-\frac{1}{2}\right)} = \frac{1}{\frac{3}{2}} = \frac{2}{3}$$

• Use the Formula for the Sum of an Infinite Geometric Series.

The sum of an infinite geometric series can be applied to nonterminating, repeating decimals.

The repeating decimal shown below has been rewritten as an infinite geometric series with first term $\frac{3}{10}$ and common ratio $\frac{1}{10}$.

$$0.33\overline{3} = 0.3 + 0.03 + 0.003 + \cdots$$
$$= \frac{3}{10} + \frac{3}{100} + \frac{3}{1000} + \cdots$$

The sum of the series can be calculated by using the Formula for the Sum of an Infinite Geometric Series.

$$S = \frac{a_1}{1-r} = \frac{\frac{3}{10}}{1-\frac{1}{10}} = \frac{\frac{3}{10}}{\frac{9}{10}} = \frac{3}{9} = \frac{1}{3}$$

• $a_1 = \frac{3}{10}, r = \frac{1}{10}$

This method can be used to find an equivalent numerical fraction for any repeating decimal.

➡ Find an equivalent fraction for $0.122\overline{2}$.

Write the decimal as an infinite geometric series. Note that the geometric series does not begin with the first term. It begins with $\frac{2}{100}$. The common ratio is $\frac{1}{10}$.

$$0.122\overline{2} = 0.1 + 0.02 + 0.002 + 0.0002 + \cdots$$
$$= \frac{1}{10} + \underbrace{\frac{2}{100} + \frac{2}{1000} + \frac{2}{10,000} + \cdots}_{\text{Geometric Series}}$$

Use the Formula for the Sum of an Infinite Geometric Series.

$$S = \frac{a_1}{1-r} = \frac{\frac{2}{100}}{1-\frac{1}{10}} = \frac{\frac{2}{100}}{\frac{9}{10}} = \frac{2}{90}$$

• $a_1 = \frac{2}{100}, r = \frac{1}{10}$

Add $\frac{1}{10}$ to the sum of the geometric series.

$$0.122\overline{2} = \frac{1}{10} + \frac{2}{90} = \frac{11}{90}$$

An equivalent fraction for $0.122\overline{2}$ is $\frac{11}{90}$.

Example 5

Find the sum of the terms of the infinite geometric sequence $2, \frac{1}{2}, \frac{1}{8}, \dots$.

Solution

$$r = \frac{a_2}{a_1} = \frac{\frac{1}{2}}{2} = \frac{1}{4}$$

$$S = \frac{a_1}{1 - r} = \frac{2}{1 - \frac{1}{4}} = \frac{2}{\frac{3}{4}} = \frac{8}{3}$$

You Try It 5

Find the sum of the terms of the infinite geometric sequence $3, -2, \frac{4}{3}, -\frac{8}{9}, \dots$.

Your solution

Example 6

Find an equivalent fraction for $0.36\overline{36}$.

Solution

$$0.36\overline{36} = 0.36 + 0.0036 + 0.000036 + \cdots$$

$$= \frac{36}{100} + \frac{36}{10,000} + \frac{36}{1,000,000} + \cdots$$

$$S = \frac{a_1}{1 - r} = \frac{\frac{36}{100}}{1 - \frac{1}{100}} = \frac{\frac{36}{100}}{\frac{99}{100}} = \frac{36}{99} = \frac{4}{11}$$

An equivalent fraction is $\frac{4}{11}$.

You Try It 6

Find an equivalent fraction for $0.6\overline{6}$.

Your solution

Example 7

Find an equivalent fraction for $0.23\overline{45}$.

Solution

$$0.23\overline{45} = 0.23 + 0.0045 + 0.000045 + \cdots$$

$$= \frac{23}{100} + \frac{45}{10,000} + \frac{45}{1,000,000} + \cdots$$

$$S = \frac{a_1}{1 - r} = \frac{\frac{45}{10,000}}{1 - \frac{1}{100}} = \frac{\frac{45}{10,000}}{\frac{99}{100}} = \frac{45}{9900}$$

$$0.23\overline{45} = \frac{23}{100} + \frac{45}{9900} = \frac{2277}{9900} + \frac{45}{9900}$$

$$= \frac{2322}{9900} = \frac{129}{550}$$

An equivalent fraction is $\frac{129}{550}$.

You Try It 7

Find an equivalent fraction for $0.34\overline{21}$.

Your solution

Solutions on p. S35

Objective D   **To solve application problems**   ◁ 12 ▷   ◎

Example 8

On the first swing, the length of the arc through which a pendulum swings is 16 in. The length of each successive swing is $\frac{7}{8}$ the length of the preceding swing. Find the length of the arc on the fifth swing. Round to the nearest tenth.

Strategy

To find the length of the arc on the fifth swing, use the Formula for the nth Term of a Geometric Sequence.

Solution

$n = 5, a_1 = 16, r = \dfrac{7}{8}$

$a_n = a_1 r^{n-1}$

$a_5 = 16\left(\dfrac{7}{8}\right)^{5-1} = 16\left(\dfrac{7}{8}\right)^4 = 16\left(\dfrac{2401}{4096}\right)$

$\qquad = \dfrac{38{,}416}{4096} \approx 9.4$

The length of the arc on the fifth swing is 9.4 in.

You Try It 8

You start a chain letter and send it to three friends. Each of the three friends sends the letter to three other friends, and the sequence is repeated. Assuming that no one breaks the chain, find how many letters will have been mailed from the first through the sixth mailings.

Your strategy

Your solution

Example 9

Assume that a mutual fund will increase at a rate of 7% per year. If $25,000 is invested on January 1, 2004, find the value of the mutual fund on December 31, 2013. Round to the nearest cent.

Strategy

To find the value of the investment, use the Formula for the nth Term of a Geometric Sequence, $a_n = a_1 r^{n-1}$.

Solution

$n = 11, a_1 = 25{,}000, r = 1.07$

$a_{11} = 25{,}000(1.07)^{11-1} = 25{,}000(1.07)^{10}$

$\qquad \approx 49{,}178.78$

The value of the investment after 10 years is $49,178.78.

You Try It 9

Assume that a mutual fund will increase at a rate of 12% per year. If $25,000 is invested on January 1, 2004, find the value of the mutual fund on December 31, 2013. Round to the nearest cent.

Your strategy

Your solution

Solutions on p. S35

12.3 Exercises

Objective A

1. Explain the difference between an arithmetic sequence and a geometric sequence.

2. Explain how to find the common ratio of a geometric sequence.

Find the indicated term of the geometric sequence.

3. $2, 8, 32, \ldots; a_9$

4. $4, 3, \dfrac{9}{4}, \ldots; a_8$

5. $6, -4, \dfrac{8}{3}, \ldots; a_7$

6. $-5, 15, -45, \ldots; a_7$

7. $-\dfrac{1}{16}, \dfrac{1}{8}, -\dfrac{1}{4}, \ldots; a_{10}$

8. $\dfrac{1}{27}, \dfrac{1}{9}, \dfrac{1}{3}, \ldots; a_7$

Find a_2 and a_3 for the geometric sequence.

9. $9, a_2, a_3, \dfrac{8}{3}, \ldots$

10. $8, a_2, a_3, \dfrac{27}{8}, \ldots$

11. $3, a_2, a_3, -\dfrac{8}{9}, \ldots$

12. $6, a_2, a_3, -48, \ldots$

13. $-3, a_2, a_3, 192, \ldots$

14. $5, a_2, a_3, 625, \ldots$

Objective B

Find the sum of the indicated number of terms of the geometric sequence.

15. $2, 6, 18, \ldots; n = 7$

16. $-4, 12, -36, \ldots; n = 7$

17. $12, 9, \dfrac{27}{4}, \ldots; n = 5$

18. $3, 3\sqrt{2}, 6, \ldots; n = 12$

Find the sum of the geometric series.

19. $\displaystyle\sum_{i=1}^{5} (2)^i$

20. $\displaystyle\sum_{n=1}^{6} \left(\dfrac{3}{2}\right)^n$

21. $\displaystyle\sum_{i=1}^{5} \left(\dfrac{1}{3}\right)^i$

22. $\displaystyle\sum_{n=1}^{6} \left(\dfrac{2}{3}\right)^n$

Objective C

Find the sum of the infinite geometric series.

23. $3 + 2 + \dfrac{4}{3} + \cdots$

24. $2 - \dfrac{1}{4} + \dfrac{1}{32} + \cdots$

25. $\dfrac{7}{10} + \dfrac{7}{100} + \dfrac{7}{1000} + \cdots$ **26.** $\dfrac{5}{100} + \dfrac{5}{10,000} + \dfrac{5}{1,000,000} + \cdots$

Find an equivalent fraction for the repeating decimal.

27. $0.88\overline{8}$ **28.** $0.55\overline{5}$ **29.** $0.22\overline{2}$ **30.** $0.99\overline{9}$

31. $0.45\overline{45}$ **32.** $0.18\overline{18}$ **33.** $0.166\overline{6}$ **34.** $0.833\overline{3}$

> **Objective D**   *Application Problems*

35. A laboratory ore sample contains 400 mg of a radioactive material with a half-life of 1 h. Find the amount of radioactive material in the sample at the beginning of the fourth hour.

36. On the first swing, the length of the arc through which a pendulum swings is 20 in. The length of each successive swing is $\dfrac{4}{5}$ the length of the preceding swing. What is the total distance the pendulum has traveled during four swings? Round to the nearest tenth of an inch.

37. To test the "bounce" of a tennis ball, the ball is dropped from a height of 10 ft. The ball bounces to 75% of its previous height with each bounce. How high does the ball bounce on the sixth bounce? Round to the nearest tenth of a foot.

38. A real estate broker estimates that a certain piece of land will increase in value at a rate of 15% each year. If the original value of the land is $25,000, what will be its value in 10 years?

39. Suppose an employee receives a wage of 1¢ on the first day of work, 2¢ the second day, 4¢ the third day, and so on in a geometric sequence. Find the total amount of money earned for working 30 days.

APPLYING THE CONCEPTS

40. State whether the following statements are true or false.
  a. The sum of an infinite geometric series is infinity.
  b. The sum of a geometric series always increases in value.
  c. $1, 8, 27, \ldots, k^3, \ldots$ is a geometric sequence.
  d. $\sqrt{1}, \sqrt{2}, \sqrt{3}, \sqrt{4}, \ldots, \sqrt{n}, \ldots$ is a geometric sequence.

41. The third term of a geometric sequence is 3, and the sixth term is $\dfrac{1}{9}$. Find the first term.

42. Given $a_n = 162, r = -3$, and $S_n = 122$ for a geometric sequence, find a_1 and n.

12.4 Binomial Expansions

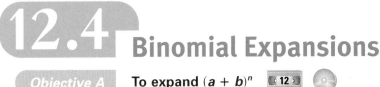

Objective A   To expand $(a + b)^n$ ⟨ 12 ⟩

By carefully observing the series for each expansion of the binomial $(a + b)^n$ shown below, it is possible to identify some interesting patterns.

$$(a + b)^1 = a + b$$
$$(a + b)^2 = a^2 + 2ab + b^2$$
$$(a + b)^3 = a^3 + 3a^2b + 3ab^2 + b^3$$
$$(a + b)^4 = a^4 + 4a^3b + 6a^2b^2 + 4ab^3 + b^4$$
$$(a + b)^5 = a^5 + 5a^4b + 10a^3b^2 + 10a^2b^3 + 5ab^4 + b^5$$

PATTERNS FOR THE VARIABLE PART

1. The first term is a^n. The exponent on a decreases by 1 for each successive term.

2. The exponent on b increases by 1 for each successive term. The last term is b^n.

3. The degree of each term is n.

➡ Write the variable parts of the terms in the expansion of $(a + b)^6$.

$a^6, a^5b, a^4b^2, a^3b^3, a^2b^4, ab^5, b^6$ • The first term is a^6. For each successive term, the exponent on a decreases by 1, and the exponent on b increases by 1. The last term is b^6.

The variable parts in the general expansion of $(a + b)^n$ are

$$a^n, a^{n-1}b, a^{n-2}b^2, \dots, a^{n-r}b^r, \dots, ab^{n-1}, b^n$$

A pattern for the coefficients of the terms of the expanded binomial can be found by writing the coefficients in a triangular array known as **Pascal's Triangle.**

Each row begins and ends with the number 1. Any other number in a row is the sum of the two closest numbers above it. For example, $4 + 6 = 10$.

For $(a + b)^1$:   1   1
For $(a + b)^2$:   1   2   1
For $(a + b)^3$:   1   3   3   1
For $(a + b)^4$:   1   4   6   4   1
For $(a + b)^5$: 1   5   10   10   5   1

➡ Write the sixth row of Pascal's Triangle.

To write the sixth row, first write the numbers of the fifth row. The first and last numbers of the sixth row are 1. Each of the other numbers of the sixth row can be obtained by finding the sum of the two closest numbers above it in the fifth row.

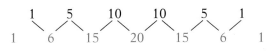

1   5   10   10   5   1
1   6   15   20   15   6   1

These numbers will be the coefficients of the terms in the expansion of $(a + b)^6$.

Using the numbers of the sixth row of Pascal's Triangle for the coefficients and using the pattern for the variable part of each term, we can write the expanded form of $(a + b)^6$ as follows:

$$(a + b)^6 = a^6 + 6a^5b + 15a^4b^2 + 20a^3b^3 + 15a^2b^4 + 6ab^5 + b^6$$

Although Pascal's Triangle can be used to find the coefficients for the expanded form of the power of any binomial, this method is inconvenient when the power of the binomial is large. An alternative method for determining these coefficients is based on the concept of **factorial.**

Point of Interest

The exclamation point was first used as a notation for factorial in 1808 in the book *Elements d'Arithmetique Universelle*, for the convenience of the printer of the book. Some English textbooks suggested using the phrase "*n*-admiration" for *n*!.

---

n Factorial

$n!$ (which is read "n factorial") is the product of the first n consecutive natural numbers. $0!$ is defined to be 1.

$$n! = n(n - 1)(n - 2) \cdots 3 \cdot 2 \cdot 1$$

$1!$, $5!$, and $7!$ are shown at the right.

$$1! = 1$$
$$5! = 5 \cdot 4 \cdot 3 \cdot 2 \cdot 1 = 120$$
$$7! = 7 \cdot 6 \cdot 5 \cdot 4 \cdot 3 \cdot 2 \cdot 1 = 5040$$

➡ Evaluate: $6!$

$$6! = 6 \cdot 5 \cdot 4 \cdot 3 \cdot 2 \cdot 1 = 720$$

• Write the factorial as a product. Then simplify.

➡ Evaluate: $\dfrac{7!}{4! \, 3!}$

$$\frac{7!}{4! \, 3!} = \frac{7 \cdot 6 \cdot 5 \cdot 4 \cdot 3 \cdot 2 \cdot 1}{(4 \cdot 3 \cdot 2 \cdot 1)(3 \cdot 2 \cdot 1)} = 35$$

• Write each factorial as a product. Then simplify.

The coefficients in a binomial expansion can be given in terms of factorials.

Note that in the expansion of $(a + b)^5$, the coefficient of a^2b^3 can be given by $\frac{5!}{2! \, 3!}$. The numerator is the factorial of the power of the binomial. The denominator is the product of the factorials of the exponents on a and b.

$$(a + b)^5 =$$
$$a^5 + 5a^4b + 10a^3b^2 + 10a^2b^3 + 5ab^4 + b^5$$
$$\frac{5!}{2! \, 3!} = \frac{5 \cdot 4 \cdot 3 \cdot 2 \cdot 1}{(2 \cdot 1)(3 \cdot 2 \cdot 1)} = 10$$

Point of Interest

Leonhard Euler (1707–1783) used the notations $\left(\dfrac{n}{r} \right)$ and $\left[\dfrac{n}{r} \right]$ for binomial coefficients around 1784. The notation $\dbinom{n}{r}$ appeared in the late 1820s.

Binomial Coefficients

The coefficients of $a^{n-r}b^r$ in the expansion of $(a + b)^n$ are given by $\dfrac{n!}{(n - r)! \, r!}$. The symbol $\dbinom{n}{r}$ is used to express this quotient.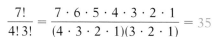

$$\binom{n}{r} = \frac{n!}{(n - r)! \, r!}$$

➡ Evaluate: $\dbinom{8}{5}$

$$\binom{8}{5} = \frac{8!}{(8 - 5)! \, 5!} = \frac{8!}{3! \, 5!}$$

• Write the quotient of the factorials.

$$= \frac{8 \cdot 7 \cdot 6 \cdot 5 \cdot 4 \cdot 3 \cdot 2 \cdot 1}{(3 \cdot 2 \cdot 1)(5 \cdot 4 \cdot 3 \cdot 2 \cdot 1)} = 56$$

• Simplify.

⇒ Evaluate: $\begin{pmatrix} 7 \\ 0 \end{pmatrix}$

$$\begin{pmatrix} 7 \\ 0 \end{pmatrix} = \frac{7!}{(7-0)!\,0!} = \frac{7!}{7!\,0!}$$

• Write the quotient of the factorials.

$$= \frac{7 \cdot 6 \cdot 5 \cdot 4 \cdot 3 \cdot 2 \cdot 1}{(7 \cdot 6 \cdot 5 \cdot 4 \cdot 3 \cdot 2 \cdot 1)(1)} = 1$$

• Recall that $0! = 1$.

Using factorials and the pattern for the variable part of each term, we can write a formula for any natural number power of a binomial.

The Binomial Expansion Formula

$$(a + b)^n = \begin{pmatrix} n \\ 0 \end{pmatrix} a^n + \begin{pmatrix} n \\ 1 \end{pmatrix} a^{n-1} b + \begin{pmatrix} n \\ 2 \end{pmatrix} a^{n-2} b^2 + \cdots +$$
$$\begin{pmatrix} n \\ r \end{pmatrix} a^{n-r} b^r + \cdots + \begin{pmatrix} n \\ n \end{pmatrix} b^n$$

The Binomial Expansion Formula is used below to expand $(a + b)^5$.

$$(a + b)^5 = \begin{pmatrix} 5 \\ 0 \end{pmatrix} a^5 + \begin{pmatrix} 5 \\ 1 \end{pmatrix} a^4 b + \begin{pmatrix} 5 \\ 2 \end{pmatrix} a^3 b^2 + \begin{pmatrix} 5 \\ 3 \end{pmatrix} a^2 b^3 + \begin{pmatrix} 5 \\ 4 \end{pmatrix} ab^4 + \begin{pmatrix} 5 \\ 5 \end{pmatrix} b^5$$
$$= a^5 + 5a^4 b + 10a^3 b^2 + 10a^2 b^3 + 5ab^4 + b^5$$

⇒ Write $(x - 2)^4$ in expanded form.

Use the Binomial Expansion Formula. Then simplify each term.

$$(x - 2)^4 = \begin{pmatrix} 4 \\ 0 \end{pmatrix} x^4 + \begin{pmatrix} 4 \\ 1 \end{pmatrix} x^3(-2) + \begin{pmatrix} 4 \\ 2 \end{pmatrix} x^2(-2)^2 + \begin{pmatrix} 4 \\ 3 \end{pmatrix} x(-2)^3 + \begin{pmatrix} 4 \\ 4 \end{pmatrix} (-2)^4$$
$$= x^4 + 4x^3(-2) + 6x^2(-2)^2 + 4x(-2)^3 + (-2)^4$$
$$= x^4 - 8x^3 + 24x^2 - 32x + 16$$

⇒ Write $(4x + 3y)^3$ in expanded form.

Use the Binomial Expansion Formula. Then simplify each term.

$$(4x + 3y)^3 = \begin{pmatrix} 3 \\ 0 \end{pmatrix} (4x)^3 + \begin{pmatrix} 3 \\ 1 \end{pmatrix} (4x)^2(3y) + \begin{pmatrix} 3 \\ 2 \end{pmatrix} (4x)(3y)^2 + \begin{pmatrix} 3 \\ 3 \end{pmatrix} (3y)^3$$
$$= 1(64x^3) + 3(16x^2)(3y) + 3(4x)(9y^2) + 1(27y^3)$$
$$= 64x^3 + 144x^2 y + 108xy^2 + 27y^3$$

The Binomial Theorem can also be used to write any term of a binomial expansion.

Note below that in the expansion of $(a + b)^5$, the exponent on b is 1 less than the term number.

$$(a + b)^5 = a^5 + 5a^4 b + 10a^3 b^2 + 10a^2 b^3 + 5ab^4 + b^5$$

The Formula for the rth Term in a Binomial Expansion

The rth term of $(a + b)^n$ is $\begin{pmatrix} n \\ r - 1 \end{pmatrix} a^{n-r+1} b^{r-1}$.

⇒ Find the 4th term in the expansion of $(x + 3)^7$.

Use the Formula for the rth Term in a Binomial Expansion. $r = 4$

$$\binom{n}{r-1} a^{n-r+1} b^{r-1}$$

$$\binom{7}{4-1} x^{7-4+1}(3)^{4-1} = \binom{7}{3} x^4 (3)^3 = 35x^4(27)$$

$$= 945x^4$$

⇒ Find the 12th term in the expansion of $(2x - 1)^{14}$.

Use the Formula for the rth Term in a Binomial Expansion. $r = 12$

$$\binom{14}{12-1}(2x)^{14-12+1}(-1)^{12-1} = \binom{14}{11}(2x)^3(-1)^{11} = 364(8x^3)(-1)$$

$$= -2912x^3$$

Example 1

Write $(2x + y)^3$ in expanded form.

Solution

$(2x + y)^3$

$$= \binom{3}{0}(2x)^3 + \binom{3}{1}(2x)^2(y) + \binom{3}{2}(2x)(y)^2 + \binom{3}{3}(y)^3$$

$$= 1(8x^3) + 3(4x^2)(y) + 3(2x)(y^2) + 1(y^3)$$

$$= 8x^3 + 12x^2y + 6xy^2 + y^3$$

You Try It 1

Write $(3m - n)^4$ in expanded form.

Your solution

Example 2

Find the first three terms in the expansion of $(x + 3)^{15}$.

Solution

$(x + 3)^{15}$

$$= \binom{15}{0}x^{15} + \binom{15}{1}x^{14}(3) + \binom{15}{2}x^{13}(3)^2 + \cdots$$

$$= 1(x^{15}) + 15x^{14}(3) + 105x^{13}(9) + \cdots$$

$$= x^{15} + 45x^{14} + 945x^{13} + \cdots$$

You Try It 2

Find the first three terms in the expansion of $(y - 2)^{10}$.

Your solution

Example 3

Find the 4th term in the expansion of $(5x - y)^6$.

Solution

$$\binom{n}{r-1} a^{n-r+1} b^{r-1}$$

$$\binom{6}{4-1}(5x)^{6-4+1}(-y)^{4-1} \quad \bullet \ n = 6, a = 5x, b = -y, \ r = 4$$

$$= \binom{6}{3}(5x)^3(-y)^3$$

$$= 20(125x^3)(-y^3)$$

$$= -2500x^3y^3$$

You Try It 3

Find the 3rd term in the expansion of $(t - 2s)^7$.

Your solution

Solutions on p. S35

12.4 Exercises

Evaluate.

1. 3!

2. 4!

3. 8!

4. 9!

5. 0!

6. 1!

7. $\dfrac{5!}{2!\,3!}$

8. $\dfrac{8!}{5!\,3!}$

9. $\dfrac{6!}{6!\,0!}$

10. $\dfrac{10!}{10!\,0!}$

11. $\dfrac{9!}{6!\,3!}$

12. $\dfrac{10!}{2!\,8!}$

Evaluate.

13. $\dbinom{7}{2}$

14. $\dbinom{8}{6}$

15. $\dbinom{10}{2}$

16. $\dbinom{9}{6}$

17. $\dbinom{9}{0}$

18. $\dbinom{10}{10}$

19. $\dbinom{6}{3}$

20. $\dbinom{7}{6}$

21. $\dbinom{11}{1}$

22. $\dbinom{13}{1}$

23. $\dbinom{4}{2}$

24. $\dbinom{8}{4}$

Write in expanded form.

25. $(x + y)^4$

26. $(r - s)^3$

27. $(x - y)^5$

28. $(y - 3)^4$

29. $(2m + 1)^4$

30. $(2x + 3y)^3$

31. $(2r - 3)^5$

32. $(x + 3y)^4$

Find the first three terms in the expansion.

33. $(a + b)^{10}$

34. $(a + b)^9$

35. $(a - b)^{11}$

36. $(a - b)^{12}$

37. $(2x + y)^8$

38. $(x + 3y)^9$

39. $(4x - 3y)^8$

40. $(2x - 5)^7$

41. $\left(x + \dfrac{1}{x}\right)^7$

42. $\left(x - \dfrac{1}{x}\right)^8$

Find the indicated term in the expansion.

43. $(2x - 1)^7$; 4th term

44. $(x + 4)^5$; 3rd term

45. $(x^2 - y^2)^6$; 2nd term

46. $(x^2 + y^2)^7$; 6th term

47. $(y - 1)^9$; 5th term

48. $(x - 2)^8$; 8th term

49. $\left(n + \dfrac{1}{n}\right)^5$; 2nd term

50. $\left(x + \dfrac{1}{2}\right)^6$; 3rd term

APPLYING THE CONCEPTS

51. State whether the following statements are true or false.
   a. $0! \cdot 4! = 0$
   b. $\dfrac{4!}{0!}$ is undefined.
   c. In the expansion of $(a + b)^8$, the exponent on a for the fifth term is 5.
   d. $n!$ is the sum of the first n positive numbers.
   e. There are n terms in the expansion of $(a + b)^n$.
   f. The sum of the exponents in each term of the expansion of $(a + b)^n$ is n.

52. Write a summary of the history of the development of Pascal's Triangle. Include some of the properties of the triangle.

Focus on Problem Solving

Forming Negations

A **statement** is a declarative sentence that is either true or false but not both true and false. Here are a few examples of statements.

>Denver is the capital of Colorado.
>The word "dog" has four letters.
>Florida is a state in the United States.
>In the year 2020, the President of the United States will be a woman.

Problem solving sometimes requires that you form the **negation** of a statement. If a statement is true, its negation must be false. If a statement is false, its negation must be true. For instance,

>"Henry Aaron broke Babe Ruth's home run record" is a true sentence. Accordingly, the sentence "Henry Aaron did not break Babe Ruth's home run record" is false.

>"Dogs have twelve paws" is a false sentence, so the sentence "Dogs do not have twelve paws" is a true sentence.

To form the negation of a statement requires a precise understanding of the statement being negated. Here are some phrases that demand special attention.

Statement	Meaning
At least seven	Greater than or equal to seven
More than seven	More than seven (does not include 7)
At most seven	Less than or equal to seven
Less than (or *fewer than*) seven	Less than seven (does not include 7)

Consider the sentence "At least three people scored over 90 on the last test." The negation of that sentence is "Fewer than three people scored over 90 on the last test."

For each of the following, write the negation of the given sentence.

1. There was at least one error on the test.
2. The temperature was less than 30 degrees.
3. The mountain is more than 5000 ft tall.
4. There are at most five vacancies for the field trip to New York.

When the word *all*, *no* (or *none*), or *some* occurs in a sentence, forming the negation requires careful thought. Consider the sentence "All roads are paved with cement." This sentence is not true because there are some roads—dirt roads, for example—that are not paved with cement. Because the sentence is false, its negation must be true. You might be tempted to write "All roads are not paved with cement" as the negation, but that sentence is not true because some roads are paved with cement. The correct negation is "*Some* roads are *not* paved with cement."

Now consider the sentence "Some fish live in aquariums." Because this sentence is true, the negation must be false. Writing "Some fish do not live in aquariums" as the negation is not correct because that sentence is true. The correct negation is "*All* fish do *not* live in aquariums."

The sentence "No houses have basements" is false because there is at least one house with a basement. Because this sentence is false, its negation must be true. The negation is "Some houses do have basements."

Statement	Negation
All A are B.	Some A are not B.
No A are B.	Some A are B.
Some A are B.	All A are not B.
Some A are not B.	All A are B.

Write the negation of each sentence.

5. All trees are tall.

6. All cats chase mice.

7. Some flowers have red blooms.

8. No golfers like tennis.

9. Some students do not like math.

10. No honest people are politicians.

11. All cars have power steering.

12. Some televisions are black and white.

Projects and Group Activities

ISBN and UPC Numbers   Every book that is cataloged for the Library of Congress must have an ISBN (International Standard Book Number). An ISBN is a 10-digit number of the form a_1-$a_2a_3a_4$-$a_5a_6a_7a_8a_9$-c. For instance, the ISBN for the Windows version of the CD-ROM containing the *American Heritage Children's Dictionary* is 0-395-73580-7. The first number, 0, indicates that the book is written in English. The next three numbers, 395, indicate the publisher (Houghton Mifflin Company). The next five numbers, 73580, identify the book (*American Heritage Children's Dictionary*), and the last digit, c, is called a *check digit*. This digit is chosen such that the following sum is divisible by 11.

$$a_1(10) + a_2(9) + a_3(8) + a_4(7) + a_5(6) + a_6(5) + a_7(4) + a_8(3) + a_9(2) + c$$

For the *American Heritage Children's Dictionary*,

$$0(10) + 3(9) + 9(8) + 5(7) + 7(6) + 3(5) + 5(4) + 8(3) + 0(2) + c$$
$$= 235 + c$$

The last digit of the ISBN is chosen as 7 because $235 + 7 = 242$ and $242 \div 11 = 22$. The value of c could be any number from 0 to 10. The number 10 is coded as an X.

One purpose of the ISBN method of coding books is to ensure that orders placed for books are accurately filled. For instance, suppose a clerk sends an order for the *American Heritage Children's Dictionary* and inadvertently enters the number 0-395-75380-7 (the 3 and 5 have been transposed). Now

$$0(10) + 3(9) + 9(8) + 5(7) + 7(6) + 5(5) + 3(4) + 8(3) + 0(2) + 7 = 244$$

and 244 is not divisible by 11. Thus an error in the ISBN has been made.

1. Determine the check digit for the book *Reader's Digest Book of Facts*. The first nine digits of the ISBN are 0-895-77692-?
2. Is 0-395-12370-4 a possible ISBN?
3. Check the ISBN of this intermediate algebra textbook.

Another coding scheme that is closely related to the ISBN is the UPC (Universal Product Code). This number is particularly useful in grocery stores. A checkout clerk passes the number by a scanner that reads the number and records the product and its price.

There are two major UPC codes: Type A consists of 12 digits and type E consists of 8 digits. For type A UPC symbols, the first six digits identify the manufacturer. The next five digits make up a product identification number given by the manufacturer to identify the product. The final digit is a check digit. The check digit is chosen such that the sum is divisible by 10.

$$a_1(13) + a_2 + a_3(13) + a_4 + a_5(13) + a_6 + a_7(13) + a_8 + a_9(13) + a_{10} + a_{11}(13) + c$$

The UPC for the TI-83 Plus scientific calculator is 0-33317-19865-8. Using the expression above,

$$\begin{aligned}0(13) + 3 + 3(13) + 3 + 1(13) + 7 + 1(13) + 9 + 8(13) + 6 + 5(13) + c \\ = 262 + c\end{aligned}$$

Because 262 + 8 is a multiple of 10, the check digit is 8.

4. What are the possible numbers for the check digit of a type A UPC?
5. Check the UPC of a product in your home.

Chapter Summary

Key Words A *sequence* is an ordered list of numbers. [p. 607]

Each of the numbers of a sequence is called a *term* of the sequence. [p. 607]

A *finite sequence* contains a finite number of terms. [p. 607]

An *infinite sequence* contains an infinite number of terms. [p. 607]

The indicated sum of the terms of a sequence is a *series*. [p. 609]

Summation notation, or *sigma notation*, is used to represent a series in compact form. In summation notation the Greek letter sigma, Σ, is used to indicate a sum. The variable used is called the *index* of the summation. [p. 609]

An *arithmetic sequence*, or *arithmetic progression*, is one in which the difference between any two consecutive terms is constant. The difference between consecutive terms is called the *common difference* of the sequence. [p. 613]

A *geometric sequence*, or *geometric progression*, is one in which each successive term of the sequence is the same nonzero constant multiple of the preceding term. The common multiple is called the *common ratio* of the sequence. [p. 619]

An *arithmetic series* is the indicated sum of the terms of an arithmetic sequence. [p. 615]

A *geometric series* is the indicated sum of the terms of a geometric sequence. [p. 621]

```
          1   1
        1   2   1
      1   3   3   1
    1   4   6   4   1
  1   5  10  10   5   1
1   6  15  20  15   6   1
```

The first six rows of the triangular array of numbers known as *Pascal's Triangle* are shown at the left. [p. 629]

n factorial, written $n!$, is the product of the first n natural numbers. [p. 630]

Essential Rules

Formula for the *n*th Term of an Arithmetic Sequence

The nth term of an arithmetic sequence with a common difference of d is given by $a_n = a_1 + (n - 1)d$. [p. 613]

Formula for the Sum of *n* Terms of an Arithmetic Series

Let a_1 be the first term of a finite arithmetic sequence, let n be the number of terms, and let a_n be the last term of the sequence. Then the sum of the series S_n is given by $S_n = \frac{n}{2}(a_1 + a_n)$. [p. 615]

Formula for the *n*th Term of a Geometric Sequence

The nth term of a geometric sequence with first term a_1 and common ratio r is given by $a_n = a_1 r^{n-1}$. [p. 620]

Formula for the Sum of *n* Terms of a Finite Geometric Series

Let a_1 be the first term of a finite geometric sequence, let n be the number of terms, and let r be the common ratio. Then the sum of the series S_n is given by $S_n = \frac{a_1(1 - r^n)}{1 - r}$. [p. 621]

Formula for the Sum of an Infinite Geometric Series

The sum of an infinite geometric series in which $|r| < 1$ and a_1 is the first term is given by $S = \frac{a_1}{1 - r}$. [p. 623]

Binomial Coefficients

The coefficients of $a^{n-r}b^r$ in the expansion of $(a + b)^n$ are given by $\binom{n}{r}$, where $\binom{n}{r} = \frac{n!}{(n - r)!\,r!}$. [p. 630]

The Binomial Expansion Formula

$$(a + b)^n = \binom{n}{0}a^n + \binom{n}{1}a^{n-1}b + \binom{n}{2}a^{n-2}b^2 + \cdots + \binom{n}{r}a^{n-r}b^r + \cdots + \binom{n}{n}b^n$$

[p. 631]

Formula for the *r*th Term in a Binomial Expansion

The rth term of $(a + b)^n$ is $\binom{n}{r - 1}a^{n-r+1}b^{r-1}$. [p. 631]

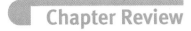

Chapter Review

1. Write $\displaystyle\sum_{i=1}^{4} 2x^{i-1}$ in expanded form.

2. Write the 10th term of the sequence whose nth term is given by the formula $a_n = 3n - 2$.

3. Find an equivalent fraction for $0.63\overline{3}$.

4. Find the sum of the infinite geometric series $\dfrac{3}{10} + \dfrac{3}{100} + \dfrac{3}{1000} + \cdots$.

5. Find the sum of the arithmetic series $\displaystyle\sum_{i=1}^{30}(4i - 1)$.

6. Find the sum of the series $\displaystyle\sum_{n=1}^{5}(3n - 2)$.

7. Find an equivalent fraction for $0.23\overline{23}$.

8. Evaluate: $\dbinom{9}{3}$

9. Find the sum of the geometric series $\displaystyle\sum_{n=1}^{5} 2(3)^n$.

10. Find the number of terms in the finite arithmetic sequence $8, 2, -4, \ldots, -118$.

11. Find the sum of the geometric series $\displaystyle\sum_{n=1}^{8}\left(\dfrac{1}{2}\right)^n$. Round to the nearest thousandth.

12. Find the 5th term in the expansion of $(3x - y)^8$.

13. Write the 5th term of the sequence whose nth term is given by the formula $a_n = \dfrac{(-1)^{2n-1} n}{n^2 + 2}$.

14. Find the sum of the first seven terms of the geometric sequence $5, 5\sqrt{5}, 25, \ldots$. Round to the nearest whole number.

15. Find the formula for the nth term of the arithmetic sequence $-7, -2, 3, \ldots$.

16. Find the 8th term in the expansion of $(x - 2y)^{11}$.

17. Find the 12th term of the geometric sequence 1, $\sqrt{3}$, 3, ….

18. Find the sum of the first 40 terms of the arithmetic sequence 11, 13, 15, ….

19. Evaluate: $\binom{12}{9}$

20. Find the sum of the series $\sum_{n=1}^{4} \dfrac{(-1)^{n-1}n}{n+1}$.

21. Write $\sum_{i=1}^{5} \dfrac{(2x)^i}{i}$ in expanded form.

22. Find the 8th term of the geometric sequence 3, 1, $\dfrac{1}{3}$, ….

23. Write $(x - 3y^2)^5$ in expanded form.

24. Find the sum of the infinite geometric series $4 - 1 + \dfrac{1}{4} - \cdots$.

25. Find the sum of the infinite geometric series $2 + \dfrac{4}{3} + \dfrac{8}{9} + \cdots$.

26. Evaluate: $\dfrac{12!}{5!\,8!}$

27. Find the 20th term of the arithmetic sequence 1, 5, 9, ….

28. The temperature of a hot water spa is 102°F. Each hour the temperature is 5% lower than during the preceding hour. Find the temperature of the spa after 8 h. Round to the nearest tenth.

29. The salary schedule for an apprentice electrician is $1200 for the first month and a $40-per-month salary increase for the next eight months. Find the total salary for the nine-month period.

30. A laboratory radioactive sample contains 200 mg of a material with a half-life of 1 h. Find the amount of radioactive material in the sample at the beginning of the seventh hour.

Chapter Test

1. Find the sum of the series $\displaystyle\sum_{n=1}^{4}(3n+1)$.

2. Evaluate: $\dbinom{9}{6}$

3. Find the 7th term of the geometric sequence $4, 4\sqrt{2}, 8, \ldots$.

4. Write the 14th term of the sequence whose nth term is given by the formula $a_n = \dfrac{8}{n+2}$.

5. Write in expanded form: $\displaystyle\sum_{i=1}^{4} 3x^i$

6. Find the sum of the first 18 terms of the arithmetic sequence $-25, -19, -13, \ldots$.

7. Find the sum of the terms of the infinite geometric sequence $4, 3, \dfrac{9}{4}, \ldots$.

8. Find the number of terms in the finite arithmetic sequence $-5, -8, -11, \ldots, -50$.

9. Find an equivalent fraction for $0.23\overline{33}$.

10. Evaluate: $\dfrac{8!}{4!\,4!}$

11. Find the 5th term of the geometric sequence $6, 2, \dfrac{2}{3}, \ldots$.

12. Find the 4th term in the expansion of $(x - 2y)^7$.

13. Find the 35th term of the arithmetic sequence $-13, -16, -19, \ldots$.

14. Find the formula for the nth term of the arithmetic sequence $12, 9, 6, \ldots$.

15. Find the sum of the first five terms of the geometric sequence $-6, 12, -24, \ldots$.

16. Find the 6th and 7th terms of the sequence whose nth term is given by the formula $a_n = \dfrac{n+1}{n}$.

17. Find the sum of the first six terms of the geometric sequence $1, \dfrac{3}{2}, \dfrac{9}{4}, \ldots$.

18. Find the sum of the first 21 terms of the arithmetic sequence $5, 12, 19, \ldots$.

19. An inventory of supplies for a fabric manufacturer indicated that 7500 yd of material were in stock on January 1. On February 1, and on the first of the month for each successive month, the manufacturer sent 550 yd of material to retail outlets. How much material was in stock after the shipment on October 1?

20. An ore sample contains 320 mg of a radioactive substance with a half-life of one day. Find the amount of radioactive material in the sample at the beginning of the fifth day.

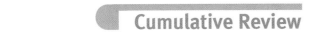

Cumulative Review

1. Graph: $3x - 2y = -4$

2. Factor: $2x^6 + 16$

3. Simplify: $\dfrac{4x^2}{x^2 + x - 2} - \dfrac{3x - 2}{x + 2}$

4. Given $f(x) = 2x^2 - 3x$, find $f(-2)$.

5. Simplify: $\sqrt{2y}(\sqrt{8xy} - \sqrt{y})$

6. Solve by using the quadratic formula:
$2x^2 - x + 7 = 0$

7. Solve: $5 - \sqrt{x} = \sqrt{x + 5}$

8. Find the equation of the circle that passes through the point $(4, 2)$ and whose center is $(-1, -1)$.

9. Solve by the addition method:
$3x - 3y = 2$
$6x - 4y = 5$

10. Evaluate the determinant:
$\begin{vmatrix} -3 & 1 \\ 4 & 2 \end{vmatrix}$

11. Solve: $2x - 1 > 3$ or $1 - 3x > 7$

12. Graph: $2x - 3y < 9$

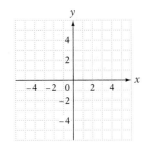

13. Write $\log_5 \sqrt{\dfrac{x}{y}}$ in expanded form.

14. Solve for x: $4^x = 8^{x-1}$

15. Write the 5th and 6th terms of the sequence whose nth term is given by the formula $a_n = n(n - 1)$.

16. Find the sum of the series $\displaystyle\sum_{n=1}^{7} (-1)^{n-1}(n + 2)$.

17. Find the 33rd term of the arithmetic sequence $-7, -10, -13, \ldots$.

18. Find the sum of the terms of the infinite geometric sequence $3, -2, \frac{4}{3}, \ldots$.

19. Find an equivalent fraction for $0.4\overline{6}$.

20. Find the 6th term in the expansion of $(2x + y)^6$.

21. How many ounces of pure water must be added to 200 oz of an 8% salt solution to make a 5% salt solution?

22. A new computer can complete a payroll in 16 min less time than it takes an older computer to complete the same payroll. Working together, both computers can complete the payroll in 15 min. How long would it take each computer working alone to complete the payroll?

23. A boat traveling with the current went 15 mi in 2 h. Against the current it took 3 h to travel the same distance. Find the rate of the boat in calm water and the rate of the current.

24. An 80-milligram sample of a radioactive material decays to 55 mg in 30 days. Use the exponential decay equation $A = A_0\left(\frac{1}{2}\right)^{t/k}$, where A is the amount of radioactive material present after time t, k is the half-life of the material, and A_0 is the original amount of radioactive material, to find the half-life of the 80-milligram sample. Round to the nearest whole number.

25. A theater-in-the-round has 62 seats in the first row, 74 seats in the second row, 86 seats in the third row, and so on in an arithmetic sequence. Find the total number of seats in the theater if there are 12 rows of seats.

26. To test the "bounce" of a ball, the ball is dropped from a height of 8 ft. The ball bounces to 80% of its previous height with each bounce. How high does the ball bounce on the fifth bounce? Round to the nearest tenth.

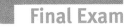

Final Exam

1. Simplify: $12 - 8[3 - (-2)]^2 \div 5 - 3$

2. Evaluate $\dfrac{a^2 - b^2}{a - b}$ when $a = 3$ and $b = -4$.

3. Simplify: $5 - 2[3x - 7(2 - x) - 5x]$

4. Solve: $\dfrac{3}{4}x - 2 = 4$

5. Solve: $\dfrac{2 - 4x}{3} - \dfrac{x - 6}{12} = \dfrac{5x - 2}{6}$

6. Solve: $8 - |5 - 3x| = 1$

7. Graph $2x - 3y = 9$ using the x- and y-intercepts.

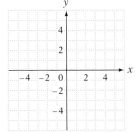

8. Find the equation of the line containing the points $(3, -2)$ and $(1, 4)$.

9. Find the equation of the line that contains the point $(-2, 1)$ and is perpendicular to the line $3x - 2y = 6$.

10. Simplify: $2a[5 - a(2 - 3a) - 2a] + 3a^2$

11. Factor: $8 - x^3y^3$

12. Factor: $x - y - x^3 + x^2y$

13. Simplify: $(2x^3 - 7x^2 + 4) \div (2x - 3)$

14. Simplify: $\dfrac{x^2 - 3x}{2x^2 - 3x - 5} \div \dfrac{4x - 12}{4x^2 - 4}$

15. Simplify: $\dfrac{x - 2}{x + 2} - \dfrac{x + 3}{x - 3}$

16. Simplify: $\dfrac{\dfrac{3}{x} + \dfrac{1}{x + 4}}{\dfrac{1}{x} + \dfrac{3}{x + 4}}$

17. Solve: $\dfrac{5}{x - 2} - \dfrac{5}{x^2 - 4} = \dfrac{1}{x + 2}$

18. Solve $a_n = a_1 + (n - 1)d$ for d.

19. Simplify: $\left(\dfrac{4x^2y^{-1}}{3x^{-1}y}\right)^{-2}\left(\dfrac{2x^{-1}y^2}{9x^{-2}y^2}\right)^3$

20. Simplify: $\left(\dfrac{3x^{2/3}y^{1/2}}{6x^2y^{4/3}}\right)^6$

21. Simplify: $x\sqrt{18x^2y^3} - y\sqrt{50x^4y}$

22. Simplify: $\dfrac{\sqrt{16x^5y^4}}{\sqrt{32xy^7}}$

23. Simplify: $\dfrac{3}{2+i}$

24. Write a quadratic equation that has integer coefficients and has solutions $-\dfrac{1}{2}$ and 2.

25. Solve by using the quadratic formula:
$2x^2 - 3x - 1 = 0$

26. Solve: $x^{2/3} - x^{1/3} - 6 = 0$

27. Solve: $\dfrac{2}{x} - \dfrac{2}{2x+3} = 1$

28. Graph: $f(x) = -x^2 + 4$

29. Graph: $\dfrac{x^2}{16} + \dfrac{y^2}{4} = 1$

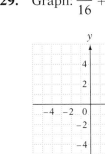

30. Find the inverse of the function
$$f(x) = \dfrac{2}{3}x - 4.$$

31. Solve by the addition method:
$3x - 2y = 1$
$5x - 3y = 3$

32. Evaluate the determinant:
$$\begin{vmatrix} 3 & 4 \\ -1 & 2 \end{vmatrix}$$

33. Solve: $x^2 - y^2 = 4$
$x + y = 1$

34. Solve: $2 - 3x < 6$ and $2x + 1 > 4$

35. Solve: $|2x + 5| < 3$

36. Graph the solution set: $3x + 2y > 6$

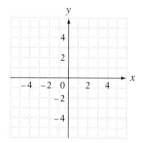

37. Graph: $f(x) = \log_2(x + 1)$

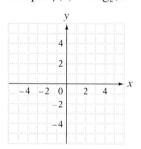

38. Write $2(\log_2 a - \log_2 b)$ as a single logarithm with a coefficient of 1.

39. Solve for x: $\log_3 x - \log_3(x - 3) = \log_3 2$

40. Write $\sum_{i=1}^{5} 2y^i$ in expanded form.

41. Find an equivalent fraction for $0.5\overline{1}$.

42. Find the 3rd term in the expansion of $(x - 2y)^9$.

43. An average score of 70–79 in a history class receives a C grade. A student has scores of 64, 58, 82, and 77 on four history tests. Find the range of scores on the fifth test that will give the student a C grade for the course.

44. A jogger and a cyclist set out at 8 A.M. from the same point headed in the same direction. The average speed of the cyclist is two and a half times the average speed of the jogger. In 2 h, the cyclist is 24 mi ahead of the jogger. How far did the cyclist ride in that time?

45. You have a total of $12,000 invested in two simple interest accounts. On one account, a money market fund, the annual simple interest rate is 8.5%. On the other account, a tax free bond fund, the annual simple interest rate is 6.4%. The total annual interest earned by the two accounts is $936. How much do you have invested in each account?

46. The length of a rectangle is 1 ft less than three times the width. The area of the rectangle is 140 ft². Find the length and width of the rectangle.

47. Three hundred shares of a utility stock earn a yearly dividend of $486. How many additional shares of the utility stock would give a total dividend income of $810?

48. An account executive traveled 45 mi by car and then an additional 1050 mi by plane. The rate of the plane was seven times the rate of the car. The total time for the trip was $3\frac{1}{4}$ h. Find the rate of the plane.

49. An object is dropped from the top of a building. Find the distance the object has fallen when its speed reaches 75 ft/s. Use the equation $v = \sqrt{64d}$, where v is the speed of the object and d is the distance in feet. Round to the nearest whole number.

50. A small plane made a trip of 660 mi in 5 h. The plane traveled the first 360 mi at a constant rate before increasing its speed by 30 mph. Then it traveled another 300 mi at the increased speed. Find the rate of the plane for the first 360 mi.

51. The intensity, L, of a light source is inversely proportional to the square of the distance, d, from the source. If the intensity is 8 foot-candles at a distance of 20 ft, what is the intensity when the distance is 4 ft?

52. A motorboat traveling with the current can go 30 mi in 2 h. Against the current, it takes 3 h to go the same distance. Find the rate of the motorboat in calm water and the rate of the current.

53. An investor deposits $4000 into an account that earns 9% annual interest compounded monthly. Use the compound interest formula $P = A(1 + i)^n$, where A is the original value of the investment, i is the interest rate per compounding period, n is the total number of compounding periods, and P is the value of the investment after n periods, to find the value of the investment after 2 years. Round to the nearest cent.

54. Assume that the average value of a home increases 6% per year. How much will a house costing $80,000 be worth in 20 years? Round to the nearest dollar.

Appendix

Guidelines for Using Graphing Calculators

Texas Instruments TI-83 and TI-83Plus

To evaluate an expression

a. Press the $\boxed{Y=}$ key. A menu showing \Y1 = through \Y7 = will be displayed vertically with a blinking cursor to the right of \Y1 =. Press $\boxed{\text{CLEAR}}$, if necessary, to delete an unwanted expression.

b. Input the expression to be evaluated. For example, to input the expression $-3a^2b - 4c$, use the following keystrokes:

$\boxed{(-)}$ 3 $\boxed{\text{ALPHA}}$ A $\boxed{X^2}$ $\boxed{\text{ALPHA}}$ B $\boxed{-}$ 4 $\boxed{\text{ALPHA}}$ C $\boxed{\text{2nd}}$ QUIT

Note the difference between the keys for a negative sign $\boxed{(-)}$ and a minus sign $\boxed{-}$.

c. Store the value of each variable that will be used in the expression. For example, to evaluate the expression above when $a = 3$, $b = -2$, and $c = -4$, use the following keystrokes:

3 $\boxed{\text{STO}\triangleright}$ $\boxed{\text{ALPHA}}$ A $\boxed{\text{ENTER}}$ $\boxed{(-)}$ 2 $\boxed{\text{STO}\triangleright}$ $\boxed{\text{ALPHA}}$ B $\boxed{\text{ENTER}}$ $\boxed{(-)}$ 4 $\boxed{\text{STO}\triangleright}$ $\boxed{\text{ALPHA}}$ C $\boxed{\text{ENTER}}$

These steps store the value of each variable.

d. Press $\boxed{\text{VARS}}$ $\boxed{\triangleright}$ $\boxed{1}$ $\boxed{1}$ $\boxed{\text{ENTER}}$. The value for the expression, Y1, for the given values is displayed; in this case, Y1 = 70.

To graph a function

a. Press the $\boxed{Y=}$ key. A menu showing \Y1 = through \Y7 = will be displayed vertically with a blinking cursor to the right of \Y1 =. Press $\boxed{\text{CLEAR}}$, if necessary, to delete an unwanted expression.

b. Input the expression for each function that is to be graphed. Press $\boxed{\text{X,T,}\theta\text{,}n}$ to input x. For example, to input $y = x^3 + 2x^2 - 5x - 6$, use the following keystrokes:

$\boxed{\text{X,T,}\theta\text{,}n}$ $\boxed{\wedge}$ 3 $\boxed{+}$ 2 $\boxed{\text{X,T,}\theta\text{,}n}$ $\boxed{X^2}$ $\boxed{-}$ 5 $\boxed{\text{X,T,}\theta\text{,}n}$ $\boxed{-}$ 6

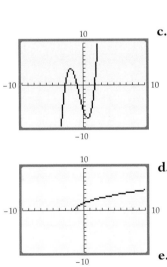

c. Set the viewing window by pressing $\boxed{\text{WINDOW}}$. Enter the values for the minimum x-value (Xmin), the maximum x-value (Xmax), the distance between tick marks on the x-axis (Xscl), the minimum y-value (Ymin), the maximum y-value (Ymax), and the distance between tick marks on the y-axis (Yscl). Now press $\boxed{\text{GRAPH}}$. For the graph shown at the left, Xmin = -10, Xmax = 10, Xscl = 1, Ymin = -10, Ymax = 10, and Yscl = 1. This is called the standard viewing window. Pressing $\boxed{\text{ZOOM}}$ $\boxed{6}$ is a quick way to set the calculator to the standard viewing window. *Note:* This will also immediately graph the function in that window.

d. Press the $\boxed{Y=}$ key. The equal sign has a black rectangle around it. This indicates that the function is active and will be graphed when the $\boxed{\text{GRAPH}}$ key is pressed. A function is deactivated by using the arrow keys. Move the cursor over the equal sign and press $\boxed{\text{ENTER}}$. When the cursor is moved to the right, the black rectangle will not be present and that equation will not be active.

e. Graphing some radical equations requires special care. To graph the equation $y = \sqrt{2x + 3}$ shown at the left, enter the following keystrokes:

$\boxed{Y=}$ $\boxed{\text{CLEAR}}$ $\boxed{\text{2nd}}$ $\sqrt{}$ 2 $\boxed{\text{X,T,}\theta\text{,}n}$ $\boxed{+}$ 3 $\boxed{)}$ $\boxed{\text{GRAPH}}$

651

To display the *x*-coordinates of rectangular coordinates as integers

a. Set the viewing window as follows: Xmin = −47, Xmax = 47, Xscl = 10, Ymin = −31, Ymax = 31, Yscl = 10.

b. Graph the function and use the TRACE feature. Press $\boxed{\text{TRACE}}$ and then move the cursor with the $\boxed{\triangleleft}$ and $\boxed{\triangleright}$ keys. The values of *x* and *y* = *f*(*x*) displayed on the bottom of the screen are the coordinates of a point on the graph.

To display the *x*-coordinates of rectangular coordinates in tenths

a. Set the viewing window as follows: $\boxed{\text{ZOOM}}$ $\boxed{4}$

b. Graph the function. Press $\boxed{\text{TRACE}}$ and then move the cursor with the $\boxed{\triangleleft}$ and $\boxed{\triangleright}$ keys. The values of *x* and *y* = *f*(*x*) displayed on the bottom of the screen are the coordinates of a point on the graph.

To evaluate a function for a given value of *x*, or to produce ordered pairs of a function

a. Input the equation; for example, input $Y_1 = 2x^3 - 3x + 2$.

b. Press $\boxed{\text{2nd}}$ QUIT.

c. To evaluate the function when *x* = 3, press $\boxed{\text{VARS}}$ $\boxed{\triangleright}$ $\boxed{1}$ $\boxed{1}$ $\boxed{(}$ $\boxed{3}$ $\boxed{)}$ $\boxed{\text{ENTER}}$. The value for the function for the given *x*-value is displayed, in this case, $Y_1(3) = 47$. An ordered pair of the function is (3, 47).

d. Repeat step **c.** to produce as many pairs as desired.

The TABLE feature can also be used to determine ordered pairs.

To use a table

a. Press $\boxed{\text{2nd}}$ TBLSET to activate the table setup menu.

b. TblStart is the beginning number for the table; ΔTbl is the difference between any two successive *x*-values in the table.

c. The portion of the table that appears as Indpnt: ⬛Auto Ask Depend: ⬛Auto Ask allows you to choose between having the calculator automatically produce the results (Auto) or having the calculator ask you for values of *x*. You can choose Ask by using the arrow keys.

d. Once a table has been set up, enter an expression for Y_1. Now select TABLE by pressing $\boxed{\text{2nd}}$ TABLE. A table showing ordered pair solutions of the equation will be displayed on the screen.

Zoom Features

To zoom in or out on a graph

Here are two methods of using ZOOM.

a. The first method uses the built-in features of the calculator. Press $\boxed{\text{TRACE}}$ and then move the cursor to a point on the graph that is of interest. Press $\boxed{\text{ZOOM}}$. The ZOOM menu will appear. Press $\boxed{2}$ $\boxed{\text{ENTER}}$ to zoom in on the graph by the amount shown under the SET FACTORS menu. The center of the new graph is the location at which you placed the cursor. Press $\boxed{\text{ZOOM}}$ $\boxed{3}$ $\boxed{\text{ENTER}}$ to zoom out on the graph by the amount under the SET FACTORS menu. (The SET FACTORS menu is accessed by pressing $\boxed{\text{ZOOM}}$ $\boxed{\triangleright}$ $\boxed{4}$.)

b. The second method uses the ZBOX option under the ZOOM menu. To use this method, press $\boxed{\text{ZOOM}}$ $\boxed{1}$. A cursor will appear on the graph. Use the arrow keys to move the cursor to a portion of the graph that is of interest. Press $\boxed{\text{ENTER}}$. Now use the arrow keys to draw a box around the portion of the graph you wish to zoom in on. Press ENTER. The portion of the graph defined by the box will be drawn.

c. Pressing ZOOM 6 resets the window to the standard viewing window.

Solving Equations

This discussion is based on the fact that the real number solutions of an equation are related to the x-intercepts of a graph. For instance, the real solutions of the equation $x^2 = x + 1$ are the x-intercepts of the graph of $f(x) = x^2 - x - 1$, which are the zeros of f.

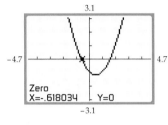

To solve $x^2 = x + 1$, rewrite the equation with all terms on one side: $x^2 - x - 1 = 0$. Think of this equation as $Y_1 = x^2 - x - 1$. The x-intercepts of the graph of Y_1 are the solutions of the equation $x^2 = x + 1$.

a. Enter $x^2 - x - 1$ into Y_1.

b. Graph the equation. You may need to adjust the viewing window so that the x-intercepts are visible.

c. Press 2nd CALC 2.

d. Move the cursor to a point on the curve that is to the left of an x-intercept. Press ENTER.

e. Move the cursor to a point on the curve that is to the right of the same x-intercept. Press ENTER ENTER.

f. The root is shown as the x-coordinate on the bottom of the screen; in this case, the root is approximately -0.618034. To find the next intercept, repeat steps **c.** through **e.** The SOLVER feature under the MATH menu can also be used to find solutions of equations.

Solving Systems of Equations in Two Variables

To solve a system of equations

To solve
$$y = x^2 - 1$$
$$\frac{1}{2}x + y = 1 \quad ,$$

a. If necessary, solve one or both of the equations for y.

b. Enter the first equation as Y_1: $Y_1 = x^2 - 1$.

c. Enter the second equation as Y_2: $Y_2 = 1 - .5x$.

d. Graph both equations. (*Note:* The point of intersection must appear on the screen. It may be necessary to adjust the viewing window so that the points of intersection are displayed.)

e. Press 2nd CALC 5.

f. Move the cursor close to the first point of intersection. Press ENTER ENTER ENTER.

g. The first point of intersection is $(-1.686141, 1.8430703)$.

h. Repeat steps **e.** and **f.** for each point of intersection.

Finding Minimum or Maximum Values of a Function

a. Enter the function into Y_1. The equation $y = x^2 - x - 1$ is used here.

b. Graph the equation. You may need to adjust the viewing window so that the maximum or minimum points are visible.

c. Press [2nd] CALC [3] to determine a minimum value or press [2nd] CALC [4] to determine a maximum value.

d. Move the cursor to a point on the curve that is to the left of the minimum (maximum). Press [ENTER].

e. Move the cursor to a point on the curve that is to the right of the minimum (maximum). Press [ENTER] [ENTER].

f. The minimum (maximum) is shown as the y-coordinate on the bottom of the screen; in this case the minimum value is -1.25.

Statistics

To calculate a linear regression equation

a. Press [STAT] to access the statistics menu. Press 1 to Edit or enter a new list of data. To delete data already in a list, press the up arrow to highlight the list name. For instance, to delete data in L_1, highlight L_1. Then press [CLEAR] and [ENTER]. Now enter each value of the independent variable in L_1. Enter each value of the dependent variable in L_2.

b. When all the data has been entered, press [STAT] [▷] 4 [ENTER]. The values of the slope and y-intercept of the linear regression equation will be displayed on the screen.

c. To graph the linear regression equation, modify step **b.** with these keystrokes: [STAT] [▷] 4 [VARS] [▷] 1 1 [ENTER]. This will store the regression equation in Y_1. Now press [GRAPH]. It may be necessary to adjust the viewing window.

d. To evaluate the regression equation for a value of x, complete step **c.** but do not graph the equation. Now press [VARS] [▷] 1 1 [(] A [)] [ENTER], where A is replaced by the number at which you want to evaluate the expression.

Proofs and Tables

Proofs of Logarithmic Properties

In each of the following proofs of logarithmic properties, it is assumed that the Properties of Exponents are true for all real number exponents.

The Logarithm Property of the Product of Two Numbers

For any positive real numbers x, y, and b, $b \neq 1$, $\log_b xy = \log_b x + \log_b y$.

Proof: Let $\log_b x = m$ and $\log_b y = n$.

Write each equation in its equivalent exponential form. $x = b^m \qquad y = b^n$

Use substitution and the Properties of Exponents. $xy = b^m b^n$

$xy = b^{m+n}$

Write the equation in its equivalent logarithmic form. $\log_b xy = m + n$

Substitute $\log_b x$ for m and $\log_b y$ for n. $\log_b xy = \log_b x + \log_b y$

The Logarithm Property of the Quotient of Two Numbers

For any positive real numbers x, y, and b, $b \neq 1$, $\log_b \dfrac{x}{y} = \log_b x - \log_b y$.

Proof: Let $\log_b x = m$ and $\log_b y = n$.

Write each equation in its equivalent exponential form. $x = b^m \qquad y = b^n$

Use substitution and the Properties of Exponents. $\dfrac{x}{y} = \dfrac{b^m}{b^n}$

$\dfrac{x}{y} = b^{m-n}$

Write the equation in its equivalent logarithmic form. $\log_b \dfrac{x}{y} = m - n$

Substitute $\log_b x$ for m and $\log_b y$ for n. $\log_b \dfrac{x}{y} = \log_b x - \log_b y$

The Logarithm Property of the Power of a Number

For any real numbers x, r, and b, $b \neq 1$, $\log_b x^r = r \log_b x$.

Proof: Let $\log_b x = m$.

Write the equation in its equivalent exponential form. $x = b^m$

Raise both sides to the r power. $x^r = (b^m)^r$

$x^r = b^{mr}$

Write the equation in its equivalent logarithmic form. $\log_b x^r = mr$

Substitute $\log_b x$ for m. $\log_b x^r = r \log_b x$

Proof of the Formula for the Sum of *n* Terms of a Geometric Series

Theorem: The sum of the first n terms of a geometric sequence whose nth term is ar^{n-1} is given by $S_n = \dfrac{a(1 - r^n)}{1 - r}$.

Proof: Let S_n represent the sum of n terms of the sequence. $S_n = a + ar + ar^2 + \cdots + ar^{n-2} + ar^{n-1}$

Multiply each side of the equation by r. $rS_n = ar + ar^2 + ar^3 + \cdots + ar^{n-1} + ar^n$

Subtract the two equations. $S_n - rS_n = a - ar^n$

Assuming $r \neq 1$, solve for S_n. $(1 - r)S_n = a(1 - r^n)$

$S_n = \dfrac{a(1 - r^n)}{1 - r}$

Proof of the Formula for the Sum of *n* Terms of an Arithmetic Series

Each term of the arithmetic sequence shown at the right was found by adding 3 to the previous term.

2, 5, 8,...,17, 20

Each term of the reverse arithmetic sequence can be found by subtracting 3 from the previous term.

20, 17, 14,....,5, 2

This idea is used in the following proof.

Theorem: The sum of the first *n* terms of an arithmetic sequence for which a_1 is the first term, *n* is the number of terms, a_n is the last term, and *d* is the common difference is given by $S_n = \dfrac{n}{2}(a_1 + a_n)$.

Proof: Let S_n represent the sum of the sequence.

$$S_n = a_1 + (a_1 + d) + (a_1 + 2d) + \cdots + a_n$$

Write the terms of the sum of the sequence in reverse order. The sum will be the same.

$$S_n = a_n + (a_n - d) + (a_n - 2d) + \cdots + a_1$$

Add the two equations.

$$2S_n = (a_1 + a_n) + (a_1 + a_n) + (a_1 + a_n) + \cdots + (a_1 + a_n)$$

Simplify the right side of the equation by using the fact that there are *n* terms in the sequence.

$$2S_n = n(a_1 + a_n)$$

Solve for S_n.

$$S_n = \dfrac{n}{2}(a_1 + a_n)$$

Table of Symbols

Symbol	Meaning
$+$	add
$-$	subtract
$\cdot, \times, (a)(b)$	multiply
$\dfrac{a}{b}, \div$	divide
$(\ )$	parentheses, a grouping symbol
$[\ ]$	brackets, a grouping symbol
π	pi, a number approximately equal to $\dfrac{22}{7}$ or 3.14
$-a$	the opposite, or additive inverse, of *a*
$\dfrac{1}{a}$	the reciprocal, or multiplicative inverse, of *a*
$=$	is equal to
$\approx$	is approximately equal to
$\neq$	is not equal to

Symbol	Meaning
$<$	is less than
$\leq$	is less than or equal to
$>$	is greater than
$\geq$	is greater than or equal to
(a, b)	an ordered pair whose first component is *a* and whose second component is *b*
$\circ$	degree (for angles)
$\sqrt{a}$	the principal square root of *a*
$\varnothing, \{\ \}$	the empty set
$|a|$	the absolute value of *a*
$\cup$	union of two sets
$\cap$	intersection of two sets
$\in$	is an element of (for sets)
$\notin$	is not an element of (for sets)

Table of Measurement Abbreviations

U.S. Customary System

Length

in.	inches
ft	feet
yd	yards
mi	miles

Capacity

oz	fluid ounces
c	cups
qt	quarts
gal	gallons

Weight

oz	ounces
lb	pounds

Area

in^2	square inches
ft^2	square feet

Metric System

Length

mm	millimeter (0.001 m)
cm	centimeter (0.01 m)
dm	decimeter (0.1 m)
m	meter
dam	decameter (10 m)
hm	hectometer (100 m)
km	kilometer (1000 m)

Capacity

ml	milliliter (0.001 L)
cl	centiliter (0.01 L)
dl	deciliter (0.1 L)
L	liter
dal	decaliter (10 L)
hl	hectoliter (100 L)
kl	kiloliter (1000 L)

Weight/Mass

mg	milligram (0.001 g)
cg	centigram (0.01 g)
dg	decigram (0.1 g)
g	gram
dag	decagram (10 g)
hg	hectogram (100 g)
kg	kilogram (1000 g)

Area

cm^2	square centimeters
m^2	square meters

Time

h	hours	min	minutes
s	seconds		

Solutions to Chapter 1 "You Try It"

SECTION 1.1

You Try It 1 Replace z by each of the elements of the set and determine whether the inequality is true.

$z > -5$
$-10 > -5$ False
$-5 > -5$ False
$6 > -5$ True
The inequality is true for 6.

You Try It 2 **a.** Replace d in $-d$ by each element of the set and determine the value of the expression.

$-d$
$-(-11) = 11$
$-(0) = 0$
$-(8) = -8$

b. Replace d in $|d|$ by each element of the set and determine the value of the expression.

$|d|$
$|-11| = 11$
$|0| = 0$
$|8| = 8$

You Try It 3 $\{-5, -4, -3, -2, -1\}$

You Try It 4 $\{x \mid x \geq 15, x \in \text{whole numbers}\}$

You Try It 5 $C \cup D = \{1, 3, 5, 7, 9, 11, 13, 17\}$

You Try It 6 $E \cap F = \varnothing$

You Try It 7 Draw a parenthesis at 0 to indicate that 0 is not in the set. Draw a solid line to the left of 0.

You Try It 8 This is the set of real numbers less than -2 *or* greater than -1.

You Try It 9 This is the set of real numbers less than 1 *and* greater than -3.

You Try It 10 **a.** The set $\{x \mid x < -1\}$ is the numbers less than -1. In interval notation, this is written $(-\infty, -1)$.

b. The set $\{x \mid -2 \leq x < 4\}$ is the numbers greater than or equal to -2 and less than 4. In interval notation, this is written $[-2, 4)$.

You Try It 11 **a.** The ∞ symbol indicates that the interval extends forever. Therefore, $(3, \infty)$ is the numbers greater than 3. In set-builder notation, this is written $\{x \mid x > 3\}$.

b. $(-4, 1]$ is the numbers greater than -4 and less than or equal to 1. In set-builder notation, this is written $\{x \mid -4 < x \leq 1\}$.

You Try It 12 Draw a bracket at -2 to show that it is in the set. The ∞ symbol indicates that the set extends forever. Draw a solid line to the right of -2.

SECTION 1.2

You Try It 1 $6 - (-8) - 10 = 6 + 8 + (-10)$
$= 14 + (-10)$
$= 4$

You Try It 2 $12(-3)|-6| = 12(-3)(6)$
$= (-36)(6)$
$= -216$

You Try It 3 $-\dfrac{28}{|-14|} = -\dfrac{28}{14}$
$= -2$

You Try It 4

$\dfrac{5}{6} - \dfrac{3}{8} - \dfrac{7}{9} = \dfrac{5}{6} + \dfrac{-3}{8} + \dfrac{-7}{9}$

$= \dfrac{5}{6} \cdot \dfrac{12}{12} + \dfrac{-3}{8} \cdot \dfrac{9}{9} + \dfrac{-7}{9} \cdot \dfrac{8}{8}$

$= \dfrac{60}{72} + \dfrac{-27}{72} + \dfrac{-56}{72}$

$= \dfrac{60 + (-27) + (-56)}{72} = \dfrac{-23}{72}$

$= -\dfrac{23}{72}$

You Try It 5

$$\frac{5}{8} \div \left(-\frac{15}{40}\right) = -\frac{5}{8} \cdot \frac{40}{15}$$
$$= -\frac{5 \cdot 40}{8 \cdot 15} = -\frac{5}{3}$$

You Try It 6 $-2^5 = -(2 \cdot 2 \cdot 2 \cdot 2 \cdot 2) = -32$
$$(-5)^2 = (-5)(-5) = 25$$

You Try It 7

$$\left(\frac{2}{3}\right)^4 = \left(\frac{2}{3}\right)\left(\frac{2}{3}\right)\left(\frac{2}{3}\right)\left(\frac{2}{3}\right)$$
$$= \frac{2 \cdot 2 \cdot 2 \cdot 2}{3 \cdot 3 \cdot 3 \cdot 3}$$
$$= \frac{16}{81}$$

You Try It 8

$$-2\left[3 - 3^3 + \frac{16 + 2}{1 - 10}\right] = -2\left[3 - 3^3 + \frac{18}{-9}\right]$$
$$= -2[3 - 3^3 + (-2)]$$
$$= -2[3 - 27 + (-2)]$$
$$= -2[-24 + (-2)]$$
$$= -2[-26]$$
$$= 52$$

You Try It 9

$$\frac{1}{3} + \frac{5}{8} \div \frac{15}{16} - \frac{7}{12} = \frac{1}{3} + \frac{5}{8} \cdot \frac{16}{15} - \frac{7}{12}$$
$$= \frac{1}{3} + \frac{2}{3} - \frac{7}{12}$$
$$= 1 - \frac{7}{12}$$
$$= \frac{5}{12}$$

You Try It 10

$$\frac{11}{12} - \frac{\dfrac{5}{4}}{2 - \dfrac{7}{2}} \cdot \frac{3}{4} = \frac{11}{12} - \frac{\dfrac{5}{4}}{-\dfrac{3}{2}} \cdot \frac{3}{4}$$
$$= \frac{11}{12} - \left[\frac{5}{4} \cdot \left(-\frac{2}{3}\right)\right] \cdot \frac{3}{4}$$
$$= \frac{11}{12} - \left(-\frac{5}{6}\right) \cdot \frac{3}{4}$$
$$= \frac{11}{12} - \left(-\frac{5}{8}\right)$$
$$= \frac{37}{24}$$

SECTION 1.3

You Try It 1 $3x + (-3x) = 0$

You Try It 2   The Associative Property of Addition

You Try It 3
$-2x^2 + 3(4xy - z)$
$-2(-3)^2 + 3[4(-3)(-1) - 2]$
$$= -2(-3)^2 + 3[12 - 2]$$
$$= -2(-3)^2 + 3(10)$$
$$= -2(9) + 3(10)$$
$$= -18 + 30$$
$$= 12$$

You Try It 4
$2x - y|4x^2 - y^2|$
$2(-2) - (-6)|4(-2)^2 - (-6)^2|$
$$= 2(-2) - (-6)|4 \cdot 4 - 36|$$
$$= 2(-2) - (-6)|16 - 36|$$
$$= 2(-2) - (-6)|-20|$$
$$= 2(-2) - (-6)(20)$$
$$= -4 - (-120)$$
$$= -4 + 120$$
$$= 116$$

You Try It 5
$9 - 2(3y - 4) + 5y = 9 - 6y + 8 + 5y$
$$= -y + 17$$

You Try It 6
$6z - 3(5 - 3z) + 5(2z - 3)$
$$= 6z - 15 + 9z + 10z - 15$$
$$= 25z - 30$$

SECTION 1.4

You Try It 1   <u>twice</u> x <u>divided by</u> the <u>difference</u> <u>between</u> x and 7

$$\frac{2x}{x - 7}$$

You Try It 2   the <u>product</u> of negative 3 and the <u>square of</u> d

$$-3d^2$$

You Try It 3   The smaller number is x.
The larger number is $16 - x$.
the <u>difference between twice the</u> smaller number and the larger number
$2x - (16 - x) = 2x - 16 + x$
$$= 3x - 16$$

You Try It 4   Let the number be x.
the <u>difference between</u> 14 and the <u>sum of</u> a number and 7
$14 - (x + 7) = 14 - x - 7$
$$= -x + 7$$

You Try It 5   the height of the Empire State
Building: x
the length of *Destiny*: $x + 56$

You Try It 7   the total credit card bill: x

the cost of restaurant meals: $\dfrac{1}{4}x$

You Try It 6   the depth of the shallow end: x
the depth of the deep end: $2x + 2$

Solutions to Chapter 2 "You Try It"

SECTION 2.1

You Try It 1

$$x + 4 = -3$$
$$x + 4 - 4 = -3 - 4$$
$$x = -7$$

The solution is -7.

You Try It 2

$$-3x = 18$$
$$\frac{-3x}{-3} = \frac{18}{-3}$$
$$x = -6$$

The solution is -6.

You Try It 3

$$6x - 5 - 3x = 14 - 5x$$
$$3x - 5 = 14 - 5x$$
$$3x - 5 + 5x = 14 - 5x + 5x$$
$$8x - 5 = 14$$
$$8x - 5 + 5 = 14 + 5$$
$$8x = 19$$
$$\frac{8x}{8} = \frac{19}{8}$$
$$x = \frac{19}{8}$$

The solution is $\dfrac{19}{8}$.

You Try It 4

$$6(5 - x) - 12 = 2x - 3(4 + x)$$
$$30 - 6x - 12 = 2x - 12 - 3x$$
$$18 - 6x = -x - 12$$
$$18 - 6x + x = -x - 12 + x$$
$$18 - 5x = -12$$
$$18 - 5x - 18 = -12 - 18$$
$$-5x = -30$$
$$\frac{-5x}{-5} = \frac{-30}{-5}$$
$$x = 6$$

The solution is 6.

You Try It 5

$$S = C - rC$$
$$S - C = -rC$$
$$\frac{S - C}{-C} = r$$
$$r = -\frac{S - C}{C} = \frac{C - S}{C}$$

SECTION 2.2

You Try It 1

Strategy   • The first number: n
The second number: $2n$
The third number: $4n - 3$
• The sum of the numbers is 81.
$n + 2n + (4n - 3) = 81$

Solution $n + 2n + (4n - 3) = 81$
$$7n - 3 = 81$$
$$7n = 84$$
$$n = 12$$
$$2n = 2(12) = 24$$
$$4n - 3 = 4(12) - 3 = 48 - 3 = 45$$

The numbers are 12, 24, and 45.

You Try It 2

Strategy   • First odd integer: n
Second odd integer: $n + 2$
Third odd integer: $n + 4$
• Three times the sum of the first
two integers is ten more than
the product of the third integer
and four.
$3[n + (n + 2)] = (n + 4)4 + 10$

Solution $3[n + (n + 2)] = (n + 4)4 + 10$
$$3[2n + 2] = 4n + 16 + 10$$
$$6n + 6 = 4n + 26$$
$$2n + 6 = 26$$
$$2n = 20$$
$$n = 10$$

Because 10 is not an odd integer,
there is no solution.

Strategy • Number of 3¢ stamps: x
Number of 10¢ stamps: $2x + 2$
Number of 15¢ stamps: $3x$

Stamps	Number	Value	Total Value
3¢	x	3	$3x$
10¢	$2x + 2$	10	$10(2x + 2)$
15¢	$3x$	15	$45x$

• The sum of the total values of each
type of stamp equals the total value
of all the stamps (156 cents).
$3x + 10(2x + 2) + 45x = 156$

Solution
$$3x + 10(2x + 2) + 45x = 156$$
$$3x + 20x + 20 + 45x = 156$$
$$68x + 20 = 156$$
$$68x = 136$$
$$x = 2$$
$$3x = 3(2) = 6$$

There are six 15¢ stamps in the
collection.

SECTION 2.3

You Try It 1

Strategy • Pounds of $3.00 hamburger: x
Pounds of $1.80 hamburger: $75 - x$
Pounds of the $2.20 mixture: 75

	Amount	Cost	Value
$3.00 hamburger	x	3.00	$3.00x$
$1.80 hamburger	$75 - x$	1.80	$1.80(75 - x)$
Mixture	75	2.20	$75(2.20)$

• The sum of the values before mixing
equals the value after mixing.
$3.00x + 1.80(75 - x) = 75(2.20)$

Solution
$$3.00x + 1.80(75 - x) = 75(2.20)$$
$$3x + 135 - 1.80x = 165$$
$$1.2x + 135 = 165$$
$$1.2x = 30$$
$$x = 25$$
$$75 - x = 75 - 25 = 50$$

The mixture must contain 25 lb of
the $3.00 hamburger and 50 lb of the
$1.80 hamburger.

You Try It 2

Strategy • Pounds of 22% hamburger: x
Pounds of 12% hamburger: $80 - x$
Pounds of 18% hamburger: 80

	Amount	Percent	Quantity
22%	x	0.22	$0.22x$
12%	$80 - x$	0.12	$0.12(80 - x)$
18%	80	0.18	$0.18(80)$

• The sum of the quantities before
mixing is equal to the quantity
after mixing.
$0.22x + 0.12(80 - x) = 0.18(80)$

Solution
$$0.22x + 0.12(80 - x) = 0.18(80)$$
$$0.22x + 9.6 - 0.12x = 14.4$$
$$0.10x + 9.6 = 14.4$$
$$0.10x = 4.8$$
$$x = 48$$
$$80 - x = 80 - 48 = 32$$

The butcher needs 48 lb of the
hamburger that is 22% fat and 32 lb
of the hamburger that is 12% fat.

You Try It 3

Strategy • Rate of the second plane: r
Rate of the first plane: $r + 30$

	Rate	Time	Distance
1st plane	$r + 30$	4	$4(r + 30)$
2nd plane	r	4	$4r$

• The total distance traveled by the
two planes is 1160 mi.
$4(r + 30) + 4r = 1160$

Solution
$$4(r + 30) + 4r = 1160$$
$$4r + 120 + 4r = 1160$$
$$8r + 120 = 1160$$
$$8r = 1040$$
$$r = 130$$
$$r + 30 = 130 + 30 = 160$$

The first plane is traveling 160 mph.
The second plane is traveling
130 mph.

SECTION 2.4

You Try It 1

$$2x - 1 < 6x + 7$$
$$-4x - 1 < 7$$
$$-4x < 8$$
$$\frac{-4x}{-4} > \frac{8}{-4}$$
$$x > -2$$
$$\{x \mid x > -2\}$$

You Try It 2

$$5x - 2 \le 4 - 3(x - 2)$$
$$5x - 2 \le 4 - 3x + 6$$
$$5x - 2 \le 10 - 3x$$
$$8x - 2 \le 10$$
$$8x \le 12$$
$$\frac{8x}{8} \le \frac{12}{8}$$
$$x \le \frac{3}{2}$$
$$\left\{x \mid x \le \frac{3}{2}\right\}$$

You Try It 3

$$-2 \le 5x + 3 \le 13$$
$$-2 - 3 \le 5x + 3 - 3 \le 13 - 3$$
$$-5 \le 5x \le 10$$
$$\frac{-5}{5} \le \frac{5x}{5} \le \frac{10}{5}$$
$$-1 \le x \le 2$$
$$\{x \mid -1 \le x \le 2\}$$

You Try It 4

$$2 - 3x > 11 \quad \text{or} \quad 5 + 2x > 7$$
$$-3x > 9 \qquad\qquad 2x > 2$$
$$x < -3 \qquad\qquad x > 1$$
$$\{x \mid x < -3\} \qquad \{x \mid x > 1\}$$

$$\{x \mid x < -3\} \cup \{x \mid x > 1\}$$
$$= \{x \mid x < -3 \text{ or } x > 1\}$$

You Try It 5

Strategy To find the maximum height, substitute the given values in the inequality $\frac{1}{2}bh < A$ and solve.

Solution

$$\frac{1}{2}bh < A$$
$$\frac{1}{2}(12)(x + 2) < 50$$
$$6(x + 2) < 50$$
$$6x + 12 < 50$$
$$6x < 38$$
$$x < \frac{19}{3}$$

The largest integer less than $\frac{19}{3}$ is 6.

$$x + 2 = 6 + 2 = 8$$

The maximum height of the triangle is 8 in.

You Try It 6

Strategy To find the range of scores, write and solve an inequality using N to represent the score on the last test.

Solution

$$80 \le \frac{72 + 94 + 83 + 70 + N}{5} \le 89$$
$$80 \le \frac{319 + N}{5} \le 89$$
$$5 \cdot 80 \le 5\left(\frac{319 + N}{5}\right) \le 5 \cdot 89$$
$$400 \le 319 + N \le 445$$
$$400 - 319 \le 319 + N - 319 \le 445 - 319$$
$$81 \le N \le 126$$

Since 100 is the maximum score, the range of scores to receive a B grade is $81 \le N \le 100$.

SECTION 2.5

You Try It 1 $|2x - 3| = 5$

$$
\begin{array}{ll}
2x - 3 = 5 & 2x - 3 = -5 \\
2x = 8 & 2x = -2 \\
x = 4 & x = -1
\end{array}
$$

The solutions are 4 and -1.

You Try It 2 $|x - 3| = -2$

There is no solution to this equation because the absolute value of a number must be nonnegative.

You Try It 3
$$5 - |3x + 5| = 3$$
$$-|3x + 5| = -2$$
$$|3x + 5| = 2$$

$$
\begin{array}{ll}
3x + 5 = 2 & 3x + 5 = -2 \\
3x = -3 & 3x = -7 \\
x = -1 & x = -\dfrac{7}{3}
\end{array}
$$

The solutions are -1 and $-\dfrac{7}{3}$.

You Try It 4 $|3x + 2| < 8$

$$-8 < 3x + 2 < 8$$
$$-8 - 2 < 3x + 2 - 2 < 8 - 2$$
$$-10 < 3x < 6$$
$$\frac{-10}{3} < \frac{3x}{3} < \frac{6}{3}$$
$$-\frac{10}{3} < x < 2$$
$$\left\{ x \,\middle|\, -\frac{10}{3} < x < 2 \right\}$$

You Try It 5 $|3x - 7| < 0$

The absolute value of a number must be nonnegative.

The solution set is the empty set.

You Try It 6 $|2x + 7| \geq -1$

The absolute value of a number must be nonnegative.

The solution set is the set of real numbers.

You Try It 7
$|5x + 3| > 8$

$$5x + 3 < -8 \quad \text{or} \quad 5x + 3 > 8$$
$$5x < -11 \qquad\qquad 5x > 5$$
$$x < -\frac{11}{5} \qquad\qquad x > 1$$

$$\left\{ x \,\middle|\, x < -\frac{11}{5} \right\} \qquad \{x \mid x > 1\}$$

$$\left\{ x \,\middle|\, x < -\frac{11}{5} \right\} \cup \{x \mid x > 1\}$$

$$= \left\{ x \,\middle|\, x < -\frac{11}{5} \text{ or } x > 1 \right\}$$

You Try It 8

Strategy

Let b represent the diameter of the bushing, T the tolerance, and d the lower and upper limits of the diameter. Solve the absolute value inequality $|d - b| \leq T$ for d.

Solution

$$|d - b| \leq T$$
$$|d - 2.55| \leq 0.003$$
$$-0.003 \leq d - 2.55 \leq 0.003$$
$$-0.003 + 2.55 \leq d - 2.55 + 2.55 \leq 0.003 + 2.55$$
$$2.547 \leq d \leq 2.553$$

The lower and upper limits of the diameter of the bushing are 2.547 in. and 2.553 in.

Solutions to Chapter 3 "You Try It"

SECTION 3.1

You Try It 1 $y = \dfrac{3x}{x + 1}$

Replace x by -2 and solve for y.

$$y = \frac{3(-2)}{-2 + 1} = \frac{-6}{-1} = 6$$

The ordered-pair solution is $(-2, 6)$.

You Try It 2

x	y
-3	2
-2	1
-1	0
0	1
1	2

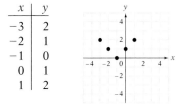

You Try It 3 $(x_1, y_1) = (5, -2) \quad (x_2, y_2) = (-4, 3)$

$$d = \sqrt{(x_1 - x_2)^2 + (y_1 - y_2)^2}$$
$$= \sqrt{[5 - (-4)]^2 + (-2 - 3)^2}$$
$$= \sqrt{9^2 + (-5)^2}$$
$$= \sqrt{81 + 25}$$
$$= \sqrt{106} \approx 10.30$$

You Try It 4 $(x_1, y_1) = (-3, -5) \quad (x_2, y_2) = (-2, 3)$

$$x_m = \frac{x_1 + x_2}{2} \qquad y_m = \frac{y_1 + y_2}{2}$$
$$= \frac{-3 + (-2)}{2} \qquad = \frac{-5 + 3}{2}$$
$$= -\frac{5}{2} \qquad\qquad = -1$$

The midpoint is $\left(-\dfrac{5}{2}, -1 \right)$.

You Try It 5

Strategy To draw a scatter diagram:
- Draw a coordinate grid with the horizontal axis representing the year and the vertical axis representing the software rental market, in billions of dollars.
- Graph the ordered pairs $(1999, 1)$, $(2000, 2)$, $(2001, 4)$, $(2002, 7)$, and $(2003, 11)$.

Solution

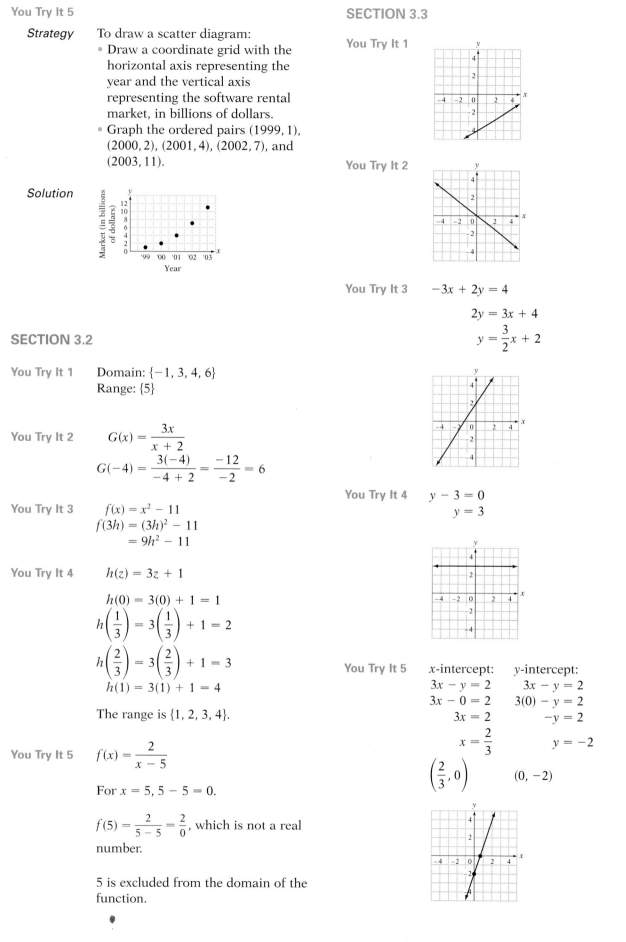

SECTION 3.2

You Try It 1 Domain: $\{-1, 3, 4, 6\}$
Range: $\{5\}$

You Try It 2
$$G(x) = \frac{3x}{x + 2}$$
$$G(-4) = \frac{3(-4)}{-4 + 2} = \frac{-12}{-2} = 6$$

You Try It 3
$$f(x) = x^2 - 11$$
$$f(3h) = (3h)^2 - 11$$
$$= 9h^2 - 11$$

You Try It 4
$$h(z) = 3z + 1$$

$$h(0) = 3(0) + 1 = 1$$
$$h\left(\frac{1}{3}\right) = 3\left(\frac{1}{3}\right) + 1 = 2$$
$$h\left(\frac{2}{3}\right) = 3\left(\frac{2}{3}\right) + 1 = 3$$
$$h(1) = 3(1) + 1 = 4$$

The range is $\{1, 2, 3, 4\}$.

You Try It 5
$$f(x) = \frac{2}{x - 5}$$

For $x = 5, 5 - 5 = 0$.

$f(5) = \dfrac{2}{5 - 5} = \dfrac{2}{0}$, which is not a real number.

5 is excluded from the domain of the function.

SECTION 3.3

You Try It 1

You Try It 2

You Try It 3
$$-3x + 2y = 4$$
$$2y = 3x + 4$$
$$y = \frac{3}{2}x + 2$$

You Try It 4
$$y - 3 = 0$$
$$y = 3$$

You Try It 5
x-intercept: y-intercept:
$$3x - y = 2 \qquad\qquad 3x - y = 2$$
$$3x - 0 = 2 \qquad\qquad 3(0) - y = 2$$
$$3x = 2 \qquad\qquad\quad -y = 2$$
$$x = \frac{2}{3} \qquad\qquad\quad y = -2$$
$$\left(\frac{2}{3}, 0\right) \qquad\qquad (0, -2)$$

You Try It 6   *x*-intercept: *y*-intercept:

$$y = \frac{1}{4}x + 1 \qquad (0, b)$$
$$b = 1$$
$$0 = \frac{1}{4}x + 1 \qquad (0, 1)$$

$$-\frac{1}{4}x = 1$$

$$x = -4$$

$$(-4, 0)$$

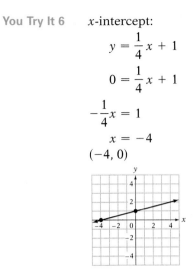

You Try It 7

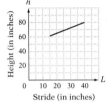

The ordered pair (32, 74) means that a person with a stride of 32 in. is 74 in. tall.

SECTION 3.4

You Try It 1   Let $P_1 = (4, -3)$ and $P_2 = (2, 7)$.

$$m = \frac{y_2 - y_1}{x_2 - x_1} = \frac{7 - (-3)}{2 - 4} = \frac{10}{-2} = -5$$

The slope is -5.

You Try It 2   Let $P_1 = (6, -1)$ and $P_2 = (6, 7)$.

$$m = \frac{y_2 - y_1}{x_2 - x_1} = \frac{7 - (-1)}{6 - 6} = \frac{8}{0}$$

Division by zero is not defined.

The slope of the line is undefined.

You Try It 3 $P_1 = (5, 25{,}000)$, $P_2 = (2, 55{,}000)$

$$m = \frac{y_2 - y_1}{x_2 - x_1}$$

$$= \frac{55{,}000 - 25{,}000}{2 - 5}$$

$$= \frac{30{,}000}{-3}$$

$$= -10{,}000$$

A slope of $-10{,}000$ means that the value of the printing press is decreasing by \$10,000 per year.

You Try It 4 $2x + 3y = 6$

$$3y = -2x + 6$$

$$y = -\frac{2}{3}x + 2$$

$$m = -\frac{2}{3} = \frac{-2}{3}$$

y-intercept $= (0, 2)$

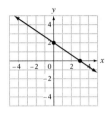

You Try It 5 $(x_1, y_1) = (-3, -2)$

$$m = 3$$

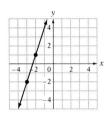

SECTION 3.5

You Try It 1 $m = -\frac{1}{3} \quad (x_1, y_1) = (-3, -2)$

$$y - y_1 = m(x - x_1)$$

$$y - (-2) = -\frac{1}{3}[x - (-3)]$$

$$y + 2 = -\frac{1}{3}(x + 3)$$

$$y + 2 = -\frac{1}{3}x - 1$$

$$y = -\frac{1}{3}x - 1 - 2$$

$$y = -\frac{1}{3}x - 3$$

The equation of the line is $y = -\frac{1}{3}x - 3$.

You Try It 2 $m = -3 \quad (x_1, y_1) = (4, -3)$

$$y - y_1 = m(x - x_1)$$
$$y - (-3) = -3(x - 4)$$
$$y + 3 = -3(x - 4)$$
$$y + 3 = -3x + 12$$
$$y = -3x + 12 - 3$$
$$y = -3x + 9$$

The equation of the line is $y = -3x + 9$.

You Try It 3   Let $(x_1, y_1) = (2, 0)$ and $(x_2, y_2) = (5, 3)$.

$$m = \frac{y_2 - y_1}{x_2 - x_1} = \frac{3 - 0}{5 - 2} = \frac{3}{3} = 1$$

$$y - y_1 = m(x - x_1)$$
$$y - 0 = 1(x - 2)$$
$$y = 1(x - 2)$$
$$y = x - 2$$

The equation of the line is $y = x - 2$.

You Try It 4   Let $(x_1, y_1) = (2, 3)$ and
$(x_2, y_2) = (-5, 3)$.

$$m = \frac{y_2 - y_1}{x_2 - x_1} = \frac{3 - 3}{-5 - 2} = \frac{0}{-7} = 0$$

The line has zero slope.
The line is a horizontal line.
All points on the line have an ordinate of 3.
The equation of the line is $y = 3$.

You Try It 5

Strategy   • Select the independent and dependent variables. The function is to predict the Celsius temperature, so that quantity is the dependent variable, y. The Fahrenheit temperature is the independent variable, x.
• From the given data, two ordered pairs are (212, 100) and (32, 0). Use these ordered pairs to determine the linear function.

Solution   Let $(x_1, y_1) = (32, 0)$ and
$(x_2, y_2) = (212, 100)$.

$$m = \frac{y_2 - y_1}{x_2 - x_1} = \frac{100 - 0}{212 - 32} = \frac{100}{180} = \frac{5}{9}$$
$$y - y_1 = m(x - x_1)$$
$$y - 0 = \frac{5}{9}(x - 32)$$
$$y = \frac{5}{9}(x - 32), \text{ or } C = \frac{5}{9}(F - 32).$$

The linear function is
$$f(F) = \frac{5}{9}(F - 32).$$

SECTION 3.6

You Try It 1 $m_1 = \dfrac{1 - (-3)}{7 - (-2)} = \dfrac{4}{9}$

$m_2 = \dfrac{-5 - 1}{6 - 4} = \dfrac{-6}{2} = -3$

$m_1 \cdot m_2 = \dfrac{4}{9} \cdot -3 = -\dfrac{4}{3}$

No, the lines are not perpendicular.

You Try It 2 $5x + 2y = 2$
$$2y = -5x + 2$$
$$y = -\frac{5}{2}x + 1$$

$$m_1 = -\frac{5}{2}$$

$5x + 2y = -6$
$$2y = -5x - 6$$
$$y = -\frac{5}{2}x - 3$$

$$m_2 = -\frac{5}{2}$$

$$m_1 = m_2 = -\frac{5}{2}$$

Yes, the lines are parallel.

You Try It 3 $x - 4y = 3$
$$-4y = -x + 3$$
$$y = \frac{1}{4}x - \frac{3}{4}$$

$$m_1 = \frac{1}{4}$$

$$m_1 \cdot m_2 = -1$$
$$\frac{1}{4} \cdot m_2 = -1$$
$$m_2 = -4$$
$$y - y_1 = m(x - x_1)$$
$$y - 2 = -4[x - (-2)]$$
$$y - 2 = -4(x + 2)$$
$$y - 2 = -4x - 8$$
$$y = -4x - 6$$

The equation of the line is
$y = -4x - 6$.

SECTION 3.7

You Try It 1 $x + 3y > 6$
$$3y > -x + 6$$
$$y > -\frac{1}{3}x + 2$$

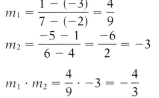

You Try It 2 $y < 2$

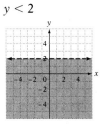

Solutions to Chapter 4 "You Try It"

SECTION 4.1

You Try It 1

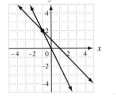

The solution is $(-1, 2)$.

You Try It 2

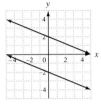

The lines are parallel and therefore do not intersect. The system of equations has no solution. The system of equations is inconsistent.

You Try It 3

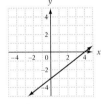

The two equations represent the same line. The system of equations is dependent. The solutions are the ordered pairs $\left(x, \dfrac{3}{4}x - 3\right)$.

You Try It 4   (1) $3x - y = 3$
(2) $6x + 3y = -4$

Solve Equation (1) for y.

$3x - y = 3$
$-y = -3x + 3$
$y = 3x - 3$

Substitute into Equation (2).

$6x + 3y = -4$
$6x + 3(3x - 3) = -4$
$6x + 9x - 9 = -4$
$15x - 9 = -4$
$15x = 5$
$x = \dfrac{5}{15} = \dfrac{1}{3}$

Substitute the value of x into Equation (1).

$3x - y = 3$
$3\left(\dfrac{1}{3}\right) - y = 3$
$1 - y = 3$
$-y = 2$
$y = -2$

The solution is $\left(\dfrac{1}{3}, -2\right)$.

You Try It 5   (1) $y = 2x - 3$
(2) $2x - 4y = 5$

$2x - 4y = 5$
$2x - 4(2x - 3) = 5$
$2x - 8x + 12 = 5$
$-6x + 12 = 5$
$-6x = -7$
$x = \dfrac{-7}{-6} = \dfrac{7}{6}$

Substitute the value of x into Equation (1).

$y = 2x - 3$
$y = 2\left(\dfrac{7}{6}\right) - 3$
$y = \dfrac{7}{3} - 3$
$y = -\dfrac{2}{3}$

The solution is $\left(\dfrac{7}{6}, -\dfrac{2}{3}\right)$.

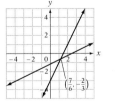

You Try It 6   (1) $6x - 3y = 6$
(2) $2x - y = 2$

Solve Equation (2) for y.

$2x - y = 2$
$-y = -2x + 2$
$y = 2x - 2$

Substitute into Equation (1).

$6x - 3y = 6$
$6x - 3(2x - 2) = 6$
$6x - 6x + 6 = 6$
$6 = 6$

The system of equations is dependent. The solutions are the ordered pairs $(x, 2x - 2)$.

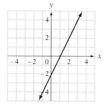

You Try It 7

Strategy   • Amount invested at 4.2%: x
Amount invested at 6%: y

	Principal	Rate	Interest
Amount at 4.2%	x	0.042	$0.042x$
Amount at 6%	y	0.06	$0.06y$

• The total investment is $13,600. The two accounts earn the same interest.

Solution $x + y = 13,600$
$0.042x = 0.06y$
$$x = \frac{10}{7}y$$

Substitute $\frac{10}{7}y$ for x in
$x + y = 13,600$ and solve for y.

$$\frac{10}{7}y + y = 13,600$$

$$\frac{17}{7}y = 13,600$$

$$y = 5600$$

$$x + 5600 = 13,600$$
$$x = 8000$$

$8000 must be invested at 4.2% and $5600 invested at 6%.

SECTION 4.2

You Try It 1   (1) $2x + 5y = 6$
(2) $3x - 2y = 6x + 2$

Write Equation (2) in the form $Ax + By = C$.

$$3x - 2y = 6x + 2$$
$$-3x - 2y = 2$$

Solve the system: $2x + 5y = 6$
$-3x - 2y = 2$

Eliminate y.
$$2(2x + 5y) = 2(6)$$
$$5(-3x - 2y) = 5(2)$$

$$4x + 10y = 12$$
$$-15x - 10y = 10$$

Add the equations.
$$-11x = 22$$
$$x = -2$$

Replace x in Equation (1).
$$2x + 5y = 6$$
$$2(-2) + 5y = 6$$
$$-4 + 5y = 6$$
$$5y = 10$$
$$y = 2$$

The solution is $(-2, 2)$.

You Try It 2 $2x + y = 5$
$4x + 2y = 6$

Eliminate y.
$$-2(2x + y) = -2(5)$$
$$4x + 2y = 6$$

$$-4x - 2y = -10$$
$$4x + 2y = 6$$

Add the equations.
$$0x + 0y = -4$$
$$0 = -4$$

This is not a true equation. The system is inconsistent and therefore has no solution.

You Try It 3   (1) $x - y + z = 6$
(2) $2x + 3y - z = 1$
(3) $x + 2y + 2z = 5$

Eliminate z. Add Equations (1) and (2).

$$x - y + z = 6$$
$$2x + 3y - z = 1$$
(4) $$3x + 2y = 7$$

Multiply Equation (2) by 2 and add to Equation (3).

$$4x + 6y - 2z = 2$$
$$x + 2y + 2z = 5$$
(5) $$5x + 8y = 7$$

Solve the system of two equations.

(4) $3x + 2y = 7$
(5) $5x + 8y = 7$

Multiply Equation (4) by -4 and add to Equation (5).
$$-12x - 8y = -28$$
$$5x + 8y = 7$$
$$-7x = -21$$
$$x = 3$$

Replace x by 3 in Equation (4).

$$3x + 2y = 7$$
$$3(3) + 2y = 7$$
$$9 + 2y = 7$$
$$2y = -2$$
$$y = -1$$

Replace x by 3 and y by -1 in Equation (1).

$$x - y + z = 6$$
$$3 - (-1) + z = 6$$
$$4 + z = 6$$
$$z = 2$$

The solution is $(3, -1, 2)$.

SECTION 4.3

You Try It 1

$$\begin{vmatrix} -1 & -4 \\ 3 & -5 \end{vmatrix} = -1(-5) - (-4)(3)$$
$$= 5 + 12 = 17$$

The value of the determinant is 17.

You Try It 2   Expand by cofactors of the first row.

$$\begin{vmatrix} 1 & 4 & -2 \\ 3 & 1 & 1 \\ 0 & -2 & 2 \end{vmatrix}$$

$$= 1\begin{vmatrix} 1 & 1 \\ -2 & 2 \end{vmatrix} - 4\begin{vmatrix} 3 & 1 \\ 0 & 2 \end{vmatrix} + (-2)\begin{vmatrix} 3 & 1 \\ 0 & -2 \end{vmatrix}$$
$$= 1(2 + 2) - 4(6 - 0) - 2(-6 - 0)$$
$$= 4 - 24 + 12$$
$$= -8$$

The value of the determinant is -8.

You Try It 3

$$\begin{vmatrix} 3 & -2 & 0 \\ 1 & 4 & 2 \\ -2 & 1 & 3 \end{vmatrix}$$

$$= 3\begin{vmatrix} 4 & 2 \\ 1 & 3 \end{vmatrix} - (-2)\begin{vmatrix} 1 & 2 \\ -2 & 3 \end{vmatrix} + 0\begin{vmatrix} 1 & 4 \\ -2 & 1 \end{vmatrix}$$
$$= 3(12 - 2) + 2(3 + 4) + 0$$
$$= 3(10) + 2(7)$$
$$= 30 + 14$$
$$= 44$$

The value of the determinant is 44.

You Try It 4

$$3x - y = 4$$
$$6x - 2y = 5$$

$$D = \begin{vmatrix} 3 & -1 \\ 6 & -2 \end{vmatrix} = 0$$

Because $D = 0$, $\dfrac{D_x}{D}$ is undefined.

Therefore, the system of equations is dependent or inconsistent.

You Try It 5

$$2x - y + z = -1$$
$$3x + 2y - z = 3$$
$$x + 3y + z = -2$$

$$D = \begin{vmatrix} 2 & -1 & 1 \\ 3 & 2 & -1 \\ 1 & 3 & 1 \end{vmatrix} = 21$$

$$D_x = \begin{vmatrix} -1 & -1 & 1 \\ 3 & 2 & -1 \\ -2 & 3 & 1 \end{vmatrix} = 9$$

$$D_y = \begin{vmatrix} 2 & -1 & 1 \\ 3 & 3 & -1 \\ 1 & -2 & 1 \end{vmatrix} = -3$$

$$D_z = \begin{vmatrix} 2 & -1 & -1 \\ 3 & 2 & 3 \\ 1 & 3 & -2 \end{vmatrix} = -42$$

$$x = \frac{D_x}{D} = \frac{9}{21} = \frac{3}{7}$$

$$y = \frac{D_y}{D} = \frac{-3}{21} = -\frac{1}{7}$$

$$z = \frac{D_z}{D} = \frac{-42}{21} = -2$$

The solution is $\left(\dfrac{3}{7}, -\dfrac{1}{7}, -2\right)$.

SECTION 4.4

You Try It 1

Strategy

• Rate of the rowing team in calm water: t
  Rate of the current: c

	Rate	Time	Distance
With current	$t + c$	2	$2(t + c)$
Against current	$t - c$	2	$2(t - c)$

• The distance traveled with the current is 18 mi. The distance traveled against the current is 10 mi.

$$2(t + c) = 18$$
$$2(t - c) = 10$$

Solution

$$2(t + c) = 18$$

$$2(t - c) = 10$$

$$\frac{1}{2} \cdot 2(t + c) = \frac{1}{2} \cdot 18$$

$$\frac{1}{2} \cdot 2(t - c) = \frac{1}{2} \cdot 10$$

$$t + c = 9$$
$$t - c = 5$$
$$2t = 14$$
$$t = 7$$

$$t + c = 9$$
$$7 + c = 9$$
$$c = 2$$

The rate of the rowing team in calm water is 7 mph. The rate of the current is 2 mph.

You Try It 2

Strategy
- Number of dimes: d
  Number of nickels: n
  Number of quarters: q
- There are 19 coins in a bank that contains only nickels, dimes, and quarters. ($n + d + q = 19$)
  The value of the coins is \$2. ($5n + 10d + 25q = 200$)
  There are twice as many nickels as dimes. ($n = 2d$)

Solution
$$
\begin{aligned}
(1) &\quad n + d + q = 19 \\
(2) &\quad n = 2d \\
(3) &\quad 5n + 10d + 25q = 200
\end{aligned}
$$

Solve the system of equations. Substitute $2d$ for n in Equation (1) and Equation (3).

$$
\begin{aligned}
2d + d + q &= 19 \\
5(2d) + 10d + 25q &= 200
\end{aligned}
$$

$$
\begin{aligned}
(4) &\quad 3d + q = 19 \\
(5) &\quad 20d + 25q = 200
\end{aligned}
$$

Solve the system of equations in two variables by multiplying Equation (4) by -25 and then adding it to Equation (5).

$$
\begin{aligned}
-75d - 25q &= -475 \\
20d + 25q &= 200 \\
\hline
-55d &= -275 \\
d &= 5
\end{aligned}
$$

Substituting the value of d into Equation (2), we have $n = 10$. Substituting the values of d and n into Equation (1), we have $q = 4$.

There are 5 dimes, 10 nickels, and 4 quarters in the bank.

SECTION 4.5

You Try It 1
Shade above the solid line $y = 2x - 3$.
Shade above the dotted line $y = -3x$.

The solution set of the system is the intersection of the solution sets of the individual inequalities.

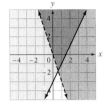

You Try It 2
$$
\begin{aligned}
3x + 4y &> 12 \\
4y &> -3x + 12 \\
y &> -\frac{3}{4}x + 3
\end{aligned}
$$

Shade above the dotted line $y = -\frac{3}{4}x + 3$.

Shade below the dotted line $y = \frac{3}{4}x - 1$.

The solution set of the system is the intersection of the solution sets of the individual inequalities.

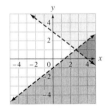

Solutions to Chapter 5 "You Try It"

SECTION 5.1

You Try It 1
$$
\begin{aligned}
(-3a^2b^4)(-2ab^3)^4 &= (-3a^2b^4)[(-2)^4a^4b^{12}] \\
&= (-3a^2b^4)(16a^4b^{12}) \\
&= -48a^6b^{16}
\end{aligned}
$$

You Try It 2
$$
\begin{aligned}
(y^{n-3})^2 &= y^{(n-3)2} \\
&= y^{2n-6}
\end{aligned}
$$

You Try It 3
$$
\begin{aligned}
[(ab^3)^3]^4 &= [a^3b^9]^4 \\
&= a^{12}b^{36}
\end{aligned}
$$

You Try It 4
$$
\begin{aligned}
\frac{20r^{-2}t^{-5}}{-16r^{-3}s^{-2}} &= -\frac{4 \cdot 5r^{-2-(-3)}s^2t^{-5}}{4 \cdot 4} \\
&= -\frac{5rs^2}{4t^5}
\end{aligned}
$$

You Try It 5
$$
\begin{aligned}
\frac{(9u^{-6}v^4)^{-1}}{(6u^{-3}v^{-2})^{-2}} &= \frac{9^{-1}u^6v^{-4}}{6^{-2}u^6v^4} \\
&= 9^{-1} \cdot 6^2u^0v^{-8} \\
&= \frac{36}{9v^8} \\
&= \frac{4}{v^8}
\end{aligned}
$$

You Try It 6

$$\frac{a^{2n+1}}{a^{n+3}} = a^{2n+1-(n+3)}$$
$$= a^{2n+1-n-3}$$
$$= a^{n-2}$$

You Try It 7 $942{,}000{,}000 = 9.42 \times 10^8$

You Try It 8 $2.7 \times 10^{-5} = 0.000027$

You Try It 9

$$\frac{5{,}600{,}000 \times 0.000000081}{900 \times 0.000000028}$$
$$= \frac{5.6 \times 10^6 \times 8.1 \times 10^{-8}}{9 \times 10^2 \times 2.8 \times 10^{-8}}$$
$$= \frac{(5.6)(8.1) \times 10^{6+(-8)-2-(-8)}}{(9)(2.8)}$$
$$= 1.8 \times 10^4 = 18{,}000$$

You Try It 10

Strategy To find the number of arithmetic operations:
- Find the reciprocal of 1×10^{-7}, which is the number of operations performed in 1 s.
- Write the number of seconds in 1 min (60) in scientific notation.
- Multiply the number of arithmetic operations per second by the number of seconds in 1 min.

Solution
$$\frac{1}{1 \times 10^{-7}} = 10^7$$
$$60 = 6 \times 10$$

$$6 \times 10 \times 10^7$$
$$6 \times 10^8$$

The computer can perform 6×10^8 operations in 1 min.

SECTION 5.2

You Try It 1
$$R(x) = -2x^4 - 5x^3 + 2x - 8$$
$$R(2) = -2(2)^4 - 5(2)^3 + 2(2) - 8$$
$$= -2(16) - 5(8) + 4 - 8$$
$$= -32 - 40 + 4 - 8$$
$$= -76$$

You Try It 2 The leading coefficient is -3; the constant term is -12; and the degree is 4.

You Try It 3
a. This is a polynomial function.
b. This is not a polynomial function. A polynomial function does not have a variable expression raised to a negative power.

c. This is not a polynomial function. A polynomial function does not have a variable expression within a radical.

You Try It 4

x	y
-4	5
-3	0
-2	-3
-1	-4
0	-3
1	0
2	5

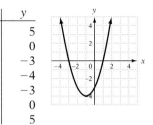

You Try It 5

x	y
-3	28
-2	9
-1	2
0	1
1	0
2	-7
3	-26

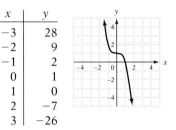

You Try It 6
$$\begin{array}{r} -3x^2 - 4x + 9 \\ -5x^2 - 7x + 1 \\ \hline -8x^2 - 11x + 10 \end{array}$$

You Try It 7 Add the additive inverse of $6x^2 + 3x - 7$ to $-5x^2 + 2x - 3$.

$$\begin{array}{r} -5x^2 + 2x - 3 \\ -6x^2 - 3x + 7 \\ \hline -11x^2 - x + 4 \end{array}$$

You Try It 8
$$S(x) = (4x^3 - 3x^2 + 2) + (-2x^2 + 2x - 3)$$
$$= 4x^3 - 5x^2 + 2x - 1$$

$$S(-1) = 4(-1)^3 - 5(-1)^2 + 2(-1) - 1$$
$$= 4(-1) - 5(1) - 2 - 1$$
$$= -4 - 5 - 2 - 1$$
$$= -12$$

You Try It 9
$$D(x) = P(x) - R(x)$$

$$D(x) = (5x^{2n} - 3x^n - 7) - (-2x^{2n} - 5x^n + 8)$$
$$= (5x^{2n} - 3x^n - 7) + (2x^{2n} + 5x^n - 8)$$
$$= 7x^{2n} + 2x^n - 15$$

SECTION 5.3

You Try It 1
$$(2b^2 - 7b - 8)(-5b) = -10b^3 + 35b^2 + 40b$$

You Try It 2 $x^2 - 2x[x - x(4x - 5) + x^2]$
$= x^2 - 2x[x - 4x^2 + 5x + x^2]$
$= x^2 - 2x[6x - 3x^2]$
$= x^2 - 12x^2 + 6x^3$
$= 6x^3 - 11x^2$

You Try It 3 $y^{n+3}(y^{n-2} - 3y^2 + 2)$
$= y^{n+3}(y^{n-2}) - (y^{n+3})(3y^2) + (y^{n+3})(2)$
$= y^{n+3+(n-2)} - 3y^{n+3+2} + 2y^{n+3}$
$= y^{2n+1} - 3y^{n+5} + 2y^{n+3}$

You Try It 4
$$\begin{array}{r} -2b^2 + 5b - 4 \\ - 3b + 2 \\ \hline -4b^2 + 10b - 8 \\ 6b^3 - 15b^2 + 12b \\ \hline 6b^3 - 19b^2 + 22b - 8 \end{array}$$

You Try It 5 $(3x - 4)(2x - 3) = 6x^2 - 9x - 8x + 12$
$= 6x^2 - 17x + 12$

You Try It 6 $(2x^n + y^n)(x^n - 4y^n)$
$= 2x^{2n} - 8x^n y^n + x^n y^n - 4y^{2n}$
$= 2x^{2n} - 7x^n y^n - 4y^{2n}$

You Try It 7 $(3x - 7)(3x + 7) = 9x^2 - 49$

You Try It 8 $(2x^n + 3)(2x^n - 3) = 4x^{2n} - 9$

You Try It 9 $(3x - 4y)^2 = 9x^2 - 24xy + 16y^2$

You Try It 10 $(2x^n - 8)^2 = 4x^{2n} - 32x^n + 64$

You Try It 11

Strategy   To find the area, replace the variables b and h in the equation $A = \frac{1}{2}bh$ by the given values and solve for A.

Solution $A = \frac{1}{2}bh$

$A = \frac{1}{2}(2x + 6)(x - 4)$

$A = (x + 3)(x - 4)$
$A = x^2 - 4x + 3x - 12$
$A = x^2 - x - 12$

The area is $(x^2 - x - 12)$ ft^2.

You Try It 12

Strategy   To find the volume, subtract the volume of the small rectangular solid from the volume of the large rectangular solid.

Large rectangular solid:
Length $= L_1 = 12x$
Width $= W_1 = 7x + 2$
Height $= H_1 = 5x - 4$
Small rectangular solid:
Length $= L_2 = 12x$
Width $= W_2 = x$
Height $= H_2 = 2x$

Solution
$V =$ Volume of large rectangular
  solid $-$ volume of small rectangular solid
$V = (L_1 \cdot W_1 \cdot H_1) - (L_2 \cdot W_2 \cdot H_2)$
$V = (12x)(7x + 2)(5x - 4) - (12x)(x)(2x)$
$= (84x^2 + 24x)(5x - 4) - (12x^2)(2x)$
$= (420x^3 - 336x^2 + 120x^2 - 96x) - 24x^3$
$= 396x^3 - 216x^2 - 96x$

The volume is $(396x^3 - 216x^2 - 96x)$ ft^3.

You Try It 13

Strategy   To find the area, replace the variable r in the equation $A = \pi r^2$ by the given value and solve for A.

Solution $A = \pi r^2$
$A = 3.14(2x + 3)^2$
$= 3.14(4x^2 + 12x + 9)$
$= 12.56x^2 + 37.68x + 28.26$

The area is
$(12.56x^2 + 37.68x + 28.26)$ cm^2.

SECTION 5.4

You Try It 1
$$\begin{array}{r} 5x - 1 \\ 3x + 4 \overline{)15x^2 + 17x - 20} \\ \underline{15x^2 + 20x} \\ -3x - 20 \\ \underline{-3x - 4} \\ -16 \end{array}$$

$$\frac{15x^2 + 17x - 20}{3x + 4} = 5x - 1 - \frac{16}{3x + 4}$$

You Try It 2
$$\begin{array}{r} x^2 + 3x - 1 \\ 3x - 1 \overline{)3x^3 + 8x^2 - 6x + 2} \\ \underline{3x^3 - x^2} \\ 9x^2 - 6x \\ \underline{9x^2 - 3x} \\ -3x + 2 \\ \underline{-3x + 1} \\ 1 \end{array}$$

$$\frac{3x^3 + 8x^2 - 6x + 2}{3x - 1} = x^2 + 3x - 1 + \frac{1}{3x - 1}$$

You Try It 3

$$
\require{enclose}
\begin{array}{r}
3x^2 - 2x + 4 \\
x^2 - 3x + 2 \enclose{longdiv}{3x^4 - 11x^3 + 16x^2 - 16x + 8}
\end{array}
$$

$$3x^4 - 9x^3 + 6x^2$$
$$-2x^3 + 10x^2 - 16x$$
$$-2x^3 + 6x^2 - 4x$$
$$4x^2 - 12x + 8$$
$$4x^2 - 12x + 8$$
$$0$$

$$\frac{3x^4 - 11x^3 + 16x^2 - 16x + 8}{x^2 - 3x + 2} = 3x^2 - 2x + 4$$

You Try It 4

$$
\begin{array}{r|rrr}
-2 & 6 & 8 & -5 \\
   &   & -12 & 8 \\
\hline
   & 6 & -4 & 3
\end{array}
$$

$$(6x^2 + 8x - 5) \div (x + 2) = 6x - 4 + \frac{3}{x + 2}$$

You Try It 5

$$
\begin{array}{r|rrrr}
2 & 5 & -12 & -8 & 16 \\
  &   & 10 & -4 & -24 \\
\hline
  & 5 & -2 & -12 & -8
\end{array}
$$

$$(5x^3 - 12x^2 - 8x + 16) \div (x - 2)$$

$$= 5x^2 - 2x - 12 - \frac{8}{x - 2}$$

You Try It 6

$$
\begin{array}{r|rrrrr}
3 & 2 & -3 & -8 & 0 & -2 \\
  &   & 6 & 9 & 3 & 9 \\
\hline
  & 2 & 3 & 1 & 3 & 7
\end{array}
$$

$$(2x^4 - 3x^3 - 8x^2 - 2) \div (x - 3)$$

$$= 2x^3 + 3x^2 + x + 3 + \frac{7}{x - 3}$$

You Try It 7

$$
\begin{array}{r|rrr}
2 & 2 & -3 & -5 \\
  &   & 4 & 2 \\
\hline
  & 2 & 1 & -3
\end{array}
$$

$$P(2) = -3$$

You Try It 8

$$
\begin{array}{r|rrrr}
-3 & 2 & -5 & 0 & 7 \\
   &   & -6 & 33 & -99 \\
\hline
   & 2 & -11 & 33 & -92
\end{array}
$$

$$P(-3) = -92$$

SECTION 5.5

You Try It 1   The GCF of $3x^3y$, $6x^2y^2$, and $3xy^3$ is $3xy$.

$$3x^3y - 6x^2y^2 - 3xy^3 = 3xy(x^2 - 2xy - y^2)$$

You Try It 2   The GCF of $6t^{2n}$ and $9t^n$ is $3t^n$.

$$6t^{2n} - 9t^n = 3t^n(2t^n - 3)$$

You Try It 3 $3(6x - 7y) - 2x^2(6x - 7y)$
$$= (6x - 7y)(3 - 2x^2)$$

You Try It 4 $4a^2 - 6a - 6ax + 9x$
$$= (4a^2 - 6a) - (6ax - 9x)$$
$$= 2a(2a - 3) - 3x(2a - 3)$$
$$= (2a - 3)(2a - 3x)$$

You Try It 5 $x^2 - x - 20 = (x + 4)(x - 5)$

You Try It 6 $x^2 + 5xy + 6y^2 = (x + 2y)(x + 3y)$

You Try It 7 $4x^2 + 15x - 4 = (x + 4)(4x - 1)$

You Try It 8 $10x^2 + 39x + 14 = (2x + 7)(5x + 2)$

You Try It 9
The GCF of $3a^3b^3$, $3a^2b^2$, and $60ab$ is $3ab$.

$$3a^3b^3 + 3a^2b^2 - 60ab = 3ab(a^2b^2 + ab - 20)$$
$$= 3ab(ab + 5)(ab - 4)$$

SECTION 5.6

You Try It 1 $x^2 - 36y^4 = x^2 - (6y^2)^2$
$$= (x + 6y^2)(x - 6y^2)$$

You Try It 2 $9x^2 + 12x + 4 = (3x + 2)^2$

You Try It 3
$(a + b)^2 - (a - b)^2$
$$= [(a + b) + (a - b)][(a + b) - (a - b)]$$
$$= (a + b + a - b)(a + b - a + b)$$
$$= (2a)(2b) = 4ab$$

You Try It 4 $a^3b^3 - 27 = (ab)^3 - 3^3$
$$= (ab - 3)(a^2b^2 + 3ab + 9)$$

You Try It 5 $8x^3 + y^3z^3 = (2x)^3 + (yz)^3$
$$= (2x + yz)(4x^2 - 2xyz + y^2z^2)$$

You Try It 6
$(x - y)^3 + (x + y)^3$
$$= [(x - y) + (x + y)]$$
$$\times [(x - y)^2 - (x - y)(x + y) + (x + y)^2]$$
$$= 2x[x^2 - 2xy + y^2 - (x^2 - y^2)$$
$$+ x^2 + 2xy + y^2]$$
$$= 2x(x^2 - 2xy + y^2 - x^2 + y^2 + x^2$$
$$+ 2xy + y^2)$$
$$= 2x(x^2 + 3y^2)$$

You Try It 7   Let $u = x^2$.

$$3x^4 + 4x^2 - 4 = 3u^2 + 4u - 4$$
$$= (u + 2)(3u - 2)$$
$$= (x^2 + 2)(3x^2 - 2)$$

You Try It 8 $18x^3 - 6x^2 - 60x = 6x(3x^2 - x - 10)$
$$= 6x(3x + 5)(x - 2)$$

You Try It 9 $4x - 4y - x^3 + x^2y$
$$= (4x - 4y) - (x^3 - x^2y)$$
$$= 4(x - y) - x^2(x - y)$$
$$= (x - y)(4 - x^2)$$
$$= (x - y)(2 + x)(2 - x)$$

You Try It 10 $x^{4n} - x^{2n}y^{2n} = x^{2n+2n} - x^{2n}y^{2n}$
$$= x^{2n}(x^{2n} - y^{2n})$$
$$= x^{2n}[(x^n)^2 - (y^n)^2]$$
$$= x^{2n}(x^n + y^n)(x^n - y^n)$$

You Try It 11 $ax^5 - ax^2y^6 = ax^2(x^3 - y^6)$
$$= ax^2(x - y^2)(x^2 + xy^2 + y^4)$$

SECTION 5.7

You Try It 1 $(x + 4)(x - 1) = 14$
$$x^2 + 3x - 4 = 14$$
$$x^2 + 3x - 18 = 0$$
$$(x + 6)(x - 3) = 0$$
$$x + 6 = 0 \qquad x - 3 = 0$$
$$x = -6 \qquad x = 3$$

The solutions are -6 and 3.

You Try It 2

Strategy   Draw a diagram. Then use the formula for the area of a triangle.

Solution $A = \dfrac{1}{2}bh$

$$54 = \dfrac{1}{2}x(x + 3)$$
$$108 = x(x + 3)$$
$$0 = x^2 + 3x - 108$$
$$0 = (x - 9)(x + 12)$$

$$x - 9 = 0 \qquad x + 12 = 0$$
$$x = 9 \qquad x = -12$$

$$x + 3 = 12$$

The base is 9 cm; the height is 12 cm.

Solutions to Chapter 6 "You Try It"

SECTION 6.1

You Try It 1 $f(x) = \dfrac{3 - 5x}{x^2 + 5x + 6}$

$$f(2) = \dfrac{3 - 5(2)}{2^2 + 5(2) + 6}$$
$$= \dfrac{3 - 10}{4 + 10 + 6}$$
$$= \dfrac{-7}{20}$$
$$= -\dfrac{7}{20}$$

You Try It 2   Set the denominator equal to zero. Then solve for x.

$$2x^2 - 7x + 3 = 0$$
$$(2x - 1)(x - 3) = 0$$
$$2x - 1 = 0 \qquad x - 3 = 0$$
$$x = \dfrac{1}{2} \qquad x = 3$$

The domain is $\left\{ x \mid x \neq \dfrac{1}{2}, 3 \right\}$.

You Try It 3 $\dfrac{6x^4 - 24x^3}{12x^3 - 48x^2} = \dfrac{6x^3(x - 4)}{12x^2(x - 4)}$

$$= \dfrac{6x^3\overset{1}{\cancel{(x - 4)}}}{12x^2\underset{1}{\cancel{(x - 4)}}} = \dfrac{x}{2}$$

You Try It 4 $\dfrac{21a^3b - 14a^3b^2}{7a^2b} = \dfrac{7a^3b(3 - 2b)}{7a^2b}$
$$= a(3 - 2b)$$

You Try It 5

$$\frac{20x - 15x^2}{15x^3 - 5x^2 - 20x} = \frac{5x(4 - 3x)}{5x(3x^2 - x - 4)}$$

$$= \frac{5x(4 - 3x)}{5x(3x - 4)(x + 1)}$$

$$= \frac{\overset{-1}{5x(4 - 3x)}}{\underset{1}{5x(3x - 4)}(x + 1)}$$

$$= -\frac{1}{x + 1}$$

You Try It 6

$$\frac{x^{2n} + x^n - 12}{x^{2n} - 3x^n} = \frac{(x^n + 4)(x^n - 3)}{x^n(x^n - 3)}$$

$$= \frac{(x^n + 4)\overset{1}{(x^n - 3)}}{x^n\underset{1}{(x^n - 3)}}$$

$$= \frac{x^n + 4}{x^n}$$

You Try It 7

$$\frac{12 + 5x - 3x^2}{x^2 + 2x - 15} \cdot \frac{2x^2 + x - 45}{3x^2 + 4x}$$

$$= \frac{(4 + 3x)(3 - x)}{(x + 5)(x - 3)} \cdot \frac{(2x - 9)(x + 5)}{x(3x + 4)}$$

$$= \frac{(4 + 3x)(3 - x)(2x - 9)(x + 5)}{(x + 5)(x - 3) \cdot x(3x + 4)}$$

$$= \frac{\overset{1}{(4 + 3x)}\overset{-1}{(3 - x)}(2x - 9)\overset{1}{(x + 5)}}{\underset{1}{(x + 5)}\underset{1}{(x - 3)} \cdot x\underset{1}{(3x + 4)}} = -\frac{2x - 9}{x}$$

You Try It 8

$$\frac{2x^2 - 13x + 20}{x^2 - 16} \cdot \frac{2x^2 + 9x + 4}{6x^2 - 7x - 5}$$

$$= \frac{(2x - 5)(x - 4)}{(x - 4)(x + 4)} \cdot \frac{(2x + 1)(x + 4)}{(3x - 5)(2x + 1)}$$

$$= \frac{(2x - 5)(x - 4) \cdot (2x + 1)(x + 4)}{(x - 4)(x + 4) \cdot (3x - 5)(2x + 1)}$$

$$= \frac{(2x - 5)\overset{1}{(x - 4)}\overset{1}{(2x + 1)}\overset{1}{(x + 4)}}{\underset{1}{(x - 4)}\underset{1}{(x + 4)}(3x - 5)\underset{1}{(2x + 1)}} = \frac{2x - 5}{3x - 5}$$

You Try It 9

$$\frac{7a^3b^7}{15x^2y^3} \div \frac{21a^5b^2}{20x^4y}$$

$$= \frac{7a^3b^7}{15x^2y^3} \cdot \frac{20x^4y}{21a^5b^2}$$

$$= \frac{7a^3b^7 \cdot 20x^4y}{15x^2y^3 \cdot 21a^5b^2}$$

$$= \frac{4b^5x^2}{9a^2y^2}$$

You Try It 10

$$\frac{6x^2 - 3xy}{10ab^4} \div \frac{16x^2y^2 - 8xy^3}{15a^2b^2}$$

$$= \frac{6x^2 - 3xy}{10ab^4} \cdot \frac{15a^2b^2}{16x^2y^2 - 8xy^3}$$

$$= \frac{3x(2x - y)}{10ab^4} \cdot \frac{15a^2b^2}{8xy^2(2x - y)}$$

$$= \frac{3x(2x - y) \cdot 15a^2b^2}{10ab^4 \cdot 8xy^2(2x - y)}$$

$$= \frac{3x\overset{1}{(2x - y)} \cdot 15a^2b^2}{10ab^4 \cdot 8xy^2\underset{1}{(2x - y)}} = \frac{9a}{16b^2y^2}$$

You Try It 11

$$\frac{6x^2 - 7x + 2}{3x^2 + x - 2} \div \frac{4x^2 - 8x + 3}{5x^2 + x - 4}$$

$$= \frac{6x^2 - 7x + 2}{3x^2 + x - 2} \cdot \frac{5x^2 + x - 4}{4x^2 - 8x + 3}$$

$$= \frac{(2x - 1)(3x - 2)}{(x + 1)(3x - 2)} \cdot \frac{(x + 1)(5x - 4)}{(2x - 1)(2x - 3)}$$

$$= \frac{(2x - 1)(3x - 2)(x + 1)(5x - 4)}{(x + 1)(3x - 2)(2x - 1)(2x - 3)}$$

$$= \frac{\overset{1}{(2x - 1)}\overset{1}{(3x - 2)}\overset{1}{(x + 1)}(5x - 4)}{\underset{1}{(x + 1)}\underset{1}{(3x - 2)}\underset{1}{(2x - 1)}(2x - 3)} = \frac{5x - 4}{2x - 3}$$

SECTION 6.2

You Try It 1

The LCM is $(2x - 5)(x + 4)$.

$$\frac{2x}{2x - 5} = \frac{2x}{2x - 5} \cdot \frac{x + 4}{x + 4} = \frac{2x^2 + 8x}{(2x - 5)(x + 4)}$$

$$\frac{3}{x + 4} = \frac{3}{x + 4} \cdot \frac{2x - 5}{2x - 5} = \frac{6x - 15}{(2x - 5)(x + 4)}$$

You Try It 2

$2x^2 - 11x + 15 = (x - 3)(2x - 5);\ x^2 - 3x = x(x - 3)$

The LCM is $x(x - 3)(2x - 5)$.

$$\frac{3x}{2x^2 - 11x + 15} = \frac{3x}{(x - 3)(2x - 5)} \cdot \frac{x}{x} = \frac{3x^2}{x(x - 3)(2x - 5)}$$

$$\frac{x - 2}{x^2 - 3x} = \frac{x - 2}{x(x - 3)} \cdot \frac{2x - 5}{2x - 5} = \frac{2x^2 - 9x + 10}{x(x - 3)(2x - 5)}$$

You Try It 3

$2x - x^2 = x(2 - x) = -x(x - 2)$;
$3x^2 - 5x - 2 = (x - 2)(3x + 1)$

The LCM is $x(x - 2)(3x + 1)$.

$$\frac{2x - 7}{2x - x^2} = -\frac{2x - 7}{x(x - 2)} \cdot \frac{3x + 1}{3x + 1} = -\frac{6x^2 - 19x - 7}{x(x - 2)(3x + 1)}$$

$$\frac{3x - 2}{3x^2 - 5x - 2} = \frac{3x - 2}{(x - 2)(3x + 1)} \cdot \frac{x}{x} = \frac{3x^2 - 2x}{x(x - 2)(3x + 1)}$$

You Try It 4

The LCM is ab.

$$\frac{2}{b} - \frac{1}{a} + \frac{4}{ab} = \frac{2}{b} \cdot \frac{a}{a} - \frac{1}{a} \cdot \frac{b}{b} + \frac{4}{ab}$$

$$= \frac{2a}{ab} - \frac{b}{ab} + \frac{4}{ab} = \frac{2a - b + 4}{ab}$$

You Try It 5

The LCM is $a(a - 5)(a + 5)$.

$$\frac{a - 3}{a^2 - 5a} + \frac{a - 9}{a^2 - 25}$$

$$= \frac{a - 3}{a(a - 5)} \cdot \frac{a + 5}{a + 5} + \frac{a - 9}{(a - 5)(a + 5)} \cdot \frac{a}{a}$$

$$= \frac{(a - 3)(a + 5) + a(a - 9)}{a(a - 5)(a + 5)}$$

$$= \frac{(a^2 + 2a - 15) + (a^2 - 9a)}{a(a - 5)(a + 5)}$$

$$= \frac{a^2 + 2a - 15 + a^2 - 9a}{a(a - 5)(a + 5)}$$

$$= \frac{2a^2 - 7a - 15}{a(a - 5)(a + 5)} = \frac{(2a + 3)(a - 5)}{a(a - 5)(a + 5)}$$

$$= \frac{(2a + 3)\overset{1}{\cancel{(a - 5)}}}{a\underset{1}{\cancel{(a - 5)}}(a + 5)} = \frac{2a + 3}{a(a + 5)}$$

You Try It 6

The LCM is $(x - 4)(x + 1)$.

$$\frac{2x}{x - 4} - \frac{x - 1}{x + 1} + \frac{2}{x^2 - 3x - 4}$$

$$= \frac{2x}{x - 4} \cdot \frac{x + 1}{x + 1} - \frac{x - 1}{x + 1} \cdot \frac{x - 4}{x - 4} + \frac{2}{(x - 4)(x + 1)}$$

$$= \frac{2x(x + 1) - (x - 1)(x - 4) + 2}{(x - 4)(x + 1)}$$

$$= \frac{(2x^2 + 2x) - (x^2 - 5x + 4) + 2}{(x - 4)(x + 1)}$$

$$= \frac{x^2 + 7x - 2}{(x - 4)(x + 1)}$$

SECTION 6.3

You Try It 1

The LCM of $x + 3$ and $x + 2$ is $(x + 3)(x + 2)$.

$$\frac{1 - \dfrac{2}{x + 3}}{1 - \dfrac{1}{x + 2}} = \frac{1 - \dfrac{2}{x + 3}}{1 - \dfrac{1}{x + 2}} \cdot \frac{(x + 3)(x + 2)}{(x + 3)(x + 2)}$$

$$= \frac{(x + 3)(x + 2) - \dfrac{2}{x + 3}(x + 3)(x + 2)}{(x + 3)(x + 2) - \dfrac{1}{x + 2}(x + 3)(x + 2)}$$

$$= \frac{x^2 + 5x + 6 - 2x - 4}{x^2 + 5x + 6 - x - 3}$$

$$= \frac{x^2 + 3x + 2}{x^2 + 4x + 3} = \frac{(x + 2)\overset{1}{\cancel{(x + 1)}}}{(x + 3)\underset{1}{\cancel{(x + 1)}}} = \frac{x + 2}{x + 3}$$

You Try It 2

The LCM is $x - 3$.

$$\frac{2x + 5 + \dfrac{14}{x - 3}}{4x + 16 + \dfrac{49}{x - 3}} = \frac{2x + 5 + \dfrac{14}{x - 3}}{4x + 16 + \dfrac{49}{x - 3}} \cdot \frac{x - 3}{x - 3}$$

$$= \frac{(2x + 5)(x - 3) + \dfrac{14}{x - 3}(x - 3)}{(4x + 16)(x - 3) + \dfrac{49}{x - 3}(x - 3)}$$

$$= \frac{2x^2 - x - 15 + 14}{4x^2 + 4x - 48 + 49} = \frac{2x^2 - x - 1}{4x^2 + 4x + 1}$$

$$= \frac{(2x + 1)(x - 1)}{(2x + 1)(2x + 1)} = \frac{\overset{1}{\cancel{(2x + 1)}}(x - 1)}{\underset{1}{\cancel{(2x + 1)}}(2x + 1)} = \frac{x - 1}{2x + 1}$$

You Try It 3

The LCM of the denominators is x.

$$2 - \frac{1}{2 - \dfrac{1}{x}} = 2 - \frac{1}{2 - \dfrac{1}{x}} \cdot \frac{x}{x}$$

$$= 2 - \frac{1 \cdot x}{2 \cdot x - \dfrac{1}{x} \cdot x} = 2 - \frac{x}{2x - 1}$$

The LCM of the denominators is $2x - 1$.

$$2 - \frac{x}{2x - 1} = 2 \cdot \frac{(2x - 1)}{(2x - 1)} - \frac{x}{2x - 1}$$

$$= \frac{4x - 2}{2x - 1} - \frac{x}{2x - 1}$$

$$= \frac{4x - 2 - x}{2x - 1} = \frac{3x - 2}{2x - 1}$$

SECTION 6.4

You Try It 1

$$\frac{5}{x - 2} = \frac{3}{4}$$

$$\frac{5}{x - 2} \cdot 4(x - 2) = \frac{3}{4} \cdot 4(x - 2)$$

$$5 \cdot 4 = 3(x - 2)$$

$$20 = 3x - 6$$

$$26 = 3x$$

$$\frac{26}{3} = x$$

The solution is $\dfrac{26}{3}$.

You Try It 2

$$\frac{5}{2x - 3} = \frac{-2}{x + 1}$$

$$\frac{5}{2x - 3}(x + 1)(2x - 3) = \frac{-2}{x + 1}(x + 1)(2x - 3)$$

$$5(x + 1) = -2(2x - 3)$$

$$5x + 5 = -4x + 6$$

$$9x + 5 = 6$$

$$9x = 1$$

$$x = \frac{1}{9}$$

The solution is $\frac{1}{9}$.

You Try It 3

Strategy　To find the cost, write and solve a proportion using x to represent the cost.

Solution

$$\frac{2}{5.80} = \frac{15}{x}$$

$$\frac{2}{5.80} \cdot x(5.80) = \frac{15}{x} \cdot x(5.80)$$

$$2x = 15(5.80)$$

$$2x = 87$$

$$x = 43.50$$

The cost of 15 lb of cashews is $43.50.

SECTION 6.5

You Try It 1

$$\frac{x}{x - 2} + x = \frac{6}{x - 2}$$

$$(x - 2)\left(\frac{x}{x - 2} + x\right) = (x - 2)\left(\frac{6}{x - 2}\right)$$

$$(x - 2)\left(\frac{x}{x - 2}\right) + (x - 2)x = (x - 2)\left(\frac{6}{x - 2}\right)$$

$$x + x^2 - 2x = 6$$

$$x^2 - x = 6$$

$$x^2 - x - 6 = 0$$

$$(x - 3)(x + 2) = 0$$

$$x = 3 \qquad x = -2$$

-2 and 3 check as solutions.

The solutions are -2 and 3.

You Try It 2

$$\frac{5}{x - 2} = 2x + \frac{3x - 1}{x - 2}$$

$$(x - 2)\left(\frac{5}{x - 2}\right) = (x - 2)\left(2x + \frac{3x - 1}{x - 2}\right)$$

$$5 = (x - 2)(2x) + 3x - 1$$

$$5 = 2x^2 - 4x + 3x - 1$$

$$0 = 2x^2 - x - 6$$

$$0 = (2x + 3)(x - 2)$$

$$2x + 3 = 0 \qquad x - 2 = 0$$

$$x = -\frac{3}{2} \qquad\qquad x = 2$$

2 does not check as a solution.

The solution is $-\frac{3}{2}$.

You Try It 3

$$\frac{P_1 V_1}{T_1} = \frac{P_2 V_2}{T_2}$$

$$T_1 T_2\left(\frac{P_1 V_1}{T_1}\right) = T_1 T_2\left(\frac{P_2 V_2}{T_2}\right)$$

$$T_2 P_1 V_1 = T_1 P_2 V_2$$

$$\frac{T_2 P_1 V_1}{P_2 V_2} = \frac{T_1 P_2 V_2}{P_2 V_2}$$

$$\frac{T_2 P_1 V_1}{P_2 V_2} = T_1$$

You Try It 4

Strategy　• Time required for the small pipe to fill the tank: x

	Rate	Time	Part
Large pipe	$\frac{1}{9}$	6	$\frac{6}{9}$
Small pipe	$\frac{1}{x}$	6	$\frac{6}{x}$

• The sum of the part of the task completed by the large pipe and the part of the task completed by the small pipe is 1.

Solution

$$\frac{6}{9} + \frac{6}{x} = 1$$

$$\frac{2}{3} + \frac{6}{x} = 1$$

$$3x\left(\frac{2}{3} + \frac{6}{x}\right) = 3x \cdot 1$$

$$2x + 18 = 3x$$

$$18 = x$$

The small pipe working alone will fill the tank in 18 h.

You Try It 5

Strategy
- Rate of the wind: r

	Distance	Rate	Time
With wind	700	$150 + r$	$\dfrac{700}{150 + r}$
Against wind	500	$150 - r$	$\dfrac{500}{150 - r}$

- The time flying with the wind equals the time flying against the wind.

Solution

$$\frac{700}{150 + r} = \frac{500}{150 - r}$$

$$(150 + r)(150 - r)\left(\frac{700}{150 + r}\right) = (150 + r)(150 - r)\left(\frac{500}{150 - r}\right)$$

$$(150 - r)700 = (150 + r)500$$

$$105{,}000 - 700r = 75{,}000 + 500r$$

$$30{,}000 = 1200r$$

$$25 = r$$

The rate of the wind is 25 mph.

SECTION 6.6

You Try It 1

Strategy   To find the distance:

- Write the basic direct variation equation, replace the variables by the given values, and solve for k.
- Write the direct variation equation, replacing k by its value. Substitute 5 for t and solve for s.

Solution $s = kt^2$
$$64 = k(2)^2$$
$$64 = k \cdot 4$$
$$16 = k$$

$$s = 16t^2 = 16(5)^2 = 400$$

The object will fall 400 ft in 5 s.

You Try It 2

Strategy   To find the resistance:

- Write the basic inverse variation equation, replace the variables by the given values, and solve for k.
- Write the inverse variation equation, replacing k by its value. Substitute 0.02 for d and solve for R.

Solution
$$R = \frac{k}{d^2}$$
$$0.5 = \frac{k}{(0.01)^2}$$
$$0.5 = \frac{k}{0.0001}$$
$$0.00005 = k$$

$$R = \frac{0.00005}{d^2} = \frac{0.00005}{(0.02)^2} = 0.125$$

The resistance is 0.125 ohm.

You Try It 3

Strategy
- Write the basic combined variation equation, replace the variables by the given values, and solve for k.
- Write the combined variation equation, replacing k by its value and substituting 4 for W, 8 for d, and 16 for L. Solve for s.

Solution

$$s = \frac{kWd^2}{L}$$ • **Combined variation equation**

$$1200 = \frac{k(2)(12)^2}{12}$$ • **Replace s by 1200, W by 12, and L by 12.**

$$1200 = 24k$$

$$50 = k$$

$$s = \frac{50Wd^2}{L}$$ • **Replace k by 50 in the combined variation equation.**

$$= \frac{50(4)8^2}{16}$$ • **Replace W by 4, d by 8, and L by 16.**

$$= 800$$

The strength of the beam is 800 lb.

Solutions to Chapter 7 "You Try It"

SECTION 7.1

You Try It 1
$$16^{-3/4} = (2^4)^{-3/4}$$
$$= 2^{-3}$$
$$= \frac{1}{2^3} = \frac{1}{8}$$

You Try It 2
$(-81)^{3/4}$

The base of the exponential expression is negative, and the denominator of the exponent is a positive even number.

Therefore, $(-81)^{3/4}$ is not a real number.

You Try It 3
$$(x^{3/4}y^{1/2}z^{-2/3})^{-4/3} = x^{-1}y^{-2/3}z^{8/9}$$
$$= \frac{z^{8/9}}{xy^{2/3}}$$

You Try It 4
$$\left(\frac{16a^{-2}b^{4/3}}{9a^4b^{-2/3}}\right)^{-1/2} = \left(\frac{2^4a^{-6}b^2}{3^2}\right)^{-1/2}$$
$$= \frac{2^{-2}a^3b^{-1}}{3^{-1}}$$
$$= \frac{3a^3}{2^2b} = \frac{3a^3}{4b}$$

You Try It 5
$$(2x^3)^{3/4} = \sqrt[4]{(2x^3)^3}$$
$$= \sqrt[4]{8x^9}$$

You Try It 6
$$-5a^{5/6} = -5(a^5)^{1/6}$$
$$= -5\sqrt[6]{a^5}$$

You Try It 7
$\sqrt[3]{3ab} = (3ab)^{1/3}$

You Try It 8
$\sqrt[4]{x^4 + y^4} = (x^4 + y^4)^{1/4}$

You Try It 9
$\sqrt[3]{-8x^{12}y^3} = -2x^4y$

You Try It 10
$-\sqrt[4]{81x^{12}y^8} = -3x^3y^2$

SECTION 7.2

You Try It 1
$$\sqrt[5]{x^7} = \sqrt[5]{x^5 \cdot x^2}$$
$$= \sqrt[5]{x^5}\,\sqrt[5]{x^2}$$
$$= x\sqrt[5]{x^2}$$

You Try It 2
$$\sqrt[3]{-64x^8y^{18}} = \sqrt[3]{-64x^6y^{18}(x^2)}$$
$$= \sqrt[3]{-64x^6y^{18}}\,\sqrt[3]{x^2}$$
$$= -4x^2y^6\sqrt[3]{x^2}$$

You Try It 3
$$3xy\sqrt[3]{81x^5y} - \sqrt[3]{192x^8y^4}$$
$$= 3xy\sqrt[3]{27x^3 \cdot 3x^2y} - \sqrt[3]{64x^6y^3 \cdot 3x^2y}$$
$$= 3xy\sqrt[3]{27x^3}\,\sqrt[3]{3x^2y} - \sqrt[3]{64x^6y^3}\,\sqrt[3]{3x^2y}$$
$$= 3xy \cdot 3x\sqrt[3]{3x^2y} - 4x^2y\sqrt[3]{3x^2y}$$
$$= 9x^2y\sqrt[3]{3x^2y} - 4x^2y\sqrt[3]{3x^2y} = 5x^2y\sqrt[3]{3x^2y}$$

You Try It 4
$$\sqrt{5b}\,(\sqrt{3b} - \sqrt{10})$$
$$= \sqrt{15b^2} - \sqrt{50b}$$
$$= \sqrt{b^2 \cdot 15} - \sqrt{25 \cdot 2b}$$
$$= \sqrt{b^2}\,\sqrt{15} - \sqrt{25}\,\sqrt{2b}$$
$$= b\sqrt{15} - 5\sqrt{2b}$$

You Try It 5
$$(2\sqrt[3]{2x} - 3)(\sqrt[3]{2x} - 5)$$
$$= 2\sqrt[3]{4x^2} - 10\sqrt[3]{2x} - 3\sqrt[3]{2x} + 15$$
$$= 2\sqrt[3]{4x^2} - 13\sqrt[3]{2x} + 15$$

You Try It 6
$$(\sqrt{a} - 3\sqrt{y})(\sqrt{a} + 3\sqrt{y})$$
$$= (\sqrt{a})^2 - (3\sqrt{y})^2$$
$$= a - 9y$$

You Try It 7
$$\frac{y}{\sqrt{3y}} = \frac{y}{\sqrt{3y}} \cdot \frac{\sqrt{3y}}{\sqrt{3y}} = \frac{y\sqrt{3y}}{\sqrt{9y^2}} = \frac{y\sqrt{3y}}{3y} = \frac{\sqrt{3y}}{3}$$

You Try It 8
$$\frac{3x}{\sqrt[3]{3x^2}} = \frac{3x}{\sqrt[3]{3x^2}} \cdot \frac{\sqrt[3]{9x}}{\sqrt[3]{9x}} = \frac{3x\sqrt[3]{9x}}{\sqrt[3]{27x^3}}$$
$$= \frac{3x\sqrt[3]{9x}}{3x} = \sqrt[3]{9x}$$

You Try It 9
$$\frac{3 + \sqrt{6}}{2 - \sqrt{6}} = \frac{3 + \sqrt{6}}{2 - \sqrt{6}} \cdot \frac{2 + \sqrt{6}}{2 + \sqrt{6}}$$
$$= \frac{6 + 3\sqrt{6} + 2\sqrt{6} + (\sqrt{6})^2}{2^2 - (\sqrt{6})^2}$$
$$= \frac{6 + 5\sqrt{6} + 6}{4 - 6} = \frac{12 + 5\sqrt{6}}{-2}$$
$$= -\frac{12 + 5\sqrt{6}}{2}$$

SECTION 7.3

You Try It 1
$\sqrt{-45} = i\sqrt{45} = i\sqrt{9 \cdot 5} = 3i\sqrt{5}$

You Try It 2
$$\sqrt{98} - \sqrt{-60} = \sqrt{98} - i\sqrt{60}$$
$$= \sqrt{49 \cdot 2} - i\sqrt{4 \cdot 15}$$
$$= 7\sqrt{2} - 2i\sqrt{15}$$

You Try It 3
$(-4 + 2i) - (6 - 8i) = -10 + 10i$

You Try It 4

$$(16 - \sqrt{-45}) - (3 + \sqrt{-20})$$
$$= (16 - i\sqrt{45}) - (3 + i\sqrt{20})$$
$$= (16 - i\sqrt{9 \cdot 5}) - (3 + i\sqrt{4 \cdot 5})$$
$$= (16 - 3i\sqrt{5}) - (3 + 2i\sqrt{5})$$
$$= 13 - 5i\sqrt{5}$$

You Try It 5

$$(3 - 2i) + (-3 + 2i) = 0 + 0i = 0$$

You Try It 6

$$(-3i)(-10i) = 30i^2 = 30(-1) = -30$$

You Try It 7

$$-\sqrt{-8} \cdot \sqrt{-5} = -i\sqrt{8} \cdot i\sqrt{5} = -i^2\sqrt{40}$$
$$= -(-1)\sqrt{40} = \sqrt{4 \cdot 10} = 2\sqrt{10}$$

You Try It 8

$$-6i(3 + 4i) = -18i - 24i^2$$
$$= -18i - 24(-1) = 24 - 18i$$

You Try It 9

$$\sqrt{-3}(\sqrt{27} - \sqrt{-6}) = i\sqrt{3}(\sqrt{27} - i\sqrt{6})$$
$$= i\sqrt{81} - i^2\sqrt{18}$$
$$= i\sqrt{81} - (-1)\sqrt{9 \cdot 2}$$
$$= 9i + 3\sqrt{2}$$
$$= 3\sqrt{2} + 9i$$

You Try It 10

$$(4 - 3i)(2 - i) = 8 - 4i - 6i + 3i^2$$
$$= 8 - 10i + 3i^2$$
$$= 8 - 10i + 3(-1)$$
$$= 5 - 10i$$

You Try It 11

$$(3 + 6i)(3 - 6i) = 3^2 + 6^2$$
$$= 9 + 36$$
$$= 45$$

You Try It 12

$$\frac{2 - 3i}{4i} = \frac{2 - 3i}{4i} \cdot \frac{i}{i}$$
$$= \frac{2i - 3i^2}{4i^2}$$
$$= \frac{2i - 3(-1)}{4(-1)}$$
$$= \frac{3 + 2i}{-4} = -\frac{3}{4} - \frac{1}{2}i$$

You Try It 13

$$\frac{2 + 5i}{3 - 2i} = \frac{2 + 5i}{3 - 2i} \cdot \frac{3 + 2i}{3 + 2i} = \frac{6 + 4i + 15i + 10i^2}{3^2 + 2^2}$$
$$= \frac{6 + 19i + 10(-1)}{13} = \frac{-4 + 19i}{13}$$
$$= -\frac{4}{13} + \frac{19}{13}i$$

SECTION 7.4

You Try It 1

$$\sqrt{x} - \sqrt{x + 5} = 1$$
$$\sqrt{x} = 1 + \sqrt{x + 5}$$
$$(\sqrt{x})^2 = (1 + \sqrt{x + 5})^2$$
$$x = 1 + 2\sqrt{x + 5} + x + 5$$
$$-6 = 2\sqrt{x + 5}$$
$$-3 = \sqrt{x + 5}$$
$$(-3)^2 = (\sqrt{x + 5})^2$$
$$9 = x + 5$$
$$4 = x$$

4 does not check as a solution. The equation has no solution.

You Try It 2

$$\sqrt[4]{x - 8} = 3$$
$$(\sqrt[4]{x - 8})^4 = 3^4$$
$$x - 8 = 81$$
$$x = 89$$

Check:
$$\sqrt[4]{x - 8} = 3$$

$$\sqrt[4]{89 - 8} \ \Big|\ 3$$
$$\sqrt[4]{81} \ \Big|\ 3$$
$$3 = 3$$

The solution is 89.

You Try It 3

Strategy To find the diagonal, use the Pythagorean Theorem. One leg is the length of the rectangle. The second leg is the width of the rectangle. The hypotenuse is the diagonal of the rectangle.

Solution
$$c^2 = a^2 + b^2$$
$$c^2 = (6)^2 + (3)^2$$
$$c^2 = 36 + 9$$
$$c^2 = 45$$
$$(c^2)^{1/2} = (45)^{1/2}$$
$$c = \sqrt{45}$$
$$c \approx 6.7$$

The diagonal is 6.7 cm.

You Try It 4

Strategy To find the height, replace d in the equation with the given value and solve for h.

Solution
$$d = \sqrt{1.5h}$$
$$5.5 = \sqrt{1.5h}$$
$$(5.5)^2 = (\sqrt{1.5h})^2$$
$$30.25 = 1.5h$$
$$20.17 \approx h$$

The periscope must be approximately 20.17 ft above the water.

You Try It 5

Strategy To find the distance, replace the variables v and a in the equation by their given values and solve for s.

Solution
$$v = \sqrt{2as}$$
$$88 = \sqrt{2 \cdot 22s}$$
$$88 = \sqrt{44s}$$
$$(88)^2 = (\sqrt{44s})^2$$
$$7744 = 44s$$
$$176 = s$$

The distance required is 176 ft.

Solutions to Chapter 8 "You Try It"

SECTION 8.1

You Try It 1
$$2x^2 = 7x - 3$$
$$2x^2 - 7x + 3 = 0$$
$$(2x - 1)(x - 3) = 0$$

$$2x - 1 = 0 \qquad x - 3 = 0$$
$$2x = 1 \qquad x = 3$$
$$x = \frac{1}{2}$$

The solutions are $\frac{1}{2}$ and 3.

You Try It 2
$$x^2 - 3ax - 4a^2 = 0$$
$$(x + a)(x - 4a) = 0$$

$$x + a = 0 \qquad x - 4a = 0$$
$$x = -a \qquad x = 4a$$

The solutions are $-a$ and $4a$.

You Try It 3
$$(x - r_1)(x - r_2) = 0$$
$$(x - 3)\left[x - \left(-\frac{1}{2}\right)\right] = 0$$
$$(x - 3)\left(x + \frac{1}{2}\right) = 0$$
$$x^2 - \frac{5}{2}x - \frac{3}{2} = 0$$
$$2\left(x^2 - \frac{5}{2}x - \frac{3}{2}\right) = 2 \cdot 0$$
$$2x^2 - 5x - 3 = 0$$

You Try It 4
$$2(x + 1)^2 - 24 = 0$$
$$2(x + 1)^2 = 24$$
$$(x + 1)^2 = 12$$

$$\sqrt{(x + 1)^2} = \sqrt{12}$$
$$x + 1 = \pm\sqrt{12} = \pm 2\sqrt{3}$$

$$x + 1 = 2\sqrt{3} \qquad x + 1 = -2\sqrt{3}$$
$$x = -1 + 2\sqrt{3} \qquad x = -1 - 2\sqrt{3}$$

The solutions are $-1 + 2\sqrt{3}$ and $-1 - 2\sqrt{3}$.

SECTION 8.2

You Try It 1
$$4x^2 - 4x - 1 = 0$$
$$4x^2 - 4x = 1$$
$$\frac{1}{4}(4x^2 - 4x) = \frac{1}{4} \cdot 1$$
$$x^2 - x = \frac{1}{4}$$

Complete the square.

$$x^2 - x + \frac{1}{4} = \frac{1}{4} + \frac{1}{4}$$
$$\left(x - \frac{1}{2}\right)^2 = \frac{2}{4}$$
$$\sqrt{\left(x - \frac{1}{2}\right)^2} = \sqrt{\frac{2}{4}}$$
$$x - \frac{1}{2} = \pm\frac{\sqrt{2}}{2}$$

$$x - \frac{1}{2} = \frac{\sqrt{2}}{2} \qquad x - \frac{1}{2} = -\frac{\sqrt{2}}{2}$$
$$x = \frac{1}{2} + \frac{\sqrt{2}}{2} \qquad x = \frac{1}{2} - \frac{\sqrt{2}}{2}$$

The solutions are $\dfrac{1 + \sqrt{2}}{2}$ and $\dfrac{1 - \sqrt{2}}{2}$.

You Try It 2

$2x^2 + x - 5 = 0$

$\qquad 2x^2 + x = 5$

$\dfrac{1}{2}(2x^2 + x) = \dfrac{1}{2} \cdot 5$

$\qquad x^2 + \dfrac{1}{2}x = \dfrac{5}{2}$

Complete the square.

$$x^2 + \dfrac{1}{2}x + \dfrac{1}{16} = \dfrac{5}{2} + \dfrac{1}{16}$$

$$\left(x + \dfrac{1}{4}\right)^2 = \dfrac{41}{16}$$

$$\sqrt{\left(x + \dfrac{1}{4}\right)^2} = \sqrt{\dfrac{41}{16}}$$

$$x + \dfrac{1}{4} = \pm\dfrac{\sqrt{41}}{4}$$

$$x + \dfrac{1}{4} = \dfrac{\sqrt{41}}{4} \qquad\qquad x + \dfrac{1}{4} = -\dfrac{\sqrt{41}}{4}$$

$$x = -\dfrac{1}{4} + \dfrac{\sqrt{41}}{4} \qquad\qquad x = -\dfrac{1}{4} - \dfrac{\sqrt{41}}{4}$$

The solutions are $\dfrac{-1 + \sqrt{41}}{4}$ and $\dfrac{-1 - \sqrt{41}}{4}$.

SECTION 8.3

You Try It 1 $x^2 - 2x + 10 = 0$

$a = 1, b = -2, c = 10$

$$x = \dfrac{-b \pm \sqrt{b^2 - 4ac}}{2a}$$

$$= \dfrac{-(-2) \pm \sqrt{(-2)^2 - 4(1)(10)}}{2 \cdot 1}$$

$$= \dfrac{2 \pm \sqrt{4 - 40}}{2} = \dfrac{2 \pm \sqrt{-36}}{2}$$

$$= \dfrac{2 \pm 6i}{2} = 1 \pm 3i$$

The solutions are $1 + 3i$ and $1 - 3i$.

You Try It 2 $\qquad 4x^2 = 4x - 1$

$4x^2 - 4x + 1 = 0$

$a = 4, b = -4, c = 1$

$$x = \dfrac{-b \pm \sqrt{b^2 - 4ac}}{2a}$$

$$= \dfrac{-(-4) \pm \sqrt{(-4)^2 - 4(4)(1)}}{2 \cdot 4}$$

$$= \dfrac{4 \pm \sqrt{16 - 16}}{8} = \dfrac{4 \pm \sqrt{0}}{8}$$

$$= \dfrac{4}{8} = \dfrac{1}{2}$$

The solution is $\dfrac{1}{2}$.

You Try It 3 $3x^2 - x - 1 = 0$

$a = 3, b = -1, c = -1$

$b^2 - 4ac =$

$(-1)^2 - 4(3)(-1) = 1 + 12 = 13$

$13 > 0$

Because the discriminant is greater than zero, the equation has two real number solutions.

SECTION 8.4

You Try It 1 $\qquad x - 5x^{1/2} + 6 = 0$

$(x^{1/2})^2 - 5(x^{1/2}) + 6 = 0$

$\qquad u^2 - 5u + 6 = 0$

$\qquad (u - 2)(u - 3) = 0$

$$u - 2 = 0 \qquad\qquad u - 3 = 0$$
$$u = 2 \qquad\qquad u = 3$$

Replace u by $x^{1/2}$.

$$x^{1/2} = 2 \qquad\qquad x^{1/2} = 3$$
$$\sqrt{x} = 2 \qquad\qquad \sqrt{x} = 3$$
$$(\sqrt{x})^2 = 2^2 \qquad\qquad (\sqrt{x})^2 = 3^2$$
$$x = 4 \qquad\qquad x = 9$$

The solutions are 4 and 9.

You Try It 2 $\sqrt{2x + 1} + x = 7$

$$\sqrt{2x + 1} = 7 - x$$
$$(\sqrt{2x + 1})^2 = (7 - x)^2$$
$$2x + 1 = 49 - 14x + x^2$$
$$0 = x^2 - 16x + 48$$
$$0 = (x - 4)(x - 12)$$

$$x - 4 = 0 \qquad\qquad x - 12 = 0$$
$$x = 4 \qquad\qquad x = 12$$

4 checks as a solution.
12 does not check as a solution.

The solution is 4.

You Try It 3 $\sqrt{2x - 1} + \sqrt{x} = 2$

Solve for one of the radical expressions.

$$\sqrt{2x - 1} = 2 - \sqrt{x}$$
$$(\sqrt{2x - 1})^2 = (2 - \sqrt{x})^2$$
$$2x - 1 = 4 - 4\sqrt{x} + x$$
$$x - 5 = -4\sqrt{x}$$

Square each side of the equation.

$$(x - 5)^2 = (-4\sqrt{x})^2$$
$$x^2 - 10x + 25 = 16x$$
$$x^2 - 26x + 25 = 0$$
$$(x - 1)(x - 25) = 0$$

$$x - 1 = 0 \qquad x - 25 = 0$$
$$x = 1 \qquad\qquad x = 25$$

1 checks as a solution.
25 does not check as a solution.

The solution is 1.

You Try It 4

$$3y + \frac{25}{3y - 2} = -8$$

$$(3y - 2)\left(3y + \frac{25}{3y - 2}\right) = (3y - 2)(-8)$$

$$(3y - 2)(3y) + (3y - 2)\left(\frac{25}{3y - 2}\right) = (3y - 2)(-8)$$

$$9y^2 - 6y + 25 = -24y + 16$$
$$9y^2 + 18y + 9 = 0$$
$$9(y^2 + 2y + 1) = 0$$
$$9(y + 1)(y + 1) = 0$$

$$y + 1 = 0 \qquad y + 1 = 0$$
$$y = -1 \qquad\quad y = -1$$

The solution is -1.

SECTION 8.5

You Try It 1

$$2x^2 - x - 10 \le 0$$
$$(2x - 5)(x + 2) \le 0$$

$$\left\{ x \mid -2 \le x \le \frac{5}{2} \right\}$$

SECTION 8.6

You Try It 1

Strategy
- This is a geometry problem.
- Width of the rectangle: w
  Length of the rectangle: $w + 3$
- Use the equation $A = L \cdot w$.

Solution
$$A = L \cdot w$$
$$54 = (w + 3)(w)$$
$$54 = w^2 + 3w$$
$$0 = w^2 + 3w - 54$$
$$0 = (w + 9)(w - 6)$$

$$w + 9 = 0 \qquad w - 6 = 0$$
$$w = -9 \qquad\quad w = 6$$

The solution -9 is not possible.

$$w + 3 = 6 + 3 = 9$$

The length is 9 m.

Solutions to Chapter 9 "You Try It"

SECTION 9.1

You Try It 1

x-coordinate of vertex:

$$-\frac{b}{2a} = -\frac{4}{2(4)} = -\frac{1}{2}$$

y-coordinate of vertex:

$$y = 4x^2 + 4x + 1$$
$$= 4\left(-\frac{1}{2}\right)^2 + 4\left(-\frac{1}{2}\right) + 1$$
$$= 1 - 2 + 1$$
$$= 0$$

Vertex: $\left(-\frac{1}{2}, 0\right)$

Axis of symmetry: $x = -\frac{1}{2}$

You Try It 2

x-coordinate of vertex:

$$-\frac{b}{2a} = -\frac{-2}{2(-1)} = -1$$

y-coordinate of vertex:

$$f(x) = -x^2 - 2x - 1$$
$$f(-1) = -(-1)^2 - 2(-1) - 1$$
$$= -1 + 2 - 1$$
$$= 0$$

Vertex: $(-1, 0)$

The domain is $\{x \mid x \in \text{real numbers}\}$. The range is $\{y \mid y \le 0\}$.

You Try It 3

$$y = x^2 + 3x + 4$$
$$0 = x^2 + 3x + 4$$

$$x = \frac{-b \pm \sqrt{b^2 - 4ac}}{2a}$$

$$= \frac{-3 \pm \sqrt{3^2 - 4(1)(4)}}{2 \cdot 1} \qquad \bullet\ a = 1,\ b = 3,\ c = 4$$

$$= \frac{-3 \pm \sqrt{-7}}{2}$$

$$= \frac{-3 \pm i\sqrt{7}}{2}$$

The equation has no real number solutions. There are no x-intercepts.

You Try It 4

$$g(x) = x^2 - x + 6$$
$$0 = x^2 - x + 6$$

$$x = \frac{-b \pm \sqrt{b^2 - 4ac}}{2a}$$

$$= \frac{-(-1) \pm \sqrt{(-1)^2 - 4(1)(6)}}{2(1)} \qquad \bullet\ a = 1,\ b = -1,\ c = 6$$

$$= \frac{1 \pm \sqrt{1 - 24}}{2}$$

$$= \frac{1 \pm \sqrt{-23}}{2} = \frac{1 \pm i\sqrt{23}}{2}$$

The zeros of the function are $\dfrac{1}{2} + \dfrac{\sqrt{23}}{2} i$ and $\dfrac{1}{2} - \dfrac{\sqrt{23}}{2} i$.

You Try It 5

$$y = x^2 - x - 6$$
$$a = 1,\ b = -1,\ c = -6$$

$$b^2 - 4ac$$
$$(-1)^2 - 4(1)(-6) = 1 + 24 = 25$$

Because the discriminant is greater than zero, the parabola has two x-intercepts.

You Try It 6

$$f(x) = -3x^2 + 4x - 1$$

$$x = -\frac{b}{2a} = -\frac{4}{2(-3)} = \frac{2}{3}$$

$$f(x) = -3x^2 + 4x - 1$$

$$f\left(\frac{2}{3}\right) = -3\left(\frac{2}{3}\right)^2 + 4\left(\frac{2}{3}\right) - 1$$

$$= -\frac{4}{3} + \frac{8}{3} - 1 = \frac{1}{3}$$

Because a is negative, the function has a maximum value.

The maximum value of the function is $\dfrac{1}{3}$.

You Try It 7

Strategy   • To find the time it takes the ball to reach its maximum height, find the t-coordinate of the vertex.

• To find the maximum height, evaluate the function at the t-coordinate of the vertex.

Solution

$$t = -\frac{b}{2a} = -\frac{64}{2(-16)} = 2$$

The ball reaches its maximum height in 2 s.

$$s(t) = -16t^2 + 64t$$
$$s(2) = -16(2)^2 + 64(2) = -64 + 128 = 64$$

The maximum height is 64 ft.

You Try It 8

Strategy

$$P = 2x + y$$
$$100 = 2x + y \qquad \bullet\ P = 100$$
$$100 - 2x = y \qquad \bullet\ \text{Solve for } y.$$

Express the area of the rectangle in terms of x.

$$A = xy$$
$$A = x(100 - 2x) \qquad \bullet\ y = 100 - 2x$$
$$A = -2x^2 + 100x$$

• To find the width, find the x-coordinate of the vertex of $f(x) = -2x^2 + 100x$.

• To find the length, replace x in $y = 100 - 2x$ by the x-coordinate of the vertex.

Solution $$x = -\frac{b}{2a} = -\frac{100}{2(-2)} = 25$$

The width is 25 ft.

$$100 - 2x = 100 - 2(25)$$
$$= 100 - 50 = 50$$

The length is 50 ft.

SECTION 9.2

You Try It 1   Any vertical line intersects the graph at most once. The graph is the graph of a function.

You Try It 2

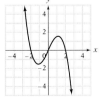

Domain: $(-\infty, \infty)$
Range: $(-\infty, \infty)$

You Try It 3

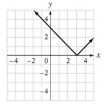

Domain: $\{x \mid x \in \text{real numbers}\}$
Range: $\{y \mid y \geq 0\}$

You Try It 4

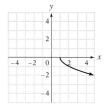

Domain: $[1, \infty)$
Range: $(-\infty, 0]$

SECTION 9.3

You Try It 1
$$
\begin{aligned}
(f + g)(-2) &= f(-2) + g(-2) \\
&= [(-2)^2 + 2(-2)] + [5(-2) - 2] \\
&= (4 - 4) + (-10 - 2) \\
&= -12
\end{aligned}
$$

$(f + g)(-2) = -12$

You Try It 2
$$
\begin{aligned}
(f \cdot g)(3) &= f(3) \cdot g(3) \\
&= (4 - 3^2) \cdot [3(3) - 4] \\
&= (4 - 9) \cdot (9 - 4) \\
&= (-5)(5) \\
&= -25
\end{aligned}
$$

$(f \cdot g)(3) = -25$

You Try It 3
$$
\begin{aligned}
\left(\frac{f}{g}\right)(4) &= \frac{f(4)}{g(4)} \\
&= \frac{4^2 - 4}{4^2 + 2 \cdot 4 + 1} \\
&= \frac{16 - 4}{16 + 8 + 1} \\
&= \frac{12}{25}
\end{aligned}
$$

$\left(\frac{f}{g}\right)(4) = \frac{12}{25}$

You Try It 4
$$
\begin{aligned}
g(x) &= x^2 \\
g(-1) &= (-1)^2 = 1
\end{aligned}
$$
$$
\begin{aligned}
f(x) &= 1 - 2x \\
f[g(-1)] &= f(1) = 1 - 2(1) = -1
\end{aligned}
$$

You Try It 5
$$
M(s) = s^3 + 1
$$
$$
\begin{aligned}
M[L(s)] &= (s + 1)^3 + 1 \\
&= s^3 + 3s^2 + 3s + 1 + 1 \\
&= s^3 + 3s^2 + 3s + 2
\end{aligned}
$$

SECTION 9.4

You Try It 1   Because any horizontal line intersects the graph at most once, the graph is the graph of a 1–1 function.

You Try It 2
$$
\begin{aligned}
f(x) &= \frac{1}{2}x + 4 \\
y &= \frac{1}{2}x + 4 \\
x &= \frac{1}{2}y + 4 \\
x - 4 &= \frac{1}{2}y \\
2x - 8 &= y \\
f^{-1}(x) &= 2x - 8
\end{aligned}
$$

The inverse of the function is given by $f^{-1}(x) = 2x - 8$.

You Try It 3
$$
\begin{aligned}
f[g(x)] &= 2\left(\frac{1}{2}x - 3\right) - 6 \\
&= x - 6 - 6 = x - 12
\end{aligned}
$$

No, $g(x)$ is not the inverse of $f(x)$.

Solutions to Chapter 10 "You Try It"

SECTION 10.1

You Try It 1

$$f(x) = \left(\frac{2}{3}\right)^x$$

$$f(3) = \left(\frac{2}{3}\right)^3 = \frac{8}{27}$$

$$f(-2) = \left(\frac{2}{3}\right)^{-2} = \left(\frac{3}{2}\right)^2 = \frac{9}{4}$$

You Try It 2 $f(x) = 2^{2x+1}$
$f(0) = 2^{2(0)+1} = 2^1 = 2$

$$f(-2) = 2^{2(-2)+1} = 2^{-3} = \frac{1}{2^3} = \frac{1}{8}$$

You Try It 3 $f(x) = e^{2x-1}$
$f(2) = e^{2 \cdot 2 - 1} = e^3 \approx 20.0855$
$f(-2) = e^{2(-2)-1} = e^{-5} \approx 0.0067$

You Try It 4

x	y
-4	4
-2	2
0	1
2	$\frac{1}{2}$
4	$\frac{1}{4}$

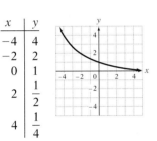

You Try It 5

x	y
-2	$\frac{5}{4}$
-1	$\frac{3}{2}$
0	2
1	3
2	5

You Try It 6

x	y
-2	6
-1	4
0	3
1	$\frac{5}{2}$
2	$\frac{9}{4}$

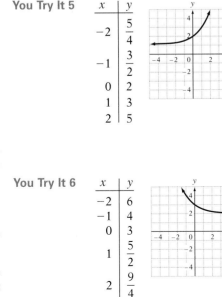

You Try It 7

x	y
-4	$\frac{9}{4}$
-2	$\frac{5}{2}$
0	3
2	4
4	6

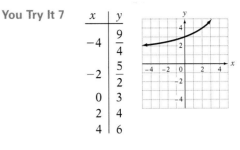

SECTION 10.2

You Try It 1 $\log_4 64 = x$
$$64 = 4^x$$
$$4^3 = 4^x$$
$$3 = x$$

$$\log_4 64 = 3$$

You Try It 2 $\log_2 x = -4$
$$2^{-4} = x$$
$$\frac{1}{2^4} = x$$
$$\frac{1}{16} = x$$

The solution is $\frac{1}{16}$.

You Try It 3 $\ln x = 3$
$$e^3 = x$$
$$20.0855 \approx x$$

You Try It 4 $\log_8 \sqrt[3]{xy^2} = \log_8 (xy^2)^{1/3} = \frac{1}{3} \log_8(xy^2)$

$$= \frac{1}{3} (\log_8 x + \log_8 y^2)$$

$$= \frac{1}{3} (\log_8 x + 2 \log_8 y)$$

$$= \frac{1}{3} \log_8 x + \frac{2}{3} \log_8 y$$

You Try It 5
$$\frac{1}{3} (\log_4 x - 2 \log_4 y + \log_4 z)$$

$$= \frac{1}{3} (\log_4 x - \log_4 y^2 + \log_4 z)$$

$$= \frac{1}{3} \left(\log_4 \frac{x}{y^2} + \log_4 z \right) = \frac{1}{3} \left(\log_4 \frac{xz}{y^2} \right)$$

$$= \log_4 \left(\frac{xz}{y^2} \right)^{1/3} = \log_4 \sqrt[3]{\frac{xz}{y^2}}$$

You Try It 6   Because $\log_b 1 = 0$, $\log_9 1 = 0$

You Try It 7 $\log_7 6.45 = \dfrac{\log 6.45}{\log 7} \approx 0.95795$

You Try It 8 $\log_3 0.834 = \dfrac{\ln 0.834}{\ln 3} \approx -0.16523$

SECTION 10.3

You Try It 1
$$f(x) = \log_2(x - 1)$$
$$y = \log_2(x - 1)$$
$$2^y = x - 1$$
$$2^y + 1 = x$$

x	y
$\dfrac{5}{4}$	-2
$\dfrac{3}{2}$	-1
2	0
3	1
5	2

You Try It 2
$$f(x) = \log_3(2x)$$
$$y = \log_3(2x)$$
$$3^y = 2x$$
$$\dfrac{3^y}{2} = x$$

x	y
$\dfrac{1}{18}$	-2
$\dfrac{1}{6}$	-1
$\dfrac{1}{2}$	0
$\dfrac{3}{2}$	1
$\dfrac{9}{2}$	2

You Try It 3
$$f(x) = -\log_3(x + 1)$$
$$y = -\log_3(x + 1)$$
$$-y = \log_3(x + 1)$$
$$3^{-y} = x + 1$$
$$3^{-y} - 1 = x$$

x	y
8	-2
2	-1
0	0
$-\dfrac{2}{3}$	1
$-\dfrac{8}{9}$	2

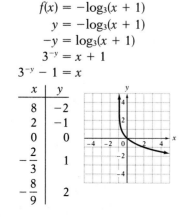

SECTION 10.4

You Try It 1
$$(1.06)^n = 1.5$$
$$\log(1.06)^n = \log 1.5$$
$$n \log 1.06 = \log 1.5$$
$$n = \dfrac{\log 1.5}{\log 1.06}$$
$$n \approx 6.9585$$

The solution is 6.9585.

You Try It 2
$$4^{3x} = 25$$
$$\log 4^{3x} = \log 25$$
$$3x \log 4 = \log 25$$
$$3x = \dfrac{\log 25}{\log 4}$$
$$3x \approx 2.3219$$
$$x \approx 0.7740$$

The solution is 0.7740.

You Try It 3 $\log_4(x^2 - 3x) = 1$

Rewrite in exponential form.

$$4^1 = x^2 - 3x$$
$$4 = x^2 - 3x$$
$$0 = x^2 - 3x - 4$$
$$0 = (x + 1)(x - 4)$$

$$x + 1 = 0 \qquad x - 4 = 0$$
$$x = -1 \qquad x = 4$$

The solutions are -1 and 4.

You Try It 4
$$\log_3 x + \log_3(x + 3) = \log_3 4$$
$$\log_3[x(x + 3)] = \log_3 4$$

Use the fact that if $\log_b u = \log_b v$,
then $u = v$.

$$x(x + 3) = 4$$
$$x^2 + 3x = 4$$
$$x^2 + 3x - 4 = 0$$
$$(x + 4)(x - 1) = 0$$

$$x + 4 = 0 \qquad x - 1 = 0$$
$$x = -4 \qquad x = 1$$

-4 does not check as a solution.
The solution is 1.

You Try It 5
$$\log_3 x + \log_3(x + 6) = 3$$
$$\log_3[x(x + 6)] = 3$$

$$x(x + 6) = 3^3$$
$$x^2 + 6x - 27 = 0$$
$$(x + 9)(x - 3) = 0$$

$$x + 9 = 0 \qquad x - 3 = 0$$
$$x = -9 \qquad x = 3$$

-9 does not check as a solution.
The solution is 3.

SECTION 10.5

You Try It 1

Strategy   To find the hydrogen ion concentration, replace pH by 2.9 in the equation $pH = -\log(H^+)$ and solve for H^+.

Solution
$$pH = -\log(H^+)$$
$$2.9 = -\log(H^+)$$
$$-2.9 = \log(H^+)$$
$$10^{-2.9} = H^+$$
$$0.00126 \approx H^+$$

The hydrogen ion concentration is approximately 0.00126.

You Try It 2

Strategy   To find the intensity, use the equation for the Richter scale magnitude of an earthquake, $M = \log\left(\dfrac{I}{I_0}\right)$. Replace M by 5.2 and solve for I.

Solution

$$M = \log\left(\frac{I}{I_0}\right)$$

$$5.2 = \log\frac{I}{I_0} \qquad \bullet \text{ Replace } M \text{ by 5.2.}$$

$$10^{5.2} = \frac{I}{I_0} \qquad \bullet \text{ Write in exponential form.}$$

$$10^{5.2}I_0 = I$$
$$158{,}489I_0 \approx I$$

The earthquake had an intensity that was approximately 158,489 times the intensity of a zero-level earthquake.

Solutions to Chapter 11 "You Try It"

SECTION 11.1

You Try It 1
$$y = x^2 + 2x + 1$$
$$-\frac{b}{2a} = -\frac{2}{2(1)} = -1$$
$$y = (-1)^2 + 2(-1) + 1$$
$$= 0$$

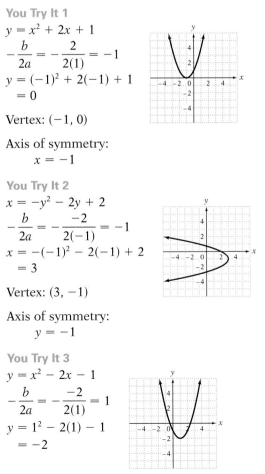

Vertex: $(-1, 0)$

Axis of symmetry:
$$x = -1$$

You Try It 2
$$x = -y^2 - 2y + 2$$
$$-\frac{b}{2a} = -\frac{-2}{2(-1)} = -1$$
$$x = -(-1)^2 - 2(-1) + 2$$
$$= 3$$

Vertex: $(3, -1)$

Axis of symmetry:
$$y = -1$$

You Try It 3
$$y = x^2 - 2x - 1$$
$$-\frac{b}{2a} = -\frac{-2}{2(1)} = 1$$
$$y = 1^2 - 2(1) - 1$$
$$= -2$$

Vertex: $(1, -2)$

Axis of symmetry:
$$x = 1$$

SECTION 11.2

You Try It 1

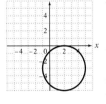

You Try It 2
$$(x - h)^2 + (y - k)^2 = r^2$$
$$(x - 2)^2 + [y - (-3)]^2 = 4^2$$
$$(x - 2)^2 + (y + 3)^2 = 16$$

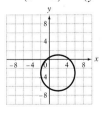

You Try It 3
$$x^2 + y^2 - 2x - 15 = 0$$
$$(x^2 - 2x) + y^2 = 15$$
$$(x^2 - 2x + 1) + y^2 = 15 + 1$$
$$(x - 1)^2 + y^2 = 16$$

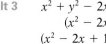

Center: (1, 0)
Radius: 4

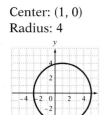

You Try It 4

$$x^2 + y^2 - 4x + 8y + 15 = 0$$
$$(x^2 - 4x) + (y^2 + 8y) = -15$$
$$(x^2 - 4x + 4) + (y^2 + 8y + 16) = -15 + 4 + 16$$
$$(x - 2)^2 + (y + 4)^2 = 5$$

Center: (2, −4)
Radius: $\sqrt{5}$

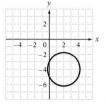

SECTION 11.3

You Try It 1
x-intercepts:
(2, 0) and (−2, 0)

y-intercepts:
(0, 5) and (0, −5)

You Try It 2
x-intercepts:
$(3\sqrt{2}, 0)$ and $(-3\sqrt{2}, 0)$

y-intercepts:
(0, 3) and (0, −3)

$$\left(3\sqrt{2} \approx 4\frac{1}{4}\right)$$

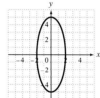

You Try It 3
Axis of symmetry:
x-axis

Vertices:
(3, 0) and (−3, 0)

Asymptotes:
$$y = \frac{5}{3}x \text{ and } y = -\frac{5}{3}x$$

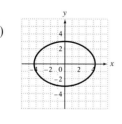

You Try It 4
Axis of symmetry:
y-axis

Vertices:
(0, 3) and (0, −3)

Asymptotes:
$y = x$ and $y = -x$

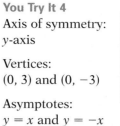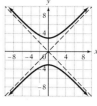

SECTION 11.4

You Try It 1   (1) $y = 2x^2 + x - 3$
(2) $y = 2x^2 - 2x + 9$

Use the substitution method.

$$y = 2x^2 + x - 3$$
$$2x^2 - 2x + 9 = 2x^2 + x - 3$$
$$-3x + 9 = -3$$
$$-3x = -12$$
$$x = 4$$

Substitute into Equation (1).

$$y = 2x^2 + x - 3$$
$$y = 2(4)^2 + 4 - 3$$
$$y = 32 + 4 - 3$$
$$y = 33$$

The solution is (4, 33).

You Try It 2
(1) $x^2 - y^2 = 10$
(2) $x^2 + y^2 = 8$

Use the addition method.

$$2x^2 = 18$$
$$x^2 = 9$$
$$x = \pm\sqrt{9} = \pm 3$$

Substitute into Equation (2).

$x^2 + y^2 = 8$	$x^2 + y^2 = 8$
$3^2 + y^2 = 8$	$(-3)^2 + y^2 = 8$
$9 + y^2 = 8$	$9 + y^2 = 8$
$y^2 = -1$	$y^2 = -1$
$y = \pm\sqrt{-1}$	$y = \pm\sqrt{-1}$

Because y is not a real number, the system of equations has no real number solution. The graphs do not intersect.

SECTION 11.5

You Try It 1 $\dfrac{x^2}{9} - \dfrac{y^2}{16} \le 1$

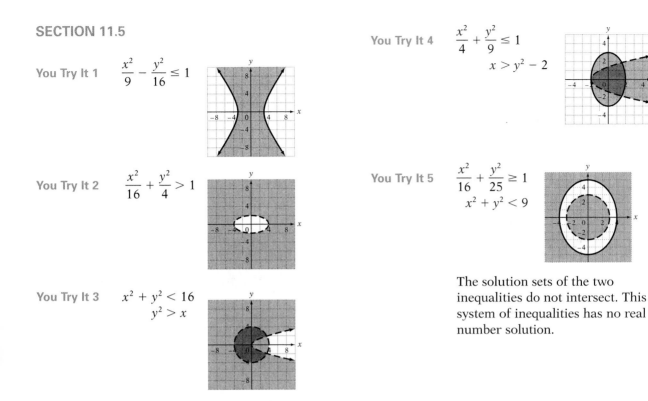

You Try It 2 $\dfrac{x^2}{16} + \dfrac{y^2}{4} > 1$

You Try It 3 $x^2 + y^2 < 16$
$y^2 > x$

You Try It 4 $\dfrac{x^2}{4} + \dfrac{y^2}{9} \le 1$
$x > y^2 - 2$

You Try It 5 $\dfrac{x^2}{16} + \dfrac{y^2}{25} \ge 1$
$x^2 + y^2 < 9$

The solution sets of the two inequalities do not intersect. This system of inequalities has no real number solution.

Solutions to Chapter 12 "You Try It"

SECTION 12.1

You Try It 1
$a_n = n(n + 1)$
$a_1 = 1(1 + 1) = 2$ The first term is 2.
$a_2 = 2(2 + 1) = 6$ The second term is 6.
$a_3 = 3(3 + 1) = 12$ The third term is 12.
$a_4 = 4(4 + 1) = 20$ The fourth term is 20.

You Try It 2
$a_n = \dfrac{1}{n(n + 2)}$

$a_6 = \dfrac{1}{6(6 + 2)} = \dfrac{1}{48}$ The sixth term is $\dfrac{1}{48}$.

$a_9 = \dfrac{1}{9(9 + 2)} = \dfrac{1}{99}$ The ninth term is $\dfrac{1}{99}$.

You Try It 3
$\displaystyle\sum_{i=3}^{6} (i^2 - 2) = (3^2 - 2) + (4^2 - 2) + (5^2 - 2)$
$+ (6^2 - 2)$
$= 7 + 14 + 23 + 34 = 78$

You Try It 4 $\displaystyle\sum_{n=1}^{4} \dfrac{x}{n} = \dfrac{x}{1} + \dfrac{x}{2} + \dfrac{x}{3} + \dfrac{x}{4}$
$= x + \dfrac{x}{2} + \dfrac{x}{3} + \dfrac{x}{4}$

SECTION 12.2

You Try It 1 $9, 3, -3, -9, \ldots$
$d = a_2 - a_1 = 3 - 9 = -6$

$a_n = a_1 + (n - 1)d$
$a_{15} = 9 + (15 - 1)(-6) = 9 + (14)(-6)$
$= 9 - 84$
$a_{15} = -75$

You Try It 2 $7, 9, 11, \ldots, 59$
$d = a_2 - a_1 = 9 - 7 = 2$

$a_n = a_1 + (n - 1)d$
$59 = 7 + (n - 1)2$
$59 = 7 + 2n - 2$
$59 = 5 + 2n$
$54 = 2n$
$27 = n$

There are 27 terms in the sequence.

You Try It 3 $-4, -2, 0, 2, 4, \ldots$

$$d = a_2 - a_1 = -2 - (-4) = 2$$

$$a_n = a_1 + (n - 1)d$$
$$a_{25} = -4 + (25 - 1)2 = -4 + (24)2$$
$$= -4 + 48$$
$$a_{25} = 44$$

$$S_n = \frac{n}{2}(a_1 + a_n)$$

$$S_{25} = \frac{25}{2}(-4 + 44) = \frac{25}{2}(40) = 25(20) = 500$$

$$S_{25} = 500$$

You Try It 4 $\displaystyle\sum_{n=1}^{18}(3n - 2)$

$$a_n = 3n - 2$$
$$a_1 = 3(1) - 2 = 1$$
$$a_{18} = 3(18) - 2 = 52$$

$$S_n = \frac{n}{2}(a_1 + a_n)$$

$$S_{18} = \frac{18}{2}(1 + 52) = 9(53) = 477$$

You Try It 5

Strategy   To find the value of the 20th-place prize:

- Write the equation for the nth-place prize.
- Find the 20th term of the sequence.

To find the total amount of prize money being awarded, use the Formula for the Sum of n Terms of an Arithmetic Sequence.

Solution
$10,000, 9700, \ldots$

$$d = a_2 - a_1 = 9700 - 10,000 = -300$$

$$a_n = a_1 + (n - 1)d$$
$$= 10,000 + (n - 1)(-300)$$
$$= 10,000 - 300n + 300$$
$$= -300n + 10,300$$

$$a_{20} = -300(20) + 10,300 = -6000 + 10,300$$
$$= 4300$$

The value of the 20th-place prize is \$4300.

$$S_n = \frac{n}{2}(a_1 + a_n)$$

$$S_{20} = \frac{20}{2}(10,000 + 4300) = 10(14,300)$$
$$= 143,000$$

The total amount of prize money being awarded is \$143,000.

SECTION 12.3

You Try It 1 $5, 2, \dfrac{4}{5}, \ldots$

$$r = \frac{a_2}{a_1} = \frac{2}{5}$$

$$a_n = a_1 r^{n-1}$$

$$a_5 = 5\left(\frac{2}{5}\right)^{5-1} = 5\left(\frac{2}{5}\right)^4 = 5\left(\frac{16}{625}\right)$$

$$a_5 = \frac{16}{125}$$

You Try It 2 $3, a_2, a_3, -192, \ldots$

$$a_n = a_1 r^{n-1}$$
$$a_4 = 3r^{4-1}$$
$$-192 = 3r^{4-1}$$
$$-192 = 3r^3$$
$$-64 = r^3$$
$$-4 = r$$

$$a_n = a_1 r^{n-1}$$
$$a_3 = 3(-4)^{3-1} = 3(-4)^2 = 3(16) = 48$$

You Try It 3 $1, -\dfrac{1}{3}, \dfrac{1}{9}, -\dfrac{1}{27}$

$$r = \frac{a_2}{a_1} = \frac{-\frac{1}{3}}{1} = -\frac{1}{3}$$

$$S_n = \frac{a_1(1 - r^n)}{1 - r}$$

$$S_4 = \frac{1\left[1 - \left(-\frac{1}{3}\right)^4\right]}{1 - \left(-\frac{1}{3}\right)} = \frac{1 - \frac{1}{81}}{\frac{4}{3}} = \frac{\frac{80}{81}}{\frac{4}{3}}$$

$$= \frac{80}{81} \cdot \frac{3}{4} = \frac{20}{27}$$

You Try It 4 $\displaystyle\sum_{n=1}^{5}\left(\frac{1}{2}\right)^n$

$$a_n = \left(\frac{1}{2}\right)^n$$

$$a_1 = \left(\frac{1}{2}\right)^1 = \frac{1}{2}$$

$$a_2 = \left(\frac{1}{2}\right)^2 = \frac{1}{4}$$

$$r = \frac{a_2}{a_1} = \frac{\frac{1}{4}}{\frac{1}{2}} = \frac{1}{4} \cdot \frac{2}{1} = \frac{1}{2}$$

$$S_n = \frac{a_1(1 - r^n)}{1 - r}$$

$$S_5 = \frac{\frac{1}{2}\left[1 - \left(\frac{1}{2}\right)^5\right]}{1 - \frac{1}{2}} = \frac{\frac{1}{2}\left(1 - \frac{1}{32}\right)}{\frac{1}{2}} = \frac{\frac{1}{2}\left(\frac{31}{32}\right)}{\frac{1}{2}}$$

$$= \frac{31}{32}$$

You Try It 5 $3, -2, \frac{4}{3}, -\frac{8}{9}, \cdots$

$$r = \frac{a_2}{a_1} = \frac{-2}{3} = -\frac{2}{3}$$

$$S = \frac{a_1}{1 - r} = \frac{3}{1 - \left(-\frac{2}{3}\right)} = \frac{3}{1 + \frac{2}{3}}$$

$$= \frac{3}{\frac{5}{3}} = \frac{9}{5}$$

You Try It 6 $0.6\overline{6} = 0.6 + 0.06 + 0.006 + \cdots$

$$= \frac{6}{10} + \frac{6}{100} + \frac{6}{1000} + \cdots$$

$$S = \frac{a_1}{1 - r} = \frac{\frac{6}{10}}{1 - \frac{1}{10}} = \frac{\frac{6}{10}}{\frac{9}{10}} = \frac{6}{9} = \frac{2}{3}$$

An equivalent fraction is $\frac{2}{3}$.

You Try It 7
$0.34\overline{21} = 0.34 + 0.0021 + 0.000021 + \cdots$

$$= \frac{34}{100} + \frac{21}{10,000} + \frac{21}{1,000,000} + \cdots$$

$$S = \frac{a_1}{1 - r} = \frac{\frac{21}{10,000}}{1 - \frac{1}{100}} = \frac{\frac{21}{10,000}}{\frac{99}{100}} = \frac{21}{9900}$$

$$0.34\overline{21} = \frac{34}{100} + \frac{21}{9900} = \frac{3366}{9900} + \frac{21}{9900}$$

$$= \frac{3387}{9900} = \frac{1129}{3300}$$

An equivalent fraction is $\frac{1129}{3300}$.

You Try It 8

Strategy   To find the total number of letters mailed, use the Formula for the Sum of n Terms of a Finite Geometric Series.

Solution $n = 6, a_1 = 3, r = 3$

$$S_n = \frac{a_1(1 - r^n)}{1 - r}$$

$$S_6 = \frac{3(1 - 3^6)}{1 - 3} = \frac{3(1 - 729)}{1 - 3}$$

$$= \frac{3(-728)}{-2} = \frac{-2184}{-2} = 1092$$

From the first through the sixth mailings, 1092 letters will have been mailed.

You Try It 9

Strategy   To find the value of the investment, use the Formula for the nth Term of a Geometric Sequence, $a_n = a_1 r^{n-1}$.

Solution $n = 11, a_1 = 25,000, r = 1.12$

$$a_{11} = 25,000(1.12)^{11-1} = 25,000(1.12)^{10}$$
$$\approx 77,646.21$$

The value of the investment after 10 years is approximately \$77,646.21.

SECTION 12.4

You Try It 1
$$(3m - n)^4 = \binom{4}{0}(3m)^4 + \binom{4}{1}(3m)^3(-n)$$

$$+ \binom{4}{2}(3m)^2(-n)^2 + \binom{4}{3}(3m)(-n)^3$$

$$+ \binom{4}{4}(-n)^4$$

$$= 1(81m^4) + 4(27m^3)(-n) + 6(9m^2)(n^2)$$
$$+ 4(3m)(-n^3) + 1(n^4)$$
$$= 81m^4 - 108m^3n + 54m^2n^2$$
$$- 12mn^3 + n^4$$

You Try It 2
$$(y - 2)^{10} = \binom{10}{0}y^{10} + \binom{10}{1}y^9(-2)$$

$$+ \binom{10}{2}y^8(-2)^2 + \cdots$$

$$= 1(y^{10}) + 10y^9(-2) + 45y^8(4) + \cdots$$
$$= y^{10} - 20y^9 + 180y^8 + \cdots$$

You Try It 3
$(t - 2s)^7$

$$\binom{n}{r - 1}a^{n-r+1}b^{r-1} \qquad \bullet \; n = 7, a = t, b = -2s, r = 3$$

$$\binom{7}{3 - 1}(t)^{7-3+1}(-2s)^{3-1} = \binom{7}{2}(t)^5(-2s)^2$$

$$= 21t^5(4s^2) = 84t^5s^2$$

Answers to Chapter 1 Odd-Numbered Exercises

PREP TEST

1. $\frac{13}{20}$ **2.** $\frac{11}{60}$ **3.** $\frac{2}{9}$ **4.** $\frac{2}{3}$ **5.** 44.405 **6.** 73.63 **7.** 7.446 **8.** 54.06 **9.** a, c, d
10. a and C; b and D; c and A; d and B

SECTION 1.1

1a. Integers: 0, −3 **b.** Rational numbers: $-\frac{15}{2}, 0, -3, 2.\overline{33}$ **c.** Irrational numbers: π, 4.232232223..., $\frac{\sqrt{5}}{4}, \sqrt{7}$
d. Real numbers: all **3.** −27 **5.** $-\frac{3}{4}$ **7.** 0 **9.** $\sqrt{33}$ **11.** 91 **13.** −3, 0 **15.** 7 **17.** −2, −1
19. 9, 0, −9 **21.** 4, 0, 4 **23.** −6, −2, 0, −1, −4 **25.** {−2, −1, 0, 1, 2, 3, 4} **27.** {2, 4, 6, 8, 10, 12}
29. {3, 6, 9, 12, 15, 18, 21, 24, 27, 30} **31.** {−35, −30, −25, −20, −15, −10, −5} **33.** $\{x \mid x > 4, x \in \text{integers}\}$
35. $\{x \mid x \geq -2\}$ **37.** $\{x \mid 0 < x < 1\}$ **39.** $\{x \mid 1 \leq x \leq 4\}$ **41.** $A \cup B = \{1, 2, 4, 6, 9\}$ **43.** $A \cup B = \{2, 3, 5, 8, 9, 10\}$
45. $A \cup B = \{-4, -2, 0, 2, 4, 8\}$ **47.** $A \cup B = \{1, 2, 3, 4, 5\}$ **49.** $A \cap B = \{6\}$ **51.** $A \cap B = \{5, 10, 20\}$
53. $A \cap B = \varnothing$ **55.** $A \cap B = \{4, 6\}$ **57.** 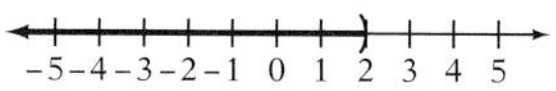

59. 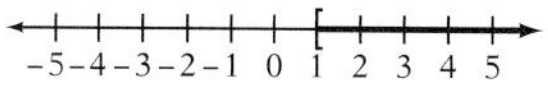

61. 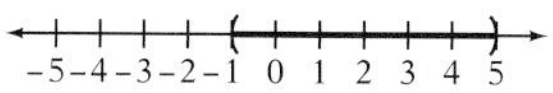

63. 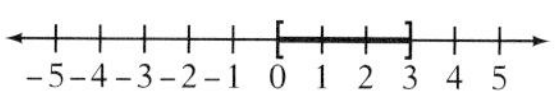

65. 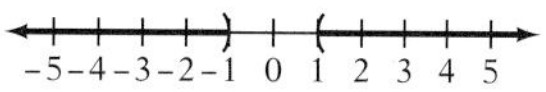

67. 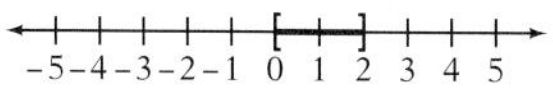

69. 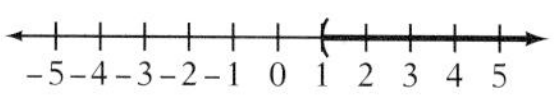

71. 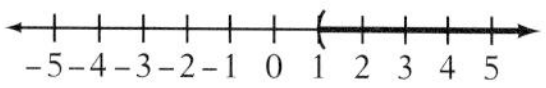

73. $\{x \mid 0 < x < 8\}$ **75.** $\{x \mid -5 \leq x \leq 7\}$ **77.** $\{x \mid -3 \leq x < 6\}$ **79.** $\{x \mid x \leq 4\}$ **81.** $\{x \mid x > 5\}$ **83.** (−2, 4)
85. [−1, 5] **87.** $(-\infty, 1)$ **89.** $[-2, \infty)$ **91.** 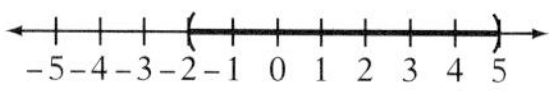

93. 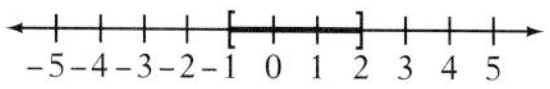

95. 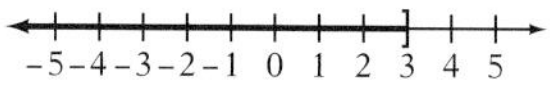

97. 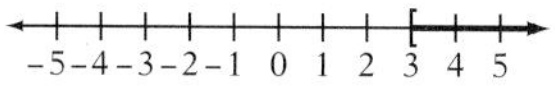

99. *A* **101.** *B* **103.** *A* **105.** *R*
107. *R* **109.** 0

SECTION 1.2

3. −30 **5.** −17 **7.** −96 **9.** −6 **11.** 5 **13.** 11,200 **15.** 98,915 **17.** 96 **19.** 23 **21.** 7 **23.** 17
25. −4 **27.** −62 **29.** −25 **31.** 7 **33.** 368 **35.** 26 **37.** −83 **41.** $\frac{43}{48}$ **43.** $-\frac{67}{45}$ **45.** $-\frac{13}{36}$
47. $\frac{11}{24}$ **49.** $\frac{13}{24}$ **51.** $-\frac{3}{56}$ **53.** $-\frac{2}{3}$ **55.** $-\frac{11}{14}$ **57.** $-\frac{1}{24}$ **59.** −12.974 **61.** −6.008 **63.** 1.9215
65. −6.02 **67.** −6.7 **69.** −1.11 **71.** −2030.44 **73.** 125 **75.** −8 **77.** −125 **79.** 324 **81.** −36
83. −72 **85.** 864 **87.** −160 **89.** 12 **91.** 2,654,208 **95.** −7 **97.** $\frac{177}{11}$ **99.** $\frac{16}{3}$ **101.** −40
103. 32 **105.** 24 **107.** $\frac{2}{15}$ **109.** $-\frac{1}{2}$ **111.** $-\frac{97}{72}$ **113.** 6.284 **115.** 5.4375 **117.** 1.891878 **119.** Zero
121. No. Zero does not have a multiplicative inverse. **123.** 9 **125.** 625 **127.** Find b^c first, then $a^{(b^c)}$.

SECTION 1.3

1. 3 **3.** 3 **5.** 0 **7.** 6 **9.** 0 **11.** 1 **13.** (2 · 3) **15.** Division Property of Zero **17.** Inverse Property of Multiplication **19.** Addition Property of Zero **21.** Division Property of Zero **23.** Distributive Property
25. Associative Property of Multiplication **27.** 10 **29.** 4 **31.** 0 **33.** $-\frac{1}{7}$ **35.** $\frac{9}{2}$ **37.** $\frac{1}{2}$ **39.** 2
41. 3 **43.** −12 **45.** 2 **47.** 6 **49.** −2 **51.** −14 **53.** 56 **55.** 256 **57.** $12x$ **59.** $-13ab$ **61.** $7x$
63. $-8a - 7b$ **65.** y **67.** $-5x + 45$ **69.** $-x - y$ **71.** $3a - 15$ **73.** $4x - 6y + 15$ **75.** $-7x + 14$
77. $48a - 75$ **79.** $-3x - 12y$ **81.** $-12x + 12y$ **83.** $33a - 25b$ **85.** $31x + 34y$ **87.** $4 - 8x - 5y$
89. $2x + 9$ **91.** Distributive Property **93.** Incorrect use of the Distributive Property. $2 + 3x = 2 + 3x$
95. Incorrect use of the Associative Property of Multiplication. $2(3y) = (2 \cdot 3)y = 6y$ **97.** Commutative Property of Addition

SECTION 1.4

1. $n - 8$ **3.** $\frac{4}{5}n$ **5.** $\frac{n}{14}$ **7.** $n - (n + 2)$; -2 **9.** $5(8n)$; $40n$ **11.** $17n - 2n$; $15n$ **13.** $n^2 - (12 + n^2)$; -12

15. $15n + (5n + 12)$; $20n + 12$ **17.** $(15 - x + 2) + 2x$; $x + 17$ **19.** $\frac{5(34 - x)}{x - 3}$ **21.** $390d$ **23.** $c + 3$

25. $10,000 - x$ **27.** Measure of angle B: x; measure of angle A: $2x$; measure of angle C: $4x$

29. The sum of twice x and 3 **31.** Twice the sum of x and 3 **33a.** $\frac{1}{2}gt^2$ **b.** ma **c.** Av^2

CHAPTER REVIEW*

1. $\{-2, -1, 0, 1, 2, 3\}$ [1.1B] **2.** $A \cap B = \{2, 3\}$ [1.1C] **3.** ⟵┼┼┼(┼┼┼┼┼┼)┼┼⟶ [1.1C] **4.** The
   −5−4−3−2−1 0 1 2 3 4 5

Associative Property of Multiplication [1.3A] **5.** -2.84 [1.2B] **6.** $\frac{7}{6}$ [1.3B] **7.** -288 [1.2C]

8. $16y - 15x + 18$ [1.3C] **9.** $\frac{3}{4}$ [1.1A] **10.** $\{x \mid x < -3\}$ [1.1B] **11.** ⟵┼┼┼┼┼┼┼┼)┼┼┼⟶ [1.1C]
   −5−4−3−2−1 0 1 2 3 4 5

12. -15 [1.2A] **13.** $-\frac{7}{30}$ [1.2B] **14.** $3 + (4 + y) = (3 + 4) + y$ [1.3A] **15.** $-\frac{5}{8}$ [1.2B] **16.** $0, 2$ [1.1A]

17. 20 [1.3B] **18.** 14 [1.2A] **19.** 52 [1.2D] **20.** ⟵┼┼(┼┼┼┼┼┼┼┼┼⟶ [1.1C]
   −5−4−3−2−1 0 1 2 3 4 5

21. $A \cup B = \{1, 2, 3, 4, 5, 6, 7, 8\}$ [1.1C] **22.** 12 [1.2A] **23.** $\{x \mid -2 \le x \le 3\}$ [1.1C] **24.** $\frac{2}{15}$ [1.2B]

25. $6x - 21y = 3(2x - 7y)$ [1.3A] **26.** ⟵┼┼┼┼┼┼(┼┼┼┼┼⟶ [1.1C] **27.** $-6x + 14$ [1.3C]
   −5−4−3−2−1 0 1 2 3 4 5

28. $-4, 0, -7$ [1.1A] **29.** The Inverse Property of Addition [1.3A] **30.** 3.1 [1.2B]

31. $12 - \frac{x + 3}{x}$; $\frac{11x - 3}{x}$ [1.4A] **32.** $2x + (40 - x) + 5$; $x + 45$ [1.4A] **33.** $3w - 3$ [1.4B]

34. $x - 47$ [1.4B]

CHAPTER TEST*

1. -30 [1.2A] **2.** $A \cap B = \{5, 7\}$ [1.1C] **3.** -72 [1.2C] **4.** ⟵┼┼┼┼┼┼┼┼)┼┼┼⟶ [1.1C]
   −5−4−3−2−1 0 1 2 3 4 5

5. $A \cap B = \{-1, 0, 1\}$ [1.1C] **6.** -5 [1.3B] **7.** 2 [1.2A] **8.** $14x + 48y$ [1.3C] **9.** 12 [1.1A]

10. -100 [1.2C] **11.** ⟵┼┼┼(┼┼┼┼┼)┼┼┼⟶ [1.1C] **12.** 12 [1.2A] **13.** $\frac{25}{36}$ [1.2B]
   −5−4−3−2−1 0 1 2 3 4 5

14. $(3 + 4) + 2 = (4 + 3) + 2$ [1.3A] **15.** $-\frac{4}{27}$ [1.2B] **16.** -5 [1.1A] **17.** 2 [1.3B] **18.** -15 [1.2A]

19. 10 [1.2D] **20.** ⟵┼┼┼┼┼┼┼┼┼┼(┼┼⟶ [1.1C] **21.** $A \cup B = \{1, 2, 3, 4, 5, 7\}$ [1.1C] **22.** $13x - y$ [1.3C]
   −5−4−3−2−1 0 1 2 3 4 5

23. 6 [1.2D] **24.** $\frac{2}{15}$ [1.2B] **25.** The Distributive Property [1.3A] **26.** ⟵┼┼┼┼┼┼┼┼┼┼┼)┼┼⟶ [1.1C]
   −5−4−3−2−1 0 1 2 3 4 5

27. -1.41 [1.2B] **28.** $A \cup B = \{-2, -1, 0, 1, 2, 3\}$ [1.1C] **29.** $13 - (n - 3)(9)$; $40 - 9n$ [1.4A]

30. $(x + 1) - 2(9 - x)$; $3x - 17$ [1.4A]

Answers to Chapter 2 Odd-Numbered Exercises

PREP TEST

1. -4 [1.2A] **2.** -6 [1.2A] **3.** 3 [1.2A] **4.** 1 [1.2B] **5.** $-\frac{1}{2}$ [1.2B] **6.** $10x - 5$ [1.3C]

7. $6x - 9$ [1.3C] **8.** $3n + 6$ [1.3C] **9.** $0.03x + 20$ [1.3C] **10.** $20 - n$ [1.4B]

*The numbers in brackets following the answers in the Chapter Review and Chapter Test are a reference to the objective that corresponds to that problem. For example, the reference [1.2A] stands for Section 1.2, Objective A. This notation will be used for all Prep Tests, Chapter Reviews, Chapter Tests, and Cumulative Reviews throughout the text.

SECTION 2.1

5. Yes   **7.** Yes   **9.** 9   **11.** -10 **13.** -2 **15.** -15 **17.** 4   **19.** $-\dfrac{2}{3}$ **21.** $\dfrac{17}{6}$ **23.** $\dfrac{1}{6}$ **25.** $\dfrac{15}{2}$

27. $-\dfrac{32}{25}$ **29.** 1   **31.** $\dfrac{15}{16}$ **33.** -64 **35.** $\dfrac{5}{6}$ **37.** $\dfrac{3}{2}$ **39.** 8   **41.** 0   **43.** 6   **45.** 6

47. The equation has no solution.   **49.** -3 **51.** 1   **53.** $\dfrac{4}{3}$ **55.** 1   **57.** $-\dfrac{4}{3}$ **59.** 6   **61.** $\dfrac{1}{2}$ **63.** -6

65. $\dfrac{5}{6}$ **67.** $-\dfrac{31}{7}$ **69.** $\dfrac{2}{3}$ **71.** $\dfrac{35}{12}$ **73.** -22 **75.** 6   **77.** 6   **79.** $\dfrac{25}{14}$ **81.** $\dfrac{11}{2}$ **83.** $-\dfrac{4}{29}$ **85.** -12

87. $\dfrac{17}{8}$ **89.** 0   **91.** $r = \dfrac{C}{2\pi}$ **93.** $h = \dfrac{2A}{b}$ **95.** $M = \dfrac{IC}{100}$ **97.** $r = \dfrac{A - P}{Pt}$ **99.** $c = 2s - a - b$

101. $h = \dfrac{S - 2\pi r^2}{2\pi r}$ **103.** $R = Pn + C$ **105.** $H = \dfrac{S - 2WL}{2W + 2L}$

SECTION 2.2

1. The number is 5.   **3.** The integers are 3 and 7.   **5.** The integers are 21 and 29.   **7.** The three numbers are 21, 44, and 58.   **9.** The integers are -20, -19, and -18.   **11.** There is no solution.   **13.** The integers are 5, 7, and 9.   **15.** There are 21 dimes in the collection.   **17.** There are 12 dimes in the bank.   **19.** There are eight 20¢ stamps and sixteen 15¢ stamps in the collection.   **21.** There are five 3¢ stamps in the collection.   **23.** There are seven 18¢ stamps in the collection.   **25.** There is no solution.

SECTION 2.3

1. The cost is $2.95 per pound.   **3.** There were 320 adult tickets sold.   **5.** The cook must use 22.5 L of imitation maple syrup.   **7.** The goldsmith used 20 oz of pure gold and 30 oz of the gold alloy.   **9.** The cost of the tea mixture is $4.11 per pound.   **11.** 23.8 gal of cranberry juice was used.   **13.** 200 lb of the 15% aluminum alloy must be used.   **15.** You should combine 2.25 oz of 70% rubbing alcohol and 1.25 oz of water.   **17.** 45 oz of pure water must be added.   **19.** 300 ml of alcohol must be added.   **21.** The result is 6% fruit juice.   **23.** 4 qt will have to be replaced with pure antifreeze.   **25.** The helicopter overtakes the car 104 mi from the starting point.   **27.** The first car is traveling 58 mph. The second car is traveling 66 mph.   **29.** The distance between the two airports is 375 mi.   **31.** The freight train is traveling 30 mph. The passenger train is traveling 48 mph.   **33.** The cyclist rode 44 mi.   **35.** 15 oz of 12-karat gold should be mixed with the 24-karat gold.   **37.** They are $3\dfrac{1}{3}$ mi apart.   **39.** The grass seed is 18 ft from the base of the 40-foot tower.

SECTION 2.4

3. a, c   **5.** $\{x \mid x < 5\}$ **7.** $\{x \mid x \le 2\}$ **9.** $\{x \mid x < -4\}$ **11.** $\{x \mid x > 3\}$ **13.** $\{x \mid x > 4\}$ **15.** $\{x \mid x \le 2\}$
17. $\{x \mid x > -2\}$ **19.** $\{x \mid x \ge 2\}$ **21.** $\{x \mid x > -2\}$ **23.** $\{x \mid x \le 3\}$ **25.** $\{x \mid x < 2\}$ **27.** $\{x \mid x < -3\}$
29. $\{x \mid x \le 5\}$ **31.** $\left\{x \mid x \ge -\dfrac{1}{2}\right\}$ **33.** $\left\{x \mid x < \dfrac{23}{16}\right\}$ **35.** $\left\{x \mid x < \dfrac{8}{3}\right\}$ **37.** $\{x \mid x > 1\}$ **39.** $\left\{x \mid x > \dfrac{14}{11}\right\}$
41. $\{x \mid x \le 1\}$ **43.** $\left\{x \mid x \le \dfrac{7}{4}\right\}$ **45.** $\left\{x \mid x \ge -\dfrac{5}{4}\right\}$ **47.** $\{x \mid x \le 2\}$ **51.** $\{x \mid -2 \le x \le 4\}$
53. $\{x \mid x < 3 \text{ or } x > 5\}$ **55.** $\{x \mid -4 < x < 2\}$ **57.** $\{x \mid x > 6 \text{ or } x < -4\}$ **59.** $\{x \mid x < -3\}$ **61.** $\{x \mid -3 < x < 2\}$
63. $\{x \mid -2 < x < 1\}$ **65.** $\{x \mid x < -2 \text{ or } x > 2\}$ **67.** $\{x \mid 2 < x < 6\}$ **69.** $\{x \mid -3 < x < -2\}$
71. $\left\{x \mid x > 5 \text{ or } x < -\dfrac{5}{3}\right\}$ **73.** The solution set is the empty set.   **75.** The solution set is the set of real numbers.   **77.** $\left\{x \mid \dfrac{17}{7} \le x \le \dfrac{45}{7}\right\}$ **79.** $\left\{x \mid -5 < x < \dfrac{17}{3}\right\}$ **81.** The solution set is the set of real numbers.
83. The solution set is the empty set.   **85.** The smallest number is -12.   **87.** The maximum width as an integer is 11 cm.
89. The TopPage plan is less expensive for more than 460 pages per month.   **91.** A call must last 7 min or less.

93. The temperature range is $32° < F < 86°$. **95.** The amount of sales is \$44,000 or more. **97.** The business writes more than 200 checks per month. **99.** The range of scores is $58 \le x \le 100$. **101.** The three consecutive even integers can be 10, 12, 14, or 12, 14, 16, or 14, 16, 18. **103.** The solution is correct.

SECTION 2.5

1. Yes **3.** Yes **5.** 7 and -7 **7.** 4 and -4 **9.** 6 and -6 **11.** 7 and -7 **13.** There is no solution.

15. There is no solution. **17.** -5 and 1 **19.** 2 and 8 **21.** 2 **23.** There is no solution. **25.** 3 and $-\dfrac{3}{2}$ **27.** $\dfrac{3}{2}$

29. There is no solution. **31.** -3 and 7 **33.** $-\dfrac{10}{3}$ and 2 **35.** 3 and 1 **37.** $\dfrac{3}{2}$ **39.** There is no solution.

41. $-\dfrac{1}{6}$ and $\dfrac{11}{6}$ **43.** -1 and $-\dfrac{1}{3}$ **45.** There is no solution. **47.** 0 and 3 **49.** There is no solution. **51.** $\dfrac{13}{3}$ and 1

53. There is no solution. **55.** $\dfrac{1}{3}$ and $\dfrac{7}{3}$ **57.** $-\dfrac{1}{2}$ **59.** $-\dfrac{7}{2}$ and $-\dfrac{1}{2}$ **61.** $\dfrac{10}{3}$ and $-\dfrac{8}{3}$ **63.** There is no solution.

65. $\{x\,|\,x > 3 \text{ or } x < -3\}$ **67.** $\{x\,|\,x > 1 \text{ or } x < -3\}$ **69.** $\{x\,|\,4 \le x \le 6\}$ **71.** $\{x\,|\,x \ge 5 \text{ or } x \le -1\}$

73. $\{x\,|\,-3 < x < 2\}$ **75.** $\left\{x\,\middle|\,x > 2 \text{ or } x < -\dfrac{14}{5}\right\}$ **77.** The solution set is the empty set. **79.** The solution set

is the set of real numbers. **81.** $\left\{x\,\middle|\,x \le -\dfrac{1}{3} \text{ or } x \ge 3\right\}$ **83.** $\left\{x\,\middle|\,-2 \le x \le \dfrac{9}{2}\right\}$ **85.** $\{x\,|\,x = 2\}$

87. $\left\{x\,\middle|\,x < -2 \text{ or } x > \dfrac{22}{9}\right\}$ **89.** $\left\{x\,\middle|\,-\dfrac{3}{2} < x < \dfrac{9}{2}\right\}$ **91.** $\left\{x\,\middle|\,x < 0 \text{ or } x > \dfrac{4}{5}\right\}$ **93.** $\{x\,|\,x > 5 \text{ or } x < 0\}$

95. Lower limit: 1.742 in.; upper limit: 1.758 in. **97.** Lower limit: 195 volts; upper limit: 245 volts **99.** Lower limit:

$9\dfrac{19}{32}$ in.; upper limit: $9\dfrac{21}{32}$ in. **101.** Lower limit: 28,420 ohms; upper limit: 29,580 ohms **103.** Lower limit: 23,750 ohms;

upper limit: 26,250 ohms **105a.** $\{x\,|\,x \ge -3\}$ **b.** $\{a\,|\,a \le 4\}$ **107a.** $\le$ **b.** $\ge$ **c.** $\ge$ **d.** $=$ **e.** $=$

CHAPTER REVIEW

1. 6 [2.1B] **2.** $\left\{x\,\middle|\,x > \dfrac{5}{3}\right\}$ [2.4A] **3.** $L = \dfrac{P - 2W}{2}$ [2.1D] **4.** -9 [2.1A] **5.** $\left\{x\,\middle|\,-3 < x < \dfrac{4}{3}\right\}$ [2.4B]

6. $-\dfrac{40}{7}$ [2.1B] **7.** $-\dfrac{2}{3}$ [2.1A] **8.** -1 and 9 [2.5A] **9.** $\{x\,|\,1 < x < 4\}$ [2.5B] **10.** $-\dfrac{9}{19}$ [2.1C]

11. $\dfrac{26}{17}$ [2.1C] **12.** $\{x\,|\,x > 2 \text{ or } x < -2\}$ [2.4B] **13.** $\left\{x\,\middle|\,x \ge 2 \text{ or } x \le \dfrac{1}{2}\right\}$ [2.5B] **14.** $C = R - Pn$ [2.1D]

15. $-\dfrac{17}{2}$ [2.1C] **16.** No solution [2.5A] **17.** The set of real numbers [2.4B] **18.** $\dfrac{5}{2}$ [2.1C]

19. 52 gal of apple juice were used. [2.3A] **20.** The amount of sales must be \$55,000 or more. [2.4C] **21.** There are 7 quarters in the collection. [2.2B] **22.** Lower limit: 2.747 in.; upper limit: 2.753 in. [2.5C] **23.** The integers are 6 and 14. [2.2A] **24.** The range of scores is $82 \le x \le 100$. [2.4C] **25.** The speed of the faster plane is 520 mph. The speed of the slower plane is 440 mph. [2.3C] **26.** 375 lb of 30% tin and 125 lb of 70% tin were used. [2.3B]

27. Lower limit: $10\dfrac{11}{32}$ in.; upper limit: $10\dfrac{13}{32}$ in. [2.5C]

CHAPTER TEST

1. -2 [2.1A] **2.** $-\dfrac{1}{8}$ [2.1A] **3.** $\dfrac{5}{6}$ [2.1A] **4.** 4 [2.1B] **5.** $\dfrac{32}{3}$ [2.1B] **6.** $-\dfrac{1}{5}$ [2.1B] **7.** 1 [2.1C]

8. $R = \dfrac{E - Ir}{I}$ [2.1D] **9.** $\dfrac{12}{7}$ [2.1C] **10.** $\{x\,|\,x \le -3\}$ [2.4A] **11.** $\{x\,|\,x > -1\}$ [2.4A] **12.** $\{x\,|\,x > -2\}$ [2.4B]

13. $\varnothing$ [2.4B] **14.** 3 and $-\dfrac{9}{5}$ [2.5A] **15.** 7 and -2 [2.5A] **16.** $\left\{x\,\middle|\,\dfrac{1}{3} \le x \le 3\right\}$ [2.5B]

17. $\left\{x\,\middle|\,x > 2 \text{ or } x < -\dfrac{1}{2}\right\}$ [2.5B] **18.** You can drive less than 120 mi. [2.4C] **19.** Lower limit: 2.648 in.; upper limit:

2.652 in. [2.5C] **20.** The integers are 4 and 11. [2.2A] **21.** There are six 24¢ stamps in the collection. [2.2B] **22.** The cost of the hamburger mixture is \$2.20 per pound. [2.3A] **23.** The total distance the jogger ran was 12 mi. [2.3C] **24.** The speed of the slower train is 60 mph. The speed of the faster train is 65 mph. [2.3C] **25.** 100 oz of pure water must be added. [2.3B]

CUMULATIVE REVIEW

1. -11 [1.2A] **2.** -108 [1.2C] **3.** 3 [1.2D] **4.** -64 [1.2D] **5.** -8 [1.3B] **6.** 1 [1.3B] **7.** Commutative Property of Addition [1.3A] **8.** $3n + (3n + 6)$; $6n + 6$ [1.4A] **9.** $B = \frac{Fc}{ev}$ [2.1D] **10.** $25y$ [1.3C] **11.** $\{-4, 0\}$ [1.1C] **12.** 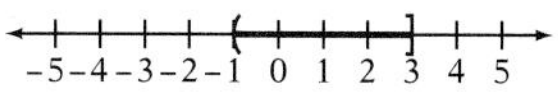 [1.1C] **13.** $y = \frac{-Ax - C}{B}$ [2.1D] **14.** $\frac{1}{2}$ [2.1A] **15.** 1 [2.1B] **16.** 24 [2.1B] **17.** 2 [2.1C] **18.** 2 [2.1C] **19.** 1 [2.1C] **20.** $-\frac{13}{5}$ [2.1C] **21.** $\{x \mid x \leq 1\}$ [2.4A] **22.** $\{x \mid -4 < x \leq 1\}$ [2.4B] **23.** -1 and 4 [2.5A] **24.** 7 and -4 [2.5A] **25.** $\left\{x \mid x < -\frac{4}{3} \text{ or } x > 2\right\}$ [2.5B] **26.** $\{x \mid -2 < x < 6\}$ [2.5B] **27.** The customer can write fewer than 50 checks. [2.4C] **28.** The first integer is -3. [2.2A] **29.** There are 15 dimes in the coin purse. [2.2B] **30.** 80 oz of pure silver were used. [2.3A] **31.** The speed of the slower plane is 220 mph. [2.3C] **32.** Lower limit: 2.449 in.; upper limit: 2.451 in. [2.5C] **33.** 3 L of 12% acid solution must be used. [2.3B]

Answers to Chapter 3 Odd-Numbered Exercises

PREP TEST

1. $-4x + 12$ [1.3C] **2.** 10 [1.2D] **3.** -2 [1.2D] **4.** 11 [1.3B] **5.** 2.5 [1.3B] **6.** 5 [1.3B] **7.** 1 [1.3B] **8.** 4 [2.1B] **9.** $y = 2x - 7$ [2.1D]

SECTION 3.1

1. 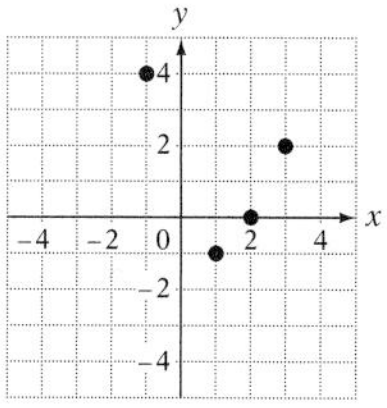

3. A is (0, 3); B is (1, 1); C is (3, -4); D is (-4, 4).

5. 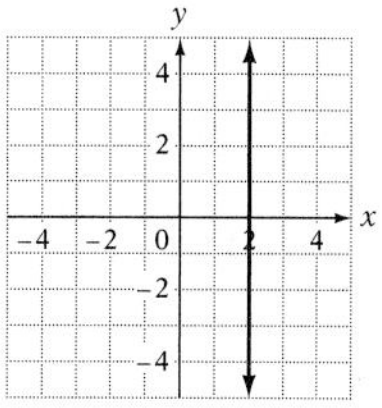

7. 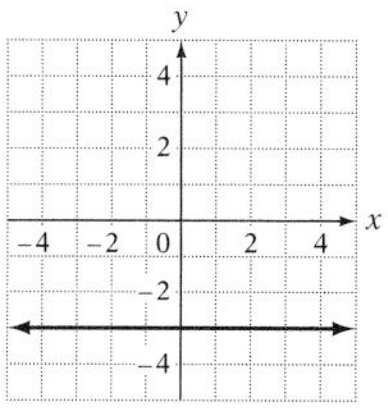

9. 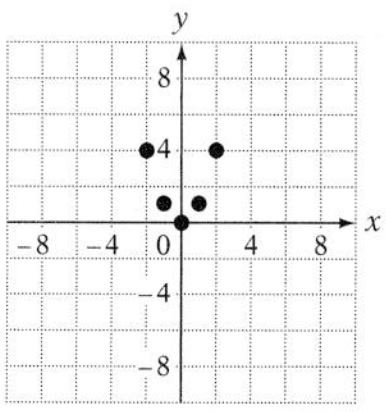

11. 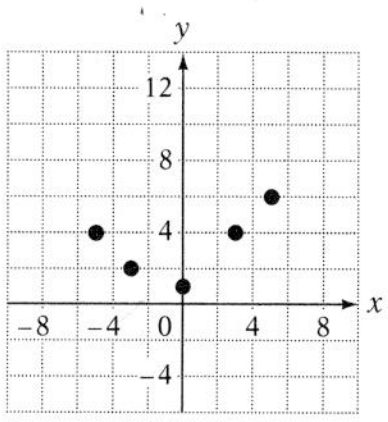

13. 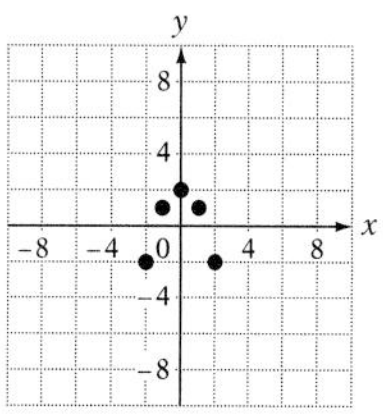

15. 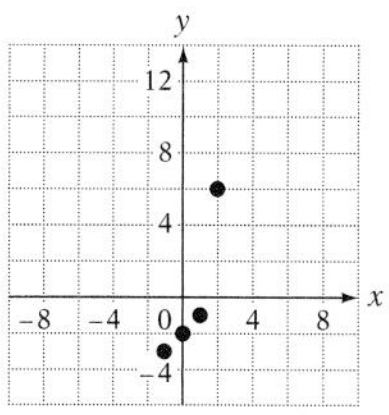

17. Length: 4.47; midpoint: (4, 3) **19.** Length: 2.24; midpoint: $\left(-1, \frac{7}{2}\right)$

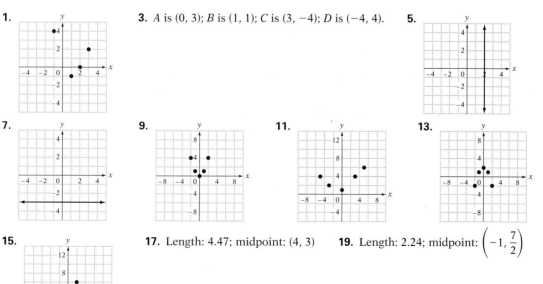

21. Length: 5.10; midpoint: $\left(-\dfrac{1}{2}, -\dfrac{9}{2}\right)$ **23.** Length: 9.90; midpoint: $\left(\dfrac{3}{2}, \dfrac{3}{2}\right)$ **25.** Length: 3; midpoint: $\left(\dfrac{7}{2}, -5\right)$

27. Length: 4.18; midpoint: $\left(\dfrac{1}{2}, \dfrac{1}{2}\right)$ **29a.** The temperature of the reaction after 20 min was 280°F.

b. The temperature was 160°F in 50 min.   **31.** 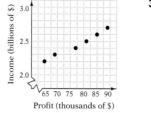   **33.**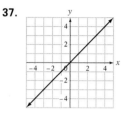

37.

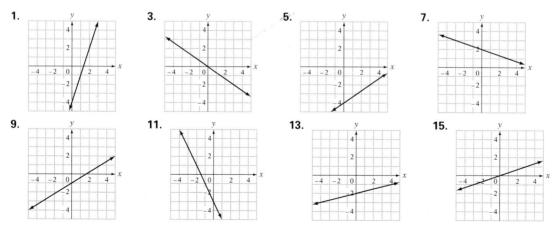

SECTION 3.2

3. Yes   **5.** Yes   **7.** No   **9.** Function   **11.** Function   **13.** Function   **15.** Not a function   **17a.** Yes

b. \$29.62   **19.** $f(3) = 11$ **21.** $f(0) = -4$ **23.** $G(0) = 4$ **25.** $G(-2) = 10$ **27.** $q(3) = 5$ **29.** $q(-2) = 0$

31. $F(4) = 24$ **33.** $F(-3) = -4$ **35.** $H(1) = 1$ **37.** $H(t) = \dfrac{3t}{t + 2}$ **39.** $s(-1) = 6$ **41.** $s(a) = a^3 - 3a + 4$

43. $P(-2 + h) - P(-2) = 4h$ **45a.** The packaging cost is \$4.75 per game.   **b.** The packaging cost is \$4.00 per game.

47a. The appraiser's fee is \$3000.   **b.** The appraiser's fee is \$950.   **49.** Domain: $\{1, 2, 3, 4, 5\}$; range: $\{1, 4, 7, 10, 13\}$

51. Domain: $\{0, 2, 4, 6\}$; range: $\{1, 2, 3, 4\}$ **53.** Domain: $\{1, 3, 5, 7, 9\}$; range: $\{0\}$ **55.** Domain: $\{-2, -1, 0, 1, 2\}$;

range: $\{0, 1, 2\}$ **57.** Domain: $\{-2, -1, 0, 1, 2\}$; range: $\{-3, 3, 6, 7, 9\}$ **59.** 1   **61.** -8 **63.** None   **65.** None

67. 0   **69.** None   **71.** None   **73.** -2 **75.** None   **77.** Range: $\{-3, 1, 5, 9\}$ **79.** Range: $\{-23, -13, -8, -3, 7\}$

81. Range: $\{0, 1, 4\}$ **83.** Range: $\{2, 14, 26, 42\}$ **85.** Range: $\left\{-5, \dfrac{5}{3}, 5\right\}$ **87.** Range: $\left\{-1, -\dfrac{1}{2}, -\dfrac{1}{3}, 1\right\}$

89. Range: $\{-38, -8, 2\}$ **93.** $\{(-2, -8), (-1, -1), (0, 0), (1, 1), (2, 8)\}$. Yes, the set defines a function.   **95.** The windmill

will produce 50.625 watts of power.   **97a.** After 5 s the parachutist is falling at a rate of 22.5 ft/s.   **b.** After 15 s the

parachutist is falling at a rate of 30 ft/s.   **99a.** After 5 h the temperature will be 64°F.   **b.** After 15 h the temperature

will be 52°F.

SECTION 3.3

1.

3.

5.

7.

9.

11.

13.

15.

17.

19. *x*-intercept: $(-4, 0)$; *y*-intercept: $(0, 2)$

21. *x*-intercept: $\left(\frac{9}{2}, 0\right)$; *y*-intercept: $(0, -3)$

23. *x*-intercept: $\left(\frac{3}{2}, 0\right)$; *y*-intercept: $(0, 3)$

25. *x*-intercept: $\left(\frac{4}{3}, 0\right)$; *y*-intercept: $(0, 2)$

27. *x*-intercept: $\left(\frac{3}{2}, 0\right)$; *y*-intercept: $(0, -2)$

29. Marlys received $165 for tutoring 15 h.

31. The roller coaster travels 550 ft in 5 s.

33. The cost of manufacturing 6000 compact disks is $33,000.

39.

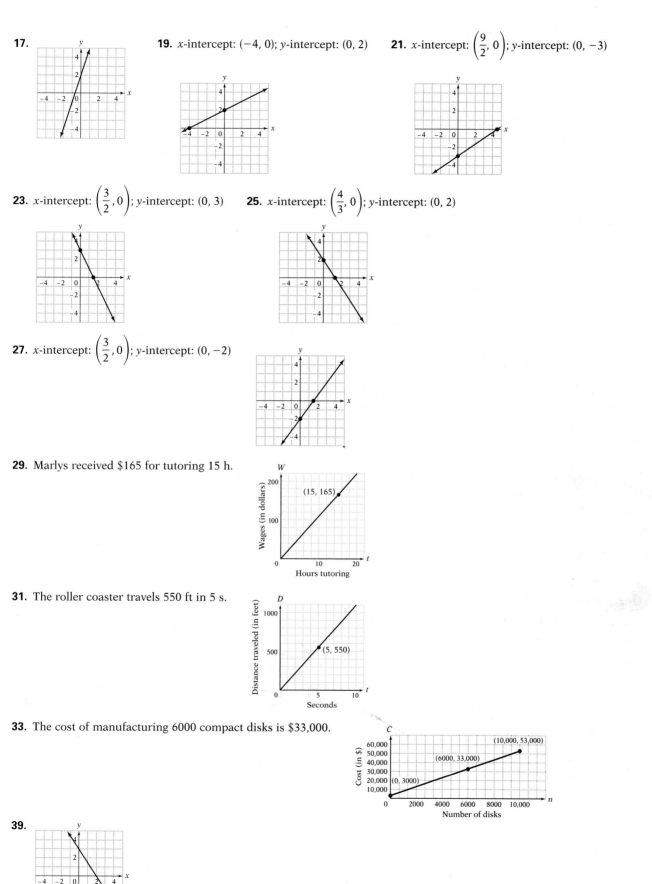

SECTION 3.4

1. -1 **3.** $\dfrac{1}{3}$ **5.** $-\dfrac{2}{3}$ **7.** $-\dfrac{3}{4}$ **9.** Undefined **11.** $\dfrac{7}{5}$ **13.** 0 **15.** $-\dfrac{1}{2}$ **17.** Undefined **19.** The slope is 40. The slope is the average speed of the motorist. **21.** The slope is -0.05. For each mile the car is driven, approximately 0.05 gal of fuel is used. **23.** The slope is approximately 385.5. The slope is the average speed of the runner in meters per minute. **25.** No

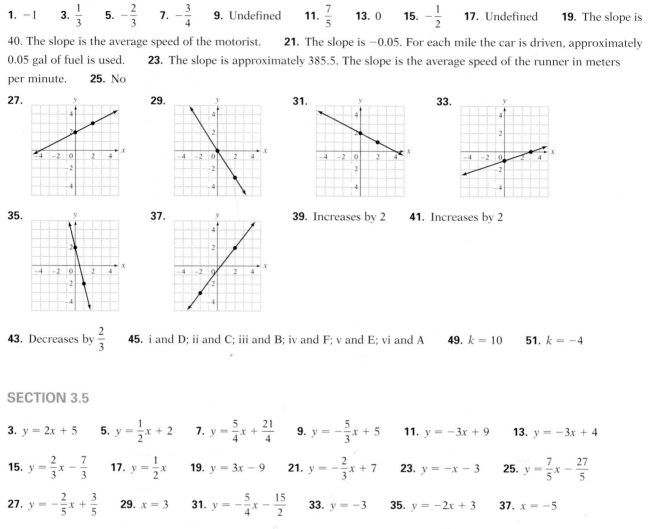

27. **29.** **31.** **33.**

35. **37.** **39.** Increases by 2 **41.** Increases by 2

43. Decreases by $\dfrac{2}{3}$ **45.** i and D; ii and C; iii and B; iv and F; v and E; vi and A **49.** $k = 10$ **51.** $k = -4$

SECTION 3.5

3. $y = 2x + 5$ **5.** $y = \dfrac{1}{2}x + 2$ **7.** $y = \dfrac{5}{4}x + \dfrac{21}{4}$ **9.** $y = -\dfrac{5}{3}x + 5$ **11.** $y = -3x + 9$ **13.** $y = -3x + 4$

15. $y = \dfrac{2}{3}x - \dfrac{7}{3}$ **17.** $y = \dfrac{1}{2}x$ **19.** $y = 3x - 9$ **21.** $y = -\dfrac{2}{3}x + 7$ **23.** $y = -x - 3$ **25.** $y = \dfrac{7}{5}x - \dfrac{27}{5}$

27. $y = -\dfrac{2}{5}x + \dfrac{3}{5}$ **29.** $x = 3$ **31.** $y = -\dfrac{5}{4}x - \dfrac{15}{2}$ **33.** $y = -3$ **35.** $y = -2x + 3$ **37.** $x = -5$

39. $y = x + 2$ **41.** $y = -2x - 3$ **43.** $y = \dfrac{2}{3}x + \dfrac{5}{3}$ **45.** $y = \dfrac{1}{3}x + \dfrac{10}{3}$ **47.** $y = \dfrac{3}{2}x - \dfrac{1}{2}$ **49.** $y = -\dfrac{3}{2}x + 3$

51. $y = -1$ **53.** $y = x - 1$ **55.** $y = -x + 1$ **57.** $y = -\dfrac{8}{3}x + \dfrac{25}{3}$ **59.** $y = \dfrac{1}{2}x - 1$ **61.** $y = -4$

63. $y = \dfrac{3}{4}x$ **65.** $y = -\dfrac{4}{3}x + \dfrac{5}{3}$ **67.** $x = -2$ **69.** $y = x - 1$ **71.** $y = \dfrac{4}{3}x + \dfrac{7}{3}$ **73.** $y = -x + 3$

75. $y = 1200x,\ 0 \le x \le 26\dfrac{2}{3}$. The height of the plane 11 min after takeoff is 13,200 ft.

77. $y = -0.032x + 16,\ 0 \le x \le 500$. After driving 150 mi, there are 11.2 gal in the tank. **83.** Answers will vary. Possible answers are $\{(0, 4), (3, 2), (9, -2)\}$. **85.** $f(4) = 0$ **87.** $y = -2x + 1$

SECTION 3.6

3. -5 **5.** $-\dfrac{1}{4}$ **7.** Yes **9.** No **11.** No **13.** Yes **15.** Yes **17.** Yes **19.** No **21.** Yes

23. $y = \dfrac{2}{3}x - \dfrac{8}{3}$ **25.** $y = \dfrac{1}{3}x - \dfrac{1}{3}$ **27.** $y = -\dfrac{5}{3}x - \dfrac{14}{3}$ **29.** $y = -2x + 15$ **31.** $\dfrac{A_1}{B_1} = \dfrac{A_2}{B_2}$ **33.** Any equation of

the form $y = 2x + b,\ b \ne -13$, or $y = -\dfrac{3}{2}x + c,\ c \ne 8$

SECTION 3.7

3. Yes **5.** No **7.**

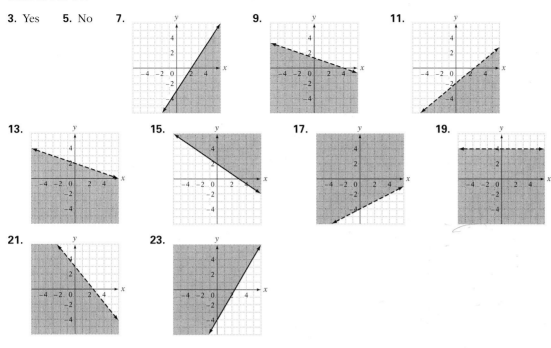

13. **15.** **17.** **19.**

21. **23.**

CHAPTER REVIEW

1. (4, 2) [3.1A] **2.** −2; 3a + 4 [3.2A] **3.** [3.1A] **4.** [3.1A]

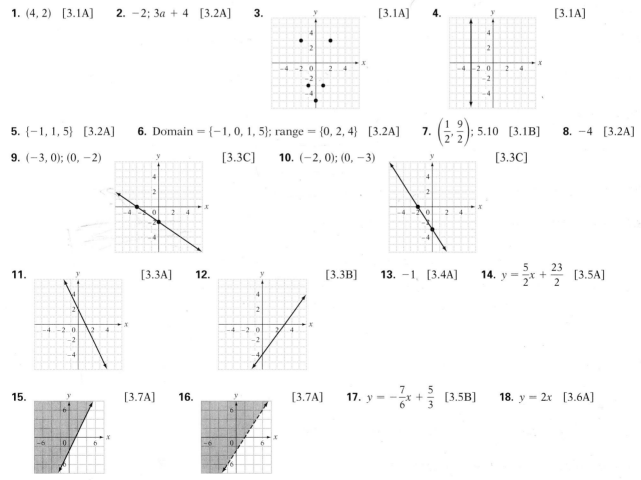

5. {−1, 1, 5} [3.2A] **6.** Domain = {−1, 0, 1, 5}; range = {0, 2, 4} [3.2A] **7.** $\left(\frac{1}{2}, \frac{9}{2}\right)$; 5.10 [3.1B] **8.** −4 [3.2A]

9. (−3, 0); (0, −2) [3.3C] **10.** (−2, 0); (0, −3) [3.3C]

11. [3.3A] **12.** [3.3B] **13.** −1 [3.4A] **14.** $y = \frac{5}{2}x + \frac{23}{2}$ [3.5A]

15. [3.7A] **16.** [3.7A] **17.** $y = -\frac{7}{6}x + \frac{5}{3}$ [3.5B] **18.** $y = 2x$ [3.6A]

19. $y = -3x + 7$ [3.6A] **20.** $y = \frac{3}{2}x + 2$ [3.6A] **21.** [3.4B]

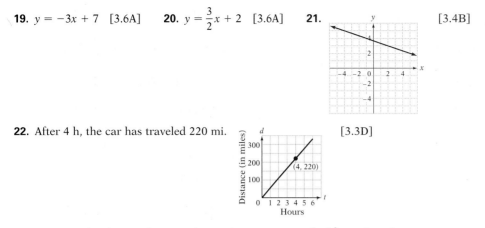

22. After 4 h, the car has traveled 220 mi. [3.3D]

23. $m = 20$; The slope is the cost of manufacturing one calculator. [3.4A]

24. $y = 80x + 25,000$; The cost to build a 2000-square-foot house is $185,000. [3.5C]

CHAPTER TEST

1. [3.1A] **2.** $(-3, 0)$ [3.1A] **3.** [3.3A] **4.** [3.3B]

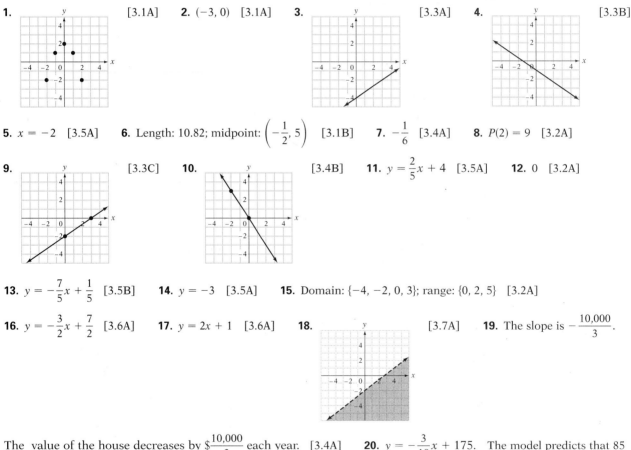

5. $x = -2$ [3.5A] **6.** Length: 10.82; midpoint: $\left(-\frac{1}{2}, 5\right)$ [3.1B] **7.** $-\frac{1}{6}$ [3.4A] **8.** $P(2) = 9$ [3.2A]

9. [3.3C] **10.** [3.4B] **11.** $y = \frac{2}{5}x + 4$ [3.5A] **12.** 0 [3.2A]

13. $y = -\frac{7}{5}x + \frac{1}{5}$ [3.5B] **14.** $y = -3$ [3.5A] **15.** Domain: $\{-4, -2, 0, 3\}$; range: $\{0, 2, 5\}$ [3.2A]

16. $y = -\frac{3}{2}x + \frac{7}{2}$ [3.6A] **17.** $y = 2x + 1$ [3.6A] **18.** [3.7A] **19.** The slope is $-\frac{10,000}{3}$.

The value of the house decreases by $\$\frac{10,000}{3}$ each year. [3.4A] **20.** $y = -\frac{3}{10}x + 175$. The model predicts that 85 students will enroll when the tuition is $300. [3.5C]

CUMULATIVE REVIEW

1. Commutative Property of Multiplication [1.3A] **2.** $\frac{9}{2}$ [2.1B] **3.** $\frac{8}{9}$ [2.1C] **4.** $-\frac{1}{14}$ [2.1C]

5. $\left\{x \mid x < -1 \text{ or } x > \frac{1}{2}\right\}$ [2.4B] **6.** $\frac{5}{2}$ and $-\frac{3}{2}$ [2.5A] **7.** $\left\{x \mid 0 < x < \frac{10}{3}\right\}$ [2.5B] **8.** 8 [1.2D]

9. -4.5 [1.3B]   **10.** 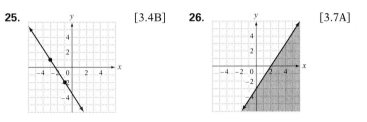 [1.1C]   **11.** $C = R - Pn$ [2.1D]   **12.** $x = 3 - \dfrac{3}{2}y$ [2.1D]

13. $\varnothing$ [2.4B]   **14.** $P(-3) = 14$ [3.2A]   **15.** $(-8, 13)$ [3.1A]   **16.** $-\dfrac{7}{4}$ [3.4A]   **17.** $y = \dfrac{3}{2}x + \dfrac{13}{2}$ [3.5A]

18. $y = -\dfrac{5}{4}x + 3$ [3.5B]   **19.** $y = -\dfrac{3}{2}x + 7$ [3.6A]   **20.** $y = -\dfrac{2}{3}x + \dfrac{8}{3}$ [3.6A]   **21.** There are 7 dimes in the

purse. [2.2B]   **22.** The speed of the first plane is 200 mph. The speed of the second plane is 400 mph.   [2.3C]

23. 20 lb of \$6 coffee and 40 lb of \$9 coffee should be used.   [2.3A]   **24.** [3.3C]

25. [3.4B]   **26.** [3.7A]

27. $y = -2500x + 15{,}000$.   The value of the truck decreases by \$2500 each year.   [3.5C]

Answers to Chapter 4 Odd-Numbered Exercises

PREP TEST

1. $6x + 5y$ [1.3C]   **2.** 7 [1.3B]   **3.** 0 [2.1B]   **4.** -3 [2.1C]   **5.** 1000 [2.1C]

6. [3.3A]   **7.** [3.3B]   **8.** [3.7A]

SECTION 4.1

1. No   **3.** Yes   **5.** Independent   **7.** Inconsistent   **9.** $(3, -1)$ **11.** $(2, 4)$

13. $(4, 3)$ **15.** $(4, -1)$ **17.** $(2, -1)$ **19.** $(4, -2)$

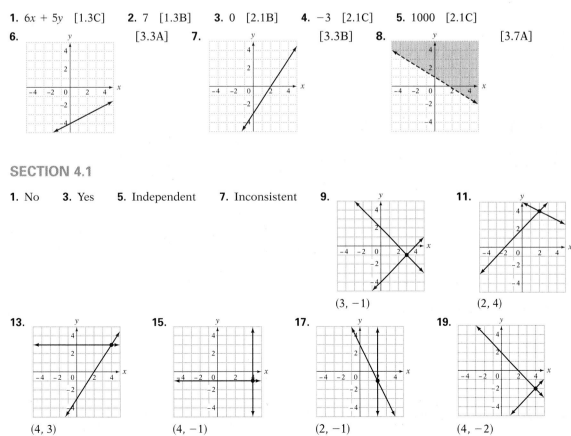

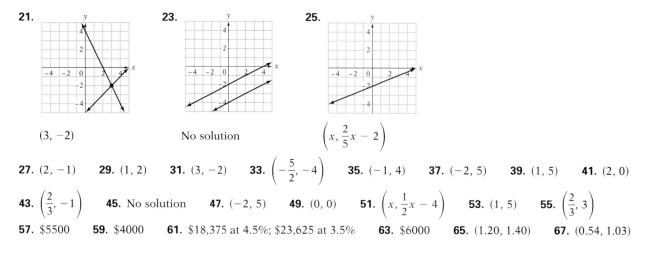

21. (3, −2)   **23.** No solution   **25.** $\left(x, \dfrac{2}{5}x - 2\right)$

27. (2, −1)   **29.** (1, 2)   **31.** (3, −2)   **33.** $\left(-\dfrac{5}{2}, -4\right)$ **35.** (−1, 4)   **37.** (−2, 5)   **39.** (1, 5)   **41.** (2, 0)

43. $\left(\dfrac{2}{3}, -1\right)$ **45.** No solution   **47.** (−2, 5)   **49.** (0, 0)   **51.** $\left(x, \dfrac{1}{2}x - 4\right)$ **53.** (1, 5)   **55.** $\left(\dfrac{2}{3}, 3\right)$

57. $5500   **59.** $4000   **61.** $18,375 at 4.5%; $23,625 at 3.5%   **63.** $6000   **65.** (1.20, 1.40)   **67.** (0.54, 1.03)

SECTION 4.2

1. (6, 1)   **3.** (1, 1)   **5.** (2, 1)   **7.** (−2, 1)   **9.** The equations are dependent. The solutions are $(x, 3x - 4)$.

11. $\left(-\dfrac{1}{2}, 2\right)$ **13.** No solution   **15.** (−1, −2)   **17.** (−5, 4)   **19.** (2, 5)   **21.** $\left(\dfrac{1}{2}, \dfrac{3}{4}\right)$ **23.** (0, 0)   **25.** (−1, 3)

27. $\left(\dfrac{2}{3}, -\dfrac{2}{3}\right)$ **29.** (−2, 3)   **31.** (2, −1)   **33.** (10, −5)   **35.** $\left(-\dfrac{1}{2}, \dfrac{2}{3}\right)$ **37.** $\left(\dfrac{5}{3}, \dfrac{1}{3}\right)$ **39.** No solution

41. (1, −1)   **43.** (2, 1, 3)   **45.** (1, −1, 2)   **47.** (1, 2, 4)   **49.** (2, −1, −2)   **51.** (−2, −1, 3)   **53.** No solution

55. (1, 4, 1)   **57.** (1, 3, 2)   **59.** (1, −1, 3)   **61.** (0, 2, 0)   **63.** (1, 5, 2)   **65.** (−2, 1, 1)   **69.** (1, −1)

71. (1, −1)   **73.** $y = \dfrac{2x}{1 - 3x}, y = \dfrac{-3x}{x + 2}$

SECTION 4.3

3. 11   **5.** 18   **7.** 0   **9.** 15   **11.** −30   **13.** 0   **15.** (3, −4)   **17.** (4, −1)   **19.** $\left(\dfrac{11}{14}, \dfrac{17}{21}\right)$ **21.** $\left(\dfrac{1}{2}, 1\right)$

23. Not independent   **25.** (−1, 0)   **27.** (1, −1, 2)   **29.** (2, −2, 3)   **31.** Not independent   **33a.** Sometimes true
b. Always true   **c.** Sometimes true   **35.** 239 ft^2 **37.** Not independent

SECTION 4.4

1. The rate of the boat is 15 mph. The rate of the current is 3 mph.   **3.** The rate of the plane is 502.5 mph. The rate of the wind is 47.5 mph.   **5.** The rate of the team is 8 km/h. The rate of the current is 2 km/h.   **7.** The rate of the plane is 180 mph. The rate of the wind is 20 mph.   **9.** The rate of the plane is 110 mph. The rate of the wind is 10 mph.

11. There are 30 nickels in the bank.   **13.** The cost of the pine was $1.10 per foot; the cost of the redwood was $3.30 per foot.   **15.** The cost of the wool carpet is $52 per yard.   **17.** The company plans to manufacture 25 mountain bikes.

19. There are 480 g of the first alloy and 40 g of the second alloy.   **21.** The cost of a Model VI computer is $4000.

23. The science museum sold 190 regular tickets, 210 student tickets, and 350 member discount tickets.

25. The investor placed $10,400 in the 8% account, $5200 in the 6% account, and $9400 in the 4% account.

27. The angles are 35° and 145°.   **29.** The oil painting is 85 years old and the watercolor is 50 years old.

SECTION 4.5

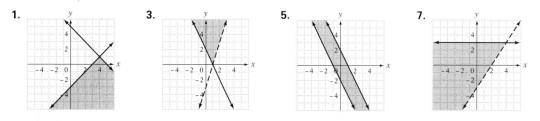

1.   **3.**   **5.**   **7.**

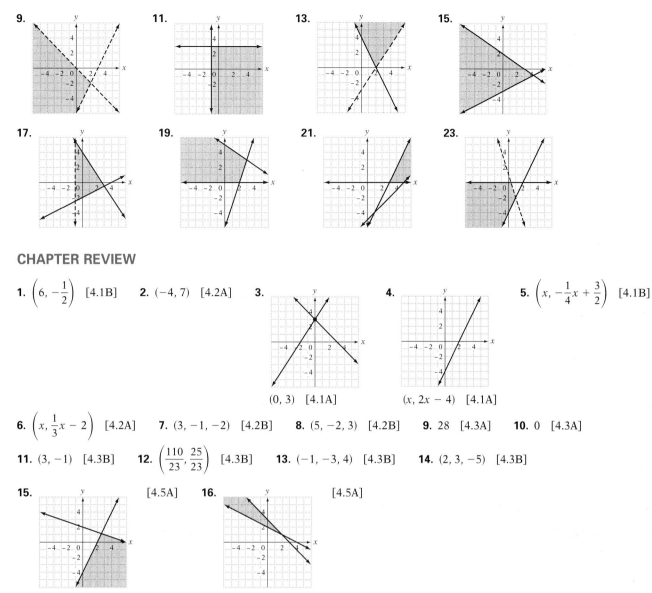

9. **11.** **13.** **15.**

17. **19.** **21.** **23.**

CHAPTER REVIEW

1. $\left(6, -\dfrac{1}{2}\right)$ [4.1B]   **2.** $(-4, 7)$ [4.2A]   **3.**   **4.**   **5.** $\left(x, -\dfrac{1}{4}x + \dfrac{3}{2}\right)$ [4.1B]

$(0, 3)$ [4.1A] $(x, 2x - 4)$ [4.1A]

6. $\left(x, \dfrac{1}{3}x - 2\right)$ [4.2A]   **7.** $(3, -1, -2)$ [4.2B]   **8.** $(5, -2, 3)$ [4.2B]   **9.** 28 [4.3A]   **10.** 0 [4.3A]

11. $(3, -1)$ [4.3B]   **12.** $\left(\dfrac{110}{23}, \dfrac{25}{23}\right)$ [4.3B]   **13.** $(-1, -3, 4)$ [4.3B]   **14.** $(2, 3, -5)$ [4.3B]

15. [4.5A]   **16.** [4.5A]

17. The rate of the cabin cruiser is 16 mph; the rate of the current is 4 mph. [4.4A]   **18.** The rate of the plane is 175 mph; the rate of the wind is 25 mph. [4.4A]   **19.** There were 100 children who attended the movie. [4.4B]
20. There was $5000 invested at 3%, and $15,000 invested at 7%. [4.1C]

CHAPTER TEST

1. $(3, 4)$ [4.1A]   **2.** $(-5, 0)$ [4.1A]

3. [4.5A]   **4.** [4.5A]   **5.** $\left(\dfrac{3}{4}, \dfrac{7}{8}\right)$ [4.1B]

6. $(-3, -4)$ [4.1B]   **7.** $(2, -1)$ [4.1B]   **8.** $(-2, 1)$ [4.2A]   **9.** No solution [4.2A]   **10.** $(1, 1)$ [4.2A]

11. No solution [4.2B]   **12.** $(2, -1, -2)$ [4.2B]   **13.** 10 [4.3A]   **14.** -32 [4.3A]   **15.** $\left(-\dfrac{1}{3}, -\dfrac{10}{3}\right)$ [4.3B]

16. $\left(\dfrac{59}{19}, -\dfrac{62}{19}\right)$ [4.3B]   **17.** $\left(\dfrac{1}{5}, -\dfrac{6}{5}, \dfrac{3}{5}\right)$ [4.3B]   **18.** The rate of the plane is 150 mph. The rate of the wind is

25 mph. [4.4A]   **19.** The cost of the cotton is $9 per yard. The cost of the wool is $14 per yard. [4.4B]   **20.** $9000

is invested at 2.7%; $6000 is invested at 5.1%. [4.1C]

CUMULATIVE REVIEW

1. $-\dfrac{11}{28}$ [2.1C]   **2.** $y = 5x - 11$ [3.5B]   **3.** $3x - 24$ [1.3C]   **4.** -4 [1.3B]   **5.** $\{x \mid x < 6\}$ [2.4B]

6. $\{x \mid -4 < x < 8\}$ [2.5B]   **7.** $\{x \mid x > 4 \text{ or } x < -1\}$ [2.5B]   **8.** $f(-3) = -98$ [3.2A]

9. Range: $\{0, 1, 5, 8, 16\}$ [3.2A]   **10.** $F(2) = 1$ [3.2A]   **11.** $f(2 + h) - f(2) = 3h$ [3.2A]

12. [1.1C]   **13.** $y = -\dfrac{2}{3}x + \dfrac{5}{3}$ [3.5A]   **14.** $y = -\dfrac{3}{2}x + \dfrac{1}{2}$ [3.6A]   **15.** 6.32 [3.1B]

16. $\left(-\dfrac{1}{2}, 4\right)$ [3.1B]   **17.** [3.4B]   **18.** [3.7A]

19. $(2, 0)$ [4.1A]   **20.** [4.5A]   **21.** $(1, 0, -1)$ [4.2A]   **22.** 3 [4.3A]

23. $(2, -3)$ [4.3B]   **24.** $(-5, -11)$ [4.1B]   **25.** There are 16 nickels in the coin purse. [2.2B]   **26.** 60 ml of
pure water must be used. [2.3B]   **27.** The rate of the wind is 12.5 mph. [4.4A]   **28.** One pound of steak costs
$6. [4.4B]   **29.** The lower limit is 10,200 ohms. The upper limit is 13,800 ohms. [2.5C]   **30.** The slope is 40. The
slope is the commission rate of the executive. [3.4A]

Answers to Chapter 5 Odd-Numbered Exercises

PREP TEST

1. $-12y$ [1.3C]   **2.** -8 [1.2C]   **3.** $3a - 8b$ [1.3C]   **4.** $11x - 2y - 2$ [1.3C]   **5.** $-x + y$ [1.3C]

6. $2 \cdot 2 \cdot 2 \cdot 5$ [1.2B]   **7.** 4 [1.2B]   **8.** -13 [1.3B]   **9.** $-\dfrac{1}{3}$ [2.1B]

SECTION 5.1

1. a^4b^4 **3.** $-18x^3y^4$ **5.** x^8y^{16} **7.** $81x^8y^{12}$ **9.** $729a^{10}b^6$ **11.** x^5y^{11} **13.** $729x^6$ **15.** $a^{18}b^{18}$ **17.** $4096x^{12}y^{12}$

19. $64a^{24}b^{18}$ **21.** x^{2n+1} **23.** y^{6n-2} **25.** a^{2n^2-6n} **27.** x^{15n+10} **29.** $-6x^5y^5z^4$ **31.** $-12a^2b^9c^2$ **33.** $-6x^4y^4z^5$

35. $-432a^7b^{11}$ **37.** $54a^{13}b^{17}$ **39.** 243   **41.** y^3 **43.** $\dfrac{a^3b^2}{4}$ **45.** $\dfrac{x}{y^4}$ **47.** $\dfrac{1}{2}$ **49.** $-\dfrac{1}{9}$ **51.** $\dfrac{1}{y^8}$ **53.** $\dfrac{1}{x^6y^{10}}$

55. $x^5 y^5$ **57.** $\dfrac{a^8}{b^9}$ **59.** $\dfrac{1}{2187a}$ **61.** $\dfrac{y^6}{x^3}$ **63.** $\dfrac{y^4}{x^3}$ **65.** $-\dfrac{1}{2x^2}$ **67.** $-\dfrac{1}{243a^5 b^{10}}$ **69.** $\dfrac{16x^8}{81y^4 z^{12}}$ **71.** $\dfrac{3a^4}{4b^3}$

73. $-\dfrac{9a}{8b^6}$ **75.** $\dfrac{16x^2}{y^6}$ **77.** $\dfrac{1}{b^{4n}}$ **79.** $-\dfrac{1}{y^{6n}}$ **81.** y^{n-2} **83.** $\dfrac{1}{y^{2n}}$ **85.** $\dfrac{x^{n-5}}{y^6}$ **87.** $\dfrac{8b^{15}}{3a^{18}}$ **89.** 4.67×10^{-6}

91. 1.7×10^{-10} **93.** 2×10^{11} **95.** 0.000000123 **97.** $8,200,000,000,000,000$ **99.** 0.039 **101.** $150,000$

103. $20,800,000$ **105.** 0.000000015 **107.** 0.0000000000178 **109.** $140,000,000$ **111.** $11,456,790$

113. 0.000008 **115.** Light travels 2.592×10^{10} km in one day. **117.** The signals traveled at a speed of

$1.08\overline{1} \times 10^7$ mi/min. **119.** The centrifuge makes one revolution in 6.6×10^{-7} s. **121.** The sun is 3.38983×10^5 times

heavier. **123.** The weight of one orchid seed is 3.225806×10^{-8} oz. **125.** The distance from the Coma cluster to Earth

is 1.6423949×10^{21} mi. **127a.** $\dfrac{8}{5}$ **b.** $\dfrac{5}{4}$

SECTION 5.2

1. $P(3) = 13$ **3.** $R(2) = 10$ **5.** $f(-1) = -11$ **7a.** The leading coefficient is -1. **b.** The constant term is 8.
c. The degree is 2. **9.** The expression is not a polynomial. **11.** The expression is not a polynomial.
13a. The leading coefficient is 3. **b.** The constant term is π. **c.** The degree is 5. **15a.** The leading coefficient is -5.
b. The constant term is 2. **c.** The degree is 3. **17a.** The leading coefficient is 14. **b.** The constant term is 14.
c. The degree is 0. **19.** **21.** **23.** **25.** $6x^2 - 6x + 5$

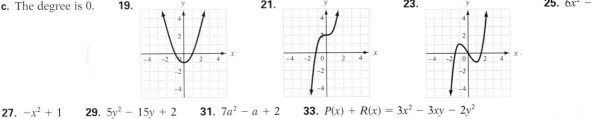

27. $-x^2 + 1$ **29.** $5y^2 - 15y + 2$ **31.** $7a^2 - a + 2$ **33.** $P(x) + R(x) = 3x^2 - 3xy - 2y^2$
35. $P(x) - R(x) = 8x^2 - 2xy + 5y^2$ **37.** $S(x) = 3x^4 - 8x^2 + 2x$; $S(2) = 20$ **39a.** $k = 8$ **b.** $k = -4$

SECTION 5.3

1. $2x^2 - 6x$ **3.** $6x^4 - 3x^3$ **5.** $6x^2 y - 9xy^2$ **7.** $x^{n+1} + x^n$ **9.** $x^{2n} + x^n y^n$ **11.** $-4b^2 + 10b$
13. $-6a^4 + 4a^3 - 6a^2$ **15.** $9b^5 - 9b^3 + 24b$ **17.** $-3y^4 - 4y^3 + 2y^2$ **19.** $-20x^2 + 15x^3 - 15x^4 - 20x^5$
21. $-2x^4 y + 6x^3 y^2 - 4x^2 y^3$ **23.** $x^{3n} + x^{2n} + x^{n+1}$ **25.** $a^{2n+1} - 3a^{n+2} + 2a^{n+1}$ **27.** $5y^2 - 11y$ **29.** $6y^2 - 31y$
31. $x^2 + 5x - 14$ **33.** $8y^2 + 2y - 21$ **35.** $8x^2 + 8xy - 30y^2$ **37.** $x^2 y^2 + xy - 12$ **39.** $2x^4 - 15x^2 + 25$
41. $10x^4 - 15x^2 y + 5y^2$ **43.** $x^{2n} - x^n - 6$ **45.** $6a^{2n} + a^n - 15$ **47.** $6a^{2n} + a^n b^n - 2b^{2n}$
49. $x^4 + 5x^3 - 3x^2 - 11x + 20$ **51.** $10a^3 - 27a^2 b + 26ab^2 - 12b^3$ **53.** $2y^5 - 10y^4 - y^3 - y^2 + 3$
55. $4x^5 - 16x^4 + 15x^3 - 4x^2 + 28x - 45$ **57.** $x^4 - 3x^3 - 6x^2 + 29x - 21$ **59.** $2a^3 + 7a^2 - 43a + 42$
61. $x^{3n} + 2x^{2n} + 2x^n + 1$ **63.** $x^{2n} - 2x^{2n} y^n + 4x^n y^n - 2x^n y^{2n} + 3y^{2n}$ **65.** $9x^2 - 4$ **67.** $36 - x^2$ **69.** $4a^2 - 9b^2$
71. $x^4 - 1$ **73.** $x^{2n} - 9$ **75.** $x^2 - 10x + 25$ **77.** $9a^2 + 30ab + 25b^2$ **79.** $x^4 - 6x^2 + 9$ **81.** $4x^4 - 12x^2 y^2 + 9y^4$
83. $a^{2n} - 2a^n b^n + b^{2n}$ **85.** $-x^2 + 2xy$ **87.** $-4xy$ **89.** $(3x^2 + 10x - 8)$ ft^2 **91.** $(x^2 + 3x)$ m^2
93. $(x^3 + 9x^2 + 27x + 27)$ cm^3 **95.** $(2x^3)$ in^3 **97.** $(78.5x^2 + 125.6x + 50.24)$ in^2 **99a.** $a^3 - b^3$ **b.** $x^3 + y^3$
101a. $k = 5$ **b.** $k = 1$ **103.** $a^2 + b^2$

SECTION 5.4

1. $x + 8$ **3.** $x^2 + 3x + 6 + \dfrac{20}{x - 3}$ **5.** $3x + 5 + \dfrac{3}{2x + 1}$ **7.** $5x + 7 + \dfrac{2}{2x - 1}$ **9.** $4x^2 + 6x + 9 + \dfrac{18}{2x - 3}$

11. $3x^2 + 1 + \dfrac{1}{2x^2 - 5}$ **13.** $x^2 - 3x - 10$ **15.** $x^2 - 2x + 1 - \dfrac{1}{x - 3}$ **17.** $2x^3 - 3x^2 + x - 4$ **19.** $2x + \dfrac{x + 2}{x^2 + 2x - 1}$

21. $x^2 + 4x + 6 + \dfrac{10x + 8}{x^2 - 2x - 1}$ **23.** $x^2 - 2x + 4 + \dfrac{-6x - 7}{x^2 + 2x + 3}$ **25.** $2x - 8$ **27.** $3x - 8$ **29.** $3x + 3 - \dfrac{1}{x - 1}$

31. $2x^2 - 3x + 9$ **33.** $4x^2 + 8x + 15 + \dfrac{12}{x - 2}$ **35.** $2x^2 - 3x + 7 - \dfrac{8}{x + 4}$ **37.** $3x^3 + 2x^2 + 12x + 19 + \dfrac{33}{x - 2}$

39. $3x^3 - x + 4 - \dfrac{2}{x + 1}$ **41.** $2x^3 + 6x^2 + 17x + 51 + \dfrac{155}{x - 3}$ **43.** $P(3) = 8$ **45.** $R(4) = 43$ **47.** $P(-2) = -39$

49. $Z(-3) = -60$ **51.** $Q(2) = 31$ **53.** $F(-3) = 178$ **55.** $P(5) = 122$ **57.** $R(-3) = 302$ **59.** $Q(2) = 0$

61. $R(-2) = -65$ **63a.** $a^2 - ab + b^2$ **b.** $x^4 - x^3y + x^2y^2 - xy^3 + y^4$ **c.** $x^5 - x^4y + x^3y^2 - x^2y^3 + xy^4 - y^5$

65a. $\dfrac{1}{2}x^2 + \dfrac{3}{4}x - 1$ **b.** $\dfrac{2}{3}x^2 + \dfrac{7}{2}x + \dfrac{10}{3} + \dfrac{64}{6x - 12}$ **c.** $2x^2 - x + 1 + \dfrac{3x - 11}{x^2 - x + 1}$ **d.** $x^2 + 6x + 17 + \dfrac{51x + 56}{x^2 - 2x - 3}$

SECTION 5.5

1. $3a(2a - 5)$ **3.** $x^2(4x - 3)$ **5.** Nonfactorable over the integers   **7.** $x(x^4 - x^2 - 1)$ **9.** $4(4x^2 - 3x + 6)$

11. $5b^2(1 - 2b + 5b^2)$ **13.** $x^n(x^n - 1)$ **15.** $x^{2n}(x^n - 1)$ **17.** $a^2(a^{2n} + 1)$ **19.** $6x^2y(2y - 3x + 4)$

21. $4a^2b^2(6a - 1 - 4b^2)$ **23.** $y^2(y^{2n} + y^n - 1)$ **25.** $(a + 2)(x - 2)$ **27.** $(x - 2)(a + b)$ **29.** $(a - 2b)(x - y)$

31. $(x + 4)(y - 2)$ **33.** $(a + b)(x - y)$ **35.** $(y - 3)(x^2 - 2)$ **37.** $(3 + y)(2 + x^2)$ **39.** $(2a + b)(x^2 - 2y)$

41. $(2b + a)(3x - 2y)$ **43.** $(y - 5)(x^n + 1)$ **45.** $(2x - 1)(x^2 + 2)$ **47.** $(x + 10)(x + 2)$ **49.** $(a + 9)(a - 8)$

51. $(a + 6)(a + 1)$ **53.** $(y - 6)(y - 12)$ **55.** $(x + 12)(x - 11)$ **57.** $(x + 10)(x + 5)$ **59.** $(b + 2)(b - 8)$

61. $(a - b)(a - 2b)$ **63.** $(a + 11b)(a - 3b)$ **65.** $(x + 3y)(x + 2y)$ **67.** $-(x - 2)(x + 1)$ **69.** $-(x - 5)(x + 1)$

71. $(x - 3)(x - 2)$ **73.** $(2x + 1)(x + 3)$ **75.** $(2y + 3)(3y - 2)$ **77.** $(3b + 7)(2b - 5)$ **79.** $(3y - 13)(y - 3)$

81. Nonfactorable over the integers   **83.** $(a + 1)(4a - 5)$ **85.** $(5x - 2)(2x - 5)$ **87.** Nonfactorable over the integers

89. $(6x - y)(x + 7y)$ **91.** $(7a - 3b)(a + 7b)$ **93.** $(6x + 5y)(3x + 2y)$ **95.** $-(5x - 3)(x + 2)$

97. Nonfactorable over the integers   **99.** $-(4b - 7)(2b + 5)$ **101.** $y^2(5y - 4)(y - 5)$ **103.** $-2x^2(5x - 2)(3x + 5)$

105. $a^2b^2(ab + 2)(ab - 5)$ **107.** $5(18a^2b^2 + 9ab + 2)$ **109.** Nonfactorable over the integers   **111.** $2a(a^2 + 5)(a^2 + 2)$

113. $3y^2(x^2 - 5)(x^2 - 8)$ **115.** $3xy(xy - 2)(xy + 4)$ **117.** $y(y^2 - 3)(y^2 - 5)$ **119.** $y(4y - 3)(4y - 1)$

121. $3(y^n - 3)(4y^n - 5)$ **123.** $x^n(x^n + 2)(x^n + 8)$ **125a.** $-9, -6, 6, 9$ **b.** $-5, -1, 1, 5$ **c.** $-7, -5, 5, 7$

d. $-9, -3, 3, 9$ **e.** $-16, -8, 8, 16$ **f.** $-5, -1, 1, 5$

SECTION 5.6

1. $4; 25x^6; 100x^4y^4$ **3.** $4z^4$ **5.** $9a^2b^3$ **7.** $(x + 4)(x - 4)$ **9.** $(2x + 1)(2x - 1)$ **11.** $(4x + 11)(4x - 11)$

13. $(1 + 3a)(1 - 3a)$ **15.** $(xy + 10)(xy - 10)$ **17.** Nonfactorable over the integers   **19.** $(5 + ab)(5 - ab)$

21. $(a^n + 1)(a^n - 1)$ **23.** $(x - 6)^2$ **25.** $(b - 1)^2$ **27.** $(4x - 5)^2$ **29.** Nonfactorable over the integers

31. Nonfactorable over the integers   **33.** $(x + 3y)^2$ **35.** $(5a - 4b)^2$ **37.** $(x^n + 3)^2$ **39.** $(x - 7)(x - 1)$

41. $(x - y + a + b)(x - y - a - b)$ **43.** $8; x^9; 27c^{15}d^{18}$ **45.** $2x^3$ **47.** $4a^2b^6$ **49.** $(x - 3)(x^2 + 3x + 9)$

51. $(2x - 1)(4x^2 + 2x + 1)$ **53.** $(x - y)(x^2 + xy + y^2)$ **55.** $(m + n)(m^2 - mn + n^2)$ **57.** $(4x + 1)(16x^2 - 4x + 1)$

59. $(3x - 2y)(9x^2 + 6xy + 4y^2)$ **61.** $(xy + 4)(x^2y^2 - 4xy + 16)$ **63.** Nonfactorable over the integers

65. Nonfactorable over the integers   **67.** $(a - 2b)(a^2 - ab + b^2)$ **69.** $(x^{2n} + y^n)(x^{4n} - x^{2n}y^n + y^{2n})$

71. $(x^n + 2)(x^{2n} - 2x^n + 4)$ **73.** $(xy - 3)(xy - 5)$ **75.** $(xy - 5)(xy - 12)$ **77.** $(x^2 - 3)(x^2 - 6)$

79. $(b^2 + 5)(b^2 - 18)$ **81.** $(x^2y^2 - 2)(x^2y^2 - 6)$ **83.** $(x^n + 1)(x^n + 2)$ **85.** $(3xy - 5)(xy - 3)$

87. $(2ab - 3)(3ab - 7)$ **89.** $(2x^2 - 15)(x^2 + 1)$ **91.** $(2x^n - 1)(x^n - 3)$ **93.** $(2a^n + 5)(3a^n + 2)$ **95.** $3(2x - 3)^2$

97. $a(3a - 1)(9a^2 + 3a + 1)$ **99.** $5(2x + 1)(2x - 1)$ **101.** $y^3(y + 11)(y - 5)$ **103.** $(4x^2 + 9)(2x + 3)(2x - 3)$

105. $2a(2 - a)(4 + 2a + a^2)$ **107.** $b^3(ab - 1)(a^2b^2 + ab + 1)$ **109.** $2x^2(2x - 5)^2$ **111.** $(x^2 + y^2)(x + y)(x - y)$

113. $(x^2 + y^2)(x^4 - x^2y^2 + y^4)$ **115.** Nonfactorable over the integers   **117.** $2a(2a - 1)(4a^2 + 2a + 1)$

119. $a^2b^2(a + 4b)(a - 12b)$ **121.** $2b^2(3a + 5b)(4a - 9b)$ **123.** $(x - 2)^2(x + 2)$ **125.** $(x + y)(x - y)(2x + 1)(2x - 1)$

127. $x(x^n + 1)^2$ **129.** $b^n(3b - 2)(b + 2)$

SECTION 5.7

1. -3 and 5 **3.** -7 and 8 **5.** $-4, 0,$ and $\dfrac{2}{3}$ **7.** -5 and 3 **9.** 1 and 3 **11.** -2 and 5 **13.** $-\dfrac{1}{2}$ and $\dfrac{3}{2}$

15. $\dfrac{1}{2}$ **17.** -3 and 3 **19.** $-\dfrac{1}{2}$ and $\dfrac{1}{2}$ **21.** -3 and 5 **23.** -1 and $\dfrac{13}{2}$ **25.** -2 and $\dfrac{4}{3}$ **27.** $-5, 0,$ and 2

29. -2 and 2 **31.** $-3, -1,$ and 3 **33.** $-2, -\dfrac{2}{3},$ and 2 **35.** $-2, -\dfrac{2}{5},$ and 2 **37.** The number is 14 or -15.

39. The length is 38 cm. The width is 10 cm.   **41.** The hypotenuse is 13 ft.   **43.** The stone will hit the bottom of the well in 2 s.   **45.** The length of the cardboard is 18 in. The width of the cardboard is 8 in.

CHAPTER REVIEW

1. $6a^2b(3a^3b - 2ab^2 + 5)$ [5.5A]   **2.** $5x + 4 + \dfrac{6}{3x - 2}$ [5.4A]   **3.** $144x^2y^{10}z^{14}$ [5.1B]

4. $(a + 2b)(2x - 3y)$ [5.5B]   **5.** $-(x - 4)(x + 3)$ [5.5C]   **6.** 1   [5.4C]   **7.** $4x^2 - 8xy + 5y^2$ [5.2B]

8. $(6x + 5)(4x + 3)$ [5.5D]   **9.** $(2x + 3y)^2$ [5.6A]   **10.** $-6a^3b^6$ [5.1A]   **11.** $(4a - 3b)(16a^2 + 12ab + 9b^2)$ [5.6B]

12. $4x^2 + 3x - 8 + \dfrac{50}{x + 6}$ [5.4A/5.4B]   **13.** -7 [5.2A]   **14.** $(x - 8)(x + 5)$ [5.5C]   **15.** $(xy + 3)(xy - 3)$ [5.6A]

16. $12x^5y^3 + 8x^3y^2 - 28x^2y^4$ [5.3A]   **17.** $(x^n - 6)^2$ [5.6A]   **18.** $\dfrac{5}{2}$ and 4   [5.7A]   **19.** $33x^2 + 24x$ [5.3A]

20. $3a^2(a^2 + 1)(a^2 - 6)$ [5.6D]   **21.** $16x^2 - 24xy + 9y^2$ [5.3C]   **22.** $x^3 + 4x^2 + 16x + 64 + \dfrac{252}{x - 4}$ [5.4A/5.4B]

23. $(3x^2 + 2)(5x^2 - 3)$ [5.6C]   **24.** $\dfrac{c^{10}}{2b^{17}}$ [5.1B]   **25.** $6x^3 - 29x^2 + 14x + 24$ [5.3B]

26. $(7x^2y^2 + 3)(3x^2y^2 + 2)$ [5.6C]   **27.** 1, 4, and -4 [5.7A]   **28.** $25a^2 - 4b^2$ [5.3C]   **29.** 0.00254   [5.1C]

30. $(3x - 2)(2x - 9)$ [5.5D]   **31.**   [5.2A]   **32a.** 3   **b.** 8   **c.** 5   [5.2A]

33. The mass of the moon is 8.103×10^{19} tons.   [5.1D]   **34.** The number is -8 or 7.   [5.7B]   **35.** The Great Galaxy of Andromeda is 1.291224×10^{19} mi from Earth.   [5.1D]   **36.** The area of the rectangle is $(10x^2 - 29x - 21)$ cm^2.   [5.3D]

CHAPTER TEST

1. $(4t + 3)^2$ [5.6A]   **2.** $-18r^2s^2 + 12rs^3 + 18rs^2$ [5.3A]   **3.** -3 [5.2A]   **4.** $(3x - 2)(9x^2 + 6x + 4)$ [5.6B]

5. $(4x - 5)(4x + 5)$ [5.6A]   **6.** $6t^5 - 8t^4 - 15t^3 + 22t^2 - 5$ [5.3B]   **7.** $35x^2 - 55x$ [5.3A]

8. $3x(2x - 3)(2x + 5)$ [5.6D]   **9.** $-1, -\dfrac{1}{6}$, and 1   [5.7A]   **10.** $2x^3 - 4x^2 + 6x - 14$ [5.2B]

11. 5.01×10^{-7} [5.1C]   **12.** $2x + 1 + \dfrac{4}{7x - 3}$ [5.4A]   **13.** $49 - 25x^2$ [5.3C]   **14.** $(2a^2 - 5)(3a^2 + 1)$ [5.6C]

15. $6a^2 - 13ab - 28b^2$ [5.3B]   **16.** $(3x^2 + 4)(x - 3)(x + 3)$ [5.6D]   **17.** $64a^7b^7$ [5.1A]   **18.** $-\dfrac{1}{3}$ and $\dfrac{1}{2}$ [5.7A]

19. -8 [5.4C]   **20.** $\dfrac{2b^7}{a^{10}}$ [5.1B]   **21.** $x^2 - 5x + 10 - \dfrac{23}{x + 3}$ [5.4A/5.4B]   **22.** $(3x - 4)(2x - 3)$ [5.5D]

23. $(3x - 2)(2x - a)$ [5.5B]   **24.** There are 6.048×10^5 s in one week.   [5.1D]   **25.** The arrow will be 64 ft above the ground after 1 s and after 2 s.   [5.7B]   **26.** The area is $(10x^2 - 3x - 1)$ ft^2.   [5.3D]

CUMULATIVE REVIEW

1. 4   [1.2D]   **2.** $-\dfrac{5}{4}$ [1.3B]   **3.** Inverse Property of Addition   [1.3A]   **4.** $-18x + 8$ [1.3C]   **5.** $-\dfrac{1}{6}$ [2.1A]

6. $-\dfrac{11}{4}$ [2.1B]   **7.** $x^2 + 3x + 9 + \dfrac{24}{x - 3}$ [5.4B]   **8.** -1 and $\dfrac{7}{3}$ [2.5A]   **9.** $P(-2) = 18$ [3.2A]

10. -2 is excluded from the domain.   [3.2A]   **11.** Range: $\{-4, -1, 8\}$ [3.2A]   **12.** The slope is $-\dfrac{1}{6}$.   [3.4A]

13. The equation of the line is $y = -\dfrac{3}{2}x + \dfrac{1}{2}$.   [3.5A]   **14.** The equation of the line is $y = \dfrac{2}{3}x + \dfrac{16}{3}$.   [3.6A]

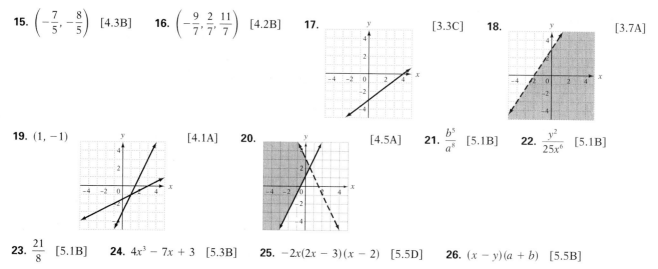

15. $\left(-\dfrac{7}{5}, -\dfrac{8}{5}\right)$ [4.3B] **16.** $\left(-\dfrac{9}{7}, \dfrac{2}{7}, \dfrac{11}{7}\right)$ [4.2B] **17.** [3.3C] **18.** [3.7A]

19. $(1, -1)$ [4.1A] **20.** [4.5A] **21.** $\dfrac{b^5}{a^8}$ [5.1B] **22.** $\dfrac{y^2}{25x^6}$ [5.1B]

23. $\dfrac{21}{8}$ [5.1B] **24.** $4x^3 - 7x + 3$ [5.3B] **25.** $-2x(2x - 3)(x - 2)$ [5.5D] **26.** $(x - y)(a + b)$ [5.5B]

27. $(x - 2)(x + 2)(x^2 + 4)$ [5.6D] **28.** $2(x - 2)(x^2 + 2x + 4)$ [5.6D] **29.** The integers are 9 and 15. [2.2A]

30. 40 oz of pure gold must be used. [2.3A] **31.** The speed of the slower cyclist is 5 mph; the speed of the faster cyclist is 7.5 mph. [2.3C] **32.** It takes the vehicle 12 h to reach the moon. [5.1D] **33.** The slope is 50. The slope represents the average speed in miles per hour. [3.4A]

Answers to Chapter 6 Odd-Numbered Exercises

PREP TEST

1. 50 [1.2B] **2.** $-\dfrac{1}{6}$ [1.2B] **3.** $-\dfrac{3}{2}$ [1.2B] **4.** $\dfrac{1}{24}$ [1.2B] **5.** $\dfrac{5}{24}$ [1.2B] **6.** $-\dfrac{2}{9}$ [1.2D]

7. $\dfrac{1}{3}$ [1.3B] **8.** -2 [2.1C] **9.** $\dfrac{10}{7}$ [2.1C] **10.** 110 mph, 130 mph [2.3C]

SECTION 6.1

3. $f(4) = 2$ **5.** $f(-2) = -2$ **7.** $f(-2) = \dfrac{1}{9}$ **9.** $f(3) = \dfrac{1}{35}$ **11.** $f(-1) = \dfrac{3}{4}$ **13.** The domain is $\{x \mid x \neq 3\}$.

15. The domain is $\{x \mid x \neq -4\}$. **17.** The domain is $\{x \mid x \neq -3\}$. **19.** The domain is $\left\{x \mid x \neq \dfrac{2}{3}, 4\right\}$.

21. The domain is $\{x \mid x \neq -3, 2\}$. **23.** The domain is $\{x \mid x \in \text{real numbers}\}$. **27.** $1 - 2x$ **29.** $3x - 1$ **31.** $2x$

33. $-\dfrac{2}{a}$ **35.** The expression is in simplest form. **37.** $x^2y^2 - 4xy + 5$ **39.** $\dfrac{x^n}{x^n - y^n}$ **41.** $\dfrac{x - 3}{x - 5}$ **43.** $\dfrac{x + 4}{x - 4}$

45. $\dfrac{a - b}{a^2 - ab + b^2}$ **47.** $\dfrac{4x^2 + 2xy + y^2}{2x + y}$ **49.** $\dfrac{(a - 2)(x + 1)}{ax}$ **51.** $\dfrac{x^2 - 3}{x^2 + 1}$ **53.** $\dfrac{2xy + 1}{3xy - 1}$ **55.** $\dfrac{a^n + 4}{a^n + 1}$ **57.** $\dfrac{a^n + b^n}{a^n - b^n}$

59. $\dfrac{a + b}{(x + 1)(x - 1)}$ **61.** $\dfrac{abx}{2}$ **63.** $\dfrac{x}{2}$ **65.** $\dfrac{y(x - 1)}{x^2(x + 1)}$ **67.** $-\dfrac{x + 5}{x - 2}$ **69.** 1 **71.** $\dfrac{x^n + 4}{x^n - 1}$

73. $\dfrac{(x + 1)(x - 1)(x - 4)}{x - 2}$ **75.** $\dfrac{x + y}{3}$ **77.** $\dfrac{4by}{3ax}$ **79.** $\dfrac{4(x - y)^2}{9x^2y}$ **81.** $\dfrac{2x - 3y}{4y^2}$ **83.** $-\dfrac{2x + 5}{2x - 5}$ **85.** $x(x - 3)$

87. -1 **89.** $(x^n + 1)^2$ **91.** $\dfrac{(x + y)(x - y)}{x^3}$ **93.** $\dfrac{x + 3}{x - 3}$ **95.** $\dfrac{5(3y^2 + 2)}{y^2}$

SECTION 6.2

1. $\frac{9y^3}{12x^2y^4}, \frac{17x}{12x^2y^4}$ 3. $\frac{2x^2-4x}{6x^2(x-2)}, \frac{3x-6}{6x^2(x-2)}$ 5. $\frac{3x-1}{2x(x-5)}, -\frac{6x^3-30x^2}{2x(x-5)}$ 7. $\frac{6x^2+9x}{(2x-3)(2x+3)}, \frac{10x^2-15x}{(2x-3)(2x+3)}$

9. $\frac{2x}{(x+3)(x-3)}, \frac{x^2+4x+3}{(x+3)(x-3)}$ 11. $\frac{6}{6(x+2y)(x-2y)}, \frac{5x+10y}{6(x+2y)(x-2y)}$ 13. $\frac{3x^2-3x}{(x+1)(x-1)^2}, \frac{5x^2+5x}{(x+1)(x-1)^2}$

15. $-\frac{x-3}{(x-2)(x^2+2x+4)}, \frac{2x-4}{(x-2)(x^2+2x+4)}$ 17. $\frac{2x^2+6x}{(x-1)(x+3)^2}, -\frac{x^2-x}{(x-1)(x+3)^2}$

19. $-\frac{12x^2-8x}{(2x-3)(2x-5)(3x-2)}, \frac{6x^2-9x}{(2x-3)(2x-5)(3x-2)}$ 21. $\frac{5}{(3x-4)(2x-3)}, -\frac{4x^2-6x}{(3x-4)(2x-3)}, \frac{3x^2-x-4}{(3x-4)(2x-3)}$

23. $\frac{2x^2+10x}{(x-3)(x+5)}, -\frac{2x-6}{(x-3)(x+5)}, -\frac{x-1}{(x-3)(x+5)}$ 25. $\frac{x-5}{(x^n+1)(x^n+2)}, \frac{2x^{n+1}+2x}{(x^n+1)(x^n+2)}$ 27. $\frac{1}{2x^2}$ 29. $\frac{1}{x+2}$

31. $\frac{12ab-9b+8a}{30a^2b^2}$ 33. $\frac{5-16b+12a}{40ab}$ 35. $\frac{7}{12x}$ 37. $\frac{2xy-8x+3y}{10x^2y^2}$ 39. $-\frac{a(2a-13)}{(a-2)(a+1)}$ 41. $\frac{5x^2-6x+10}{(2x-5)(5x-2)}$

43. $\frac{a}{b(a-b)}$ 45. $\frac{a^2+18a-9}{a(a-3)}$ 47. $\frac{17x^2+20x-25}{x(6x-5)}$ 49. $\frac{6}{(x+3)(x-3)^2}$ 51. $\frac{2(x-1)}{(x+2)^2}$ 53. $-\frac{5x^2-17x+8}{(x+4)(x-2)}$

55. $\frac{3x^n+2}{(x^n+1)(x^n-1)}$ 57. 1   59. $\frac{x^2-52x+160}{4(x+3)(x-3)}$ 61. $\frac{3x-1}{4x+1}$ 63. $\frac{2(5x-3)}{(x+3)(x+4)(x-3)}$ 65. $\frac{x-2}{x+3}$

67. 1   69. $\frac{x+1}{2x-1}$ 71. $\frac{1}{2x-1}$ 73. $\frac{1}{x^2+4}$ 75. $\frac{3-a}{3a}$ 77. $-\frac{2x^2+5x-2}{(x+2)(x+1)}$ 79. $\frac{b-a}{b+2a}$ 81. $\frac{2}{x+2}$

83a. $\frac{b+6}{3(b-2)}$ b. $\frac{6(x-1)}{x(2x+1)}$ 85a. $f(4) = \frac{2}{3}; g(4) = 4; S(4) = 4\frac{2}{3}$; yes; $S(a) = f(a) + g(a)$

SECTION 6.3

3. $\frac{5}{23}$ 5. $\frac{2}{5}$ 7. $\frac{x}{x-1}$ 9. $-\frac{a}{a+2}$ 11. $-\frac{a-1}{a+1}$ 13. $\frac{2}{5}$ 15. $-\frac{1}{2}$ 17. $\frac{x^2-x-1}{x^2+x+1}$ 19. $\frac{x+2}{x-1}$

21. $-\frac{x+4}{2x+3}$ 23. $\frac{a(a^2+a+1)}{a^2+1}$ 25. $\frac{x+1}{x-4}$ 27. $\frac{(x-1)(x+1)}{x^2+1}$ 29. $-\frac{1}{x(x+h)}$ 31. $\frac{x-2}{x-3}$ 33. $\frac{x-3}{x+4}$

35. $\frac{3x+1}{x-5}$ 37. $-\frac{2a+2}{7a-4}$ 39. $\frac{x+y}{x-y}$ 41. $-\frac{2x}{x^2+1}$ 43. $-\frac{a^2}{1-2a}$ 45. $-\frac{3x+2}{x-2}$ 47. $\frac{3n^2+12n+8}{n(n+2)(n+4)}$

49a. $\frac{2}{2+x}$ b. -1 c. $\frac{x-3}{x-2}$

SECTION 6.4

3. 9   5. $\frac{15}{2}$ 7. 3   9. $\frac{10}{3}$ 11. -2 13. $\frac{2}{5}$ 15. -4 17. $\frac{3}{4}$ 19. $-\frac{6}{5}$ 21. -2 23. There are 56.75 g of protein in a 454-gram box of pasta.   25. There would be approximately 60,000 squawfish in the area.   27. You would expect 504 of the computers to have defective CD-ROM drives.   29. The dimensions of the room are 18 ft by 24 ft.   31. The 210-pound adult would require 0.75 additional ounces of medication.   33. A person would need to walk 21.54 mi to lose one pound.

SECTION 6.5

1. -5 3. $\frac{5}{2}$ 5. -1 7. 8   9. -3 11. -4 13. The equation has no solution.   15. -1 and $-\frac{5}{3}$

17. $-\frac{1}{3}$ and 5   19. $-\frac{11}{3}$ 21. The equation has no solution.   23. $P_2 = \frac{P_1V_1T_2}{T_1V_2}$ 25. $b = \frac{af}{a-f}$ 27. It would take the experienced bricklayer 14 h to do the job.   29. It would have taken the slower machine 160 min to send the fax.   31. With all three machines working, it would take 4 h to fill the bottles.   33. With both pipes open, it would take 90 min to empty the tank.   35. With both clowns working, it would take 120 min to have 76 balloons.   37. With both clerks

working, it would take 2400 s to address the envelopes. **39.** The rate of the runner is 8 mph. **41.** The tortoise was running at 0.4 ft/s. The hare was running at 72 ft/s. **43.** The rate of the cyclist is 10 mph. **45.** The rate of the current is 2 mph. **47.** The rate of the wind is 75 mph. **49.** The rate of the wind is 60.80 mph. **51.** The rate of the current is 2 mph. **53a.** 3 **b.** 0 and y **c.** $3y$ and $\frac{3}{2}y$ **55.** The bus usually travels 60 mph.

SECTION 6.6

1. The profit is \$80,000. **3.** The pressure is 6.75 lb/in². **5.** The object will fall 1600 ft. **7.** The ball will roll 54 ft in 3 s. **9.** The length is 10 ft when the width is 4 ft. **11.** The gear will make 30 revolutions/min. **13.** The current will be 7.5 amps. **15.** The intensity of light will be 48 foot-candles. **17.** y is doubled. **19.** inversely **21.** inversely

CHAPTER REVIEW

1. $\frac{1}{a-1}$ [6.1B] **2.** $-\frac{5x^2-17x-9}{(x-3)(x+2)}$ [6.2B] **3.** 4 [6.1A] **4.** 4 [6.4A] **5.** $-\frac{9x^2+16}{24x}$ [6.3A]

6. $\frac{16x^2+4x}{(4x-1)(4x+1)}, \frac{12x^2-7x+1}{(4x-1)(4x+1)}$ [6.2A] **7.** $r=\frac{S-a}{S}$ [6.5A] **8.** $\frac{2}{21}$ [6.1A] **9.** $\frac{2}{5}$ [6.4A]

10. The domain is $\{x\,|\,x\neq -3,2\}$. [6.1A] **11.** $\frac{3x^2-1}{3x^2+1}$ [6.1A] **12.** The domain is $\{x\,|\,x\neq 3\}$. [6.1A]

13. x [6.1B] **14.** $\frac{4x-1}{x+2}$ [6.2B] **15.** $N=\frac{S}{1-Q}$ [6.5A] **16.** The equation has no solution. [6.5A]

17. $\frac{x}{x-3}$ [6.3A] **18.** $\frac{x^2-7x+12}{(x-5)(x-4)}; \frac{x}{(x-5)(x-4)}; -\frac{x-5}{(x-5)(x-4)}$ [6.2A] **19.** 10 [6.5A] **20.** $\frac{1}{x}$ [6.1C]

21. $\frac{3x+2}{x}$ [6.1C] **22.** $\frac{9x^2+x+4}{(3x-1)(x-2)}$ [6.2B] **23.** $\frac{x^2+3x+9}{x+3}$ [6.1A] **24.** $\frac{21a+40b}{24a^2b^4}$ [6.2B] **25.** The rate of the helicopter is 180 mph. [6.5C] **26.** It will take the student 375 min to read 150 pages. [6.4B] **27.** The current in the circuit is 2 amps. [6.6A] **28.** 48 mi would be represented by 12 in. [6.4B] **29.** The stopping distance of the car is 61.2 ft. [6.6A] **30.** It would take the apprentice 104 min working alone. [6.5B]

CHAPTER TEST

1. 2 [6.4A] **2.** $\frac{x+2}{x-1}$ [6.1C] **3.** $\frac{x^2-9x+3}{(x+2)(x-3)}$ [6.2B] **4.** $\frac{x^2-2x-3}{(x+3)(x-3)(x-2)}; \frac{2x^2-4x}{(x+3)(x-3)(x-2)}$ [6.2A]

5. The equation has no solution. [6.5A] **6.** $\frac{v(v+2)}{2v-1}$ [6.1A] **7.** $\frac{6(x-2)}{5}$ [6.1B] **8.** The domain is $\{x\,|\,x\neq -3,3\}$. [6.1A] **9.** $\frac{x-4}{x+3}$ [6.3A] **10.** $\frac{x+4}{x+2}$ [6.3A] **11.** $\frac{x+1}{3x-4}$ [6.1C] **12.** 1 and 2 [6.5A] **13.** $-\frac{2(a-2)}{3a+2}$ [6.1A]

14. $t=\frac{4r}{r-2}$ [6.5A] **15.** $f(-1)=2$ [6.1A] **16.** $\frac{-x^2-5x+2}{(x+1)(x+4)(x-1)}$ [6.2B] **17.** -3 and 5 [6.4A]

18. The rate of the cyclist is 10 mph. [6.5C] **19.** The resistance of the cable is 0.4 ohm. [6.6A] **20.** The designer needs 14 rolls of wallpaper. [6.4B] **21.** It would take both landscapers 10 min to till the soil. [6.5B] **22.** The resistance is 1.6 ohms. [6.6A]

CUMULATIVE REVIEW

1. $\frac{36}{5}$ [1.2D] **2.** $\frac{15}{2}$ [2.1C] **3.** 7 and 1 [2.5A] **4.** The domain is $\{x\,|\,x\neq 3\}$. [6.1A] **5.** $P(-2)=\frac{3}{7}$ [6.1A]

6. 3.5×10^{-8} [5.1C] **7.** $\frac{a^5}{b^6}$ [5.1B] **8.** $\{x\,|\,x\leq 4\}$ [2.4A] **9.** $-4a^4+6a^3-2a^2$ [5.3A]

10. $(2x^n-1)(x^n+2)$ [5.6C] **11.** $(xy-3)(x^2y^2+3xy+9)$ [5.6B] **12.** $x+3y$ [6.1A] **13.** The equation of the line is $y=\frac{3}{2}x+2$. [3.6A] **14.** $\frac{3}{2}$ and $-\frac{3}{4}$ [5.7A] **15.** $4x^2+10x+10+\frac{21}{x-2}$ [5.4A/5.4B] **16.** $\frac{3xy}{x-y}$ [6.1C]

17. $\dfrac{x^3y - 2x^2y}{2x^2(x+1)(x-2)}; \dfrac{2}{2x^2(x+1)(x-2)}$ [6.2A]   **18.** $-\dfrac{x}{(3x+2)(x+1)}$ [6.2B]   **19.**

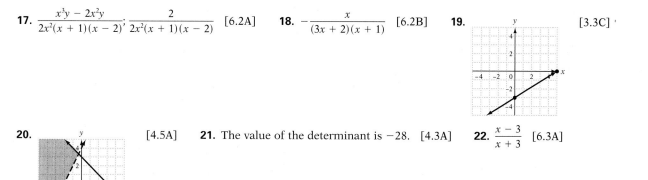

[3.3C]

20.   [4.5A]   **21.** The value of the determinant is -28. [4.3A]   **22.** $\dfrac{x-3}{x+3}$ [6.3A]

23. $(1, 1, 1)$ [4.2B/4.3B]   **24.** $\left\{x \mid x < -\dfrac{2}{3} \text{ or } x > 2\right\}$ [2.5B]   **25.** 9 [6.4A]   **26.** 13 [6.5A]   **27.** $r = \dfrac{E - IR}{I}$ [2.1D]

28. $\dfrac{x}{x-1}$ [5.1B]   **29.** The integers are 5 and 10. [2.2A]   **30.** The mixture must contain 50 lb of almonds. [2.3A]

31. The number of people expected to vote is 75,000. [6.4B]   **32.** The new computer would complete the job in 14 min. [6.5B]   **33.** The rate of the wind is 60 mph. [6.5C]   **34.** The first person has made 12 laps and is at the starting point. The second person has made 20 laps and is at the starting point. [2.3C]

Answers to Chapter 7 Odd-Numbered Exercises

PREP TEST

1. 16 [1.2B]   **2.** 32 [1.2C]   **3.** 9 [1.2B]   **4.** $\dfrac{1}{12}$ [1.2B]   **5.** $-5x - 1$ [1.3C]   **6.** $\dfrac{xy^5}{4}$ [5.1B]

7. $9x^2 - 12x + 4$ [5.3C]   **8.** $-12x^2 + 14x + 10$ [5.3B]   **9.** $36x^2 - 1$ [5.3C]   **10.** $-1, 15$ [5.7A]

SECTION 7.1

1. 2   **3.** 27   **5.** $\dfrac{1}{9}$ **7.** 4   **9.** $(-25)^{5/2}$ is not a real number.   **11.** $\dfrac{343}{125}$ **13.** x **15.** $y^{1/2}$ **17.** $x^{1/12}$

19. $a^{7/12}$ **21.** $\dfrac{1}{a}$ **23.** $\dfrac{1}{y}$ **25.** $y^{3/2}$ **27.** $\dfrac{1}{x}$ **29.** $\dfrac{1}{x^4}$ **31.** a **33.** $x^{3/10}$ **35.** a^3 **37.** $\dfrac{1}{x^{1/2}}$ **39.** $y^{1/9}$

41. x^4y **43.** $x^6y^3z^9$ **45.** $\dfrac{x}{y^2}$ **47.** $\dfrac{x^{3/2}}{y^{1/4}}$ **49.** $\dfrac{x^2}{y^8}$ **51.** $\dfrac{1}{x^{11/12}}$ **53.** $\dfrac{1}{y^{5/2}}$ **55.** $\dfrac{1}{b^{7/8}}$ **57.** a^5b^{13} **59.** $\dfrac{m^2}{4n^{3/2}}$

61. $\dfrac{y^{17/2}}{x^3}$ **63.** $\dfrac{16b^2}{a^{1/3}}$ **65.** $y^2 - y$ **67.** $a - a^2$ **69.** x^{4n} **71.** $x^{3n/2}$ **73.** $y^{3n/2}$ **75.** x^{2n^2} **77.** $x^{2n}y^n$

79. $x^{4n}y^{2n}$ **81.** $\sqrt[4]{3}$ **83.** $\sqrt{a^3}$ **85.** $\sqrt{32t^5}$ **87.** $-2\sqrt[3]{x^2}$ **89.** $\sqrt[3]{a^4b^2}$ **91.** $\sqrt[5]{a^6b^{12}}$ **93.** $\sqrt[4]{(4x-3)^3}$

95. $\dfrac{1}{\sqrt[3]{x^2}}$ **97.** $14^{1/2}$ **99.** $x^{1/3}$ **101.** $x^{4/3}$ **103.** $b^{3/5}$ **105.** $(2x^2)^{1/3}$ **107.** $-(3x^5)^{1/2}$ **109.** $3xy^{2/3}$

111. $(a^2 - 2)^{1/2}$ **113.** x^8 **115.** $-x^4$ **117.** xy^3 **119.** $-x^5y$ **121.** $4a^2b^6$ **123.** $\sqrt{-16x^4y^2}$ is not a real number.
125. $3x^3$ **127.** $-4x^3y^4$ **129.** $-x^2y^3$ **131.** x^4y^2 **133.** $3xy^5$ **135.** $2ab^2$ **137a.** False; 2   **b.** True   **c.** True
d. False; $(a^n + b^n)^{1/n}$ **e.** False; $a + 2a^{1/2}b^{1/2} + b$ **f.** False; $a^{n/m}$

SECTION 7.2

1. $x^2yz^2\sqrt{yz}$ **3.** $2ab^4\sqrt{2a}$ **5.** $3xyz^2\sqrt{5yz}$ **7.** $\sqrt{-9x^3}$ is not a real number.   **9.** $a^5b^2\sqrt[3]{ab^2}$ **11.** $-5y\sqrt[3]{x^2y}$
13. $abc^2\sqrt[3]{ab^2}$ **15.** $2x^2y\sqrt[4]{xy}$ **17.** $-6\sqrt{x}$ **19.** $-2\sqrt{2}$ **21.** $3\sqrt{2b} + 5\sqrt{3b}$ **23.** $-2xy\sqrt{2y}$

25. $6ab^2\sqrt{3ab} + 3ab\sqrt{3ab}$ **27.** $-\sqrt[3]{2}$ **29.** $8b\sqrt[3]{2b^2}$ **31.** $3a\sqrt[4]{2a}$ **33.** $17\sqrt{2} - 15\sqrt{5}$ **35.** $5b\sqrt{b}$
37. $-8xy\sqrt{2x} + 2xy\sqrt{xy}$ **39.** $2y\sqrt[3]{2x}$ **41.** $-4ab\sqrt[4]{2b}$ **43.** 16 **45.** $2\sqrt[3]{4}$ **47.** $xy^3\sqrt{x}$ **49.** $8xy\sqrt{x}$
51. $2x^2y\sqrt[3]{2}$ **53.** $2ab\sqrt[4]{3a^2b}$ **55.** 6 **57.** $x - \sqrt{2x}$ **59.** $4x - 8\sqrt{x}$ **61.** $x - 6\sqrt{x} + 9$ **63.** $84 + 16\sqrt{5}$
65. $672x^2y^2$ **67.** $4a^3b^3\sqrt[3]{a}$ **69.** $-8\sqrt{5}$ **71.** $x - y^2$ **73.** $12x - y$ **77.** $y\sqrt{5y}$ **79.** $b\sqrt{13b}$ **81.** $\dfrac{\sqrt{2}}{2}$
83. $\dfrac{2\sqrt{3y}}{3y}$ **85.** $\dfrac{3\sqrt{3a}}{a}$ **87.** $\dfrac{\sqrt{2y}}{2}$ **89.** $\dfrac{5\sqrt{3}}{3}$ **91.** $\dfrac{5\sqrt[3]{9y^2}}{3y}$ **93.** $\dfrac{b\sqrt{2a}}{2a^2}$ **95.** $\dfrac{\sqrt{15x}}{5x}$ **97.** $2 + 2\sqrt{2}$
99. $-\dfrac{12 + 4\sqrt{2}}{7}$ **101.** $-\dfrac{10 + 5\sqrt{7}}{3}$ **103.** $-\dfrac{7\sqrt{x} + 21}{x - 9}$ **105.** $-\sqrt{6} + 3 - 2\sqrt{2} + 2\sqrt{3}$ **107.** $-\dfrac{17 + 5\sqrt{5}}{4}$
109. $\dfrac{8a - 10\sqrt{ab} + 3b}{16a - 9b}$ **111.** $\dfrac{3 - 7\sqrt{y} + 2y}{1 - 4y}$ **113a.** False; $\sqrt[6]{432}$ **b.** True **c.** False; $\sqrt[3]{x} \cdot \sqrt[3]{x} = x^{2/3}$
d. False; $\sqrt{x} + \sqrt{y}$ **e.** False; $\sqrt[2]{2} + \sqrt[3]{3}$ **f.** True **115.** $\sqrt[4]{a + b}$

SECTION 7.3

3. $2i$ **5.** $7i\sqrt{2}$ **7.** $3i\sqrt{3}$ **9.** $4 + 2i$ **11.** $2\sqrt{3} - 3i\sqrt{2}$ **13.** $4\sqrt{10} - 7i\sqrt{3}$ **15.** $8 - i$ **17.** $-8 + 4i$
19. $6 - 6i$ **21.** $19 - 7i\sqrt{2}$ **23.** $6\sqrt{2} - 3i\sqrt{2}$ **25.** 63 **27.** -4 **29.** $-3\sqrt{2}$ **31.** $-4 + 12i$ **33.** $-2 + 4i$
35. $17 - i$ **37.** $8 + 27i$ **39.** 1 **41.** 1 **43.** $-3i$ **45.** $\dfrac{3}{4} + \dfrac{1}{2}i$ **47.** $\dfrac{10}{13} - \dfrac{2}{13}i$ **49.** $\dfrac{4}{5} + \dfrac{2}{5}i$ **51.** $-i$
53. $-\dfrac{\sqrt{5}}{5} + \dfrac{2\sqrt{5}}{5}i$ **55.** $\dfrac{3}{10} - \dfrac{11}{10}i$ **57.** $\dfrac{6}{5} + \dfrac{7}{5}i$ **59a.** Yes **b.** Yes

SECTION 7.4

1. -2 **3.** 9 **5.** 7 **7.** $\dfrac{13}{3}$ **9.** 35 **11.** -7 **13.** -12 **15.** 9 **17.** 2 **19.** 2 **21.** 1 **23.** The
object will fall 576 ft. **25.** The HDTV screen is 7.15 in. wider. **27.** The length of the pendulum is 7.30 ft. **29.** 2
31a. $A = \dfrac{s^2 + s^2\sqrt{5}}{2}$ **b.** $\dfrac{1 + \sqrt{5}}{2}$ **33.** The length of the side labeled x is $\sqrt{6}$.

CHAPTER REVIEW

1. $20x^2y^2$ [7.1A] **2.** 7 [7.4A] **3.** $39 - 2i$ [7.3C] **4.** $7x^{2/3}y$ [7.1B] **5.** $6\sqrt{3} - 13$ [7.2C] **6.** -2 [7.4A]
7. $\dfrac{1}{x^5}$ [7.1A] **8.** $\dfrac{8\sqrt{3y}}{3y}$ [7.2D] **9.** $-2a^2b^4$ [7.1C] **10.** $2a^2b\sqrt{2b}$ [7.2B]
11. $\dfrac{x\sqrt{x} - x\sqrt{2} + 2\sqrt{x} - 2\sqrt{2}}{x - 2}$ [7.2D] **12.** $\dfrac{2}{3} - \dfrac{5}{3}i$ [7.3D] **13.** $3ab^3\sqrt{2a}$ [7.2A] **14.** $-4\sqrt{2} + 8i\sqrt{2}$ [7.3B]
15. $5x^3y^3\sqrt[3]{2x^2y}$ [7.2B] **16.** $4xy^2\sqrt[3]{x^2}$ [7.2C] **17.** $7 + 3i$ [7.3C] **18.** $3\sqrt[4]{x^3}$ [7.1B] **19.** $-2ab^2\sqrt[5]{2a^3b^2}$ [7.2A]
20. $-2 + 7i$ [7.3D] **21.** $-6\sqrt{2}$ [7.3C] **22.** 30 [7.4A] **23.** $3a^2b^3$ [7.1C] **24.** $5i\sqrt{2}$ [7.3A]
25. $-12 + 10i$ [7.3B] **26.** $31 - 10\sqrt{6}$ [7.2C] **27.** $6x^2\sqrt{3y}$ [7.2B] **28.** The amount of power generated is
120 watts. [7.4B] **29.** The distance required to reach a velocity of 88 ft/s is 242 ft. [7.4B] **30.** The bottom of the
ladder is 6.63 ft from the building. [7.4B]

CHAPTER TEST

1. $\dfrac{1}{2}x^{3/4}$ [7.1B] **2.** $-2x^2y\sqrt[3]{2x}$ [7.2B] **3.** $3\sqrt[5]{y^2}$ [7.1B] **4.** $18 + 16i$ [7.3C] **5.** $4x + 4\sqrt{xy} + y$ [7.2C]
6. $r^{1/6}$ [7.1A] **7.** 4 [7.4A] **8.** $2xy^2$ [7.1C] **9.** $-4x\sqrt{3}$ [7.2C] **10.** $-3 + 2i$ [7.3B] **11.** $4x^2y^3\sqrt{2y}$ [7.2A]
12. $14 + 10\sqrt{3}$ [7.2C] **13.** $10 + 2i$ [7.3C] **14.** 2 [7.2D] **15.** $8a\sqrt{2a}$ [7.2B] **16.** $2a - \sqrt{ab} - 15b$ [7.2C]
17. $\dfrac{64x^3}{y^6}$ [7.1A] **18.** $\dfrac{x + \sqrt{xy}}{x - y}$ [7.2D] **19.** $-\dfrac{4}{5} + \dfrac{7}{5}i$ [7.3D] **20.** -3 [7.4A] **21.** $\dfrac{b^3}{8a^6}$ [7.1A]
22. $3abc^2\sqrt[3]{ac}$ [7.2A] **23.** $\dfrac{4x^2}{y}$ [7.2D] **24.** -4 [7.3C] **25.** The object has fallen 576 ft. [7.4B]

CUMULATIVE REVIEW

1. Distributive Property [1.3A]   **2.** 34 [3.2A]   **3.** $\dfrac{3}{2}$ [2.1B]   **4.** $\dfrac{2}{3}$ [2.1C]   **5.** $\dfrac{1}{3}$ and $\dfrac{7}{3}$ [2.5A]

6. $\{x \mid x > 1\}$ [2.4A]   **7.** $\{x \mid -6 \le x \le 3\}$ [2.5B]   **8.** $(9x + y)(9x - y)$ [5.6A]   **9.** $x(x^2 + 3)(x + 1)(x - 1)$ [5.6D]

10. The equation of the line is $y = \dfrac{1}{3}x + \dfrac{7}{3}$. [3.5B]   **11.** The value of the determinant is 3. [4.3A]

12. $C = R - nP$ [2.1D]   **13.** $2x^2y^2$ [5.1B]   **14.** $\dfrac{x(2x - 3)}{y(x + 4)}$ [6.1B]   **15.** $-x\sqrt{10x}$ [7.2B]   **16.** $-\dfrac{1}{2}$ and 3 [6.5A]

17.

$m = \dfrac{3}{2}, b = 3$ [3.3B]   **18.**

[3.7A]   **19.** $-\dfrac{1}{5} + \dfrac{3}{5}i$ [7.3D]

20. -20 [7.4A]   **21.** $\dfrac{x^2 + 12x - 12}{(2x - 3)(x + 4)}$ [6.2B]   **22.** $\left(\dfrac{17}{4}, \dfrac{9}{2}\right)$ [4.3B]   **23.** There are nineteen 18¢ stamps in the collection. [2.2B]   **24.** The rate of the plane was 250 mph. [2.3C]   **25.** Light travels to Earth from the moon in 1.25 s. [5.1D]   **26.** The periscope would have to be 32.7 ft above the surface of the water. [7.4B]   **27.** The slope is 0.08. The slope indicates that the annual income is 8% of the investment. [3.4A]

Answers to Chapter 8 Odd-Numbered Exercises

PREP TEST

1. $3\sqrt{2}$ [7.2A]   **2.** Not a real number [7.2A]   **3.** $\dfrac{2x - 1}{x - 1}$ [6.2B]   **4.** 8 [1.3B]   **5.** yes [5.6A]   **6.** $(2x - 1)^2$ [5.6A]   **7.** $(3x + 2)(3x - 2)$ [5.6A]   **8.**

[1.1C]   **9.** -3 and 5 [5.7A]   **10.** 4 [6.4A]

SECTION 8.1

3. $2x^2 - 4x - 5 = 0; a = 2, b = -4, c = -5$ **5.** $4x^2 - 5x + 6 = 0; a = 4, b = -5, c = 6$ **7.** 0 and 4   **9.** 5 and -5

11. 3 and -2 **13.** 3   **15.** 0 and 2   **17.** 5 and -2 **19.** 2 and 5   **21.** 6 and $-\dfrac{3}{2}$ **23.** 2 and $\dfrac{1}{4}$ **25.** $\dfrac{1}{3}$ and -4

27. $\dfrac{9}{2}$ and $-\dfrac{2}{3}$ **29.** $\dfrac{1}{4}$ and -4 **31.** 9 and -2 **33.** -2 and $-\dfrac{3}{4}$ **35.** 2 and -5 **37.** -4 and $-\dfrac{3}{2}$ **39.** $2b$ and $7b$

41. $7c$ and $-c$ **43.** $-\dfrac{b}{2}$ and $-b$ **45.** $4a$ and $\dfrac{2a}{3}$ **47.** $-\dfrac{a}{3}$ and $3a$ **49.** $-\dfrac{3y}{2}$ and $-\dfrac{y}{2}$ **51.** $-\dfrac{a}{2}$ and $-\dfrac{4a}{3}$

53. $x^2 - 7x + 10 = 0$ **55.** $x^2 + 6x + 8 = 0$ **57.** $x^2 - 5x - 6 = 0$ **59.** $x^2 - 9 = 0$ **61.** $x^2 - 8x + 16 = 0$

63. $x^2 - 5x = 0$ **65.** $x^2 - 3x = 0$ **67.** $2x^2 - 7x + 3 = 0$ **69.** $4x^2 - 5x - 6 = 0$ **71.** $3x^2 + 11x + 10 = 0$

73. $9x^2 - 4 = 0$ **75.** $6x^2 - 5x + 1 = 0$ **77.** $10x^2 - 7x - 6 = 0$ **79.** $8x^2 + 6x + 1 = 0$ **81.** $50x^2 - 25x - 3 = 0$

83. 7 and -7 **85.** $2i$ and $-2i$ **87.** 2 and -2 **89.** $\dfrac{9}{2}$ and $-\dfrac{9}{2}$ **91.** $7i$ and $-7i$ **93.** $4\sqrt{3}$ and $-4\sqrt{3}$

95. $5\sqrt{3}$ and $-5\sqrt{3}$ **97.** $3i\sqrt{2}$ and $-3i\sqrt{2}$ **99.** 7 and -5 **101.** 0 and -6 **103.** -7 and 3   **105.** 1 and 0

107. $-5 + \sqrt{6}$ and $-5 - \sqrt{6}$ **109.** $3 + 3i\sqrt{5}$ and $3 - 3i\sqrt{5}$ **111.** $-\dfrac{2 - 9\sqrt{2}}{3}$ and $-\dfrac{2 + 9\sqrt{2}}{3}$ **113.** $x^2 - 2 = 0$

115. $x^2 + 1 = 0$ **117.** $x^2 - 8 = 0$ **119.** $x^2 + 2 = 0$ **121.** $\dfrac{3b}{a}$ and $-\dfrac{3b}{a}$ **123.** $-a + 2$ and $-a - 2$ **125.** $-\dfrac{1}{2}$

SECTION 8.2

1. 5 and -1 **3.** -9 and 1 **5.** 3 **7.** $-2 + \sqrt{11}$ and $-2 - \sqrt{11}$ **9.** $3 + \sqrt{2}$ and $3 - \sqrt{2}$ **11.** $1 + i$ and $1 - i$

13. 8 and -3 **15.** 4 and -9 **17.** $\dfrac{3 + \sqrt{5}}{2}$ and $\dfrac{3 - \sqrt{5}}{2}$ **19.** $\dfrac{1 + \sqrt{5}}{2}$ and $\dfrac{1 - \sqrt{5}}{2}$ **21.** $3 + \sqrt{13}$ and $3 - \sqrt{13}$

23. 5 and 3 **25.** $2 + 3i$ and $2 - 3i$ **27.** $-3 + 2i$ and $-3 - 2i$ **29.** $1 + 3\sqrt{2}$ and $1 - 3\sqrt{2}$

31. $\dfrac{1 + \sqrt{17}}{2}$ and $\dfrac{1 - \sqrt{17}}{2}$ **33.** $1 + 2i\sqrt{3}$ and $1 - 2i\sqrt{3}$ **35.** $\dfrac{1}{2} + i$ and $\dfrac{1}{2} - i$ **37.** $\dfrac{1}{3} + \dfrac{1}{3}i$ and $\dfrac{1}{3} - \dfrac{1}{3}i$

39. $\dfrac{2 + \sqrt{14}}{2}$ and $\dfrac{2 - \sqrt{14}}{2}$ **41.** 1 and $-\dfrac{3}{2}$ **43.** $1 + \sqrt{5}$ and $1 - \sqrt{5}$ **45.** $\dfrac{1}{2}$ and 5 **47.** $2 + \sqrt{5}$ and $2 - \sqrt{5}$

49. 1.236 and -3.236 **51.** 1.707 and 0.293 **53.** 0.309 and -0.809 **55.** $-a$ and $2a$ **57.** $-5a$ and $2a$
59. No. The ball will have gone only 197.2 ft when it hits the ground.

SECTION 8.3

3. 5 and -2 **5.** 4 and -9 **7.** $4 + 2\sqrt{22}$ and $4 - 2\sqrt{22}$ **9.** 3 and -8 **11.** $\dfrac{-5 + \sqrt{33}}{4}$ and $\dfrac{-5 - \sqrt{33}}{4}$

13. $\dfrac{3}{2}$ and $-\dfrac{1}{4}$ **15.** $7 + 3\sqrt{5}$ and $7 - 3\sqrt{5}$ **17.** $\dfrac{1 + \sqrt{3}}{2}$ and $\dfrac{1 - \sqrt{3}}{2}$ **19.** $-1 + i$ and $-1 - i$

21. $1 + 2i$ and $1 - 2i$ **23.** $2 + 3i$ and $2 - 3i$ **25.** $\dfrac{1 + \sqrt{11}}{2}$ and $\dfrac{1 - \sqrt{11}}{2}$ **27.** $-\dfrac{3}{2} + \dfrac{1}{2}i$ and $-\dfrac{3}{2} - \dfrac{1}{2}i$

29. $\dfrac{3}{4} + \dfrac{3\sqrt{3}}{4}i$ and $\dfrac{3}{4} - \dfrac{3\sqrt{3}}{4}i$ **31.** 7.606 and 0.394 **33.** 0.236 and -4.236 **35.** 1.851 and -1.351

37. Two complex number solutions **39.** One real number solution **41.** Two real number solutions **43.** $\{p \,|\, p < 9\}$
45. $\{p \,|\, p > 1\}$ **47.** i and $-2i$

SECTION 8.4

1. 3, -3, 2, and -2 **3.** $\sqrt{2}$, $-\sqrt{2}$, 2, and -2 **5.** 1 and 4 **7.** 16 **9.** $2i$, $-2i$, 1, and -1 **11.** $4i$, $-4i$, 2, and -2

13. 16 **15.** 1 and 512 **17.** $\dfrac{2}{3}$, $-\dfrac{2}{3}$, 1, and -1 **19.** 3 **21.** 9 **23.** 2 and -1 **25.** 0 and 2 **27.** 2 and $-\dfrac{1}{2}$

29. -2 **31.** 1 **33.** 1 **35.** -3 **37.** 10 and -1 **39.** $-\dfrac{1}{2} + \dfrac{\sqrt{7}}{2}i$ and $-\dfrac{1}{2} - \dfrac{\sqrt{7}}{2}i$ **41.** 1 and -3

43. 0 and -1 **45.** $\dfrac{1}{2}$ and $-\dfrac{1}{3}$ **47.** $-\dfrac{2}{3}$ and 6 **49.** $\dfrac{4}{3}$ and 3 **51.** $-\dfrac{1}{4}$ and 3 **53.** 9 and 36 **55.** $\sqrt{5}$ or $-\sqrt{5}$

SECTION 8.5

3. $\{x \,|\, x < -2 \text{ or } x > 4\}$ **5.** $\{x \,|\, x \le 1 \text{ or } x \ge 2\}$

7. $\{x \,|\, -3 < x < 4\}$ **9.** $\{x \,|\, x < -2 \text{ or } 1 < x < 3\}$

11. $\{x \,|\, -4 \le x \le 1 \text{ or } x \ge 2\}$ **13.** $\{x \,|\, x < -2 \text{ or } x > 4\}$

15. $\{x \,|\, -1 < x \le 3\}$ **17.** $\{x \,|\, x \le -2 \text{ or } 1 \le x < 3\}$

19. $\{x \,|\, x > 4 \text{ or } x < -4\}$ **21.** $\{x \,|\, -3 \le x \le 12\}$ **23.** $\left\{x \,\middle|\, \dfrac{1}{2} < x < \dfrac{3}{2}\right\}$ **25.** $\left\{x \,\middle|\, x < 1 \text{ or } x > \dfrac{5}{2}\right\}$

27. $\{x \,|\, x < -1 \text{ or } 1 < x \le 2\}$ **29.** $\left\{x \,\middle|\, \dfrac{1}{2} < x \le 1\right\}$ **31.** $\{x \,|\, 2 < x \le 3\}$ **33.** $\{x \,|\, x > 5 \text{ or } -4 < x < -1\}$

35. **37.** **39.**

SECTION 8.6

1. The height is 3 cm; the base is 14 cm.   **3.** The dimensions of Colorado are approximately 272 mi by 383 mi.
5. The maximum safe speed is 33 mph.   **7.** The time for the projectile to return to Earth is 12.5 s.
9. The maximum speed is 72.5 km/h.   **11.** The smaller pipe requires 12 min. The larger pipe requires 6 min.
13. The rate of the wind is 108 mph.   **15.** The rowing rate of the guide is 6 mph.   **17.** The radius of the cone is 1.5 in.

CHAPTER REVIEW

1. 0 and $\dfrac{3}{2}$ [8.1A]   **2.** $\dfrac{c}{2}$ and $-2c$ [8.1A]   **3.** $4\sqrt{3}$ and $-4\sqrt{3}$ [8.1C]   **4.** $-\dfrac{1}{2} + 2i$ and $-\dfrac{1}{2} - 2i$ [8.1C]

5. -3 and -1 [8.2A]   **6.** $\dfrac{7 + 2\sqrt{7}}{7}$ and $\dfrac{7 - 2\sqrt{7}}{7}$ [8.2A]   **7.** $\dfrac{3}{4}$ and $\dfrac{4}{3}$ [8.3A]   **8.** $\dfrac{1}{2} + \dfrac{\sqrt{31}}{2}i$ and $\dfrac{1}{2} - \dfrac{\sqrt{31}}{2}i$ [8.3A]

9. $x^2 + 3x = 0$ [8.1B]   **10.** $12x^2 - x - 6 = 0$ [8.1B]   **11.** $1 + i\sqrt{7}$ and $1 - i\sqrt{7}$ [8.2A]   **12.** $2i$ and $-2i$ [8.2A]

13. $\dfrac{11 + \sqrt{73}}{6}$ and $\dfrac{11 - \sqrt{73}}{6}$ [8.3A]   **14.** Two real number solutions [8.3A]   **15.** $\left\{ x \mid -3 < x < \dfrac{5}{2} \right\}$ [8.5A]

16. $\left\{ x \mid x \le -4 \text{ or } -\dfrac{3}{2} \le x \le 2 \right\}$ [8.5A]   **17.** 27 and -64 [8.4A]   **18.** $\dfrac{5}{4}$ [8.4B]   **19.** -1 and 3 [8.4C]

20. -1 [8.4C]   **21.** [8.5A]   **22.** [8.5A]   **23.** 4 [8.4B]
$\left\{ x \mid x < \dfrac{3}{2} \text{ or } x \ge 2 \right\}$ $\left\{ x \mid x \le -3 \text{ or } \dfrac{1}{2} \le x < 4 \right\}$

24. 5 [8.4B]   **25.** $\dfrac{-3 + \sqrt{249}}{10}$ and $\dfrac{-3 - \sqrt{249}}{10}$ [8.4C]   **26.** $\dfrac{-11 + \sqrt{129}}{2}$ and $\dfrac{-11 - \sqrt{129}}{2}$ [8.4C]   **27.** The length
is 12 cm; the width is 5 cm. [8.6A]   **28.** The integers are 2, 4, and 6 or -6, -4, and -2. [8.6A]   **29.** It takes the
computer 12 min. [8.6A]   **30.** The first car is traveling 40 mph; the second car is traveling 50 mph. [8.6A]

CHAPTER TEST

1. $\dfrac{2}{3}$ and -4 [8.1A]   **2.** $\dfrac{3}{2}$ and $-\dfrac{2}{3}$ [8.1A]   **3.** $x^2 - 9 = 0$ [8.1B]   **4.** $2x^2 + 7x - 4 = 0$ [8.1B]

5. $2 + 2\sqrt{2}$ and $2 - 2\sqrt{2}$ [8.1C]   **6.** $3 + \sqrt{11}$ and $3 - \sqrt{11}$ [8.2A]   **7.** $\dfrac{3 + \sqrt{15}}{3}$ and $\dfrac{3 - \sqrt{15}}{3}$ [8.2A]

8. $\dfrac{1 + \sqrt{3}}{2}$ and $\dfrac{1 - \sqrt{3}}{2}$ [8.3A]   **9.** $-2 + 2i\sqrt{2}$ and $-2 - 2i\sqrt{2}$ [8.3A]   **10.** Two real number solutions [8.3A]

11. Two complex number solutions [8.3A]   **12.** $\dfrac{1}{4}$ [8.4A]   **13.** $1, -1, \sqrt{3}$, and $-\sqrt{3}$ [8.4A]   **14.** 4 [8.4B]

15. No solution [8.4B]   **16.** 2 and -9 [8.4C]   **17.** $\{ x \mid x < -4 \text{ or } 2 < x < 4 \}$ [8.5A]

18. $\left\{ x \mid -4 < x \le \dfrac{3}{2} \right\}$ [8.5A]   **19.** The base is 15 ft. The height is 4 ft. [8.6A]   **20.** The
rowing rate in calm water is 4 mph. [8.6A]

CUMULATIVE REVIEW

1. 14 [1.3B]   **2.** -28 [2.1C]   **3.** $-\dfrac{3}{2}$ [3.4A]   **4.** $y = x + 1$ [3.6A]   **5.** $-3xy(x^2 - 2xy + 3y^2)$ [5.5A]

6. $(2x - 5)(3x + 4)$ [5.5D]   **7.** $(x + y)(a^n - 2)$ [5.5B]   **8.** $x^2 - 3x - 4 - \dfrac{6}{3x - 4}$ [5.4A]   **9.** $\dfrac{x}{2}$ [6.1B]

10. $2\sqrt{5}$ [3.1B]   **11.** $b = \dfrac{2S - an}{n}$ [2.1D]   **12.** $-8 - 14i$ [7.3C]   **13.** $1 - a$ [7.1A]   **14.** $\dfrac{x\sqrt[3]{4y^2}}{2y}$ [7.2D]

15. $-\dfrac{3}{2}$ and -1 [8.4C]   **16.** $-\dfrac{1}{2}$ [6.5A]   **17.** $2, -2, \sqrt{2}$, and $-\sqrt{2}$ [8.4A]   **18.** 0 and 1 [8.4B]

19. $\left\{x \,\middle|\, -2 < x < \dfrac{10}{3}\right\}$ [2.5B]   **20.** $\left(\dfrac{5}{2}, 0\right)$; $(0, -3)$ [3.3C]   **21.** [4.5A]   **22.** $(1, -1, 2)$ [4.2B]

23. $f(-2) = -\dfrac{7}{3}$ [6.1A]   **24.** $\{x \,|\, x \neq -3, 5\}$ [6.1A]   **25.** $\{x \,|\, x < -3 \text{ or } 0 < x < 2\}$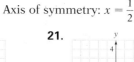

[8.5A]   **26.** $\{x \,|\, -3 < x \leq 1 \text{ or } x \geq 5\}$ 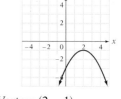 [8.5A]   **27.** Lower limit: $9\dfrac{23}{64}$ in.; upper limit:

$9\dfrac{25}{64}$ in. [2.5C]   **28.** The area of the triangle is $(x^2 + 6x - 16)$ ft². [5.3D]   **29.** Two complex number solutions [8.3A]

30. The slope is $-\dfrac{25{,}000}{3}$. The slope indicates that the building decreases $\dfrac{\$25{,}000}{3}$ in value each year. [3.4A]

Answers to Chapter 9 Odd-Numbered Exercises

PREP TEST

1. 1 [1.3B]   **2.** -7 [1.3B]   **3.** 30 [3.2A]   **4.** $h^2 + 4h - 1$ [3.2A]   **5.** $-\dfrac{2}{3}$, 3 [8.1A]

6. $2 - \sqrt{3}$, $2 + \sqrt{3}$ [8.2A, 8.3A]   **7.** $y = \dfrac{1}{2}x - 2$ [2.1D]   **8.** Domain: $\{-2, 3, 4, 6\}$; Range: $\{4, 5, 6\}$; Yes [3.2A]

9. 8 [3.2A]   **10.** 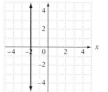 [3.3B]

SECTION 9.1

5. -5 **7.** $x = 7$ **9.**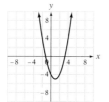

Vertex: $(1, -5)$
Axis of symmetry: $x = 1$

11.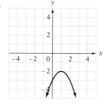

Vertex: $(1, -2)$
Axis of symmetry: $x = 1$

13.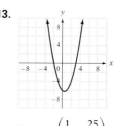

Vertex: $\left(\dfrac{1}{2}, -\dfrac{25}{4}\right)$

Axis of symmetry: $x = \dfrac{1}{2}$

15.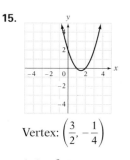

Vertex: $\left(\dfrac{3}{2}, -\dfrac{1}{4}\right)$

Axis of symmetry: $x = \dfrac{3}{2}$

17.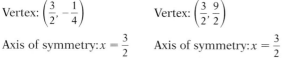

Vertex: $\left(\dfrac{3}{2}, \dfrac{9}{2}\right)$

Axis of symmetry: $x = \dfrac{3}{2}$

19.

Vertex: $(0, -1)$

Axis of symmetry: $x = 0$

21.

Vertex: $(2, -1)$

Axis of symmetry: $x = 2$

23.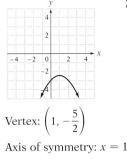

Vertex: $\left(1, -\dfrac{5}{2}\right)$

Axis of symmetry: $x = 1$

25. The domain is $\{x \mid x \in \text{real numbers}\}$. The range is $\{y \mid y \geq -5\}$.

27. The domain is $\{x \mid x \in \text{real numbers}\}$. The range is $\{y \mid y \leq 0\}$.   **29.** The domain is $\{x \mid x \in \text{real numbers}\}$. The range is $\{y \mid y \geq -7\}$.   **33.** $(3, 0)$ and $(-3, 0)$ **35.** $(0, 0)$ and $(-2, 0)$ **37.** $(4, 0)$ and $(-2, 0)$ **39.** $\left(-\dfrac{1}{2}, 0\right)$ and $(3, 0)$

41. $\left(-2 + \sqrt{7}, 0\right)$ and $\left(-2 - \sqrt{7}, 0\right)$ **43.** No x-intercepts   **45.** 3   **47.** $\dfrac{3 + \sqrt{41}}{2}$ and $\dfrac{3 - \sqrt{41}}{2}$ **49.** 0 and $\dfrac{4}{3}$

51. $i\sqrt{2}$ and $-i\sqrt{2}$ **53.** $\dfrac{1}{6} + \dfrac{\sqrt{47}}{6}i$ and $\dfrac{1}{6} - \dfrac{\sqrt{47}}{6}i$ **55.** $\dfrac{1 + \sqrt{41}}{4}$ and $\dfrac{1 - \sqrt{41}}{4}$ **57.** Two x-intercepts

59. One x-intercept   **61.** No x-intercepts   **63.** Two x-intercepts   **65.** No x-intercepts   **67.** Two x-intercepts
69. $(-4, 0)$ and $(5, 0)$ **73a.** Minimum value   **b.** Maximum value   **c.** Minimum value   **75.** Minimum value: -2

77. Maximum value: -3 **79.** Minimum value: $\dfrac{9}{8}$ **81.** Minimum value: $-\dfrac{11}{4}$ **83.** Maximum value: $\dfrac{9}{4}$

85. Maximum value: $-\dfrac{1}{12}$ **87.** b   **89.** The function has a maximum value at -5.   **91.** The diver will attain a maximum height above the water of 13.1 m.   **93.** A price of \$250 will give the maximum revenue.   **95.** The minimum height of the cable above the bridge is 24.36 ft.   **97.** The maximum height of the water is 141.6 ft.   **99.** A car could be traveling at a speed of 20 mph and still stop at a stop sign 44 ft away.   **101.** The two numbers are 7 and -7.

103. The other root is $-\dfrac{3}{2}$.   **105.** The roots are -2 and 3.

SECTION 9.2

1. Yes   **3.** No   **5.** Yes

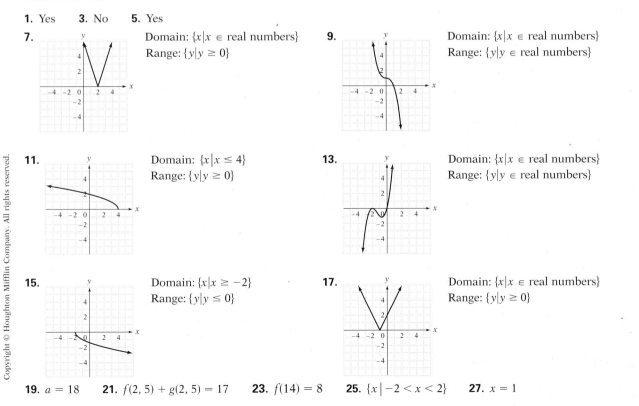

7.   Domain: $\{x \mid x \in \text{real numbers}\}$
Range: $\{y \mid y \geq 0\}$

9.   Domain: $\{x \mid x \in \text{real numbers}\}$
Range: $\{y \mid y \in \text{real numbers}\}$

11.   Domain: $\{x \mid x \leq 4\}$
Range: $\{y \mid y \geq 0\}$

13.   Domain: $\{x \mid x \in \text{real numbers}\}$
Range: $\{y \mid y \in \text{real numbers}\}$

15.   Domain: $\{x \mid x \geq -2\}$
Range: $\{y \mid y \leq 0\}$

17.   Domain: $\{x \mid x \in \text{real numbers}\}$
Range: $\{y \mid y \geq 0\}$

19. $a = 18$ **21.** $f(2, 5) + g(2, 5) = 17$ **23.** $f(14) = 8$ **25.** $\{x \mid -2 < x < 2\}$ **27.** $x = 1$

SECTION 9.3

1. $f(2) - g(2) = 5$ **3.** $f(0) + g(0) = 1$ **5.** $(f \cdot g)(2) = 0$ **7.** $\left(\dfrac{f}{g}\right)(4) = -\dfrac{29}{4}$ **9.** $\left(\dfrac{g}{f}\right)(-3) = \dfrac{2}{3}$ **11.** $f(1) + g(1) = 2$

13. $f(4) - g(4) = 39$ **15.** $(f \cdot g)(1) = -8$ **17.** $\left(\dfrac{f}{g}\right)(-3) = -\dfrac{4}{5}$ **19.** $f(2) - g(2) = -2$ **21.** $\left(\dfrac{f}{g}\right)(-2) = 7$

23. $g[f(0)] = -13$ **25.** $g[f(-2)] = -29$ **27.** $g[f(x)] = 8x - 13$ **29.** $f[h(0)] = 4$ **31.** $f[h(-1)] = 3$

33. $f[h(x)] = x + 4$ **35.** $h[g(0)] = 1$ **37.** $h[g(-2)] = 5$ **39.** $h[g(x)] = x^2 + 1$ **41.** $h[f(0)] = 5$

43. $h[f(-2)] = 11$ **45.** $h[f(x)] = 3x^2 + 3x + 5$ **47.** $f[g(-1)] = -3$ **49.** $g[f(-1)] = -27$

51. $g[f(x)] = x^3 - 6x^2 + 12x - 8$ **53.** $g(3 + h) - g(3) = 6h + h^2$ **55.** $\dfrac{g(1 + h) - g(1)}{h} = 2 + h$

57. $\dfrac{g(a + h) - g(a)}{h} = 2a + h$ **59.** $g(h[f(1)]) = -1$ **61.** $f(h[g(0)]) = -6$ **63.** $g(f[h(x)]) = 6x - 13$

SECTION 9.4

3. Yes   **5.** No   **7.** Yes   **9.** No   **11.** No   **13.** No   **17.** $\{(0, 1), (3, 2), (8, 3), (15, 4)\}$ **19.** No inverse

21. $\{(-2, 0), (5, -1), (3, 3), (6, -4)\}$ **23.** No inverse   **25.** $f^{-1}(x) = \dfrac{1}{4}x + 2$ **27.** $f^{-1}(x) = \dfrac{1}{2}x - 2$

29. $f^{-1}(x) = 2x + 2$ **31.** $f^{-1}(x) = -\dfrac{1}{2}x + 1$ **33.** $f^{-1}(x) = \dfrac{3}{2}x - 6$ **35.** $f^{-1}(x) = -3x + 3$ **37.** $f^{-1}(x) = \dfrac{1}{2}x + \dfrac{5}{2}$

39. $f^{-1}(x) = \dfrac{1}{5}x + \dfrac{2}{5}$ **41.** $f^{-1}(x) = \dfrac{1}{6}x + \dfrac{1}{2}$ **43.** $f^{-1}(0) = \dfrac{5}{3}$ **45.** $f^{-1}(4) = 3$ **47.** Yes; Yes   **49.** Yes   **51.** No

53. Yes   **55.** Yes   **57.**   **59.**   **61.**

65. $f^{-1}(-3) = 5$ **67.** $f^{-1}(1) = 0$ **69.** $f^{-1}(7) = -4$

CHAPTER REVIEW

1. Yes   [9.2A]   **2.** Yes   [9.4A]   **3.**   **4.**

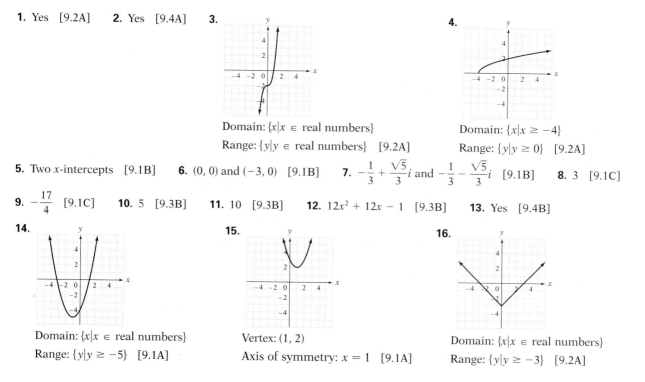

Domain: $\{x | x \in \text{real numbers}\}$
Range: $\{y | y \in \text{real numbers}\}$ [9.2A]

Domain: $\{x | x \geq -4\}$
Range: $\{y | y \geq 0\}$ [9.2A]

5. Two x-intercepts   [9.1B]   **6.** $(0, 0)$ and $(-3, 0)$ [9.1B]   **7.** $-\dfrac{1}{3} + \dfrac{\sqrt{5}}{3}i$ and $-\dfrac{1}{3} - \dfrac{\sqrt{5}}{3}i$ [9.1B]   **8.** 3   [9.1C]

9. $-\dfrac{17}{4}$ [9.1C]   **10.** 5   [9.3B]   **11.** 10   [9.3B]   **12.** $12x^2 + 12x - 1$ [9.3B]   **13.** Yes   [9.4B]

14.   **15.**   **16.**

Domain: $\{x | x \in \text{real numbers}\}$
Range: $\{y | y \geq -5\}$ [9.1A]

Vertex: $(1, 2)$
Axis of symmetry: $x = 1$ [9.1A]

Domain: $\{x | x \in \text{real numbers}\}$
Range: $\{y | y \geq -3\}$ [9.2A]

17. No [9.4A]   **18.** 7 [9.3A]   **19.** −9 [9.3A]   **20.** 70 [9.3A]   **21.** $\dfrac{12}{7}$ [9.3A]   **22.** $6x^2 + 3x - 16$ [9.3B]

23. $f^{-1}(x) = -\dfrac{1}{6}x + \dfrac{2}{3}$ [9.4B]   **24.** $f^{-1}(x) = \dfrac{3}{2}x + 18$ [9.4B]   **25.** $f^{-1}(x) = 2x - 16$ [9.4B]   **26.** The dimensions of the rectangle are 7 ft by 7 ft. [9.1D]

CHAPTER TEST

1. The zeros of the function are $\dfrac{3}{4} + \dfrac{\sqrt{23}}{4}i$ and $\dfrac{3}{4} - \dfrac{\sqrt{23}}{4}i$. [9.1B]   **2.** The x-intercepts are $\dfrac{-3 + \sqrt{41}}{2}$ and

$\dfrac{-3 - \sqrt{41}}{2}$. [9.1B]   **3.** The parabola has two real zeros. [9.1B]   **4.** $(f - g)(2) = -2$ [9.3A]

5. $(f \cdot g)(-3) = 234$ [9.3A]   **6.** $\left(\dfrac{f}{g}\right)(-2) = -\dfrac{13}{2}$ [9.3A]   **7.** $(f - g)(-4) = -5$ [9.3A]   **8.** $f[g(3)] = 5$ [9.3B]

9. $f[g(x)] = 2x^2 - 4x - 5$ [9.3B]   **10.** The maximum value of the function is 9. [9.1C]   **11.** $f^{-1}(x) = \dfrac{1}{4}x + \dfrac{1}{2}$ [9.4B]

12. $f^{-1}(x) = 4x + 16$ [9.4B]   **13.** {(6, 2), (5, 3), (4, 4), (3, 5)} [9.4B]   **14.** Yes [9.4B]

15. The maximum product is 100. [9.1D]   **16.**

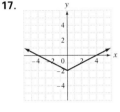

Domain: $\{x \mid x \le 3\}$; range: $\{y \mid y \le 0\}$ [9.2A]

17.

Domain: $\{x \mid x \in \text{real numbers}\}$; range: $\{y \mid y \ge -2\}$ [9.2A]

18.

Domain: $\{x \mid x \in \text{real numbers}\}$; range: $\{y \mid y \in \text{real numbers}\}$ [9.2A]

19. No [9.4A]   **20.** The dimensions are 50 cm by 50 cm. The maximum area is 2500 cm². [9.1D]

CUMULATIVE REVIEW

1. $-\dfrac{23}{4}$ [1.3B]   **2.** 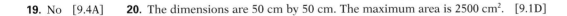 [1.1C]   **3.** 3 [2.1C]   **4.** $\{x \mid x < -2 \text{ or } x > 3\}$ [2.4B]

5. The set of all real numbers [2.5B]   **6.** $-\dfrac{a^{10}}{12b^4}$ [5.1B]   **7.** $2x^3 - 4x^2 - 17x + 4$ [5.3B]

8. $(a + 2)(a - 2)(a^2 + 2)$ [5.6D]   **9.** $xy(x - 2y)(x + 3y)$ [5.6D]   **10.** −3 and 8 [5.7A]

11. $\{x \mid x < -3 \text{ or } x > 5\}$ [8.5A]   **12.** $\dfrac{5}{2x - 1}$ [6.2B]   **13.** −2 [6.5A]   **14.** $-3 - 2i$ [7.3D]

15.

Vertex: (0, 0); axis of symmetry: $x = 0$ [9.1A]

16. 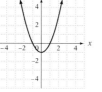 [3.7A]   **17.** $y = -2x - 2$ [3.5B]   **18.** $y = -\dfrac{3}{2}x - \dfrac{7}{2}$ [3.6A]

19. $\dfrac{1}{2} + \dfrac{\sqrt{3}}{6}i$ and $\dfrac{1}{2} - \dfrac{\sqrt{3}}{6}i$ [8.3A]   **20.** 3 [8.4B]   **21.** -3 [9.1C]   **22.** $\{1, 2, 4, 5\}$ [3.2A]   **23.** Yes [3.2A]

24. 2 [7.4A]   **25.** $g[h(2)] = 10$ [9.3B]   **26.** $f^{-1}(x) = -\dfrac{1}{3}x + 3$ [9.4B]   **27.** The cost per pound is \$3.96. [2.3A]

28. There must be 25 lb of the 80% copper alloy. [2.3B]   **29.** There must be an additional 4.5 oz of insecticide. [6.4B]

30. It would take the larger pipe 4 min. [8.6A]   **31.** The spring will stretch 24 in. [6.6A]   **32.** The frequency in the pipe is 80 vibrations/min. [6.6A]

Answers to Chapter 10 Odd-Numbered Exercises

PREP TEST

1. $\dfrac{1}{9}$ [5.1B]   **2.** 16 [5.1B]   **3.** -3 [5.1B]   **4.** 0; 108 [3.2A]   **5.** -6 [2.1B]   **6.** $-2, 8$ [8.1A]

7. 6326.60 [1.3B]   **8.** 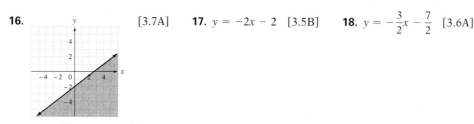 [9.1A]

SECTION 10.1

3. c   **5a.** $f(2) = 9$ **b.** $f(0) = 1$ **c.** $f(-2) = \dfrac{1}{9}$ **7a.** $g(3) = 16$ **b.** $g(1) = 4$ **c.** $g(-3) = \dfrac{1}{4}$

9a. $P(0) = 1$ **b.** $P\left(\dfrac{3}{2}\right) = \dfrac{1}{8}$ **c.** $P(-2) = 16$ **11a.** $G(4) = 7.3891$ **b.** $G(-2) = 0.3679$ **c.** $G\left(\dfrac{1}{2}\right) = 1.2840$

13a. $H(-1) = 54.5982$ **b.** $H(3) = 1$ **c.** $H(5) = 0.1353$ **15a.** $F(2) = 16$ **b.** $F(-2) = 16$ **c.** $F\left(\dfrac{3}{4}\right) = 1.4768$

17a. $f(-2) = 0.1353$ **b.** $f(2) = 0.1353$ **c.** $f(-3) = 0.0111$ **19.**   **21.**

23.   **25.**   **27.**   **29.**

31. b and d   **33.** $(0, 1)$ **35.** There is no x-intercept. The y-intercept is $(0, 1)$.

37. 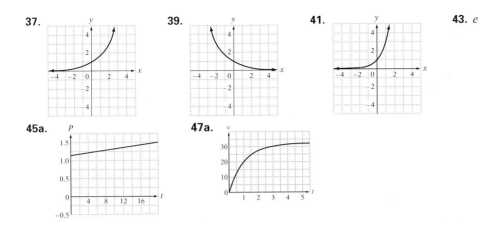 **39.** **41.** **43.** e

45a. 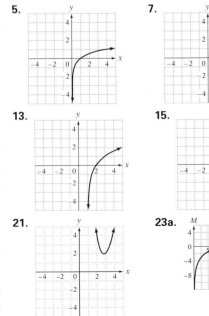 **47a.**

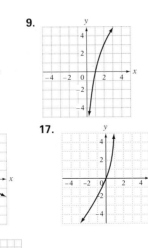

SECTION 10.2

3. $\log_5 25 = 2$ **5.** $\log_4 \dfrac{1}{16} = -2$ **7.** $\log_{10} x = y$ **9.** $\log_a w = x$ **11.** $3^2 = 9$ **13.** $10^{-2} = 0.01$ **15.** $e^y = x$

17. $b^y = u$ **19.** $\log_3 81 = 4$ **21.** $\log_2 128 = 7$ **23.** $\log 100 = 2$ **25.** $\ln e^3 = 3$ **27.** $\log_8 1 = 0$

29. $\log_5 625 = 4$ **31.** 9 **33.** 64 **35.** $\dfrac{1}{7}$ **37.** 1 **39.** 316.23 **41.** 0.02 **43.** 7.39 **45.** 0.61

49. $\log_3(x^3 y^2)$ **51.** $\ln\left(\dfrac{x^4}{y^2}\right)$ **53.** $\log_7 x^3$ **55.** $\ln(x^3 y^4)$ **57.** $\log_4(x^2 y^2)$ **59.** $\log_3\left(\dfrac{x^2 z^2}{y}\right)$ **61.** $\ln\left(\dfrac{x}{y^2 z}\right)$

63. $\log_6 \sqrt{\dfrac{x}{y}}$ **65.** $\log_4\left(\dfrac{s^2 r^2}{t^4}\right)$ **67.** $\ln\left(\dfrac{x}{y^2 z^2}\right)$ **69.** $\log_2\left(\dfrac{t^3 v^2}{r^2}\right)$ **71.** $\log_4 \sqrt{\dfrac{x^3 z}{y^2}}$ **73.** $\ln\sqrt{\dfrac{x}{y^3}}$ **75.** $\log_2\left(\dfrac{\sqrt{xz}}{\sqrt[3]{y^2}}\right)$

77. $\log_8 x + \log_8 z$ **79.** $5\log_3 x$ **81.** $\log_b r - \log_b s$ **83.** $2\log_3 x + 6\log_3 y$ **85.** $3\log_7 u - 4\log_7 v$

87. $2\log_2 r + 2\log_2 s$ **89.** $2\ln x + \ln y + \ln z$ **91.** $\log_5 x + 2\log_5 y - 4\log_5 z$ **93.** $2\log_8 x - \log_8 y - 2\log_8 z$

95. $\dfrac{3}{2}\log_4 x + \dfrac{1}{2}\log_4 y$ **97.** $\dfrac{3}{2}\log_7 x - \dfrac{1}{2}\log_7 y$ **99.** $\log_3 t - \dfrac{1}{2}\log_3 x$ **101.** 0.8451 **103.** -0.2218 **105.** 1.3863

107. 1.0415 **109.** 0.8617 **111.** 2.1133 **113.** -0.6309 **115.** 0.2727 **117.** 1.6826 **119.** 1.9266

121. 0.6752 **123.** 2.6125 **125a.** False **b.** True **c.** False **d.** True **e.** False **f.** True

SECTION 10.3

5. **7.** **9.** **11.**

13. **15.** **17.** **19.**

21. **23a.**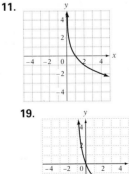

SECTION 10.4

3. $-\dfrac{1}{3}$ **5.** -3 **7.** 1   **9.** 6   **11.** $\dfrac{4}{7}$ **13.** $\dfrac{9}{8}$ **15.** 1.1133   **17.** 1.0986   **19.** 1.3222   **21.** -2.8074

23. 3.5850   **25.** 1.1309   **29.** $\dfrac{11}{2}$ **31.** -4 and 2   **33.** $\dfrac{5}{3}$ **35.** $\dfrac{1}{2}$ **37.** $\dfrac{9}{2}$ **39.** 17.5327   **41.** 3   **43.** 2

45. No solution   **47.** -0.55 **49.** 1.38   **51.** No solution   **53.** 0.09   **55a.**

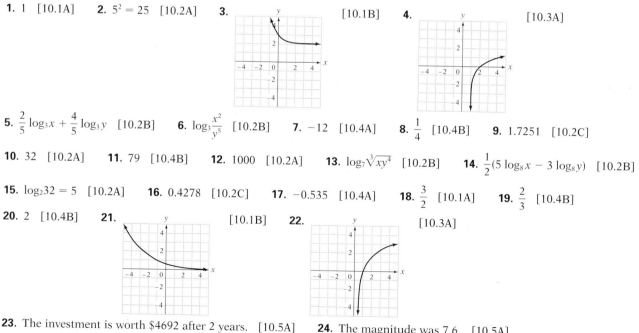

b. It will take the object approximately 2.64 s to fall 100 ft.

SECTION 10.5

1. The value of the investment after 2 years is \$1172.   **3.** The investment will be worth \$15,000 in 18 years.
5a. The technetium level will be 21.2 mg after 3 h.   **b.** It will take 3.5 h for the technetium level to reach 20 mg.
7. The half-life of promethium-147 is 2.5 years.   **9a.** The approximate pressure at 40 km above Earth's surface is 0.098
newtons per square centimeter.   **b.** The approximate pressure on Earth's surface is 10.13 newtons per square centimeter.
c. The atmospheric pressure decreases as you rise about Earth's surface.   **11.** The pH of milk is 6.4.
13. At a depth of 2.5 m, the percent of light will be 75% of the light at the surface of the pool.
15. Normal conversation is 65 decibels.   **17.** The average time of a Major League baseball game increased 22.5 min
during the years 1981 to 1999.   **19.** A thickness of 0.4 cm is needed.   **21.** The Richter scale magnitude of the earthquake
is 6.8.   **23.** The intensity of the earthquake is $794{,}328{,}235I_0$.   **25.** The magnitude of the earthquake is 5.3.
27. The magnitude of the earthquake is 5.6.   **29.** The value of the investment after 5 years is \$3210.06.

CHAPTER REVIEW

1. 1   [10.1A]   **2.** $5^2 = 25$ [10.2A]   **3.**   [10.1B]   **4.**   [10.3A]

5. $\dfrac{2}{5}\log_3 x + \dfrac{4}{5}\log_3 y$ [10.2B]   **6.** $\log_3 \dfrac{x^2}{y^5}$ [10.2B]   **7.** -12 [10.4A]   **8.** $\dfrac{1}{4}$ [10.4B]   **9.** 1.7251   [10.2C]

10. 32   [10.2A]   **11.** 79   [10.4B]   **12.** 1000   [10.2A]   **13.** $\log_7 \sqrt[3]{xy^4}$ [10.2B]   **14.** $\dfrac{1}{2}(5\log_8 x - 3\log_8 y)$ [10.2B]

15. $\log_2 32 = 5$ [10.2A]   **16.** 0.4278   [10.2C]   **17.** -0.535 [10.4A]   **18.** $\dfrac{3}{2}$ [10.1A]   **19.** $\dfrac{2}{3}$ [10.4B]

20. 2   [10.4B]   **21.**   [10.1B]   **22.**   [10.3A]

23. The investment is worth \$4692 after 2 years.   [10.5A]   **24.** The magnitude was 7.6.   [10.5A]
25. The half-life is 27 days.   [10.5A]   **26.** The sound at a busy street corner is 107 decibels.   [10.5A]

CHAPTER TEST

1. $f(0) = 1$ [10.1A]   **2.** $f(-2) = \dfrac{1}{3}$ [10.1A]   **3.**   [10.1B]   **4.**   [10.1B]

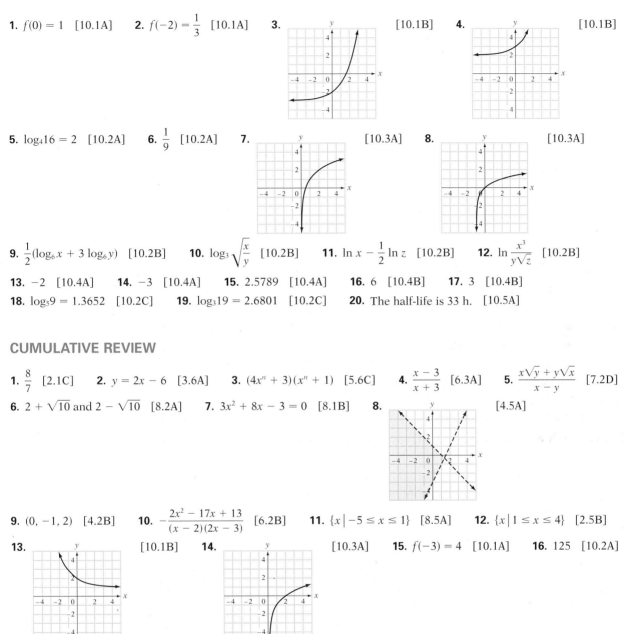

5. $\log_4 16 = 2$ [10.2A]   **6.** $\dfrac{1}{9}$ [10.2A]   **7.**   [10.3A]   **8.**   [10.3A]

9. $\dfrac{1}{2}(\log_6 x + 3\log_6 y)$ [10.2B]   **10.** $\log_3 \sqrt{\dfrac{x}{y}}$ [10.2B]   **11.** $\ln x - \dfrac{1}{2}\ln z$ [10.2B]   **12.** $\ln \dfrac{x^3}{y\sqrt{z}}$ [10.2B]

13. -2 [10.4A]   **14.** -3 [10.4A]   **15.** 2.5789 [10.4A]   **16.** 6 [10.4B]   **17.** 3 [10.4B]

18. $\log_5 9 = 1.3652$ [10.2C]   **19.** $\log_3 19 = 2.6801$ [10.2C]   **20.** The half-life is 33 h.   [10.5A]

CUMULATIVE REVIEW

1. $\dfrac{8}{7}$ [2.1C]   **2.** $y = 2x - 6$ [3.6A]   **3.** $(4x^n + 3)(x^n + 1)$ [5.6C]   **4.** $\dfrac{x-3}{x+3}$ [6.3A]   **5.** $\dfrac{x\sqrt{y} + y\sqrt{x}}{x - y}$ [7.2D]

6. $2 + \sqrt{10}$ and $2 - \sqrt{10}$ [8.2A]   **7.** $3x^2 + 8x - 3 = 0$ [8.1B]   **8.**   [4.5A]

9. $(0, -1, 2)$ [4.2B]   **10.** $-\dfrac{2x^2 - 17x + 13}{(x - 2)(2x - 3)}$ [6.2B]   **11.** $\{x \mid -5 \le x \le 1\}$ [8.5A]   **12.** $\{x \mid 1 \le x \le 4\}$ [2.5B]

13.   [10.1B]   **14.**   [10.3A]   **15.** $f(-3) = 4$ [10.1A]   **16.** 125 [10.2A]

17. $\log_b \dfrac{x^3}{y^5}$ [10.2B]   **18.** $\log_3 7 = 1.7712$ [10.2C]   **19.** 2 [10.4A]   **20.** $\dfrac{1}{2}$ [10.4B]   **21.** The customer can write 49 checks.   [2.4C]   **22.** The mixture costs \$3.10/lb.   [2.3A]   **23.** The rate of the wind is 25 mph.   [6.5C]   **24.** The force will stretch the spring 10.2 in.   [6.6A]   **25.** The redwood costs 40¢ per foot. The fir costs 25¢ per foot.   [4.4B]   **26.** The investment should double in 8 years.   [10.5A]

Answers to Chapter 11 Odd-Numbered Exercises

PREP TEST

1. 7.21 [3.1B]   **2.** $x^2 - 8x + 16; (x - 4)^2$ [8.2A]   **3.** $0; \pm 4$ [8.1C]   **4.** $(1, -1)$ [4.1B]   **5.** $(1, -5)$ [4.2A]

6. $x = 2$; $(2, -2)$ [9.1A]   **7.** 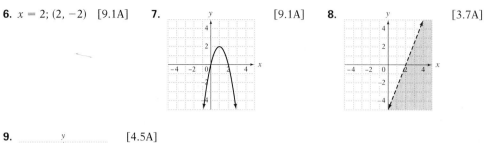   [9.1A]   **8.**   [3.7A]

9. 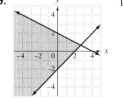   [4.5A]

SECTION 11.1

1a. The axis of symmetry is a vertical line.   **b.** The parabola opens up.   **3a.** The axis of symmetry is a horizontal line.

b. The parabola opens right.   **5a.** The axis of symmetry is a horizontal line.   **b.** The parabola opens left.

7.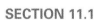

Vertex: $\left(-\dfrac{25}{4}, \dfrac{3}{2}\right)$

Axis of symmetry: $y = \dfrac{3}{2}$

9.

Vertex: $(0, 2)$

Axis of symmetry: $x = 0$

11.

Vertex: $(-1, 0)$

Axis of symmetry: $y = 0$

13.

Vertex: $(-1, 2)$

Axis of symmetry: $y = 2$

15.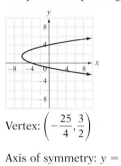

Vertex: $\left(-1, -\dfrac{7}{2}\right)$

Axis of symmetry: $x = -1$

17.

Vertex: $\left(-\dfrac{25}{4}, \dfrac{1}{2}\right)$

Axis of symmetry: $y = \dfrac{1}{2}$

19.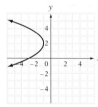

Vertex: $(-1, -7)$

Axis of symmetry: $x = -1$

21.

Vertex: $\left(\dfrac{1}{4}, -\dfrac{25}{8}\right)$

Axis of symmetry: $x = \dfrac{1}{4}$

23.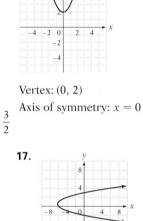

Vertex: $\left(-\dfrac{5}{2}, -\dfrac{1}{4}\right)$

Axis of symmetry: $x = -\dfrac{5}{2}$

25. $\left(0, \dfrac{1}{8}\right)$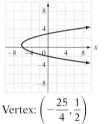

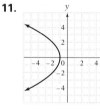

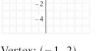

SECTION 11.2

1.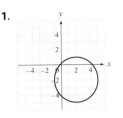

3.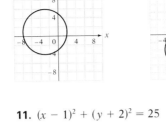

5.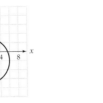

7. $(x - 2)^2 + (y + 1)^2 = 4$

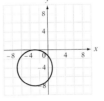

9. $(x + 1)^2 + (y - 1)^2 = 5$ **11.** $(x - 1)^2 + (y + 2)^2 = 25$ **13.** $(x + 3)^2 + (y + 4)^2 = 16$ **15.** $(x - 1)^2 + (y + 1)^2 = 25$

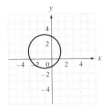

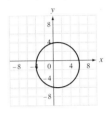

17. $(x - 4)^2 + y^2 = 16$ **19.** $r = 6\sqrt{3}$ in.

SECTION 11.3

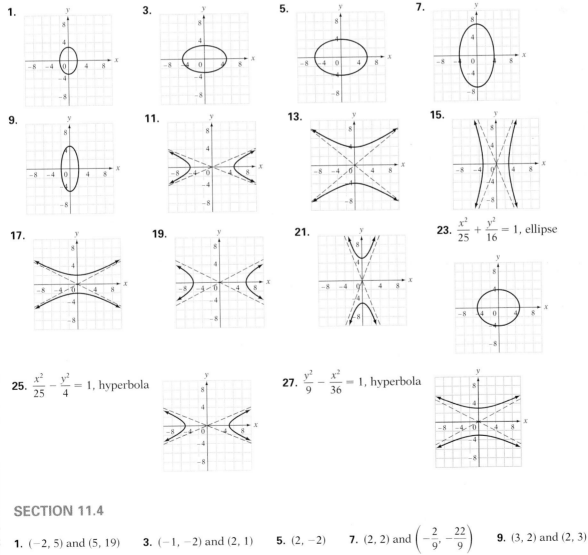

23. $\dfrac{x^2}{25} + \dfrac{y^2}{16} = 1$, ellipse

25. $\dfrac{x^2}{25} - \dfrac{y^2}{4} = 1$, hyperbola

27. $\dfrac{y^2}{9} - \dfrac{x^2}{36} = 1$, hyperbola

SECTION 11.4

1. $(-2, 5)$ and $(5, 19)$ **3.** $(-1, -2)$ and $(2, 1)$ **5.** $(2, -2)$ **7.** $(2, 2)$ and $\left(-\dfrac{2}{9}, -\dfrac{22}{9}\right)$ **9.** $(3, 2)$ and $(2, 3)$

11. $\left(\dfrac{\sqrt{3}}{2}, 3\right)$ and $\left(-\dfrac{\sqrt{3}}{2}, 3\right)$ **13.** The system of equations has no real number solution.

15. $(1, 2), (1, -2), (-1, 2),$ and $(-1, -2)$ **17.** $(3, 2), (3, -2), (-3, 2),$ and $(-3, -2)$ **19.** $(2, 7)$ and $(-2, 11)$

21. $(3, \sqrt{2}), (3, -\sqrt{2}), (-3, \sqrt{2}),$ and $(-3, -\sqrt{2})$ **23.** The system of equations has no real number solution.

25. $(\sqrt{2}, 3), (\sqrt{2}, -3), (-\sqrt{2}, 3),$ and $(-\sqrt{2}, -3)$ **27.** $(2, -1)$ and $(8, 11)$ **29.** $(1, 0)$ **31.** $(2, -1)$ and $(1, -2)$

33. $\left(-\dfrac{1}{2}, \dfrac{9}{4}\right)$ and $(-4, 4)$ **35.** $(1.000, 2.000)$ **37.** $(1.755, 0.811), (0.505, -0.986)$ **39.** $(5.562, -1.562), (0.013, 3.987)$

SECTION 11.5

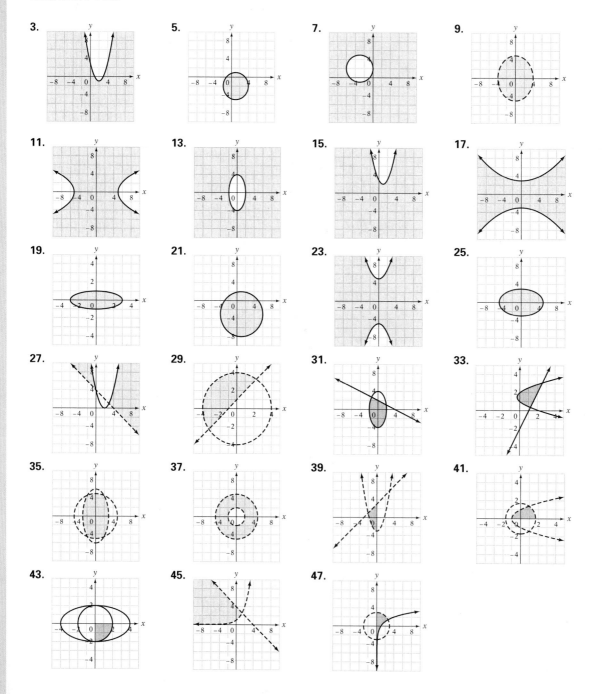

CHAPTER REVIEW

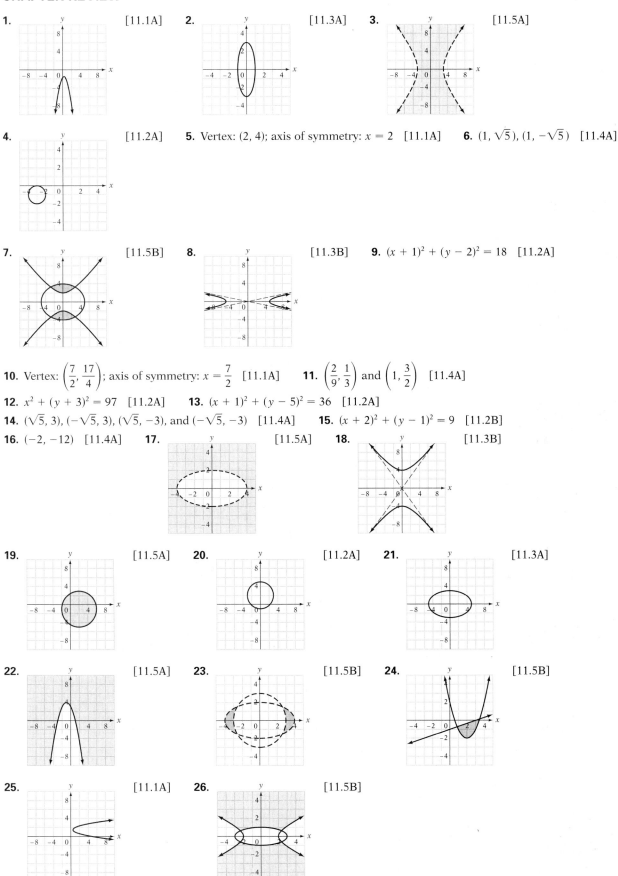

1. [11.1A]

2. [11.3A]

3. [11.5A]

4. [11.2A]

5. Vertex: (2, 4); axis of symmetry: $x = 2$ [11.1A]

6. $(1, \sqrt{5}), (1, -\sqrt{5})$ [11.4A]

7. [11.5B]

8. [11.3B]

9. $(x + 1)^2 + (y - 2)^2 = 18$ [11.2A]

10. Vertex: $\left(\dfrac{7}{2}, \dfrac{17}{4}\right)$; axis of symmetry: $x = \dfrac{7}{2}$ [11.1A]

11. $\left(\dfrac{2}{9}, \dfrac{1}{3}\right)$ and $\left(1, \dfrac{3}{2}\right)$ [11.4A]

12. $x^2 + (y + 3)^2 = 97$ [11.2A]

13. $(x + 1)^2 + (y - 5)^2 = 36$ [11.2A]

14. $(\sqrt{5}, 3), (-\sqrt{5}, 3), (\sqrt{5}, -3)$, and $(-\sqrt{5}, -3)$ [11.4A]

15. $(x + 2)^2 + (y - 1)^2 = 9$ [11.2B]

16. $(-2, -12)$ [11.4A]

17. [11.5A]

18. [11.3B]

19. [11.5A]

20. [11.2A]

21. [11.3A]

22. [11.5A]

23. [11.5B]

24. [11.5B]

25. [11.1A]

26. [11.5B]

CHAPTER TEST

1. $(x + 3)^2 + (y + 3)^2 = 16$ [11.2A]   **2.** $(x + 2)^2 + (y - 1)^2 = 32$ [11.2A]   **3.**

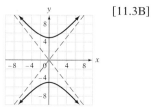 [11.3B]

4. 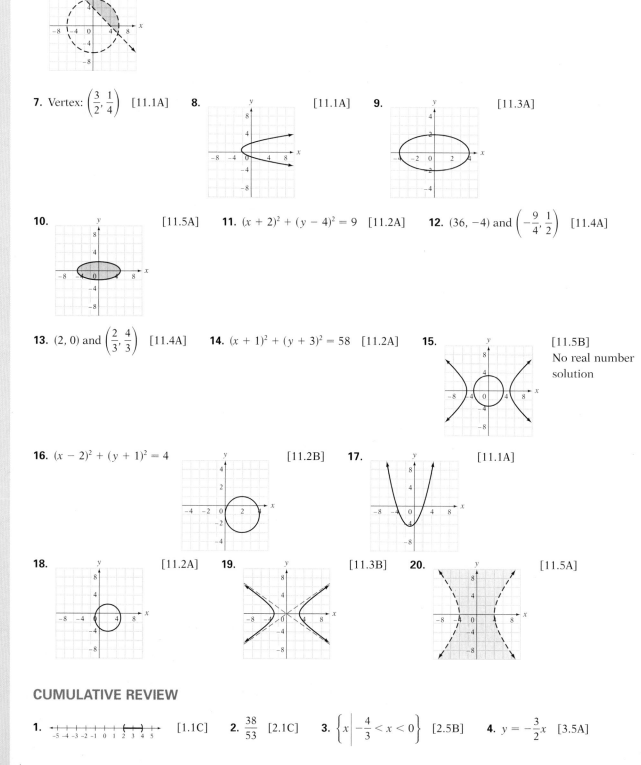 [11.5B]   **5.** The axis of symmetry is $x = 3$.   [11.1A]   **6.** $(5, 1), (-5, 1), (5, -1)$, and $(-5, -1)$ [11.4A]

7. Vertex: $\left(\dfrac{3}{2}, \dfrac{1}{4}\right)$ [11.1A]   **8.** [11.1A]   **9.** [11.3A]

10. [11.5A]   **11.** $(x + 2)^2 + (y - 4)^2 = 9$ [11.2A]   **12.** $(36, -4)$ and $\left(-\dfrac{9}{4}, \dfrac{1}{2}\right)$ [11.4A]

13. $(2, 0)$ and $\left(\dfrac{2}{3}, \dfrac{4}{3}\right)$ [11.4A]   **14.** $(x + 1)^2 + (y + 3)^2 = 58$ [11.2A]   **15.** [11.5B] No real number solution

16. $(x - 2)^2 + (y + 1)^2 = 4$ [11.2B]   **17.** [11.1A]

18. [11.2A]   **19.** [11.3B]   **20.** [11.5A]

CUMULATIVE REVIEW

1. [1.1C]   **2.** $\dfrac{38}{53}$ [2.1C]   **3.** $\left\{ x \middle| -\dfrac{4}{3} < x < 0 \right\}$ [2.5B]   **4.** $y = -\dfrac{3}{2}x$ [3.5A]

5. $y = x - 6$ [3.6A]   **6.** $x^{4n} + 2x^{3n} - 3x^{2n+1}$ [5.3A]   **7.** $(x - y - 1)(x^2 - 2x + 1 + xy - y + y^2)$ [5.6B]

8. $\{x \mid -4 < x \le 3\}$ [8.5A]   **9.** $\dfrac{x}{x + y}$ [6.1A]   **10.** $\dfrac{x - 5}{3x - 2}$ [6.2B]   **11.** $\dfrac{5}{2}$ [6.5A]

12. 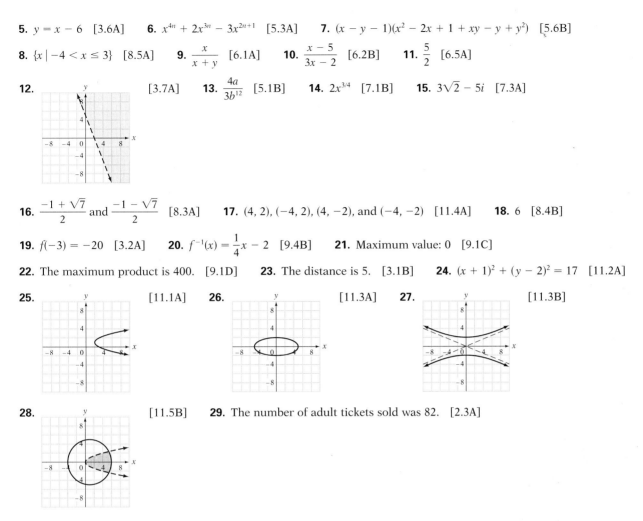   [3.7A]   **13.** $\dfrac{4a}{3b^{12}}$ [5.1B]   **14.** $2x^{3/4}$ [7.1B]   **15.** $3\sqrt{2} - 5i$ [7.3A]

16. $\dfrac{-1 + \sqrt{7}}{2}$ and $\dfrac{-1 - \sqrt{7}}{2}$ [8.3A]   **17.** $(4, 2), (-4, 2), (4, -2),$ and $(-4, -2)$ [11.4A]   **18.** 6 [8.4B]

19. $f(-3) = -20$ [3.2A]   **20.** $f^{-1}(x) = \dfrac{1}{4}x - 2$ [9.4B]   **21.** Maximum value: 0 [9.1C]

22. The maximum product is 400. [9.1D]   **23.** The distance is 5. [3.1B]   **24.** $(x + 1)^2 + (y - 2)^2 = 17$ [11.2A]

25.   [11.1A]   **26.**   [11.3A]   **27.**   [11.3B]

28.   [11.5B]   **29.** The number of adult tickets sold was 82. [2.3A]

30. The rate of the motorcycle is 60 mph. [6.5C]   **31.** The rowing rate in calm water is 4.5 mph. [6.5C]
32. The gear will make 18 revolutions/min. [6.6A]

Answers to Chapter 12 Odd-Numbered Exercises

PREP TEST

1. 12 [1.2D]   **2.** $\dfrac{3}{4}$ [3.2A]   **3.** 18 [1.3B]   **4.** 96 [1.3B]   **5.** -410 [1.3B]   **6.** $\dfrac{4}{9}$ [1.2D]

7. $x^2 + 2xy + y^2$ [5.3C]   **8.** $x^3 + 3x^2y + 3xy^2 + y^3$ [5.3B, 5.3C]

SECTION 12.1

3. 8   **5.** 2, 3, 4, 5   **7.** 3, 5, 7, 9   **9.** 0, -2, -4, -6 **11.** 2, 4, 8, 16   **13.** 2, 5, 10, 17   **15.** 0, $\dfrac{7}{2}$, $\dfrac{26}{3}$, $\dfrac{63}{4}$

17. 40   **19.** 110   **21.** 225   **23.** $\dfrac{1}{256}$ **25.** 380   **27.** $-\dfrac{1}{36}$ **29.** 45   **31.** 20   **33.** 91   **35.** 0   **37.** 432

39. $\dfrac{5}{6}$ **41.** $2x + 2x^2 + 2x^3 + 2x^4 + 2x^5$ **43.** $x + \dfrac{x^2}{2} + \dfrac{x^3}{3} + \dfrac{x^4}{4} + \dfrac{x^5}{5}$ **45.** $x^2 + x^4 + x^6 + x^8 + x^{10}$

47. $x + \dfrac{x^2}{4} + \dfrac{x^3}{9} + \dfrac{x^4}{16}$ **49.** $\dfrac{1}{x} + \dfrac{1}{x^2} + \dfrac{1}{x^3} + \dfrac{1}{x^4} + \dfrac{1}{x^5}$ **51.** $\displaystyle\sum_{i=1}^{n} \dfrac{1}{i}$

SECTION 12.2

1. $a_{15} = 141$ **3.** $a_{15} = 50$ **5.** $a_{31} = 17$ **7.** $a_{12} = 12\frac{1}{2}$ **9.** $a_{40} = -109$ **11.** $a_n = 3n - 2$ **13.** $a_n = -3n + 6$

15. $a_n = -2.5n + 9.5$ **17.** 20 **19.** 20 **21.** 13 **23.** 20 **25.** 2500 **27.** -760 **29.** 189 **31.** 345

33. $\dfrac{187}{2}$ **35.** -180 **37.** It will take 11 weeks. **39.** The object will fall 144 ft during the fifth second.

41. The apprentice will receive $3600 during the eighth month. The apprentice will receive $20,400 for the eight months.
43. log 8

SECTION 12.3

3. $a_9 = 131,072$ **5.** $a_7 = \dfrac{128}{243}$ **7.** $a_{10} = 32$ **9.** $a_2 = 6, a_3 = 4$ **11.** $a_2 = -2, a_3 = \dfrac{4}{3}$ **13.** $a_2 = 12, a_3 = -48$

15. 2186 **17.** $\dfrac{2343}{64}$ **19.** 62 **21.** $\dfrac{121}{243}$ **23.** 9 **25.** $\dfrac{7}{9}$ **27.** $\dfrac{8}{9}$ **29.** $\dfrac{2}{9}$ **31.** $\dfrac{5}{11}$ **33.** $\dfrac{1}{6}$ **35.** The

amount of radioactive material is 50 mg. **37.** The ball will bounce 1.8 ft. **39.** The total amount earned is
$10,737,418.23. **41.** 27

SECTION 12.4

1. 6 **3.** 40,320 **5.** 1 **7.** 10 **9.** 1 **11.** 84 **13.** 21 **15.** 45 **17.** 1 **19.** 20 **21.** 11 **23.** 6
25. $x^4 + 4x^3y + 6x^2y^2 + 4xy^3 + y^4$ **27.** $x^5 - 5x^4y + 10x^3y^2 - 10x^2y^3 + 5xy^4 - y^5$ **29.** $16m^4 + 32m^3 + 24m^2 + 8m + 1$
31. $32r^5 - 240r^4 + 720r^3 - 1080r^2 + 810r - 243$ **33.** $a^{10} + 10a^9b + 45a^8b^2$ **35.** $a^{11} - 11a^{10}b + 55a^9b^2$
37. $256x^8 + 1024x^7y + 1792x^6y^2$ **39.** $65,536x^8 - 393,216x^7y + 1,032,192x^6y^2$ **41.** $x^7 + 7x^5 + 21x^3$ **43.** $-560x^4$
45. $-6x^{10}y^2$ **47.** $126y^5$ **49.** $5n^3$ **51a.** False **b.** False **c.** False **d.** False **e.** False **f.** True

CHAPTER REVIEW

1. $2 + 2x + 2x^2 + 2x^3$ [12.1B] **2.** 28 [12.1A] **3.** $\dfrac{19}{30}$ [12.3C] **4.** $\dfrac{1}{3}$ [12.3C] **5.** 1830 [12.2B]

6. 35 [12.2B] **7.** $\dfrac{23}{99}$ [12.3C] **8.** 84 [12.4A] **9.** 726 [12.3B] **10.** 22 terms [12.2A] **11.** 0.996 [12.3B]

12. $5670x^4y^4$ [12.4A] **13.** $-\dfrac{5}{27}$ [12.1A] **14.** 1127 [12.3B] **15.** $5n - 12$ [12.2A] **16.** $-42,240x^4y^7$ [12.4A]

17. $243\sqrt{3}$ [12.3A] **18.** 2000 [12.2B] **19.** 220 [12.4A] **20.** $-\dfrac{13}{60}$ [12.1B]

21. $2x + 2x^2 + \dfrac{8x^3}{3} + 4x^4 + \dfrac{32x^5}{5}$ [12.1B] **22.** $\dfrac{1}{729}$ [12.3A]

23. $x^5 - 15x^4y^2 + 90x^3y^4 - 270x^2y^6 + 405xy^8 - 243y^{10}$ [12.4A] **24.** $\dfrac{16}{5}$ [12.3C] **25.** 6 [12.3C] **26.** 99 [12.4A]

27. 77 [12.2A] **28.** The temperature of the spa is 67.7°F. [12.3D] **29.** The total salary is $12,240. [12.2C]
30. The amount of radioactive material is 3.125 mg. [12.3D]

CHAPTER TEST

1. 34 [12.2B] **2.** 84 [12.4A] **3.** 32 [12.3A] **4.** $\dfrac{1}{2}$ [12.1A] **5.** $3x + 3x^2 + 3x^3 + 3x^4$ [12.1B]

6. 468 [12.2B] **7.** 16 [12.3C] **8.** 16 [12.2A] **9.** $\dfrac{7}{30}$ [12.3C] **10.** 70 [12.4A] **11.** $\dfrac{2}{27}$ [12.3A]

12. $-280x^4y^3$ [12.4A] **13.** -115 [12.2A] **14.** $a_n = -3n + 15$ [12.2A] **15.** -66 [12.3B] **16.** $\dfrac{7}{6}; \dfrac{8}{7}$ [12.1A]

17. $\dfrac{665}{32}$ [12.3B] **18.** 1575 [12.2B] **19.** There were 2550 yd of material in stock after the shipment on October 1.

[12.2C] **20.** There will be 20 mg of radioactive material in the sample at the beginning of the fifth day. [12.3D]

CUMULATIVE REVIEW

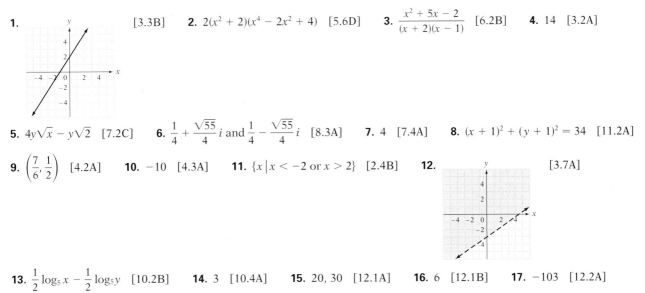

1. [3.3B]   **2.** $2(x^2 + 2)(x^4 - 2x^2 + 4)$ [5.6D]   **3.** $\dfrac{x^2 + 5x - 2}{(x + 2)(x - 1)}$ [6.2B]   **4.** 14 [3.2A]

5. $4y\sqrt{x} - y\sqrt{2}$ [7.2C]   **6.** $\dfrac{1}{4} + \dfrac{\sqrt{55}}{4}i$ and $\dfrac{1}{4} - \dfrac{\sqrt{55}}{4}i$ [8.3A]   **7.** 4 [7.4A]   **8.** $(x + 1)^2 + (y + 1)^2 = 34$ [11.2A]

9. $\left(\dfrac{7}{6}, \dfrac{1}{2}\right)$ [4.2A]   **10.** -10 [4.3A]   **11.** $\{x \mid x < -2 \text{ or } x > 2\}$ [2.4B]   **12.** [3.7A]

13. $\dfrac{1}{2}\log_5 x - \dfrac{1}{2}\log_5 y$ [10.2B]   **14.** 3 [10.4A]   **15.** 20, 30 [12.1A]   **16.** 6 [12.1B]   **17.** -103 [12.2A]

18. $\dfrac{9}{5}$ [12.3C]   **19.** $\dfrac{7}{15}$ [12.3C]   **20.** $12xy^5$ [12.4A]   **21.** There must be 120 oz of pure water added. [2.3B]

22. The new computer takes 24 min. The old computer takes 40 min. [6.5B]   **23.** The rate of the boat is 6.25 mph. The rate of the current is 1.25 mph. [6.5C]   **24.** The half-life of the sample is 55 days. [10.5A]   **25.** There are 1536 seats. [12.2C]   **26.** The ball bounces 2.6 ft on the fifth bounce. [12.3D]

FINAL EXAM

1. -31 [1.2D]   **2.** -1 [1.3B]   **3.** $33 - 10x$ [1.3C]   **4.** 8 [2.1B]   **5.** $\dfrac{2}{3}$ [2.1C]   **6.** 4 and $-\dfrac{2}{3}$ [2.5A]

7. [3.3C]   **8.** $y = -3x + 7$ [3.5B]   **9.** $y = -\dfrac{2}{3}x - \dfrac{1}{3}$ [3.6A]   **10.** $6a^3 - 5a^2 + 10a$ [5.3A]

x-intercept: $\left(\dfrac{9}{2}, 0\right)$

y-intercept: $(0, -3)$

11. $(2 - xy)(4 + 2xy + x^2y^2)$ [5.6B]   **12.** $(x - y)(1 + x)(1 - x)$ [5.6D]   **13.** $x^2 - 2x - 3 - \dfrac{5}{2x - 3}$ [5.4A]

14. $\dfrac{x(x - 1)}{2x - 5}$ [6.1C]   **15.** $\dfrac{-10x}{(x + 2)(x - 3)}$ [6.2B]   **16.** $\dfrac{x + 3}{x + 1}$ [6.3A]   **17.** $-\dfrac{7}{4}$ [6.5A]   **18.** $d = \dfrac{a_n - a_1}{n - 1}$ [2.1D]

19. $\dfrac{y^4}{162x^3}$ [5.1B]   **20.** $\dfrac{1}{64x^8y^5}$ [7.1A]   **21.** $-2x^2y\sqrt{2y}$ [7.2B]   **22.** $\dfrac{x^2\sqrt{2y}}{2y^2}$ [7.2D]   **23.** $\dfrac{6}{5} - \dfrac{3}{5}i$ [7.3D]

24. $2x^2 - 3x - 2 = 0$ [8.1B]   **25.** $\dfrac{3 + \sqrt{17}}{4}$ and $\dfrac{3 - \sqrt{17}}{4}$ [8.3A]   **26.** 27 and -8 [8.4A]   **27.** $\dfrac{3}{2}$ and -2 [8.4C]

28. [11.1A]   **29.** [11.3A]   **30.** $f^{-1}(x) = \dfrac{3}{2}x + 6$ [9.4B]   **31.** $(3, 4)$ [4.2A]

32. 10 [4.3A]   **33.** $\left(\dfrac{5}{2}, -\dfrac{3}{2}\right)$ [11.4A]   **34.** $\left\{x \mid x > \dfrac{3}{2}\right\}$ [2.4B]   **35.** $\{x \mid -4 < x < -1\}$ [2.5B]

36. 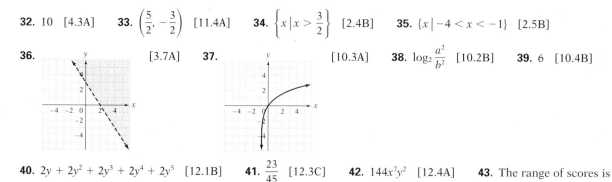 [3.7A]   **37.** [10.3A]   **38.** $\log_2 \dfrac{a^2}{b^2}$ [10.2B]   **39.** 6 [10.4B]

40. $2y + 2y^2 + 2y^3 + 2y^4 + 2y^5$ [12.1B]   **41.** $\dfrac{23}{45}$ [12.3C]   **42.** $144x^7y^2$ [12.4A]   **43.** The range of scores is $69 \le x \le 100$. [2.4C]   **44.** The cyclist rode 40 mi. [2.3C]   **45.** There is $8000 invested at 8.5% and $4000 invested at 6.4%. [4.1C]   **46.** The length is 20 ft. The width is 7 ft. [8.6A]   **47.** An additional 200 shares are needed. [6.4B]   **48.** The rate of the plane was 420 mph. [4.4B]   **49.** The object has fallen 88 ft when the speed reaches 75 ft/s. [7.4B]   **50.** The rate of the plane for the first 360 mi was 120 mph. [8.6A]   **51.** The intensity is 200 foot-candles. [6.6A]   **52.** The rate of the boat is 12.5 mph. The rate of the current is 2.5 mph. [4.4A]   **53.** The value of the investment after 2 years would be $4785.65. [10.5A]   **54.** The house will be worth $256,571 in 20 years. [12.3D]

Glossary

abscissa The first number of an ordered pair; it measures a horizontal distance and is also called the first coordinate of an ordered pair. (Sec. 3.1)

absolute value of a number The distance of the number from zero on the number line. (Sec. 1.1, 2.5)

absolute value equation An equation containing the absolute-value symbol. (Sec. 2.5)

addition method An algebraic method of finding an exact solution of a system of linear equations. (Sec. 4.2)

additive inverses Numbers that are the same distance from zero on the number line but lie on different sides of zero; also called opposites. (Sec. 1.1)

analytic geometry Geometry in which a coordinate system is used to study relationships between variables. (Sec. 3.1)

antilogarithm If $\log_b M = N$, then the antilogarithm, base b, of N is M. (Sec. 10.2)

arithmetic progression A sequence in which the difference between any two consecutive terms is constant; also called an arithmetic sequence. (Sec. 12.2)

arithmetic sequence A sequence in which the difference between any two consecutive terms is constant; also called an arithmetic progression. (Sec. 12.2)

arithmetic series The indicated sum of the terms of an arithmetic sequence. (Sec. 12.2)

asymptotes The two straight lines that a hyperbola "approaches." (Sec. 11.3)

axes The two number lines that form a rectangular coordinate system; also called coordinate axes. (Sec. 3.1)

axis of symmetry of a parabola A line of symmetry that passes through the vertex of the parabola and is parallel to the y-axis for an equation of the form $y = ax^2 + bx + c$ or parallel to the x-axis for an equation of the form $x = ay^2 + by + c$. (Sec. 9.1, 11.1)

base In an exponential expression, the number that is taken as a factor as many times as indicated by the exponent. (Sec. 1.2)

binomial A polynomial of two terms. (Sec. 5.2)

center of a circle The central point that is equidistant from all the points that make up a circle. (Sec. 11.2)

center of an ellipse The intersection of the two axes of symmetry of the ellipse. (Sec. 11.3)

characteristic The integer part of a common logarithm. (Sec. 10.2)

circle The set of all points (x, y) in the plane that are a fixed distance from a given point (h, k) called the center. (Sec. 11.2)

clearing denominators Removing denominators from an equation that contains fractions by multiplying each side of the equation by the LCM of the denominators. (Sec. 6.5)

closed interval In set-builder notation, an interval that contains its endpoints. (Sec. 1.1)

coefficient The number part of a variable term. (Sec. 1.3)

cofactor of an element of a matrix $(-1)^{i+j}$ times the minor of that element, where i is the row number of the element and j is its column number. (Sec. 4.3)

combined variation A variation in which two or more types of variation occur at the same time. (Sec. 6.6)

combining like terms Using the Distributive Property to add the coefficients of like variable terms; adding like terms of a variable expression. (Sec. 1.3)

common difference of a sequence The difference between any two consecutive terms in an arithmetic sequence. (Sec. 12.2)

common logarithms Logarithms to the base 10. (Sec. 10.2)

common ratio of a sequence In a geometric sequence, each successive term of the sequence is the same nonzero constant multiple of the preceding term. This common multiple is called the common ratio of the sequence. (Sec. 12.3)

completing the square Adding to a binomial the constant term that makes it a perfect-square trinomial. (Sec. 8.2)

complex fraction A fraction whose numerator or denominator contains one or more fractions. (Sec. 1.2, 6.3)

complex number A number of the form $a + bi$, where a and b are real numbers and $i = \sqrt{-1}$. (Sec. 7.3)

composite number A number that has whole-number factors besides 1 and itself. For example, 12 is a composite number. (Sec. 1.1)

composition of functions The operation on two functions f and g denoted by $f \circ g$. The value of the composition of f and g is given by $(f \circ g)(x) = f[g(x)]$. (Sec. 9.3)

compound inequality Two inequalities joined with a connective word such as "and" or "or." (Sec. 2.4)

compound interest Interest that is computed not only on the original principal but also on the interest already earned. (Sec. 10.5)

conditional equation An equation that is true if the variable it contains is replaced by the proper value. $x + 2 = 5$ is a conditional equation. (Sec. 2.1)

conic section A curve that can be constructed from the intersection of a plane and a right circular cone. The four conic sections are the parabola, hyperbola, ellipse, and circle. (Sec. 11.1)

conjugates Binomial expressions that differ only in the sign of a term. The expressions $a + b$ and $a - b$ are conjugates. (Sec. 7.2)

consecutive even integers Even integers that follow one another in order. (Sec. 2.2)

consecutive integers Integers that follow one another in order. (Sec. 2.2)

consecutive odd integers Odd integers that follow one another in order. (Sec. 2.2)

constant function A function given by $f(x) = b$, where b is a constant. Its graph is a horizontal line passing through $(0, b)$. (Sec. 3.3)

constant of proportionality k in a variation equation; also called the constant of variation. (Sec. 6.6)

constant of variation k in a variation equation; also called the constant of proportionality. (Sec. 6.6)

constant term A term that contains no variable part. (Sec. 1.3)

contradiction An equation in which any replacement for the variable will result in a false equation. $x = x + 1$ is a contradiction. (Sec. 2.1)

coordinate axes The two number lines that form a rectangular coordinate system; also called axes. (Sec. 3.1)

coordinates of a point The numbers in the ordered pair that is associated with the point. (Sec. 3.1)

cube root of a perfect cube One of the three equal factors of the perfect cube. (Sec. 5.6)

cubic function A third-degree polynomial function. (Sec. 5.2)

degree of a monomial The sum of the exponents of the variables. (Sec. 5.1)

degree of a polynomial The greatest of the degrees of any of its terms. (Sec. 5.2)

dependent system of equations A system of equations whose graphs coincide. (Sec. 4.1)

dependent variable A variable whose value depends on that of another variable known as the independent variable. (Sec. 3.2)

descending order The terms of a polynomial in one variable are arranged in descending order when the exponents of the variable decrease from left to right. (Sec. 5.2)

determinant A number associated with a square matrix. (Sec. 4.3)

direct variation A special function that can be expressed as the equation $y = kx$, where k is a constant called the constant of variation or the constant of proportionality. (Sec. 6.6)

discriminant For an equation of the form $ax^2 + bx + c = 0$, the quantity $b^2 - 4ac$ is called the discriminant. (Sec. 8.3)

domain The set of the first coordinates of all the ordered pairs of a function. (Sec. 1.1, 3.2)

double root When a quadratic equation has two solutions that are the same number, the solution is called a double root of the equation. (Sec. 8.1)

element of a matrix A number in a matrix. (Sec. 4.3)

elements of a set The objects in the set. (Sec. 1.1)

ellipse An oval shape that is one of the conic sections. (Sec. 11.3)

empty set The set that contains no elements. (Sec. 1.1)

equation A statement of the equality of two mathematical expressions. (Sec. 2.1)

equivalent equations Equations that have the same solution. (Sec. 2.1)

evaluating a function Determining $f(x)$ for a given value of x. (Sec. 3.2)

even integer An integer that is divisible by 2. (Sec. 2.2)

expanding by cofactors A technique for finding the value of a 3×3 or larger determinant. (Sec. 4.3)

exponent In an exponential expression, the raised number that indicates how many times the factor, or base, occurs in the multiplication. (Sec. 1.2)

exponential equation An equation in which the variable occurs in the exponent. (Sec. 10.4)

exponential form The expression 2^6 is in exponential form. Compare *factored form*. (Sec. 1.2)

exponential function The exponential function with base b is defined by $f(x) = b^x$, where b is a positive real number not equal to one. (Sec. 10.1)

extraneous solution When each side of an equation is raised to an even power, the resulting equation may have a solution that is not a solution of the original equation. Such a solution is called an extraneous solution. (Sec. 7.4)

factored form The multiplication $2 \cdot 2 \cdot 2 \cdot 2 \cdot 2 \cdot 2$ is in factored form. Compare *exponential form*. (Sec. 1.2)

factoring a polynomial Writing the polynomial as a product of other polynomials. (Sec. 5.5)

factoring a quadratic trinomial Expressing the trinomial as the product of two binomials. (Sec. 5.5)

finite sequence A sequence that contains a finite number of terms. (Sec. 12.1)

finite set A set for which all the elements can be listed. (Sec. 1.1)

first coordinate of an ordered pair The first number of the ordered pair; it measures a horizontal distance and is also called the abscissa. (Sec. 3.1)

first-degree equation An equation in which all variables have an exponent of 1. (Sec. 2.1)

FOIL A method of finding the product of two binomials. The letters stand for First, Outer, Inner, and Last. (Sec. 5.3)

formula A literal equation that states a rule about measurement. (Sec. 2.1)

function A relation in which no two ordered pairs that have the same first coordinate have different second coordinates. (Sec. 3.2)

functional notation Notation used for those equations that define functions. The letter f is commonly used to name a function. (Sec. 3.2)

general term of a sequence In the sequence $a_1, a_2, a_3, \ldots, a_n, \ldots$ the general term of the sequence is a_n. (Sec. 12.1)

geometric progression A sequence in which each successive term of the sequence is the same nonzero constant multiple of the preceding term; also called a geometric sequence. (Sec. 12.3)

geometric sequence A sequence in which each successive term of the sequence is the same nonzero constant multiple of the preceding term; also called a geometric progression. (Sec. 12.3)

geometric series The indicated sum of the terms of a geometric sequence. (Sec. 12.3)

graph of a function A graph of the ordered pairs that belong to the function. (Sec. 3.3)

graph of a quadratic inequality in two variables A region of the plane that is bounded by one of the conic sections. (Sec. 11.5)

graph of a real number A heavy dot placed directly above the number on the number line. (Sec. 1.1)

graph of an ordered pair The dot drawn at the coordinates of the point in the plane. (Sec. 3.1)

graphing a point in the plane Placing a dot at the location given by the ordered pair; also called plotting a point in the plane. (Sec. 3.1)

greater than A number that lies to the right of another number on the number line is said to be greater than that number. (Sec. 1.1)

greatest common factor (GCF) The GCF of two or more numbers is the largest integer that divides evenly into all of them. (Sec. 1.2)

half-open interval In set-builder notation, an interval that contains one of its endpoints. (Sec. 1.1)

half-plane The solution set of a linear inequality in two variables. (Sec. 3.7)

hyperbola A conic section formed by the intersection of a cone and a plane perpendicular to the base of the cone. (Sec. 11.3)

hypotenuse In a right triangle, the side opposite the $90°$ angle. (Sec. 7.4)

identity An equation in which any replacement for the variable will result in a true equation. $x + 2 = x + 2$ is an identity. (Sec. 2.1)

imaginary number A number of the form ai, where a is a real number and $i = \sqrt{-1}$. (Sec. 7.3)

imaginary part of a complex number For the complex number $a + bi$, b is the imaginary part. (Sec. 7.3)

inconsistent system of equations A system of equations that has no solution. (Sec. 4.1)

independent system of equations A system of equations whose graphs intersect at only one point. (Sec. 4.1)

independent variable A variable whose value determines that of another variable known as the dependent variable. (Sec. 3.2)

index In the expression $\sqrt[n]{a}$, n is the index of the radical. (Sec. 7.1)

infinite geometric series The indicated sum of the terms of an infinite geometric sequence. (Sec. 12.3)

infinite sequence A sequence that contains an infinite number of terms. (Sec. 12.1)

infinite set A set in which the list of elements continues without end. (Sec. 1.1)

integers The numbers $\ldots, -3, -2, -1, 0, 1, 2, 3, \ldots$. (Sec. 1.1)

intersection of two sets The set that contains all elements that are common to both of the sets. (Sec. 1.1)

interval notation A type of set-builder notation in which the property that distinguishes the elements of the set is their location within a specified interval. (Sec. 1.1)

inverse of a function The set of ordered pairs formed by reversing the coordinates of each ordered pair of the function. (Sec. 9.4)

inverse variation A function that can be expressed as the equation $y = \dfrac{k}{x}$, where k is a constant. (Sec. 6.6)

irrational number The decimal representation of an irrational number never terminates or repeats and can only be approximated. (Sec. 1.1)

joint variation A variation in which a variable varies directly as the product of two or more variables. A joint variation can be expressed as the equation $z = kxy$, where k is a constant. (Sec. 6.6)

leading coefficient In a polynomial, the coefficient of the variable with the largest exponent. (Sec. 5.2)

least common multiple (LCM) The LCM of two or more numbers is the smallest number that is a multiple of each of those numbers. (Sec. 1.2)

least common multiple of two polynomials The simplest polynomial of least degree that contains the factors of each polynomial. (Sec. 6.2)

leg In a right triangle, one of the two sides that are not opposite the $90°$ angle. (Sec. 7.4)

less than A number that lies to the left of another number on the number line is said to be less than that number. (Sec. 1.1)

like terms Terms of a variable expression that have the same variable part. Having no variable part, constant terms are like terms. (Sec. 1.3)

linear equation in three variables An equation of the form $Ax + By + Cz = D$, where A, B, and C are coefficients of the variables and D is a constant. (Sec. 4.2)

linear equation in two variables An equation of the form $y = mx + b$ or $Ax + By = C$. (Sec. 3.3)

linear function A function that can be expressed in the form $f(x) = mx + b$. Its graph is a straight line. (Sec. 3.3)

linear inequality in two variables An inequality of the form $y > mx + b$ or $Ax + By > C$. (The symbol $>$ could be replaced by $\geq$, $<$, or $\leq$.) (Sec. 3.7)

literal equation An equation that contains more than one variable. (Sec. 2.1)

logarithm For b greater than zero and not equal to 1, the statement $y = \log_b x$ (the logarithm of x to the base b) is equivalent to $x = b^y$. (Sec. 10.2)

mantissa The decimal part of a common logarithm. (Sec. 10.2)

matrix A rectangular array of numbers. (Sec. 4.3)

minor of an element The minor of an element in a 3×3 determinant is the 2×2 determinant obtained by eliminating the row and column that contain that element. (Sec. 4.3)

monomial A number, a variable, or a product of a number and variables; a polynomial of one term. (Sec. 5.1, 5.2)

multiplicative inverse The multiplicative inverse of a nonzero real number a is $\dfrac{1}{a}$; also called the reciprocal. (Sec. 1.2)

n factorial The product of the first n natural numbers; n factorial is written $n!$ (Sec. 12.4)

natural exponential function The function defined by $f(x) = e^x$, where $e \approx 2.71828$. (Sec. 10.1)

natural logarithm When e (the base of the natural exponential function) is used as the base of a logarithm, the logarithm is referred to as the natural logarithm and is abbreviated $\ln x$. (Sec. 10.2)

natural numbers The numbers $1, 2, 3, \ldots$; also called the positive integers. (Sec. 1.1)

negative integers The numbers $\ldots, -3, -2, -1$. (Sec. 1.1)

negative slope The slope of a line that slants downward to the right. (Sec. 3.4)

nonfactorable over the integers A polynomial is nonfactorable over the integers if it does not factor using only integers. (Sec. 5.5)

nonlinear system of equations A system of equations in which one or more of the equations are not linear equations. (Sec. 11.4)

nth root of a A number b such that $b^n = a$. The nth root of a can be written $a^{1/n}$ or $\sqrt[n]{a}$. (Sec. 7.1)

null set The set that contains no elements. (Sec. 1.1)

numerical coefficient The number part of a variable term. (Sec. 1.3)

odd integer An integer that is not divisible by 2. (Sec. 2.2)

one-to-one function In a one-to-one function, given any y, there is only one x that can be paired with the given y. (Sec. 9.4)

open interval In set-builder notation, an interval that does not contain its endpoints. (Sec. 1.1)

opposites Numbers that are the same distance from zero on the number line but lie on different sides of zero; also called additive inverses. (Sec. 1.1)

order $m \times n$ A matrix of m rows and n columns is of order $m \times n$. (Sec. 4.3)

Order of Operations Agreement Rules that specify the order in which we perform operations in simplifying numerical expressions. (Sec. 1.2)

ordered pair A pair of numbers expressed in the form (a, b) and used to locate a point in the plane determined by a rectangular coordinate system. (Sec. 3.1)

ordinate The second number of an ordered pair; it measures a vertical distance and is also called the second coordinate of an ordered pair. (Sec. 3.1)

origin The point of intersection of the two number lines that form a rectangular coordinate system. (Sec. 3.1)

parabola The graph of a quadratic function is called a parabola. (Sec. 9.1)

parallel lines Lines that have the same slope and thus do not intersect. (Sec. 3.6)

Pascal's Triangle A pattern for the coefficients of the terms of the expansion of the binomial $(a + b)^n$ can be formed by writing the coefficients in a triangular array known as Pascal's Triangle. (Sec. 12.4)

perfect cube The product of the same three factors. (Sec. 5.6)

perfect square The product of a term and itself. (Sec. 5.6)

perpendicular lines Lines that intersect at right angles. The slopes of perpendicular lines are negative reciprocals of each other. (Sec. 3.6)

plotting a point in the plane Placing a dot at the location given by the ordered pair; also called graphing a point in the plane. (Sec. 3.1)

point-slope formula The equation $y - y_1 = m(x - x_1)$, where m is the slope of a line and (x_1, y_1) is a point on the line. (Sec. 3.5)

polynomial A variable expression in which the terms are monomials. (Sec. 5.2)

positive integers The numbers $1, 2, 3, \ldots$; also called the natural numbers. (Sec. 1.1)

positive slope The slope of a line that slants upward to the right. (Sec. 3.4)

prime number A number whose only whole-number factors are 1 and itself. For example, 17 is a prime number. (Sec. 1.1)

prime polynomial A polynomial that is nonfactorable over the integers. (Sec. 5.5)

principal square root The positive square root of a number. (Sec. 7.1)

proportion An equation that states the equality of two ratios or rates. (Sec. 6.4)

Pythagorean Theorem The square of the hypotenuse of a right triangle is equal to the sum of the squares of the two legs. (Sec. 7.4)

quadrant One of the four regions into which a rectangular coordinate system divides the plane. (Sec. 3.1)

quadratic equation An equation of the form $ax^2 + bx + c = 0$, where a and b are coefficients, c is a constant, and $a \neq 0$; also called a second-degree equation. (Sec. 5.7, 8.1)

quadratic formula A general formula, derived by applying the method of completing the square to the standard form of a quadratic equation, used to solve quadratic equations. (Sec. 8.3)

quadratic function A function that can be expressed by the equation $f(x) = ax^2 + bx + c$, where a is not equal to zero. (Sec. 5.2, 9.1)

quadratic inequality An inequality that can be written in the form $ax^2 + bx + c < 0$ or $ax^2 + bx + c > 0$, where a is not equal to zero. The symbols $\leq$ and $\geq$ can also be used. (Sec. 8.5)

quadratic trinomial A trinomial of the form $ax^2 + bx + c$, where a and b are nonzero coefficients and c is a nonzero constant. (Sec. 5.5)

radical In a radical expression, the symbol $\sqrt{}$. (Sec. 7.1)

radical equation An equation that contains a variable expression in a radicand. (Sec. 7.4)

radicand The expression under a radical sign. (Sec. 7.1)

radius The fixed distance, from the center of a circle, of all points that make up the circle. (Sec. 11.2)

range The set of the second coordinates of all the ordered pairs of a function. (Sec. 3.2)

rate The quotient of two quantities that have different units. (Sec. 6.4)

rate of work That part of a task that is completed in one unit of time. (Sec. 6.5)

ratio The quotient of two quantities that have the same unit. (Sec. 6.4)

rational expression A fraction in which the numerator or denominator is a polynomial. (Sec. 6.1)

rational function A function that is written in terms of a rational expression. (Sec. 6.1)

rational number A number of the form $\frac{a}{b}$, where a and b are integers and b is not equal to zero. (Sec. 1.1)

rationalizing the denominator The procedure used to remove a radical from the denominator of a fraction. (Sec. 7.2)

real numbers The rational numbers and the irrational numbers taken together. (Sec. 1.1)

real part of a complex number For the complex number $a + bi$, a is the real part. (Sec. 7.3)

reciprocal The reciprocal of a nonzero real number a is $\frac{1}{a}$; also called the multiplicative inverse. (Sec. 1.2)

reciprocal of a rational expression The rational expression with the numerator and denominator interchanged. (Sec. 6.1)

rectangular coordinate system A coordinate system formed by two number lines, one horizontal and one vertical, that intersect at the zero point of each line. (Sec. 3.1)

relation A set of ordered pairs. (Sec. 3.2)

repeating decimal A decimal formed when dividing the numerator of its fractional counterpart by the denominator results in a decimal part wherein one or more digits repeat infinitely. (Sec. 1.1)

root(s) of an equation The replacement value(s) of the variable that will make the equation true; also called the solution(s) of the equation. (Sec. 2.1)

roster method A method of designating a set by enclosing a list of its elements in braces. (Sec. 1.1)

scatter diagram A graph of ordered-pair data. (Sec. 3.1)

scientific notation Notation in which a number is expressed as the product of a number between 1 and 10 and a power of 10. (Sec. 5.1)

second coordinate of an ordered pair The second number of the ordered pair; it measures a vertical distance and is also called the ordinate. (Sec. 3.1)

second-degree equation An equation of the form $ax^2 + bx + c = 0$, where a and b are coefficients, c is a constant, and $a \neq 0$; also called a quadratic equation. (Sec. 8.1)

sequence An ordered list of numbers. (Sec. 12.1)

series The indicated sum of the terms of a sequence. (Sec. 12.1)

set A collection of objects. (Sec. 1.1)

set-builder notation A method of designating a set that makes use of a variable and a certain property that only elements of that set possess. (Sec. 1.1)

sigma notation Notation used to represent a series in a compact form; also called summation notation. (Sec. 12.1)

simplest form of a rational expression A rational expression is in simplest form when the numerator and denominator have no common factors. (Sec. 6.1)

slope A measure of the slant, or tilt, of a line. The symbol for slope is m. (Sec. 3.4)

slope-intercept form of a straight line The equation $y = mx + b$, where m is the slope of the line and $(0, b)$ is the y-intercept. (Sec. 3.4)

solution of a system of equations in three variables An ordered triple that is a solution of each equation of the system. (Sec. 4.2)

solution of a system of equations in two variables An ordered pair that is a solution of each equation of the system. (Sec. 4.1)

solution of an equation in three variables An ordered triple (x, y, z) whose coordinates make the equation a true statement. (Sec. 4.2)

solution of an equation in two variables An ordered pair whose coordinates make the equation a true statement. (Sec. 3.1)

solution set of a system of inequalities The intersection of the solution sets of the individual inequalities. (Sec. 4.5)

solution set of an inequality A set of numbers, each element of which, when substituted for the variable, results in a true inequality. (Sec. 2.4)

solution(s) of an equation The replacement value(s) of the variable that will make the equation true; also called the root(s) of the equation. (Sec. 2.1)

solving an equation Finding a root, or solution, of the equation. (Sec. 2.1)

square matrix A matrix that has the same number of rows as columns. (Sec. 4.3)

square root of a perfect square One of the two equal factors of the perfect square. (Sec. 5.6)

standard form of a quadratic equation A quadratic equation is in standard form when the polynomial is in descending order and equal to zero. (Sec. 8.1)

substitution method An algebraic method of finding an exact solution of a system of linear equations. (Sec. 4.1)

summation notation Notation used to represent a series in a compact form; also called sigma notation. (Sec. 12.1)

synthetic division A shorter method of dividing a polynomial by a binomial of the form $x - a$. This method uses only the coefficients of the variable terms. (Sec. 5.4)

system of equations Two or more equations considered together. (Sec. 4.1)

system of inequalities Two or more inequalities considered together. (Sec. 4.5)

term of a sequence A number in a sequence. (Sec. 12.1)

terminating decimal A decimal formed when dividing the numerator of its fractional counterpart by the denominator results in a remainder of zero. (Sec. 1.1)

terms of a variable expression The addends of the expression. (Sec. 1.3)

tolerance of a component The acceptable amount by which the component may vary from a given measurement. (Sec. 2.5)

trinomial A polynomial of three terms. (Sec. 5.2)

undefined slope The slope of a vertical line is undefined. (Sec. 3.4)

uniform motion The motion of an object whose speed and direction do not change. (Sec. 2.3, 6.5)

union of two sets The set that contains all elements that belong to either of the sets. (Sec. 1.1)

value mixture problem A problem that involves combining two ingredients that have different prices into a single blend. (Sec. 2.3)

value of a function The value of the dependent variable for a given value of the independent variable. (Sec. 3.2)

variable A letter used to stand for a quantity that is unknown or that can change. (Sec. 1.1)

variable expression An expression that contains one or more variables. (Sec. 1.3)

variable term A term composed of a numerical coefficient and a variable part. When the numerical coefficient is 1 or -1, the 1 is usually not written. (Sec. 1.3)

vertex of a parabola The point on the parabola with the smallest y-coordinate or the largest y-coordinate. (Sec. 9.1)

whole numbers The numbers 0, 1, 2, 3, 4, 5,... (Sec. 1.1)

x-coordinate The abscissa in an xy-coordinate system. (Sec. 3.1)

x-intercept The point at which a graph crosses the x-axis. (Sec. 3.3)

xy-coordinate system A rectangular coordinate system in which the horizontal axis is labeled x and the vertical axis is labeled y. (Sec. 3.1)

y-coordinate The ordinate in an xy-coordinate system. (Sec. 3.1)

y-intercept The point at which a graph crosses the y-axis. (Sec. 3.3)

zero slope The slope of a horizontal line. (Sec. 3.4)

Index